Formulas/Equations

Distance Formula   If $P_1 = (x_1, y_1)$ and $P_2 = (x_2, y_2)$, the distance from

$$d(P_1, P_2) = \sqrt{(x_2 - x_1)^2 + (y}$$

Equation of a Circle   The equation of a circle of radius r with center at (h, k) is

$$(x - h)^2 + (y - k)^2 = r^2$$

Slope Formula   The slope m of the line containing the points $P_1 = (x_1, y_1)$ and $P_2 = (x_2, y_2)$ is

$$m = \frac{y_2 - y_1}{x_2 - x_1} \qquad \text{if } x_1 \neq x_2$$

$$m \text{ is undefined} \qquad \text{if } x_1 = x_2$$

Point–Slope Equation of a Line   The equation of a line with slope m containing the point (x_1, y_1) is

$$y - y_1 = m(x - x_1)$$

Slope-Intercept Equation of a Line   The equation of a line with slope m and y-intercept b is

$$y = mx + b$$

Quadratic Formula   The solutions of the equation $ax^2 + bx + c = 0$, $a \neq 0$, are

$$x = \frac{-b \pm \sqrt{b^2 - 4ac}}{2a}$$

If $b^2 - 4ac > 0$, there are two real unequal solutions.

If $b^2 - 4ac = 0$, there is a repeated real solution.

If $b^2 - 4ac < 0$, there are two complex solutions.

Geometry Formulas

Circle $r = $ Radius, $A = $ Area, $C = $ Circumference

$$A = \pi r^2 \qquad C = 2\pi r$$

Triangle $b = $ Base, $h = $ Altitude (Height), $A = $ Area

$$A = \tfrac{1}{2}bh$$

Rectangle $l = $ Length, $w = $ Width, $A = $ Area, $P = $ Perimeter

$$A = lw \qquad P = 2l + 2w$$

Rectangular Box $l = $ Length, $w = $ Width, $h = $ Height, $V = $ Volume

$$V = lwh$$

Sphere $r = $ Radius, $V = $ Volume, $S = $ Surface area

$$V = \tfrac{4}{3}\pi r^3 \qquad S = 4\pi r^2$$

$61.35

Trigonometry

Enhanced with Graphing Utilities

Michael Sullivan
Chicago State University

Michael Sullivan, III
South Suburban College

 Prentice Hall, Upper Saddle River, NJ 07458

Library of Congress Cataloging-in-Publication Data

Sullivan, Michael, 1942–
   Trigonometry: enhanced with graphing utilities / Michael
Sullivan, Michael Sullivan, III.
     p. cm.
   Includes index.
   ISBN 0–13–456401–4
   1. Trigonometry. 2. Graphic calculators. I. Sullivan, Michael,
1967–   . II. Title.
QA531.S86 1996
516′.24—dc20   96-22996
                                         CIP

Acquisitions Editor: Sally Denlow
Editorial Assistant: Joanne Wendelken
Editor-in Chief of Development: Ray Mullaney
Director of Production and Manufacturing: David W. Riccardi
Managing Editor: Linda Behrens
Production Editor: Robert C. Walters
Marketing Manager: Jolene L. Howard
Marketing Assistant: William Paquin
Copy Editor: William Thomas
Interior Designer: Rosemarie Votta
Cover Designer: Rosemarie Votta
Creative Director: Paula Maylahn
Art Director: Amy Rosen
Assistant to Art Director: Rod Hernandez
Manufacturing Manager: Alan Fischer
Photo Researcher: Kathy Ringrose
Photo Editor: Lorinda Morris-Nantz
Supplements Editor: Audra J. Walsh

© 1996 by Prentice-Hall, Inc.
Simon & Schuster/A Viacom Company
Upper Saddle River, NJ 07458

Printed in the United States of America

10 9 8 7 6 5 4 3 2 1

ISBN 0-13-456401-4

PRENTICE-HALL INTERNATIONAL (UK) LIMITED, LONDON
PRENTICE-HALL OF AUSTRALIA PTY. LIMITED, SYDNEY
PRENTICE-HALL CANADA INC. TORONTO
PRENTICE-HALL HISPANOAMERICANA, S.A., MEXICO
PRENTICE-HALL OF INDIA PRIVATE LIMITED, NEW DELHI
PRENTICE-HALL OF JAPAN, INC., TOKYO
SIMON & SCHUSTER ASIA PTE. LTD., SINGAPORE
EDITORA PRENTICE-HALL DO BRASIL, LTDA., RIO DE JANEIRO

For Shannon, Patrick, and Ryan (Murphy)
and Michael S. (Sullivan)
The Next Generation

Contents

Preface to the Instructor   ix
Preface to the Student   xvii

CHAPTER 1   FUNCTIONS AND GRAPHS   1

1.1   Rectangular Coordinates; Graphing Utilities   2
1.2   Graphs of Equations   14
1.3   The Straight Line; Circles   35
1.4   Functions   57
1.5   More about Functions   73
1.6   Graphing Techniques   94
1.7   One-to-One Functions; Inverse Functions   107
         Chapter Review   118

CHAPTER 2   TRIGONOMETRIC FUNCTIONS   125

2.1   Angles and Their Measure   126
2.2   Trigonometric Functions: Unit Circle Approach   136
2.3   Properties of the Trigonometric Functions   152
2.4   Right Triangle Trigonometry   163
2.5   Graphs of the Trigonometric Functions   173
2.6   The Inverse Trigonometric Functions   183
         Chapter Review   200

CHAPTER 3   ANALYTIC TRIGONOMETRY   205

3.1   Trigonometric Identities   206
3.2   Sum and Difference Formulas   211
3.3   Double-angle and Half-angle Formulas   221

3.4 Product-to-Sum and Sum-to-Product Formulas   230
3.5 Trigonometric Equations   234
     Chapter Review   246

CHAPTER 4 APPLICATIONS OF TRIGONOMETRIC FUNCTIONS   249

4.1 Solving Right Triangles   250
4.2 The Law of Sines   259
4.3 The Law of Cosines   270
4.4 The Area of a Triangle   278
4.5 Sinusoidal Graphs; Simple Harmonic Motion   283
4.6 Damped Vibrations   298
     Chapter Review   303

CHAPTER 5 POLAR COORDINATES; VECTORS   311

5.1 Polar Coordinates   312
5.2 Polar Equations and Graphs   319
5.3 The Complex Plane; Demoivre's Theorem   340
5.4 Vectors   348
5.5 The Dot Product   359
     Chapter Review   367

CHAPTER 6 ANALYTIC GEOMETRY   371

6.1 Conics   372
6.2 The Parabola   373
6.3 The Ellipse   386
6.4 The Hyperbola   401
6.5 Rotation of Axes; General Form of a Conic   418
6.6 Polar Equations of Conics   426
6.7 Plane Curves and Parametric Equations   433
     Chapter Review   442

CHAPTER 7 EXPONENTIAL AND LOGARITHMIC FUNCTIONS   447

7.1 Exponential Functions   448
7.2 Logarithmic Functions   461
7.3 Properties of Logarithms   471
7.4 Logarithmic and Exponential Equations   478
7.5 Compound Interest   486
7.6 Growth and Decay   496
7.7 Nonlinear Curve Fitting   502
7.8 Logarithmic Scales   517
     Chapter Review   521

APPENDIX   527

1. Topics from Algebra and Geometry   527
2. Solving Equations   539
3. Completing the Square   543
4. Complex Numbers   546
5. Linear Curve Fitting   552

ANSWERS   AN1

INDEX   I1

Preface

Times certainly are changing. This text represents two generations of mathematics learning and teaching in the Sullivan family.

As the father, I have been teaching at an urban public university for over 30 years. Many things have changed during those years. . . . Students today have many varied needs—some students have very little mathematical background and a fear of the subject, while others are extremely motivated and have a strong educational background. Many are trying to balance family, work and school. All of these added pressures enter into the classroom. To be able to reach these students and excite them about the subject I love is truly wonderful. Technology, too, is changing. For me, as with many of my colleagues, the move has been from slide rules to arithmetic/scientific calculators to graphing calculators and computer algebra systems. As a result, our careers have been exciting, challenging, and, most importantly, filled with a wide range of learning experiences.

As the son, I grew up with teaching and learning all around me. Having completed dual advanced degrees in economics and mathematics has impacted the way I think about college courses . . . applications and skills blend together. Unlike my father, technology was in every aspect of my college career. Now as a mathematics teacher at a community college, I have the opportunity to see the strengths of the traditional foundations of mathematics combined with the exploratory nature of technology and the power of real world applications. For many of my students, the visual nature of mathematics through technology compels them to become more active in the classroom. Being able to use real world data, do experiments, and work in groups has made learning interesting again to them.

Welcome to Trigonometry: Enhanced with Graphing Utilities. We hope you enjoy using it as much as we do.

Michael Sullivan

Michael Sullivan, III

Our Philosophy on Technology

MAA, AMATYC, and NCTM all recommend the appropriate use of technology in the classroom and the mathematics laboratory. Appropriate is the difficult word to define. Our text assumes immediate access to graphing utilities by both students and instructors. We fully utilize the graphers ability to promote visualization, exploration, foreshadowing of concepts, data analysis, curve fitting and most of all, excitement in the classroom. We have not, however, thrown out all hand work. By blending together technology and traditional methods of problem solving, students are empowered with knowledge and ability that they will use throughout their lives.

How we integrate technology through the text is subtle—no big boxes, no icons. You will see both TI-82 screens and traditional line art where appropriate. Many examples are solved in two ways: with a graphing solution and with an algebraic solution. Many exercises are multi-tasked, beginning with questions that require the use of algebra or trigonometry and ending with questions that require not just a graph, but also an analysis of the graph. We encourage students to distinguish between solving a given problem using the full power of the technology vs. recognizing when the technology is limited or the simple nature of the problem makes a solution by hand the better choice.

One very effective use of a graphing utility is to develop the ability of students to recognize patterns. With a graphing utility, students can quickly and effectively start to recognize patterns; they can actually 'see' the mathematics, resulting in a better conceptual understanding of the concept at hand. Throughout the text, we have students graph related functions on the same screen, asking them leading questions to help them recognize patterns.

Another strong reason to use the technologies available is to move away from contrived and 'simple' numbers. Students can work with real information involving 'messy' numbers in a way that traditionally they could not do. Without taking away from the conceptual understanding of the objective at hand, students can analyze and interpret real data. Using this premise, we have included problems using real world data, as well as CBL exercises where students can actively collect their own data and manipulate it.

Technology can also be used to introduce ideas and solve problems that go beyond the traditional limitations of trigonometry. For example, many concepts and problems typically covered in a calculus class can be investigated and solved using graphing utilities. Determining where a function is increasing and decreasing and where its local maxima and minima occur are topics that can now be introduced, discussed, and solved in Trigonometry, thanks to the technology. Curve fitting is another topic that can be introduced, discussed, and solved, enabling students to relate algebra skills with data analysis.

These are a few of the uses made of technology in *Trigonometry: Enhanced with Graphing Utilities*. A graphing utility is required for the student to use this text. If you or your school would like to use graphing utilities, but not require them, *Trigonometry,* 4th Edition, by Michael Sullivan, which provides an optional use of graphing utilities, is also available. To receive an examination copy, contact your Prentice Hall representative.

About This Book

Content The content of this book is not different than that of a traditional Trigonometry—but the emphasis and the approach is! For example, we approach the idea of Horizontal Shifts by first exploring the graphs of several functions using a graphing utility and then drawing a conclusion about what is happening. Also, many Examples and Exercises found in this text are ones that cannot be handled using only traditional methods.

Organization With the use of fully integrated technology, the order of presentation of certain topics shifts.

To emphasize the focus of the book, Chapter 1 begins with graphing. In recognition of the fact that some students may not be fully prepared to start here, an Appendix has been carefully designed to provide necessary review. The sections of the Appendix that apply are referenced at the beginning of a Chapter and also within the body of the chapter. Also found in the Appendix is a section on Linear Curve Fitting that demonstrates the power of a graphing utility to solve applied problems in this area.

Chapter 1 treats the graphs of certain key equations, including the line and the circle. Chapter 1 also develops the concept of a function, the graphs of functions, and properties of functions.

In Chapter 2, Trigonometric Functions, the trigonometric functions are introduced using the unit circle approach. Later it is shown that right triangle trigonometry is a special case. The graphs of the trigonometric functions are then discussed. The chapter concludes with the inverse trigonometric functions.

Chapter 3, Analytic Trigonometry, deals with trigonometric identities and trigonometric equations. Chapter 4, Applications of Trigonometric Functions, provides applications of trigonometry to solving triangles, both right and oblique. Also included is a discussion of sinusoidal graphs, harmonic motion, and damped vibrations.

Chapter 5, Polar Coordinates; Vectors, introduces polar coordinates, with an emphasis on graphing. Also included are vectors and the dot product. Chapter 6, Analytic Geometry, discusses the conics, including rotation of axes and polar forms. Parametric equations are also included.

Chapter 7, Exponential and Logarithmic Functions, also places emphasis on the analysis of graphs, but, in addition, contains an entire section devoted to Nonlinear Curve Fitting.

Examples Recognizing that many students learn through working problems by hand as well as through the use of a graphing utility, we have provided both traditional and technological examples of solving problems. Examples are worked out in appropriate detail, starting with simple, reasonable problems and working gradually up to more challenging ones. There are no magic steps and we encourage and often show the final step of checking the result. Many of the examples involve applications that will be seen in calculus or in other disciplines. At the end of many examples there are "Now Work" suggestions which refer students to an odd numbered problem in section exercises which is similar to the worked out example shown. This allows students immediate feedback on whether they understand the concept clearly before moving on.

Exercises The exercises in the text are mostly of three types: visual—where students are asked to draw conclusions about a graph; technological—where the students will fully utilize the power of their graphing utility; and open-ended—where critical thinking, writing, research or collaborative effort is required. The text contains over 3200 tested and true exercises, over 500 are applied problems. Exercise sets begin with problems designed to build confidence, continue with problems which relate to worked out examples in the text and conclude with problems that are more challenging. Many of the problems, especially those at the beginning are visual in character—such as showing a graph and asking for conclusions.

Answers are given in the back of the text for all the odd-numbered problems. Fully worked out, step by step solutions for the odd-numbered problems are found in the **Student's Solutions Manual,** while the even solutions are equally worked out in the **Instructor's Solutions Manual.**

Illustrations The design uses color effectively and functionally to help the student identify definitions, theorems, formulas and procedures. Included in the Preface to the Student is an overview of how these colorized elements can be helpful when studying. Many illustrations have been included to provide a dynamic realism to selected examples and exercises. All the graphing utility and line art has been computer generated for consistency and accuracy. For the purpose of clarity, all line art utilizes two colors. With over 800 pieces of art throughout the text we aim to help the student to visualize mathematics and show how to use it to solve mathematical problems.

Applications Every opportunity has been taken to present understandable, realistic applications consistent with the abilities of the student, drawing from such sources as tax rate tables, the Guinness Book of World Records, Government publications and newspaper articles. For added interest, some of the applied exercises have been adapted from textbooks the students may be using in other courses (such as economics, chemistry, physics, etc.)

Communication and Problem Solving The recommendations of the NCTM, AMATYC and MAA support the inclusion of writing, verbalizing, research and critical thinking in mathematics. Throughout the text there are many ways we encourage these activities. In the exercises, these kind of problems are designated by a ▨ . All of these problems not only ask the student to write, discuss or to do active problem solving, but they also drive forward the concepts of the section.

Collaborative Projects In each chapter, a full page "Mission Possible" has been devoted to collaborative learning. These multi-tasked projects, written by Hester Lewellen, one of the co-authors of the University of Chicago High School Mathematics Project, will help your students work together to solve some unusual problems. Some of the projects require utilizing the graphing utility but all require critical thinking and communication. Suggested solutions to the collaborative projects are found in the Instructor's Solutions Manual.

Calculus Many of your students will be going on to calculus after this course. To encourage students and to let them preview calculus, we have included many examples and exercises that foreshadow topics found in calculus.

CBL Projects In key places in the text, CBL (Calculator Based Laboratory) experiments are given in the exercises. Each demonstrates real world applications of the topics just covered. Students are asked to perform experiments and collect and analyze the data obtained.

Instructor's Supplementary Aids

Instructor's Solutions Manual Contains complete step-by-step worked out solutions to all the even numbered exercises in the textbook. Also included are strategies for using the collaborative learning projects found in each chapter. Transparency masters which duplicate important illustrations in the text can be found as well.

Video Review A new videotape series which has been created to accompany the Sullivan texts include a half hour review of the most important topics in each chapter. Each segment uses both traditional and technological ways of solving mathematical problems. Entertaining and educational, these videos provide an alternative process which can add to your students' success in this course. Also included is a graphing calculator video tutorial which walks one through the most common uses of the various calculators. Written by mathematics teachers, these videos concentrate on those topics that 'get them every time'.

Written Test Item File Features six tests per chapter plus four forms of a final examination, prepared and ready to be photocopied. Of these tests, three are multiple choice and three are free response.

Prentice Hall Custom Test (Computerized Testing Generator - Mac and Windows) PH Custom Test is a fully networkable, easy to use test generator. Instructors may select questions by objective, section, chapter or use a ready-made test for each chapter. As the questions are algorithmically generated, an instructor can create up to 99 versions of each question while keeping the problem type and objective constant. PH Custom Test allows for on-line testing and offers a gradebook feature that not only organizes test grades but can be used for any other classroom grades or information. Instructors can download questions into word processing programs or create their own problems and insert them into PH Custom Test. Graphics and mathematical symbols are integrated into the program and other graphics from programs such as Mathematica, Maple, Derive or Matlab can be imported into the program.

Student's Supplementary Aids

Student's Solutions Manual Contains complete step-by-step worked out solutions to all the odd numbered exercises in the textbook. This is terrific for getting instant feedback for your students.

Prentice Hall Tutorial Program (Software tutorial both Mac and Windows) This computerized tutorial program is based on the highly successful Prentice Hall testing program. Utilizing the same algorithms programmed for the test generator, students are able to pretest their abilities, receive a diagnostic recommendation for further study, work through a step-by-step tutorial—complete with a graphing utility built into the program, and tutorial tests. The software is fully networkable and can be used with the gradebook feature in the Prentice Hall Custom Test.

Visual Precalculus   A software package for IBM compatible computers which consists of two parts. Part One contains routines to graph and evaluate functions, graph conic sections, investigate series, carry out synthetic division and illuminate important concepts with animation. Part Two contains routines to solve triangles, graph systems of linear equations and inequalities, evaluate matrix expressions, apply Gaussian elimination to reduce or invert matrices and graphically solve linear programming problems. Those routines will provide additional insights to the material covered within the text.

X(PLORE)   A powerful (yet inexpensive) fully programmable symbolic and numeric mathematical processor for IBM and Macintosh computers. This program will allow your students to evaluate expressions, graph curves, solve equations and use matrices. This software package may also be used for calculus or differential equations.

New York Times Supplement   A free newspaper from Prentice Hall and the New York Times which includes interesting and current articles on mathematics in the world around us. Great for getting students to talk and write about mathematics. This supplement is created new each year.

For any of the above supplements, please contact your Prentice Hall representative.

Acknowledgments

Textbooks are written by an author, but evolve from an idea into final form through the efforts of many people. Special thanks to Don Dellen, who first suggested this book and the other books in this series.

There are many people we would like to thank for their input, encouragement, patience and support. They have our deepest thanks and appreciation. We apologize for any omissions . . .

James Africh, College of Du Page
Steve Agronsky, Cal Poly State University
Dave Anderson, South Suburban College
Joby Milo Anthony, University of Central Florida
James E. Arnold, University of Wisconsin, Milwaukee
Agnes Azzolino, Middlesex County College
Wilson P. Banks, Illinois State University
Dale R. Bedgood, East Texas State University
Beth Beno, South Suburban College
William H. Beyer, University of Akron
Richelle Blair, Lakeland Community College
Trudy Bratten, Grossmont College
William J. Cable, University of Wisconsin—Stevens Point
Lois Calamia, Brookdale Community College
Roger Carlsen, Moraine Valley Community College
John Collado, South Suburban College
Denise Corbett, East Carolina University
Theodore C. Coskey, South Seattle Community College
John Davenport, East Texas State University
Duane E. Deal, Ball State University
Vivian Dennis, Eastfield College
Karen R. Dougan, University of Florida
Louise Dyson, Clark College
Paul D. East, Lexington Community College
Don Edmondson, University of Texas, Austin
Christopher Ennis, University of Minnesota
Garret J. Etgen, University of Houston
W. A. Ferguson, University of Illinois, Urbana/Champaign

Iris B. Fetts, Clemson University
Mason Flake, student at Edison Community College
Merle Friel, Humboldt State University
Richard A. Fritz, Moraine Valley Community College
Carolyn Funk, South Suburban College
Dewey Furness, Ricke College
Wayne Gibson, Rancho Santiago College
Joan Goliday, Santa Fe Community College
Frederic Gooding, Goucher College
Ken Gurganus, University of North Carolina
James E. Hall, University of Wisconsin, Madison
Judy Hall, West Virginia University
Edward R. Hancock, DeVry Institute of Technology
Brother Herron, Brother Rice High School
Kim Hughes, California State College, San Bernardino
Ron Jamison, Brigham Young University
Richard A. Jensen, Manatee Community College
Sandra G. Johnson, St. Cloud State University
Moana H. Karsteter, Tallahassee Community College
Arthur Kaufman, College of Staten Island
Thomas Kearns, North Kentucky University
Keith Kuchar, Manatee Community College
Tor Kwembe, Chicago State University
Linda J. Kyle, Tarrant County Jr. College
H. E. Lacey, Texas A & M University
Christopher Lattin, Oakton Community College
Adele LeGere, Oakton Community College
Stanley Lukawecki, Clemson University
Virginia McCarthy, Iowa State University
James McCollow, DeVry Institute of Technology
Laurence Maher, North Texas State University
James Maxwell, Oklahoma State University, Stillwater
Carolyn Meitler, Concordia University
Eldon Miller, University of Mississippi
James Miller, West Virginia University
Michael Miller, Iowa State University
Jane Murphy, Middlesex Community College
Bill Naegele, South Suburban College
James Nymann, University of Texas, El Paso
Sharon O'Donnell, Chicago State University
Seth F. Oppenheimer, Mississippi State University
E. James Peake, Iowa State University
Thomas Radin, San Joaquin Delta College
Ken A. Rager, Metropolitan State College
Elsi Reinhardt, Truckee Meadows Community College
Jane Ringwald, Iowa State University
Stephen Rodi, Austin Community College
Howard L. Rolf, Baylor University
Edward Rozema, University of Tennessee at Chattanooga
Dennis C. Runde, Manatee Community College
John Sanders, Chicago State University
Susan Sandmeyer, Jamestown Community College
A.K. Shamma, University of West Florida
Martin Sherry, Lower Columbia College
Timothy Sipka, Alma College
John Spellman, Southwest Texas State University
Becky Stamper, Western Kentucky University
Neil Stephens, Hinsdale South High School
Diane Tesar, South Suburban College
Tommy Thompson, Brookhaven College
Richard J. Tondra, Iowa State University
Marvel Townsend, University of Florida
Jim Trudnowski, Carroll College
Richard G. Vinson, University of Southern Alabama
Darlene Whitkenack, Northern Illinois University
Chris Wilson, West Virginia University
Carlton Woods, Auburn University
George Zazi, Chicago State University

Recognition and thanks are due particularly to the following individuals for their valuable assistance in the preparation of this edition: Jerome Grant for his support and commitment; Sally Denlow, for her genuine interest and insightful direction; Bob Walters for his organizational skill as production supervisor; Jolene Howard and Evan Hanby for their innovative marketing efforts; Ray Mullaney for his specific editorial comments; the entire Prentice-Hall sales staff for their confidence; and to Katy Murphy for checking the answers to all the exercises.

Preface

TO THE STUDENT

As you begin your study of trigonometry, you might feel overwhelmed by the number of theorems, definitions, procedures and equations that confront you. You may even wonder whether you can learn all this material in a single course. For many of you, this may be your last mathematics course, while for others, just the first in a series of many. Don't worry—either way, this text was written with you in mind.

This text was designed to help you—the student, master the terminology and basic concepts of trigonometry. These aims have helped to shape every aspect of the book. Many learning aids are built into the format of the text to make your study of this material easier and more rewarding. This book is meant to be a "machine for learning," one that can help you to focus your efforts and get the most from the time and energy you invest.

This book requires that you have access to a graphing utility: a graphing calculator or a computer software package that has a graphing component. Be sure you have some familiarity with the device you are using before the course begins.

Here are some hints we give our students at the beginning of the course:

1. Take advantage of the feature PREPARING FOR THIS CHAPTER. At the beginning of each chapter, we have prepared a list of topics to review. Be sure to take the time to do this. It will help you proceed quicker and more confidently through the chapter.
2. Read the material in the book before the lecture. Knowing what to expect and what is in the book, you can take fewer notes and spend more time listening and understanding the lecture.
3. After each lecture, rewrite your notes as you re-read the book, jotting down any additional facts that seem helpful. Be sure to do the Now Work Problem x as you proceed through a section. After completing a section, be sure to do the assigned problems. Answers to the Odd ones are in the back of the book.
4. If you are confused about something, visit your instructor during office hours immediately, before you fall behind. Bring your attempted solutions to problems with you to show your instructor where you are having trouble.
5. To prepare for an exam, review your notes. Then proceed through the Chapter Review. It contains a capsule summary of all the important material of the chapter. If you are uncertain of any concept, go back into the chapter and study it further. Be sure to do the Review Exercises for practice.

Remember the two "golden rules" of trigonometry:

1. DON'T GET BEHIND! The course moves too fast, and it's hard to catch up.
2. WORK LOTS OF PROBLEMS. Everyone needs to practice, and problems show where you need more work. If you can't solve the homework problems without help, you won't be able to do them on exams.

We encourage you to examine the following overview for some hints on how to use this text.

Best Wishes!

Michael Sullivan
Michael Sullivan, III

OVERVIEW

Chapter 1

PREPARING FOR THIS CHAPTER

Before getting started on this chapter, review the following concepts:
Topics from Algebra and Geometry (Appendix, Section 1)
Solving equations (Appendix, Section 2)
Completing the square (Appendix, Section 3)

FUNCTIONS AND GRAPHS

1.1 Rectangular Coordinates; Graphing Utilities
1.2 Graphs of Equations
1.3 The Straight Line; Circles
1.4 Functions
1.5 More about Functions
1.6 Graphing Techniques
1.7 One-to-One Functions; Inverse Functions
Chapter Review

Preview Getting from an Island to Town

An island is 2 miles from the nearest point P on a straight shoreline. A town is 12 miles down the shore from P.
(a) If a person can row a boat at an average speed of 3 miles per hour and the same person can walk 5 miles per hour, express the time T it takes to go from the island to town as a function of the distance x from P to where the person lands the boat.
(b) What is the domain of T?
(c) How long will it take to travel from the island to town if the person lands the boat 4 miles from P?
(d) How long will it take if the person lands the boat 8 miles from P?
(e) Use a graphing utility to graph the function T = T(x).
(f) Use the TRACE function to see how the time T varies as x changes from 0 to 12.
(g) What value of x results in the least time?
   [Example 9 in Section 1.4] ■

Each chapter begins with a list of concepts to review. Refresh your memory by turning to the pages listed.

A section by section outline is also provided. This is a good way to organize your notes for studying.

Highlighting an application from the chapter, you are given a "Preview" of a coming use of the mathematics introduced in this chapter.

Most chapters open with a brief historical discussion of "where this material came from". It is helpful to understand how others created and used these ideas to solve their everyday problems.

New terms appear in boldface type where they are defined.

Perhaps the most central idea in mathematics is the notion of a *function*. This important chapter deals with what a function is, how to graph functions, how to perform operations on functions, and how functions are used in applications.

The word *function* apparently was introduced by René Descartes in 1637. For him, a function simply meant any positive integral power of a variable x. Gottfried Wilhelm von Leibniz (1646–1716), who always emphasized the geometric side of mathematics, used the word function to denote any quantity associated with a curve, such as the coordinates of a point on the curve. Leonhard Euler (1707–1783) employed the word to mean any equation or formula involving variables and constants. His idea of a function is similar to the one most often used today in courses that precede calculus. Later, the use of functions in investigating heat flow equations led to a very broad definition, due to Lejeune Dirichlet (1805–1859), which describes a function as a rule or correspondence between two sets. It is his definition that we use here.

1.1

Rectangular Coordinates; Graphing Utilities

FIGURE 1

FIGURE 2

We locate a point on the real number line by assigning it a single real number, called the *coordinate of the point*. For work in a two-dimensional plane, we locate points by using two numbers.

We begin with two real number lines located in the same plane: one horizontal and the other vertical. We call the horizontal line the **x-axis**, the vertical line the **y-axis**, and the point of intersection the **origin O**. We assign coordinates to every point on these number lines as shown in Figure 1, using a convenient scale. In mathematics, we usually use the same scale on each axis; in applications, a different scale is often used on each axis.

The origin O has a value of 0 on both the x-axis and the y-axis. We follow the usual convention that points on the x-axis to the right of O are associated with positive real numbers, and those to the left of O are associated with negative real numbers. Those on the y-axis above O are associated with positive real numbers, and those below O are associated with negative real numbers. In Figure 1, the x-axis and y-axis are labeled as x and y, respectively, and we have used an arrow at the end of each axis to denote the positive direction.

The coordinate system described here is called a **rectangular**, or **Cartesian*** **coordinate system.** The plane formed by the x-axis and y-axis is sometimes called the **xy-plane**, and the x-axis and y-axis are referred to as the **coordinate axes.**

Any point P in the xy-plane can then be located by using an **ordered pair** (x, y) of real numbers. Let x denote the signed distance of P from the y-axis (*signed* in the sense that, if P is to the right of the y-axis, then x > 0, and if P is to the left of the y-axis, then x < 0); and let y denote the signed distance of P from the x-axis. The ordered pair (x, y), also called the **coordinates** of P, then gives us enough information to locate the point P in the plane.

For example, to locate the point whose coordinates are (−3, 1), go 3 units along the x-axis to the left of O and then go straight up 1 unit. We **plot** this point by placing a dot at this location. See Figure 2, in which the points with coordinates (−3, 1), (−2, −3), (3, −2), and (3, 2) are plotted.

The origin O has coordinates (0, 0). Any point on the x-axis has coordinates of the form (x, 0), and any point on the y-axis has coordinates of the form (0, y).

If (x, y) are the coordinates of a point P, then x is called the **x-coordinate**, or **abscissa**, of P and y is the **y-coordinate**, or **ordinate**, of P. We identify the

*Named after René Descartes (1596–1650), a French mathematician, philosopher, and theologian.

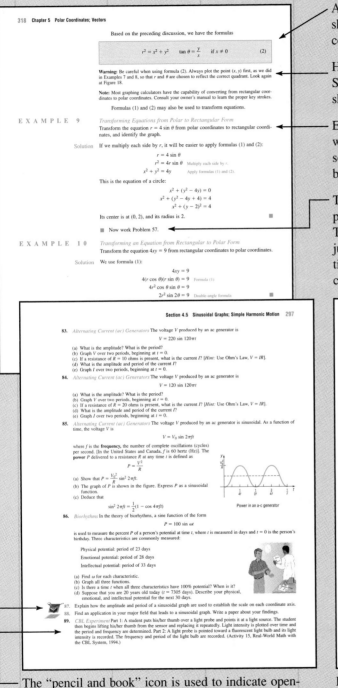

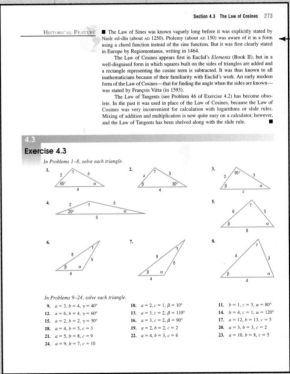

All important formulas are enclosed by a box and shown in color. This is done to alert you to important concepts.

Hints or warnings are offered where appropriate. Sometimes there are short cuts or pitfalls that students should know about—we've included them.

Examples are easy to locate and are titled to tell you what concept they highlight. Most examples work the solution first with the graphing utility and then algebraically.

The Now Work Problem xx feature asks you to do a particular problem before you go on in the section. This is to ensure that you have mastered the material just presented before going on. By following the practice of doing these Now Work Problems, you will gain confidence and save time.

The "pencil and book" icon is used to indicate open-ended questions for discussion, writing, group or research projects.

CBL Experiments are provided at key places in the exercises.

Historical Features place the mathematics you are learning in a historical context. By learning how others have used similar concepts you will understand how they may be used in your own life.

Theorem
Laws of Exponents

If s, t, a, and b are real numbers with $a > 0$ and $b > 0$, then

$$a^s \cdot a^t = a^{s+t} \qquad (a^s)^t = a^{st} \qquad (ab)^s = a^s \cdot b^s$$
$$1^s = 1 \qquad a^{-s} = \frac{1}{a^s} = \left(\frac{1}{a}\right)^s \qquad a^0 = 1 \qquad (1)$$

We are now ready for the following definition.

Exponential Function

An **exponential function** is a function of the form

$$f(x) = a^x$$

where a is a positive real number and $a \neq 1$. The domain of f is the set of all real numbers.

We exclude the base $a = 1$, because this function is simply the constant function $f(x) = 1^x = 1$. We also need to exclude bases that are negative, because, otherwise, we would have to exclude many values of x from the domain, such as $x = \frac{1}{2}$, $x = \frac{3}{4}$, and so on. [Recall that $(-2)^{1/2}$, $(-3)^{3/4}$, and so on, are not defined in the system of real numbers.]

Graphs of Exponential Functions

First, we graph the exponential function $y = 2^x$.

EXAMPLE 2   *Graphing an Exponential Function*

Graph the exponential function: $f(x) = 2^x$

Solution   The domain of $f(x) = 2^x$ consists of all real numbers. We begin by locating some points on the graph of $f(x) = 2^x$, as listed in Table 1.

Since $2^x > 0$ for all x, the range of f is $(0, \infty)$. From this, we conclude that the graph has no x-intercepts, and, in fact, the graph will lie above the x-axis. As Table 1 indicates, the y-intercept is 1. Table 1 also indicates that as $x \to -\infty$ the value of $f(x) = 2^x$ gets closer and closer to 0. Thus, the x-axis is a horizontal asymptote to the graph as $x \to -\infty$. Look again at Table 1. As $x \to \infty$, $f(x) = 2^x$ grows very quickly, causing the graph of $f(x) = 2^x$ to rise very rapidly. Thus, it is apparent that f is an increasing function and, hence, is one-to-one.

Figure 2 shows the graph of $f(x) = 2^x$. Notice that all the conclusions given earlier are confirmed by the graph.

Major definitions appear in large type enclosed within a color screen. These are important vocabulary items for you to know.

Important Procedures and STEPS are noted in the left column and are separated from the body of the text by two horizontal color rules. You will need to know these procedures and steps to do your homework problems and prepare for exams.

---

16   Chapter 1   Functions and Graphs

becomes evident; then we connect these points with a smooth curve following the suggested pattern. Shortly, we shall investigate various techniques that will enable us to graph an equation by hand without plotting so many points. Other times, we shall graph equations using a graphing utility.

Using a Graphing Utility to Graph Equations

From Examples 1 and 2, we see that a graph can be obtained by plotting points in a rectangle coordinate system and connecting them. Graphing utilities perform these same steps when graphing an equation. For example, the TI-82 determines 95 evenly spaced input values*, uses the equation to determine the output values, plots these points on the screen and finally, (if in the connected mode), draws a line between consecutive points.

Most graphing utilities require the following steps in order to obtain the graph of an equation:

STEP 1:   Solve the equation for y in terms of x.
STEP 2:   Select the viewing rectangle.
STEP 3:   Get into the graphing mode of your graphing utility. The screen will usually display $y =$, prompting you to enter the expression involving x that you found in STEP 1. (Consult your manual for the correct way to enter the expression; for example, $y = x^2$ might be entered as x^2 or as $x*x$ or as $x\,x^y\,2$).
STEP 4:   Execute.

EXAMPLE 3   *Graphing an Equation on a Graphing Utility*

Graph the equation: $6x^2 + 3y = 24$

Solution   STEP 1:   We solve for y in terms of x.

$$6x^2 + 3y = 24$$
$$3y = -6x^2 + 24$$
$$y = -2x^2 + 8$$

STEP 2:   Select a viewing rectangle. We will use the one given next.

$$\text{Xmin} = -5$$
$$\text{Xmax} = 5$$
$$\text{Xscl} = 1$$
$$\text{Ymin} = -10$$
$$\text{Ymax} = 20$$
$$\text{Yscl} = 2$$

STEP 3:   From the graphing mode, enter the expression $-2x^2 + 8$ after the prompt $y =$.

STEP 4:   Execute

The screen should look like Figure 20.

*These input values depend on the values of Xmin and Xmax. For example, if Xmin $= -10$ and Xmax $= 10$, then the first input value will be -10 and the next input value will be $-10 + (10 - (-10))/94 = -9.7872$, and so on.

---

38   Chapter 1   Functions and Graphs

Figure 58 shows the graph of the line $y = x$ on a square screen using the viewing rectangle given in Example 1(b). Notice that the line now bisects the first and third quadrants. Compare this illustration to Figure 57.

FIGURE 58

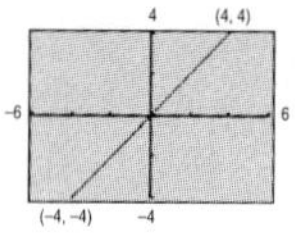

FIGURE 59

Seeing the Concept: On the same square screen, graph the following equations:

$y = 0$ Slope of line is 0.
$y = \frac{1}{4}x$ Slope of line is $\frac{1}{4}$.
$y = \frac{1}{2}x$ Slope of line is $\frac{1}{2}$.
$y = x$ Slope of line is 1.
$y = 2x$ Slope of line is 2.
$y = 6x$ Slope of line is 6.

See Figure 59.

FIGURE 60

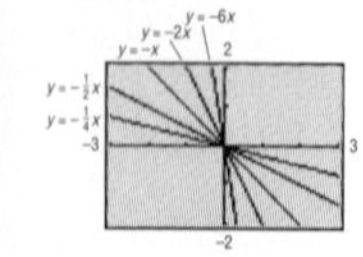

Seeing the Concept: On the same square screen, graph the following equations:

$y = 0$ Slope of line is 0.
$y = -\frac{1}{4}x$ Slope of line is $-\frac{1}{4}$.
$y = -\frac{1}{2}x$ Slope of line is $-\frac{1}{2}$.
$y = -x$ Slope of line is -1.
$y = -2x$ Slope of line is -2.
$y = -6x$ Slope of line is -6.

See Figure 60.

The next example illustrates how the slope of a line can be used to graph the line.

EXAMPLE 4   *Graphing a Line Given a Point and a Slope*

Draw a graph of the line that passes through the point (3, 2) and has a slope of:
(a) $\frac{3}{4}$ (b) $-\frac{4}{5}$

Solution   (a) Slope = Rise/Run. The fact that the slope is $\frac{3}{4}$ means that for every horizontal movement (run) of 4 units to the right there will be a vertical movement (rise) of 3 units. If we start at the given point (3, 2) and move 4 units to the right and 3 units up, we reach the point (7, 5). By drawing the line through this point and the point (3, 2), we have the graph. See Figure 61.

"Seeing the Concept." Use your graphing utility to recognize patterns in the algebra.

The Chapter Review is for your use in checking your understanding of the chapter materials.

"Things to Know" is the best place to start. Check your understanding of the concepts listed there.

The formulas found throughout the chapter are listed here.

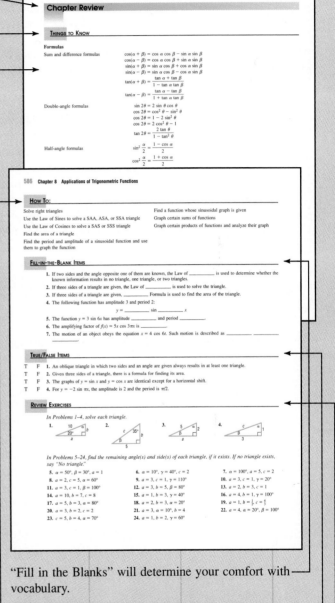

Chapter Review

THINGS TO KNOW

Formulas

Sum and difference formulas	$\cos(\alpha + \beta) = \cos \alpha \cos \beta - \sin \alpha \sin \beta$
	$\cos(\alpha - \beta) = \cos \alpha \cos \beta + \sin \alpha \sin \beta$
	$\sin(\alpha + \beta) = \sin \alpha \cos \beta + \cos \alpha \sin \beta$
	$\sin(\alpha - \beta) = \sin \alpha \cos \beta - \cos \alpha \sin \beta$
	$\tan(\alpha + \beta) = \dfrac{\tan \alpha + \tan \beta}{1 - \tan \alpha \tan \beta}$
	$\tan(\alpha - \beta) = \dfrac{\tan \alpha - \tan \beta}{1 + \tan \alpha \tan \beta}$
Double-angle formulas	$\sin 2\theta = 2 \sin \theta \cos \theta$
	$\cos 2\theta = \cos^2 \theta - \sin^2 \theta$
	$\cos 2\theta = 1 - 2 \sin^2 \theta$
	$\cos 2\theta = 2 \cos^2 \theta - 1$
	$\tan 2\theta = \dfrac{2 \tan \theta}{1 - \tan^2 \theta}$
Half-angle formulas	$\sin^2 \dfrac{\alpha}{2} = \dfrac{1 - \cos \alpha}{2}$
	$\cos^2 \dfrac{\alpha}{2} = \dfrac{1 + \cos \alpha}{2}$

HOW TO:

Solve right triangles	Find a function whose sinusoidal graph is given
Use the Law of Sines to solve a SAA, ASA, or SSA triangle	Graph certain sums of functions
Use the Law of Cosines to solve a SAS or SSS triangle	Graph certain products of functions and analyze their graph
Find the area of a triangle	
Find the period and amplitude of a sinusoidal function and use them to graph the function	

FILL-IN-THE-BLANK ITEMS

1. If two sides and the angle opposite one of them are known, the Law of _____ is used to determine whether the known information results in no triangle, one triangle, or two triangles.
2. If three sides of a triangle are given, the Law of _____ is used to solve the triangle.
3. If three sides of a triangle are given, _____ Formula is used to find the area of the triangle.
4. The following function has amplitude 3 and period 2:
$$y = \underline{\quad} \sin \underline{\quad} x$$
5. The function $y = 3 \sin 6x$ has amplitude _____ and period _____.
6. The amplifying factor of $f(x) = 5x \cos 3\pi x$ is _____.
7. The motion of an object obeys the equation $x = 4 \cos 6t$. Such motion is described as _____ _____.

TRUE/FALSE ITEMS

T F 1. An oblique triangle in which two sides and an angle are given always results in at least one triangle.
T F 2. Given three sides of a triangle, there is a formula for finding its area.
T F 3. The graphs of $y = \sin x$ and $y = \cos x$ are identical except for a horizontal shift.
T F 4. For $y = -2 \sin \pi x$, the amplitude is 2 and the period is $\pi/2$.

REVIEW EXERCISES

In Problems 1–4, solve each triangle.

1. 2. 3. 4.

In Problems 5–24, find the remaining angle(s) and side(s) of each triangle, if it exists. If no triangle exists, say "No triangle."

5. $\alpha = 50°$, $\beta = 30°$, $a = 1$
6. $\alpha = 10°$, $\gamma = 40°$, $c = 2$
7. $\alpha = 100°$, $a = 5$, $c = 2$
8. $a = 2$, $c = 5$, $\alpha = 60°$
9. $a = 3$, $c = 1$, $\gamma = 110°$
10. $a = 3$, $c = 1$, $\gamma = 20°$
11. $a = 3$, $c = 1$, $\beta = 100°$
12. $a = 3$, $b = 5$, $\beta = 80°$
13. $a = 2$, $b = 3$, $c = 1$
14. $a = 10$, $b = 7$, $c = 8$
15. $a = 1$, $b = 3$, $\gamma = 40°$
16. $a = 4$, $b = 1$, $\gamma = 100°$
17. $a = 5$, $b = 3$, $\alpha = 80°$
18. $a = 2$, $b = 3$, $\alpha = 20°$
19. $a = 1$, $b = \frac{1}{2}$, $c = \frac{4}{3}$
20. $a = 3$, $b = 2$, $c = 2$
21. $a = 3$, $\alpha = 10°$, $b = 4$
22. $a = 4$, $\alpha = 20°$, $\beta = 100°$
23. $c = 5$, $b = 4$, $\alpha = 70°$
24. $a = 1$, $b = 2$, $\gamma = 60°$

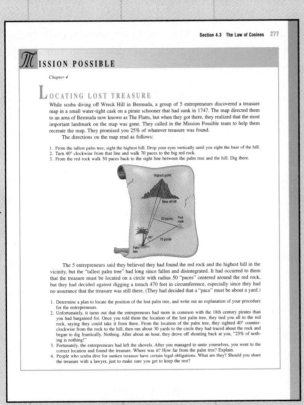

$\mathcal{M}$ISSION POSSIBLE

Chapter 4

LOCATING LOST TREASURE

While scuba diving off Wreck Hill in Bermuda, a group of 5 entrepreneurs discovered a treasure map in a small water-tight cask on a pirate schooner that had sunk in 1747. The map directed them to an area of Bermuda now known as The Flatts, but when they got there, they realized that the most important landmark on the map was gone. They called in the Mission Possible team to help them recreate the map. They promised you 25% of whatever treasure was found.

The directions on the map read as follows:

1. From the tallest palm tree, sight the highest hill. Drop your eyes vertically until you sight the base of the hill.
2. Turn 40° clockwise from that line and walk 70 paces to the big red rock.
3. From the red rock walk 50 paces back to the sight line between the palm tree and the hill. Dig there.

The 5 entrepreneurs said they believed they had found the red rock and the highest hill in the vicinity, but the "tallest palm tree" had long since fallen and disintegrated. It had occurred to them that the treasure must be located on a circle with radius 50 "paces" centered around the red rock, but they had decided against digging a trench 470 feet in circumference, especially since they had no assurance that the treasure was still there. (They had decided that a "pace" must be about a yard.)

1. Determine a plan to locate the position of the lost palm tree, and write out an explanation of your procedure for the entrepreneurs.
2. Unfortunately, it turns out that the entrepreneurs had more in common with the 18th century pirates than you had bargained for. Once you told them the location of the lost palm tree, they tied you all to the red rock, saying they could take it from there. From the location of the palm tree, they sighted 40° counterclockwise from the rock to the hill, then ran about 30 yards to the circle they had traced about the rock and began to dig frantically. Nothing. After about an hour, they drove off shouting back at you, "25% of nothing is nothing!"
3. Fortunately, the entrepreneurs had left the shovels. After you managed to untie yourselves, you went to the correct location and found the treasure. Where was it? How far from the palm tree? Explain.
4. People who scuba dive for sunken treasure have certain legal obligations. What are they? Should you share the treasure with a lawyer, just to make sure you get to keep the rest?

Mission Possible: In the "real world" colleagues often collaborate to solve more difficult problems—or problems that may have more than one answer. Every chapter includes a "Mission Possible" for you and your classmates to work together. All of these projects require verbal communication or written answers. Good communication skills are very important to becoming successful—no matter what your future holds.

Demonstrate to yourself that you know "How to" deal with the concepts listed.

"Fill in the Blanks" will determine your comfort with vocabulary.

"True/False" is a stickler for knowing definitions!

The "Review Exercises" provide a comprehensive supply of exercises using all the concepts contained in the chapter.

Student's Supplementary Aids

You may find yourself seeking out extra help with this course. Many students have found the following items to be useful in becoming successful in trigonometry. Your college bookstore should have these items available, but if not, they can order them for you.

Student's Solutions Manual Contains complete step-by-step worked out solutions to all the odd numbered exercises in the textbook. This is terrific for getting instant feedback on whether you are proceeding correctly while solving problems.

Visual Precalculus A software package for IBM compatible computers which consists of two parts. Part One contains routines to graph and evaluate functions, graph conic sections, investigate series, carry out synthetic division, and illuminate important concepts with animation. Part Two contains routines to solve triangles, graph systems of linear equations and inequalities, evaluate matrix expressions, apply Gaussian elimination to reduce or invert matrices and graphically solve linear programming problems. These routines will provide additional insights into the material covered within the text.

X (Plore) A powerful (yet inexpensive) fully programmable symbolic and numeric mathematical processor for IBM and Macintosh computers. This program will allow you to evaluate expressions, graph curves, solve equations and matrices. This software package may also be used for calculus or differential equations.

New York Times Supplement A free newspaper from Prentice Hall and the New York Times which includes interesting and current articles on mathematics in the world around us. Great for getting together to talk and write about mathematics! This supplement is created new each year.

Photo Credits

Chapter 1	Port of Soller	PedroColl/The Stock Market
Chapter 2	Cadillac Mountain, Acadia National Park	Larry Ulrich/Tony Stone Images
Chapter 3	Shotput	Focus on Sports
Chapter 4	Gibbs Hill Lighthouse	Marvullo/The Stock Market
Chapter 5	Corporate Jet	Joe Towers/The Stock Market
Chapter 6	Satellite Station	Four by Five/Superstock
Chapter 7	Sobriety Testing	Bachmann/Photrl

PREPARING FOR THIS CHAPTER

Before getting started on this chapter, review the following concepts:

Topics from Algebra and Geometry (Appendix, Section 1)
Solving equations (Appendix, Section 2)
Completing the square (Appendix, Section 3)

FUNCTIONS AND GRAPHS

1.1 Rectangular Coordinates; Graphing Utilities
1.2 Graphs of Equations
1.3 The Straight Line; Circles
1.4 Functions
1.5 More about Functions
1.6 Graphing Techniques
1.7 One-to-One Functions; Inverse Functions
Chapter Review

Preview   Getting from an Island to Town

An island is 2 miles from the nearest point P on a straight shoreline. A town is 12 miles down the shore from P.

(a) If a person can row a boat at an average speed of 3 miles per hour and the same person can walk 5 miles per hour, express the time T it takes to go from the island to town as a function of the distance x from P to where the person lands the boat.

(b) What is the domain of T?

(c) How long will it take to travel from the island to town if the person lands the boat 4 miles from P?

(d) How long will it take if the person lands the boat 8 miles from P?

(e) Use a graphing utility to graph the function T = T(x).

(f) Use the TRACE function to see how the time T varies as x changes from 0 to 12.

(g) What value of x results in the least time?
[Example 9 in Section 1.4] ■

erhaps the most central idea in mathematics is the notion of a *function*. This important chapter deals with what a function is, how to graph functions, how to perform operations on functions, and how functions are used in applications.

The word *function* apparently was introduced by René Descartes in 1637. For him, a function simply meant any positive integral power of a variable *x*. Gottfried Wilhelm von Leibniz (1646–1716), who always emphasized the geometric side of mathematics, used the word function to denote any quantity associated with a curve, such as the coordinates of a point on the curve. Leonhard Euler (1707–1783) employed the word to mean any equation or formula involving variables and constants. His idea of a function is similar to the one most often used today in courses that precede calculus. Later, the use of functions in investigating heat flow equations led to a very broad definition, due to Lejeune Dirichlet (1805–1859), which describes a function as a rule or correspondence between two sets. It is his definition that we use here.

1.1

Rectangular Coordinates; Graphing Utilities

FIGURE 1

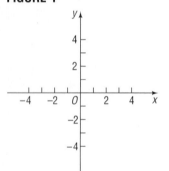

FIGURE 2

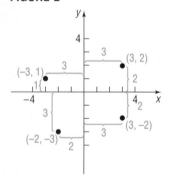

We locate a point on the real number line by assigning it a single real number, called the *coordinate of the point.* For work in a two-dimensional plane, we locate points by using two numbers.

We begin with two real number lines located in the same plane: one horizontal and the other vertical. We call the horizontal line the **x-axis,** the vertical line the **y-axis,** and the point of intersection the **origin O.** We assign coordinates to every point on these number lines as shown in Figure 1, using a convenient scale. In mathematics, we usually use the same scale on each axis; in applications, a different scale is often used on each axis.

The origin *O* has a value of 0 on both the *x*-axis and the *y*-axis. We follow the usual convention that points on the *x*-axis to the right of *O* are associated with positive real numbers, and those to the left of *O* are associated with negative real numbers. Those on the *y*-axis above *O* are associated with positive real numbers, and those below *O* are associated with negative real numbers. In Figure 1, the *x*-axis and *y*-axis are labeled as *x* and *y,* respectively, and we have used an arrow at the end of each axis to denote the positive direction.

The coordinate system described here is called a **rectangular,** or **Cartesian* coordinate system.** The plane formed by the *x*-axis and *y*-axis is sometimes called the **xy-plane,** and the *x*-axis and *y*-axis are referred to as the **coordinate axes.**

Any point *P* in the *xy*-plane can then be located by using an **ordered pair** (x, y) of real numbers. Let *x* denote the signed distance of *P* from the *y*-axis (*signed* in the sense that, if *P* is to the right of the *y*-axis, then $x > 0$, and if *P* is to the left of the *y*-axis, then $x < 0$); and let *y* denote the signed distance of *P* from the *x*-axis. The ordered pair (x, y), also called the **coordinates** of *P,* then gives us enough information to locate the point *P* in the plane.

For example, to locate the point whose coordinates are $(-3, 1)$, go 3 units along the *x*-axis to the left of *O* and then go straight up 1 unit. We **plot** this point by placing a dot at this location. See Figure 2, in which the points with coordinates $(-3, 1)$, $(-2, -3)$, $(3, -2)$, and $(3, 2)$ are plotted.

The origin has coordinates $(0, 0)$. Any point on the *x*-axis has coordinates of the form $(x, 0)$, and any point on the *y*-axis has coordinates of the form $(0, y)$.

If (x, y) are the coordinates of a point *P,* then *x* is called the **x-coordinate,** or **abscissa,** of *P* and *y* is the **y-coordinate,** or **ordinate,** of *P.* We identify the

*Named after René Descartes (1596–1650), a French mathematician, philosopher, and theologian.

FIGURE 3

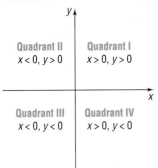

point P by its coordinates (x, y) by writing $P = (x, y)$. Usually, we will simply say "the point (x, y)" rather than "the point whose coordinates are (x, y)."

The coordinate axes divide the xy-plane into four sections, called **quadrants,** as shown in Figure 3. In quadrant I, both the x-coordinate and the y-coordinate of all points are positive; in quadrant II, x is negative and y is positive; in quadrant III, both x and y are negative; and in quadrant IV, x is positive and y is negative. Points on the coordinate axes belong to no quadrant.

Graphing Utilities

All graphing utilities, that is, all graphing calculators and all computer software graphing packages, graph equations by plotting points on a screen. The screen itself actually consists of small rectangles, called **pixels.** The more pixels the screen has, the better the resolution. Most graphing calculators have 48 pixels per square inch; most computer screens have 32 to 108 pixels per square inch. When a point to be plotted lies inside a pixel, the pixel is turned on (lights up). Thus, the graph of an equation is a collection of pixels. Figure 4 shows how the graph of $y = 2x$ looks on a T1-82 graphing calculator.

FIGURE 4
$y = 2x$

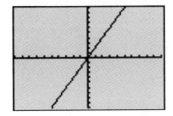

The screen of a graphing utility will display the coordinate axes of a rectangular coordinate system. However, you must set the scale on each axis. You must also include the smallest and largest values of x and y that you want included in the graph. This is called **setting the RANGE** and it gives the **viewing rectangle** or **window.**

Figure 5 illustrates a typical viewing rectangle.

FIGURE 5

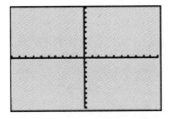

To select the viewing rectangle, we must give values to the following expressions:

Xmin: the smallest value of x

Xmax: the largest value of x

Xscl: the number of units per tick mark on the x-axis

Ymin: the smallest value of y

Ymax: the largest value of y

Yscl: the number of units per tick mark on the y-axis

Figure 6 illustrates these settings for a typical screen.

FIGURE 6

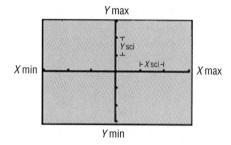

If the scale used on each axis is known, we can determine the minimum and maximum values of x and y shown on the screen by counting the tick marks. Look again at Figure 5. For a scale of 1 on each axis, the minimum and maximum values of x are -10 and 10, respectively; the minimum and maximum values of y are also -10 and 10. If the scale is 2 on each axis, then the minimum and maximum values of x are -20 and 20, respectively; the minimum and maximum value of y are -20 and 20, respectively.

Conversely, if we know the minimum and maximum values of x and y, we can determine the scales being used by counting the tick marks displayed. We shall follow the practice of showing the minimum and maximum values of x and y in our illustrations so that you will know how the RANGE was set. See Figure 7.

FIGURE 7

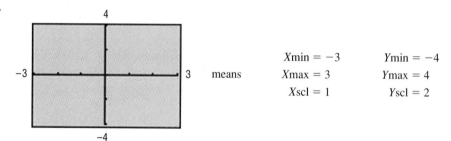

means

$$Xmin = -3 \qquad Ymin = -4$$
$$Xmax = 3 \qquad Ymax = 4$$
$$Xscl = 1 \qquad Yscl = 2$$

E X A M P L E   1   *Finding the Coordinates of a Point Shown on a Graphing Utility Screen*

Find the coordinates of the point shown in Figure 8.

FIGURE 8

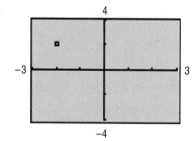

Solution   First we note that the range setting used in Figure 8 is

$$X\text{min} = -3 \qquad X\text{scl} = 1 \qquad Y\text{max} = 4$$
$$X\text{max} = 3 \qquad Y\text{min} = -4 \qquad Y\text{scl} = 2$$

The point shown is 2 tic units to the left on the horizontal axis (scale = 1) and 1 tic up on the vertical (scale = 2). Thus, the coordinates of the point shown are $(-2, 2)$. ■

■ Now work Problems 5 and 15.

Distance between Points

If the same units of measurement, such as inches, centimeters, and so on, are used for both the x-axis and the y-axis, then all distances in the xy-plane can be measured using this unit of measurement.

E X A M P L E   2   *Finding the Distance between Two Points*

Find the distance d between the points $(1, 3)$ and $(5, 6)$.

Solution   First we plot the points $(1, 3)$ and $(5, 6)$ as shown in Figure 9(a). Then we draw a horizontal line from $(1, 3)$ to $(5, 3)$ and a vertical line from $(5, 3)$ to $(5, 6)$, forming a right triangle, as in Figure 9(b). One leg of the triangle is of length 4 and the other is of length 3. By the Pythagorean Theorem (see the Appendix, Section 1), the square of the distance d we seek is

$$d^2 = 4^2 + 3^2 = 16 + 9 = 25$$
$$d = 5$$

FIGURE 9

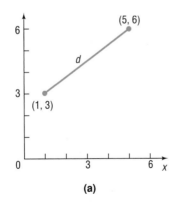

(a)

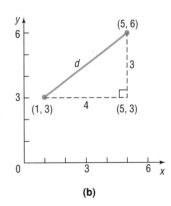

(b)

■

The **distance formula** provides a straightforward method for computing the distance between two points.

Theorem   The distance between two points $P_1 = (x_1, y_1)$ and $P_2 = (x_2, y_2)$, denoted by $d(P_1, P_2)$, is

Distance Formula

$$d(P_1, P_2) = \sqrt{(x_2 - x_1)^2 + (y_2 - y_1)^2} \qquad (1)$$

■

That is, to compute the distance between two points, find the difference of the x-coordinates, square it, and add this to the square of the difference of the y-coordinates. The square root of this sum is the distance. See Figure 10.

FIGURE 10

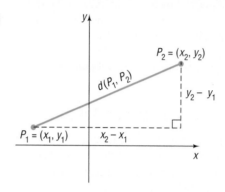

Proof of the Distance Formula   Let (x_1, y_1) denote the coordinates of point P_1, and let (x_2, y_2) denote the coordinates of point P_2. Assume that the line joining P_1 and P_2 is neither horizontal nor vertical. Refer to Figure 11(a). The coordinates of P_3 are (x_2, y_1). The horizontal distance from P_1 to P_3 is the absolute value of the difference of the x-coordinates, $|x_2 - x_1|$. The vertical distance from P_3 to P_2 is the absolute value of the difference of the y-coordinates, $|y_2 - y_1|$. See Figure 11(b). The distance $d(P_1, P_2)$ that we seek is the length of the hypotenuse of the right triangle, so, by the Pythagorean Theorem, it follows that

$$[d(P_1, P_2)]^2 = |x_2 - x_1|^2 + |y_2 - y_1|^2$$
$$= (x_2 - x_1)^2 + (y_2 - y_1)^2$$
$$d(P_1, P_2) = \sqrt{(x_2 - x_1)^2 + (y_2 - y_1)^2}$$

FIGURE 11

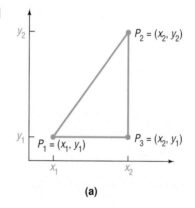

(a)

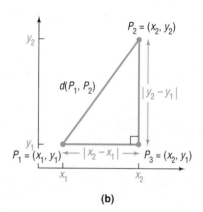

(b)

Now, if the line joining P_1 and P_2 is horizontal, then the y-coordinate of P_1 equals the y-coordinate of P_2; that is, $y_1 = y_2$. Refer to Figure 12(a). In this case, the distance formula (1) still works, because, for $y_1 = y_2$, it reduces to

$$d(P_1, P_2) = \sqrt{(x_2 - x_1)^2 + 0^2} = \sqrt{(x_2 - x_1)^2} = |x_2 - x_1|$$

A similar argument holds if the line joining P_1 and P_2 is vertical. See Figure 12(b). Thus, the distance formula is valid in all cases.

FIGURE 12

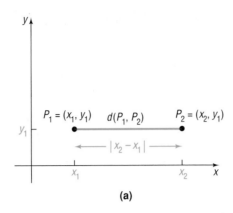

(a)

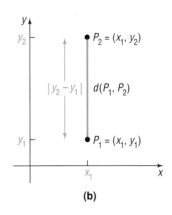

(b)

EXAMPLE 3

Finding the Length of a Line Segment

Find the length of the line segment shown in Figure 13.

FIGURE 13

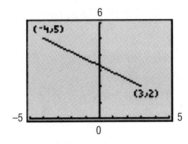

Solution

The length of the line segment is the distance between the points $(-4, 5)$ and $(3, 2)$. Using the distance formula (1), the length d is

$$d = \sqrt{[3 - (-4)]^2 + (2 - 5)^2} = \sqrt{7^2 + (-3)^2}$$
$$= \sqrt{49 + 9} = \sqrt{58} \approx 7.62$$

Now work Problem 25.

The distance between two points $P_1 = (x_1, y_1)$ and $P_2 = (x_2, y_2)$ is never a negative number. Furthermore, the distance between two points is 0 only when the points are identical, that is, when $x_1 = x_2$ and $y_1 = y_2$. Also, because $(x_2 - x_1)^2 = (x_1 - x_2)^2$ and $(y_2 - y_1)^2 = (y_1 - y_2)^2$, it makes no difference whether the distance is computed from P_1 to P_2 or from P_2 to P_1; that is, $d(P_1, P_2) = d(P_2, P_1)$.

The use of rectangular coordinates enables us to translate geometry problems into algebra problems, and vice versa. The next example shows how algebra (the distance formula) can be used to solve geometry problems.

EXAMPLE 4

Using Algebra to Solve Geometry Problems

Consider the three points $A = (-2, 1)$, $B = (2, 3)$ and $C = (3, 1)$.

(a) Plot each point and form the triangle ABC.

(b) Find the length of each side of the triangle.

(c) Verify that the triangle is a right triangle.

(d) Find the area of the triangle.

FIGURE 14

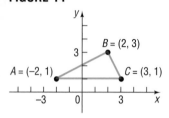

Solution

(a) Points A, B, C, and triangle ABC are plotted in Figure 14.

(b) $d(A, B) = \sqrt{[2 - (-2)]^2 + (3 - 1)^2} = \sqrt{16 + 4} = \sqrt{20} = 2\sqrt{5}$

$d(B, C) = \sqrt{(3 - 2)^2 + (1 - 3)^2} = \sqrt{1 + 4} = \sqrt{5}$

$d(A, C) = \sqrt{[3 - (-2)]^2 + (1 - 1)^2} = \sqrt{5^2 + 0^2} = 5$

(c) To show that the triangle is a right triangle, we need to show that the sum of the squares of the lengths of two of the sides equals the square of the length of the third side. (Why is this sufficient?) Looking at Figure 14, it seems reasonable to conjecture that the right angle is at vertex B. Thus, we shall check to see whether

$$[d(A, B)]^2 + [d(B, C)]^2 = [d(A, C)]^2$$

We find that

$$[d(A, B)]^2 + [d(B, C)]^2 = (2\sqrt{5})^2 + (\sqrt{5})^2$$
$$= 20 + 5 = 25 = [d(A, C)]^2$$

so it follows from the converse of the Pythagorean Theorem that triangle ABC is a right triangle.

(d) Because the right angle is at B, the sides AB and BC form the base and altitude of the triangle. Its area is therefore

$$\text{Area} = \frac{1}{2}(\text{Base})(\text{Altitude}) = \frac{1}{2}(2\sqrt{5})(\sqrt{5}) = 5 \text{ square units} \qquad \blacksquare$$

■   Now work Problem 43.

Midpoint Formula

FIGURE 15

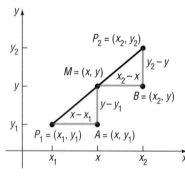

We now derive a formula for the coordinates of the **midpoint of a line segment.** Let $P_1 = (x_1, y_1)$ and $P_2 = (x_2, y_2)$ be the endpoints of a line segment, and let $M = (x, y)$ be the point on the line segment that is the same distance from P_1 as it is from P_2. See Figure 15. The triangles P_1AM and MBP_2 are congruent.* [Do you see why? Angle AP_1M = Angle BMP_2,† Angle P_1MA = Angle MP_2B, and $d(P_1, M) = d(M, P_2)$ is given. Thus, we have Angle–Side–Angle.] Hence, corresponding sides are equal in length. That is,

$$x - x_1 = x_2 - x \quad \text{and} \quad y - y_1 = y_2 - y$$
$$2x = x_1 + x_2 \qquad\qquad 2y = y_1 + y_2$$
$$x = \frac{x_1 + x_2}{2} \qquad\qquad y = \frac{y_1 + y_2}{2}$$

---

*The following statement is a postulate from geometry. Two triangles are congruent if their sides are the same length (SSS), or if two sides and the included angle are the same (SAS), or if two angles and the included side are the same (ASA).

†Another postulate from geometry states that the transversal $\overline{P_1P_2}$ forms equal corresponding angles with the parallel lines $\overline{P_1A}$ and $\overline{MB}$.

Theorem   The midpoint (x, y) of the line segment from $P_1 = (x_1, y_1)$ to $P_2 = (x_2, y_2)$ is

Midpoint Formula

$$(x, y) = \left(\frac{x_1 + x_2}{2}, \frac{y_1 + y_2}{2} \right) \qquad (2)$$

◼

Thus, to find the midpoint of a line segment, we average the x-coordinates and the y-coordinates of the endpoints.

E X A M P L E   5   *Finding the Midpoint of a Line Segment*

Find the midpoint of a line segment from $P_1 = (-5, 3)$ to $P_2 = (3, 1)$. Plot the points P_1 and P_2 and their midpoint. Check your answer.

Solution   We apply the midpoint formula (2) using $x_1 = -5$, $x_2 = 3$, $y_1 = 3$, and $y_2 = 1$. Then the coordinates (x, y) of the midpoint M are

$$x = \frac{x_1 + x_2}{2} = \frac{-5 + 3}{2} = -1 \quad \text{and} \quad y = \frac{y_1 + y_2}{2} = \frac{3 + 1}{2} = 2$$

FIGURE 16

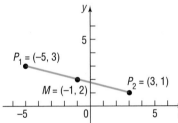

That is, $M = (-1, 2)$. See Figure 16.

Check: Because M is the midpoint, we check the answer by verifying that $d(P_1, M) = d(M, P_2)$:

$$d(P_1, M) = \sqrt{[-1 - (-5)]^2 + (2 - 3)^2} = \sqrt{16 + 1} = \sqrt{17}$$
$$d(M, P_2) = \sqrt{[3 - (-1)]^2 + (1 - 2)^2} = \sqrt{16 + 1} = \sqrt{17}$$

◼

◼   Now work Problem 53.

1.1

Exercise 1.1

In Problems 1 and 2, plot each point in the xy-plane. Tell in which quadrant or on what coordinate axis each point lies.

1. (a) $A = (-3, 2)$ (b) $B = (6, 0)$ (c) $C = (-2, -2)$
   (d) $D = (6, 5)$ (e) $E = (0, -3)$ (f) $F = (6, -3)$

2. (a) $A = (1, 4)$ (b) $B = (-3, -4)$ (c) $C = (-3, 4)$
   (d) $D = (4, 1)$ (e) $E = (0, 1)$ (f) $F = (-3, 0)$

3. Plot the points $(2, 0)$, $(2, -3)$, $(2, 4)$, $(2, 1)$, and $(2, -1)$. Describe the set of all points of the form $(2, y)$, where y is a real number.

4. Plot the points $(0, 3)$, $(1, 3)$, $(-2, 3)$, $(5, 3)$ and $(-4, 3)$. Describe the set of all points of the form $(x, 3)$, where x is a real number.

In Problems 5–8, determine the coordinates of the points shown. Tell in which quadrant each point lies.

5.
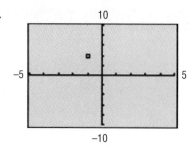

6.
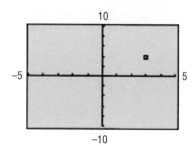

7.
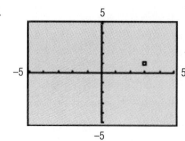

8.
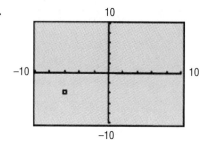

In Problems 9–14, select a RANGE setting so that each of the given points will lie within the viewing rectangle.

9. $(-10, 5), (3, -2), (4, -1)$

10. $(5, 0), (6, 8), (-2, -3)$

11. $(40, 20), (-20, -80), (10, 40)$

12. $(-80, 60), (20, -30), (-20, -40)$

13. $(0, 0), (100, 5), (5, 150)$

14. $(0, -1), (100, 50), (-10, 30)$

In Problems 15–24, determine the RANGE settings used for each viewing rectangle.

15.
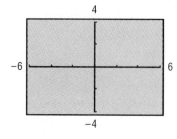

16.
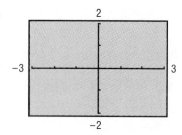

17.
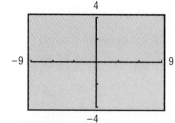

18.
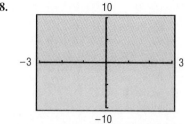

19.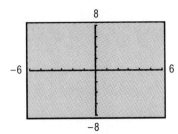

20.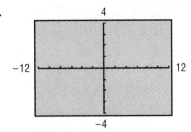

21.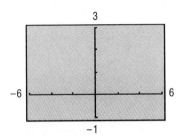

22.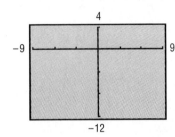

23.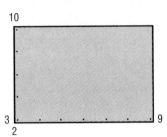

24.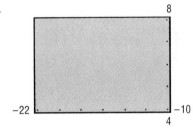

In Problems 25–38, find the distance $d(P_1, P_2)$ between the points P_1 and P_2.

25.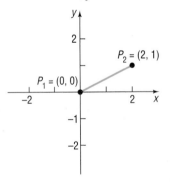

26.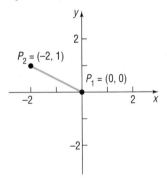

27.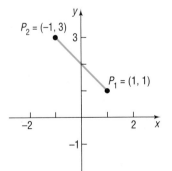

28.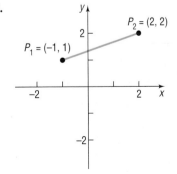

29. $P_1 = (3, -4)$; $P_2 = (5, 4)$

30. $P_1 = (-1, 0)$; $P_2 = (2, 4)$

31. $P_1 = (-3, 2)$; $P_2 = (6, 0)$

32. $P_1 = (2, -3)$; $P_2 = (4, 2)$

33. $P_1 = (4, -3)$; $P_2 = (6, 4)$

34. $P_1 = (-4, -3)$; $P_2 = (6, 2)$

35. $P_1 = (-0.2, 0.3)$; $P_2 = (2.3, 1.1)$

36. $P_1 = (1.2, 2.3)$; $P_2 = (-0.3, 1.1)$

37. $P_1 = (a, b)$; $P_2 = (0, 0)$

38. $P_1 = (a, a)$; $P_2 = (0, 0)$

In Problems 39–42, find the length of the line segment.

39.

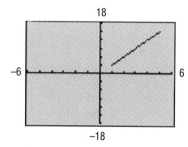

40.

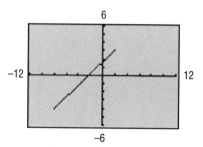

41.

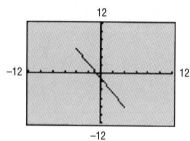

42.

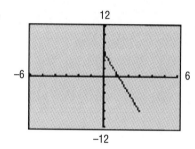

In Problems 43–48, plot each point and form the triangle ABC. Verify that the triangle is a right triangle. Find its area.

43. $A = (-2, 5)$; $B = (1, 3)$; $C = (-1, 0)$

44. $A = (-2, 5)$; $B = (12, 3)$; $C = (10, -11)$

45. $A = (-5, 3)$; $B = (6, 0)$; $C = (5, 5)$

46. $A = (-6, 3)$; $B = (3, -5)$; $C = (-1, 5)$

47. $A = (4, -3)$; $B = (0, -3)$; $C = (4, 2)$

48. $A = (4, -3)$; $B = (4, 1)$; $C = (2, 1)$

49.   Find all points having an x-coordinate of 2 whose distance from the point $(-2, -1)$ is 5.

50.   Find all points having a y-coordinate of -3 whose distance from the point $(1, 2)$ is 13.

51.   Find all points on the x-axis that are 5 units from the point $(4, -3)$.

52.   Find all points on the y-axis that are 5 units from the point $(4, 4)$.

In Problems 53–62, find the midpoint of the line segment joining the points P_1 and P_2.

53. $P_1 = (5, -4)$; $P_2 = (3, 2)$

54. $P_1 = (-1, 0)$; $P_2 = (2, 4)$

55. $P_1 = (-3, 2)$; $P_2 = (6, 0)$

56. $P_1 = (2, -3)$; $P_2 = (4, 2)$

57. $P_1 = (4, -3)$; $P_2 = (6, 1)$

58. $P_1 = (-4, -3)$; $P_2 = (2, 2)$

59. $P_1 = (-0.2, 0.3)$; $P_2 = (2.3, 1.1)$

60. $P_1 = (1.2, 2.3)$; $P_2 = (-0.3, 1.1)$

61. $P_1 = (a, b)$; $P_2 = (0, 0)$

62. $P_1 = (a, a)$; $P_2 = (0, 0)$

63. The **medians** of a triangle are the line segments from each vertex to the midpoint of the opposite side (see the figure in the margin). Find the lengths of the medians of the triangle with vertices at $A = (0, 0)$, $B = (0, 6)$, and $C = (8, 0)$.

64. An **equilateral triangle** is one in which all three sides are of equal length. If two vertices of an equilateral triangle are $(0, 4)$ and $(0, 0)$, find the third vertex. How many of these triangles are possible?

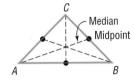

*In Problems 65–68, find the length of each side of the triangle determined by the three points P_1, P_2, and P_3. State whether the triangle is an isosceles triangle, a right triangle, neither of these, or both. (An **isosceles triangle** is one in which at least two of the sides are of equal length.)*

65. $P_1 = (2, 1)$; $P_2 = (-4, 1)$; $P_3 = (-4, -3)$

66. $P_1 = (-1, 4)$; $P_2 = (6, 2)$; $P_3 = (4, -5)$

67. $P_1 = (-2, -1)$; $P_2 = (0, 7)$; $P_3 = (3, 2)$

68. $P_1 = (7, 2)$; $P_2 = (-4, 0)$; $P_3 = (4, 6)$

69. If r is a real number, prove that the point $P = (x, y)$ that divides the line segment from $P_1 = (x_1, y_1)$ to $P_2 = (x_2, y_2)$ in the ratio r, that is,

$$\frac{d(P_1, P)}{d(P_1, P_2)} = r$$

has coordinates

$$x = x_1 + r(x_2 - x_1) \quad \text{and} \quad y = y_1 + r(y_2 - y_1)$$

[*Hint:* Use similar triangles.*]

Problems 70–74 use the result of Problem 69.

70. Verify that the midpoint of a line segment divides the line segment from $P_1 = (x_1, y_1)$ to $P_2 = (x_2, y_2)$ in the ratio $r = \frac{1}{2}$.

71. What point P divides the line segment from P_1 to P_2 in the ratio $r = 1$?

72. What point P divides the line segment from P_1 to P_2 in the ratio $r = 0$?

73. Find the point P on the line joining $P_1 = (2, 4)$ and $P_2 = (5, 6)$ that is twice as far from P_1 as P_2 is from P_1 and that lies on the same side of P_1 as P_2 does.

74. Find the point P on the line joining $P_1 = (0, 4)$ and $P_2 = (-4, 1)$ that is three times as far from P_1 as P_2 is from P_1.

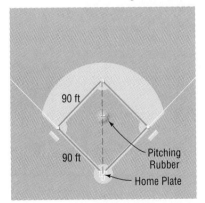

75. *Baseball* A major league baseball "diamond" is actually a square, 90 feet on a side (see the figure). What is the distance directly from home plate to second base (the diagonal of the square)?

76. *Little League Baseball* The layout of a Little League playing field is a square, 60 feet on a side.† How far is it directly from home plate to second base (the diagonal of the square)?

77. *Baseball* Refer to Problem 75. Overlay a rectangular coordinate system on a major league baseball diamond so that the origin is at home plate, the positive x-axis lies in the direction from home plate to first base, and the positive y-axis lies in the direction from home plate to third base.
(a) What are the coordinates of first base, second base, and third base? Use feet as the unit of measurement.
(b) If the right fielder is located at $(310, 15)$, how far is it from there to second base?
(c) If the center fielder is located at $(300, 300)$, how far is it from there to third base?

78. *Little League Baseball* Refer to Problem 76. Overlay a rectangular coordinate system on a Little League baseball diamond so that the origin is at home plate, the positive x-axis lies in the direction from home plate to first base, and the positive y-axis lies in the direction from home plate to third base.
(a) What are the coordinates of first base, second base, and third base? Use feet as the unit of measurement.
(b) If the right fielder is located at $(180, 20)$, how far is it from there to second base?
(c) If the center fielder is located at $(220, 220)$, how far is it from there to third base?

*Two triangles are **similar** if they have the same angles. In similar triangles, corresponding sides are in proportion.
†Source: Little League Baseball, Official Regulations and Playing Rules, 1991.

79. *Geometry*   Find the midpoint of each diagonal of a square with side of length *s*. Draw the conclusion that the diagonals of a square intersect at their midpoints. [*Hint:* Use (0, 0), (0, *s*), (*s*, 0), and (*s*, *s*) as the vertices of the square.]

80. *Geometry*   Verify that the points (0, 0), (*a*, 0), and (*a*/2, $\sqrt{3}a/2$) are the vertices of an equilateral triangle. Then show that the midpoints of the three sides are the vertices of a second equilateral triangle. (See Problem 64.)

81. An automobile and a truck leave an intersection at the same time. The automobile heads east at an average speed of 30 miles per hour, while the truck heads south at an average speed of 40 miles per hour. Find an expression for their distance apart *d* (in miles) at the end of *t* hours.

82. A hot air balloon, headed due east at an average speed of 15 miles per hour and at a constant altitude of 100 feet, passes over an intersection (see the figure). Find an expression for its distance *d* (measured in feet) from the intersection *t* seconds later.

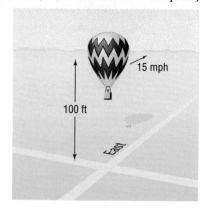

15 mph

100 ft

East

1.2

Graphs of Equations

Illustrations play an important role in helping us to visualize the relationships that exist between two variable quantities. For example, Figure 17 shows the variation in the price of oil from January to October 1991. Such illustrations are usually referred to as *graphs*. The **graph of an equation** in two variables *x* and *y* consists of the set of points in the *xy*-plane whose coordinates (*x*, *y*) satisfy the equation. Let's look at some examples.

FIGURE 17
Oil futures prices
Source: Dow Jones News Retrieval

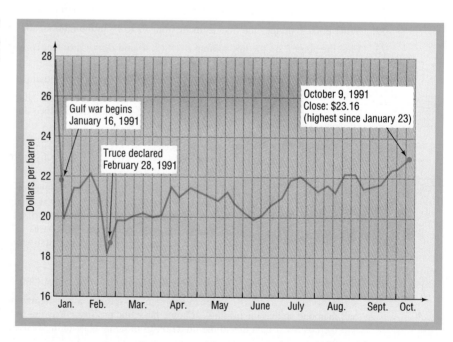

Dollars per barrel

Gulf war begins
January 16, 1991

Truce declared
February 28, 1991

October 9, 1991
Close: $23.16
(highest since January 23)

Jan.   Feb.   Mar.   Apr.   May   June   July   Aug.   Sept.   Oct.

E X A M P L E   1   *Graphing an Equation by Hand*

Graph the equation: *y* = 2*x* + 5

FIGURE 18
$y = 2x + 5$

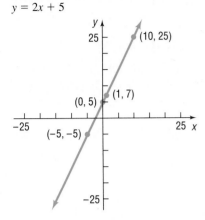

Solution   We want to find all points (x, y) that satisfy the equation. To locate some of these points (and thus get an idea of the pattern of the graph), we assign some numbers to x and find corresponding values for y:

IF	THEN	POINT ON GRAPH
$x = 0$	$y = 2(0) + 5 = 5$	$(0, 5)$
$x = 1$	$y = 2(1) + 5 = 7$	$(1, 7)$
$x = -5$	$y = 2(-5) + 5 = -5$	$(-5, -5)$
$x = 10$	$y = 2(10) + 5 = 25$	$(10, 25)$

By plotting these points and then connecting them, we obtain the graph of the equation (a straight line), as shown in Figure 18. ■

E X A M P L E   2   *Graphing an Equation by Hand*

Graph the equation: $y = x^2$

Solution   Table 1 provides several points on the graph. In Figure 19 we plot these points and connect them with a smooth curve to obtain the graph (a *parabola*).

TABLE 1

x	$y = x^2$	(x, y)
-4	16	$(-4, 16)$
-3	9	$(-3, 9)$
-2	4	$(-2, 4)$
-1	1	$(-1, 1)$
0	0	$(0, 0)$
1	1	$(1, 1)$
2	4	$(2, 4)$
3	9	$(3, 9)$
4	16	$(4, 16)$

FIGURE 19
$y = x^2$

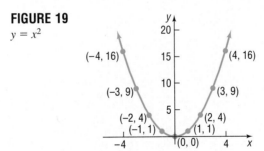

The graphs of the equations shown in Figures 18 and 19 do not show all points. For example, in Figure 18, the point $(20, 45)$ is a part of the graph of $y = 2x + 5$, but it is not shown. Since the graph of $y = 2x + 5$ could be extended out as far as we please, we use arrows to indicate that the pattern shown continues. Thus, it is important when illustrating a graph to present enough of the graph so that any viewer of the illustration will "see" the rest of it as an obvious continuation of what is actually there. This is referred to as a **complete graph.**

So, one way to obtain a complete graph of an equation is to plot a sufficient number of points on the graph until a pattern becomes evident. Then these points are connected with a smooth curve following the suggested pattern. But how many points are sufficient? Sometimes knowledge about the equation tells us. For example, we will learn in the next section that, if an equation is of the form $y = mx + b$, then its graph is a straight line. In this case, two points would suffice to obtain the graph.

One of the purposes of this book is to investigate the properties of equations in order to decide whether a graph is complete. Sometimes we shall graph equations by hand by plotting a sufficient number of points on the graph until a pattern becomes evident; then we connect these points with a smooth curve following the suggested pattern. Shortly, we shall investigate various techniques that will enable us to graph an equation by hand without plotting so many points. Other times, we shall graph equations using a graphing utility.

Using a Graphing Utility to Graph Equations

From Examples 1 and 2, we see that a graph can be obtained by plotting points in a rectangle coordinate system and connecting them. Graphing utilities perform these same steps when graphing an equation. For example, the T1-82 determines 95 evenly spaced input values*, uses the equation to determine the output values, plots these points on the screen and finally, (if in the connected mode), draws a line between consecutive points.

Most graphing utilities require the following steps in order to obtain the graph of an equation:

STEP 1:   Solve the equation for y in terms of x.
STEP 2:   Select the viewing rectangle.
STEP 3:   Get into the graphing mode of your graphing utility. The screen will usually display $y =$, prompting you to enter the expression involving x that you found in STEP 1. (Consult your manual for the correct way to enter the expression; for example, $y = x^2$ might be entered as x^2 or as $x*x$ or as $x\,x^Y\,2$).
STEP 4:   Execute.

E X A M P L E   3   *Graphing an Equation on a Graphing Utility*

Graph the equation: $6x^2 + 3y = 24$

Solution   STEP 1:   We solve for y in terms of x.

$$6x^2 + 3y = 24$$
$$3y = -6x^2 + 24$$
$$y = -2x^2 + 8$$

STEP 2:   Select a viewing rectangle. We will use the one given next.

$$\text{Xmin} = -5$$
$$\text{Xmax} = 5$$
$$\text{Xscl} = 1$$
$$\text{Ymin} = -10$$
$$\text{Ymax} = 20$$
$$\text{Yscl} = 2$$

STEP 3:   From the graphing mode, enter the expression $-2x^2 + 8$ after the prompt $y =$.

STEP 4:   Execute

The screen should look like Figure 20.

*These input values depend on the values of Xmin and Xmax. For example, if Xmin $= -10$ and Xmax $= 10$, then the first input value will be -10 and the next input value will be $-10 + (10 - (-10))/94 = -9.7872$, and so on.

FIGURE 20

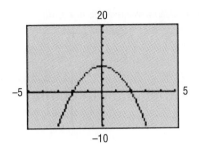

Now that we have seen the graph of $y = -2x^2 + 8$, we might want to graph it again using the settings given below:

$$\text{Xmin} = -5$$
$$\text{Xmax} = 5$$
$$\text{Xscl} = 1$$
$$\text{Ymin} = -10$$
$$\text{Ymax} = 10$$
$$\text{Yscl} = 1$$

Figure 21 shows the graph obtained. Notice the improvement over Figure 20.

FIGURE 21

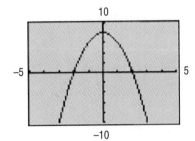

On some graphing utilities, you can scroll the graph up or down and left or right to see more of it, instead of setting a new viewing rectangle. With scrolling, the scale stays the same as it was originally; only the min/max values change.

■   Now work Problems 11(a) and 11(b).

E X A M P L E   4

Graphing an Equation

Graph the equation: $y = x^3$

Solution   Using a graphing utility, we obtain Figure 22.

FIGURE 22

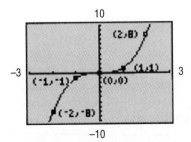

To graph by hand, we use the equation to obtain several points on the graph (Table 2) and then draw the graph on paper. See Figure 23.

TABLE 2

x	$y = x^3$	(x, y)
-3	-27	$(-3, -27)$
-2	-8	$(-2, -8)$
-1	-1	$(-1, -1)$
0	0	$(0, 0)$
1	1	$(1, 1)$
2	8	$(2, 8)$
3	27	$(3, 27)$

FIGURE 23

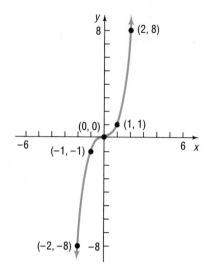

Table 2 can also be obtained using a graphing utility. (Check your manual to see how.) On a graphing utility, Table 2 takes the form shown in Table 3.

TABLE 3

X	Y1	
-3	-27	
-2	-8	
-1	-1	
0	0	
1	1	
2	8	
3	27	

Y1⊟X^3

E X A M P L E   5   *Graphing an Equation*

Graph the equation: $x = y^2$

Solution   To graph an equation using a graphing utility, we must write the equation in the form y = expression involving x. So to graph $x = y^2$, we must solve for y:

$$x = y^2$$
$$y^2 = x$$
$$y = \pm\sqrt{x}$$

Thus, to graph $x = y^2$, we need to graph both $y = \sqrt{x}$ and $y = -\sqrt{x}$ on the same screen. Figure 24 shows the result. Table 4 shows various values of y for a given value of x when $y_1 = \sqrt{x}$ and $y_2 = -\sqrt{x}$. Notice when $x < 0$, we get an error. Can you explain why?

TABLE 4

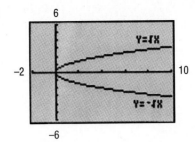

FIGURE 24

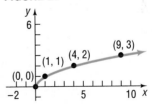

To graph by hand, we use the equation to obtain several points on the graph (Table 5) and then draw the graph on paper. See Figure 25.

TABLE 5

y	$x = y^2$	(x, y)
-3	9	$(9, -3)$
-2	4	$(4, -2)$
-1	1	$(1, -1)$
0	0	$(0, 0)$
1	1	$(1, 1)$
2	4	$(4, 2)$
3	9	$(9, 3)$
4	16	$(16, 4)$

FIGURE 25

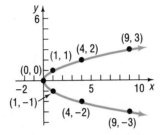

Look again at either Figure 24 or Figure 25. If we restrict y so that $y \geq 0$, the equation $x = y^2$, $y \geq 0$, may be written equivalently as $y = \sqrt{x}$. The portion of the graph of $x = y^2$ in quadrant I is therefore the graph of $y = \sqrt{x}$. See Figure 26 and Figure 27.

FIGURE 26

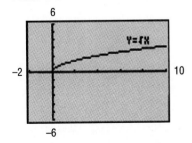

FIGURE 27

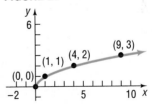

■ Now work Problem 47.

Obtaining a Complete Graph Using a Graphing Utility

Often the first choice for the RANGE setting does not show a complete graph. To obtain a complete graph requires adjusting the RANGE. One way to do this is to use ZOOM-OUT.

E X A M P L E   6   *Using ZOOM-OUT to Obtain a Complete Graph*

Graph the equation $y = x^3 - 11x^2 - 190x + 200$ using the following settings for the viewing rectangle

$$\begin{aligned}
&\text{Xmin:} \ -12 \\
&\text{Xmax:} \ 12 \\
&\text{Xscl:} \ \ \ 4 \\
&\text{Ymin:} \ -200 \\
&\text{Ymax:} \ 200 \\
&\text{Yscl:} \ \ \ 100
\end{aligned}$$

Solution   Figure 28 shows the graph.

FIGURE 28

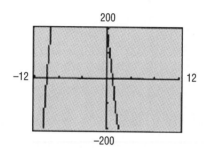

Notice how ragged the graph looks. The *y*-scale is clearly not adequate. The graph is not complete. We can use the ZOOM-OUT function to help obtain a complete graph.

After the first ZOOM-OUT, we obtain Figure 29.

FIGURE 29

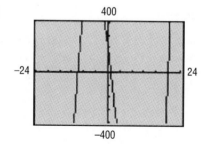

Notice that the min/max settings have been doubled, while the scales remained the same.* However, we still cannot see the top or bottom portion of the graph. ZOOM-OUT again to obtain the graph in Figure 30.

---

*On some graphing utilities, the default factor for the ZOOM-OUT function is 4, meaning that the original min/max setting will be multiplied by 4. Also, you can set the factor yourself, if you want; furthermore, the factor need not be the same for *x* and *y*.

Activate the ZOOM-IN function. The result is shown in Figure 40.

FIGURE 40

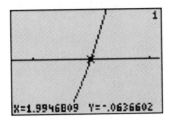

Move the cursor up. The result is shown in Figure 41.

FIGURE 41

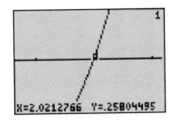

Notice the improvement. Now we know that the x-intercept lies between 1.9946809 and 2.0212766. ZOOM-IN again. The result is shown in Figure 42(a). Figure 42(b) is the result of moving the cursor down.

FIGURE 42

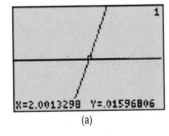
(a)

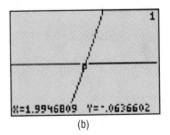
(b)

Now we know that the x-intercept lies between 1.9946809 and 2.0013298.

■

Most graphing utilities have a BOX function that allows you to box in a specific part of the graph of an equation.

EXAMPLE 11

Using the BOX Function

Graph the equation $y = x^3 - 8$ and use the BOX function to improve on the approximation found in Example 9 for the x-intercept.

Solution   Using the viewing rectangle of Example 9, graph the equation. Activate the BOX function. (With some graphing utilities, this requires positioning the cursor at one corner of the box and then tracing out the sides of the box to the diagonal corner.) See Figure 43.

FIGURE 43

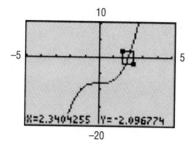

Once executed, the box becomes the viewing rectangle. The result is shown in Figure 44.

FIGURE 44

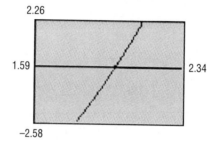

Now you can TRACE to get the approximations shown in Figure 45.

FIGURE 45

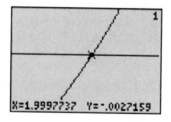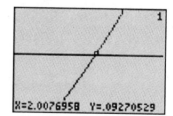

Now we know the x-intercept lies between 1.9997737 and 2.0076958.   ■

Symmetry

We have just seen the role intercepts play in obtaining key points on the graph of an equation. Another helpful tool for graphing equations involves *symmetry*, particularly symmetry with respect to the x-axis, the y-axis, and the origin.

Symmetry with Respect
to the x-Axis

A graph is said to be **symmetric with respect to the x-axis** if, for every point (x, y) on the graph, the point (x, −y) is also on the graph.

Figure 46 illustrates the definition. Notice that when a graph is symmetric with respect to the *x*-axis the part of the graph above the *x*-axis is a reflection or mirror image of the part below it, and vice versa.

FIGURE 46

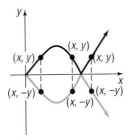

Symmetric with respect to
the *x*-axis

E X A M P L E   1 2   *Points Symmetric with Respect to the x-Axis*

If a graph is symmetric with respect to the *x*-axis and the point (3, 2) is on the graph, then the point (3, −2) is also on the graph. ∎

Symmetry with Respect
to the *y*-Axis

A graph is said to be **symmetric with respect to the y-axis** if, for every point (*x*, *y*) on the graph, the point (−*x*, *y*) is also on the graph.

Figure 47 illustrates the definition. Notice that when a graph is symmetric with respect to the *y*-axis the part of the graph to the right of the *y*-axis is a reflection of the part to the left of it, and vice versa.

FIGURE 47

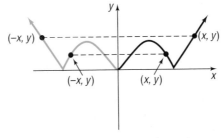

Symmetric with respect to the *y*-axis

E X A M P L E   1 3   *Points Symmetric with Respect to the y-Axis*

If a graph is symmetric with respect to the *y*-axis and the point (5, 8) is on the graph, then the point (−5, 8) is also on the graph. ∎

Symmetry with Respect
to the Origin

A graph is said to be **symmetric with respect to the origin** if, for every point (*x*, *y*) on the graph, the point (−*x*, −*y*) is also on the graph.

Figure 48 illustrates the definition. Notice that symmetry with respect to the origin may be viewed in two ways:

1. As a reflection about the y-axis, followed by a reflection about the x-axis.
2. As a projection along a line through the origin so that the distances from the origin are equal.

FIGURE 48

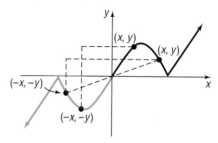

Symmetric with respect to
the origin

E X A M P L E   1 4   *Points Symmetric with Respect to the Origin*

If a graph is symmetric with respect to the origin and the point $(4, 2)$ is on the graph, then the point $(-4, -2)$ is also on the graph.   ■

■   Now work Problems 1 and 31(b).

When the graph of an equation is symmetric with respect to a coordinate axis or the origin, the number of points that you need to plot in order to see the pattern is reduced. For example, if the graph of an equation is symmetric with respect to the y-axis, then, once points to the right of the y-axis are plotted, an equal number of points on the graph can be obtained by reflecting them about the y-axis. Thus, before we graph an equation, we first want to determine whether it has any symmetry. The following tests are used for that purpose.

Tests for Symmetry   To test the graph of an equation for symmetry with respect to the:

x-Axis   Replace y by $-y$ in the equation. If an equivalent equation results, the graph of the equation is symmetric with respect to the x-axis.

y-Axis   Replace x by $-x$ in the equation. If an equivalent equation results, the graph of the equation is symmetric with respect to the y-axis.

Origin   Replace x by $-x$ and y by $-y$ in the equation. If an equivalent equation results, the graph of the equation is symmetric with respect to the origin.

Let's look at an equation we have already graphed to see how these tests are used.

E X A M P L E   1 5   *Testing Equations for Symmetry*

(a) To test the graph of the equation $x = y^2$ for symmetry with respect to the x-axis, we replace y by $-y$ in the equation, as follows:

$$x = y^2 \qquad \text{Original equation}$$
$$x = (-y)^2 \qquad \text{Replace } y \text{ by } -y$$
$$x = y^2 \qquad \text{Simplify.}$$

When we replace y by $-y$, the result is the same equation. Thus, the graph is symmetric with respect to the x-axis.

(b) To test the graph of the equation $x = y^2$ for symmetry with respect to the y-axis, we replace x by $-x$ in the equation:

$$x = y^2 \quad \text{Original equation}$$
$$-x = y^2 \quad \text{Replace } x \text{ by } -x.$$

Because we arrive at the equation $-x = y^2$, which is not equivalent to the original equation, we conclude that the graph is not symmetric with respect to the y-axis.

(c) To test for symmetry with respect to the origin, we replace x by $-x$ and y by $-y$:

$$x = y^2 \quad \text{Original equation}$$
$$-x = (-y)^2 \quad \text{Replace } x \text{ by } -x \text{ and } y \text{ by } -y.$$
$$-x = y^2 \quad \text{Simplify.}$$

The resulting equation, $-x = y^2$, is not equivalent to the original equation. We conclude that the graph is not symmetric with respect to the origin. ■

Figure 49(a) illustrates the graph of $x = y^2$. In forming a table of points on the graph of $x = y^2$, we can restrict ourselves to points whose y-coordinates are positive. Once these are plotted and connected, a reflection about the x-axis (because of the symmetry) provides the rest of the graph.

Figures 49(b) and 49(c) illustrate two other equations we graphed earlier. Notice how the existence of symmetry reduces the number of points we need to plot.

FIGURE 49

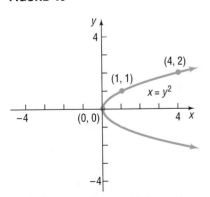

(a) Symmetry with respect to the x-axis

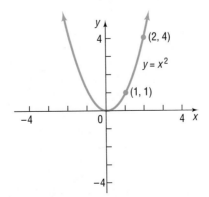

(b) Symmetry with respect to the y-axis

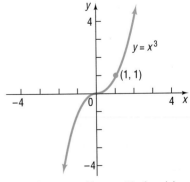

(c) Symmetry with respect to the origin

■ Now work Problem 89.

E X A M P L E   1 6   *Graphing the Equation y = 1/x*

Consider the equation $y = 1/x$

(a) Graph this equation using a graphing utility. Set the viewing rectangle as

RANGE

$\text{XMin} = -3$

$\text{XMax} = 3$

$\text{XScl} = 1$

$\text{YMin} = -4$

$\text{YMax} = 4$

$\text{YScl} = 1$

(b) Use algebra to find any intercepts and test for symmetry.

(c) Draw the graph by hand.

Solution   (a) Figure 50 illustrates the graph.

FIGURE 50

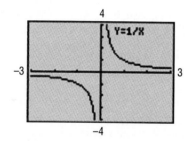

We infer from the graph that there are no intercepts; we may also infer that symmetry with respect to the origin is a possibility. The TRACE function on a graphing utility can provide further evidence of symmetry with respect to the origin. Using TRACE we observe that for any (x, y) ordered pair, the ordered pair $(-x, -y)$ is also a point on the graph. For example, the points $(0.95744681, 1.0444444)$ and $(-0.95744681, -1.0444444)$ both lie on the graph.

(b) We check for intercepts first. If we let $x = 0$, we obtain a 0 denominator, which is not allowed. Hence, there is no y-intercept. If we let $y = 0$, we get the equation $1/x = 0$, which has no solution. Hence, there is no x-intercept. Thus, the graph of $y = 1/x$ does not cross the coordinate axes.

Next we check for symmetry:

x-Axis: Replacing y by $-y$ yields $-y = 1/x$, which is not equivalent to $y = 1/x$.

y-Axis: Replacing x by $-x$ yields $y = -1/x$, which is not equivalent to $y = 1/x$.

Origin: Replacing x by $-x$ and y by $-y$ yields $-y = -1/x$, which is equivalent to $y = 1/x$.

The graph is symmetric with respect to the origin. This confirms the inferences drawn in part (a) of the solution.

(c) We can use the equation to obtain some points on the graph. Because of symmetry, we need only find points (x, y) for which x is positive. Also from the equation $y = 1/x$ we infer that if x is a large and positive number, then $y = 1/x$ is a positive number close to 0. We also infer that if x is a positive number close to 0, then $y = 1/x$ is a large and positive number. Armed with this information, we can graph the equation. Figure 51 illustrates some of these points and the graph of $y = 1/x$. Observe how the absence of intercepts and the existence of symmetry with respect to the origin were utilized.

TABLE 7

x	$y = 1/x$	(x, y)
$\frac{1}{10}$	10	$(\frac{1}{10}, 10)$
$\frac{1}{3}$	3	$(\frac{1}{3}, 3)$
$\frac{1}{2}$	2	$(\frac{1}{2}, 2)$
1	1	$(1, 1)$
2	$\frac{1}{2}$	$(2, \frac{1}{2})$
3	$\frac{1}{3}$	$(3, \frac{1}{3})$
10	$\frac{1}{10}$	$(10, \frac{1}{10})$

FIGURE 51

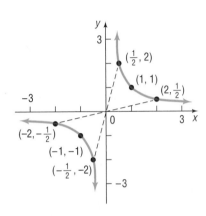

1.2

Exercise 1.2

In Problems 1–10, plot each point. Then plot the point that is symmetric to it with respect to:
(a) The x-axis   (b) The y-axis   (c) The origin

1. $(3, 4)$ **2.** $(5, 3)$ **3.** $(-2, 1)$ **4.** $(4, -2)$ **5.** $(1, 1)$

6. $(-1, -1)$ **7.** $(-3, -4)$ **8.** $(4, 0)$ **9.** $(0, -3)$ **10.** $(-3, 0)$

In Problems 11–30, graph each equation using the following RANGE settings:

(a) Xmin $= -5$	(b) Xmin $= -10$	(c) Xmin $= -10$	(d) Xmin $= -5$
Xmax $= 5$	Xmax $= 10$	Xmax $= 10$	Xmax $= 5$
Xscl $= 1$	Xscl $= 1$	Xscl $= 2$	Xscl $= 1$
Ymin $= -4$	Ymin $= -8$	Ymin $= -8$	Ymin $= -20$
Ymax $= 4$	Ymax $= 8$	Ymax $= 8$	Ymax $= 20$
Yscl $= 1$	Yscl $= 1$	Yscl $= 2$	Yscl $= 5$

11. $y = x + 2$ **12.** $y = x - 2$ **13.** $y = -x + 2$ **14.** $y = -x - 2$

15. $y = 2x + 2$ **16.** $y = 2x - 2$ **17.** $y = -2x + 2$ **18.** $y = -2x - 2$

19. $y = x^2 + 2$ **20.** $y = x^2 - 2$ **21.** $y = -x^2 + 2$ **22.** $y = -x^2 - 2$

23. $y = 2x^2 + 2$ **24.** $y = 2x^2 - 2$ **25.** $y = -2x^2 + 2$ **26.** $y = -2x^2 - 2$

27. $3x + 2y = 6$ **28.** $3x - 2y = 6$ **29.** $-3x + 2y = 6$ **30.** $-3x - 2y = 6$

In Problems 31–46, the graph of an equation is given.
(a) List the intercepts of the graph.
(b) Based on the graph, tell whether the graph is symmetric with respect to the x-axis, y-axis, and/or origin.

31.

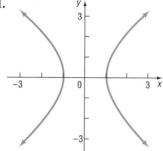

32.

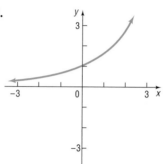

33.

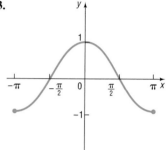

34.

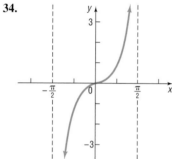

35.

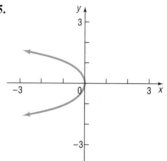

36.

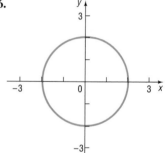

37.

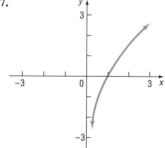

38.

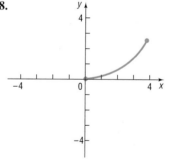

39.

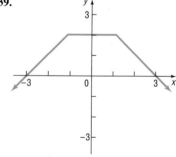

40.

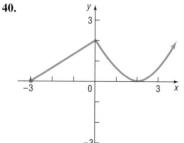

41.

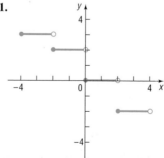

42.

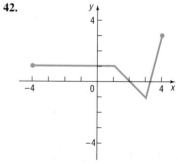

43.

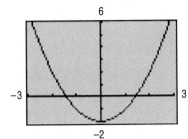

44.

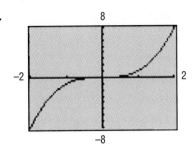

45.

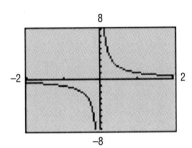

46.

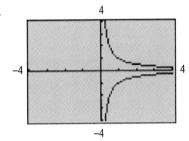

In Problems 47–52, tell whether the given points are on the graph of the equation.

47. Equation: $y = x^4 + \sqrt{x}$
Points: $(0, 0)$; $(1, 1)$; $(-1, 0)$

48. Equation: $y = x^3 - 2\sqrt{x}$
Points: $(0, 0)$; $(1, 1)$; $(1, -1)$

49. Equation: $y^2 = x^2 + 4$
Points: $(0, 2)$; $(2, 0)$; $(-2, 0)$

50. Equation: $y^3 = x + 1$
Points: $(1, 2)$; $(0, 1)$; $(-1, 0)$

51. Equation: $x^2 + y^2 = 4$
Points: $(0, 2)$; $(-2, 2)$; $(\sqrt{2}, \sqrt{2})$

52. Equation: $x^2 + 4y^2 = 4$
Points: $(0, 1)$; $(2, 0)$; $(2, \frac{1}{2})$

53. If $(a, 2)$ is a point on the graph of $y = 5x + 4$, what is a?

54. If $(2, b)$ is a point on the graph of $y = x^2 + 3x$, what is b?

55. If (a, b) is a point on the graph of $2x + 3y = 6$, write an equation that relates a to b.

56. If $(2, 0)$ and $(0, 5)$ are points on the graph of $y = mx + b$, what are m and b?

In Problems 57–76, use a graphing utility to graph each equation. State the viewing rectangle used and draw the graph by hand.

57. $3x + 5y = 75$

58. $3x - 5y = 75$

59. $3x + 5y = -75$

60. $3x - 5y = -75$

61. $y = (x - 10)^2$

62. $y = (x + 10)^2$

63. $y = x^2 - 100$

64. $y = x^2 + 100$

65. $x^2 + y^2 = 100$

66. $x^2 + y^2 = 164$

67. $3x^2 + y^2 = 900$

68. $4x^2 + y^2 = 1600$

69. $x^2 + 3y^2 = 900$

70. $x^2 + 4y^2 = 1600$

71. $y = x^2 - 10x$

72. $y = x^2 + 10x$

73. $y = x^2 - 18x$

74. $y = x^2 + 18x$

75. $y = x^2 - 36x$

76. $y = x^2 + 36x$

In Problems 77–80, use the graph on the right.

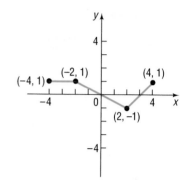

77. Extend the graph to make it symmetric with respect to the *x*-axis.

78. Extend the graph to make it symmetric with respect to the *y*-axis.

79. Extend the graph to make it symmetric with respect to the origin.

80. Extend the graph to make it symmetric with respect to the *x*-axis, *y*-axis, and origin.

In Problems 81–84, use the graph on the right.

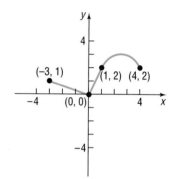

81. Extend the graph to make it symmetric with respect to the *x*-axis.

82. Extend the graph to make it symmetric with respect to the *y*-axis.

83. Extend the graph to make it symmetric with respect to the origin.

84. Extend the graph to make it symmetric with respect to the *x*-axis, *y*-axis, and origin.

In Problems 85–98, list the intercepts and test for symmetry. Graph each equation using a graphing utility.

85. $x^2 = y$

86. $y^2 = x$

87. $y = 3x$

88. $y = -5x$

89. $x^2 + y - 9 = 0$

90. $y^2 - x - 4 = 0$

91. $9x^2 + 4y^2 = 36$

92. $4x^2 + y^2 = 4$

93. $y = x^3 - 27$

94. $y = x^4 - 1$

95. $y = x^2 - 3x - 4$

96. $y = x^2 + 4$

97. $y = \dfrac{x}{x^2 + 9}$

98. $y = \dfrac{x^2 - 4}{x}$

In Problem 99, you may use a graphing utility, but it is not required.

99. (a) Graph $y = \sqrt{x^2}$, $y = x$, $y = |x|$, and $y = (\sqrt{x})^2$, noting which graphs are the same.
  (b) Explain why the graphs of $y = \sqrt{x^2}$ and $y = |x|$ are the same.
  (c) Explain why the graphs of $y = x$ and $y = (\sqrt{x})^2$ are not the same.
  (d) Explain why the graphs of $y = \sqrt{x^2}$ and $y = x$ are not the same.

100. Make up an equation with the intercepts $(2, 0)$, $(4, 0)$, and $(0, 1)$. Compare your equation with a friend's equation. Comment on any similarities.

101. An equation is being tested for symmetry with respect to the *x*-axis, the *y*-axis, and the origin. Explain why, if two of these symmetries are present, then the remaining one must also be present.

102. Draw a graph that contains the points $(-2, -1)$, $(0, 1)$, $(1, 3)$, and $(3, 5)$. Compare your graph with those of other students. Are most of the graphs almost straight lines? How many are "curved"? Discuss the various ways these points might be connected.

1.3

The Straight Line; Circles

In this section we study a certain type of equation that contains two variables, called a *linear equation,* and its graph, a *straight line.*

Slope of a Line

FIGURE 52

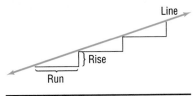

Consider the staircase illustrated in Figure 52. Each step contains exactly the same horizontal **run** and the same vertical **rise**. The ratio of the rise to the run, called the *slope,* is a numerical measure of the steepness of the staircase. For example, if the run is increased and the rise remains the same, the staircase becomes less steep. If the run is kept the same, but the rise is increased, the staircase becomes more steep. This important characteristic of a line is best defined using rectangular coordinates.

Slope of a Line

Let $P = (x_1, y_1)$ and $Q = (x_2, y_2)$ be two distinct points with $x_1 \neq x_2$. The **slope m** of the nonvertical line L containing P and Q is defined by the formula

$$m = \frac{y_2 - y_1}{x_2 - x_1} \qquad x_1 \neq x_2 \tag{1}$$

If $x_1 = x_2$, L is a **vertical line** and the slope m of L is **undefined** (since this results in division by 0).

Figure 53(a) provides an illustration of the slope of a nonvertical line; Figure 53(b) illustrates a vertical line.

FIGURE 53

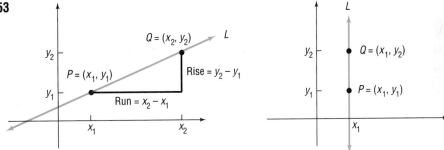

(a) Slope of L is $m = \dfrac{y_2 - y_1}{x_2 - x_1}$

(b) Slope is undefined; L is vertical

As Figure 53(a) illustrates, the slope m of a nonvertical line may be viewed as

$$m = \frac{y_2 - y_1}{x_2 - x_1} = \frac{\text{Rise}}{\text{Run}}$$

We can also express the slope m of a nonvertical line as

$$m = \frac{y_2 - y_1}{x_2 - x_1} = \frac{\text{Change in } y}{\text{Change in } x} = \frac{\Delta y}{\Delta x}$$

That is, the slope m of a nonvertical line L is the ratio of the change in the y-coordinates from P to Q, $\Delta y = y_2 - y_1$, to the change in the x-coordinates from P to Q, $\Delta x = x_2 - x_1$.

Two comments about computing the slope of a nonvertical line may prove helpful:

1. Any two distinct points on the line can be used to compute the slope of the line. (See Figure 54 for justification.)

FIGURE 54

Triangles ABC and PQR are similar (equal angles). Hence, ratios of corresponding sides are proportional. Thus:

Slope using P and $Q = \dfrac{y_2 - y_1}{x_2 - x_1}$

$= $ Slope using A and $B = \dfrac{d(B, C)}{d(A, C)}$

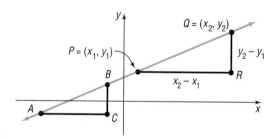

2. The slope of a line may be computed from $P = (x_1, y_1)$ to $Q = (x_2, y_2)$ or from Q to P, because

$$\frac{y_2 - y_1}{x_2 - x_1} = \frac{y_1 - y_2}{x_1 - x_2}$$

E X A M P L E   1   *Finding the Slope of a Line Joining Two Points*

The slope m of the line joining the points $(1, 2)$ and $(5, -3)$ may be computed as

$$m = \frac{-3 - 2}{5 - 1} = \frac{-5}{4} \quad \text{or as} \quad m = \frac{2 - (-3)}{1 - 5} = \frac{5}{-4} = \frac{-5}{4} \quad \blacksquare$$

■   Now work Problem 3.

To get a better idea of the meaning of the slope m of a line L, consider the following example.

E X A M P L E   2   *Finding the Slopes of Various Lines Containing the Same Point (2, 3)*

Compute the slopes of the lines L_1, L_2, L_3, and L_4 containing the following pairs of points. Graph all four lines on the same set of coordinate axes.

$$L_1: \quad P = (2, 3) \qquad Q_1 = (-1, -2)$$
$$L_2: \quad P = (2, 3) \qquad Q_2 = (3, -1)$$
$$L_3: \quad P = (2, 3) \qquad Q_3 = (5, 3)$$
$$L_4: \quad P = (2, 3) \qquad Q_4 = (2, 5)$$

Solution   Let m_1, m_2, m_3, and m_4 denote the slopes of the lines L_1, L_2, L_3, and L_4, respectively. Then

$$m_1 = \frac{-2 - 3}{-1 - 2} = \frac{-5}{-3} = \frac{5}{3} \quad \text{A rise of 5 divided by a run of 3}$$

$$m_2 = \frac{-1 - 3}{3 - 2} = \frac{-4}{1} = -4$$

$$m_3 = \frac{3 - 3}{5 - 2} = \frac{0}{3} = 0$$

m_4 is undefined

The graphs of these lines are given in Figure 55.

FIGURE 65

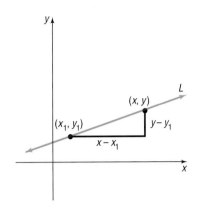

Theorem	An equation of a nonvertical line of slope m that passes through the point (x_1, y_1) is

**Point–Slope Form
of an Equation of a Line**

$$y - y_1 = m(x - x_1) \qquad (2)$$

E X A M P L E   6

Using the Point–Slope Form of a Line

An equation of the line with slope 4 and passing through the point $(1, 2)$ can be found by using the point–slope form with $m = 4$, $x_1 = 1$, and $y_1 = 2$:

$$y - y_1 = m(x - x_1)$$
$$y - 2 = 4(x - 1)$$
$$y = 4x - 2$$

See Figure 66.

FIGURE 66
$y = 4x - 2$

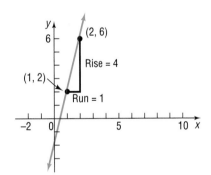 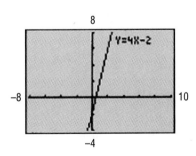

E X A M P L E   7

Finding the Equation of a Horizontal Line

Find an equation of the horizontal line passing through the point $(3, 2)$.

Solution

The slope of a horizontal line is 0. To get an equation, we use the point–slope form with $m = 0$, $x_1 = 3$, and $y_1 = 2$:

$$y - y_1 = m(x - x_1)$$
$$y - 2 = 0 \cdot (x - 3)$$
$$y - 2 = 0$$
$$y = 2$$

See Figure 67 for the graph.

FIGURE 67
$y = 2$

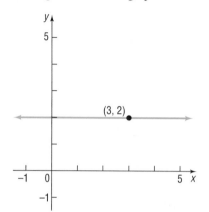

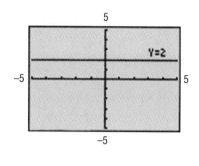

As suggested by Example 7, we have the following result:

Theorem   A horizontal line is given by an equation of the form

Equation of a Horizontal Line

$$y = b$$

where b is the y-intercept.

E X A M P L E   8   *Finding an Equation of a Line Given Two Points*

Find an equation of the line L passing through the points $(2, 3)$ and $(-4, 5)$. Graph the line L.

Solution   Since two points are given, we first compute the slope of the line:

$$m = \frac{5 - 3}{-4 - 2} = \frac{2}{-6} = \frac{-1}{3}$$

We use the point $(2, 3)$ and the fact that the slope $m = -\dfrac{1}{3}$ to get the point–slope form of the equation of the line:

$$y - 3 = -\frac{1}{3}(x - 2)$$

See Figure 68 for the graph.

FIGURE 68
$y - 3 = -\frac{1}{3}(x - 2)$

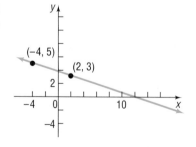

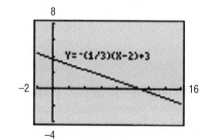

In the solution to Example 8, we could have used the other point, $(-4, 5)$, instead of the point $(2, 3)$. The equation that results, although it looks different, is equivalent to the equation we obtained in the example. (Try it for yourself.)

Another form of the equation of the line in Example 8 can be obtained by multiplying both sides of the point–slope equation by 3 and collecting terms:

$$y - 3 = -\frac{1}{3}(x - 2)$$

$$3(y - 3) = 3\left(-\frac{1}{3}\right)(x - 2) \quad \text{Multiply by 3.}$$

$$3y - 9 = -1(x - 2)$$

$$3y - 9 = -x + 2$$

$$x + 3y - 11 = 0$$

General Form of an Equation of a Line

The equation of a line L is in **general form** when it is written as

$$Ax + By + C = 0 \tag{3}$$

where A, B, and C are three real numbers and A and B are not both 0.

■ Now work Problem 31.

Every line has an equation that is equivalent to an equation written in general form. For example, a vertical line whose equation is

$$x = a$$

can be written in the general form

$$1 \cdot x + 0 \cdot y - a = 0 \quad A = 1, B = 0, C = -a$$

A horizontal line whose equation is

$$y = b$$

can be written in the general form

$$0 \cdot x + 1 \cdot y - b = 0 \quad A = 0, B = 1, C = -b$$

Lines that are neither vertical nor horizontal have general equations of the form

$$Ax + By + C = 0 \quad A \neq 0 \text{ and } B \neq 0$$

Because the equation of every line can be written in general form, any equation equivalent to (3) is called a **linear equation.**

The next example illustrates one way of graphing a linear equation.

E X A M P L E   9   *Finding the Intercepts of a Line*

Find the intercepts of the line $2x + 3y - 6 = 0$. Graph this line.

Solution   To find the point at which the graph crosses the x-axis, that is, to find the x-intercept, we need to find the number x for which $y = 0$. Thus, we let $y = 0$ to get

$$2x + 3(0) - 6 = 0$$
$$2x - 6 = 0$$
$$x = 3$$

The x-intercept is 3. To find the y-intercept, we let $x = 0$ and solve for y:

$$2(0) + 3y - 6 = 0$$
$$3y - 6 = 0$$
$$y = 2$$

The y-intercept is 2.

To graph the line by hand, we use the intercepts. Since the x-intercept is 3 and the y-intercept is 2, we know two points on the line: $(3, 0)$ and $(0, 2)$. Because two points determine a unique line, we do not need further information to graph the line. See Figure 69.

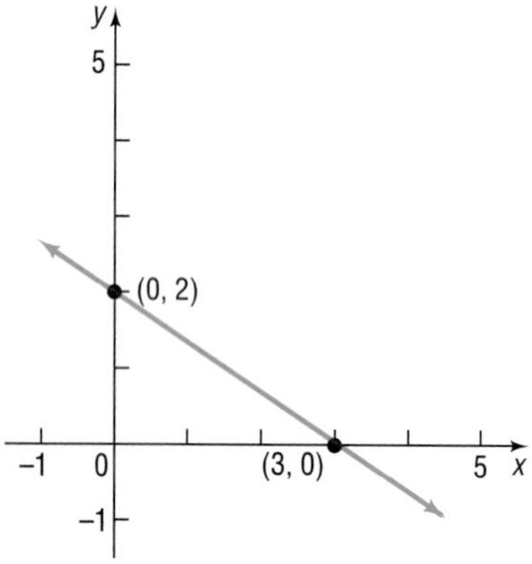

To graph the line using a graphing utility, we need to solve for y.

$$2x + 3y - 6 = 0$$
$$3y = -2x + 6$$
$$y = -\tfrac{2}{3}x + 2$$

See Figure 70 for the graph.

FIGURE 69
$2x + 3y - 6 = 0$

(graph showing line through points $(0, 2)$ and $(3, 0)$, with axes marked -1, 0, 5 on x-axis and 5, -1 on y-axis)

FIGURE 70

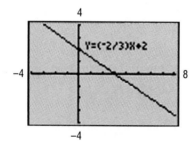

Another useful equation of a line is obtained when the slope m and y-intercept b are known. In this event, we know both the slope m of the line and a point $(0, b)$ on the line; thus, we may use the point–slope form, equation (2), to obtain the following equation:

$$y - b = m(x - 0) \quad \text{or} \quad y = mx + b$$

Theorem   An equation of a line L with slope m and y-intercept b is

Slope–Intercept Form
of an Equation of a Line

$$y = mx + b \qquad (4)$$

Seeing the Concept: To see the role the slope m plays, graph the following lines on the same square screen

$$y = 2$$
$$y = x + 2$$
$$y = -x + 2$$
$$y = 3x + 2$$
$$y = -3x + 2$$

See Figure 71.

FIGURE 71
$y = mx + 2$

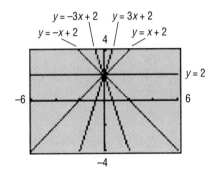

What do you conclude about the lines $y = mx + 2$?

Seeing the Concept: To see the role of the y-intercept b, graph the following lines on the same square screen.

$$y = 2x$$
$$y = 2x + 1$$
$$y = 2x - 1$$
$$y = 2x + 4$$
$$y = 2x - 4$$

FIGURE 72
$y = 2x + b$

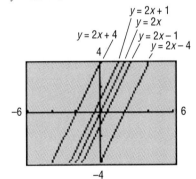

See Figure 72. What do you conclude about the lines $y = 2x + b$?

When the equation of a line is written in slope–intercept form, it is easy to find the slope m and y-intercept b of the line. For example, suppose the equation of a line is

$$y = -2x + 3$$

Compare it to $y = mx + b$:

$$y = -2x + 3$$
$$\uparrow \qquad \uparrow$$
$$y = \quad mx \ + b$$

The slope of this line is -2 and its y-intercept is 3.
Let's look at another example.

E X A M P L E   1 0   *Finding the Slope and y-Intercept of a Line*

Find the slope m and y-intercept b of the line $2x + 4y - 8 = 0$. Graph the line.

Solution   To obtain the slope and y-intercept, we transform the equation into its slope–intercept form. Thus, we need to solve for y:

$$2x + 4y - 8 = 0$$
$$4y = -2x + 8$$
$$y = -\tfrac{1}{2}x + 2$$

The coefficient of x, $-\tfrac{1}{2}$, is the slope, and the y-intercept is 2. Figure 73 shows the graph using a graphing utility.

FIGURE 73

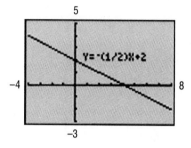

We can graph the line by hand in two ways:

1. Use the fact that the y-intercept is 2 and the slope is $-\tfrac{1}{2}$. Then, starting at the point $(0, 2)$, go to the right 2 units and then down 1 unit to the point $(2, 1)$. See Figure 74.

Or:

2. Locate the intercepts. Because the y-intercept is 2, we know one intercept is $(0, 2)$. To obtain the x-intercept, let $y = 0$ and solve for x. When $y = 0$, we have

$$2x + 4 \cdot 0 - 8 = 0$$
$$2x - 8 = 0$$
$$x = 4$$

Thus, the intercepts are $(4, 0)$ and $(0, 2)$. See Figure 75.

FIGURE 74
$2x + 4y - 8 = 0$

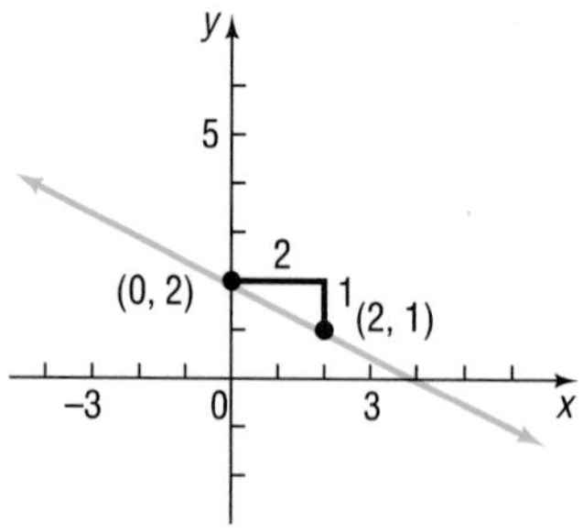

FIGURE 75
$2x + 4y - 8 = 0$

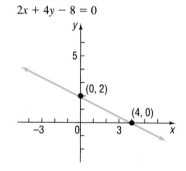

■   Now work Problem 47.

Circles

One advantage of a coordinate system is that it enables us to translate a geometric statement into an algebraic statement, and vice versa. Consider, for example, the following geometric statement that defines a circle.

Circle

> A **circle** is a set of points in the *xy*-plane that are a fixed distance *r* from a fixed point (*h*, *k*). The fixed distance *r* is called the **radius,** and the fixed point (*h*, *k*) is called the **center** of the circle.

FIGURE 76

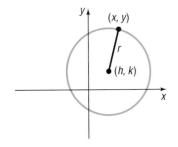

Figure 76 shows the graph of a circle. Is there an equation having this graph? If so, what is the equation? To find the equation, we let (*x*, *y*) represent the coordinates of any point on a circle with radius *r* and center (*h*, *k*). Then the distance between the points (*x*, *y*) and (*h*, *k*) must always equal *r*. That is, by the distance formula,

$$\sqrt{(x - h)^2 + (y - k)^2} = r$$

or, equivalently,

$$(x - h)^2 + (y - k)^2 = r^2$$

The **standard form of an equation of a circle** with radius *r* and center (*h*, *k*) is

Standard Form
of an Equation of a Circle

$$(x - h)^2 + (y - k)^2 = r^2 \qquad (5)$$

Conversely, by reversing the steps, we conclude: The graph of any equation of the form of equation (5) is that of a circle with radius *r* and center (*h*, *k*).

E X A M P L E   1 1

Graphing a Circle

Graph the equation: $(x + 3)^2 + (y - 2)^2 = 16$

Solution

The graph of the equation is a circle. To graph a circle on a graphing utility we must write the equation in the form *y* = expression involving *x**. Thus, we must solve for *y* in the equation

$$(x + 3)^2 + (y - 2)^2 = 16$$
$$(y - 2)^2 = 16 - (x + 3)^2$$
$$y - 2 = \pm\sqrt{16 - (x + 3)^2}$$
$$y = 2 \pm \sqrt{16 - (x + 3)^2}$$

To graph the circle, we first graph the top half

$$y = 2 + \sqrt{16 - (x + 3)^2}$$

*Some graphing utilities (e.g. TI-82, TI-85) have a CIRCLE function which allows the user to enter only the coordinates of the center of the circle and its radius to graph the circle.

and then graph the bottom half

$$y = 2 - \sqrt{16 - (x + 3)^2}$$

Also, be sure to use a square screen. Otherwise the circle will appear distorted. Figure 77 shows the graph.

FIGURE 77

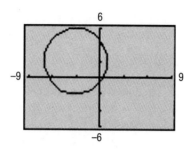

FIGURE 78

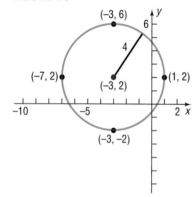

To graph the equation by hand, we first compare the given equation to the standard form of the equation of a circle. The comparison yields information about the circle:

$$(x + 3)^2 + (y - 2)^2 = 16$$
$$(x - (-3))^2 + (y - 2)^2 = 4^2$$
$$(x - h)^2 + (y - k)^2 = r^2$$

We see that $h = -3$, $k = 2$, and $r = 4$. Hence, the circle has center $(-3, 2)$ and a radius of 4 units. To graph this circle, we first plot the center $(-3, 2)$. Since the radius is 4, we can locate four points on the circle by going out 4 units to the left, to the right, up, and down from the center. These four points can then be used as guides to obtain the graph. See Figure 78. ∎

■ Now work Problem 75.

E X A M P L E   1 2   *Writing the Standard Form of the Equation of a Circle*

Write the standard form of the equation of the circle with radius 3 and center $(1, -2)$.

Solution   Using the form of equation (5) and substituting the values $r = 3$, $h = 1$, and $k = -2$, we have

$$(x - h)^2 + (y - k)^2 = r^2$$
$$(x - 1)^2 + (y + 2)^2 = 9$$ ∎

■ Now work Problem 59.

The standard form of an equation of a circle of radius r with center at the origin $(0, 0)$ is

$$x^2 + y^2 = r^2$$

If the radius $r = 1$, the circle whose center is at the origin is called the **unit circle** and has the equation

$$x^2 + y^2 = 1$$

See Figure 79.

FIGURE 79
Unit circle $x^2 + y^2 = 1$

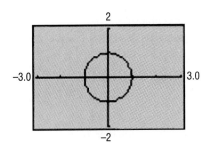
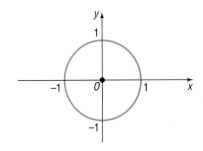

If we eliminate the parentheses from the standard form of the equation of the circle obtained in Example 12, we get

$$(x - 1)^2 + (y + 2)^2 = 9$$
$$x^2 - 2x + 1 + y^2 + 4y + 4 = 9$$

which we find, upon simplifying, is equivalent to

$$x^2 + y^2 - 2x + 4y - 4 = 0$$

By completing the squares on both the x- and y-terms, it can be shown that any equation of the form

$$x^2 + y^2 + ax + by + c = 0$$

has a graph that is a circle, or a point, or has no graph at all. For example, the graph of the equation $x^2 + y^2 = 0$ is the single point $(0, 0)$. The equation $x^2 + y^2 + 5 = 0$, or $x^2 + y^2 = -5$, has no graph, because sums of squares of real numbers are never negative. When its graph is a circle, the equation

General Form
of the Equation of a Circle

$$x^2 + y^2 + ax + by + c = 0$$

is referred to as the **general form of the equation of a circle.**

The next example shows how to transform an equation in the general form to an equivalent equation in standard form. As we said earlier, the idea is to use the method of completing the square on both the x- and y-terms. See the Appendix, Section 3, for a discussion of completing the square.

E X A M P L E   1 3   *Graphing a Circle Whose Equation Is in General Form*
Graph the equation: $x^2 + y^2 + 4x - 6y + 12 = 0$

Solution　We rearrange the equation as follows:

$$(x^2 + 4x) + (y^2 - 6y) = -12$$

Next, we complete the square of each expression in parentheses. Remember that any number added on the left also must be added on the right:

$$(x^2 + 4x + 4) + (y^2 - 6y + 9) = -12 + 4 + 9$$
$$(x + 2)^2 + (y - 3)^2 = 1$$

We recognize this equation as the standard form of the equation of a circle with radius 1 and center $(-2, 3)$.

To graph the equation using a graphing utility, we need to solve for y:

$$(y - 3)^2 = 1 - (x + 2)^2$$
$$y - 3 = \pm\sqrt{1 - (x + 2)^2}$$
$$y = 3 \pm \sqrt{1 - (x + 2)^2}$$

To graph the equation by hand, we use the center $(-2, 3)$ and the radius of 1. Figure 80 illustrates the graph.

FIGURE 80

$$x^2 + y^2 + 4x - 6y + 12 = 0$$

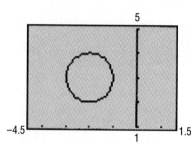
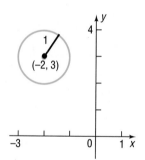

■　Now work Problem 77.

E X A M P L E   1 4 　*Finding the General Equation of a Circle*

Find the general equation of the circle whose center is $(1, -2)$ and whose graph contains the point $(4, 2)$.

FIGURE 81

$$x^2 + y^2 - 2x + 4y - 20 = 0$$

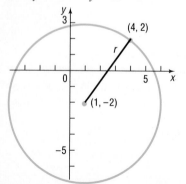

Solution　See Figure 81. To find the equation of a circle, we need to know its center and its radius. Here, we know that the center is $(1, -2)$. Since the point $(4, 2)$ is on the graph, the radius r will equal the distance from $(4, 2)$ to the center $(1, -2)$. Thus,

$$r = \sqrt{(4 - 1)^2 + [2 - (-2)]^2}$$
$$= \sqrt{9 + 16} = 5$$

The standard form of the equation of the circle is

$$(x - 1)^2 + (y + 2)^2 = 25$$

Eliminating the parentheses and rearranging terms, we get the general equation

$$x^2 + y^2 - 2x + 4y - 20 = 0$$

■

Overview

The preceding discussion about lines and circles dealt with two main types of problems that can be generalized as follows:

1. Given an equation, classify it and graph it.
2. Given a graph, or information about a graph, find its equation.

   This text deals mainly with the first type of problem. We shall study various equations, classify them, and graph them. The second type of problem is usually more difficult to solve than the first. However, we shall tackle such problems when it is practical to do so. See the Appendix, Section 5, for a discussion of Linear Curve Fitting.

1.3

Exercise 1.3

In Problems 1–4, find the slope of the line.

1.

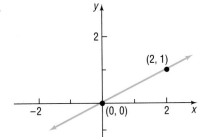

2.

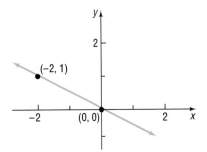

3.

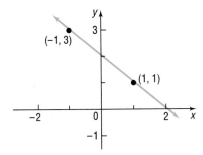

4.

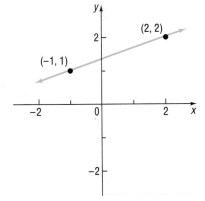

In Problems 5–14, plot each pair of points and determine the slope of the line containing them. By hand, graph the line.

5. $(2, 3); (4, 0)$

6. $(4, 2); (3, 4)$

7. $(-2, 3); (2, 1)$

8. $(-1, 1); (2, 3)$

9. $(-3, -1); (2, -1)$

10. $(4, 2); (-5, 2)$

11. $(-1, 2); (-1, -2)$

12. $(2, 0); (2, 2)$

13. $(\sqrt{2}, 3); (1, \sqrt{3})$

14. $(-2\sqrt{2}, 0); (4, \sqrt{5})$

In Problems 15–22, graph by hand the line passing through the point P and having slope m.

15. $P = (1, 2)$; $m = 3$

16. $P = (2, 1)$; $m = 4$

17. $P = (2, 4)$; $m = \frac{-3}{4}$

18. $P = (1, 3)$; $m = \frac{-2}{5}$

19. $P = (-1, 3)$; $m = 0$

20. $P = (2, -4)$; $m = 0$

21. $P = (0, 3)$; slope undefined

22. $P = (-2, 0)$; slope undefined

In Problems 23–26, find an equation of each line. Express your answer using either the general form or the slope–intercept form of the equation of a line, whichever you prefer.

23.

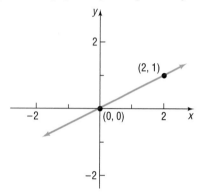

24.

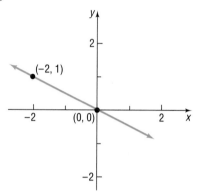

25.

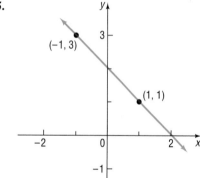

26.

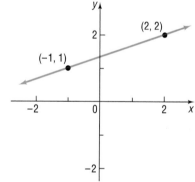

In Problems 27–38, find an equation for the line with the given properties. Express your answer using either the general form or the slope–intercept form of the equation of a line, whichever you prefer.

27. Slope = 3; passing through $(-2, 3)$

28. Slope = 2; passing through $(4, -3)$

29. Slope = $-\frac{2}{3}$; passing through $(1, -1)$

30. Slope = $\frac{1}{2}$; passing through $(3, 1)$

31. Passing through $(1, 3)$ and $(-1, 2)$

32. Passing through $(-3, 4)$ and $(2, 5)$

33. Slope = -3; y-intercept = 3

34. Slope = -2; y-intercept = -2

35. x-intercept = 2; y-intercept = -1

36. x-intercept = -4; y-intercept = 4

37. Slope undefined; passing through $(2, 4)$

38. Slope undefined; passing through $(3, 8)$

In Problems 39–58, find the slope and y-intercept of each line. By hand, graph the line. Check your graph using a graphing utility.

39. $y = 2x + 3$

40. $y = -3x + 4$

41. $\frac{1}{2}y = x - 1$

42. $\frac{1}{3}x + y = 2$

43. $y = \frac{1}{2}x + 2$

44. $y = 2x + \frac{1}{2}$

45. $x + 2y = 4$

46. $-x + 3y = 6$

47. $2x - 3y = 6$ **48.** $3x + 2y = 6$ **49.** $x + y = 1$ **50.** $x - y = 2$

51. $x = -4$ **52.** $y = -1$ **53.** $y = 5$ **54.** $x = 2$

55. $y - x = 0$ **56.** $x + y = 0$ **57.** $2y - 3x = 0$ **58.** $3x + 2y = 0$

In Problems 59–62, find the center and radius of each circle. Write the standard form of the equation.

59. 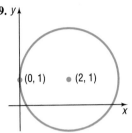   **60.** 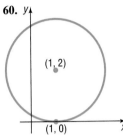   **61.** 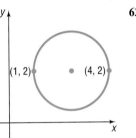   **62.**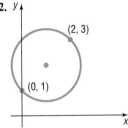

In Problems 63–72, write the standard form of the equation and the general form of the equation of each circle of radius r and center (h, k). By hand, graph each circle.

63. $r = 1$; $(h, k) = (1, -1)$ **64.** $r = 2$; $(h, k) = (-2, 1)$

65. $r = 2$; $(h, k) = (0, 2)$ **66.** $r = 3$; $(h, k) = (1, 0)$

67. $r = 5$; $(h, k) = (4, -3)$ **68.** $r = 4$; $(h, k) = (2, -3)$

69. $r = 2$; $(h, k) = (0, 0)$ **70.** $r = 3$; $(h, k) = (0, 0)$

71. $r = \frac{1}{2}$; $(h, k) = \left(\frac{1}{2}, 0\right)$ **72.** $r = \frac{1}{2}$; $(h, k) = \left(0, -\frac{1}{2}\right)$

In Problems 73–82, find the center (h, k) and radius r of each circle. By hand, graph each circle.

73. $x^2 + y^2 = 4$ **74.** $x^2 + (y - 1)^2 = 1$

75. $(x - 3)^2 + y^2 = 4$ **76.** $(x + 1)^2 + (y - 1)^2 = 2$

77. $x^2 + y^2 + 4x - 4y - 1 = 0$ **78.** $x^2 + y^2 - 6x + 2y + 9 = 0$

79. $x^2 + y^2 - x + 2y + 1 = 0$ **80.** $x^2 + y^2 + x + y - \frac{1}{2} = 0$

81. $2x^2 + 2y^2 - 12x + 8y - 24 = 0$ **82.** $2x^2 + 2y^2 + 8x + 7 = 0$

In Problems 83–88, find the general form of the equation of each circle.

83. Center at the origin and containing the point $(-2, 3)$ **86.** Center $(-3, 1)$ and tangent to the y-axis

84. Center $(1, 0)$ and containing the point $(-3, 2)$ **87.** With endpoints of a diameter at $(1, 4)$ and $(-3, 2)$

85. Center $(2, 3)$ and tangent to the x-axis   **88.** With endpoints of a diameter at $(4, 3)$ and $(0, 1)$

In Problems 89–92, match each graph with the correct equation:
(a) $y = x$ (b) $y = 2x$ (c) $y = x/2$ (d) $y = 4x$

89.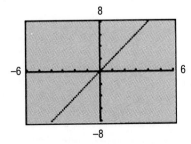

90.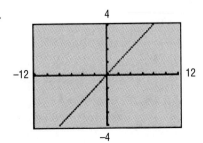

91.

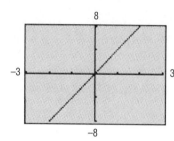

92.

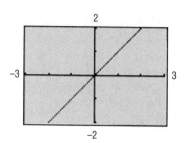

In Problems 93–98, write an equation of each line. Express your answer using either the general form or the slope–intercept form of the equation of a line, whichever you prefer.

93.

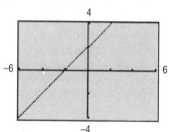

94.

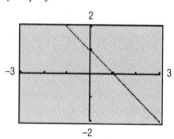

95.

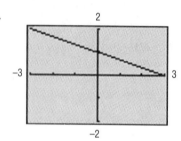

96.

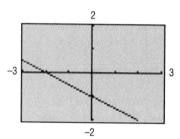

97.

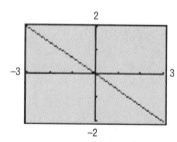

98.

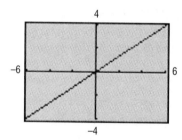

In Problems 99–102, match each graph with the correct equation.
(a) $(x - 3)^2 + (y + 3)^2 = 9$ *(c)* $(x - 1)^2 + (y + 2)^2 = 4$
(b) $(x + 1)^2 + (y - 2)^2 = 4$ *(d)* $(x + 3)^2 + (y - 3)^2 = 9$

99.

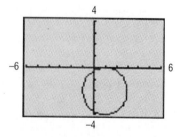

100.

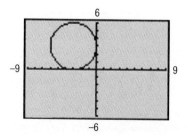

101.

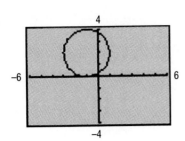

102.

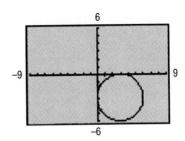

In Problems 103–106, find the standard form of the equation of each circle.

103.

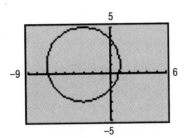

104.

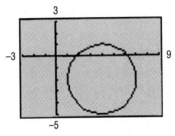

105.

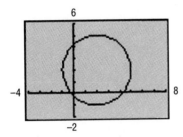

106.

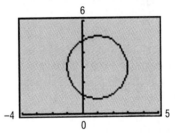

107. Find an equation of the x-axis.

108. Find an equation of the y-axis.

109. *Measuring Temperature* The relationship between Celsius (°C) and Fahrenheit (°F) degrees for measuring temperature is linear. Find an equation relating °C and °F if 0°C corresponds to 32°F and 100°C corresponds to 212°F. Use the equation to find the Celsius measure of 70°F.

110. *Measuring Temperature* The Kelvin (K) scale for measuring temperature is obtained by adding 273 to the Celsius temperature.
   (a) Write an equation relating K and °C.
   (b) Write an equation relating K and °F (see Problem 109).

111. *Business: Computing Profit* Each Sunday, a newspaper agency sells x copies of a certain newspaper for $1.00 per copy. The cost to the agency of each newspaper is $0.50. The agency pays a fixed cost for storage, delivery, and so on, of $100 per Sunday.
   (a) Write an equation that relates the profit P, in dollars, to the number x of copies sold. Graph this equation.
   (b) What is the profit to the agency if 1000 copies are sold?
   (c) What is the profit to the agency if 5000 copies are sold?

112. *Business: Computing Profit* Repeat Problem 111 if the cost to the agency is $0.45 per copy and the fixed cost is $125 per Sunday.

113. *Cost of Electricity* In 1991, Commonwealth Edison Company supplied electricity in the summer months to residential customers for a monthly customer charge of $9.06 plus 10.819¢ per kilowatt-hour supplied in the month.* Write an equation that relates the monthly charge C, in dollars, to the number x of kilowatt-hours in the month. Graph this equation. What is the monthly charge for using 300 kilowatt-hours? For using 900 kilowatt-hours?

Source: Commonwealth Edison Co., Chicago, Illinois, 1991.

114. *Weather Satellites* Earth is represented on a map of a portion of the solar system so that its surface is the circle with equation $x^2 + y^2 + 2x + 4y - 4091 = 0$. A weather satellite circles 0.6 unit above Earth with the center of its circular orbit at the center of Earth. Find the equation for the orbit of the satellite on this map.

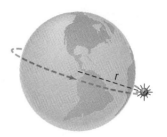

115. The **tangent line** to a circle may be defined as the line that intersects the circle in a single point, called the **point of tangency** (see the figure). If the equation of the circle is $x^2 + y^2 = r^2$ and the equation of the tangent line is $y = mx + b$, show that:
  (a) $r^2(1 + m^2) = b^2$ [*Hint:* the quadratic equation $x^2 + (mx + b)^2 = r^2$ has exactly one solution.]
  (b) The point of tangency is $(-r^2m/b, r^2/b)$.
  (c) The tangent line is perpendicular to the line containing the center of the circle and the point of tangency.

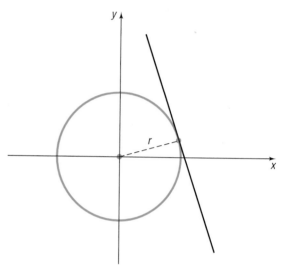

116. The Greek method for finding the equation of the tangent line to a circle used the fact that at any point on a circle the line containing the radius and the tangent line are perpendicular (see Problem 115). Use this method to find an equation of the tangent line to the circle $x^2 + y^2 = 9$ at the point $(1, 2\sqrt{2})$.

117. Use the Greek method described in Problem 116 to find an equation of the tangent line to the circle $x^2 + y^2 - 4x + 6y + 4 = 0$ at the point $(3, 2\sqrt{2} - 3)$.

118. Refer to Problem 115. The line $x - 2y + 4 = 0$ is tangent to a circle at $(0, 2)$. The line $y = 2x - 7$ is tangent to the same circle at $(3, -1)$. Find the center of the circle.

119. Find an equation of the line containing the centers of the two circles
$$x^2 + y^2 - 4x + 6y + 4 = 0 \quad \text{and} \quad x^2 + y^2 + 6x + 4y + 9 = 0$$

120. If a circle of radius 2 is made to roll along the x-axis, what is an equation for the path of the center of the circle?

121. The equation $2x - y + C = 0$ defines a **family of lines,** one line for each value of C. On one set of coordinate axes, graph the members of the family when $C = -4$, $C = 0$, and $C = 2$. Can you draw a conclusion from the graph about each member of the family?

122. Rework Problem 121 for the family of lines $Cx + y + 4 = 0$.

123. Which form of the equation of a line do you prefer to use? Justify your position with an example that shows that your choice is better than another. Have reasons.

124. Can every line be written in slope–intercept form? Explain.

125. Does every line have two distinct intercepts? Explain. Are there lines that have no intercepts? Explain.

126. What can you say about two lines that have equal slopes and equal y-intercepts?

127. What can you say about two lines with the same x-intercept and the same y-intercept? Assume that the x-intercept is not 0.

128. If two lines have the same slope, but different x-intercepts, can they have the same y-intercept?

129. If two lines have the same y-intercept, but different slopes, can they have the same x-intercept? What is the only way this can happen?

130. The accepted symbol used to denote the slope of a line is the letter m. Investigate the origin of this symbolism. Begin by consulting a French dictionary and looking up the French word *monter.* Write a brief essay on your findings.

131. The term *grade* is used to describe the inclination of a road. How does this term relate to the notion of slope of a line? Is a 4% grade very steep? Investigate the grades of some mountainous roads and determine their slopes. Write a brief essay on your findings.

132. *Carpentry* Carpenters use the term pitch to describe the steepness of staircases and roofs. How does pitch relate to slope? Investigate typical pitches used for stairs and for roofs. Write a brief essay on your findings.

1.4

Functions

In many applications, a correspondence often exists between two sets of numbers. For example, the revenue R resulting from the sale of x items selling for $10 each is $R = 10x$ dollars. If we know how many items have been sold, then we can calculate the revenue by using the rule $R = 10x$. This rule is an example of a *function*.

As another example, if an object is dropped from a height of 64 feet above the ground, the distance s (in feet) of the object from the ground after t seconds is given (approximately) by the formula $s = 64 - 16t^2$. When $t = 0$ seconds, the object is $s = 64$ feet above the ground. After 1 second, the object is $s = 64 - 16(1)^2 = 48$ feet above the ground. After 2 seconds, the object strikes the ground. The formula $s = 64 - 16t^2$ provides a way of finding the distance s when the time t ($0 \leq t \leq 2$) is prescribed. There is a correspondence between each time t in the interval $0 \leq t \leq 2$ and the distance s. We say that the distance s is a *function* of the time t because:

1. There is a correspondence between the set of times and the set of distances.
2. There is exactly one distance s obtained for a prescribed time t in the interval $0 \leq t \leq 2$.

Let's now look at the definition of a function.

Definition of Function

Function

> Let X and Y be two nonempty sets of real numbers.* A **function** from X into Y is a rule or a correspondence that associates with each element of X a unique element of Y. The set X is called the **domain** of the function. For each element x in X, the corresponding element y in Y is called the **value** of the function at x, or the **image** of x. The set of all images of the elements of the domain is called the **range** of the function.

*The two sets X and Y can also be sets of complex numbers, and then we have defined a complex function. In the broad definition (due to Lejeune Dirichlet), X and Y can be any two sets.

FIGURE 82

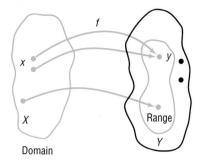

Domain

When we select a viewing rectangle to graph a function, the values of Xmin, Xmax give the domain we wish to view, while Ymin, Ymax give the range we wish to view. These settings usually do not represent the actual domain and range of the function.

Warning: Do not confuse the two meanings given for the word *range*. When used in connection with a function, it means the set of all images of the elements of the domain of the function. When the word RANGE is used in connection with a graphing utility, it means the settings used for the viewing rectangle.

Refer to Figure 82. Since there may be some elements in Y that are not the image of some x in X, it follows that the range of a function may be a subset of Y.

The rule (or correspondence) referred to in the definition of a function is most often given as an equation in two variables, usually denoted x and y.

EXAMPLE 1

Example of a Function

Consider the function defined by the equation

$$y = 2x - 5 \qquad 1 \le x \le 6$$

The domain $1 \le x \le 6$ specifies that the number x is restricted to the real numbers from 1 to 6, inclusive. The rule $y = 2x - 5$ specifies that the number x is to be multiplied by 2 and then 5 is to be subtracted from the result to get y. For example, the value of the function at $x = \frac{3}{2}$ (that is, the image of $x = \frac{3}{2}$) is $y = 2 \cdot \frac{3}{2} - 5 = -2$. ■

Functions are often denoted by letters such as f, F, g, G, and so on. If f is a function, then for each number x in its domain the corresponding image in the range is designated by the symbol $f(x)$, read as "f of x" or as "f at x." We refer to $f(x)$ as the **value of f at the number x.** Thus, $f(x)$ is the number that results when x is given and the rule for f is applied; $f(x)$ does *not* mean "f times x." For example, the function given in Example 1 may be written as $f(x) = 2x - 5$, $1 \le x \le 6$.

Figure 83 illustrates some other functions. Note that for each of the functions illustrated, to each x in the domain, there is one value in the range.

FIGURE 83

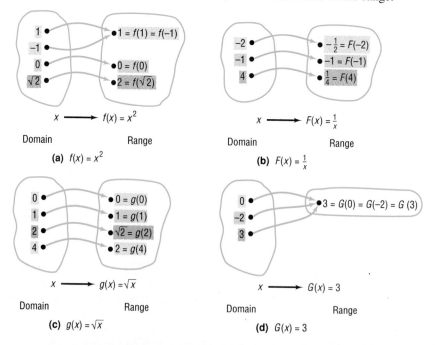

E X A M P L E 2   *Finding Values of a Function*

For the function

$$f(x) = x^2 + 3x - 4 \qquad -5 \leq x \leq 5$$

find the value of f at:

(a) $x = 0$ (b) $x = 1$ (c) $x = -4$ (d) $x = 5$

Solution   (a) The value of f at $x = 0$ is found by replacing x by 0 in the stated rule. Thus,

$$f(0) = 0^2 + 3(0) - 4 = -4$$

(b) $f(1) = 1^2 + 3(1) - 4 = 1 + 3 - 4 = 0$
(c) $f(-4) = (-4)^2 + 3(-4) - 4 = 16 - 12 - 4 = 0$
(d) $f(5) = 5^2 + 3(5) - 4 = 25 + 15 - 4 = 36$ ■

■   Now work Problem 3.

In general, when the rule that defines a function f is given by an equation in x and y, we say that the function f is given **implicitly.** If it is possible to solve the equation for y in terms of x, then we write $y = f(x)$ and say that the function is given **explicitly.** In fact, we usually write "the function $y = f(x)$" when we mean "the function f defined by the equation $y = f(x)$." Although this usage is not entirely correct, it is rather common and should not cause any confusion. For example:

IMPLICIT FORM   **EXPLICIT FORM**

$3x + y = 5$ $y = f(x) = -3x + 5$

$x^2 - y = 6$ $y = f(x) = x^2 - 6$

$xy = 4$ $y = f(x) = 4/x$

Not all equations in x and y define a function $y = f(x)$. If an equation is solved for y and two or more values of y can be obtained for a given x, then the equation does not define a function $y = f(x)$. For example, consider the equation $x^2 + y^2 = 1$, which defines a circle. If we solve for y, we obtain $y = \pm\sqrt{1 - x^2}$ so that two values of y will result for numbers x between -1 and 1. Thus, $x^2 + y^2 = 1$ does not define a function.

The explicit form of a function is the form required by a graphing calculator. Now do you see why it is necessary to graph a circle in two "pieces"?

We list below a summary of some important facts to remember about a function f.

Summary of Important Facts about Functions

1. $f(x)$ is the image of x, or the value of f at x, when the rule f is applied to an x in the domain.
2. To each x in the domain of f, there is one and only one image $f(x)$ in the range.
3. f is the symbol we use to denote the function. It is symbolic of the domain and the rule we use to get from an x in the domain to $f(x)$ in the range.

Function Keys

Most graphing utilities have special keys that enable you to find the value of certain commonly used functions. For example, you should be able to find the square function, $f(x) = x^2$; the square root function, $f(x) = \sqrt{x}$; the reciprocal function, $f(x) = 1/x = x^{-1}$; and many others that will be discussed later in this book (such as $\ln x$, $\log x$, and so on). Verify the results of Example 3 on your graphing utility.

E X A M P L E   3   *Finding Values of a Function on a Calculator*

(a) $f(x) = x^2$; $f(1.234) = 1.522756$

(b) $F(x) = 1/x$; $F(1.234) = 0.8103727715$

(c) $g(x) = \sqrt{x}$; $g(1.234) = 1.110855526$ ∎

Domain of a Function

Often, the domain of a function f is not specified; instead, only a rule or equation defining the function is given. In such cases, we agree that the domain of f is the largest set of real numbers for which the rule makes sense or, more precisely, for which the value $f(x)$ is a real number. Thus, the domain of f is the same as the domain of the variable x in the expression $f(x)$.

E X A M P L E   4   *Finding the Domain of a Function*

Find the domain of each of the following functions:

(a) $f(x) = \dfrac{3x}{x^2 - 4}$ (b) $g(x) = \sqrt{4 - 3x}$

Solution   (a) The rule f tells us to divide $3x$ by $x^2 - 4$. Since division by 0 is not allowed, the denominator $x^2 - 4$ can never be 0. Thus, x can never equal 2 or -2. The domain of the function f is $\{x | x \ne -2, x \ne 2\}$.

(b) The rule g tells us to take the square root of $4 - 3x$. But only nonnegative numbers have real square roots. Hence, we require that

$$4 - 3x \ge 0$$
$$-3x \ge -4$$
$$x \le \tfrac{4}{3}$$

The domain of g is $\{x | -\infty < x \le \tfrac{4}{3}\}$ or the interval $(-\infty, \tfrac{4}{3}]$. ∎

We can use a graphing utility to estimate the domain of a function. For example, Figure 84 shows the graph of $y = \sqrt{4 - 3x}$. We can approximate the

FIGURE 84

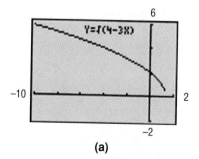

(a)

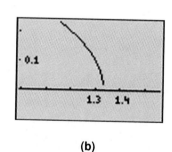

(b)

domain by noting that the domain consists of all numbers x less than or equal to the number where the graph begins. By utilizing the TRACE and ZOOM functions, experiment to see how close you can come to $x = \frac{4}{3}$, the largest value of x in the domain.

■   Now work Problem 41.

If x is in the domain of a function f, we shall say that **f is defined at x,** or **$f(x)$ exists.** If x is not in the domain of f, we say that **f is not defined at x, or $f(x)$ does not exist.** For example, if $f(x) = x/(x^2 - 1)$, then $f(0)$ exists, but $f(1)$ and $f(-1)$ do not exist. (Do you see why?)

We have not said much about finding the range of a function. The reason is that when a function is defined by an equation, it is often difficult to find the range. Therefore, we shall usually be content to find just the domain of a function when only the rule for the function is given. We shall express the domain of a function using interval notation, set notation, or words, whichever is most convenient.

When we use functions in applications, the domain may be restricted by physical or geometric considerations. For example, the domain of the function f defined by $f(x) = x^2$ is the set of all real numbers. However, if f is used as the rule for obtaining the area of a square when the length x of a side is known, then we must restrict the domain of f to the positive real numbers, since the length of a side can never be 0 or negative.

Independent Variable; Dependent Variable

Consider a function $y = f(x)$. The variable x is called the **independent variable,** because it can be assigned any of the permissible numbers from the domain. The variable y is called the **dependent variable,** because its value depends on x.

Any symbol can be used to represent the independent and dependent variables. For example, if f is the *cube function,* then f can be defined by $f(x) = x^3$ or $f(t) = t^3$ or $f(z) = z^3$. All three rules are identical: each tells us to cube the independent variable. In practice, the symbols used for the independent and dependent variables are based on common usage.

EXAMPLE 5

Construction Cost

The cost per square foot to build a house is $110. Express the cost C as a function of x, the number of square feet. What is the cost to build a 2000 square foot house?

Solution

The cost C of building a house containing x square feet is $110x$ dollars. A function expressing this relationship is

$$C(x) = 110x$$

where x is the independent variable and C is the dependent variable. In this setting, the domain is $\{x \mid x > 0\}$ since a house cannot have 0 or negative square feet. The cost to build a 2000 square foot house is

$$C(2000) = 110(2000) = \$220,000$$ ■

It is worth observing that in the solution to Example 5 we used the symbol C in two ways; it is used to name the function, and it is used to symbolize the dependent variable. This double use is common in applications and should not cause any difficulty.

E X A M P L E   6   *Area of a Circle*

Express the area of a circle as a function of its radius.

Solution   We know that the formula for the area A of a circle of radius r is $A = \pi r^2$. If we use r to represent the independent variable and A to represent the dependent variable, the function expressing this relationship is

$$A(r) = \pi r^2$$

In this setting, the domain is $\{r | r > 0\}$. (Do you see why?)

■   Now work Problem 61.

The Graph of a Function

In applications, a graph often demonstrates more clearly the relationship between two variables than, say, an equation or table would. For example, Figure 85 shows the price per share (vertical axis) of McDonald's Corp. stock at the end of each week from Nov. 4, 1994 to Jan. 27, 1995 (horizontal axis). We can see from the graph that the price of the stock was falling over the few days preceding Nov. 25 and was rising over the days from Jan. 20 through Jan. 27. The graph also shows that the lowest price during this period occurred on Dec. 9 while the highest occurred on Jan. 27. Equations and tables, on the other hand, usually require some calculations and interpretation before this kind of information can be "seen."

FIGURE 85
Weekly closing prices of McDonald's
Corp. stock

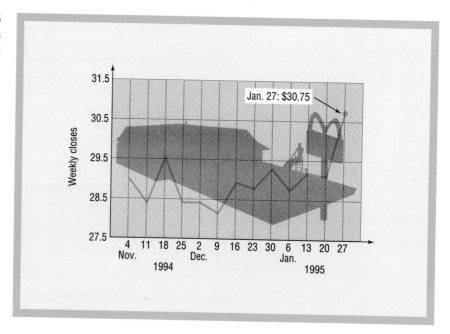

Look again at Figure 85. The graph shows that, for each time on the horizontal axis, there is only one price on the vertical axis. Thus, the graph represents a function, although the exact rule for getting from time to price is not given.

When the rule that defines a function f is given by an equation in x and y, the **graph of f** is the graph of the equation, that is, the set of points (x, y) in the *xy*-plane that satisfies the equation.

Not every collection of points in the *xy*-plane represents the graph of a function. Remember, for a function *f*, each number *x* in the domain of *f* has one and only one image *f(x)*. Thus, the graph of a function *f* cannot contain two points with the same *x*-coordinate and different *y*-coordinates. Therefore, the graph of a function must satisfy the following **vertical-line test:**

Theorem
Vertical-Line Test

A set of points in the *xy*-plane is the graph of a function if and only if a vertical line intersects the graph in at most one point. ∎

It follows that, if any vertical line intersects a graph at more than one point, the graph is not the graph of a function.

E X A M P L E   7

Identifying the Graph of a Function

Which of the graphs in Figure 86 are graphs of functions?

FIGURE 86

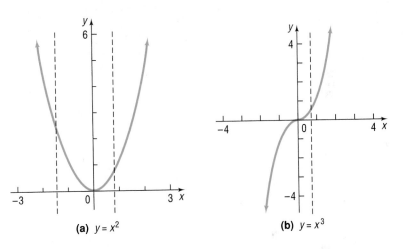

(a) $y = x^2$ **(b)** $y = x^3$

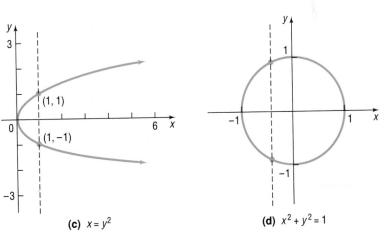

(c) $x = y^2$ **(d)** $x^2 + y^2 = 1$

Solution

The graphs in Figures 86(a) and 86(b) are graphs of functions, because a vertical line intersects each graph in at most one point. The graphs in Figures 86(c) and 86(d) are not graphs of functions, because some vertical line intersects each graph in more than one point. ∎

Ordered Pairs

The preceding discussion provides an alternative way to think of a function. We may consider a function f as a set of **ordered pairs** (x, y) or $(x, f(x))$, in which no two pairs have the same first element. The set of all first elements is the domain of the function, and the set of all second elements is its range. Thus, there is associated with each element x in the domain a unique element y in the range. An example is the set of all ordered pairs (x, y) such that $y = x^2$. Some of the pairs in this set are

$$(2, 2^2) = (2, 4) \qquad (0, 0^2) = (0, 0)$$

$$(-2, (-2)^2) = (-2, 4) \qquad \left(\frac{1}{2}, \left(\frac{1}{2}\right)^2\right) = \left(\frac{1}{2}, \frac{1}{4}\right)$$

In this set, no two pairs have the same *first* element (although there are pairs that have the same *second* element). This set is the *square function*, which associates with each real number x the number x^2. Look again at Figure 86(a).

On the other hand, the ordered pairs (x, y) for which $y^2 = x$ do not represent a function, because there are ordered pairs with the same first element but different second elements. For example, $(1, 1)$ and $(1, -1)$ are ordered pairs obeying the relationship $y^2 = x$ with the same first element but different second elements. Look again at Figure 86(c).

The next example illustrates how to determine the domain and range of a function if its graph is given.

E X A M P L E   8   *Obtaining Information from the Graph of a Function*

Let f be the function whose graph is given in Figure 87. Some points on the graph are labeled.

(a) What is the value of the function when $x = -6$, $x = -4$, $x = 0$, and $x = 6$?

(b) What is the domain of f?

(c) What is the range of f?

(d) List the intercepts. (Recall that these are the points, if any, where the graph crosses the coordinate axes.)

FIGURE 87

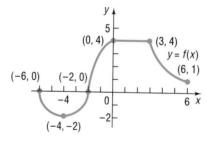

Solution   (a) Since $(-6, 0)$ is on the graph of f, the y-coordinate 0 must be the value of f at the x-coordinate -6; that is, $f(-6) = 0$. In a similar way, we find that when $x = -4$ then $y = -2$, or $f(-4) = -2$; when $x = 0$, then $y = 4$, or $f(0) = 4$; and when $x = 6$, then $y = 1$, or $f(6) = 1$.

(b) To determine the domain of f, we notice that the points on the graph of f all have x-coordinates between -6 and 6, inclusive; and, for each number x between -6 and 6, there is a point $(x, f(x))$ on the graph. Thus, the domain of f is $\{x | -6 \leq x \leq 6\}$, or the interval $[-6, 6]$.

(c) The points on the graph all have *y*-coordinates between −2 and 4, inclusive; and, for each such number *y*, there is at least one number *x* in the domain. Hence, the range of *f* is $\{y| -2 \le y \le 4\}$, or the interval $[-2, 4]$.

(d) The intercepts are $(-6, 0)$, $(-2, 0)$, and $(0, 4)$.   ■

   When the graph of a function is given, its domain may be viewed as the shadow created by the graph on the *x*-axis by vertical beams of light. Its range can be viewed as the shadow created by the graph on the *y*-axis by horizontal beams of light. Try this technique with the graph given in Figure 87.

   ■   Now work Problems 25 and 27.

E X A M P L E   9

Getting from an Island to Town

An island is 2 miles from the nearest point *P* on a straight shoreline. A town is 12 miles down the shore from *P*.

(a) If a person can row a boat at an average speed of 3 miles per hour and the same person can walk 5 miles per hour, express the time *T* it takes to go from the island to town as a function of the distance *x* from *P* to where the person lands the boat. See Figure 88.

(b) What is the domain of *T*?

(c) How long will it take to travel from the island to town if the person lands the boat 4 miles from *P*?

(d) How long will it take if the person lands the boat 8 miles from *P*?

(e) Use a graphing utility to graph the function $T = T(x)$.

(f) Use the TRACE function to see how the time *T* varies as *x* changes from 0 to 12.

(g) What value of *x* results in the least time?

Solution   (a) Figure 88 illustrates the situation. The distance d_1 from the island to the landing point satisfies the equation

$$d_1^2 = 4 + x^2$$

FIGURE 88

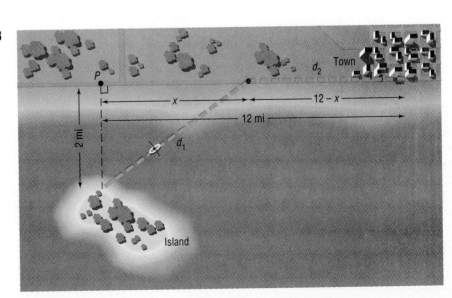

Since the average speed of the boat is 3 miles per hour, the time t_1 it takes to cover the distance d_1 is

$$d_1 = 3t_1$$

Thus,

$$t_1 = \frac{d_1}{3} = \frac{\sqrt{4 + x^2}}{3}$$

The distance d_2 from the landing point to town is $12 - x$, and the time t_2 it takes to cover this distance at an average walking speed of 5 miles per hour obey the equation

$$d_2 = 5t_2$$

Thus,

$$t_2 = \frac{d_2}{5} = \frac{12 - x}{5}$$

The total time T of the trip is $t_1 + t_2$. Thus,

$$T(x) = \frac{\sqrt{4 + x^2}}{3} + \frac{12 - x}{5}$$

(b) Since x equals the distance from P to where the boat lands, it follows that the domain of T is $0 \le x \le 12$.

FIGURE 89

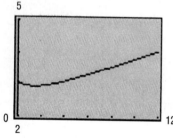

(c) If the boat is landed 4 miles from P, then $x = 4$. The time T the trip takes is

$$T(4) = \frac{\sqrt{20}}{3} + \frac{8}{5} \approx 3.09 \text{ hours}$$

(d) If the boat is landed 8 miles from P, then $x = 8$. The time T the trip takes is

$$T(8) = \frac{\sqrt{68}}{3} + \frac{4}{5} \approx 3.55 \text{ hours}$$

(e) See Figure 89.

(f) As x varies from 0 to 12, the time T varies from about 2.93 to about 4.05.

(g) Using the TRACE function, for x approximately 1.53 miles, the time T is least, about 2.93 hours. See Figure 90(a).

NOTE: Most graphing utilities have a function minimum command that determines the minimum value of a function for a specified domain. Consult your manual. Using this command, we find that x is approximately 1.50 miles with the time T about 2.93 hours. See Figure 90(b).

FIGURE 90

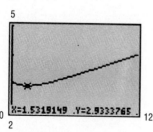

(a) Using TRACE

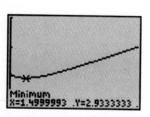

(b) Using minimum on a TI-82

Now work Problem 63.

Summary

We list here some of the important vocabulary introduced in this section, with a brief description of each term.

Function	A rule or correspondence between two sets of real numbers so that each number x in the first set, the domain, has corresponding to it exactly one number y in the second set.
	A set of ordered pairs (x, y) or $(x, f(x))$ in which no two pairs have the same first element.
	The range is the set of y values of the function for the x values in the domain.
	A function f may be defined implicitly by an equation involving x and y or explicitly by writing $y = f(x)$.
Unspecified domain	If a function f is defined by an equation and no domain is specified, then the domain will be taken to be the largest set of real numbers for which the rule defines a real number.
Function notation	$y = f(x)$
	f is a symbol for the rule that defines the function.
	x is the independent variable.
	y is the dependent variable.
	$f(x)$ is the value of the function at x, or the image of x.
Graph of a function	The collection of points (x, y) that satisfies the equation $y = f(x)$.
	A collection of points is the graph of a function provided vertical lines intersect the graph in at most one point (vertical-line test).

1.4

Exercise 1.4

In Problems 1–8, find the following values for each function:

(a) f(0)   (b) f(1)   (c) f(−1)   (d) f(3)

1. $f(x) = -3x^2 + 2x - 4$

2. $f(x) = 2x^2 + x - 1$

3. $f(x) = \dfrac{x}{x^2 + 1}$

4. $f(x) = \dfrac{x^2 - 1}{x + 4}$

5. $f(x) = |x| + 4$

6. $f(x) = \sqrt{x^2 + x}$

7. $f(x) = \dfrac{2x + 1}{3x - 5}$

8. $f(x) = 1 - \dfrac{1}{(x + 2)^2}$

In Problems 9–20, use the graph of the function f *given in the figure.*

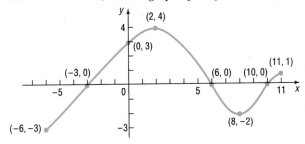

9. Find $f(0)$ and $f(-6)$.

10. Find $f(6)$ and $f(11)$.

11. Is $f(2)$ positive or negative?

12. Is $f(8)$ positive or negative?

13. For what numbers x is $f(x) = 0$?

14. For what numbers x is $f(x) > 0$?

15. What is the domain of f?

16. What is the range of f?

17. What are the x-intercepts?

18. What are the y-intercepts?

19. How often does the line $y = \frac{1}{2}$ intersect the graph?

20. How often does the line $y = 3$ intersect the graph?

In Problems 21–24, answer the questions about the given function.

21. $f(x) = \dfrac{x + 2}{x - 6}$

    (a) Is the point $(3, 14)$ on the graph of f?

    (b) If $x = 4$, what is $f(x)$?

    (c) If $f(x) = 2$, what is x?

    (d) What is the domain of f?

22. $f(x) = \dfrac{x^2 + 2}{x + 4}$

    (a) Is the point $(1, \frac{3}{5})$ on the graph of f?

    (b) If $x = 0$, what is $f(x)$?

    (c) If $f(x) = \frac{1}{2}$, what is x?

    (d) What is the domain of f?

23. $f(x) = \dfrac{2x^2}{x^4 + 1}$

    (a) Is the point $(-1, 1)$ on the graph of f?

    (b) If $x = 2$, what is $f(x)$?

    (c) If $f(x) = 1$, what is x?

    (d) What is the domain of f?

24. $f(x) = \dfrac{2x}{x - 2}$

    (a) Is the point $(\frac{1}{2}, -\frac{2}{3})$ on the graph of f?

    (b) If $x = 4$, what is $f(x)$?

    (c) If $f(x) = 1$, what is x?

    (d) What is the domain of f?

In Problems 25–36, determine whether the graph is that of a function by using the vertical-line test. If it is, use the graph to find:

(a) Its domain and range

(b) The intercepts, if any

(c) Any symmetry with respect to the x-axis, y-axis, or origin

25.

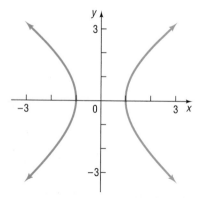

26.

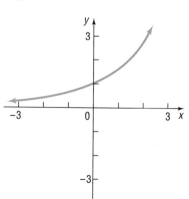

1.5

More about Functions

Function Notation

The independent variable of a function is sometimes called the **argument** of the function. Thinking of the independent variable as an argument sometimes can make it easier to apply the rule of the function. For example, if f is the function defined by $f(x) = x^3$, then f is the rule that tells us to cube the argument. Thus, $f(2)$ means to cube 2, $f(a)$ means to cube the number a, and $f(x + h)$ means to cube the quantity $x + h$.

E X A M P L E   1

Finding Values of a Function

For the function G defined by $G(x) = 2x^2 - 3x$, evaluate:

(a) $G(3)$ (b) $G(x) + G(3)$ (c) $G(-x)$

(d) $-G(x)$ (e) $G(x + 3)$

Solution

(a) We replace x by 3 in the rule for G to get

$$G(3) = 2(3)^2 - 3(3) = 18 - 9 = 9$$

(b) $G(x) + G(3) = (2x^2 - 3x) + (9) = 2x^2 - 3x + 9$

(c) We replace x by $-x$ in the rule for G:

$$G(-x) = 2(-x)^2 - 3(-x) = 2x^2 + 3x$$

(d) $-G(x) = -(2x^2 - 3x) = -2x^2 + 3x$

(e) $G(x + 3) = 2(x + 3)^2 - 3(x + 3)$ Notice the use of parentheses here.

$$= 2(x^2 + 6x + 9) - 3x - 9$$
$$= 2x^2 + 12x + 18 - 3x - 9$$
$$= 2x^2 + 9x + 9$$

Notice in this example that $G(x + 3) \neq G(x) + G(3)$.

■ Now work Problem 31.

Example 1 illustrates certain uses of **function notation.** Let's look at another use.

E X A M P L E   2

Using Function Notation

For the function $f(x) = x^2 + 1$, find: $\dfrac{f(x) - f(1)}{x - 1}, x \neq 1$.

Solution

First, we find $f(1)$:

$$f(1) = (1)^2 + 1 = 2$$

Then

$$\frac{f(x) - f(1)}{x - 1} = \frac{(x^2 + 1) - (2)}{x - 1} = \frac{x^2 - 1}{x - 1} = \frac{(x + 1)(x - 1)}{x - 1} = x + 1 \quad ■$$

Expressions like the one we worked with in Example 2 occur frequently in calculus.

Difference Quotient

For a number c in the domain of a function f, the expression

$$\frac{f(x) - f(c)}{x - c} \qquad x \neq c \qquad (1)$$

is called the **difference quotient** of f at c.

The difference quotient of a function has an important geometric interpretation. Look at the graph of $y = f(x)$ in Figure 91. We have labeled two points on the graph: $(c, f(c))$ and $(x, f(x))$. The slope of the line containing these two points is

$$\frac{f(x) - f(c)}{x - c}$$

This line is called a **secant line.** Thus, the difference quotient of a function equals the slope of a secant line containing two points on its graph.

FIGURE 91

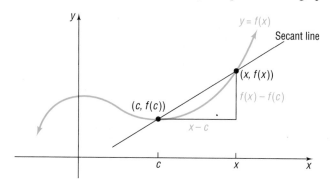

E X A M P L E   3   *Finding a Difference Quotient*

Find the difference quotient of $f(x) = 2x^2 - x + 1$ at $c = 2$.

Solution   From expression (1), we seek

$$\frac{f(x) - f(2)}{x - 2} \qquad x \neq 2$$

We begin by finding $f(2)$:

$$f(2) = 2(2)^2 - (2) + 1 = 8 - 2 + 1 = 7$$

Then the difference quotient of f at 2 is

$$\frac{f(x) - f(2)}{x - 2} = \frac{2x^2 - x + 1 - 7}{x - 2} = \frac{2x^2 - x - 6}{x - 2} = \frac{(2x + 3)(x - 2)}{x - 2} = 2x + 3$$

■   Now work Problem 43.

Increasing and Decreasing Functions

Consider the graph given in Figure 92. If you look from left to right along the graph of this function, you will notice that parts of the graph are rising, parts are falling, and parts are horizontal. In such cases, the function is described as *increasing, decreasing,* and *constant,* respectively.

FIGURE 92

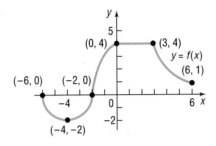

EXAMPLE   4   *Determining Where a Function Is Increasing, Decreasing, or Constant*

Where is the function in Figure 92 increasing? Where is it decreasing? Where is it constant?

Solution   To answer the question of where a function is increasing, where it is decreasing, and where it is constant, we use inequalities involving the independent variable x or we use intervals of x-coordinates. The graph in Figure 92 is rising (increasing) from the point $(-4, -2)$ to the point $(0, 4)$, so we conclude that it is increasing on the interval $[-4, 0]$ (or for $-4 \leq x \leq 0$). The graph is falling (decreasing) from the point $(-6, 0)$ to the point $(-4, -2)$ and from the point $(3, 4)$ to the point $(6, 1)$. We conclude that the graph is decreasing on the intervals $[-6, -4]$ and $[3, 6]$ (or for $-6 \leq x \leq -4$ and $3 \leq x \leq 6$). The graph is constant on the interval $[0, 3]$ (or for $0 \leq x \leq 3$). ■

More precise definitions follow.

Increasing Function
> A function f is **increasing** on an interval I if, for any choice of x_1 and x_2 in I, with $x_1 < x_2$, we have $f(x_1) < f(x_2)$.

Decreasing Function
> A function f is **decreasing** on an interval I if, for any choice of x_1 and x_2 in I, with $x_1 < x_2$, we have $f(x_1) > f(x_2)$.

Constant Function
> A function f is **constant** on an interval I if, for all choices of x in I, the values $f(x)$ are equal.

Thus, the graph of an increasing function goes up from left to right, the graph of a decreasing function goes down from left to right, and the graph of a constant function remains at a fixed height. Figure 93 illustrates the definitions.

FIGURE 93

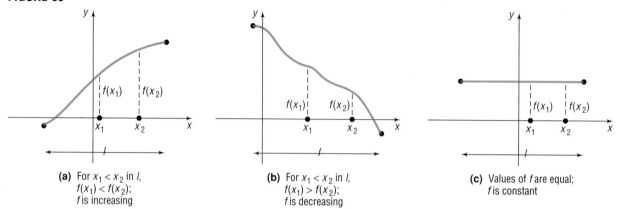

(a) For $x_1 < x_2$ in I,
$f(x_1) < f(x_2)$;
f is increasing

(b) For $x_1 < x_2$ in I,
$f(x_1) > f(x_2)$;
f is decreasing

(c) Values of f are equal;
f is constant

E X A M P L E   5   *Using a Graphing Utility to Determine Where A Function Is Increasing and Decreasing*

Use a graphing utility to graph the function $f(x) = x^3 - 2x^2 + 3$ for $-1 \leq x \leq 2$. Determine where f is increasing and where it is decreasing.

Solution   Figure 94 shows the graph of f. Use the TRACE function, beginning at the point $(-1, 0)$. As we proceed along the graph, the values of y increase until we cross the y-axis (where $y = 3$). We infer that f is increasing on the interval $[-1, 0]$.

As we continue along the graph to the right of the y-axis, the values of f decrease until we reach $x \approx 1.33$ (where $y \approx 1.815$). We infer that f is decreasing on the interval $[0, 1.33]$.

As we continue, the values of y increase from $x = 1.33$ to $x = 2$. We infer that f is increasing on the interval $[1.33, 2]$.

FIGURE 94

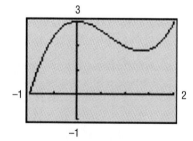

Local Maximum; Local Minimum

When the graph of a function is increasing to the left of $x = c$ and decreasing to the right of $x = c$, then at c the value of f is largest. This value is called a *local maximum* of f.

When the graph of a function is decreasing to the left of $x = c$ and is increasing to the right of $x = c$, then at c the value of f is the smallest. This value is called a *local minimum* of f.

Thus, if f has a local maximum at c, then the value of f at c is greater than or equal to the values of f near c. If f has a local minimum at c, then the value of

f at c is less than or equal to the values of f near c. The word *local* is used to suggest that it is only near c that the value $f(c)$ is largest or smallest. See Figure 95.

FIGURE 95

f has local maximum at x_1 and x_3;
f has a local minimum at x_2.

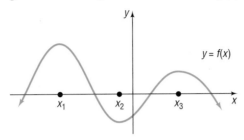

A function f has a **local maximum at c** if there is an interval I containing c so that, for all x in I, $f(x) < f(c)$. We call $f(c)$ a **local maximum of f**.

A function f has a **local minimum at c** if there is an interval I containing c so that, for all x in I, $f(x) > f(c)$. We call $f(c)$ a **local minimum of f**.

For example, in Figure 94, f has a local maximum at 0 and a local minimum at 1.33. The local maximum is $f(0) = 3$; the local minimum is $f(1.33) = 1.815$.

To locate the exact value at which a function f has a local maximum or a local minimum usually requires calculus. However, a graphing utility may be used to approximate these values.

E X A M P L E   6   *Using a Graphing Utility to Locate Local Maxima and Minima*

Use a graphing utility to graph $f(x) = 6x^3 - 12x + 5$ for $-2 \leq x \leq 2$. Determine where f has a local maximum and where f has a local minimum.

Solution   Figure 96 shows the graph of f on the interval $[-2, 2]$.

Using the TRACE function, we find the local maximum to occur at approximately -0.80 and the local minimum to occur at 0.80, correct to two decimal places. The local maximum is 11.53 and the local minimum is -1.53, correct to two decimal places. See Figures 96(a) and (b).

FIGURE 96

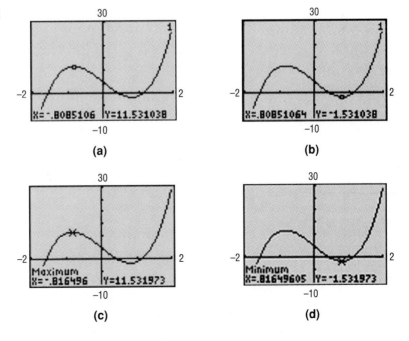

(a)

(b)

(c)

(d)

Most graphing utilities have a function that finds the maximum or minimum point of a graph within a given domain. Using this function, we find the maximum occurs at $(-0.81, 11.53)$ and the minimum occurs at $(0.81, -1.53)$, correct to two decimal places, for $-2 \le x \le 2$, as seen in Figures 96(c) and (d). ■

■   Now work Problem 103.

Even and Odd Functions

Even Function

A function f is **even** if for every number x in its domain the number $-x$ is also in the domain and

$$f(-x) = f(x)$$

Odd Function

A function f is **odd** if for every number x in its domain the number $-x$ is also in the domain and

$$f(-x) = -f(x)$$

Refer to Section 1.2, where the tests for symmetry are listed. The following results are then evident:

Theorem   A function is even if and only if its graph is symmetric with respect to the y-axis. A function is odd if and only if its graph is symmetric with respect to the origin. ■

E X A M P L E   7   *Determining Even and Odd Functions from the Graph*

Determine whether each graph given in Figure 97 is the graph of an even function, an odd function, or a function that is neither even nor odd.

FIGURE 97

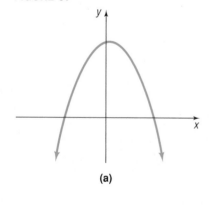

(a)

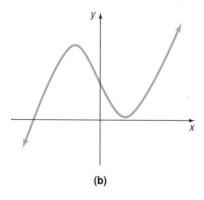

(b)

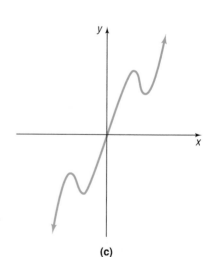

(c)

Solution   The graph in Figure 97(a) is that of an even function, because the graph is symmetric with respect to the y-axis. The function whose graph is given in Figure 97(b) is neither even nor odd, because the graph is neither symmetric with respect to the y-axis nor symmetric with respect to the origin. The function whose graph is given in Figure 97(c) is odd, because its graph is symmetric with respect to the origin. ■

■   Now work Problem 9.

A graphing utility can be used to conjecture whether a function is even, odd, or neither. As stated, a function is even if $f(-x) = f(x)$. This condition implies that when an even function contains the point (x, y) it must also contain the point $(-x, y)$. Therefore, if TRACE indicates that both the point (x, y) and the point $(-x, y)$ are on the graph for every x, then we would conjecture that the function is even.*

In addition, a function is odd if $f(-x) = -f(x)$. This condition implies that an odd function contains the points $(-x, -y)$ and (x, y). TRACE could be used in the same way to conjecture that the function is odd.*

In the next example, we show how to verify whether a function is even, odd, or neither.

E X A M P L E   8   *Identifying Even and Odd Functions*

Determine whether each of the following functions is even, odd, or neither. Then determine whether the graph is symmetric with respect to the y-axis or with respect to the origin.

(a) $f(x) = x^2 - 5$ (b) $g(x) = x^3 - 1$

(c) $h(x) = 5x^3 - x$ (d) $F(x) = |x|$

(a) Graphing Solution   Graph the function. Use TRACE to determine different pairs of points (x, y) and $(-x, y)$. For example, the point $(1.9574468, -1.168402)$ is on the graph. See Figure 98(a). Now move the cursor to -1.9574468 and determine the corresponding y-coordinate. See Figure 98(b). Since $(1.9574468, -1.168402)$ and $(-1.9574468, -1.168402)$ both lie on the graph, we have evidence that the function is even. Repeating this procedure yields similar results. Therefore, we conjecture that the function is even.

FIGURE 98

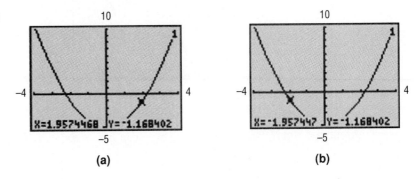

(a)   (b)

*—Xmin and Xmax must be equal for this to work.

Algebraic Solution   We replace x by $-x$ in $f(x) = x^2 - 5$. Then

$$f(-x) = (-x)^2 - 5 = x^2 - 5$$

Since $f(-x) = f(x)$, we conclude that f is an even function, and the graph is symmetric with respect to the y-axis.

(b) Graphing Solution   Graph the function. Using TRACE, the point (1.787234, 4.7087929) is on the graph. See Figure 99(a). Now move the cursor to -1.787234 and determine the corresponding y-coordinate, -6.708793. See Figure 99(b). We conjecture that the function is neither even nor odd since (x, y) does not equal $(-x, y)$ (so it is not even) and (x, y) also does not equal $(-x, -y)$ (so it is not odd).

FIGURE 99

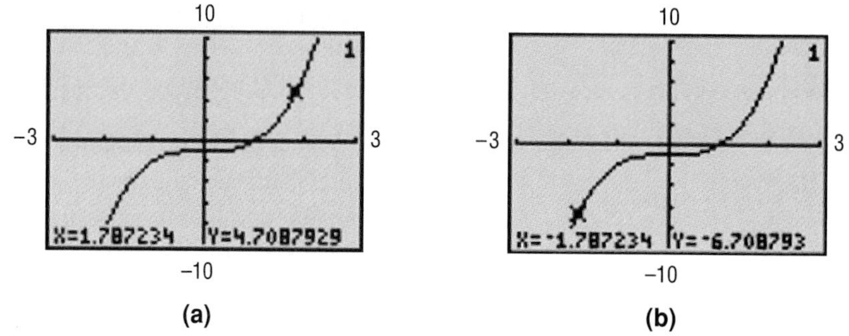

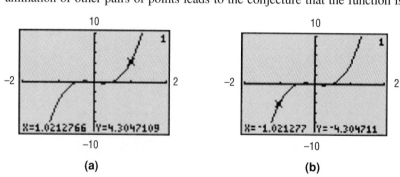

(a)   (b)

Algebraic Solution   We replace x by $-x$. Then

$$g(-x) = (-x)^3 - 1 = -x^3 - 1$$

Since $g(-x) \neq g(x)$ and $g(-x) \neq -g(x)$, we conclude that g is neither even nor odd. The graph is not symmetric with respect to the y-axis nor with respect to the origin.

(c) Graphing Solution   Graph the function. Using TRACE, the point (1.0212766, 4.3047109) is on the graph. See Figure 100(a). We move the cursor to -1.021277 and determine the corresponding y-coordinate, -4.304711. See Figure 100(b). Since (x, y) equals $(-x, -y)$, correct to five decimal places we have evidence that the function is odd. An examination of other pairs of points leads to the conjecture that the function is odd.

FIGURE 100

(a)   (b)

Algebraic Solution   We replace x by $-x$ in $h(x) = 5x^3 - x$. Then

$$h(-x) = 5(-x)^3 - (-x) = -5x^3 + x$$

Since $h(-x) = -h(x)$, h is an odd function, and the graph of h is symmetric with respect to the origin.

(d) Graphing Solution

Graph the function. Using TRACE, the point $(-5.744681, 5.7446809)$ is on the graph. See Figure 101(a). We move the cursor to 5.7446809 and determine the corresponding y-coordinate, 5.7446809. See Figure 101(b). Since (x, y) equals $(-x, y)$, we have evidence that the function is even. Repeating this procedure yields similar results. We conjecture that the function is even.

FIGURE 101

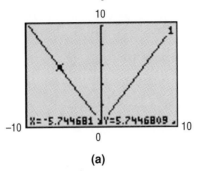
(a)

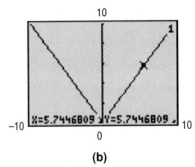
(b)

Algebraic Solution

We replace x by $-x$ in $F(x) = |x|$. Then

$$F(-x) = |-x| = |x|$$

Since $F(-x) = F(x)$, F is an even function, and the graph of F is symmetric with respect to the y-axis. ∎

■   Now work Problem 53.

Important Functions

We now give names to some of the functions we have encountered. In going through this list, pay special attention to the characteristics of each function, particularly to the shape of each graph.

Linear Function

$$f(x) = mx + b \qquad m \text{ and } b \text{ are real numbers}$$

The domain of the **linear function** f consists of all real numbers. The graph of this function is a nonvertical straight line with slope m and y-intercept b. A linear function is increasing if $m > 0$, decreasing if $m < 0$, and constant if $m = 0$.

Constant Function

$$f(x) = b \qquad b \text{ a real number}$$

See Figure 102.

FIGURE 102

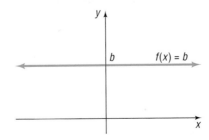

A **constant function** is a special linear function ($m = 0$). Its domain is the set of all real numbers; its range is the set consisting of a single number b. Its graph is a horizontal line whose y-intercept is b. The constant function is an even function whose graph is constant over its domain.

Identity Function

$$f(x) = x$$

See Figure 103.

FIGURE 103

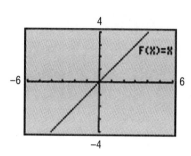 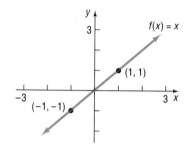

The **identity function** is also a special linear function. Its domain and its range are the set of all real numbers. Its graph is a line whose slope is $m = 1$ and whose y-intercept is 0. The line consists of all points for which the x-coordinate equals the y-coordinate. The identity function is an odd function that is increasing over its domain. Note that the graph bisects quadrants I and III.

Square Function

$$f(x) = x^2$$

See Figure 104.

FIGURE 104

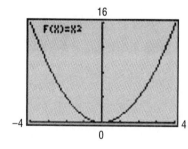 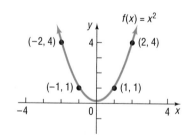

The domain of the **square function** f is the set of all real numbers; its range is the set of nonnegative real numbers. The graph of this function is a parabola, whose intercept is at $(0, 0)$. The square function is an even function that is decreasing on the interval $(-\infty, 0]$ and increasing on the interval $[0, \infty)$.

Cube Function

$$f(x) = x^3$$

See Figure 105.

FIGURE 105

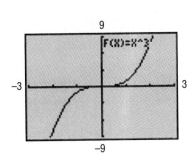

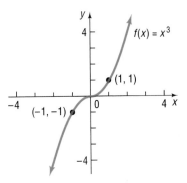

The domain and range of the **cube function** are the set of all real numbers. The intercept of the graph is at (0, 0). The cube function is odd and is increasing on the interval $(-\infty, \infty)$.

Square Root Function

$$f(x) = \sqrt{x}$$

See Figure 106.

FIGURE 106

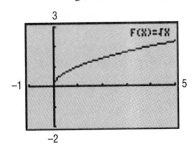

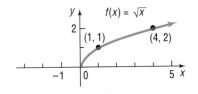

The domain and range of the **square root function** are the set of nonnegative real numbers. The intercept of the graph is at (0, 0). The square root function is neither even nor odd and is increasing on the interval $[0, \infty)$.

Reciprocal Function

$$f(x) = \frac{1}{x}$$

Refer to Example 16, p. 30, for a discussion of the equation $y = 1/x$. See Figure 107.

FIGURE 107

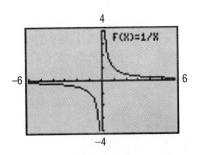

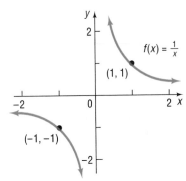

The domain and range of the **reciprocal function** are the set of all nonzero real numbers. The graph has no intercepts. The reciprocal function is decreasing on the intervals $(-\infty, 0)$ and $(0, \infty)$ and is an odd function.

Absolute Value Function

$$f(x) = |x|$$

See Figure 108.

FIGURE 108

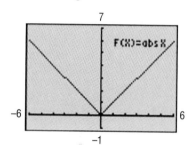
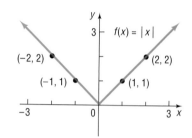

The domain of the **absolute value function** is the set of all real numbers; its range is the set of nonnegative real numbers. The intercept of the graph is at $(0, 0)$. If $x \geq 0$, then $f(x) = x$ and the graph of f is part of the line $y = x$; if $x < 0$, then $f(x) = -x$ and the graph of f is part of the line $y = -x$. The absolute value function is an even function; it is decreasing on the interval $(-\infty, 0]$ and increasing on the interval $[0, \infty)$.

Comment: If your utility has no built-in absolute value function, you can still graph $f(x) = |x|$ by using the fact that $|x| = \sqrt{(x^2)}$.

The symbol $[[x]]$, read as **"bracket x,"** stands for the largest integer less than or equal to x. For example,

$$[[1]] = 1 \qquad [[2.5]] = 2 \qquad [[\tfrac{1}{2}]] = 0 \qquad [[-\tfrac{3}{4}]] = -1 \qquad [[\pi]] = 3$$

This type of correspondence occurs frequently enough in mathematics that we give it a name.

Greatest-integer Function

$$f(x) = [[x]] = \text{Greatest integer less than or equal to } x$$

We obtain the graph of $f(x) = [[x]]$ by plotting several points. See Table 8. For values of x, $-1 \leq x < 0$, the value of $f(x) = [[x]]$ is -1; for values of x, $0 \leq x < 1$, the value of f is 0. See Figure 109 for the graph.

TABLE 8

X	Y1
-1	-1
-.75	-1
-.5	-1
-.25	-1
0	0
.25	0
.5	0

Y1☐int X

FIGURE 109

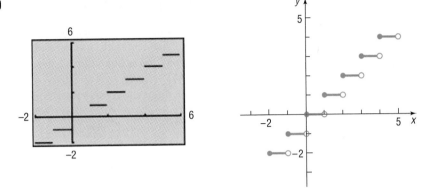

The domain of the **greatest integer function** is the set of all real numbers; its range is the set of integers. The y-intercept of the graph is at 0. The x-intercepts lie in the interval $[0, 1)$. The greatest-integer function is neither even nor odd. It is constant on every interval of the form $[k, k + 1)$, for k an integer. In Figure 109, we use a solid dot to indicate, for example, that at $x = 1$ the value of f is $f(1) = 1$; we use an open circle to illustrate that the function does not assume the value of 0 at $x = 1$.

From the graph of the greatest-integer function, we can see why it is also called a **step function.** At $x = 0$, $x = \pm 1$, $x = \pm 2$, and so on, this function exhibits what is called a *discontinuity;* that is, at integer values, the graph suddenly "steps" from one value to another without taking on any of the intermediate values. For example, to the immediate left of $x = 3$, the y-coordinates are 2, and to the immediate right of $x = 3$, the y-coordinates are 3.

Comment: When graphing a function, you can choose either the **connected mode,** in which points plotted on the screen are connected, making the graph appear without any breaks, or the **dot mode,** in which only the points plotted appear. When graphing the greatest integer function with a graphing utility, it is necessary to be in the **dot mode.** This is to prevent the utility from "connecting the dots" when $f(x)$ changes from one integer value to the next. See Figure 110.

FIGURE 110

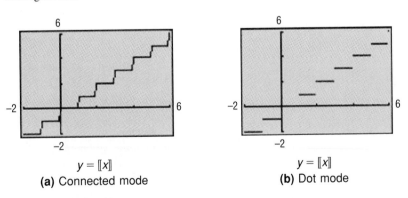

$y = [\![x]\!]$
(a) Connected mode

$y = [\![x]\!]$
(b) Dot mode

The functions we have discussed so far are basic. Whenever you encounter one of them, you should see a mental picture of its graph. For example, if you encounter the function $f(x) = x^2$, you should see in your mind's eye a picture like Figure 104.

■ Now work Problem 69.

Piecewise Defined Functions

Sometimes, a function is defined by a rule consisting of two or more equations. The choice of which equation to use depends on the value of the independent variable x. For example, the absolute value function $f(x) = |x|$ is actually defined by two equations: $f(x) = x$ if $x \geq 0$ and $f(x) = -x$ if $x < 0$. For convenience, we generally combine these equations into one expression as

$$f(x) = |x| = \begin{cases} x & \text{if } x \geq 0 \\ -x & \text{if } x < 0 \end{cases}$$

When functions are defined by more than one equation, they are called **piecewise defined** functions.

Let's look at another example of a piecewise defined function.

EXAMPLE 9   *Analyzing a Piecewise Defined Function*

For the following function f,

$$f(x) = \begin{cases} -x + 1 & \text{if } -1 \leq x < 1 \\ 2 & \text{if } x = 1 \\ x^2 & \text{if } x > 1 \end{cases}$$

(a) Find $f(0)$, $f(1)$, and $f(2)$.   (b) Determine the domain of f.

(c) Graph f.   (d) Use the graph to find the range of f.

Solution   (a) To find $f(0)$, we observe that when $x = 0$ the equation for f is given by $f(x) = -x + 1$. So we have

$$f(0) = -0 + 1 = 1$$

When $x = 1$, the equation for f is $f(x) = 2$. Thus

$$f(1) = 2$$

When $x = 2$, the equation for f is $f(x) = x^2$. So

$$f(2) = 2^2 = 4$$

(b) To find the domain of f, we look at its definition. We conclude that the domain of f is $\{x | x \geq -1\}$, or $[-1, \infty)$.

(c) On a graphing utility, the procedure for graphing a piecewise defined function varies depending on the particular utility. In general, you need to enter each piece as a function with a restricted domain.

To graph the middle piece [the point $(1, 2)$], use the $\boxed{\begin{smallmatrix}\text{STAT}\\\text{PLOT}\end{smallmatrix}}$ function. See Figure 111(a).

In graphing piecewise defined functions, it is usually better to be in the dot mode, since such functions may have breaks.

To graph f on paper, we graph "each piece." Thus, we first graph the line $y = -x + 1$ and keep only the part for which $-1 \leq x < 1$. Then we plot the point $(1, 2)$, because when $x = 1$, $f(x) = 2$. Finally, we graph the parabola $y = x^2$ and keep only the part for which $x > 1$. See Figure 111(b).

FIGURE 111

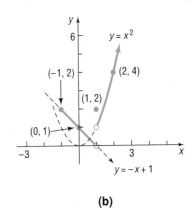

(a)

(b)

(d) From the graph, we conclude that the range of f is $\{y|y > 0\}$, or $(0, \infty)$. ∎

■ Now work Problem 83.

E X A M P L E   1 0   *Cost of Electricity*

In the winter, Commonwealth Edison Company supplies electricity to residences for a monthly customer charge of \$9.06 plus 10.819¢ per kilowatt-hour (kWhr) for the first 400 kWhr supplied in the month, and 7.093¢ per kWhr for all usage over 400 kWhr in the month.*

(a) What is the charge for using 300 kWhr in a month?

(b) What is the charge for using 700 kWhr in a month?

(c) If C is the monthly charge for x kWhr, express C as a function of x.

Solution   (a) For 300 kWhr, the charge is \$9.06 plus 10.819¢ = \$0.10819 per kWhr. Thus,

$$\text{Charge} = \$9.06 + \$0.10819(300) = \$41.52$$

(b) For 700 kWhr, the charge is \$9.06 plus 10.819¢ for the first 400 kWhr plus 7.093¢ for the 300 kWhr in excess of 400. Thus,

$$\text{Charge} = \$9.06 + \$0.10819(400) + \$0.07093(300) = \$73.62$$

(c) If $0 \le x \le 400$, the monthly charge C (in dollars) can be found by multiplying x times \$0.10819 and adding the monthly customer charge of \$9.06. Thus, if $0 \le x \le 400$, then $C(x) = 0.10819x + 9.06$. For $x > 400$, the charge is $0.10819(400) + 9.06 + 0.07093(x - 400)$, since $x - 400$ equals the usage in excess of 400 kWhr, which costs \$0.07093 per kWhr. Thus, if $x > 400$, then

$$C(x) = 0.10819(400) + 9.06 + 0.07093(x - 400)$$
$$= 52.336 + 0.07093(x - 400)$$
$$= 0.07093x + 23.964$$

The rule for computing C follows two rules:

$$C(x) = \begin{cases} 0.10819x + 9.06 & \text{if } 0 \le x \le 400 \\ 0.07093x + 23.964 & \text{if } x > 400. \end{cases}$$

*Source: Commonwealth Edison Co., Chicago, Illinois, 1991.

See Figure 112 for the graph.

FIGURE 112

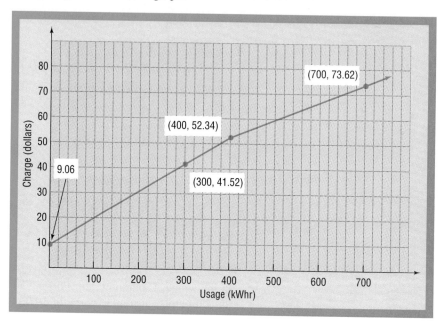

1.5

Exercise 1.5

In Problems 1–8, match each graph to the function listed whose graph most resembles the one given.

A. *Constant function*   B. *Linear function*
C. *Square function*   D. *Cube function*
E. *Square root function*   F. *Reciprocal function*
G. *Absolute value function*   H. *Greatest integer function*

1.

2.

3.

4.

5.

6.

7.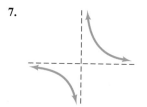

8.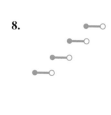

In Problems 9–24, the graph of a function is given. Use the graph to find:

(a) Its domain and range

(b) The intervals on which it is increasing, decreasing, or constant

(c) Whether it is even, odd, or neither

(d) The intercepts, if any

9.

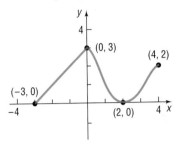

10.

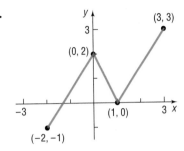

11.

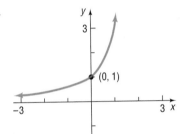

12.

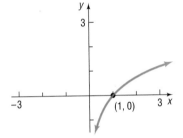

13.

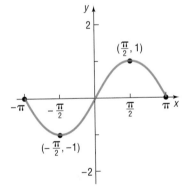

14.

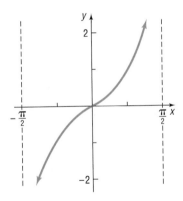

15.

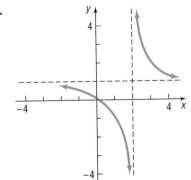

16.

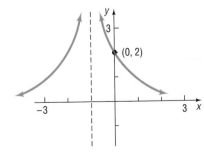

17.

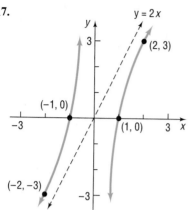

18.

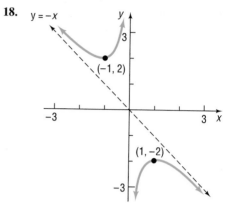

19.

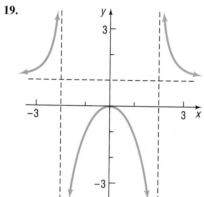

20.

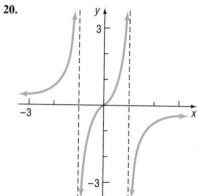

For Problems 21–24, assume that the graph shown is complete.

21.

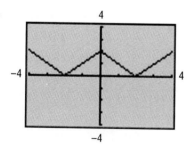

22.

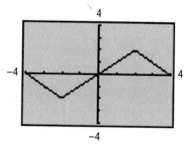

23.

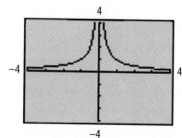

24.

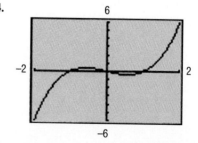

25. If $f(x) = [[2x]]$, find: (a) $f(1.2)$ (b) $f(1.6)$ (c) $f(-1.8)$

26. If $f(x) = [[x/2]]$, find: (a) $f(1.2)$ (b) $f(1.6)$ (c) $f(-1.8)$

27. If

$$f(x) = \begin{cases} x^2 & \text{if } x < 0 \\ 2 & \text{if } x = 0 \\ 2x + 1 & \text{if } x > 0 \end{cases}$$

find: (a) $f(-2)$ (b) $f(0)$ (c) $f(2)$

28. If

$$f(x) = \begin{cases} x^3 & \text{if } x < 0 \\ 3x + 2 & \text{if } x \geq 0 \end{cases}$$

find: (a) $f(-1)$ (b) $f(0)$ (c) $f(1)$

In Problems 29–40, find the following for each function:

   (a) $f(-x)$ (b) $-f(x)$ (c) $f(2x)$ (d) $f(x - 3)$ (e) $f(1/x)$ (f) $1/f(x)$

29. $f(x) = 2x + 5$ **30.** $f(x) = 3 - x$ **31.** $f(x) = 2x^2 - 4$ **32.** $f(x) = x^3 + 1$

33. $f(x) = x^3 - 3x$ **34.** $f(x) = x^2 + x$ **35.** $f(x) = \dfrac{x}{x^2 + 1}$ **36.** $f(x) = \dfrac{x^2}{x^2 + 1}$

37. $f(x) = |x|$ **38.** $f(x) = \dfrac{1}{x}$ **39.** $f(x) = 1 + \dfrac{1}{x}$ **40.** $f(x) = 4 + \dfrac{2}{x}$

In Problems 41–52, find the difference quotient,

$$\frac{f(x) - f(1)}{x - 1} \qquad x \neq 1$$

for each function. Be sure to simplify.

41. $f(x) = 3x$ **42.** $f(x) = -2x$ **43.** $f(x) = 1 - 3x$ **44.** $f(x) = x^2 + 1$

45. $f(x) = 3x^2 - 2x$ **46.** $f(x) = 4x - 2x^2$ **47.** $f(x) = x^3 - x$ **48.** $f(x) = x^3 + x$

49. $f(x) = \dfrac{2}{x + 1}$ **50.** $f(x) = \dfrac{1}{x^2}$ **51.** $f(x) = \sqrt{x}$ **52.** $f(x) = \sqrt{x + 3}$

In Problems 53–64, tell whether each function is even, odd, or neither without drawing a graph. Graph each function and use TRACE to verify your results.

53. $f(x) = 4x^3$ **54.** $f(x) = 2x^4 - x^2$ **55.** $g(x) = 2x^2 - 5$ **56.** $h(x) = 3x^3 + 2$

57. $F(x) = \sqrt[3]{x}$ **58.** $G(x) = \sqrt{x}$ **59.** $f(x) = x + |x|$ **60.** $f(x) = \sqrt[3]{2x^2 + 1}$

61. $g(x) = \dfrac{1}{x^2}$ **62.** $h(x) = \dfrac{x}{x^2 - 1}$ **63.** $h(x) = \dfrac{x^3}{3x^2 - 9}$ **64.** $F(x) = \dfrac{x}{|x|}$

65. How many x-intercepts can a function defined on an interval have if it is increasing on that interval? Explain.

66. How many y-intercepts can a function have? Explain.

In Problems 67–92:

   (a) *Find the domain of each function.*   (b) *Locate any intercepts.*

   (c) *Graph each function by hand.*   (d) *Based on the graph, find the range.*

   (e) *Verify your results using a graphing utility.*

67. $f(x) = 3x - 3$ **68.** $f(x) = 4 - 2x$ **69.** $g(x) = x^2 - 4$

70. $g(x) = x^2 + 4$ **71.** $h(x) = -x^2$ **72.** $F(x) = 2x^2$

73. $f(x) = \sqrt{x - 2}$ **74.** $g(x) = \sqrt{x} + 2$ **75.** $h(x) = \sqrt{2 - x}$

76. $F(x) = -\sqrt{x}$ **77.** $f(x) = |x| + 3$ **78.** $g(x) = |x + 3|$

79. $h(x) = -|x|$ **80.** $F(x) = |3 - x|$

81. $f(x) = \begin{cases} 2x & \text{if } x \neq 0 \\ 0 & \text{if } x = 0 \end{cases}$ **82.** $f(x) = \begin{cases} 3x & \text{if } x \neq 0 \\ 4 & \text{if } x = 0 \end{cases}$

83. $f(x) = \begin{cases} 1 + x & \text{if } x < 0 \\ x^2 & \text{if } x \geq 0 \end{cases}$ **84.** $f(x) = \begin{cases} 1/x & \text{if } x < 0 \\ \sqrt{x} & \text{if } x \geq 0 \end{cases}$

85. $f(x) = \begin{cases} |x| & \text{if } -2 \leq x < 0 \\ 1 & \text{if } x = 0 \\ x^3 & \text{if } x > 0 \end{cases}$ **86.** $f(x) = \begin{cases} 3 + x & \text{if } -3 \leq x < 0 \\ 3 & \text{if } x = 0 \\ \sqrt{x} & \text{if } x > 0 \end{cases}$

87. $g(x) = \begin{cases} 1 & \text{if } x \text{ is an integer} \\ -1 & \text{if } x \text{ is not an integer} \end{cases}$ **88.** $g(x) = \begin{cases} x & \text{if } x \geq 1 \\ 1 & \text{if } x < 1 \end{cases}$

89. $h(x) = 2[[x]]$ **90.** $f(x) = [[2x]]$

91. $F(x) = \begin{cases} 4 - x^2 & \text{if } |x| \leq 2 \\ x^2 - 4 & \text{if } |x| > 2 \end{cases}$ **92.** $G(x) = |x^2 - 4|$

Problems 93–96, require the following definition. Secant Line *The slope of the secant line containing the two points $(x, f(x))$ and $(x + h, f(x + h))$ on the graph of a function $y = f(x)$ may be given as*

$$\frac{f(x + h) - f(x)}{h}$$

In Problems 93–96, express the slope of the secant line of each function in terms of x and h. Be sure to simplify your answer.

93. $f(x) = 2x + 5$ **94.** $f(x) = -3x + 2$

95. $f(x) = x^2 + 2x$ **96.** $f(x) = 1/x$

In Problems 97–100, the graph of a piecewise-defined function is given. Write a definition for each function.

97.

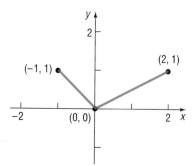

98.

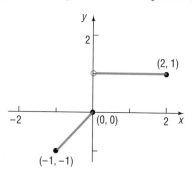

99.

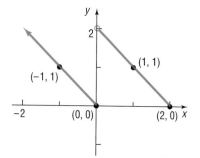

100.

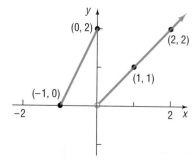

In Problems 101 and 102, decide whether each function is even. Give a reason.

101. $f(x) = \begin{cases} x^2 + 4 & \text{if } x \neq 2 \\ 6 & \text{if } x = 2 \end{cases}$

102. $f(x) = \begin{cases} x^2 + 4 & \text{if } x \neq 2 \\ 5 & \text{if } x = 2 \end{cases}$

In Problems 103–106, use a graphing utility to graph each function over the indicated interval. Determine where the function is increasing and where it is decreasing. Approximate any local maxima and local minima.

103. $f(x) = x^3 - 3x + 2$ $[-2, 2]$

104. $f(x) = x^3 - 3x^2 + 5$ $[-1, 3]$

105. $f(x) = x^5 - x^3$ $[-2, 2]$

106. $f(x) = x^4 - x^2$ $[-2, 2]$

107. Graph $y = x^2$. Then on the same screen graph $y = x^2 + 2$, followed by $y = x^2 + 4$, followed by $y = x^2 - 2$. What pattern do you observe? Can you predict the graph of $y = x^2 - 4$? Of $y = x^2 + 5$?

108. Graph $y = x^2$. Then on the same screen graph $y = (x - 2)^2$, followed by $y = (x - 4)^2$, followed by $y = (x + 2)^2$. What pattern do you observe? Can you predict the graph of $y = (x + 4)^2$? Of $y = (x - 5)^2$?

109. Graph $y = |x|$. Then on the same screen graph $y = 2|x|$, followed by $y = 4|x|$, followed by $y = \frac{1}{2}|x|$. What pattern do you observe? Can you predict the graph of $y = \frac{1}{4}|x|$? Of $y = 5|x|$?

110. Graph $y = x^2$. Then on the same screen graph $y = -x^2$. What pattern do you observe? Now try $y = |x|$ and $y = -|x|$. What do you conclude?

111. Graph $y = \sqrt{x}$. Then on the same screen graph $y = \sqrt{-x}$. What pattern do you observe? Now try $y = 2x + 1$ and $y = 2(-x) + 1$. What do you conclude?

112. Graph $y = x^3$. Then on the same screen graph $y = (x - 1)^3 + 2$. Could you have predicted the result?

113. Graph $y = x^2$, $y = x^4$, and $y = x^6$ on the same screen. What do you notice is the same about each graph? What do you notice that is different?

114. Graph $y = x^3$, $y = x^5$, and $y = x^7$ on the same screen. What do you notice is the same about each graph? What do you notice that is different?

115. *Cost of Natural Gas* On November 12, 1991, the Peoples Gas Light and Coke Company had the following rate schedule* for natural gas usage in single-family residences:

Monthly service charge	$7.00
Per therm service charge, 1st 90 therms	$0.21054/therm
Over 90 therms	$0.11242/therm
Gas charge	$0.26341/therm

(a) What is the charge for using 50 therms in a month?
(b) What is the charge for using 500 therms in a month?
(c) Construct a function that relates the monthly charge C for x therms of gas.
(d) Graph this function.

116. *Cost of Natural Gas* On November 19, 1991, Northern Illinois Gas Company had the following rate schedule† for natural gas usage in single-family residences:

Monthly customer charge	$4.00
Distribution charge, 1st 50 therms	$0.1402/therm
Over 50 therms	$0.0547/therm
Gas supply charge	$0.2406/therm

(a) What is the charge for using 40 therms in a month?
(b) What is the charge for using 202 therms in a month?
(c) Construct a function that gives the monthly charge C for x therms of gas.
(d) Graph this function.

*Source: The Peoples Gas Light and Coke Company, Chicago, Illinois.
†Source: Northern Illinois Gas Company, Naperville, Illinois.

117. Let f denote any function with the property that, whenever x is in its domain, then so is $-x$. Define the functions $E(x)$ and $O(x)$ to be

$$E(x) = \frac{1}{2}[f(x) + f(-x)] \qquad O(x) = \frac{1}{2}[f(x) - f(-x)]$$

(a) Show that $E(x)$ is an even function.   (b) Show that $O(x)$ is an odd function.
(c) Show that $f(x) = E(x) + O(x)$.
(d) Draw the conclusion that any such function f can be written as the sum of an even function and an odd function.

118. Let f and g be two functions defined on the same interval $[a, b]$. Suppose that we define two functions min (f, g) and max (f, g) as follows:

$$\min(f, g)(x) = \begin{cases} f(x) & \text{if } f(x) \le g(x) \\ g(x) & \text{if } f(x) > g(x) \end{cases} \qquad \max(f, g)(x) = \begin{cases} g(x) & \text{if } f(x) \le g(x) \\ f(x) & \text{if } f(x) > g(x) \end{cases}$$

Show that

$$\min(f, g)(x) = \frac{f(x) + g(x)}{2} - \frac{|f(x) - g(x)|}{2}$$

Develop a similar formula for max (f, g).

119. Consider the equation

$$y = \begin{cases} 1 & \text{if } x \text{ is rational} \\ 0 & \text{if } x \text{ is irrational} \end{cases}$$

Is this a function? What is its domain? What is its range? What is its y-intercept, if any? What are its x-intercepts, if any? Is it even, odd, or neither? How would you describe its graph?

120. Define some functions that pass through $(0, 0)$ and $(1, 1)$ and are increasing for $x \ge 0$. Begin your list with $y = \sqrt{x}$, $y = x$, and $y = x^2$. Can you propose a general result about such functions?

121. Can you think of a function that is both even and odd?

1.6

Graphing Techniques

At this stage, if you were asked to graph any of the functions defined by $y = x^2$, $y = x^3$, $y = x$, $y = \sqrt{x}$, $y = |x|$, or $y = 1/x$, your response should be, "Yes, I recognize these functions and know the general shapes of their graphs." (If this is not your answer, review the previous section and Figures 102 through 110.)

Sometimes, we are asked to graph a function that is "almost" like one we already know how to graph. In this section, we look at some of these functions and develop techniques for graphing them.

Vertical Shifts

E X A M P L E   1   *Vertical Shifts*

On the same screen, graph each of the following functions:

$$f(x) = x^2$$
$$f(x) = x^2 + 1$$
$$f(x) = x^2 + 2$$
$$f(x) = x^2 - 1$$
$$f(x) = x^2 - 2$$

Solution   Figure 113 illustrates the graphs. You should have observed a general pattern. With $y = x^2$ on the screen, the graph of $y = x^2 + 1$ is identical to that of $y = x^2$, except that it is shifted vertically up 1 unit. Similarly, $y = x^2 + 2$ is identical to that of $y = x^2$, except that it is shifted vertically up 2 units. The graph of $y = x^2 - 1$ is identical to that of $y = x^2$, except that it is shifted vertically down 1 unit.

FIGURE 113

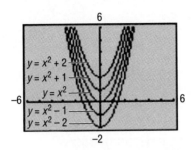

We are led to the following conclusion: If a real number c is added to the right side of a function $y = f(x)$, the graph of the new function $y = f(x) + c$ is the graph of f **shifted vertically** up (if $c > 0$) or down (if $c < 0$). Let's look at another example.

E X A M P L E   2

Vertical Shift Down

Use the graph of $f(x) = x^2$ to obtain the graph of $h(x) = x^2 - 4$.

Solution   Table 9 lists some points on the graphs of $f = Y_1$ and $h = Y_2$. The graph of h is identical to that of f, except that it is shifted down 4 units. See Figure 114.

TABLE 9

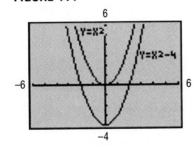

FIGURE 114

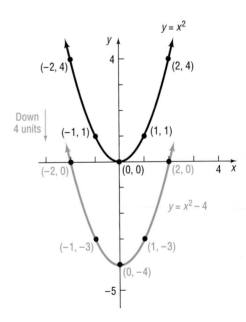

Horizontal Shifts

E X A M P L E   3

On the same screen, graph each of the following functions:

$$Y = x^2$$
$$Y = (x - 1)^2$$
$$Y = (x - 3)^2$$
$$Y = (x + 2)^2$$

Solution   Figure 115 illustrates the graphs.

FIGURE 115

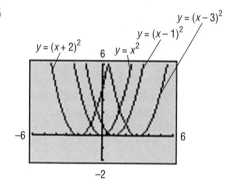

You should have observed the following pattern. With the graph of $y = x^2$ on the screen, the graph of $y = (x - 1)^2$ is identical to that of $y = x^2$, except it is shifted horizontally to the right 1 unit. Similarly, the graph of $y = (x - 3)^2$ is identical to that of $y = x^2$, except it is shifted horizontally to the right 3 units. Finally, the graph of $y = (x + 2)^2$ is identical to that of $y = x^2$, except it is shifted horizontally to the left 2 units. ■

We are led to the following conclusion.

If a real number c is added to the argument x of a function f, the graph of the new function $g(x) = f(x + c)$ is the graph of f shifted horizontally left (if $c > 0$) or right (if $c < 0$).

■ Now work Problem 31.

Vertical and horizontal shifts are sometimes combined.

E X A M P L E   4

Combining Vertical and Horizontal Shifts

Graph the function: $f(x) = (x - 1)^3 + 3$

Solution   We graph f in steps. First, we note that the rule for f is basically a cube function. Thus, we begin with the graph of $y = x^3$. See Figure 116(a). Next, to get the graph of $y = (x - 1)^3$, we shift the graph of $y = x^3$ horizontally 1 unit to the right. See Figure 116(b). Finally, to get the graph of $y = (x - 1)^3 + 3$, we shift the graph of $y = (x - 1)^3$ vertically up 3 units. See Figure 116(c). Note the three points that have been plotted on each graph. Using key points such as these can be helpful in keeping track of just what is taking place.

Notice that the values for Y_2 in Table 12 are larger than the values of Y_1 for every x, except $x = 0$. Therefore, the graph of Y_2 is horizontally *compressed*. So, $f(kx)$ will have a graph that is horizontally *compressed* [that is, its graph will be steeper than $f(x)$] when $k > 1$. Likewise, the values of Y_3 in Table 13 are smaller than the values of Y_1 for every x, except $x = 0$. Therefore, the graph of Y_3 will be horizontally *stretched*. So, $f(kx)$ will have a graph that is horizontally *stretched* [that is, its graph will be flatter than $f(x)$] when $0 < k < 1$.

TABLE 12

X	Y₁	Y₂
-3	6	30
-2	2	12
-1	0	2
0	0	0
1	2	6
2	6	20
3	12	42

Y₂ = 4X² + 2X

TABLE 13

X	Y₁	Y₃
-3	6	.75
-2	2	0
-1	0	-.25
0	0	0
1	2	.75
2	6	2
3	12	3.75

Y₃ = .25X² + .5X

Reflections about the *x*-Axis and the *y*-Axis

E X A M P L E   7

Reflection about the x-Axis

(a) Graph $y = x^2$ followed by $y = -x^2$.

(b) Graph $y = |x|$ followed by $y = -|x|$.

(c) Graph $y = x^2 - 4$ followed by $y = -(x^2 - 4) = -x^2 + 4$.

Solution   See Tables 14(a), (b), and (c) and Figures 119(a), (b), and (c). In each instance, the second graph is the reflection about the *x*-axis of the first graph.

TABLE 14

X	Y₁	Y₂
-3	9	-9
-2	4	-4
-1	1	-1
0	0	0
1	1	-1
2	4	-4
3	9	-9

Y₂ = -X²

(a)

X	Y₁	Y₂
-3	3	-3
-2	2	-2
-1	1	-1
0	0	0
1	1	-1
2	2	-2
3	3	-3

Y₂ = -abs X

(b)

X	Y₁	Y₂
-3	5	-5
-2	0	0
-1	-3	3
0	-4	4
1	-3	3
2	0	0
3	5	-5

Y₂ = -X² + 4

(c)

FIGURE 119

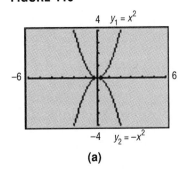

(a)

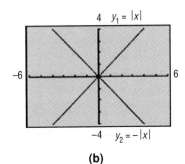

(b)

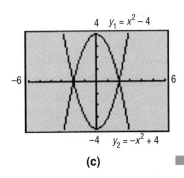

(c)

When the right side of the equation $y = f(x)$ is multiplied by -1, the graph of the new function $y = -f(x)$ is the **reflection about the x-axis** of the graph of the function $y = f(x)$.

■ Now work Problem 39.

E X A M P L E   8

(a) Graph $y = \sqrt{x}$ followed by $y = \sqrt{-x}$.

(b) Graph $y = x + 1$ followed by $y = -x + 1$.

(c) Graph $y = x^4 + x$ followed by $y = (-x)^4 + (-x) = x^4 - x$.

Solution See Tables 15(a), (b), and (c) and Figures 120(a), (b), and (c). In each instance, the second graph is the reflection about the y-axis of the first graph.

TABLE 15

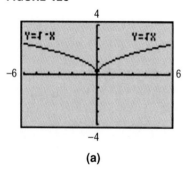

(a)

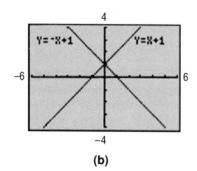

(b)

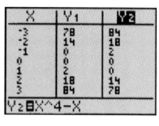

(c)

FIGURE 120

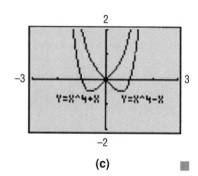

(a)   (b)   (c)   ■

When the graph of the function $y = f(x)$ is known, the graph of the new function $y = f(-x)$ is the **reflection about the y-axis** of the graph of the function $y = f(x)$.

Summary of Graphing Techniques

Table 16 summarizes the graphing procedures we have just discussed.

TABLE 16

To Graph:	Draw the Graph of f and:
Vertical shifts	
$y = f(x) + c, \quad c > 0$	Raise the graph of f by c units.
$y = f(x) - c, \quad c > 0$	Lower the graph of f by c units.
Horizontal shifts	
$y = f(x + c), \quad c > 0$	Shift the graph of f to the left c units.
$y = f(x - c), \quad c > 0$	Shift the graph of f to the right c units.
Compressing or stretching	
$y = kf(x), \quad k > 0$	Compress or stretch the graph of f by a factor of k.
$y = f(kx), \quad k > 0$	
Reflection about the x-axis	
$y = -f(x)$	Reflect the graph of f about the x-axis.
Reflection about the y-axis	
$y = f(-x)$	Reflect the graph of f about the y-axis.

The examples that follow combine some of the procedures outlined in this section to get the required graph.

E X A M P L E   9   *Combining Graphing Procedures*

Graph the function: $f(x) = \dfrac{3}{x-2} + 1$

Solution   We use the following steps to obtain the graph of f:

STEP 1: $y = \dfrac{1}{x}$ Reciprocal function

STEP 2: $y = \dfrac{3}{x}$ Vertical stretch of the graph of $y = \dfrac{1}{x}$ by a factor of 3

STEP 3: $y = \dfrac{3}{x-2}$ Horizontal shift to the right 2 units; replace x by $x-2$

STEP 4: $y = \dfrac{3}{x-2} + 1$ Vertical shift up 1 unit; add 1

See Figure 121.

FIGURE 121

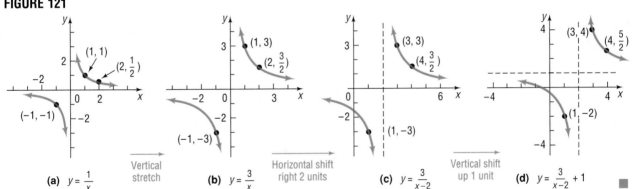

(a) $y = \dfrac{1}{x}$ Vertical stretch   (b) $y = \dfrac{3}{x}$ Horizontal shift right 2 units   (c) $y = \dfrac{3}{x-2}$ Vertical shift up 1 unit   (d) $y = \dfrac{3}{x-2} + 1$

There are other orderings of the steps shown in Example 9 that would also result in the graph of f. For example, try this one;

STEP 1: $y = \dfrac{1}{x}$ Reciprocal function

STEP 2: $y = \dfrac{1}{x-2}$ Horizontal shift to the right 2 units; replace x by $x-2$

STEP 3: $y = \dfrac{3}{x-2}$ Vertical stretch of the graph of $y = \dfrac{1}{x-2}$ by factor of 3

STEP 4: $y = \dfrac{3}{x-2} + 1$ Vertical shift up 1 unit; add 1

■   Now work Problem 45.

E X A M P L E   1 0   *Combining Graphing Procedures*

Graph the function: $f(x) = \sqrt{1-x} + 2$

Solution   We use the following steps to get the graph of $y = \sqrt{1 - x} + 2$:

STEP 1: $y = \sqrt{x}$ Square root function
STEP 2: $y = \sqrt{x + 1}$ Replace x by $x + 1$; horizontal shift left 1 unit
STEP 3: $y = \sqrt{-x + 1} = \sqrt{1 - x}$ Replace x by $-x$; reflect about y-axis
STEP 4: $y = \sqrt{1 - x} + 2$ Vertical shift up 2 units

See Figure 122.

FIGURE 122

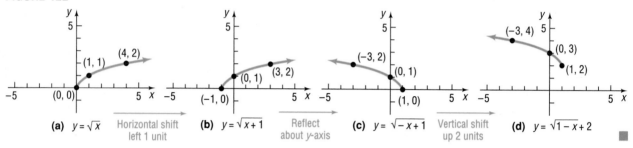

(a) $y = \sqrt{x}$ Horizontal shift left 1 unit
(b) $y = \sqrt{x + 1}$ Reflect about y-axis
(c) $y = \sqrt{-x + 1}$ Vertical shift up 2 units
(d) $y = \sqrt{1 - x} + 2$

1.6

Exercise 1.6

In Problems 1–12, match each graph to one of the following functions:

A. $y = x^2 + 2$ B. $y = -x^2 + 2$ C. $y = |x| + 2$ D. $y = -|x| + 2$

E. $y = (x - 2)^2$ F. $y = -(x + 2)^2$ G. $y = |x - 2|$ H. $y = -|x + 2|$

I. $y = 2x^2$ J. $y = -2x^2$ K. $y = 2|x|$ L. $y = -2|x|$

1.

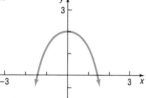

2.

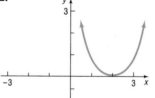

3.

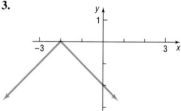

4.

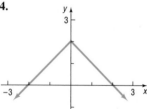

5.

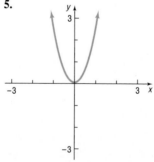

6.

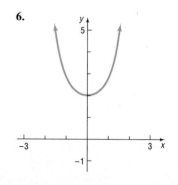

7.

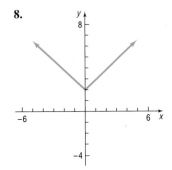

8.

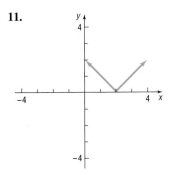

9.

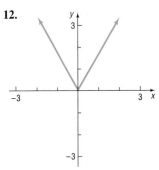

10.

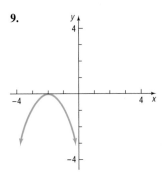

11.

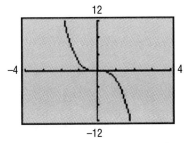

12.

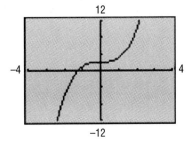

In Problems 13–16, match each graph to one of the following functions:

A. $y = 2x^3$ *B.* $y = (x + 2)^3$ *C.* $y = -2x^3$ *D.* $y = x^3 + 2$

13.

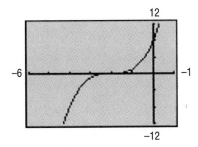

14.

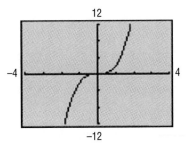

15.

16.

In Problems 17–24, use the function $f(x) = x^3$. Write the function whose graph is the graph of $y = x^3$, but is:

17. Shifted to the right 4 units

18. Shifted to the left 4 units

19. Shifted up 4 units

20. Shifted down 4 units

21. Reflected about the y-axis

22. Reflected about the x-axis

23. Vertically stretched by a factor of 4

24. Horizontally stretched by a factor of 4

In Problems 25–54, graph each function by hand using the techniques of shifting, compressing, stretching, and/or reflecting. Start with the graph of the basic function (for example, $y = x^2$) and show all stages. Verify your answer by using a graphing utility.

25. $f(x) = x^2 - 1$

26. $f(x) = x^2 + 4$

27. $g(x) = x^3 + 1$

28. $g(x) = x^3 - 1$

29. $h(x) = \sqrt{x - 2}$

30. $h(x) = \sqrt{x + 1}$

31. $f(x) = (x - 1)^3$

32. $f(x) = (x + 2)^3$

33. $g(x) = 4\sqrt{x}$

34. $g(x) = \frac{1}{2}\sqrt{x}$

35. $h(x) = \dfrac{1}{2x}$

36. $h(x) = \dfrac{4}{x}$

37. $f(x) = -|x|$

38. $f(x) = -\sqrt{x}$

39. $g(x) = -\dfrac{1}{x}$

40. $g(x) = -x^3$

41. $h(x) = [[-x]]$

42. $h(x) = \dfrac{1}{-x}$

43. $f(x) = (x + 1)^2 - 3$

44. $f(x) = (x - 2)^2 + 1$

45. $g(x) = \sqrt{x - 2} + 1$

46. $g(x) = |x + 1| - 3$

47. $h(x) = \sqrt{-x} - 2$

48. $h(x) = \dfrac{4}{x} + 2$

49. $f(x) = (x + 1)^3 - 1$

50. $f(x) = 4\sqrt{x - 1}$

51. $g(x) = 2|1 - x|$

52. $g(x) = 4\sqrt{2 - x}$

53. $h(x) = 2[[x - 1]]$

54. $h(x) = -x^3 + 2$

In Problems 55–60, the graph of a function f is illustrated. Use the graph of f as the first step toward graphing each of the following functions:

(a) $F(x) = f(x) + 3$

(b) $G(x) = f(x + 2)$

(c) $P(x) = -f(x)$

(d) $Q(x) = \frac{1}{2}f(x)$

(e) $g(x) = f(-x)$

(f) $h(x) = 3f(x)$

55.

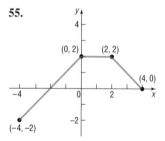

56.

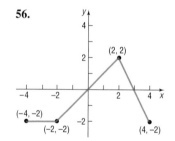

57.

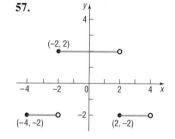

58.

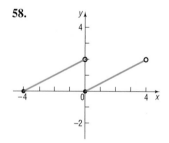

59.

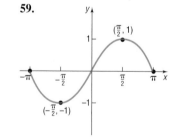

60.

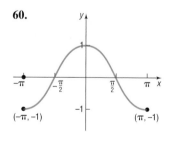

61. (a) Use a graphing utility to graph $y = x + 1$ and $y = |x + 1|$.
(b) Graph $y = 4 - x^2$ and $y = |4 - x^2|$.
(c) Graph $y = x^3 + x$ and $y = |x^3 + x|$.
(d) What do you conclude about the relationship between the graphs of $y = f(x)$ and $y = |f(x)|$?

62. (a) Use a graphing utility to graph $y = x + 1$ and $y = |x| + 1$.
(b) Graph $y = 4 - x^2$ and $y = 4 - |x|^2$.
(c) Graph $y = x^3 + x$ and $y = |x|^3 + |x|$.
(d) What do you conclude about the relationship between the graphs of $y = f(x)$ and $y = f(|x|)$?

63. The graph of a function f is illustrated in the figure.

    (a) Draw the entire graph of $y = |f(x)|$.

    (b) Draw the entire graph of $y = f(|x|)$.

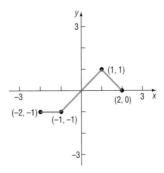

64. Repeat Problem 63 for the graph shown.

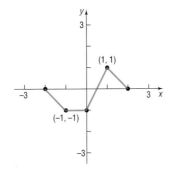

In Problems 65–70, complete the square of each quadratic expression. Then graph each function by hand using the technique of shifting. Verify your results using a graphing utility.

65. $f(x) = x^2 + 2x$ **66.** $f(x) = x^2 - 6x$ **67.** $f(x) = x^2 - 8x + 1$

68. $f(x) = x^2 + 4x + 2$ **69.** $f(x) = x^2 + x + 1$ **70.** $f(x) = x^2 - x + 1$

71. The equation $y = (x - c)^2$ defines a *family of parabolas,* one parabola for each value of c. On one set of coordinate axes, graph the members of the family for $c = 0$, $c = 3$, $c = -2$.

72. Repeat Problem 71 for the family of parabolas $y = x^2 + c$.

73. *Temperature Measurements* The relationship between the Celsius (°C) and Fahrenheit (°F) scales for measuring temperature is given by the equation

$$F = \frac{9}{5}C + 32$$

The relationship between the Celsius (°C) and Kelvin (K) scales is $K = C + 273$. Graph the equation $F = \frac{9}{5}C + 32$ using degrees Fahrenheit on the y-axis and degrees Celsius on the x-axis. Use the techniques introduced in this section to obtain the graph showing the relationship between Kelvin and Fahrenheit temperatures.

74. *Period of a Pendulum* The period T (in seconds) of a simple pendulum is a function of its length l (in feet) defined by the equation

$$T = 2\pi\sqrt{\frac{l}{g}}$$

where $g \approx 32.2$ feet per second per second is the acceleration of gravity.

    (a) Use a graphing utility to graph the function $T = T(l)$.

    (b) Now graph the functions $T = T(l + 1)$, $T = T(l + 2)$, and $T = T(l + 3)$.

    (c) Discuss how adding to the length l changes the period T.

    (d) Now graph the functions $T = T(2l)$, $T = T(3l)$, and $T = T(4l)$.

    (e) Discuss how multiplying the length l by factors of 2, 3, and 4 changes the period T.

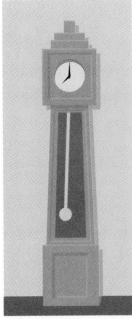

1.7

One-to-One Functions; Inverse Functions

Suppose that (x_1, y_1) and (x_2, y_2) are any two *distinct* points on the graph of a function $y = f(x)$. Then it follows that $x_1 \neq x_2$. For some functions, it also happens that the y-coordinates of distinct points are always unequal. Such functions are called *one-to-one* functions. See Figure 123.

FIGURE 123

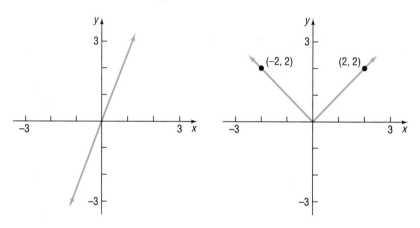

(a) $f(x) = 2x$
One-to-one:
Every distinct point has
a different y-coordinate

(b) $g(x) = |x|$
Not one-to-one:
The distinct points $(-2, 2)$ and $(2, 2)$
have the same y-coordinate

One-to-One Function

A function f is said to be **one-to-one** if, for any choice of numbers x_1 and x_2, $x_1 \neq x_2$, in the domain of f, then $f(x_1) \neq f(x_2)$.

In other words, if f is a one-to-one function, then for each x in the domain of f, there is exactly one y in the range, and no y in the range is the image of more than one x in the domain. See Figure 124.

FIGURE 124

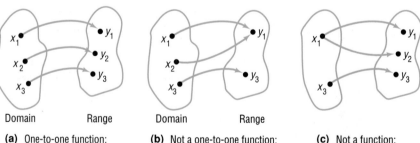

(a) One-to-one function:
Each x in the domain has
one and only one image
in the range

(b) Not a one-to-one function:
y_1 is the image of both
x_1 and x_2

(c) Not a function:
x_1 has two images,
y_1 and y_2

As Figure 125 illustrates, if the graph of a function f is known, there is a simple test, called the **horizontal-line test,** to determine whether f is one-to-one.

Theorem
Horizontal Line Test

If horizontal lines intersect the graph of a function f in at most one point, then f is one-to-one. ∎

and the point (b, a) on f^{-1} is shown in Figure 130. The line joining (a, b) and (b, a) is perpendicular to the line $y = x$ and is bisected by the line $y = x$. (Do you see why?) It follows that the point (b, a) on f^{-1} is the reflection about the line $y = x$ of the point (a, b) on f.

FIGURE 130

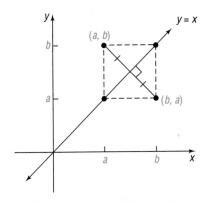

Theorem   The graph of a function f and the graph of its inverse f^{-1} are symmetric with respect to the line $y = x$. ■

Figure 131 illustrates this result. Notice that, once the graph of f is known, the graph of f^{-1} may be obtained by folding the paper along the line $y = x$.

FIGURE 131

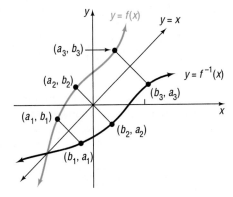

E X A M P L E   3   *Graphing the Inverse Function*

The graph in Figure 132(a) is that of a one-to-one function $y = f(x)$. Draw the graph of its inverse.

Solution   We begin by adding the graph of $y = x$ to Figure 132(a). Since the points $(-2, -1)$, $(-1, 0)$, and $(2, 1)$ are on the graph of f, we know that the points $(-1, -2)$, $(0, -1)$, and $(1, 2)$ must be on the graph of f^{-1}. Keeping in mind that the graph of f^{-1} is the reflection about the line $y = x$ of the graph of f, we can draw f^{-1}. See Figure 132(b).

FIGURE 132

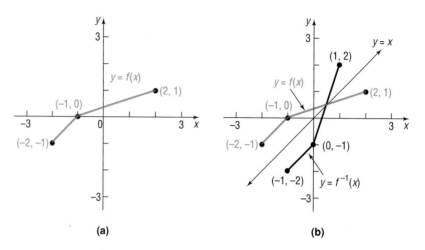

(a)   (b)

■   Now work Problem 7.

Finding the Inverse Function

The fact that the graph of a one-to-one function f and its inverse are symmetric with respect to the line $y = x$ tells us more. It says that we can obtain f^{-1} by interchanging the roles of x and y in f. Look again at Figure 131. That is, if f is defined by the equation

$$y = f(x)$$

then f^{-1} is defined by the equation

$$x = f(y)$$

Be careful! The equation $x = f(y)$ defines f^{-1} implicitly. If we can solve this equation for y, we will have the explicit form of f^{-1}, that is,

$$y = f^{-1}(x)$$

Let's use this procedure to find the inverse of $f(x) = 2x + 3$. (Since f is a linear function and is increasing, we know that f is one-to-one.)

EXAMPLE   4   *Finding the Inverse Function*

Find the inverse of $f(x) = 2x + 3$. Also find the domain and range of f and f^{-1}. Graph f and f^{-1} on the same coordinate axes.

Solution   In the equation $y = 2x + 3$, interchange the variables x and y. The result,

$$x = 2y + 3$$

is an equation that defines the inverse f^{-1} implicitly. Solving for y, we obtain

$$2y + 3 = x$$
$$2y = x - 3$$
$$y = \tfrac{1}{2}(x - 3)$$

The explicit form of the inverse f^{-1} is therefore

$$f^{-1}(x) = \tfrac{1}{2}(x - 3)$$

which we verified in Example 2(c).

In Problems 35–46, the function f is one-to-one. Find its inverse and check your answer. State the domain and range of f and f^{-1}. Using a graphing utility, graph f, f^{-1}, and $y = x$ on the same square screen.

35. $f(x) = \dfrac{2}{3 + x}$

36. $f(x) = \dfrac{4}{2 - x}$

37. $f(x) = (x + 2)^2, \quad x \geq -2$

38. $f(x) = (x - 1)^2, \quad x \geq 1$

39. $f(x) = \dfrac{2x}{x - 1}$

40. $f(x) = \dfrac{3x + 1}{x}$

41. $f(x) = \dfrac{3x + 4}{2x - 3}$

42. $f(x) = \dfrac{2x - 3}{x + 4}$

43. $f(x) = \dfrac{2x + 3}{x + 2}$

44. $f(x) = \dfrac{-3x - 4}{x - 2}$

45. $f(x) = 2\sqrt[3]{x}$

46. $f(x) = \dfrac{4}{\sqrt{x}}$

47. Find the inverse of the linear function $f(x) = mx + b, m \neq 0$.

48. Find the inverse of the function $f(x) = \sqrt{r^2 - x^2}, 0 \leq x \leq r$.

49. Can an even function be one-to-one? Explain.

50. Is every odd function one-to-one? Explain.

51. A function f has an inverse. If the graph of f lies in quadrant I, in which quadrant does the graph of f^{-1} lie?

52. A function f has an inverse. If the graph of f lies in quadrant II, in which quadrant does the graph of f^{-1} lie?

53. The function $f(x) = |x|$ is not one-to-one. Find a suitable restriction on the domain of f so that the new function that results is one-to-one. Then find the inverse of f.

54. The function $f(x) = x^4$ is not one-to-one. Find a suitable restriction on the domain of f so that the new function that results is one-to-one. Then find the inverse of f.

55. *Temperature Conversion* To convert from x degrees Celsius to y degrees Fahrenheit, we use the formula $y = f(x) = \frac{9}{5}x + 32$. To convert from x degrees Fahrenheit to y degrees Celsius, we use the formula $y = g(x) = \frac{5}{9}(x - 32)$. Show that f and g are inverse functions.

56. *Demand for Corn* The demand for corn obeys the equation $p(x) = 300 - 50x$, where p is the price per bushel (in dollars) and x is the number of bushels produced, in millions. Express the production amount x as a function of the price p.

57. *Period of a Pendulum* The period T (in seconds) of a simple pendulum is a function of its length l (in feet), given by $T(l) = 2\pi\sqrt{l/g}$, where $g \approx 32.2$ feet per second per second is the acceleration of gravity. Express the length l as a function of the period T.

58. Give an example of a function whose domain is the set of real numbers and that is neither increasing nor decreasing on its domain, but is one-to-one. [*Hint:* Use a piecewise defined function.]

59. Given

$$f(x) = \frac{ax + b}{cx + d}$$

find $f^{-1}(x)$. If $c \neq 0$, under what conditions on $a, b, c,$ and d is $f = f^{-1}$?

60. We said earlier that finding the range of a function f is not easy. However, if f is one-to-one, we can find its range by finding the domain of the inverse function f^{-1}. Use this technique to find the range of each of the following one-to-one functions:

(a) $f(x) = \dfrac{2x + 5}{x - 3}$

(b) $g(x) = 4 - \dfrac{2}{x}$

(c) $F(x) = \dfrac{3}{4 - x}$

For Problems 61–66, write a program that will graph the inverse of a function $y = f(x)$. Then graph the function f and its inverse on the same screen. Compare your answers with those of Problems 23–28.

61. $f(x) = 3x$

62. $f(x) = -4x$

63. $f(x) = 4x + 2$

64. $f(x) = 1 - 3x$

65. $f(x) = x^3 - 1$

66. $f(x) = x^3 + 1$

67. If the graph of a function and its inverse intersect, where must this necessarily occur? Can they intersect anywhere else? Must they intersect?

68. Can a one-to-one function and its inverse be equal? What must be true about the graph of *f* for this to happen? Give some examples to support your conclusion.

69. Draw the graph of a one-to-one function that contains the points $(-2, -3)$, $(0, 0)$, and $(1, 5)$. Now draw the graph of its inverse. Compare your graph to those of other students. Discuss any similarities. What differences do you see?

Chapter Review

THINGS TO KNOW

Formulas

Distance formula	$d = \sqrt{(x_2 - x_1)^2 + (y_2 - y_1)^2}$
Midpoint formula	$(x, y) = \left(\dfrac{x_1 + x_2}{2}, \dfrac{y_1 + y_2}{2} \right)$
Slope	$m = \dfrac{y_2 - y_1}{x_2 - x_1}$, if $x_1 \neq x_2$; undefined if $x_1 = x_2$

Equations

Vertical line	$x = a$
Horizontal line	$y = b$
Point–slope form of the equation of a line	$y - y_1 = m(x - x_1)$; m is the slope of the line, (x_1, y_1) is a point on the line
General form of the equation of a line	$Ax + By + C = 0$, A, B not both 0
Slope–intercept form of the equation of a line	$y = mx + b$; m is the slope of the line, b is the y-intercept
Standard form of the equation of a circle	$(x - h)^2 + (y - k)^2 = r^2$; r is the radius of the circle, (h, k) is the center of the circle
General form of the equation of a circle	$x^2 + y^2 + ax + by + c = 0$
Equation of the unit circle	$x^2 + y^2 = 1$

Function

A rule or correspondence between two sets of real numbers so that each number x in the first set, the domain, has corresponding to it exactly one number y in the second set. The range is the set of y values of the function for the x values in the domain. x is the independent variable; y is the dependent variable.
A function f may be defined implicitly by an equation involving x and y or explicitly by writing $y = f(x)$.
A function can also be characterized as a set of ordered pairs (x, y) or $(x, f(x))$ in which no two pairs have the same first element.

Function notation

$y = f(x)$
f is a symbol for the function or rule that defines the function.
x is the argument, or independent variable.
y is the dependent variable.
$f(x)$ is the value of the function at x, or the image of x.

Domain

If unspecified, the domain of a function f is the largest set of real numbers for which the rule defines a real number.

Vertical line test

A set of points in the plane is the graph of a function if and only if every vertical line intersects the graph in at most one point.

Local maximum

A function f has a local maximum at c if there is an interval I containing c so that for all x in I, $f(x) < f(c)$.

Local minimum

A function f has a local minimum at c if there is an interval I containing c so that for all x in I, $f(x) > f(c)$.

Even function f

$f(-x) = f(x)$ for every x in the domain ($-x$ must also be in the domain).

Odd function f

$f(-x) = -f(x)$ for every x in the domain ($-x$ must also be in the domain).

One-to-one function f

If $x_1 \neq x_2$, then $f(x_1) \neq f(x_2)$ for any choice of x_1 and x_2 in the domain.

Horizontal line test

If horizontal lines intersect the graph of a function f in at most one point, then f is one-to-one.

Inverse function f^{-1} of f

Domain of f = Range of f^{-1}; Range of f = Domain of f^{-1}
$f^{-1}(f(x)) = x$ and $f(f^{-1}(x)) = x$.
Graphs of f and f^{-1} are symmetric with respect to the line $y = x$.

IMPORTANT FUNCTIONS

Linear function

$f(x) = mx + b$ Graph is a straight line with slope m and y-intercept b.

Constant function

$f(x) = b$ Graph is a horizontal line with y-intercept b (see Figure 102).

Identity function

$f(x) = x$ Graph is a straight line with slope 1 and y-intercept 0 (see Figure 103).

Square function

$f(x) = x^2$ Graph is a parabola with intercept at (0, 0) (see Figure 104).

Cube function

$f(x) = x^3$ See Figure 105.

Square root function

$f(x) = \sqrt{x}$ See Figure 106.

Reciprocal function

$f(x) = 1/x$ See Figure 107.

Absolute value function

$f(x) = |x|$ See Figure 108.

How To:

Use the distance formula

Graph equations by plotting points

Find the intercepts of a graph

Test an equation for symmetry

Find the slope and intercepts of a line, given the equation

Graph lines by hand

Graph equations using a graphing utility

Obtain the equation of a line

Obtain the equation of a circle

Find the center and radius of a circle, given the equation

Graph circles by hand

Find the domain and range of a function from its graph

Find the domain of a function given its equation

Determine whether a function is even or odd without graphing it

Graph certain functions by shifting, compressing, stretching, and/or reflecting (see Table 16)

Use a graphing utility to determine where the graph of a function is increasing or decreasing

Find the inverse of certain one-to-one functions (see the procedure given on page 113)

Graph f^{-1} given the graph of f

Use a graphing utility to find the local maxima and local minima of a function

Fill-In-The-Blank Items

1. If (x, y) are the coordinates of a point P in the xy-plane, then x is called the _____ of P and y is the _____ of P.

2. If three points $P, Q,$ and R all lie on a line and if $d(P, Q) = d(Q, R)$, then Q is called the _____ of the line segment from P to R.

3. If for every point (x, y) on a graph, the point $(-x, y)$ is also on the graph, then the graph is symmetric with respect to the _____.

4. The set of points in the xy-plane that are a fixed distance from a fixed point is called a(n) _____. The fixed distance is called the _____; the fixed point is called the _____.

5. The slope of a vertical line is _____; the slope of a horizontal line is _____.

6. If f is a function defined by the equation $y = f(x)$, then x is called the _____ variable and y is the _____ variable.

7. A set of points in the xy-plane is the graph of a function if and only if no _____ line contains more than one point of the set.

8. A(n) _____ function f is one for which $f(-x) = f(x)$ for every x in the domain of f; a(n) _____ function f is one for which $f(-x) = -f(x)$ for every x in the domain of f.

9. Suppose that the graph of a function f is known. Then the graph of $y = f(x - 2)$ may be obtained by a(n) _____ shift of the graph of f to the _____ a distance of 2 units.

10. If every horizontal line intersects the graph of a function f at no more than one point, then f is a(n) _____ function.

11. If f^{-1} denotes the inverse of a function f, then the graphs of f and f^{-1} are symmetric with respect to the line _____.

True/False Items

T   F   1. The distance between two points is sometimes a negative number.

T   F   2. The graph of the equation $y = x^4 + x^2 + 1$ is symmetric with respect to the y-axis.

T   F   3. Vertical lines have undefined slope.

T   F   4. The slope of the line $2y = 3x + 5$ is 3.

T   F   5. The radius of the circle $x^2 + y^2 = 9$ is 3.

T   F   **6.** Vertical lines intersect the graph of a function in no more than one point.

T   F   **7.** The y-intercept of the graph of the function $y = f(x)$ whose domain is all real numbers is $f(0)$.

T   F   **8.** Even functions have graphs that are symmetric with respect to the origin.

T   F   **9.** The graph of $y = f(-x)$ is the reflection about the y-axis of the graph of $y = f(x)$.

T   F   **10.** If f and g are inverse functions, then the domain of f is the same as the domain of g.

T   F   **11.** If f and g are inverse functions, then their graphs are symmetric with respect to the line $y = x$.

REVIEW EXERCISES

In Problems 1–10, find an equation of the line having the given characteristics. Express your answer using either the general form or the slope–intercept form of the equation of a line, whichever you prefer.

1. Slope $= -2$; passing through $(3, -1)$

2. Slope $= 0$; passing through $(-5, 4)$

3. Slope undefined; passing through $(-3, 4)$

4. x-intercept $= 2$; passing through $(4, -5)$

5. y-intercept $= -2$; passing through $(5, -3)$

6. Passing through $(3, -4)$ and $(2, 1)$

7. Slope $= \frac{2}{3}$; passing through $(-5, 3)$

8. Slope $= -1$; passing through $(1, -3)$

9. x-intercept $= 7$; y-intercept $= -7$

10. x-intercept $= 7$; no y-intercept

In Problems 11–16, graph each line using a graphing utility. Use BOX and TRACE to find the x-intercept and y-intercept. Also draw each graph by hand, labeling any intercepts.

11. $4x - 5y + 20 = 0$

12. $3x + 4y - 12 = 0$

13. $\frac{1}{2}x - \frac{1}{3}y + \frac{1}{6} = 0$

14. $-\frac{3}{4}x + \frac{1}{2}y = 0$

15. $\sqrt{2}x + \sqrt{3}y = \sqrt{6}$

16. $\frac{x}{3} + \frac{y}{4} = 1$

In Problems 17–20, find the center and radius of each circle. Graph each circle using a graphing utility.

17. $x^2 + y^2 - 2x + 4y - 4 = 0$

18. $x^2 + y^2 + 4x - 4y - 1 = 0$

19. $3x^2 + 3y^2 - 6x + 12y = 0$

20. $2x^2 + 2y^2 - 4x = 0$

21. Find the slope of the line containing the points $(7, 4)$ and $(-3, 2)$. What is the distance between these points? What is their midpoint?

22. Find the slope of the line containing the points $(2, 5)$ and $(6, -3)$. What is the distance between these points? What is their midpoint?

In Problems 23–30, list the intercepts and test for symmetry.

23. $2x = 3y^2$

24. $y = 5x$

25. $x^2 + 4y^2 = 16$

26. $9x^2 - y^2 = 9$

27. $y = x^4 + 2x^2 + 1$

28. $y = x^3 - x$

29. $x^2 + x + y^2 + 2y = 0$

30. $x^2 + 4x + y^2 - 2y = 0$

31. Given that f is a linear function, $f(4) = -5$, and $f(0) = 3$, write the equation that defines f.

32. Given that g is a linear function with slope $= -4$ and $g(-2) = 2$, write the equation that defines g.

33. A function f is defined by

$$f(x) = \frac{Ax + 5}{6x - 2}$$

If $f(1) = 4$, find A.

34. A function g is defined by

$$g(x) = \frac{A}{x} + \frac{8}{x^2}$$

If $g(-1) = 0$, find A.

35. (a) Tell which of the following graphs are graphs of functions.
(b) Tell which are graphs of one-to-one functions.

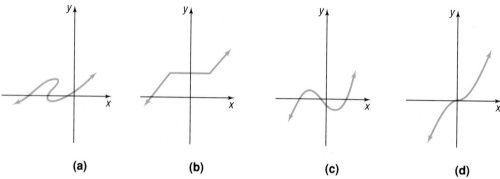

(a)   (b)   (c)   (d)

36. Use the graph of the function f shown to find:
(a) The domain and range of f
(b) The intervals on which f is increasing
(c) The intervals on which f is constant
(d) The intercepts of f

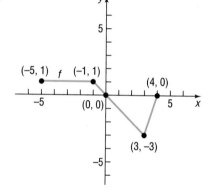

In Problems 37–42, find the following for each function:
(a) $f(-x)$ (b) $-f(x)$ (c) $f(x + 2)$ (d) $f(x - 2)$

37. $f(x) = \dfrac{3x}{x^2 - 4}$ **38.** $f(x) = \dfrac{x^2}{x + 2}$ **39.** $f(x) = \sqrt{x^2 - 4}$ **40.** $f(x) = |x^2 - 4|$

41. $f(x) = \dfrac{x^2 - 4}{x^2}$ **42.** $f(x) = \dfrac{x^3}{x^2 - 4}$

In Problems 43–48, determine whether the given function is even, odd, or neither without drawing a graph.

43. $f(x) = x^3 - 4x$ **44.** $g(x) = \dfrac{4 + x^2}{1 + x^4}$ **45.** $h(x) = \dfrac{1}{x^4} + \dfrac{1}{x^2} + 1$ **46.** $F(x) = \sqrt{1 - x^3}$

47. $G(x) = 1 - x + x^3$ **48.** $H(x) = 1 + x + x^2$

In Problems 49–60, find the domain of each function.

49. $f(x) = \dfrac{x}{x^2 - 9}$ **50.** $f(x) = \dfrac{3x^2}{x - 2}$ **51.** $f(x) = \sqrt{2 - x}$ **52.** $f(x) = \sqrt{x + 2}$

53. $h(x) = \dfrac{\sqrt{x}}{|x|}$

54. $g(x) = \dfrac{|x|}{x}$

55. $f(x) = \dfrac{x}{x^2 + 2x - 3}$

56. $F(x) = \dfrac{1}{x^2 - 3x - 4}$

57. $G(x) = \begin{cases} |x| & \text{if } -1 \le x \le 1 \\ 1/x & \text{if } x > 1 \end{cases}$

58. $H(x) = \begin{cases} 1/x & \text{if } 0 < x < 4 \\ x - 4 & \text{if } 4 \le x \le 8 \end{cases}$

59. $f(x) = \begin{cases} 1/(x-2) & \text{if } x > 2 \\ 0 & \text{if } x = 2 \\ 3x & \text{if } 0 \le x < 2 \end{cases}$

60. $g(x) = \begin{cases} |1 - x| & \text{if } x < 1 \\ 3 & \text{if } x = 1 \\ x + 1 & \text{if } 1 < x \le 3 \end{cases}$

In Problems 61–80:
(a) Find the domain of each function.
(b) Locate any intercepts.
(c) Graph each function.
(d) Based on the graph, find the range.
(e) Use a graphing utility to verify your answers.

61. $F(x) = |x| - 4$

62. $f(x) = |x| + 4$

63. $g(x) = -|x|$

64. $g(x) = \frac{1}{2}|x|$

65. $h(x) = \sqrt{x - 1}$

66. $h(x) = \sqrt{x} - 1$

67. $f(x) = \sqrt{1 - x}$

68. $f(x) = -\sqrt{x}$

69. $F(x) = \begin{cases} x^2 + 4 & \text{if } x < 0 \\ 4 - x^2 & \text{if } x \ge 0 \end{cases}$

70. $H(x) = \begin{cases} |1 - x| & \text{if } 0 \le x \le 2 \\ |x - 1| & \text{if } x > 2 \end{cases}$

71. $h(x) = (x - 1)^2 + 2$

72. $h(x) = (x + 2)^2 - 3$

73. $g(x) = (x - 1)^3 + 1$

74. $g(x) = (x + 2)^3 - 8$

75. $f(x) = \begin{cases} 2\sqrt{x} & \text{if } x \ge 4 \\ x & \text{if } 0 < x < 4 \end{cases}$

76. $f(x) = \begin{cases} 3|x| & \text{if } x < 0 \\ \sqrt{1 - x} & \text{if } 0 \le x \le 1 \end{cases}$

77. $g(x) = \dfrac{1}{x - 1} + 1$

78. $g(x) = \dfrac{1}{x + 2} - 2$

79. $h(x) = [\![-x]\!]$

80. $h(x) = -[\![x]\!]$

In Problems 81–86, the function f is one-to-one. Find the inverse of each function and check your answer.
Find the domain and range of f and f^{-1}. Use a graphing utility to graph f, f^{-1}, and $y = x$ on the same
square screen.

81. $f(x) = \dfrac{2x + 3}{5x - 2}$

82. $f(x) = \dfrac{2 - x}{3 + x}$

83. $f(x) = \dfrac{1}{x - 1}$

84. $f(x) = \sqrt{x - 2}$

85. $f(x) = \dfrac{3}{x^{1/3}}$

86. $f(x) = x^{1/3} + 1$

87. For the graph of the function f
shown:
(a) Draw the graph of $y = f(-x)$.
(b) Draw the graph of $y = -f(x)$.
(c) Draw the graph of $y = f(x + 2)$.
(d) Draw the graph of $y = f(x) + 2$.
(e) Draw the graph of $y = f(2 - x)$.
(f) Draw the graph of f^{-1}.

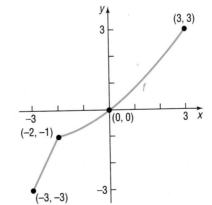

88. Repeat Problem 87 for the graph of the function g shown here.

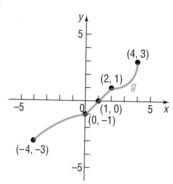

89. *Temperature Conversion* The temperature T of the air is approximately a linear function of the altitude h for altitudes within 10,000 meters of the surface of Earth. If the surface temperature is 30°C and the temperature at 10,000 meters is 5°C, find the function $T = T(h)$.

90. *Speed as a Function of Time* The speed v (in feet per second) of a car is a linear function of the time t (in seconds) for $10 \le t \le 30$. If after each second the speed of the car has increased by 5 feet per second, and if after 20 seconds the speed is 80 feet per second, how fast is the car going after 30 seconds? Find the function $v = v(t)$.

91. Show that the points $A = (3, 4)$, $B = (1, 1)$, and $C = (-2, 3)$ are the vertices of an isosceles triangle.

92. Show that the points $A = (-2, 0)$, $B = (-4, 4)$, and $C = (8, 5)$ are the vertices of a right triangle in two ways:
  (a) By using the converse of the Pythagorean Theorem
  (b) By using the slopes of the lines joining the vertices

93. Show that the points $A = (2, 5)$, $B = (6, 1)$, and $C = (8, -1)$ lie on a straight line by using slopes.

94. Show that the points $A = (1, 5)$, $B = (2, 4)$, and $C = (-3, 5)$ lie on a circle with center $(-1, 2)$. What is the radius of this circle?

95. The endpoints of the diameter of a circle are $(-3, 2)$ and $(5, -6)$. Find the center and radius of the circle. Write the general equation of this circle.

96. Find two numbers y such that the distance from $(-3, 2)$ to $(5, y)$ is 10.

97. Make up four problems you might be asked to do given the two points $(-3, 4)$ and $(6, 1)$. Each problem should involve a different concept. Be sure your directions are clearly stated.

98. Describe each of the following graphs. Give justification.
  (a) $x = 0$ (b) $y = 0$ (c) $x + y = 0$ (d) $xy = 0$ (e) $x^2 + y^2 = 0$

99. Graph $y = x$ and $y = 0.99x + 0.01$ on the same screen. It appears the graphs are identical. Explain why this happens. Provide a way to correct this situation.

100. Why does the graph of $y = x/6$, a straight line, appear to consist of a collection of tiny horizontal line segments?

PREPARING FOR THIS CHAPTER

Before getting started on this chapter, review the following concepts:

Pythagorean Theorem (Appendix, Section 1)
Unit circle (p. 49)
Functions (Sections 1.4 and 1.5)
Inverse functions (pp. 109–115)

TRIGONOMETRIC FUNCTIONS

2.1 Angles and Their Measure
2.2 Trigonometric Functions: Unit Circle Approach
2.3 Properties of the Trigonometric Functions
2.4 Right Triangle Trigonometry
2.5 Graphs of the Trigonometric Functions
2.6 The Inverse Trigonometric Functions
    Chapter Review

Preview   Being the First to See the Rising Sun

Cadillac Mountain, elevation 1530 feet, is located in Acadia National Park, Maine, and is the highest peak on the east coast of the United States. It is said that a person standing on the summit will be the first person in the United States to see the rays of the rising Sun. How much sooner would a person atop Cadillac Mountain see the first rays than a person standing below at sea level? [Hint: Consult the figure on page 199. When the person at D sees the first rays of the Sun, the person at P does not. The person at P sees the first rays of the Sun only after Earth has rotated so that P is at location Q. Compute the length of arc s subtended by the central angle θ. Then use the fact that in 24 hours a length of 2π(3960) miles is subtended, and find the time it takes to subtend the length s.]
[Problem 79, Exercise 2.6] ■

Trigonometry was developed by Greek astronomers who regarded the sky as the inside of a sphere, so it was natural that triangles on a sphere were investigated early (by Menelaus of Alexandria about AD 100) and that triangles in the plane were studied much later. The first book containing a systematic treatment of plane and spherical trigonometry was written by the Persian astronomer Nasîr ed-dîn (about AD 1250).

Regiomontanus (1436–1476) is the man most responsible for moving trigonometry from astronomy into mathematics. His work was improved by Copernicus (1473–1543) and Copernicus's student Rhaeticus (1514–1576). Rhaeticus's book was the first to define the six trigonometric functions as ratios of sides of triangles, although he did not give the functions their present names. Credit for this is due to Thomas Fincke (1583), but Fincke's notation was by no means universally accepted at the time. The notation was finally stabilized by the textbooks of Leonhard Euler (1707–1783).

Trigonometry has since evolved from its use by surveyors, navigators, and engineers to present applications involving ocean tides, the rise and fall of food supplies in certain ecologies, brain wave patterns, and many other phenomena.

There are two widely accepted approaches to the development of the *trigonometric functions:* One uses circles, especially the *unit circle;* the other uses *right triangles.* In this book, we introduce trigonometric functions using the unit circle. In Section 2.4, we shall show that right triangle trigonometry is a special case of the unit circle approach.

2.1

Angles and Their Measure

A **ray,** or **half-line,** is that portion of a line that starts at a point V on the line and extends indefinitely in one direction. The starting point V of a ray is called its **vertex.** See Figure 1.

If two rays are drawn with a common vertex, they form an **angle.** We call one of the rays of an angle the **initial side** and the other the **terminal side.** The angle that is formed is identified by showing the direction and amount of rotation from the initial side to the terminal side. If the rotation is in the counterclockwise direction, the angle is **positive;** if the rotation is clockwise, the angle is **negative.** See Figure 2. Lowercase Greek letters, such as α (alpha), β (beta), γ (gamma), θ (theta), and so on, will be used to denote angles. Notice in Figure 2(a) that the angle α is positive because the direction of the rotation from the initial side to the terminal side is counterclockwise. The angle β in Figure 2(b) is negative because the rotation is clockwise. The angle γ in Figure 2(c) is positive. Notice that the angle α in Figure 2(a) and the angle γ have the same initial side and the same terminal side. However, α and γ are unequal because the amount of rotation required to go from the initial side to the terminal side is greater for angle γ than for angle α.

FIGURE 1

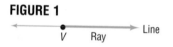

FIGURE 2

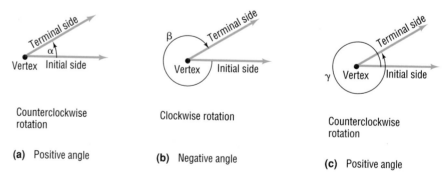

(a) Positive angle **(b)** Negative angle **(c)** Positive angle

An angle θ is said to be in **standard position** if its vertex is at the origin of a rectangular coordinate system and its initial side coincides with the positive x-axis. See Figure 3.

FIGURE 3

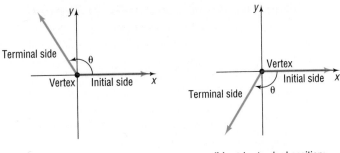

(a) θ in standard position; θ positive

(b) θ in standard position; θ negative

When an angle θ is in standard position, the terminal side either will lie in a quadrant, in which case we say θ **lies in that quadrant,** or it will lie on the x-axis or the y-axis, in which case we say θ is a **quadrantal angle.** For example, the angle θ in Figure 4(a) lies in quadrant II, the angle θ in Figure 4(b) lies in quadrant IV, and the angle θ in Figure 4(c) is a quadrantal angle.

FIGURE 4

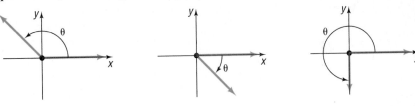

(a) θ lies in quadrant II

(b) θ lies in quadrant IV

(c) θ is a quadrantal angle

We measure angles by determining the amount of rotation needed for the initial side to become coincident with the terminal side. There are two commonly used measures for angles: *degrees* and *radians*.

Degrees

The angle formed by rotating the initial side exactly once in the counterclockwise direction until it coincides with itself (1 revolution) is said to measure 360 degrees, abbreviated 360°. Thus **one degree, 1°,** is $\frac{1}{360}$ revolution. A **right angle** is an angle of 90°, or $\frac{1}{4}$ revolution; a **straight angle** is an angle of 180°, or $\frac{1}{2}$ revolution. See Figure 5. As Figure 5(b) shows, it is customary to indicate a right angle by using the symbol ⌐.

FIGURE 5

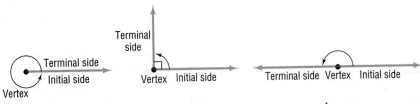

(a) 1 revolution counterclockwise, 360°

(b) $\frac{1}{4}$ revolution counterclockwise, 90°

(c) $\frac{1}{2}$ revolution counterclockwise, 180°

E X A M P L E   1   *Drawing an Angle*

Draw each angle:

(a) 45°   (b) −90°   (c) 225°   (d) 405°

Solution   (a) An angle of 45° is $\frac{1}{2}$ of a right angle. See Figure 6.

FIGURE 6

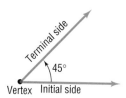

(b) An angle of −90° is $\frac{1}{4}$ revolution in the clockwise direction. See Figure 7.

FIGURE 7

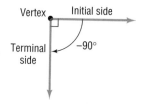

(c) An angle of 225° consists of a rotation through 180° followed by a rotation through 45°. See Figure 8.

FIGURE 8

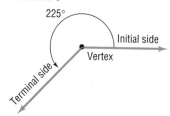

(d) An angle of 405° consists of 1 revolution (360°) followed by a rotation through 45°. See Figure 9.

FIGURE 9

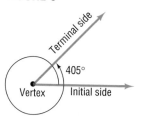

■ Now work Problem 1.

Although subdivisions of a degree may be obtained by using decimals, we also may use the notion of *minutes* and *seconds*. **One minute,** denoted by **1′**, is defined as $\frac{1}{60}$ degree. **One second,** denoted by **1″**, is defined as $\frac{1}{60}$ minute or, equivalently, $\frac{1}{3600}$ degree. An angle of, say, 30 degrees, 40 minutes, 10 seconds is written compactly as 30°40′10″. To summarize:

$$1 \text{ counterclockwise revolution} = 360°$$
$$60' = 1° \qquad 60'' = 1' \tag{1}$$

Because calculators use decimals, it is sometimes necessary to convert from the degree, minute, second notation (D°M′S″) to a decimal form, and vice versa. Check your calculator; it should be capable of doing the conversion for you.

Before getting started, though, you must set the mode to degrees, because there are two common ways to measure angles: degree mode and radian mode. (We will define radians shortly.) Usually, a menu is used to change from one mode to another. Check your instruction manual to find out how you particular calculator works.

EXAMPLE   2   *Converting between Degrees, Minutes, Seconds, and Decimal Forms*

(a) Convert 50°6′21″ to a decimal in degrees.

(b) Convert 21.256° to the D°M′S″ form.

Solution   (a) Figure 10 shows the solution using a TI-82 graphing calculator.

FIGURE 10

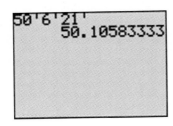

(b) Figure 11 shows the solution using a TI-82 graphing calculator.

FIGURE 11

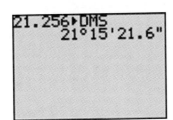

■ Now work Problems 57 and 63.

In many applications, such as describing the exact location of a star or the precise position of a boat at sea, angles measured in degrees, minutes, and even seconds are used. For calculation purposes, these are transformed to decimal form. In many other applications, especially those in calculus, angles are measured using *radians*.

Radians

Consider a circle of radius r. Construct an angle whose vertex is at the center of this circle, called a **central angle,** and whose rays subtend an arc on the circle whose length equals r. See Figure 12. The measure of such an angle is **1 radian.**

Now consider a circle and two central angles, θ and θ_1. Suppose that these central angles subtend arcs of lengths s and s_1, respectively, as shown in Figure 13. From geometry, we know that the ratio of the measures of the angles equals the ratio of the corresponding lengths of the arcs subtended by those angles; that is,

FIGURE 12

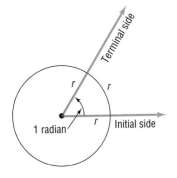

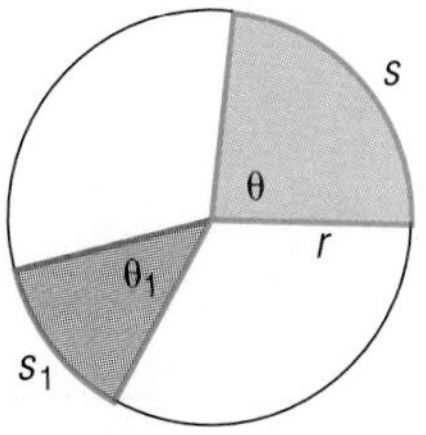

$$\frac{\theta}{\theta_1} = \frac{s}{s_1} \tag{2}$$

Suppose that θ and θ_1 are measured in radians, and suppose that $\theta_1 = 1$ radian. Refer again to Figure 13. Then the amount of arc s_1 subtended by the central angle θ_1 equals the radius r of the circle. Thus, $s_1 = r$, so formula (2) reduces to

FIGURE 13

$$\frac{\theta}{\theta_1} = \frac{s}{s_1}$$

$$\frac{\theta}{1} = \frac{s}{r} \quad \text{or} \quad s = r\theta \tag{3}$$

Theorem
Arc Length

For a circle of radius r, a central angle of θ radians subtends an arc whose length s is

$$s = r\theta \qquad (4)$$

Note: Formulas must be consistent with regard to the units used. In equation (4), we write

$$s = r\theta$$

To see the units, however, we must go back to equation (3) and write

$$\frac{\theta \text{ radians}}{1 \text{ radian}} = \frac{s \text{ length units}}{r \text{ length units}}$$

$$s \text{ length units} = (r \text{ length units})\frac{\theta \text{ radians}}{1 \text{ radian}}$$

Since the radians cancel, we are left with

$$s \text{ length units} = (r \text{ length units})\theta \quad s = r\theta$$

where θ appears to be "dimensionless" but, in fact, is measured in radians. Thus, in using the formula $s = r\theta$, the dimension of radians for θ is usually omitted, and any convenient unit of length (such as inches or meters) may be used for s and r.

E X A M P L E   3

Finding the Length of Arc of a Circle

Find the length of the arc of a circle of radius 2 meters subtended by a central angle of 0.25 radian.

Solution

We use equation (4) with $r = 2$ meters and $\theta = 0.25$. The length s of the arc is

$$s = r\theta = 2(0.25) = 0.5 \text{ meter}$$

◾   Now work Problem 33.

FIGURE 14
1 revolution $= 2\pi$ radians

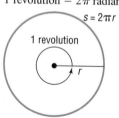

Relationship between Degrees and Radians

Consider a circle of radius r. A central angle of 1 revolution will subtend an arc equal to the circumference of the circle (Figure 14). Because the circumference of a circle equals $2\pi r$, we use $s = 2\pi r$ in equation (4) to find that, for an angle θ of 1 revolution,

$$s = r\theta$$
$$2\pi r = r\theta$$
$$\theta = 2\pi \text{ radians}$$

Thus,

$$1 \text{ revolution} = 2\pi \text{ radians} \tag{5}$$

so that

$$360° = 2\pi \text{ radians}$$

or

$$180° = \pi \text{ radians} \tag{6}$$

Divide both sides of equation (6) by 180. Then

$$1 \text{ degree} = \frac{\pi}{180} \text{ radian}$$

Divide both sides of (6) by π. Then

$$\frac{180}{\pi} \text{ degrees} = 1 \text{ radian}$$

Thus, we have the following two conversion formulas:

$$1 \text{ degree} = \frac{\pi}{180} \text{ radian} \qquad 1 \text{ radian} = \frac{180}{\pi} \text{ degrees} \tag{7}$$

E X A M P L E   4

Converting from Degrees to Radians

Convert each angle in degrees to radians:

(a) 60°   (b) 150°   (c) −45°   (d) 90°

Solution   (a) $60° = 60 \cdot 1 \text{ degree} = 60 \cdot \dfrac{\pi}{180} \text{ radian} = \dfrac{\pi}{3} \text{ radians}$

(b) $150° = 150 \cdot \dfrac{\pi}{180} \text{ radian} = \dfrac{5\pi}{6} \text{ radians}$

(c) $-45° = -45 \cdot \dfrac{\pi}{180} \text{ radian} = -\dfrac{\pi}{4} \text{ radian}$

(d) $90° = 90 \cdot \dfrac{\pi}{180} \text{ radian} = \dfrac{\pi}{2} \text{ radians}$ ■

■   Now work Problem 13.

Example 4 illustrates that angles that are fractions of a revolution are expressed in radian measure as fractional multiples of π, rather than as decimals. Thus, a right angle, as in Example 4(d), is left in the form $\pi/2$ radians, which is exact, rather than using the approximation $\pi/2 \approx 3.1416/2 = 1.5708$ radians.

E X A M P L E   5   *Converting Radians to Degrees*

Convert each angle in radians to degrees.

(a) $\dfrac{\pi}{6}$ radian   (b) $\dfrac{3\pi}{2}$ radians   (c) $-\dfrac{3\pi}{4}$ radians   (d) $\dfrac{7\pi}{3}$ radians

Solution   (a) $\dfrac{\pi}{6}$ radian $= \dfrac{\pi}{6} \cdot 1$ radian $= \dfrac{\pi}{6} \cdot \dfrac{180}{\pi}$ degrees $= 30°$

(b) $\dfrac{3\pi}{2}$ radians $= \dfrac{3\pi}{2} \cdot \dfrac{180}{\pi}$ degrees $= 270°$

(c) $-\dfrac{3\pi}{4}$ radians $= -\dfrac{3\pi}{4} \cdot \dfrac{180}{\pi}$ degrees $= -135°$

(d) $\dfrac{7\pi}{3}$ radians $= \dfrac{7\pi}{3} \cdot \dfrac{180}{\pi}$ degrees $= 420°$ ■

■   Now work Problem 23.

Table 1 lists the degree and radian measures of some commonly encountered angles. You should learn to feel equally comfortable using degree or radian measure for these angles.

TABLE 1

DEGREES	0°	30°	45°	60°	90°	120°	135°	150°	180°
RADIANS	0	$\dfrac{\pi}{6}$	$\dfrac{\pi}{4}$	$\dfrac{\pi}{3}$	$\dfrac{\pi}{2}$	$\dfrac{2\pi}{3}$	$\dfrac{3\pi}{4}$	$\dfrac{5\pi}{6}$	π
DEGREES	210°	225°	240°	270°	300°	315°	330°	360°	
RADIANS	$\dfrac{7\pi}{6}$	$\dfrac{5\pi}{4}$	$\dfrac{4\pi}{3}$	$\dfrac{3\pi}{2}$	$\dfrac{5\pi}{3}$	$\dfrac{7\pi}{4}$	$\dfrac{11\pi}{6}$	2π	

E X A M P L E   6   *Finding the Length of Arc of a Circle*

Find the length of the arc of a circle of radius $r = 3$ feet subtended by a central angle of $30°$.

Solution   We use equation (4), but first we must convert the central angle of $30°$ to radians. Since $30° = \pi/6$ radian, we use $\theta = \pi/6$ and $r = 3$ feet in equation (4). The length of the arc is

$$s = r\theta = 3 \cdot \frac{\pi}{6} = \frac{\pi}{2} \approx \frac{3.14}{2} = 1.57 \text{ feet}$$ ■

When an angle is measured in degrees, the degree symbol always will be shown. However, when an angle is measured in radians, we will follow the usual practice and omit the word *radians*. Thus, if the measure of an angle is given as $\pi/6$, it is understood to mean $\pi/6$ radian.

Circular Motion

We have already defined the average speed of an object as the distance traveled divided by the elapsed time. Suppose that an object moves along a circle of radius

r at a constant speed. If s is the distance traveled in time t along this circle, then the **linear speed** v of the object is defined as

$$v = \frac{s}{t} \qquad (8)$$

FIGURE 15

$v = \dfrac{s}{t}, \ \omega = \dfrac{\theta}{t}$

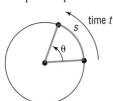

As this object travels along the circle, suppose that θ (measured in radians) is the central angle swept out in time t (See Figure 15.). Then the **angular speed** ω (the Greek letter omega) of this object is the angle (measured in radians) swept out divided by the elapsed time; that is,

$$\omega = \frac{\theta}{t} \qquad (9)$$

Angular speed is the way the speed of a phonograph record is described. For example, a 45 rpm (revolutions per minute) record is one that rotates at an angular speed of

$$\frac{45 \ \text{revolutions}}{\text{Minute}} = \frac{45 \ \text{revolutions}}{\text{Minute}} \cdot \frac{2\pi \ \text{radians}}{\text{Revolution}} = \frac{90\pi \ \text{radians}}{\text{Minute}}$$

There is an important relationship between linear speed and angular speed. In the formula $s = r\theta$, divide each side by t:

$$\frac{s}{t} = r\frac{\theta}{t}$$

Then, using equations (8) and (9), we obtain

$$v = r\omega \qquad (10)$$

When using equation (10), remember that $v = s/t$ (the linear speed) has the dimensions of length per unit of time (such as feet per second or miles per hour), r (the radius of the circular motion) has the same length dimension as s, and ω (the angular speed) has the dimensions of radians per unit of time. As noted earlier, we leave the radian dimension off the numerical value of the angular speed ω so that both sides of the equation will be dimensionally consistent (with "length per unit of time"). If the angular speed is given in terms of *revolutions* per unit of time (as is often the case), be sure to convert it to *radians* per unit of time before attempting to use equation (10).

E X A M P L E   7   *Finding Linear Speed*

Find the linear speed of a $33\frac{1}{3}$ rpm record at the point where the needle is 3 inches from the spindle (center of the record).

Solution   Look at Figure 16. The point P is traveling along a circle of radius $r = 3$ inches. The angular speed ω of the record is

FIGURE 16

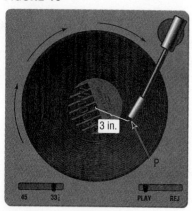

$$\omega = \frac{33\frac{1}{3} \text{ revolutions}}{\text{Minute}} = \frac{100 \text{ revolutions}}{3 \text{ minutes}} \cdot \frac{2\pi \text{ radians}}{\text{Revolution}}$$

$$= \frac{200\pi \text{ radians}}{3 \text{ minutes}}$$

From equation (10), the linear speed v of the point P is

$$v = r\omega = 3 \text{ inches} \cdot \frac{200\pi \text{ radians}}{3 \text{ minutes}} = \frac{200\pi \text{ inches}}{\text{Minute}} \approx \frac{628 \text{ inches}}{\text{Minute}}$$

■   Now work Problem 71.

2.1

Exercise 2.1

In Problems 1–12, draw each angle.

1. 30°   **2.** 60°   **3.** 135°   **4.** −120°   **5.** 450°   **6.** 540°

7. $3\pi/4$ **8.** $4\pi/3$ **9.** $-\pi/6$ **10.** $-2\pi/3$ **11.** $16\pi/3$ **12.** $21\pi/4$

In Problems 13–22, convert each angle in degrees to radians. Express your answer as a multiple of π.

13. 30°   **14.** 120°   **15.** 240°   **16.** 330°   **17.** −60°

18. −30°   **19.** 180°   **20.** 270°   **21.** 135°   **22.** −225°

In Problems 23–32, convert each angle in radians to degrees.

23. $\pi/3$ **24.** $5\pi/6$ **25.** $-5\pi/4$ **26.** $-2\pi/3$ **27.** $\pi/2$

28. 4π **29.** $\pi/12$ **30.** $5\pi/12$ **31.** $2\pi/3$ **32.** $5\pi/4$

In Problems 33–40, s denotes the length of arc of a circle of radius r subtended by the central angle θ. Find the missing quantity.

33. $r = 10$ meters, $\theta = \frac{1}{2}$ radian, $s = ?$ **34.** $r = 6$ feet, $\theta = 2$ radians, $s = ?$

35. $\theta = \frac{1}{3}$ radian, $s = 2$ feet, $r = ?$ **36.** $\theta = \frac{1}{4}$ radian, $s = 6$ centimeters, $r = ?$

37. $r = 5$ miles, $s = 3$ miles, $\theta = ?$ **38.** $r = 6$ meters, $s = 8$ meters, $\theta = ?$

39. $r = 2$ inches, $\theta = 30°$, $s = ?$ **40.** $r = 3$ meters, $\theta = 120°$, $s = ?$

In Problems 41–48, convert each angle in degrees to radians. Express your answer in decimal form, rounded to two decimal places.

41. 17°   **42.** 73°   **43.** −40°   **44.** −51°   **45.** 125°   **46.** 200°   **47.** 340°   **48.** 350°

In Problems 49–56, convert each angle in radians to degrees. Express your answer in decimal form, rounded to two decimal places.

49. 3.14   **50.** π **51.** 10.25   **52.** 0.75   **53.** 2   **54.** 3   **55.** 6.32   **56.** $\sqrt{2}$

In Problems 57–62, convert each angle to a decimal in degrees. Round off your answer to two decimal places.

57. 40°10′25″   **58.** 61°42′21″   **59.** 1°2′3″   **60.** 73°40′40″   **61.** 9°9′9″   **62.** 98°22′45″

In Problems 63–68, convert each angle to D°M'S" *form. Round off your answer to the nearest second.*

63. 40.32°   **64.** 61.24°   **65.** 18.255°   **66.** 29.411°   **67.** 19.99°   **68.** 44.01°

69. *Minute Hand of a Clock* The minute hand of a clock is 6 inches long. How far does the tip of the minute hand move in 15 minutes? How far does it move in 25 minutes?

70. *Movement of a Pendulum* A pendulum swings through an angle of 20° each second. If the pendulum is 40 inches long, how far does its tip move each second?

71. An object is traveling around a circle with a radius of 5 centimeters. If in 20 seconds a central angle of $\frac{1}{3}$ radian is swept out, what is the angular speed of the object? What is its linear speed?

72. An object is traveling around a circle with a radius of 2 meters. If in 20 seconds the object travels 5 meters, what is its angular speed? What is its linear speed?

73. *Bicycle Wheels* The diameter of each wheel of a bicycle is 26 inches. If you are traveling at a speed of 35 miles per hour on this bicycle, through how many revolutions per minute are the wheels turning?

74. *Car Wheels* The radius of each wheel of a car is 15". If the wheels are turning at the rate of 3 revolutions per second, how fast is the car moving? Express your answer in inches per second and in miles per hour.

75. *Windshield Wipers* The windshield wiper of a car is 18 inches long. How many inches will the tip of the wiper trace out in $\frac{1}{3}$ revolution?

76. *Windshield Wipers* The windshield wiper of a car is 18 inches long. If it takes 1 second to trace out $\frac{1}{3}$ revolution, how fast is the tip of the wiper moving?

77. *Speed of the Moon* The mean distance of the Moon from Earth is 2.39×10^5 miles. Assuming that the orbit of the Moon around Earth is circular and that 1 revolution takes 27.3 days, find the linear speed of the Moon. Express your answer in miles per hour.

78. *Speed of the Earth* The mean distance of Earth from the Sun is 9.29×10^7 miles. Assuming that the orbit of Earth around the Sun is circular and that 1 revolution takes 365 days, find the linear speed of Earth. Express your answer in miles per hour.

79. *Pulleys* Two pulleys, one with radius 2 inches and the other with radius 8 inches, are connected by a belt. (See the figure.) If the 2 inch pulley is caused to rotate at 3 revolutions per minute, determine the revolutions per minute of the 8 inch pulley. [*Hint:* The linear speeds of the pulleys, that is, the speed of the belt, are the same.]

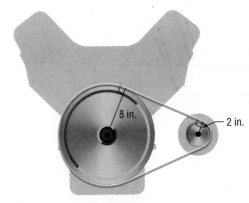

80. *Pulleys* Two pulleys, one with radius r_1 and the other with radius r_2, are connected by a belt. The pulley with radius r_1 rotates at ω_1 revolutions per minute, whereas the pulley with radius r_2 rotates at ω_2 revolutions per minute. Show that $r_1/r_2 = \omega_2/\omega_1$.

81. *Computing the Speed of River Current* To approximate the speed of the current of a river, a circular paddle wheel with radius 4 feet is lowered into the water. If the current causes the wheel to rotate at a speed of 10 revolutions per minute, what is the speed of the current? Express your answer in miles per hour.

82. *Spin Balancing Tires* A spin balancer rotates the wheel of a car at 480 revolutions per minute. If the diameter of the wheel is 26 inches, what road speed is being tested? Express your answer in miles per hour. At how many revolutions per minute should the balancer be set to test a road speed of 80 miles per hour?

83. *The Cable Cars of San Francisco* At the Cable Car Museum you can see the four cable lines that are used to pull cable cars up and down the hills of San Francisco. Each cable travels at a speed of 9.55 mi/hr, caused by a rotating wheel whose diameter is 8.5 feet. How fast is the wheel rotating? Express your answer in revolutions per minute.

84. *Difference in Time of Sun Rise* Naples, Florida, is approximately 90 miles due west of Ft. Lauderdale. How much sooner would a person in Ft. Lauderdale first see the rising Sun than a person in Naples? [*Hint:* Consult the figure. When a person at Q sees the first rays of the Sun, a person at P is still in the dark. The person at P sees the first rays after Earth has rotated so that P is at the location Q. Now use the fact that in 24 hours a length of arc of $2\pi(3960)$ miles is subtended.]

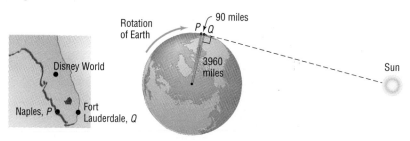

85. *Keeping up with the Sun* How fast would you have to travel on the surface of Earth to keep up with the Sun (that is, so that the Sun would appear to remain in the same position in the sky)?

86. *Nautical Miles* A **nautical mile** equals the length of arc subtended by a central angle of 1 minute on a great circle* on the surface of Earth. (See the figure.) If the radius of Earth is taken as 3960 miles, express 1 nautical mile in terms of ordinary, or **statute,** miles (5280 feet).

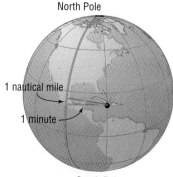

87. Do you prefer to measure angles using degrees or radians? Provide justification and a rationale for your choice.

88. Discuss why ships and airplanes use nautical miles to measure distance. Explain the difference between a nautical mile and a statute mile.

89. Investigate the way speed bicycles work. In particular, explain the differences and similarities between 10-speed and 18-speed derailleurs. Be sure to include a discussion of linear speed and angular speed.

2.2

Trigonometric Functions: Unit Circle Approach

We are now ready to introduce the trigonometric functions. As we said earlier, the approach taken uses the unit circle.

The Unit Circle

Recall that the **unit circle** is a circle whose radius is 1 and whose center is at the origin of a rectangular coordinate system. Because the radius r of the unit circle

*Any circle drawn on the surface of Earth that divides Earth into two equal hemispheres.

is 1, we see from the formula $s = r\theta$ that on the unit circle a central angle of θ radians subtends an arc whose length s is

$$s = \theta$$

FIGURE 17
Unit circle: $x^2 + y^2 = 1$

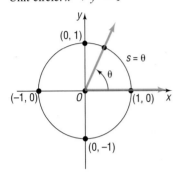

See Figure 17. Thus, on the unit circle, the length measure of the arc s equals the radian measure of the central angle θ. In other words, on the unit circle, the real number used to measure an angle θ in radians corresponds exactly with the real number used to measure the length of the arc subtended by that angle.

   For example suppose that $r = 1$ foot. Then, if $\theta = 3$ radians, $s = 3$ feet; if $\theta = 8.2$ radians, then $s = 8.2$ feet; and so on.

   Now, let t be any real number and let θ be the angle, in standard position, equal to t radians. Let P be the point on the unit circle that is also on the terminal side of θ. If $t \geq 0$, this point P is reached by moving *counterclockwise* along the unit circle, starting at $(1, 0)$, for a length of arc equal to t units. See Figure 18(a). If $t < 0$, this point P is reached by moving *clockwise* along the unit circle, starting at $(1, 0)$, for a length of arc equal to $|t|$ units. See Figure 18(b).

FIGURE 18

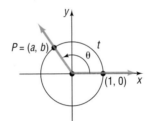

(a) $\theta = t$ radians; length of arc from $(1, 0)$ to P is t units, $t \geq 0$

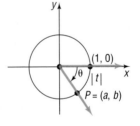

(b) $\theta = t$ radians; length of arc from $(1, 0)$ to P is $|t|$ units, $t < 0$

   Thus, to each real number t there corresponds a unique point $P = (a, b)$ on the unit circle. This is the important idea here. No matter what real number t is chosen, there corresponds a unique point P on the unit circle. We use the coordinates of the point $P = (a, b)$ on the unit circle corresponding to the real number t to define the **six trigonometric functions.**

Trigonometric Functions

Let t be a real number and let $P = (a, b)$ be the point on the unit circle that corresponds to t.

Sine Function

The **sine function** associates with t the y-coordinate of P and is denoted by

$$\sin t = b$$

Cosine Function	The **cosine function** associates with t the x-coordinate of P and is denoted by

$$\cos t = a$$

Tangent Function	If $a \neq 0$, the **tangent function** is defined as

$$\tan t = \frac{b}{a}$$

Cosecant Function	If $b \neq 0$, the **cosecant function** is defined as

$$\csc t = \frac{1}{b}$$

Secant Function	If $a \neq 0$, the **secant function** is defined as

$$\sec t = \frac{1}{a}$$

Cotangent Function	If $b \neq 0$, the **cotangent function** is defined as

$$\cot t = \frac{a}{b}$$

Notice in these definitions that if $a = 0$, that is, if the point $P = (0, b)$ is on the y-axis, then the tangent function and the secant function are undefined. Also, if $b = 0$, that is, if the point $P = (a, 0)$ is on the x-axis, then the cosecant function and the cotangent function are undefined.

Because we use the unit circle in these definitions of the trigonometric functions, they are also sometimes referred to as **circular functions.**

E X A M P L E   1   *Finding the Value of the Six Trigonometric Functions*

Let t be a real number and let $P = (-\frac{1}{2}, \sqrt{3}/2)$ be the point on the unit circle that corresponds to t. See Figure 19. Then

FIGURE 19

$\theta = t$ radians

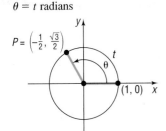

$$\sin t = \frac{\sqrt{3}}{2} \qquad \cos t = -\frac{1}{2} \qquad \tan t = \frac{\sqrt{3}/2}{-\frac{1}{2}} = -\sqrt{3}$$

$$\csc t = \frac{1}{\sqrt{3}/2} = \frac{2\sqrt{3}}{3} \qquad \sec t = \frac{1}{-\frac{1}{2}} = -2 \qquad \cot t = \frac{-\frac{1}{2}}{\sqrt{3}/2} = \frac{-\sqrt{3}}{3}$$

■

Trigonometric Functions of Angles

Let P be the point on the unit circle corresponding to the real number t. Then the angle θ, in standard position and measured in radians, whose terminal side is the ray from the origin through P is

$$\theta = t \text{ radians}$$

FIGURE 20

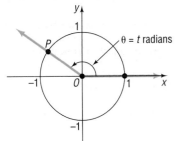

$\theta = t$ radians

See Figure 20.

Thus, on the unit circle, the measure of the angle θ in radians equals the value of the real number t. As a result, we can say that

$$\sin t = \sin \theta$$

Real number $\theta = t$ radians

and so on. We now can define the trigonometric functions of the angle θ.

> If $\theta = t$ radians, the **six trigonometric functions of the angle θ** are defined as
>
> $$\sin \theta = \sin t \qquad \cos \theta = \cos t \qquad \tan \theta = \tan t$$
> $$\csc \theta = \csc t \qquad \sec \theta = \sec t \qquad \cot \theta = \cot t$$

Even though the distinction between trigonometric functions of real numbers and trigonometric functions of angles is important, it is customary to refer to trigonometric functions of real numbers and trigonometric functions of angles collectively as *the trigonometric functions*. We shall follow this practice from now on.

If an angle θ is measured in degrees, we shall use the degree symbol when writing a trigonometric function of θ, as, for example, in $\sin 30°$ and $\tan 45°$. If an angle θ is measured in radians, then no symbol is used when writing a trigonometric function of θ, as, for example, in $\cos \pi$ and $\sec \pi/3$.

Finally, since the values of the trigonometric functions of an angle θ are determined by the coordinates of the point $P = (a, b)$ on the unit circle corresponding to θ, the units used to measure the angle θ are irrelevant. For example, it does not matter whether we write $\theta = \pi/2$ radians or $\theta = 90°$. The point on the unit circle corresponding to this angle is $P = (0, 1)$. Hence,

$$\sin \frac{\pi}{2} = \sin 90° = 1 \quad \text{and} \quad \cos \frac{\pi}{2} = \cos 90° = 0$$

Evaluating the Trigonometric Functions

To find the exact value of a trigonometric function of an angle θ requires that we locate the corresponding point P on the unit circle. In fact, though, any circle whose center is at the origin can be used.

Let θ be any nonquadrantal angle placed in standard position. Let $P^* = (a^*, b^*)$ be the point where the terminal side of θ intersects the unit circle. See Figure 21.

FIGURE 21

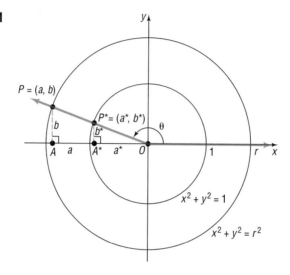

Let $P = (a, b)$ be any point on the terminal side of θ. Suppose that r is the distance from the origin to P. Then P is on the circle $x^2 + y^2 = r^2$. Refer again to Figure 21. Notice that the triangles OA^*P^* and OAP are similar; thus, ratios of corresponding sides are equal:

$$\frac{b^*}{1} = \frac{b}{r} \qquad \frac{a^*}{1} = \frac{a}{r} \qquad \frac{b^*}{a^*} = \frac{b}{a}$$

$$\frac{1}{b^*} = \frac{r}{b} \qquad \frac{1}{a^*} = \frac{r}{a} \qquad \frac{a^*}{b^*} = \frac{a}{b}$$

These results lead us to formulate the following theorem:

Theorem   For an angle θ in standard position, let $P = (a, b)$ be any point on the terminal side of θ. Let r equal the distance from the origin to P. Then

$$\sin \theta = \frac{b}{r} \qquad\qquad \cos \theta = \frac{a}{r} \qquad\qquad \tan \theta = \frac{b}{a}, \quad a \neq 0$$

$$\csc \theta = \frac{r}{b}, \quad b \neq 0 \qquad \sec \theta = \frac{r}{a}, \quad a \neq 0 \qquad \cot \theta = \frac{a}{b}, \quad b \neq 0$$

■

E X A M P L E   2   *Finding the Exact Value of the Six Trigonometric Functions*

Find the exact value of each of the six trigonometric functions of an angle θ if $(4, -3)$ is a point on its terminal side.

Solution   Figure 22 illustrates the situation for θ a positive angle. For the point $(a, b) = (4, -3)$, we have $a = 4$ and $b = -3$. Then $r = \sqrt{a^2 + b^2} = \sqrt{16 + 9} = 5$. Thus,

FIGURE 22

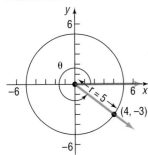

$$\sin \theta = \frac{b}{r} = -\frac{3}{5} \qquad \cos \theta = \frac{a}{r} = \frac{4}{5} \qquad \tan \theta = \frac{b}{a} = -\frac{3}{4}$$

$$\csc \theta = \frac{r}{b} = -\frac{5}{3} \qquad \sec \theta = \frac{r}{a} = \frac{5}{4} \qquad \cot \theta = \frac{a}{b} = -\frac{4}{3} \qquad \blacksquare$$

■   Now work Problem 1.

To evaluate the trigonometric functions of a given angle θ in standard position, we need to find the coordinates of any point on the terminal side of that angle. This is not always so easy to do. In the examples that follow, we will evaluate the trigonometric functions of certain angles for which this process is relatively easy. A calculator will need to be used to evaluate trigonometric functions of most angles.

E X A M P L E   3   *Finding the Exact Value of the Six Trigonometric Functions of Quadrantal Angles*

Find the exact value of each of the trigonometric functions at

(a) $\theta = 0 = 0°$ (b) $\theta = \pi/2 = 90°$ (c) $\theta = \pi = 180°$

(d) $\theta = 3\pi/2 = 270°$

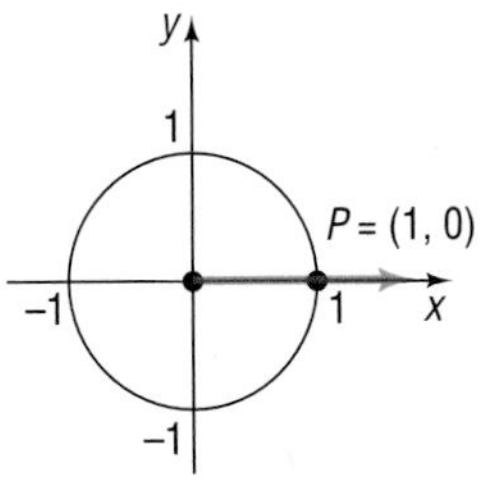

Solution   (a) The point $P = (1, 0)$ is on the terminal side of $\theta = 0 = 0°$ and is a distance of 1 unit from the origin. See Figure 23. Thus,

FIGURE 23

$\theta = 0 = 0°$

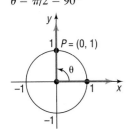

$$\sin 0 = \sin 0° = \frac{0}{1} = 0 \qquad \cos 0 = \cos 0° = \frac{1}{1} = 1$$

$$\tan 0 = \tan 0° = \frac{0}{1} = 0 \qquad \sec 0 = \sec 0° = \frac{1}{1} = 1$$

Since the y-coordinate of P is 0, $\csc 0$ and $\cot 0$ are not defined.

FIGURE 24

$\theta = \pi/2 = 90°$

(b) The point $P = (0, 1)$ is on the terminal side of $\theta = \pi/2 = 90°$ and is a distance of 1 unit from the origin. See Figure 24. Thus,

$$\sin \frac{\pi}{2} = \sin 90° = \frac{1}{1} = 1 \qquad \cos \frac{\pi}{2} = \cos 90° = \frac{0}{1} = 0$$

$$\csc \frac{\pi}{2} = \csc 90° = \frac{1}{1} = 1 \qquad \cot \frac{\pi}{2} = \cot 90° = \frac{0}{1} = 0$$

Since the x-coordinate of P is 0, $\tan \pi/2$ and $\sec \pi/2$ are not defined.

FIGURE 25

$\theta = \pi = 180°$

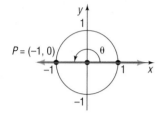

(c) The point $P = (-1, 0)$ is on the terminal side of $\theta = \pi = 180°$ and is a distance of 1 unit from the origin. See Figure 25. Thus,

$$\sin \pi = \sin 180° = \frac{0}{1} = 0 \qquad \cos \pi = \cos 180° = \frac{-1}{1} = -1$$

$$\tan \pi = \tan 180° = \frac{0}{1} = 0 \qquad \sec \pi = \sec 180° = \frac{1}{-1} = -1$$

Since the y-coordinate of P is 0, $\csc \pi$ and $\cot \pi$ are not defined.

FIGURE 26

$\theta = 3\pi/2 = 270°$

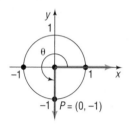

(d) The point $P = (0, -1)$ is on the terminal side of $\theta = 3\pi/2 = 270°$ and is a distance of 1 unit from the origin. See Figure 26. Thus,

$$\sin \frac{3\pi}{2} = \sin 270° = \frac{-1}{1} = -1 \qquad \cos \frac{3\pi}{2} = \cos 270° = \frac{0}{1} = 0$$

$$\csc \frac{3\pi}{2} = \csc 270° = \frac{1}{-1} = -1 \qquad \cot \frac{3\pi}{2} = \cot 270° = \frac{0}{-1} = 0$$

Since the x-coordinate of P is 0, $\tan 3\pi/2$ and $\sec 3\pi/2$ are not defined. ■

Note: The results obtained in Example 3 are the same whether we pick a point on the unit circle or a point on any circle whose center is at the origin. For example, $P = (0, 5)$ is a point on the terminal side of $\theta = \pi/2 = 90°$ and is a distance of 5 units from the origin. Using this point, $\sin \theta = \frac{5}{5} = 1$, $\cos \theta = \frac{0}{5} = 0$, and so on, as in Example 3(b).

Table 2 summarizes the values of the trigonometric functions found in Example 3.

TABLE 2　QUADRANTAL ANGLES

θ (RADIANS)	θ (DEGREES)	$\sin \theta$	$\cos \theta$	$\tan \theta$	$\csc \theta$	$\sec \theta$	$\cot \theta$
0	0°	0	1	0	Not defined	1	Not defined
$\pi/2$	90°	1	0	Not defined	1	Not defined	0
π	180°	0	-1	0	Not defined	-1	Not defined
$3\pi/2$	270°	-1	0	Not defined	-1	Not defined	0

■　Now work Problem 39.

E X A M P L E　4　　*Finding Exact Values of Trigonometric Functions* ($\theta = 45°$, $\theta = -45°$)

Find the exact value of each of the trigonometric functions at:

(a)　$\theta = \pi/4 = 45°$　　(b)　$\theta = -\pi/4 = -45°$

FIGURE 27

$\theta = \pi/4 = 45°$

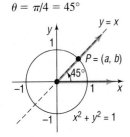

Solution (a) We seek the coordinates of a point $P = (a, b)$ on the terminal side of $\theta = \pi/4 = 45°$. See Figure 27. First, we observe that P lies on the line $y = x$. (Do you see why? Since $\theta = 45° = \frac{1}{2} \cdot 90°$, P must lie on the line that bisects quadrant I.) Suppose that P also lies on the unit circle so that P is a distance of 1 unit from the origin. Then it follows that

$$a^2 + b^2 = 1 \quad a = b, a > 0, b > 0$$
$$a^2 + a^2 = 1$$
$$2a^2 = 1$$
$$a = \frac{1}{\sqrt{2}} = \frac{\sqrt{2}}{2}, \qquad b = \frac{\sqrt{2}}{2}$$

Thus,

$$\sin\frac{\pi}{4} = \sin 45° = \frac{\sqrt{2}}{2} \qquad \cos\frac{\pi}{4} = \cos 45° = \frac{\sqrt{2}}{2} \qquad \tan\frac{\pi}{4} = \tan 45° = \frac{\sqrt{2}/2}{\sqrt{2}/2} = 1$$

$$\csc\frac{\pi}{4} = \csc 45° = \frac{1}{\sqrt{2}/2} = \sqrt{2} \qquad \sec\frac{\pi}{4} = \sec 45° = \frac{1}{\sqrt{2}/2} = \sqrt{2} \qquad \cot\frac{\pi}{4} = \cot 45° = \frac{\sqrt{2}/2}{\sqrt{2}/2} = 1$$

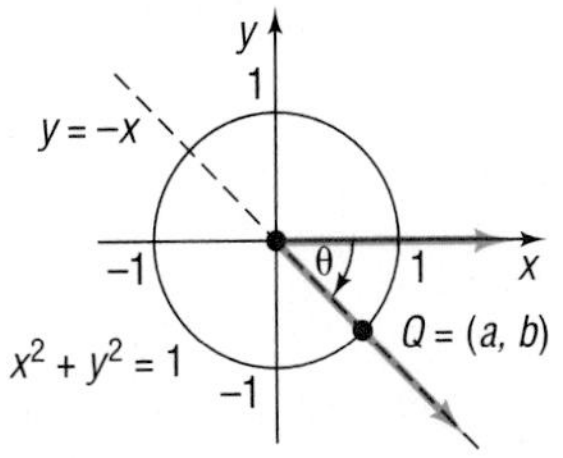

(b) We seek the coordinates of a point $Q = (a, b)$ on the terminal side of $\theta = -\pi/4 = -45°$. See Figure 28. First, we notice that Q lies on the line $y = -x$. If Q also lies on the unit circle $x^2 + y^2 = 1$, then

FIGURE 28

$\theta = -\pi/4 = -45°$

$$a^2 + b^2 = 1 \quad b = -a, a > 0$$
$$a^2 + (-a)^2 = 1$$
$$2a^2 = 1$$
$$a = \frac{1}{\sqrt{2}} = \frac{\sqrt{2}}{2}, \qquad b = -a = -\frac{\sqrt{2}}{2}$$

Thus,

$$\sin\left(-\frac{\pi}{4}\right) = \sin(-45°) \qquad \cos\left(-\frac{\pi}{4}\right) = \cos(-45°) \qquad \tan\left(-\frac{\pi}{4}\right) = \tan(-45°)$$

$$= -\frac{\sqrt{2}}{2} \qquad\qquad = \frac{\sqrt{2}}{2} \qquad\qquad = \frac{-\sqrt{2}/2}{\sqrt{2}/2} = -1$$

$$\csc\left(-\frac{\pi}{4}\right) = \csc(-45°) \qquad \sec\left(-\frac{\pi}{4}\right) = \sec(-45°) \qquad \cot\left(-\frac{\pi}{4}\right) = \cot(-45°)$$

$$= \frac{1}{-\sqrt{2}/2} = -\sqrt{2} \qquad = \frac{1}{\sqrt{2}/2} = \sqrt{2} \qquad = \frac{\sqrt{2}/2}{-\sqrt{2}/2} = -1$$

■

In solving Example 4(b), we could have located the point Q by using the coordinates of $P = (\sqrt{2}/2, \sqrt{2}/2)$ from Example 4(a), noting the symmetry of Q and P with respect to the x-axis.

E X A M P L E   5   *Finding the Exact Value of a Trigonometric Expression*

Find the exact value of each expression:

(a) $\sin 45° \cos 180°$ (b) $\tan\frac{\pi}{4} - \sin\frac{3\pi}{2}$

FIGURE 29

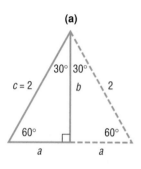

(a)

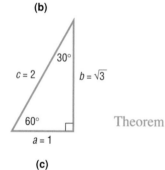

(b)

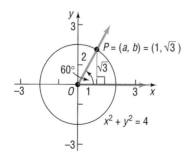

(c)

Solution

(a) $\sin 45° \cos 180° = \dfrac{\sqrt{2}}{2} \cdot (-1) = \dfrac{-\sqrt{2}}{2}$

From Example 4(a) ⌐   ⌐ From Table 2

(b) $\tan \dfrac{\pi}{4} - \sin \dfrac{3\pi}{2} = 1 - (-1) = 2$

From Example 4(a) ⌐   ⌐ From Table 2 ■

■ Now work Problems 15 and 37.

Trigonometric Functions of 30° and 60°

Consider a right triangle in which one of the angles is 30°. It then follows that the other angle is 60°. Figure 29(a) illustrates such a triangle with hypotenuse of length 2. Our problem is to determine a and b.

We begin by placing next to this triangle another triangle congruent to the first, as shown in Figure 29(b). Notice that we now have a triangle whose angles are each 60°. This triangle is therefore equilateral, so each side is of length 2. In particular, the base is $2a = 2$, and so $a = 1$. By the Pythagorean Theorem, b satisfies the equation $a^2 + b^2 = c^2$, so we have

$$a^2 + b^2 = c^2$$
$$1^2 + b^2 = 2^2$$
$$b^2 = 4 - 1 = 3$$
$$b = \sqrt{3}$$

This results in Figure 29(c) and leads to the following theorem:

Theorem

In a 30, 60, 90 degree right triangle, the length of the side opposite the 30° angle is half the length of the hypotenuse. The length of the adjacent side is $\sqrt{3}/2$ times the length of the hypotenuse. ■

E X A M P L E   6

Finding Exact Values of the Trigonometric Functions ($\theta = 60°$)

Find the exact value of each of the trigonometric functions of $\pi/3 = 60°$.

Solution

We reposition the triangle in Figure 29(c) in a rectangular coordinate system. We seek the coordinates of the point $P = (a, b)$ on the terminal side of $\theta = \pi/3 = 60°$, a distance of 2 units from the origin. See Figure 30.

FIGURE 30

Based on the preceding theorem, it follows that $a = 1$ and $b = \sqrt{3}$. Since $r = 2$, we have

$$\sin \frac{\pi}{3} = \sin 60° = \frac{\sqrt{3}}{2} \qquad \cos \frac{\pi}{3} = \cos 60° = \frac{1}{2} \qquad \tan \frac{\pi}{3} = \tan 60° = \frac{\sqrt{3}}{1} = \sqrt{3}$$

$$\csc \frac{\pi}{3} = \csc 60° = \frac{2}{\sqrt{3}} = \frac{2\sqrt{3}}{3} \qquad \sec \frac{\pi}{3} = \sec 60° = \frac{2}{1} = 2 \qquad \cot \frac{\pi}{3} = \cot 60° = \frac{1}{\sqrt{3}} = \frac{\sqrt{3}}{3} \qquad \blacksquare$$

E X A M P L E   7   *Finding Exact Values of the Trigonometric Functions ($\theta = 30°$)*

Find the exact value of each of the trigonometric functions of $\pi/6 = 30°$.

Solution   Again, we reposition the triangle in Figure 29(c) in a rectangular coordinate system. We seek the coordinates of the point $P = (a, b)$ on the terminal side of $\theta = \pi/6 = 30°$, a distance of 2 units from the origin. See Figure 31. Based on the preceding theorem, it follows that $a = \sqrt{3}$ and $b = 1$. Since $r = 2$, we have

$$\sin \frac{\pi}{6} = \sin 30° = \frac{1}{2} \qquad \cos \frac{\pi}{6} = \cos 30° = \frac{\sqrt{3}}{2} \qquad \tan \frac{\pi}{6} = \tan 30° = \frac{1}{\sqrt{3}} = \frac{\sqrt{3}}{3}$$

$$\csc \frac{\pi}{6} = \csc 30° = \frac{2}{1} = 2 \qquad \sec \frac{\pi}{6} = \sec 30° = \frac{2}{\sqrt{3}} = \frac{2\sqrt{3}}{3} \qquad \cot \frac{\pi}{6} = \cot 30° = \frac{\sqrt{3}}{1} = \sqrt{3}$$

FIGURE 31

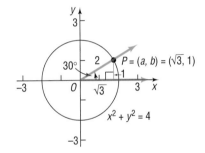

Table 3 summarizes the information just derived for $\pi/6$ (30°), $\pi/4$ (45°), and $\pi/3$ (60°). Until you memorize the entries in Table 3, you should draw an appropriate diagram to determine the values given in the table.

TABLE 3

θ (RADIANS)	θ (DEGREES)	$\sin \theta$	$\cos \theta$	$\tan \theta$	$\csc \theta$	$\sec \theta$	$\cot \theta$
$\pi/6$	30°	$\frac{1}{2}$	$\sqrt{3}/2$	$\sqrt{3}/3$	2	$2\sqrt{3}/3$	$\sqrt{3}$
$\pi/4$	45°	$\sqrt{2}/2$	$\sqrt{2}/2$	1	$\sqrt{2}$	$\sqrt{2}$	1
$\pi/3$	60°	$\sqrt{3}/2$	$\frac{1}{2}$	$\sqrt{3}$	$2\sqrt{3}/3$	2	$\sqrt{3}/3$

■   Now work Problems 21 and 31.

E X A M P L E   8   *Constructing a Rain Gutter*

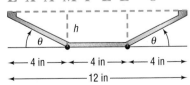

FIGURE 32

A rain gutter is to be constructed of aluminum sheets 12 inches wide. After marking off a length of 4 inches from each edge, this length is bent up at an angle θ. See Figure 32. The area A of the opening can be expressed as a function of θ:

$$A(\theta) = 16 \sin \theta \, (\cos \theta + 1)$$

Find the area A of the opening for $\theta = 30°$, $\theta = 45°$, and $\theta = 60°$.

Solution For $\theta = 30°$: $A(30°) = 16 \sin 30° (\cos 30° + 1)$

$$= 16\left(\frac{1}{2}\right)\left(\frac{\sqrt{3}}{2} + 1\right) = 4\sqrt{3} + 8$$

The area of the opening for $\theta = 30°$ is about 14.9 square inches.

For $\theta = 45°$: $A(45°) = 16 \sin 45° (\cos 45° + 1)$

$$= 16\left(\frac{\sqrt{2}}{2}\right)\left(\frac{\sqrt{2}}{2} + 1\right) = 8 + 8\sqrt{2}$$

The area of the opening for $\theta = 45°$ is about 19.3 square inches.

For $\theta = 60°$: $A(60°) = 16 \sin 60° (\cos 60° + 1)$

$$= 16\left(\frac{\sqrt{3}}{2}\right)(\tfrac{1}{2} + 1) = 12\sqrt{3}$$

The area of the opening for $\theta = 60°$ is about 20.8 square inches. ■

Using a Calculator to Find Values of Trigonometric Functions

Before getting started, you must first decide whether to enter the angle in the calculator using radians or degrees and then set the calculator to the correct MODE. Your calculator probably only has the keys marked $\boxed{\sin}$, $\boxed{\cos}$, and $\boxed{\tan}$. To find the values of the remaining three trigonometric functions, secant, cosecant, and cotangent, we use the facts that

$$\sec \theta = \frac{1}{\cos \theta} \qquad \csc \theta = \frac{1}{\sin \theta} \qquad \cot \theta = \frac{1}{\tan \theta}$$

These facts are a direct consequence of the theorem stated on page 140.

E X A M P L E 9 *Using a Calculator to Approximate the Value of a Trigonometric Function*

Use a calculator to find the value of

(a) $\cos 48°$ (b) $\csc 21°$ (c) $\tan \dfrac{\pi}{12}$

Express your answer rounded to two decimal places.

Solution (a) First, we set the MODE to receive degrees. See Figure 33(a). Figure 33(b) shows the solution using a TI-82 graphing calculator.

FIGURE 33

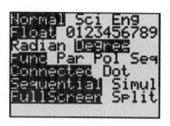

(a)

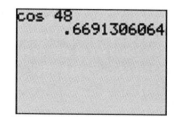

(b)

Thus,

$$\cos 48° \approx 0.67$$

rounded to two decimal places.

(b) Most calculators do not have a $\boxed{\text{CSC}}$ key. The manufacturers assume that the user knows some trigonometry. Thus, to find the value of csc 21°, we use the fact that csc 21° = 1/(sin 21°). Figure 34 shows the solution using a TI-82 graphing calculator.

FIGURE 34

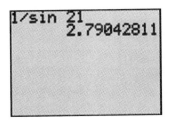

Thus,

$$\csc 21° \approx 2.79$$

rounded to two decimal places.

(c) Set the mode to receive radians. Figure 35 shows the solution using a TI-82 graphing calculator.

FIGURE 35

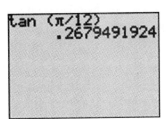

Thus,

$$\tan \frac{\pi}{12} \approx 0.27$$

rounded to two decimal places.   ■

Note: In Figure 35, the parentheses are required. Without them, the calculator will evaluate tan π first, then divide by 12.

■ Now work Problem 47.

Summary

For the quadrantal angles and the angles 30°, 45°, 60°, and their integral multiples, we can find the exact value of each trigonometric function by using the geometric features of these angles and symmetry.

Figure 36 shows points on the unit circle that are on the terminal sides of some angles that are integral multiples of $\pi/6$ (30°), $\pi/4$ (45°), and $\pi/3$ (60°).

FIGURE 36

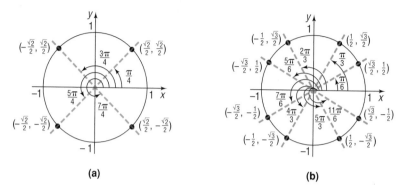

(a)

(b)

Notice the relationship between integral multiples of an angle and symmetry. For example, in Figure 36(a), the point on the unit circle corresponding to $7\pi/4$ is the reflection about the x-axis of the point corresponding to $\pi/4$. The point on the unit circle corresponding to $5\pi/4$ is the reflection about the origin of the point corresponding to $\pi/4$. Thus, using Figure 36(a), we see that

$$\sin\frac{7\pi}{4} = -\frac{\sqrt{2}}{2} \qquad \cos\left(-\frac{\pi}{4}\right) = \frac{\sqrt{2}}{2} \qquad \tan\frac{5\pi}{4} = \frac{-\sqrt{2}/2}{-\sqrt{2}/2} = 1$$

Using Figure 36(b), we see that

$$\sin\frac{11\pi}{6} = -\frac{1}{2} \qquad \cos\frac{7\pi}{6} = -\frac{\sqrt{3}}{2} \qquad \tan\frac{4\pi}{3} = \frac{-\sqrt{3}/2}{-\frac{1}{2}} = \sqrt{3}$$

For most other angles, besides the quadrantal angles and those listed in Figure 36, we can only approximate the value of each trigonometric function using a calculator.

■ The name *sine* for the sine function is due to a medieval confusion. The name comes from the Sanskrit word *jıva* (meaning chord), first used in India by Āryabhata the Elder (AD 510). He really meant half-chord, but abbreviated it. This was brought into Arabic as *jıba*, which was meaningless. Because the proper Arabic word *jaib* would be written the same way (short vowels are not written out in Arabic), *jıba* was pronounced as *jaib*, which meant bosom or hollow, and *jaib* remains as the Arabic word for sine to this day. Scholars translating the Arabic works into Latin found that the word *sinus* also meant bosom or hollow, and from *sinus* we get the word *sine*.

The name *tangent*, due to Thomas Fincke (1583), can be understood by looking at Figure 37. The line segment $\overline{DC}$ is tangent to the circle at C. If $d(O, B) = d(O, C) = 1$, then the length of the line segment $\overline{DC}$ is

FIGURE 37

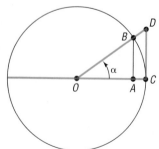

$$d(D, C) = \frac{d(D, C)}{1} = \frac{d(D, C)}{d(O, C)} = \tan\alpha$$

The old name for the tangent is *umbra versa* (meaning turned shadow), referring to the use of the tangent in solving height problems with shadows.

The names of the remaining functions came about as follows. If α and β are complementary angles, then $\cos\alpha = \sin\beta$. Because β is the complement of α, it was natural to write the cosine of α as *sin co α*. Probably for reasons involving ease of pronunciation, the *co* migrated to the front, and then cosine received a three-letter abbreviation to match sin, sec, and tan. The two other cofunctions were similarly treated, except that the long forms *cotan* and *cosec* survive to this day in some countries. ■

2.2

Exercise 2.2

In Problems 1–10, a point on the terminal side of an angle θ is given. Find the exact value of each of the six trigonometric functions of θ.

1. $(-3, 4)$ **2.** $(5, -12)$ **3.** $(2, -3)$ **4.** $(-1, -2)$ **5.** $(-2, -2)$

6. $(1, -1)$ **7.** $(-3, -2)$ **8.** $(2, 2)$ **9.** $(\frac{1}{3}, -\frac{1}{4})$ **10.** $(-0.3, -0.4)$

In Problems 11–30, find the exact value of each expression. Do not use a calculator.

11. $\sin 45° + \cos 60°$ **12.** $\sin 30° - \cos 45°$ **13.** $\sin 90° + \tan 45°$

14. $\cos 180° - \sin 180°$ **15.** $\sin 45° \cos 45°$ **16.** $\tan 45° \cos 30°$

17. $\csc 45° \tan 60°$ **18.** $\sec 30° \cot 45°$ **19.** $4 \sin 90° - 3 \tan 180°$

20. $5 \cos 90° - 8 \sin 270°$ **21.** $2 \sin \dfrac{\pi}{3} - 3 \tan \dfrac{\pi}{6}$ **22.** $2 \sin \dfrac{\pi}{4} + 3 \tan \dfrac{\pi}{4}$

23. $\sin \dfrac{\pi}{4} - \cos \dfrac{\pi}{4}$ **24.** $\tan \dfrac{\pi}{3} + \cos \dfrac{\pi}{3}$ **25.** $2 \sec \dfrac{\pi}{4} + 4 \cot \dfrac{\pi}{3}$

26. $3 \csc \dfrac{\pi}{3} + \cot \dfrac{\pi}{4}$ **27.** $\tan \pi - \cos 0$ **28.** $\sin \dfrac{3\pi}{2} + \tan \pi$

29. $\csc \dfrac{\pi}{2} + \cot \dfrac{\pi}{2}$ **30.** $\sec \pi - \csc \dfrac{\pi}{2}$

In Problems 31–46, find the exact value of each of the six trigonometric functions of the given angle. If any are not defined, say "not defined." Do not use a calculator.

31. $2\pi/3$ **32.** $3\pi/4$ **33.** $150°$ **34.** $330°$

35. $-\pi/6$ **36.** $-\pi/3$ **37.** $225°$ **38.** $210°$

39. $5\pi/2$ **40.** 3π **41.** $-180°$ **42.** $-270°$

43. $3\pi/2$ **44.** $-\pi$ **45.** $450°$ **46.** $-90°$

In Problems 47–70, use a calculator to find the approximate value of each expression rounded to two decimal places.

47. $\sin 28°$ **48.** $\cos 14°$ **49.** $\tan 21°$ **50.** $\sin 15°$

51. $\sec 41°$ **52.** $\csc 55°$ **53.** $\cot 70°$ **54.** $\tan 80°$

55. $\sin \dfrac{\pi}{10}$ **56.** $\cos \dfrac{\pi}{8}$ **57.** $\tan \dfrac{5\pi}{12}$ **58.** $\sin \dfrac{3\pi}{10}$

59. $\sec \dfrac{\pi}{12}$ **60.** $\csc \dfrac{5\pi}{13}$ **61.** $\cot \dfrac{\pi}{18}$ **62.** $\sin \dfrac{\pi}{18}$

63. $\sin 1$ **64.** $\tan 1$ **65.** $\sin 1°$ **66.** $\tan 1°$

67. $\cos 21.5°$ **68.** $\cos 35.2°$ **69.** $\tan 0.3$ **70.** $\tan 0.1$

In Problems 71–82, find the exact value of each expression if θ = 60°. Do not use a calculator.

71. $\sin \theta$ **72.** $\cos \theta$ **73.** $\sin \dfrac{\theta}{2}$ **74.** $\cos \dfrac{\theta}{2}$

75. $(\sin \theta)^2$ **76.** $(\cos \theta)^2$ **77.** $\sin 2\theta$ **78.** $\cos 2\theta$

79. $2 \sin \theta$ **80.** $2 \cos \theta$ **81.** $\dfrac{\sin \theta}{2}$ **82.** $\dfrac{\cos \theta}{2}$

83. Find the exact value of $\sin 45° + \sin 135° + \sin 225° + \sin 315°$.

84. Find the exact value of $\tan 60° + \tan 150°$.

85. If $\sin \theta = 0.1$, find $\sin(\theta + \pi)$.

86. If $\cos \theta = 0.3$, find $\cos(\theta + \pi)$.

87. If $\tan \theta = 3$, find $\tan(\theta + \pi)$.

88. If $\cot \theta = -2$, find $\cot(\theta + \pi)$.

89. If $\sin \theta = \frac{1}{5}$, find $\csc \theta$.

90. If $\cos \theta = \frac{2}{3}$, find $\sec \theta$.

The path of a projectile fired at an inclination θ to the horizontal with initial speed v_0 is a parabola (see the figure). The range R of the projectile, that is, the horizontal distance that the projectile travels, is found by using the formula

$$R = \frac{v_0^2 \sin 2\theta}{g}$$

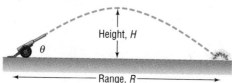

v_0 = Intial speed

Height, H

θ

Range, R

where $g \approx 32.2$ feet per second per second ≈ 9.8 meters per second per second is the acceleration due to gravity. The maximum height H of the projectile is

$$H = \frac{v_0^2 \sin^2\theta}{2g}$$

In Problems 91–94, find the range R and maximum height H.

91. The projectile is fired at an angle of 45° to the horizontal with an initial speed of 100 feet per second.

92. The projectile is fired at an angle of 30° to the horizontal with an initial speed of 150 meters per second.

93. The projectile is fired at an angle of 25° to the horizontal with an initial speed of 500 meters per second.

94. The projectile is fired at an angle of 50° to the horizontal with an initial speed of 200 feet per second.

95. *Inclined Plane* If friction is ignored, the time t (in seconds) required for a block to slide down an inclined plane (see the figure) is given by the formula

$$t = \sqrt{\frac{2a}{g \sin \theta \cos \theta}}$$

where a is the length (in feet) of the base and $g \approx 32$ feet per second per second is the acceleration of gravity. How long does it take a block to slide down an inclined plane with base $a = 10$ feet when
(a) $\theta = 30°$?   (b) $\theta = 45°$?   (c) $\theta = 60°$?

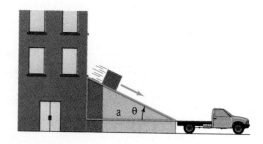

96. *Piston Engines* In a certain piston engine, the distance x (in meters) from the center of the drive shaft to the head of the piston is given by

$$x = \cos \theta + \sqrt{16 + 0.5 \cos 2\theta}$$

where θ is the angle between the crank and the path of the piston head (see the figure). Find x when $\theta = 30°$ and when $\theta = 45°$.

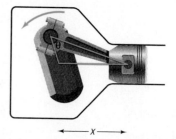

x

97. *Calculating the Time of a Trip* Two oceanfront homes are located 8 miles apart on a straight stretch of beach, each a distance of 1 mile from a paved road that parallels the ocean. Sally can walk 8 miles per hour along the paved road, but only 3 miles per hour in the sand that separates the road from the ocean. Because of a river directly between the two houses, it is necessary to walk in the sand to the road, continue on the road, and then walk directly back in the sand to get from one house to the other. See the illustration. The time T to get from one house to the other as a function of the angle θ shown in the illustration is

$$T(\theta) = 1 + \frac{2}{3 \sin \theta} - \frac{1}{4 \tan \theta}, \ 0° < \theta < 90°$$

(a) Calculate the time T for $\theta = 30°$. How long is Sally on the paved road?
(b) Calculate the time T for $\theta = 45°$. How long is Sally on the paved road?
(c) Calculate the time T for $\theta = 60°$. How long is Sally on the paved road?
(d) Calculate the time T for $\theta = 90°$. Describe the path taken. Why can't the formula for T be used?

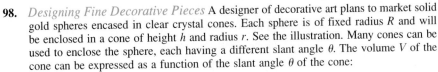

98. *Designing Fine Decorative Pieces* A designer of decorative art plans to market solid gold spheres encased in clear crystal cones. Each sphere is of fixed radius R and will be enclosed in a cone of height h and radius r. See the illustration. Many cones can be used to enclose the sphere, each having a different slant angle θ. The volume V of the cone can be expressed as a function of the slant angle θ of the cone:

$$V(\theta) = \frac{1}{3}\pi R^3 \frac{(1 + \sec \theta)^3}{\tan^2 \theta}, \ 0° < \theta < 90°$$

What volume V is required to enclose a sphere of radius 2 centimeters in a cone whose slant angle θ is 30°? 45°? 60°?

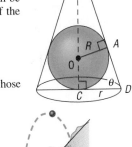

99. *Projectile Motion* An object is propelled upward at an angle θ, $45° < \theta < 90°$, to the horizontal with an initial velocity of v_0 feet per second from the base of a plane that makes an angle of 45° with the horizontal. See the illustration. If air resistance is ignored, the distance R it travels up the inclined plane is given by

$$R = \frac{v_0^2 \sqrt{2}}{32}(\sin 2\theta - \cos 2\theta - 1)$$

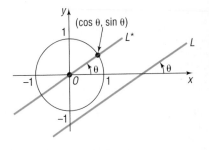

(a) Find the distance R that the object travels along the inclined plane if the initial velocity is 32 feet per second and $\theta = 60°$.
(b) Graph $R = R(\theta)$ if the initial velocity is 32 ft/sec.
(c) What value of θ makes R largest?

100. If θ $(0 < \theta < \pi)$ is the angle between a horizontal ray directed to the right (say, the positive x-axis) and a nonhorizontal, nonvertical line L, show that the slope m of L equals $\tan \theta$. The angle θ is called the **inclination** of L. [*Hint:* See the illustration, where we have drawn the line L^* parallel to L and passing through the origin. Use the fact that L^* intersects the unit circle at the point $(\cos \theta, \sin \theta)$.]

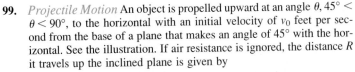

101. Write a brief paragraph that explains how to quickly compute the trigonometric functions of 30°, 45°, and 60°.

102. Write a brief paragraph that explains how to quickly compute the trigonometric functions of 0°, 90°, 180°, and 270°.

103. How would you explain the meaning of the sine function to a fellow student who has just completed college algebra?

2.3

Properties of the Trigonometric Functions

FIGURE 38

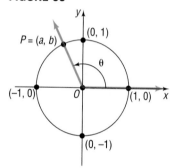

Domain and Range of the Trigonometric Functions

Let θ be an angle in standard position, and let $P = (a, b)$ be a point on the terminal side of θ. Suppose, for convenience, that P also lies on the unit circle. See Figure 38. Then, by definition,

$$\sin \theta = b \qquad \cos \theta = a \qquad \tan \theta = \frac{b}{a}, \quad a \neq 0$$

$$\csc \theta = \frac{1}{b}, \quad b \neq 0 \qquad \sec \theta = \frac{1}{a}, \quad a \neq 0 \qquad \cot \theta = \frac{a}{b}, \quad b \neq 0$$

For $\sin \theta$ and $\cos \theta$, θ can be any angle, so it follows that the domain of the sine function and cosine function is the set of all real numbers.

> The domain of the sine function is the set of all real numbers.
> The domain of the cosine function is the set of all real numbers.

If $a = 0$, then the tangent function and the secant function are not defined. Thus, for the tangent function and secant function the x-coordinate of $P = (a, b)$ cannot be 0. On the unit circle, there are two such points, $(0, 1)$ and $(0, -1)$. These two points correspond to the angles $\pi/2$ (90°) and $3\pi/2$ (270°) or, more generally, to any angle that is an odd multiple of $\pi/2$ (90°), such as $\pi/2$ (90°), $3\pi/2$ (270°), $5\pi/2$ (450°), $-\pi/2$ (−90°), $-3\pi/2$ (−270°), and so on. Such angles must therefore be excluded from the domain of the tangent function and secant function.

> The domain of the tangent function is the set of all real numbers, except odd multiples of $\pi/2$ (90°).
>
> The domain of the secant function is the set of all real numbers, except odd multiples of $\pi/2$ (90°).

If $b = 0$, then the cotangent function and the cosecant function are not defined. Thus, for the cotangent function and cosecant function the y-coordinate of $P = (a, b)$ cannot be 0. On the unit circle, there are two such points, $(1, 0)$ and $(-1, 0)$. These two points correspond to the angles 0 (0°) and π (180°) or, more generally, to any angle that is an integral multiple of π (180°), such as 0 (0°), π (180°), 2π (360°), 3π (540°), $-\pi$ (−180°), and so on. Such angles must therefore be excluded from the domain of the cotangent function and cosecant function.

Quotient Identities

$$\tan \theta = \frac{\sin \theta}{\cos \theta} \qquad \cot \theta = \frac{\cos \theta}{\sin \theta} \tag{3}$$

The proofs of formulas (2) and (3) follow from the definitions of the trigonometric functions. (See Problems 99 and 100.)

Seeing the Concept To see the identity $\tan \theta = (\sin \theta)/(\cos \theta)$, graph $y = \tan x$ and $y = (\sin x)/(\cos x)$ on the same screen.

If $\sin \theta$ and $\cos \theta$ are known, formulas (2) and (3) make it easy to find the values of the remaining trigonometric functions.

E X A M P L E   3   *Finding Exact Values Using Identities When Sine and Cosine Are Given*

Given $\sin \theta = 1/\sqrt{5}$ and $\cos \theta = 2/\sqrt{5}$, find the exact values of the four remaining trigonometric functions of θ.

Solution   Based on a quotient identity from formula (3), we have

$$\tan \theta = \frac{\sin \theta}{\cos \theta} = \frac{1/\sqrt{5}}{2/\sqrt{5}} = \frac{1}{2}$$

Then we use the reciprocal identities from formula (2) to get

$$\csc \theta = \frac{1}{\sin \theta} = \frac{1}{1/\sqrt{5}} = \sqrt{5} \qquad \sec \theta = \frac{1}{\cos \theta} = \frac{1}{2/\sqrt{5}} = \frac{\sqrt{5}}{2} \qquad \cot \theta = \frac{1}{\tan \theta} = \frac{1}{\frac{1}{2}} = 2$$

◼

◼ Now work Problem 25.

The equation of the unit circle is $x^2 + y^2 = 1$. Thus, if $P = (a, b)$ is the point on the terminal side of an angle θ and if P lies on the unit circle, then

$$b^2 + a^2 = 1$$

But $b = \sin \theta$ and $a = \cos \theta$. Thus,

$$(\sin \theta)^2 + (\cos \theta)^2 = 1 \tag{4}$$

It is customary to write $\sin^2 \theta$ instead of $(\sin \theta)^2$, $\cos^2 \theta$ instead of $(\cos \theta)^2$, and so on. With this notation, we can rewrite equation (4) as

$$\sin^2 \theta + \cos^2 \theta = 1 \tag{5}$$

If $\cos \theta \neq 0$, we can divide each side of equation (5) by $\cos^2 \theta$:

$$\frac{\sin^2 \theta}{\cos^2 \theta} + 1 = \frac{1}{\cos^2 \theta}$$

$$\left(\frac{\sin \theta}{\cos \theta}\right)^2 + 1 = \left(\frac{1}{\cos \theta}\right)^2$$

Now use formulas (2) and (3) to get

$$\tan^2 \theta + 1 = \sec^2 \theta \tag{6}$$

Similarly, if $\sin \theta \neq 0$, we can divide equation (5) by $\sin^2 \theta$ and use formulas (2) and (3) to get the result:

$$1 + \cot^2 \theta = \csc^2 \theta \tag{7}$$

Collectively, the identities in equations (5), (6), and (7) are referred to as the **Pythagorean identities.**

Let's pause here to summarize the fundamental identities.

Fundamental Identities

$$\tan \theta = \frac{\sin \theta}{\cos \theta} \qquad \cot \theta = \frac{\cos \theta}{\sin \theta}$$

$$\cot \theta = \frac{1}{\tan \theta} \qquad \sec \theta = \frac{1}{\cos \theta} \qquad \csc \theta = \frac{1}{\sin \theta}$$

$$\sin^2 \theta + \cos^2 \theta = 1 \qquad \tan^2 \theta + 1 = \sec^2 \theta \qquad 1 + \cot^2 \theta = \csc^2 \theta$$

The Pythagorean identity

$$\sin^2 \theta + \cos^2 \theta = 1$$

can be solved for $\sin \theta$ in terms of $\cos \theta$ (or vice versa) as follows:

$$\sin^2 \theta = 1 - \cos^2 \theta$$
$$\sin \theta = \pm\sqrt{1 - \cos^2 \theta}$$

where the $+$ sign is used if $\sin \theta > 0$ and the $-$ sign is used if $\sin \theta < 0$.

EXAMPLE 4   *Finding Exact Values Given One Value and the Sign of Another*

Given that $\sin \theta = \frac{1}{3}$ and $\cos \theta < 0$, find the exact value of each of the remaining five trigonometric functions.

Solution   We solve this problem in two ways: the first way uses the definition of the trigonometric functions; the second method uses the fundamental identities.

Solution 1   *Using the Definition*

Suppose that $P = (a, b)$ is a point on the terminal side of θ that lies a distance of $r = 3$ units from the origin. Since $\sin \theta > 0$ and $\cos \theta < 0$, the point P lies in quadrant II. See Figure 44. (Do you see why we chose 3 units? Notice that $\sin \theta = \frac{1}{3} = b/r$. The choice $r = 3$ will make our calculations easy.) With this choice, $b = 1$ and $r = 3$. Since $\cos \theta = a/r < 0$, it follows that $a < 0$. Thus,

FIGURE 44

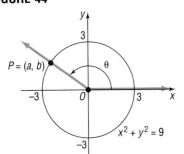

$$a^2 + b^2 = r^2 \qquad b = 1, r = 3, a < 0$$
$$a^2 + 1^2 = 3^2$$
$$a^2 = 8$$
$$a = -2\sqrt{2}$$

Thus,

$$\cos \theta = \frac{a}{r} = \frac{-2\sqrt{2}}{3} \qquad \tan \theta = \frac{b}{a} = \frac{1}{-2\sqrt{2}} = \frac{-\sqrt{2}}{4}$$

$$\csc \theta = \frac{r}{b} = \frac{3}{1} = 3 \qquad \sec \theta = \frac{r}{a} = \frac{3}{-2\sqrt{2}} = \frac{-3\sqrt{2}}{4} \qquad \cot \theta = \frac{a}{b} = \frac{-2\sqrt{2}}{1} = -2\sqrt{2}$$

■

Solution 2   *Using Identities*

First, we solve equation (5) for $\cos \theta$:

$$\sin^2 \theta + \cos^2 \theta = 1$$
$$\cos^2 \theta = 1 - \sin^2 \theta$$
$$\cos \theta = \pm\sqrt{1 - \sin^2 \theta}$$

Because $\cos \theta < 0$, we choose the minus sign:

$$\cos \theta = -\sqrt{1 - \sin^2 \theta} = -\sqrt{1 - \frac{1}{9}} = -\sqrt{\frac{8}{9}} = -\frac{2\sqrt{2}}{3}$$
$$\underset{\uparrow}{}$$
$$\sin \theta = \frac{1}{3}$$

Now we know the values of $\sin \theta$ and $\cos \theta$, so we can use formulas (2) and (3) to get

$$\tan \theta = \frac{\sin \theta}{\cos \theta} = \frac{\frac{1}{3}}{-2\sqrt{2}/3} = \frac{1}{-2\sqrt{2}} = \frac{-\sqrt{2}}{4} \qquad \cot \theta = \frac{1}{\tan \theta} = -2\sqrt{2}$$

$$\sec \theta = \frac{1}{\cos \theta} = \frac{1}{-2\sqrt{2}/3} = \frac{-3}{2\sqrt{2}} = \frac{-3\sqrt{2}}{4} \qquad \csc \theta = \frac{1}{\sin \theta} = \frac{1}{\frac{1}{3}} = 3$$

■

■   Now work Problem 33.

Even–Odd Properties

Recall that a function f is even if $f(-\theta) = f(\theta)$ for all θ in the domain of f; a function f is odd if $f(-\theta) = -f(\theta)$ for all θ in the domain of f. We will now show that the trigonometric functions sine, tangent, cotangent, and cosecant are odd functions, whereas the functions cosine and secant are even functions.

Theorem
Even–Odd Properties

$$\sin(-\theta) = -\sin \theta \qquad \cos(-\theta) = \cos \theta \qquad \tan(-\theta) = -\tan \theta$$
$$\csc(-\theta) = -\csc \theta \qquad \sec(-\theta) = \sec \theta \qquad \cot(-\theta) = -\cot \theta$$

FIGURE 45

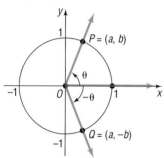

Proof   Let $P = (a, b)$ be the point on the terminal side of the angle θ that is on the unit circle. (See Figure 45.) The point Q on the terminal side of the angle $-\theta$ that is on the unit circle will have coordinates $(a, -b)$. Using the definition for the trigonometric functions, we have

$$\sin \theta = b \qquad \cos \theta = a \qquad \sin(-\theta) = -b \qquad \cos(-\theta) = a$$

so that

$$\sin(-\theta) = -\sin \theta \qquad \cos(-\theta) = \cos \theta$$

Now, using these results and some of the fundamental identities, we have

$$\tan(-\theta) = \frac{\sin(-\theta)}{\cos(-\theta)} = \frac{-\sin \theta}{\cos \theta} = -\tan \theta \qquad \cot(-\theta) = \frac{1}{\tan(-\theta)} = \frac{1}{-\tan \theta} = -\cot \theta$$

$$\sec(-\theta) = \frac{1}{\cos(-\theta)} = \frac{1}{\cos \theta} = \sec \theta \qquad \csc(-\theta) = \frac{1}{\sin(-\theta)} = \frac{1}{-\sin \theta} = -\csc \theta$$

■

Seeing the Concept To see that the cosine function is even, graph $y = \cos x$ and then $y = \cos(-x)$. Clear the screen. Now graph $y = -\sin x$ and $y = \sin(-x)$.

E X A M P L E   5   *Finding Exact Values Using Even–Odd Properties*

Find the exact value of

(a) $\sin(-45°)$ (b) $\cos(-\pi)$ (c) $\cot(-3\pi/2)$ (d) $\tan(-37\pi/4)$

Solution   (a) $\sin(-45°) = -\sin 45° = -\dfrac{\sqrt{2}}{2}$ (b) $\cos(-\pi) = \cos \pi = -1$
$\qquad\qquad\quad$ ↑ $\qquad\qquad\qquad\qquad\qquad\qquad\qquad\qquad\qquad$ ↑
$\qquad\qquad\quad$ Odd function $\qquad\qquad\qquad\qquad\qquad\qquad\qquad$ Even function

(c) $\cot\left(-\dfrac{3\pi}{2}\right) = -\cot\dfrac{3\pi}{2} = 0$
$\qquad\qquad\qquad$ ↑
$\qquad\qquad\qquad$ Odd function

(d) $\tan\left(-\dfrac{37\pi}{4}\right) = -\tan\dfrac{37\pi}{4} = -\tan\left(\dfrac{\pi}{4} + 9\pi\right) = -\tan\dfrac{\pi}{4} = -1$
$\qquad\qquad\qquad\quad$ ↑ $\qquad\qquad\qquad\qquad\qquad\qquad\qquad$ ↑
$\qquad\qquad\qquad\quad$ Odd function $\qquad\qquad\qquad\qquad$ Period is π

■

■ Now work Problem 49.

Exercise 2.3

In Problems 1–16, use the fact that the trigonometric functions are periodic to find the exact value of each expression. Do not use a calculator.

1. $\sin 405°$ **2.** $\cos 420°$ **3.** $\tan 405°$ **4.** $\sin 390°$ **5.** $\csc 450°$ **6.** $\sec 540°$

7. $\cot 390°$ **8.** $\sec 420°$ **9.** $\cos\dfrac{33\pi}{4}$ **10.** $\sin\dfrac{9\pi}{4}$ **11.** $\tan 21\pi$ **12.** $\csc\dfrac{9\pi}{2}$

13. $\sec\dfrac{17\pi}{4}$ **14.** $\cot\dfrac{17\pi}{4}$ **15.** $\tan\dfrac{19\pi}{6}$ **16.** $\sec\dfrac{25\pi}{6}$

In Problems 17–24, name the quadrant in which the angle θ lies.

17. $\sin \theta > 0$, $\cos \theta < 0$ **18.** $\sin \theta < 0$, $\cos \theta > 0$ **19.** $\sin \theta < 0$, $\tan \theta < 0$

20. $\cos \theta > 0$, $\tan \theta > 0$ **21.** $\cos \theta > 0$, $\tan \theta < 0$ **22.** $\cos \theta < 0$, $\tan \theta > 0$

23. $\sec \theta < 0$, $\sin \theta > 0$ **24.** $\csc \theta > 0$, $\cos \theta < 0$

In Problems 25–32, sin θ and cos θ are given. Find the exact value of each of the four remaining trigonometric functions.

25. $\sin\theta = 2/\sqrt{5}, \quad \cos\theta = 1/\sqrt{5}$

26. $\sin\theta = -1/\sqrt{5}, \quad \cos\theta = -2/\sqrt{5}$

27. $\sin\theta = \frac{1}{2}, \quad \cos\theta = \sqrt{3}/2$

28. $\sin\theta = \sqrt{3}/2, \quad \cos\theta = \frac{1}{2}$

29. $\sin\theta = -\frac{1}{3}, \quad \cos\theta = 2\sqrt{2}/3$

30. $\sin\theta = 2\sqrt{2}/3, \quad \cos\theta = -\frac{1}{3}$

31. $\sin\theta = 0.2588, \quad \cos\theta = 0.9659$

32. $\sin\theta = 0.6428, \quad \cos\theta = 0.7660$

In Problems 33–48, find the exact value of each of the remaining trigonometric functions of θ.

33. $\sin\theta = \frac{12}{13}, \quad 90° < \theta < 180°$

34. $\cos\theta = \frac{3}{5}, \quad 270° < \theta < 360°$

35. $\cos\theta = -\frac{4}{5}, \quad \pi < \theta < 3\pi/2$

36. $\sin\theta = -\frac{5}{13}, \quad \pi < \theta < 3\pi/2$

37. $\sin\theta = \frac{5}{13}, \quad \cos\theta < 0$

38. $\cos\theta = \frac{4}{5}, \quad \sin\theta < 0$

39. $\cos\theta = -\frac{1}{3}, \quad \csc\theta > 0$

40. $\sin\theta = -\frac{2}{3}, \quad \sec\theta > 0$

41. $\sin\theta = \frac{2}{3}, \quad \tan\theta < 0$

42. $\cos\theta = -\frac{1}{4}, \quad \tan\theta > 0$

43. $\sec\theta = 2, \quad \sin\theta < 0$

44. $\csc\theta = 3, \quad \cot\theta < 0$

45. $\tan\theta = \frac{3}{4}, \quad \sin\theta < 0$

46. $\cot\theta = \frac{4}{3}, \quad \cos\theta < 0$

47. $\tan\theta = -\frac{1}{3}, \quad \sin\theta > 0$

48. $\sec\theta = -2, \quad \tan\theta > 0$

In Problems 49–66, use the even–odd properties to find the exact value of each expression. Do not use a calculator.

49. $\sin(-60°)$

50. $\cos(-30°)$

51. $\tan(-30°)$

52. $\sin(-135°)$

53. $\sec(-60°)$

54. $\csc(-30°)$

55. $\sin(-90°)$

56. $\cos(-270°)$

57. $\tan\left(-\dfrac{\pi}{4}\right)$

58. $\sin(-\pi)$

59. $\cos\left(-\dfrac{\pi}{4}\right)$

60. $\sin\left(-\dfrac{\pi}{3}\right)$

61. $\tan(-\pi)$

62. $\sin\left(-\dfrac{3\pi}{2}\right)$

63. $\csc\left(-\dfrac{\pi}{4}\right)$

64. $\sec(-\pi)$

65. $\sec\left(-\dfrac{\pi}{6}\right)$

66. $\csc\left(-\dfrac{\pi}{3}\right)$

In Problems 67–78, find the exact value of each expression. Do not use a calculator.

67. $\sin(-\pi) + \cos 5\pi$

68. $\tan\left(-\dfrac{5\pi}{4}\right) - \cot\dfrac{7\pi}{2}$

69. $\sec(-\pi) + \csc\left(-\dfrac{\pi}{2}\right)$

70. $\tan(-6\pi) + \cos\dfrac{9\pi}{4}$

71. $\sin\left(-\dfrac{9\pi}{4}\right) - \tan\left(-\dfrac{9\pi}{4}\right)$

72. $\cos\left(-\dfrac{17\pi}{4}\right) - \sin\left(-\dfrac{3\pi}{2}\right)$

73. $\sin^2 40° + \cos^2 40°$

74. $\sec^2 18° - \tan^2 18°$

75. $\sin 80° \csc 80°$

76. $\tan 10° \cot 10°$

77. $\tan 40° - \dfrac{\sin 40°}{\cos 40°}$

78. $\cot 20° - \dfrac{\cos 20°}{\sin 20°}$

79. If $\sin\theta = 0.3$, find the value of $\sin\theta + \sin(\theta + 2\pi) + \sin(\theta + 4\pi)$.

80. If $\cos\theta = 0.2$, find the value of $\cos\theta + \cos(\theta + 2\pi) + \cos(\theta + 4\pi)$.

81. If $\tan\theta = 3$, find the value of $\tan\theta + \tan(\theta + \pi) + \tan(\theta + 2\pi)$.

82. If $\cot\theta = -2$, find the value of $\cot\theta + \cot(\theta - \pi) + \cot(\theta - 2\pi)$.

83. *Calculating the Time of a Trip* From a parking lot, you want to walk to a house on the ocean. The house is located 1500 feet down a paved path that parallels the ocean, which is 500 feet away. See the illustration. Along the path you can walk 300 feet per minute, but in the sand on the beach you can only walk 100 feet per minute.

The time T to get from the parking lot to the beachhouse can be expressed as a function of the angle θ shown in the illustration is

$$T(\theta) = 5 - \frac{5}{3\tan\theta} + \frac{5}{\sin\theta}, \ 0 < \theta < \frac{\pi}{2}$$

Calculate the time T if you walk directly from the parking lot to the house. [*Hint:* $\tan\theta = 500/1500$.]

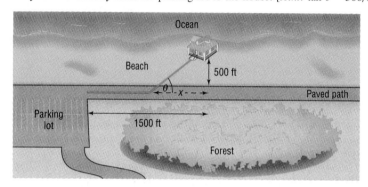

84. *Calculating the Time of a Trip* Two oceanfront homes are located 8 miles apart on a straight stretch of beach, each a distance of 1 mile from a paved road that parallels the ocean. Sally can walk 8 miles per hour along the paved road, but only 3 miles per hour in the sand that separates the road from the ocean. Because of a river directly between the two houses, it is necessary to walk in the sand to the road, continue on the road, and then walk directly back in the sand to get from one house to the other. See the illustration. The time T to get from one house to the other as a function of the angle θ shown in the illustration is

$$T(\theta) = 1 + \frac{2}{3\sin\theta} - \frac{1}{4\tan\theta}, \ 0 < \theta < \frac{\pi}{2}$$

(a) Calculate the time T for $\tan\theta = 1/4$.
(b) Describe the path taken.
(c) Explain why θ must be larger than $14°$.

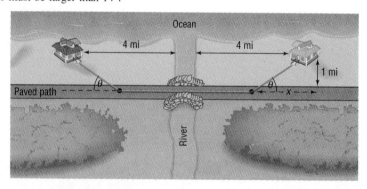

85. For what numbers θ is $f(\theta) = \tan\theta$ not defined? **86.** For what numbers θ is $f(\theta) = \cot\theta$ not defined?

87. For what numbers θ is $f(\theta) = \sec\theta$ not defined? **88.** For what numbers θ is $f(\theta) = \csc\theta$ not defined?

89. What is the value of $\sin k\pi$, where k is any integer? **90.** What is the value of $\cos k\pi$, where k is any integer?

91. Show that the range of the tangent function is the set of all real numbers.

92. Show that the range of the cotangent function is the set of all real numbers.

93. Show that the period of $f(\theta) = \sin\theta$ is 2π. [*Hint:* Assume that $0 < p < 2\pi$ exists so that $\sin(\theta + p) = \sin\theta$ for all θ. Let $\theta = 0$ to find p. Then let $\theta = \pi/2$ to obtain a contradiction.]

94. Show that the period of $f(\theta) = \cos\theta$ is 2π.

95. Show that the period of $f(\theta) = \sec\theta$ is 2π.

96. Show that the period of $f(\theta) = \csc\theta$ is 2π.

97. Show that the period of $f(\theta) = \tan\theta$ is π.

98. Show that the period of $f(\theta) = \cot\theta$ is π.

99. Prove the reciprocal identities given in formula (2).

100. Prove the quotient identities given in formula (3).

101. Establish the identity: $(\sin\theta\cos\phi)^2 + (\sin\theta\sin\phi)^2 + \cos^2\theta = 1$

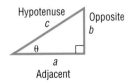

102. Write down five characteristics of the tangent function. Explain the meaning of each.

103. Describe your understanding of the meaning of a periodic function.

2.4

Right Triangle Trigonometry

A triangle in which one angle is a right angle (90°) is called a **right triangle.** Recall that the side opposite the right angle is called the **hypotenuse,** and the remaining two sides are called the **legs** of the triangle. In Figure 46(a), we have labeled the hypotenuse as c, to indicate its length is c units, and, in a like manner, we have labeled the legs as a and b. Because the triangle is a right triangle, the Pythagorean Theorem tells us that

$$a^2 + b^2 = c^2$$

FIGURE 46

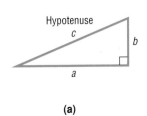

(a)

(b)

Now, suppose that θ is an **acute angle;** that is, $0° < \theta < 90°$ (if θ is measured in degrees) or $0 < \theta < \pi/2$ (if θ is measured in radians). Place θ in standard position, and let $P = (a, b)$ be any point except the origin O on the terminal side of θ. Form a right triangle by dropping the perpendicular from P to the x-axis, as shown in Figure 46(b).

By referring to the lengths of the sides of the triangle by the names hypotenuse (c), opposite (b), and adjacent (a), as indicated in Figure 47, we can express the trigonometric functions of θ as ratios of the sides of a right triangle:

FIGURE 47

Hypotenuse
c

Opposite
b

θ

a

Adjacent

$$\sin\theta = \frac{\text{Opposite}}{\text{Hypotenuse}} = \frac{b}{c} \qquad \cos\theta = \frac{\text{Adjacent}}{\text{Hypotenuse}} = \frac{a}{c}$$

$$\tan\theta = \frac{\text{Opposite}}{\text{Adjacent}} = \frac{b}{a} \qquad \csc\theta = \frac{\text{Hypotenuse}}{\text{Opposite}} = \frac{c}{b} \qquad (1)$$

$$\sec\theta = \frac{\text{Hypotenuse}}{\text{Adjacent}} = \frac{c}{a} \qquad \cot\theta = \frac{\text{Adjacent}}{\text{Opposite}} = \frac{a}{b}$$

Notice that each of the trigonometric functions of the acute angle θ is positive.

E X A M P L E   1   *Finding the Value of Trigonometric Functions from a Right Triangle*

Find the exact value of each of the six trigonometric functions of the angle θ in Figure 48.

Solution   We see in Figure 48 that the two given sides of the triangle are

$$c = \text{Hypotenuse} = 5 \qquad a = \text{Adjacent} = 3$$

FIGURE 48

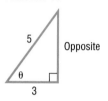

To find the length of the opposite side, we use the Pythagorean Theorem:

$$(\text{Adjacent})^2 + (\text{Opposite})^2 = (\text{Hypotenuse})^2$$
$$3^2 + (\text{Opposite})^2 = 5^2$$
$$(\text{Opposite})^2 = 25 - 9 = 16$$
$$\text{Opposite} = 4$$

Now that we know the lengths of the three sides, we use the ratios in (1) to find the value of each of the six trigonometric functions:

$$\sin \theta = \frac{\text{Opposite}}{\text{Hypotenuse}} = \frac{4}{5} \qquad \cos \theta = \frac{\text{Adjacent}}{\text{Hypotenuse}} = \frac{3}{5} \qquad \tan \theta = \frac{\text{Opposite}}{\text{Adjacent}} = \frac{4}{3}$$

$$\csc \theta = \frac{\text{Hypotenuse}}{\text{Opposite}} = \frac{5}{4} \qquad \sec \theta = \frac{\text{Hypotenuse}}{\text{Adjacent}} = \frac{5}{3} \qquad \cot \theta = \frac{\text{Adjacent}}{\text{Opposite}} = \frac{3}{4}$$

■

■   Now work Problem 1.

Thus, the values of the trigonometric functions of an acute angle are ratios of the lengths of the sides of a right triangle. This way of viewing the trigonometric functions leads to many applications and, in fact, was the point of view used by early mathematicians (before calculus) in studying the subject of trigonometry.

Complementary Angles: Cofunctions

Two acute angles are called **complementary** if their sum is a right angle. Because the sum of the angles of any triangle is 180°, it follows that, for a right triangle, the two acute angles are complementary.

Refer now to Figure 49; we have labeled the angle opposite side b as β and the angle opposite side a as α. Notice that side b is adjacent to angle α and side a is adjacent to angle β. As a result,

FIGURE 49

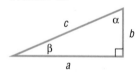

$$\sin \beta = \frac{b}{c} = \cos \alpha \qquad \cos \beta = \frac{a}{c} = \sin \alpha \qquad \tan \beta = \frac{b}{a} = \cot \alpha$$

$$\csc \beta = \frac{c}{b} = \sec \alpha \qquad \sec \beta = \frac{c}{a} = \csc \alpha \qquad \cot \beta = \frac{a}{b} = \tan \alpha$$

(2)

Because of these relationships, the functions sine and cosine, tangent and cotangent, and secant and cosecant are called **cofunctions** of each other. The identities (2) may be expressed in words as follows:

Complementary Angle Theorem   Cofunctions of complementary angles are equal.   ■

Examples of this theorem are given next:

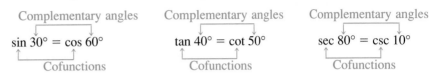

If θ is an acute angle measured in degrees, the angle $90° - \theta$ (or $\pi/2 - \theta$, if θ is in radians) is the angle complementary to θ. Table 6 restates the preceding theorem on cofunctions.

TABLE 6

θ (DEGREES)	θ (RADIANS)
$\sin \theta = \cos(90° - \theta)$	$\sin \theta = \cos(\pi/2 - \theta)$
$\cos \theta = \sin(90° - \theta)$	$\cos \theta = \sin(\pi/2 - \theta)$
$\tan \theta = \cot(90° - \theta)$	$\tan \theta = \cot(\pi/2 - \theta)$
$\csc \theta = \sec(90° - \theta)$	$\csc \theta = \sec(\pi/2 - \theta)$
$\sec \theta = \csc(90° - \theta)$	$\sec \theta = \csc(\pi/2 - \theta)$
$\cot \theta = \tan(90° - \theta)$	$\cot \theta = \tan(\pi/2 - \theta)$

Although the angle θ in Table 6 is acute, we will see later that these results are valid for any angle θ.

Seeing the Concept Graph $y = \sin x$ and $y = \cos(90° - x)$. Be sure that the mode is set to degrees.

E X A M P L E   2   *Using the Complementary Angle Theorem*

(a) $\sin 62° = \cos(90° - 62°) = \cos 28°$ (b) $\tan \dfrac{\pi}{12} = \cot\left(\dfrac{\pi}{2} - \dfrac{\pi}{12}\right) = \cot \dfrac{5\pi}{12}$

(c) $\cos \dfrac{\pi}{4} = \sin\left(\dfrac{\pi}{2} - \dfrac{\pi}{4}\right) = \sin \dfrac{\pi}{4}$ (d) $\csc \dfrac{\pi}{6} = \sec\left(\dfrac{\pi}{2} - \dfrac{\pi}{6}\right) = \sec \dfrac{\pi}{3}$

◼

◼   Now work Problem 57(a).

Reference Angle

Next, we concentrate on angles that lie in a quadrant. Once we know in which quadrant an angle lies, we know the sign of each value of the trigonometric functions of that angle. The use of a certain reference angle may help us to evaluate the trigonometric functions of such an angle.

Reference Angle

> Let θ denote a nonacute angle that lies in a quadrant. The acute angle formed by the terminal side of θ and either the positive x-axis or the negative x-axis is called the **reference angle** for θ.

Figure 50 illustrates the reference angle for some general angles θ. Note that a reference angle is always an acute angle, that is, an angle whose measure is between 0° and 90°.

Although formulas can be given for calculating reference angles, usually it is easier to find the reference angle for a given angle by making a quick sketch of the angle.

FIGURE 50

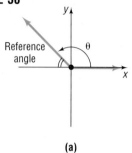

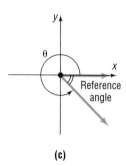

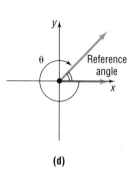

(a)   (b)   (c)   (d)

E X A M P L E   3   *Finding Reference Angles*

Find the reference angle for each of the following angles:

(a) 150°   (b) −45°   (c) 9π/4   (d) −5π/6

Solution   (a) Refer to Figure 51. The reference   (b) Refer to Figure 52. The reference
angle for 150° is 30°.   angle for −45° is 45°.

FIGURE 51   **FIGURE 52**

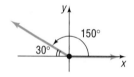

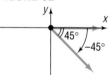

(c) Refer to Figure 53. The reference   (d) Refer to Figure 54. The reference
angle for 9π/4 is π/4.   angle for −5π/6 is π/6.

FIGURE 53   **FIGURE 54**

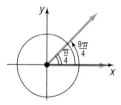

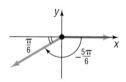

■   Now work Problem 11.

The advantage of using reference angles is that, except for the correct sign,
the values of the trigonometric functions of a general angle θ equal the values of
the trigonometric functions of its reference angle.

Theorem   If θ is an angle that lies in a quadrant and if α is its reference angle, then
Reference Angles

$$
\begin{array}{lll}
\sin \theta = \pm \sin \alpha & \cos \theta = \pm \cos \alpha & \tan \theta = \pm \tan \alpha \\
\csc \theta = \pm \csc \alpha & \sec \theta = \pm \sec \alpha & \cot \theta = \pm \cot \alpha
\end{array}
\tag{3}
$$

where the + or − sign depends on the quadrant in which θ lies.   ■

FIGURE 55

$\sin \theta = b/c$, $\sin \alpha = b/c$;
$\cos \theta = a/c$, $\cos \alpha = |a|/c$

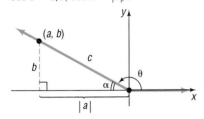

For example, suppose that θ lies in quadrant II and α is its reference angle. See Figure 55. If (a, b) is a point on the terminal side of θ and if $c = \sqrt{a^2 + b^2}$, we have

$$\sin \theta = \frac{b}{c} = \sin \alpha \qquad \cos \theta = \frac{a}{c} = \frac{-|a|}{c} = -\cos \alpha$$

and so on.

The next example illustrates how the theorem on reference angles is used.

E X A M P L E   4   *Using Reference Angles to Find the Value of Trigonometric Functions*

Find the exact value of each of the following trigonometric functions using reference angles:

(a) $\sin 135°$ (b) $\cos 240°$ (c) $\cos \dfrac{5\pi}{6}$ (d) $\tan\left(-\dfrac{\pi}{3}\right)$

Solution   (a) Refer to Figure 56. The angle $135°$ is in quadrant II, where the sine function is positive. The reference angle for $135°$ is $45°$. Thus,

$$\sin 135° = \sin 45° = \frac{\sqrt{2}}{2}$$

FIGURE 56

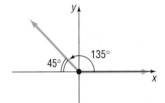

(b) Refer to Figure 57. The angle $240°$ is in quadrant III, where the cosine function is negative. The reference angle for $240°$ is $60°$. Thus,

$$\cos 240° = -\cos 60° = -\tfrac{1}{2}$$

FIGURE 57

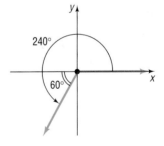

(c) Refer to Figure 58. The angle $5\pi/6$ is in quadrant II, where the cosine function is negative. The reference angle for $5\pi/6$ is $\pi/6$. Thus,

$$\cos \frac{5\pi}{6} = -\cos \frac{\pi}{6} = -\frac{\sqrt{3}}{2}$$

FIGURE 58

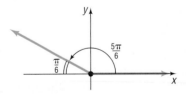

(d) Refer to Figure 59. The angle $-\pi/3$ is in quadrant IV, where the tangent function is negative. The reference angle for $-\pi/3$ is $\pi/3$. Thus,

$$\tan\left(-\frac{\pi}{3}\right) = -\tan \frac{\pi}{3} = -\sqrt{3}$$

FIGURE 59

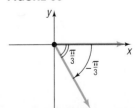

■ Now work Problems 27 and 45.

E X A M P L E 5

Using Right Triangles to Find Values When One Value Is Known

Given that $\cos \theta = -\frac{2}{3}$, $\pi/2 < \theta < \pi$, find the exact value of each of the remaining trigonometric functions.

Solution *Using Right Triangles* We have already discussed two ways of solving this type of problem: using the definition of the trigonometric functions and using identities. (Refer to Example 4 in Section 2.3). Here we present a third method: using right triangles.

The angle θ lies in quadrant II, so we know that $\sin \theta$ and $\csc \theta$ are positive, whereas the other trigonometric functions are negative. If α is the reference angle for θ, then $\cos \alpha = \frac{2}{3}$. The values of the remaining trigonometric functions of the angle α can be found by drawing the appropriate triangle. We use Figure 60 to obtain

FIGURE 60

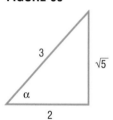

$$\sin \alpha = \frac{\sqrt{5}}{3} \qquad \cos \alpha = \frac{2}{3} \qquad \tan \alpha = \frac{\sqrt{5}}{2}$$

$$\csc \alpha = \frac{3}{\sqrt{5}} = \frac{3\sqrt{5}}{5} \qquad \sec \alpha = \frac{3}{2} \qquad \cot \alpha = \frac{2}{\sqrt{5}} = \frac{2\sqrt{5}}{5}$$

Now, we assign the appropriate sign to each of these values to find the values of the trigonometric functions of θ:

$$\sin \theta = \frac{\sqrt{5}}{3} \qquad \cos \theta = -\frac{2}{3} \qquad \tan \theta = -\frac{\sqrt{5}}{2}$$

$$\csc \theta = \frac{3\sqrt{5}}{5} \qquad \sec \theta = -\frac{3}{2} \qquad \cot \theta = -\frac{2\sqrt{5}}{5}$$ ■

E X A M P L E 6

Constructing a Rain Gutter

FIGURE 61(a)

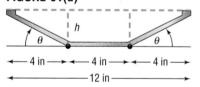

A rain gutter is to be constructed of aluminum sheets 12 inches wide. After marking off a length of 4 inches from each edge, this length is bent up at an angle θ. See Figure 61(a).

(a) Express the area A of the opening as a function of θ.

(b) Graph $A = A(\theta)$. Find the angle θ that makes A largest. (This bend will allow the most water to flow through the gutter).

Solution (a) The area A of the opening is the sum of the areas of two right triangles and one rectangle:

$$A = 2 \cdot \frac{1}{2} \cdot h\sqrt{16 - h^2} + 4h = (4 \sin \theta)(4 \cos \theta) + 4(4 \sin \theta)$$
$$A(\theta) = 16 \sin \theta(\cos \theta + 1)$$

(b) See Figure 61(b). The angle θ that makes A largest is 60°.

FIGURE 61(b)

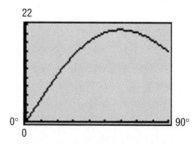

2.4

Exercise 2.4

In Problems 1–10, find the exact value of the six trigonometric functions of the angle θ in each figure.

1.
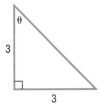

2.
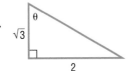

3.
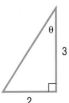

4.

5.
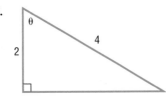

6.
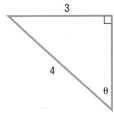

7.

8.
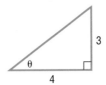

9.
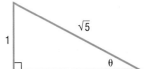

10.
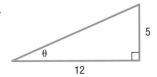

In Problems 11–26, find the reference angle of each angle.

11. $-30°$	**12.** $60°$	**13.** $120°$	**14.** $300°$	**15.** $210°$
16. $330°$	**17.** $5\pi/4$	**18.** $5\pi/6$	**19.** $8\pi/3$	**20.** $7\pi/4$
21. $-135°$	**22.** $-240°$	**23.** $-2\pi/3$	**24.** $-7\pi/6$	**25.** $420°$
26. $480°$				

In Problems 27–56, find the exact value of each expression. Do not use a calculator.

27. $\sin 150°$	**28.** $\cos 210°$	**29.** $\cos 315°$	**30.** $\sin 120°$
31. $\sec 240°$	**32.** $\csc 300°$	**33.** $\cot 330°$	**34.** $\tan 225°$
35. $\sin \dfrac{3\pi}{4}$	**36.** $\cos \dfrac{2\pi}{3}$	**37.** $\cot \dfrac{7\pi}{6}$	**38.** $\csc \dfrac{7\pi}{4}$
39. $\cos(-60°)$	**40.** $\tan(-120°)$	**41.** $\sin\left(-\dfrac{2\pi}{3}\right)$	**42.** $\cot\left(-\dfrac{\pi}{6}\right)$

43. $\tan \dfrac{14\pi}{3}$

44. $\sec \dfrac{11\pi}{4}$

45. $\csc(-315°)$

46. $\sec(-225°)$

47. $\sin 38° - \cos 52°$

48. $\tan 12° - \cot 78°$

49. $\dfrac{\cos 10°}{\sin 80°}$

50. $\dfrac{\cos 40°}{\sin 50°}$

51. $1 - \cos^2 20° - \cos^2 70°$

52. $1 + \tan^2 5° - \csc^2 85°$

53. $\tan 20° - \dfrac{\cos 70°}{\cos 20°}$

54. $\cot 40° - \dfrac{\sin 50°}{\sin 40°}$

55. $\cos 35° \sin 55° + \sin 35° \cos 55°$

56. $\sec 35° \csc 55° - \tan 35° \cot 55°$

57. If $\sin \theta = \frac{1}{3}$, find the exact value of: (a) $\cos(90° - \theta)$ (b) $\cos^2 \theta$ (c) $\csc \theta$ (d) $\sec\left(\dfrac{\pi}{2} - \theta\right)$

58. If $\sin \theta = 0.2$, find the exact value of: (a) $\cos\left(\dfrac{\pi}{2} - \theta\right)$ (b) $\cos^2 \theta$ (c) $\sec(90° - \theta)$ (d) $\csc \theta$

59. If $\tan \theta = 4$, find the exact value of: (a) $\sec^2 \theta$ (b) $\cot \theta$ (c) $\cot\left(\dfrac{\pi}{2} - \theta\right)$ (d) $\csc^2 \theta$

60. If $\sec \theta = 3$, find the exact value of: (a) $\cos \theta$ (b) $\tan^2 \theta$ (c) $\csc(90° - \theta)$ (d) $\sin^2 \theta$

61. If $\csc \theta = 4$, find the exact value of: (a) $\sin \theta$ (b) $\cot^2 \theta$ (c) $\sec(90° - \theta)$ (d) $\sec^2 \theta$

62. If $\cot \theta = 2$, find the exact value of: (a) $\tan \theta$ (b) $\csc^2 \theta$ (c) $\tan\left(\dfrac{\pi}{2} - \theta\right)$ (d) $\sec^2 \theta$

63. If $\sin \theta = 0.3$, find the exact value of $\sin \theta + \cos\left(\dfrac{\pi}{2} - \theta\right)$.

64. If $\tan \theta = 4$ find the exact value of $\tan \theta + \tan\left(\dfrac{\pi}{2} - \theta\right)$.

65. Find the exact value of: $\sin 1° + \sin 2° + \sin 3° + \cdots + \sin 358° + \sin 359°$.

66. Find the exact value of: $\cos 1° + \cos 2° + \cos 3° + \cdots + \cos 358° + \cos 359°$.

67. Find the acute angle θ that satisfies the equation: $\sin \theta = \cos(2\theta + 30°)$

68. Find the acute angle θ that satisfies the equation: $\tan \theta = \cot(\theta + 45°)$

69. *Calculating the Time of a Trip* Two oceanfront homes are located 8 miles apart on a straight stretch of beach, each a distance of 1 mile from a paved road that parallels the ocean. Sally can walk 8 miles per hour along the paved road, but only 3 miles per hour in the sand that separates the road from the ocean. Because of a river directly between the two houses, it is necessary to walk in the sand to the road, continue on the road, and then walk directly back in the sand to get from one house to the other. See the illustration.

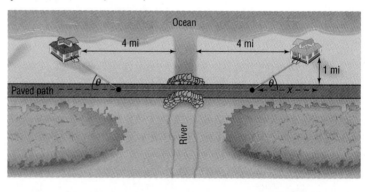

(a) Express the time T to get from one house to the other as a function of the angle θ shown in the illustration.

(b) Graph $T = T(\theta)$. What angle θ results in the least time? What is the least time? How long is Sally on the paved road?

70. *Designing Fine Decorative Pieces* A designer of decorative art plans to market solid gold spheres encased in clear crystal cones. Each sphere is of fixed radius R and will be enclosed in a cone of height h and radius r. See the illustration. Many cones can be used to enclose the sphere, each having a different slant angle θ.

(a) Express the volume V of the cone as a function of the slant angle θ of the cone. [*Hint*: The volume V of a cone of height h and radius r is $V = \frac{1}{3}\pi r^2 h$.]

(b) What slant angle θ should be used for the volume V of the cone to be a minimum? (This choice minimizes the amount of crystal required and gives maximum emphasis to the gold sphere).

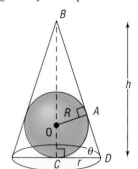

71. *Calculating the Time of a Trip* From a parking lot, you want to walk to a house on the ocean. The house is located 1500 feet down a paved path that parallels the ocean, which is 500 feet away. See the illustration. Along the path you can walk 300 feet per minute, but in the sand on the beach you can only walk 100 feet per minute.

(a) Calculate the time T if you walk 1500 feet along the paved path and then 500 feet in the sand to the house.

(b) Calculate the time T if you walk in the sand first for 500 feet and then walk along the beach for 1500 feet to the house.

(c) Express the time T to get from the parking lot to the beachhouse as a function of the angle θ shown in the illustration.

(d) Calculate the time T if you walk 1000 feet along the paved path and then walk directly to the house.

(e) Graph $T = T(\theta)$. For what angle θ is T least? What is the least time? What is x for this angle?

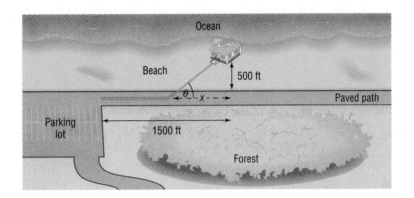

72. *Carrying a Ladder around a Corner* A ladder of length L is carried horizontally around a corner from a hall 3 feet wide into a hall 4 feet wide. See the illustration. Find the length L of the ladder as a function of the angle θ shown in the illustration.

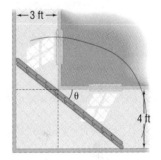

73. Give three examples that will show a fellow student how to use the Theorem on Reference Angles. Give them to the fellow student and ask for a critique.

74. Refer to Example 4, Section 2.3, and Example 5, Section 2.4. Which of the three methods of solution do you prefer? Which is your least favorite? Are there situations in which one method is sometimes best and others in which another method is best? Give reasons.

75. Use a graphing utility set in radian mode to complete the following table. What can you conclude about the ratio $(\sin \theta)/\theta$ as θ approaches 0?

θ	0.5	0.4	0.2	0.1	0.01	0.001	0.0001	0.00001
$\sin \theta$								
$\dfrac{\sin \theta}{\theta}$								

76. Use a graphing utility set in radian mode to complete the following table. What can you conclude about the ratio $(\cos \theta - 1)/\theta$ as θ approaches 0?

θ	0.5	0.4	0.2	0.1	0.01	0.001	0.0001	0.00001
$\cos \theta - 1$								
$\dfrac{\cos \theta - 1}{\theta}$								

77. Suppose that the angle θ is a central angle of a circle of radius 1 (see the figure). Show that

(a) Angle $OAC = \dfrac{\theta}{2}$ (b) $|CD| = \sin\theta$ and $|OD| = \cos\theta$

(c) $\tan\dfrac{\theta}{2} = \dfrac{\sin\theta}{1 + \cos\theta}$

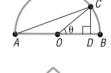

78. Show that the area of an isosceles triangle is $A = a^2 \sin\theta\cos\theta$, where a is the length of one of the two equal sides and θ is the measure of one of the two equal angles (see the figure).

79. Let $n > 0$ be any real number, and let θ be any angle for which $0 < \theta < \pi/(1 + n)$. Then we can construct a triangle with the angles θ and $n\theta$ and included side of length 1 (do you see why?) and place it on the unit circle as illustrated. Now, drop the perpendicular from C to $D = (x, 0)$, and show that

$$x = \frac{\tan n\theta}{\tan\theta + \tan n\theta}$$

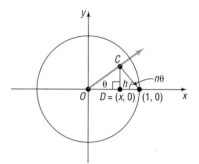

80. Refer to the accompanying figure. The smaller circle, whose radius is a, is tangent to the larger circle, whose radius is b. The ray OA contains a diameter of each circle, and the ray OB is tangent to each circle. Show that

$$\cos\theta = \frac{\sqrt{ab}}{\dfrac{a + b}{2}}$$

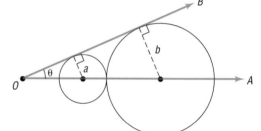

(that is, $\cos\theta$ equals the ratio of the geometric mean of a and b to the arithmetic mean of a and b). [*Hint:* First show that $\sin\theta = (b - a)/(b + a)$.]

81. Refer to the figure. If $|OA| = 1$, show that

(a) Area $\triangle OAC = \frac{1}{2}\sin\alpha\cos\alpha$ (b) area $\triangle OCB = \frac{1}{2}|OB|^2 \sin\beta\cos\beta$

(c) Area $\triangle OAB = \frac{1}{2}|OB|\sin(\alpha + \beta)$ (d) $|OB| = \dfrac{\cos\alpha}{\cos\beta}$

(e) $\sin(\alpha + \beta) = \sin\alpha\cos\beta + \cos\alpha\sin\beta$

[*Hint:*   Area $\triangle OAB =$ Area $\triangle OAC +$ Area $\triangle OCB$.]

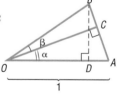

82. Refer to the accompanying figure, where a unit circle is drawn. The line DB is tangent to the circle.

(a) Express the area of $\triangle OBC$ in terms of $\sin\theta$ and $\cos\theta$.

(b) Express the area of $\triangle OBD$ in terms of $\sin\theta$ and $\cos\theta$.

(c) The area of the sector of the circle $\widehat{OBC}$ is $\frac{1}{2}\theta$, where θ is measured in radians. Use the results of parts (a) and (b) and the fact that

$$\text{Area } \triangle OBC < \text{Area } \widehat{OBC} < \text{Area } \triangle OBD$$

to show that

$$1 < \frac{\theta}{\sin\theta} < \frac{1}{\cos\theta}$$

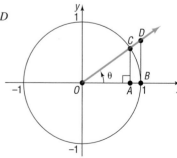

83. If $\cos\alpha = \tan\beta$ and $\cos\beta = \tan\alpha$, where α and β are acute angles, show that

$$\sin\alpha = \sin\beta = \sqrt{\frac{3 - \sqrt{5}}{2}}$$

84. If θ is an acute angle, explain why $\sec\theta > 1$.

85. If θ is an acute angle, explain why $0 < \sin\theta < 1$.

2.5

Graphs of the Trigonometric Functions

We have discussed properties of the trigonometric functions $f(\theta) = \sin \theta$, $f(\theta) = \cos \theta$, and so on. In this section, we shall use the traditional symbols x to represent the independent variable (or argument) and y for the dependent variable (or value at x) for each function. Thus, we write the six trigonometric functions as

$y = \sin x$	$y = \cos x$	$y = \tan x$
$y = \csc x$	$y = \sec x$	$y = \cot x$

Our purpose in this section is to graph each of these functions. Unless indicated otherwise, we shall use radian measure throughout for the independent variable x.

The Graph of $y = \sin x$

Since the sine function has period 2π, we need to graph $y = \sin x$ only on the interval $[0, 2\pi]$. The remainder of the graph will consist of repetitions of this portion of the graph.

We begin by constructing Table 7, which lists some points on the graph of $y = \sin x$, $0 \le x \le 2\pi$. As the table shows, the graph of $y = \sin x$, $0 \le x \le 2\pi$, begins at the origin. As x increases from 0 to $\pi/2$, the value of $y = \sin x$ increases from 0 to 1; as x increases from $\pi/2$ to π to $3\pi/2$, the value of y decreases from 1 to 0 to -1; as x increases from $3\pi/2$ to 2π, the value of y increases from -1 to 0. Based on this, we set the viewing rectangle as shown in Figure 62(a) and graph $y = \sin x$, $0 \le x \le 2\pi$. See Figure 62(b). Figure 62(c) shows the graph drawn by hand.

TABLE 7

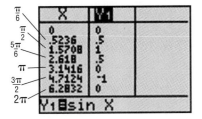

FIGURE 62

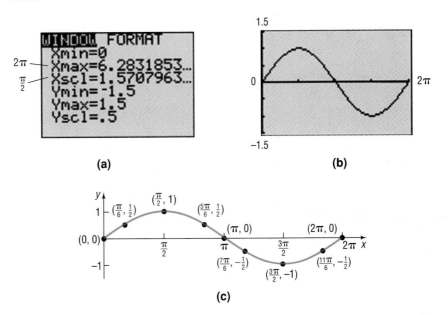

(a)   (b)

(c)

The graph in Figure 62 is one period of the graph of $y = \sin x$. To obtain a more complete graph of $y = \sin x$, we repeat this period in each direction, as shown in Figure 63.

FIGURE 63

$y = \sin x$,
$-\infty < x < \infty$

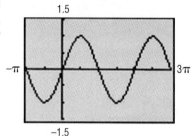

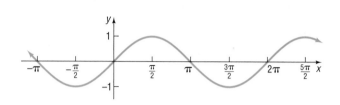

The graph of $y = \sin x$ illustrates some of the facts we already know about the sine function:

Characteristics
of the Sine Function

1. The domain is the set of all real numbers.
2. The range consists of all real numbers from -1 to 1, inclusive.
3. The sine function is an odd function, as the symmetry of the graph with respect to the origin indicates.
4. The sine function is periodic, with period 2π.
5. The x-intercepts are $\ldots, -2\pi, -\pi, 0, \pi, 2\pi, 3\pi, \ldots$; the y-intercept is 0.
6. The maximum value is 1 and occurs at $x = \ldots, -3\pi/2, \pi/2, 5\pi/2, 9\pi/2, \ldots$; the minimum value is -1 and occurs at $x = \ldots, -\pi/2, 3\pi/2, 7\pi/2, 11\pi/2, \ldots$.

▪ Now work Problems 1, 3, and 5.

The graphing techniques introduced in Chapter 1 may be used to graph functions that are variations of the sine function.

E X A M P L E   1   *Graphing Variations of y = sin x Using Shifts, Reflections, and the Like*

Use the graph of $y = \sin x$ to graph $y = -\sin x + 2$.

Solution   Figure 64 illustrates the steps.

FIGURE 64

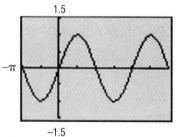

(a) $y = \sin x$

Reflection
about x-axis

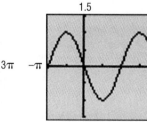

(b) $y = -\sin x$

Vertical shift
up 2 units

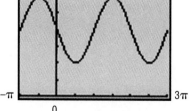

(c) $y = -\sin x + 2$

▪

E X A M P L E   2

Graphing Variations of y = sin x Using Shifts, Reflections, and the Like

Use the graph of $y = \sin x$ to graph $y = \sin\left(x - \dfrac{\pi}{4}\right)$.

Solution   Figure 65 illustrates the steps.

FIGURE 65

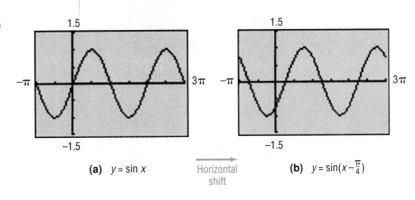

(a) $y = \sin x$ Horizontal shift   (b) $y = \sin\left(x - \dfrac{\pi}{4}\right)$

■   Now work Problem 31.

The Graph of $y = \cos x$

TABLE 8

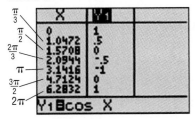

The cosine function also has period 2π. Thus, we proceed as we did with the sine function by constructing Table 8, which lists some points on the graph of $y = \cos x$, $0 \le x \le 2\pi$. As the table shows, the graph of $y = \cos x$, $0 \le x \le 2\pi$, begins at the point $(0, 1)$. As x increases from 0 to $\pi/2$ to π, the value of y decreases from 1 to 0 to -1; as x increases from π to $3\pi/2$ to 2π, the value of y increases from -1 to 0 to 1. As before, we set the viewing rectangle as shown in Figure 66(a) and graph $y = \cos x$, $0 \le x \le 2\pi$. See Figure 66(b). Figure 66(c) shows the graph drawn by hand.

FIGURE 66
$y = \cos x$, $0 \le x \le 2\pi$

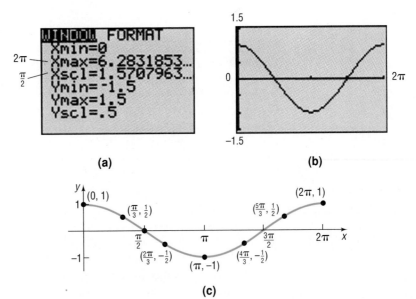

(a)

(b)

(c)

A more complete graph of $y = \cos x$ is obtained by repeating this period in each direction, as shown in Figure 67.

FIGURE 67

$y = \cos x, \; -\infty < x < \infty$

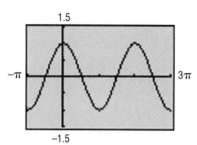

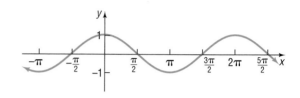

The graph of $y = \cos x$ illustrates some of the facts we already know about the cosine function:

Characteristics of the Cosine Function	

Characteristics of the Cosine Function

1. The domain is the set of all real numbers.
2. The range consists of all real numbers from -1 to 1, inclusive.
3. The cosine function is an even function, as the symmetry of the graph with respect to the y-axis indicates.
4. The cosine function is periodic, with period 2π.
5. The x-intercepts are $\ldots$, $-3\pi/2$, $-\pi/2$, $\pi/2$, $3\pi/2$, $5\pi/2$, $\ldots$; the y-intercept is 1.
6. The maximum value is 1 and occurs at $x = \ldots$, -2π, 0, 2π, 4π, 6π, $\ldots$; the minimum value is -1 and occurs at $x = \ldots$, $-\pi$, π, 3π, 5π, $\ldots$.

Again, the graphing techniques from Chapter 1 may be used to graph variations of the cosine function.

E X A M P L E   3   *Graphing Variations of $y = \cos x$ Using Shifts, Reflections, and the Like*

Use the graph of $y = \cos x$ to graph $y = 2 \cos x$

Solution   Figure 68 illustrates the graph, which is a vertical stretch of the graph of $y = \cos x$.

FIGURE 68

$y = 2 \cos x$

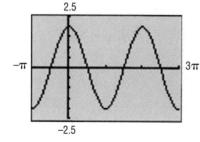

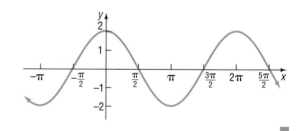

E X A M P L E   4   *Graphing Variations of $y = \cos x$ Using Shifts, Reflections, and the Like*

Use the graph of $y = \cos x$ to graph $y = \cos 3x$.

Solution   Figure 69 illustrates the graph, which is a horizontal compression of the graph of $y = \cos x$. Notice that, due to this compression, the period of $y = \cos 3x$ is $2\pi/3$, whereas the period of $y = \cos x$ is 2π.

FIGURE 72

$y = \tan\left(x + \dfrac{\pi}{4}\right)$

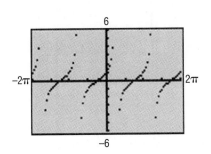

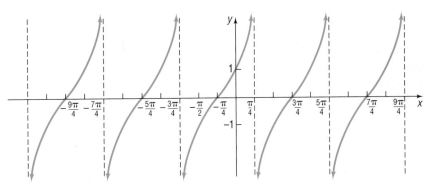

■   Now work Problem 51.

The Graphs of $y = \csc x$, $y = \sec x$, and $y = \cot x$

The cosecant and secant functions, sometimes referred to as **reciprocal functions,** are graphed by making use of the reciprocal identities

$$\csc x = \frac{1}{\sin x} \quad \text{and} \quad \sec x = \frac{1}{\cos x}$$

For example, the value of the cosecant function $y = \csc x$ at a given number x equals the reciprocal of the corresponding value of the sine function, provided the value of the sine function is not 0. If the value of $\sin x$ is 0, then, at such numbers x, the cosecant function is not defined. In fact, the graph of the cosecant function has vertical asymptotes at integral multiples of π. Figure 73 shows the graph drawn by hand and Figure 74 shows the graph using a graphing utility.

FIGURE 73   **FIGURE 74**

$y = \csc x$, $-\infty < x < \infty$, x not equal to
integral multiples of π, $|y| \geq 1$

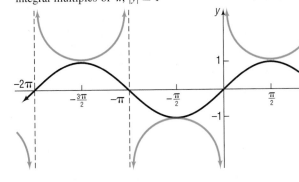

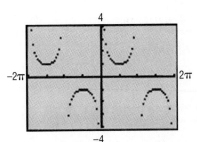

E X A M P L E   6   *Graphing Variations of* $y = \csc x$ *Using Shifts, Reflections, and the Like*

Graph $y = 2 \csc(x - \pi/2)$, $-\pi \leq x \leq \pi$.

Solution   Figure 75 shows the required steps.

FIGURE 75

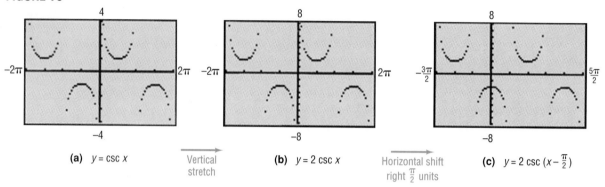

(a) $y = \csc x$ Vertical stretch   (b) $y = 2 \csc x$ Horizontal shift right $\frac{\pi}{2}$ units   (c) $y = 2 \csc \left(x - \frac{\pi}{2}\right)$ ■

Using the idea of reciprocals, we can similarly obtain the graph of $y = \sec x$. See Figure 76 and Figure 77.

FIGURE 76

$y = \sec x$, $-\infty < x < \infty$, x not equal to odd multiples of $\pi/2$, $|y| \geq 1$

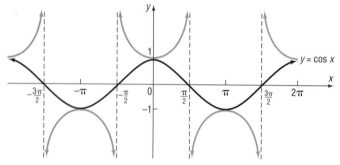

FIGURE 77

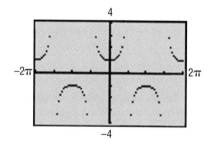

TABLE 11

x	$y = \cot x$	(x, y)
$\pi/6$	$\sqrt{3}$	$(\pi/6, \sqrt{3})$
$\pi/4$	1	$(\pi/4, 1)$
$\pi/3$	$\sqrt{3}/3$	$(\pi/3, \sqrt{3}/3)$
$\pi/2$	0	$(\pi/2, 0)$
$2\pi/3$	$-\sqrt{3}/3$	$(2\pi/3, -\sqrt{3}/3)$
$3\pi/4$	-1	$(3\pi/4, -1)$
$5\pi/6$	$-\sqrt{3}$	$(5\pi/6, -\sqrt{3})$

FIGURE 78

$y = \cot x$, $-\infty < x < \infty$, x not equal to integral multiples of π, $-\infty < y < \infty$

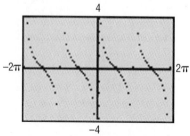

We obtain the graph of $y = \cot x$ as we did the graph of $y = \tan x$. The period of $y = \cot x$ is π. Because the cotangent function is not defined for integral multiples of π, we shall concentrate on the interval $(0, \pi)$. Table 11 lists some points on the graph of $y = \cot x$, $0 < x < \pi$. As x approaches 0, but remains greater than 0, the value of $\cos x$ will be close to 1 and the value of $\sin x$ will be positive and close to 0. Hence, the ratio $(\cos x)/(\sin x) = \cot x$ will be positive and large, so, as x approaches 0, $\cot x$ approaches ∞. Similarly, as x approaches π, but remains less than π, the value of $\cos x$ will be close to -1 and the value of $\sin x$ will be positive and close to 0. Hence, the ratio $(\cos x)/(\sin x) = \cot x$ will be negative and will approach $-\infty$ as x approaches π. Figure 78 shows the graph.

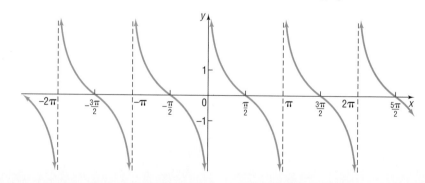

2.5

Exercise 2.5

In Problems 1–20, refer to the graphs to answer each question, if necessary.

1. What is the y-intercept of $y = \sin x$?

2. What is the y-intercept of $y = \cos x$?

3. For what numbers x, $-\pi \le x \le \pi$ is the graph of $y = \sin x$ increasing?

4. For what numbers x, $-\pi \le x \le \pi$ is the graph of $y = \cos x$ decreasing?

5. What is the largest value of $y = \sin x$?

6. What is the smallest value of $y = \cos x$?

7. For what numbers x, $0 \le x \le 2\pi$, does $\sin x = 0$?

8. For what numbers x, $0 \le x \le 2\pi$, does $\cos x = 0$?

9. For what numbers x, $-2\pi \le x \le 2\pi$, does $\sin x = 1$? What about $\sin x = -1$?

10. For what numbers x, $-2\pi \le x \le 2\pi$, does $\cos x = 1$? What about $\cos x = -1$?

11. What is the y-intercept of $y = \tan x$?

12. What is the y-intercept of $y = \cot x$?

13. What is the y-intercept of $y = \sec x$?

14. What is the y-intercept of $y = \csc x$?

15. For what numbers x, $-2\pi \le x \le 2\pi$, does $\sec x = 1$? What about $\sec x = -1$?

16. For what numbers x, $-2\pi \le x \le 2\pi$, does $\csc x = 1$? What about $\csc x = -1$?

17. For what numbers x, $-2\pi \le x \le 2\pi$, does the graph of $y = \sec x$ have vertical asymptotes?

18. For what numbers x, $-2\pi \le x \le 2\pi$, does the graph of $y = \csc x$ have vertical asymptotes?

19. For what numbers x, $-2\pi \le x \le 2\pi$, does the graph of $y = \tan x$ have vertical asymptotes?

20. For what numbers x, $-2\pi \le x \le 2\pi$, does the graph of $y = \cot x$ have vertical asymptotes?

In Problems 21 and 22, match the graph to a function. Three answers are possible.

A. $y = -\sin x$ B. $y = -\cos x$ C. $y = \sin\left(x - \dfrac{\pi}{2}\right)$

D. $y = -\cos\left(x - \dfrac{\pi}{2}\right)$ E. $y = \sin(x + \pi)$ F. $y = \cos(x + \pi)$

21.
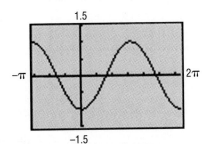

22.
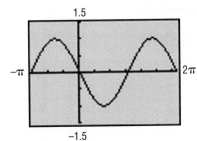

In Problems 23–26, match each function to its graph.

A. $y = \sin 2x$ B. $y = \sin 4x$ C. $y = \cos 2x$ D. $y = \cos 4x$

23.

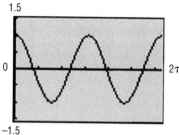

24.

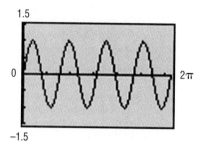

25.

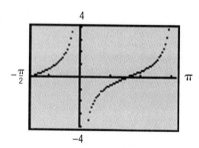

26.

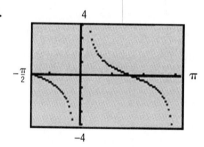

In Problems 27–30, match each function to its graph.

A. $y = -\tan x$ B. $y = \tan\left(x + \dfrac{\pi}{2}\right)$ C. $y = \tan(x + \pi)$ D. $y = -\tan\left(x - \dfrac{\pi}{2}\right)$

27.

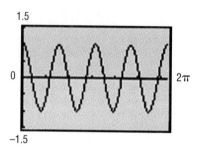

28.

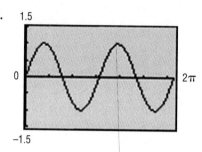

29.

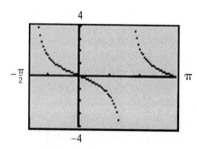

30.

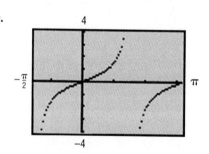

In Problems 31–62, show the stages required to graph each function.

31. $y = 3 \sin x$

32. $y = 4 \cos x$

33. $y = \cos\left(x + \dfrac{\pi}{4}\right)$

34. $y = \sin(x - \pi)$

35. $y = \sin x - 1$

36. $y = \cos x + 1$

37. $y = -2 \sin x$

38. $y = -3 \cos x$

39. $y = \sin \pi x$

40. $y = \cos \dfrac{\pi}{2} x$

41. $y = 2 \sin x + 2$

42. $y = 3 \cos x + 3$

43. $y = -2 \cos\left(x - \dfrac{\pi}{2}\right)$

44. $y = -3 \sin\left(x + \dfrac{\pi}{2}\right)$

45. $y = 3 \sin(\pi - x)$

46. $y = 2 \cos(\pi - x)$

47. $y = -\sec x$

48. $y = -\cot x$

49. $y = \sec\left(x - \dfrac{\pi}{2}\right)$

50. $y = \csc(x - \pi)$

51. $y = \tan(x - \pi)$

52. $y = \cot(x - \pi)$

53. $y = 3 \tan 2x$

54. $y = 4 \tan \frac{1}{2} x$

55. $y = \sec 2x$

56. $y = \csc \frac{1}{2} x$

57. $y = \cot \pi x$

58. $y = \cot 2x$

59. $y = -3 \tan 4x$

60. $y = -3 \tan 2x$

61. $y = 2 \sec \frac{1}{2} x$

62. $y = 2 \sec 3x$

63. *Carrying a Ladder Around a Corner* A ladder of length L is carried horizontally around a corner from a hall 3 feet wide into a hall 4 feet wide. See the illustration.

(a) Show that the length L of the ladder as a function of the angle θ is

$$L = 3 \sec \theta + 4 \csc \theta$$

(b) Graph L, $0 < \theta < \dfrac{\pi}{2}$.

(c) Where is L the least?

(d) What is the length of the largest ladder that can be carried around the corner? Why is this also the least value of L?

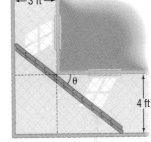

64. Graph $y = \sin x$, $y = 2 \sin x$, $y = \frac{1}{2} \sin x$, and $y = 8 \sin x$. What do you conclude about the graph of $y = A \sin x$, $A > 0$?

65. Graph $y = \sin x$, $y = \sin 2x$, $y = \sin 4x$, and $y = \sin \frac{1}{2} x$. What do you conclude about the graph of $y = \sin \omega x$?

66. Graph $y = \sin x$, $y = \sin[x - (\pi/3)]$, $y = \sin[x - (\pi/4)]$, and $y = \sin[x - (\pi/6)]$. What do you conclude about the graph of $y = \sin(x - \phi)$, $\phi > 0$?

67. Graph

$$y = \sin x \quad \text{and} \quad y = \cos\left(x - \dfrac{\pi}{2}\right)$$

Do you think that $\sin x = \cos\left(x - \dfrac{\pi}{2}\right)$?

68. Graph

$$y = \tan x \quad \text{and} \quad y = -\cot\left(x + \dfrac{\pi}{2}\right)$$

Do you think that $\tan x = -\cot\left(x + \dfrac{\pi}{2}\right)$?

2.6

The Inverse Trigonometric Functions

In Section 1.7 we discussed inverse functions, and we noted that if a function is one-to-one it will have an inverse. We also observed that if a function is not one-to-one, it may be possible to restrict its domain in some suitable manner such that the restricted function is one-to-one. In this section, we use these ideas to define inverse trigonometric functions. (You may wish to review Section 1.7 at this time.) We begin with the inverse of the sine function.

The Inverse Sine Function

In Figure 79, we reproduce the graph of $y = \sin x$. Because every horizontal line $y = b$, where b is between -1 and 1, intersects the graph of $y = \sin x$ infinitely many times, it follows from the horizontal-line test that the function $y = \sin x$ is not one-to-one.

FIGURE 79
$y = \sin x, \ -\infty < x < \infty, \ -1 \le y \le 1$

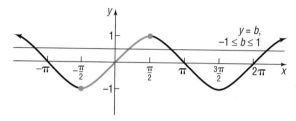

However, if we restrict the domain of $y = \sin x$ to the interval $[-\pi/2, \pi/2]$, the restricted function

$$y = \sin x, \qquad -\frac{\pi}{2} \le x \le \frac{\pi}{2}$$

is one-to-one and hence, will have an inverse.* See Figure 80.

FIGURE 80
$y = \sin x, \ -\pi/2 \le x \le \pi/2, \ -1 \le y \le 1$

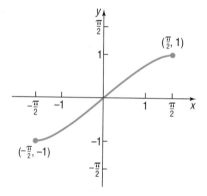

The inverse function is called the **inverse sine** of x and is symbolized by $y = \sin^{-1} x$. Thus,

Inverse Sine Function

$$y = \sin^{-1} x \quad \text{means} \quad x = \sin y \tag{1}$$

$$\text{where } -\frac{\pi}{2} \le y \le \frac{\pi}{2} \quad \text{and} \quad -1 \le x \le 1$$

Because $y = \sin^{-1} x$ means $x = \sin y$, we read $y = \sin^{-1} x$ as "y is the angle whose sine equals x." Alternatively, we can say that "y is the inverse sine of x." Be careful about the notation used. The superscript -1 that appears in $y = \sin^{-1} x$ is not an exponent, but is reminiscent of the symbolism f^{-1} used to denote the inverse of a function f. (To avoid this notation, some books use the notation $y = \arcsin x$ instead of $y = \sin^{-1} x$.)

---

*Although there are many other ways to restrict the domain and obtain a one-to-one function, mathematicians have agreed on a consistent use of the interval $[-\pi/2, \pi/2]$ in order to define the inverse of $y = \sin x$.

𝓜 ISSION POSSIBLE

Chapter 2

IDENTIFYING THE MOUNTAINS OF THE HAWAIIN ISLANDS SEEN FROM OAHU

A wealthy tourist with a strong desire for the perfect scrapbook has called in your consulting firm for help in labeling a photograph he took on a clear day from the southeast shore of Oahu. He shows you the photograph in which you see mostly sea and sky. But on the horizon are three mountain peaks, equally spaced and apparently all of the same height. The tourist tells you that the T-shirt vendor at the beach informed him that the three mountain peaks are the volcanoes of Lanai, Maui, and the Big Island (Hawaii). The vendor even added, "You almost never see the Big Island from here."

The tourist thinks his photograph may have some special value. He has some misgivings, however. He is wondering if, in fact, it is ever really possible to see the Big Island from Oahu. He wants accurate labels on his photos, so he asks you for help.

You realize that some trigonometry is needed as well as a good atlas and encyclopedia. After doing some research you discover the following facts:

The heights of the peaks vary. Lanaihale, the tallest peak on Lanai, is only about 3370 feet above sea level. Maui's Haleakala is 10,023 feet above sea level. Mauna Kea on the Big Island is the highest of all, 13,796 feet above sea level. (If measured from its base on the ocean floor, it would qualify as the tallest mountain on Earth.)

Measuring from the southeast corner of Oahu, the distance to Lanai is about 65 miles, to Maui it's about 110 miles, and to the Big Island it is about 190 miles. These distances, measured along the surface of the Earth, represent the length of arc that originates at sea level on Oahu and terminates in a imaginary location directly below the peak of each mountain at what would be sea level.

1. These volcanic peaks are all different heights. If your wealthy client took a photo of three mountain peaks that all look the same height, how could they possibly represent these three volcanoes?
2. The radius of the earth is approximately 3960 miles. Based on that figure, what is the circumference of the earth? How is the distance between islands related to the circumference of the earth?
3. To determine which of these mountain peaks would actually be visible from Oahu, you need to consider that the tourist standing on the shore and looking "straight out" would have a line of sight tangent to the surface of the earth at that point. Make a rough sketch of the right triangle formed by the tourist's line of sight, the radius from the center of the earth to the tourist, and a line from the center of the earth that passes through Mauna Kea. Can you determine the angle formed at the center of the earth?
4. What would the length of the hypotenuse of the triangle be? Can you tell from that whether or not Mauna Kea would be visible from Oahu?
5. Repeat this procedure with the information about Maui and Lanai. Will they be visible from Oahu?
6. There is another island off Oahu that the T-shirt vendor did not mention. Molokai is about 40 miles from Oahu, and its highest peak, "Kamakou," is 4961 feet above sea level. Would Kamakou be visible from Oahu?
7. Your team should now be prepared to name the three volcanic peaks in the photograph. How do you decide which one is which?

Based on the general discussion of functions and their inverses (Section 1.7), we have the following results:

$$\sin^{-1}(\sin u) = u \qquad \text{where} \qquad -\frac{\pi}{2} \le u \le \frac{\pi}{2}$$

$$\sin(\sin^{-1} v) = v \qquad \text{where} \qquad -1 \le v \le 1$$

Let's examine the function $y = \sin^{-1} x$. Its domain is $-1 \le x \le 1$, and its range is $-\pi/2 \le y \le \pi/2$. Its graph can be obtained by reflecting the restricted portion of the graph of $y = \sin x$ about the line $y = x$, as shown in Figure 81(a). Figure 81(b) shows the graph using a graphing utility.

FIGURE 81

$y = \sin^{-1} x,\ -1 \le x \le 1,$
$-\pi/2 \le y \le \pi/2$

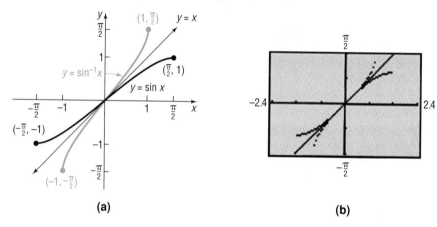

(a)   (b)

For some numbers x it is possible to find the exact value of $y = \sin^{-1} x$.

E X A M P L E   1

Finding the Exact Value of an Inverse Sine Function

Find the exact value of $\sin^{-1} 1$.

Solution   Let $\theta = \sin^{-1} 1$. Then we seek the angle θ, $-\pi/2 \le \theta \le \pi/2$, whose sine equals 1:

$$\theta = \sin^{-1} 1, \qquad -\frac{\pi}{2} \le \theta \le \frac{\pi}{2}$$

$$\sin \theta = 1, \qquad -\frac{\pi}{2} \le \theta \le \frac{\pi}{2} \quad \text{By definition}$$

From Figure 82 we see that the only angle θ within the interval $[-\pi/2, \pi/2]$ whose sine is 1 is $\pi/2$. (Note that $\sin(5\pi/2)$ also equals 1, but $5\pi/2$ lies outside the interval $[-\pi/2, \pi/2]$ and hence is not admissible.)

FIGURE 82

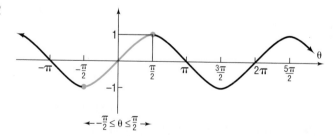

We conclude that

$$\theta = \frac{\pi}{2}$$

So

$$\sin^{-1} 1 = \frac{\pi}{2}$$ ■

■   Now work Problem 1.

E X A M P L E   2   *Finding the Exact Value of an Inverse Sine Function*

Find the exact value of $\sin^{-1}(\sqrt{3}/2)$.

Solution   Let $\theta = \sin^{-1}(\sqrt{3}/2)$. Then we seek the angle θ, $-\pi/2 \le \theta \le \pi/2$, whose sine equals $\sqrt{3}/2$:

$$\theta = \sin^{-1} \frac{\sqrt{3}}{2}, \qquad -\frac{\pi}{2} \le \theta \le \frac{\pi}{2}$$

$$\sin\theta = \frac{\sqrt{3}}{2}, \qquad -\frac{\pi}{2} \le \theta \le \frac{\pi}{2} \quad \text{By definition}$$

From Figure 83, we see that the only angle θ within the interval $[-\pi/2, \pi/2]$ whose sine is $\sqrt{3}/2$ is $\pi/3$. (Note that $\sin(2\pi/3)$ also equals $\sqrt{3}/2$, but $2\pi/3$ lies outside the interval $[-\pi/2, \pi/2]$ and hence is not admissible.)

FIGURE 83

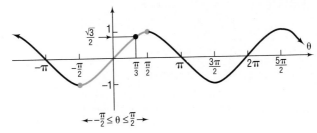

We conclude that

$$\theta = \frac{\pi}{3}$$

So

$$\sin^{-1} \frac{\sqrt{3}}{2} = \frac{\pi}{3}$$ ■

E X A M P L E   3   *Finding the Exact Value of an Inverse Sine Function*

Find the exact value of $\sin^{-1}(-\frac{1}{2})$.

Solution   Let $\theta = \sin^{-1}(-\frac{1}{2})$. Then we seek the angle θ, $-\pi/2 \le \theta \le \pi/2$, whose sine equals $-\frac{1}{2}$:

$$\theta = \sin^{-1}\left(-\frac{1}{2}\right), \qquad -\frac{\pi}{2} \le \theta \le \frac{\pi}{2}$$

$$\sin\theta = -\frac{1}{2}, \qquad -\frac{\pi}{2} \le \theta \le \frac{\pi}{2}$$

(Refer to Figure 83, if necessary.) The only angle within the interval $[-\pi/2, \pi/2]$ whose sine is $-\frac{1}{2}$ is $-\pi/6$. Thus,

$$\theta = -\frac{\pi}{6}$$

So

$$\sin^{-1}\left(-\frac{1}{2}\right) = -\frac{\pi}{6}$$ ∎

■   Now work Problem 3.

For most numbers x, the value $y = \sin^{-1} x$ must be approximated.

E X A M P L E   4   *Finding an Approximate Value of an Inverse Sine Function*

Find the approximate value of

(a) $\sin^{-1}\frac{1}{3}$ (b) $\sin^{-1}(-\frac{1}{4})$

Express the answer rounded to two decimal places.

Solution   Because we want the angle measured in radians, we first set the mode to radians.

(a) Figure 84 shows the solution using a TI-82 graphing calculator.

FIGURE 84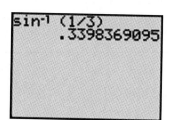

Thus, $\sin^{-1}\dfrac{1}{3} = 0.34$ rounded to two decimal places.

(b) Figure 85 shows the solution using a TI-82 graphing calculator.

FIGURE 85

Thus, $\sin^{-1}\left(-\dfrac{1}{4}\right) = -0.25$ rounded to two decimal places.   ∎

■   Now work Problem 13.

The Inverse Cosine Function

In Figure 86 we reproduce the graph of $y = \cos x$. Because every horizontal line $y = b$, where b is between -1 and 1, intersects the graph of $y = \cos x$ infinitely many times, it follows that the cosine function is not one-to-one.

FIGURE 86
$y = \cos x,\ -\infty < x < \infty,\ -1 \leq y \leq 1$

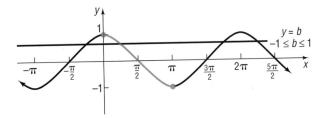

However, if we restrict the domain of $y = \cos x$ to the interval $[0, \pi]$, the restricted function

$$y = \cos x, \qquad 0 \leq x \leq \pi$$

is one-to-one and hence will have an inverse.* See Figure 87.

FIGURE 87
$y = \cos x,\ 0 \leq x \leq \pi,\ -1 \leq y \leq 1$

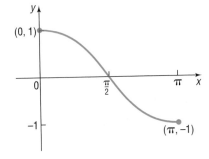

The inverse function is called the **inverse cosine** of x and is symbolized by $y = \cos^{-1} x$ (or by $y = \arccos x$). Thus,

Inverse Cosine Function

$$y = \cos^{-1} x \quad \text{means} \quad x = \cos y \tag{2}$$
$$\text{where} \quad 0 \leq y \leq \pi \quad \text{and} \quad -1 \leq x \leq 1$$

Here, y is the angle whose cosine is x. The domain of the function $y = \cos^{-1} x$ is $-1 \leq x \leq 1$, and its range is $0 \leq y \leq \pi$. The graph of $y = \cos^{-1} x$ can be obtained by reflecting the restricted portion of the graph of $y = \cos x$ about the line $y = x$, as shown in Figure 88(a). Figure 88(b) shows the graph using a graphing utility.

*This is the generally accepted restriction to define the inverse.

FIGURE 88

$y = \cos^{-1} x, \; -1 \le x \le 1, \; 0 \le y \le \pi$

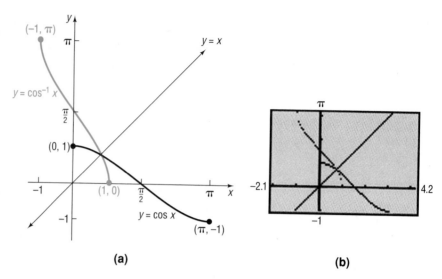

(a)

(b)

In general,

$$\cos^{-1}(\cos u) = u, \qquad \text{where} \quad 0 \le u \le \pi$$
$$\cos(\cos^{-1} v) = v, \qquad \text{where} \quad -1 \le v \le 1$$

E X A M P L E   5

Finding the Exact Value of an Inverse Cosine Function

Find the exact value of $\cos^{-1} 0$.

Solution   Let $\theta = \cos^{-1} 0$. Then we seek the angle θ, $0 \le \theta \le \pi$, whose cosine equals 0:

$$\theta = \cos^{-1} 0, \qquad 0 \le \theta \le \pi$$
$$\cos \theta = 0, \qquad 0 \le \theta \le \pi$$

From Figure 89, we see that the only angle θ within the interval $[0, \pi]$ whose cosine is 0 is $\pi/2$. (Note that $\cos(3\pi/2)$ also equals 0, but $3\pi/2$ lies outside the interval $[0, \pi]$ and hence is not admissible.)

FIGURE 89

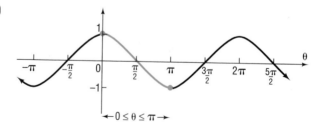

We conclude that

$$\theta = \frac{\pi}{2}$$

So

$$\cos^{-1} 0 = \frac{\pi}{2}$$

■

E X A M P L E   6

Finding the Exact Value of an Inverse Cosine Function
Find the exact value of $\cos^{-1}(\sqrt{2}/2)$.

Solution   Let $\theta = \cos^{-1}(\sqrt{2}/2)$. Then we seek the angle θ, $0 \leq \theta \leq \pi$, whose cosine equals $\sqrt{2}/2$:

$$\theta = \cos^{-1}\frac{\sqrt{2}}{2}, \qquad 0 \leq \theta \leq \pi$$

$$\cos\theta = \frac{\sqrt{2}}{2}, \qquad 0 \leq \theta \leq \pi$$

From Figure 90, we see that the only angle θ within the interval $[0, \pi]$ whose cosine is $\sqrt{2}/2$ is $\pi/4$.

FIGURE 90

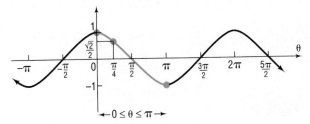

We conclude that

$$\theta = \frac{\pi}{4}$$

So

$$\cos^{-1}\frac{\sqrt{2}}{2} = \frac{\pi}{4}$$

∎

E X A M P L E   7

Finding the Exact Value of an Inverse Cosine Function
Find the exact value of $\cos^{-1}(-\frac{1}{2})$.

Solution   Let $\theta = \cos^{-1}(-\frac{1}{2})$. Then we seek the angle θ, $0 \leq \theta \leq \pi$, whose cosine equals $-\frac{1}{2}$:

$$\theta = \cos^{-1}(-\tfrac{1}{2}), \qquad 0 \leq \theta \leq \pi$$

$$\cos\theta = -\tfrac{1}{2}, \qquad 0 \leq \theta \leq \pi$$

(Refer to Figure 90, if necessary.) The only angle within the interval $[0, \pi]$ whose cosine is $-\frac{1}{2}$ is $2\pi/3$. Thus,

$$\theta = \frac{2\pi}{3}$$

So

$$\cos^{-1}\left(-\frac{1}{2}\right) = \frac{2\pi}{3}$$

∎

The Inverse Tangent Function

In Figure 91 we reproduce the graph of $y = \tan x$. Because every horizontal line intersects the graph infinitely many times, it follows that the tangent function is not one-to-one.

FIGURE 91

$y = \tan x$, $-\infty < x < \infty$, x not equal to odd multiples of $\pi/2$, $-\infty < y < \infty$

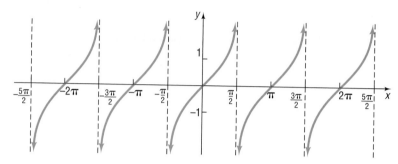

However, if we restrict the domain of $y = \tan x$ to the interval $(-\pi/2, \pi/2)$,* the restricted function

$$y = \tan x, \qquad -\frac{\pi}{2} < x < \frac{\pi}{2}$$

is one-to-one and hence has an inverse. See Figure 92.

FIGURE 92

$y = \tan x$, $-\pi/2 < x < \pi/2$, $-\infty < y < \infty$

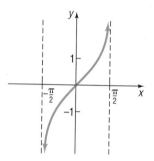

The inverse function is called the **inverse tangent** of x and is symbolized by $y = \tan^{-1} x$ (or by $y = \arctan x$). Thus,

Inverse Tangent Function

$$y = \tan^{-1} x \quad \text{means} \quad x = \tan y \tag{3}$$

$$\text{where} \quad -\frac{\pi}{2} < y < \frac{\pi}{2} \quad \text{and} \quad -\infty < x < \infty$$

Here, y is the angle whose tangent is x. The domain of the function $y = \tan^{-1} x$ is $-\infty < x < \infty$, and its range is $-\pi/2 < y < \pi/2$. The graph of $y = \tan^{-1} x$ can be obtained by reflecting the restricted portion of the graph of $y = \tan x$ about the line $y = x$, as shown in Figure 93(a). Figure 93(b) shows the graph using a graphing utility.

*This is the generally accepted restriction.

Solution   Let $\theta = \tan^{-1} v$ so that $\tan \theta = v$, $-\pi/2 < \theta < \pi/2$. There are two possibilities: either $-\pi/2 < \theta < 0$ or $0 \leq \theta < \pi/2$. If $0 \leq \theta < \pi/2$, then $\tan \theta = v \geq 0$. Based on Figure 98, we conclude that

$$\sin(\tan^{-1} v) = \sin \theta = \frac{v}{\sqrt{1 + v^2}}$$

If $-\pi/2 < \theta < 0$, then $\tan \theta = v < 0$. Based on Figure 99, we conclude that

$$\sin(\tan^{-1} v) = \sin \theta = \frac{v}{\sqrt{1 + v^2}}$$

FIGURE 98
$v > 0$

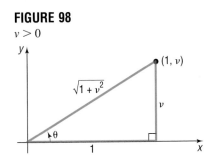

FIGURE 99
$v < 0$

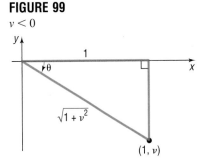

An alternative solution to Example 17 uses the fundamental identities. If $\theta = \tan^{-1} v$, then $\tan \theta = v$ and $-\pi/2 < \theta < \pi/2$, as before. As a result, we know that $\sec \theta > 0$. Thus,

$$\sin(\tan^{-1} v) = \sin \theta = \cos \theta \tan \theta = \frac{\tan \theta}{\sec \theta} = \frac{\tan \theta}{\sqrt{1 + \tan^2 \theta}} = \frac{v}{\sqrt{1 + v^2}}$$

$$\tan \theta = \frac{\sin \theta}{\cos \theta} \qquad \begin{array}{l} \sec^2 \theta = 1 + \tan^2 \theta \\ \sec \theta > 0 \end{array}$$

Now work Problem 57.

The Remaining Inverse Trigonometric Functions

The inverse cotangent, inverse secant, and inverse cosecant functions are defined as follows:

$$y = \cot^{-1} x \quad \text{means} \quad x = \cot y \qquad (4)$$
$$\text{where} \quad -\infty < x < \infty \quad \text{and} \quad 0 < y < \pi$$

$$y = \sec^{-1} x \quad \text{means} \quad x = \sec y \qquad (5)$$
$$\text{where} \quad |x| \geq 1 \quad \text{and} \quad 0 \leq y \leq \pi, \ y \neq \frac{\pi}{2}$$

$$y = \csc^{-1} x \quad \text{means} \quad x = \csc y \qquad (6)$$
$$\text{where} \quad |x| \geq 1 \quad \text{and} \quad \frac{-\pi}{2} \leq y \leq \frac{\pi}{2}, \ y \neq 0$$

Most calculators do not have keys for evaluating these inverse trigonometric functions. The easiest way to evaluate them is to convert to an inverse trigonometric function whose range is the same as the one to be evaluated. In this regard, notice that $y = \cot^{-1} x$ and $y = \sec^{-1} x$ (except where undefined) each have the same range as $y = \cos^{-1} x$, while $y = \csc^{-1} x$ (except where undefined) has the same range as $y = \sin^{-1} x$.

E X A M P L E   1 8   *Approximating the Value of Inverse Trigonometric Functions*

Use a calculator to approximate each expression rounded to two decimal places:

(a) $\sec^{-1} 3$ (b) $\csc^{-1}(-4)$ (c) $\cot^{-1} \frac{1}{2}$ (d) $\cot^{-1}(-2)$

Solution   First, set your calculator to radian mode.

(a) Let $\theta = \sec^{-1} 3$. Then $\sec \theta = 3$ and $0 \le \theta \le \pi$, $\theta \ne \pi/2$. Thus, $\cos \theta = \frac{1}{3}$ and

$$\sec^{-1} 3 = \theta = \cos^{-1} \tfrac{1}{3} \approx 1.23$$

Use a calculator.

(b) Let $\theta = \csc^{-1}(-4)$. Then $\csc \theta = -4$, $-\pi/2 \le \theta \le \pi/2$, $\theta \ne 0$. Thus, $\sin \theta = -\frac{1}{4}$ and

$$\csc^{-1}(-4) = \theta = \sin^{-1}(-\tfrac{1}{4}) \approx -0.25$$

FIGURE 100
$\cot \theta = \frac{1}{2}, 0 < \theta < \pi$

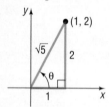

(c) Let $\theta = \cot^{-1} \frac{1}{2}$. Then $\cot \theta = \frac{1}{2}, 0 < \theta < \pi$. From these facts we know that θ is in quadrant I. We draw Figure 100 to help us to find $\cos \theta$. Thus, $\cos \theta = 1/\sqrt{5}$, $0 < \theta < \pi/2$, and

$$\cot^{-1} \frac{1}{2} = \theta = \cos^{-1} \frac{1}{\sqrt{5}} \approx 1.11$$

FIGURE 101
$\cot \theta = -2, 0 < \theta < \pi$

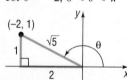

(d) Let $\theta = \cot^{-1}(-2)$. Then $\cot \theta = -2, 0 < \theta < \pi$. From these facts we know that θ lies in quadrant II. We draw Figure 101 to help us to find $\cos \theta$. Thus, $\cos \theta = -2/\sqrt{5}, \pi/2 < \theta < \pi$, and

$$\cot^{-1}(-2) = \theta = \cos^{-1}\left(-\frac{2}{\sqrt{5}}\right) \approx 2.68$$

■ Now work Problem 67.

2.6

Exercise 2.6

In Problems 1–12, find the exact value of each expression.

1. $\sin^{-1} 0$

2. $\cos^{-1} 1$

3. $\sin^{-1}(-1)$

4. $\cos^{-1}(-1)$

5. $\tan^{-1} 0$

6. $\tan^{-1}(-1)$

7. $\sin^{-1} \dfrac{\sqrt{2}}{2}$

8. $\tan^{-1} \dfrac{\sqrt{3}}{3}$

9. $\tan^{-1} \sqrt{3}$

10. $\sin^{-1}\left(-\dfrac{\sqrt{3}}{2}\right)$

11. $\cos^{-1}\left(-\dfrac{\sqrt{3}}{2}\right)$

12. $\sin^{-1}\left(-\dfrac{\sqrt{2}}{2}\right)$

In Problems 13–24, use a calculator to find the value of each expression rounded to two decimal places.

13. $\sin^{-1} 0.1$

14. $\cos^{-1} 0.6$

15. $\tan^{-1} 5$

16. $\tan^{-1} 0.2$

17. $\cos^{-1} \frac{7}{8}$

18. $\sin^{-1} \frac{1}{8}$

19. $\tan^{-1}(-0.4)$

20. $\tan^{-1}(-3)$

21. $\sin^{-1}(-0.12)$

22. $\cos^{-1}(-0.44)$

23. $\cos^{-1} \dfrac{\sqrt{2}}{3}$

24. $\sin^{-1} \dfrac{\sqrt{3}}{5}$

be *known* (just as you know your name rather than have it memorized). In fact, the use made of a basic identity is often a minor variation of the form listed here. For example, we might want to use $\sin^2 \theta = 1 - \cos^2 \theta$ instead of $\sin^2 \theta + \cos^2 \theta = 1$. For this reason, among others, you need to know these relationships and be quite comfortable with variations of them.

In the examples that follow, the directions will read "Establish the identity. . . ." As you will see, this is accomplished by starting with one side of the given equation (usually the one containing the more complicated expression) and, using appropriate basic identities and algebraic manipulations, arriving at the other side. The selection of an appropriate basic identity to obtain the desired result is learned only through experience and lots of practice.

E X A M P L E   1   *Establishing an Identity*

Establish the identity: $\sec \theta \cdot \sin \theta = \tan \theta$

Solution   We start with the left side, because it contains the more complicated expression, and apply a reciprocal identity:

$$\sec \theta \cdot \sin \theta = \frac{1}{\cos \theta} \cdot \sin \theta = \frac{\sin \theta}{\cos \theta} = \tan \theta$$

Having arrived at the right side, the identity is established.   ■

Comment: A graphing utility can be used to provide evidence of an identity. For example, if we graph $Y_1 = \sec \theta \cdot \sin \theta$ and $Y_2 = \tan \theta$, the graphs appear to be the same. This provides evidence that $Y_1 = Y_2$. However, it does not prove their equality. A graphing utility *cannot be used to establish an identity*—identities must be established algebraically.

■   Now work Problem 1.

E X A M P L E   2   *Establishing an Identity*

Establish the identity: $\sin^2(-\theta) + \cos^2(-\theta) = 1$

Solution   We begin with the left side and apply even–odd identities:

$$\begin{aligned}
\sin^2(-\theta) + \cos^2(-\theta) &= [\sin(-\theta)]^2 + [\cos(-\theta)]^2 \\
&= (-\sin \theta)^2 + (\cos \theta)^2 \\
&= (\sin \theta)^2 + (\cos \theta)^2 \\
&= 1
\end{aligned}$$

■

E X A M P L E   3   *Establishing an Identity*

Establish the identity: $\dfrac{\sin^2(-\theta) - \cos^2(-\theta)}{\sin(-\theta) - \cos(-\theta)} = \cos \theta - \sin \theta$

Solution   We begin with two observations: The left side appears to contain the more complicated expression. Also, the left side contains expressions with the argument $-\theta$, whereas the right side contains expressions with the argument θ. We decide, therefore, to start with the left side and apply even–odd identities:

$$\frac{\sin^2(-\theta) - \cos^2(-\theta)}{\sin(-\theta) - \cos(-\theta)} = \frac{[\sin(-\theta)]^2 - [\cos(-\theta)]^2}{\sin(-\theta) - \cos(-\theta)}$$

$$= \frac{(-\sin\theta)^2 - (\cos\theta)^2}{-\sin\theta - \cos\theta} \qquad \text{Even–odd identities}$$

$$= \frac{(\sin\theta)^2 - (\cos\theta)^2}{-\sin\theta - \cos\theta} \qquad \text{Simplify.}$$

$$= \frac{(\sin\theta - \cos\theta)(\sin\theta + \cos\theta)}{-(\sin\theta + \cos\theta)} \qquad \text{Factor.}$$

$$= \cos\theta - \sin\theta \qquad \text{Cancel and simplify.} \quad\blacksquare$$

E X A M P L E   4   *Establishing an Identity*

Establish the identity: $\dfrac{1 + \tan\theta}{1 + \cot\theta} = \tan\theta$

Solution

$$\frac{1 + \tan\theta}{1 + \cot\theta} = \frac{1 + \tan\theta}{1 + \dfrac{1}{\tan\theta}} = \frac{1 + \tan\theta}{\dfrac{\tan\theta + 1}{\tan\theta}} = \frac{\tan\theta(1 + \tan\theta)}{\tan\theta + 1} = \tan\theta \quad\blacksquare$$

■   Now work Problem 9.

When sums or differences of quotients appear, it is usually best to rewrite them as a single quotient, especially if the other side of the identity consists of only one term.

E X A M P L E   5   *Establishing an Identity*

Establish the identity: $\dfrac{\sin\theta}{1 + \cos\theta} + \dfrac{1 + \cos\theta}{\sin\theta} = 2\csc\theta$

Solution   The left side is more complicated, so we start with it and proceed to add:

$$\frac{\sin\theta}{1 + \cos\theta} + \frac{1 + \cos\theta}{\sin\theta} = \frac{\sin^2\theta + (1 + \cos\theta)^2}{(1 + \cos\theta)(\sin\theta)}$$

$$= \frac{\sin^2\theta + 1 + 2\cos\theta + \cos^2\theta}{(1 + \cos\theta)(\sin\theta)}$$

$$= \frac{(\sin^2\theta + \cos^2\theta) + 1 + 2\cos\theta}{(1 + \cos\theta)(\sin\theta)}$$

$$= \frac{2 + 2\cos\theta}{(1 + \cos\theta)(\sin\theta)}$$

$$= \frac{2(1 + \cos\theta)}{(1 + \cos\theta)(\sin\theta)}$$

$$= \frac{2}{\sin\theta}$$

$$= 2\csc\theta \qquad\qquad\blacksquare$$

E X A M P L E   6   *Establishing an Identity*

Establish the identity: $\dfrac{1}{\cos\theta} - \dfrac{\cos\theta}{1+\sin\theta} = \tan\theta$

Solution

$$\frac{1}{\cos\theta} - \frac{\cos\theta}{1+\sin\theta} = \frac{1+\sin\theta - \cos^2\theta}{\cos\theta(1+\sin\theta)}$$

$$= \frac{\sin\theta + (1-\cos^2\theta)}{\cos\theta(1+\sin\theta)}$$

$$= \frac{\sin\theta + \sin^2\theta}{\cos\theta(1+\sin\theta)} \qquad 1-\cos^2\theta = \sin^2\theta$$

$$= \frac{\sin\theta(1+\sin\theta)}{\cos\theta(1+\sin\theta)}$$

$$= \tan\theta \qquad\qquad\qquad\qquad\qquad ■$$

■ Now work Problem 23.

Sometimes, multiplying the numerator and denominator by an appropriate factor will result in a simplification.

E X A M P L E   7   *Establishing an Identity*

Establish the identity: $\dfrac{1-\sin\theta}{\cos\theta} = \dfrac{\cos\theta}{1+\sin\theta}$

Solution   We start with the left side and multiply the numerator and the denominator by $1+\sin\theta$. (Alternatively, we could multiply the numerator and denominator of the right side by $1-\sin\theta$.)

$$\frac{1-\sin\theta}{\cos\theta} = \frac{1-\sin\theta}{\cos\theta}\cdot\frac{1+\sin\theta}{1+\sin\theta}$$

$$= \frac{1-\sin^2\theta}{\cos\theta(1+\sin\theta)}$$

$$= \frac{\cos^2\theta}{\cos\theta(1+\sin\theta)}$$

$$= \frac{\cos\theta}{1+\sin\theta} \qquad\qquad\qquad ■$$

■ Now work Problem 35.

Although a lot of practice is the only real way to learn how to establish identities, the following guidelines should prove helpful:

Guidelines
for Establishing
Identities

1. It is almost always preferable to start with the side containing the more complicated expression.
2. Rewrite sums or differences of quotients as a single quotient.
3. Sometimes rewriting one side in terms of sines and cosines only will help.
4. Always keep your goal in mind. As you manipulate one side of the expression, you must keep in mind the form of the expression on the other side.

Warning: Be careful not to handle identities to be established as if they were equations. You *cannot* establish an identity by such methods as adding the same expression to each side and obtaining a true statement. This practice is not allowed, because the original statement is precisely the one that you are trying to establish. You do not know until it has been established that it is, in fact, true.

3.1

Exercise 3.1

In Problems 1–80, establish each identity.

1. $\csc \theta \cdot \cos \theta = \cot \theta$

2. $\csc \theta \cdot \tan \theta = \sec \theta$

3. $1 + \tan^2(-\theta) = \sec^2 \theta$

4. $1 + \cot^2(-\theta) = \csc^2 \theta$

5. $\cos \theta(\tan \theta + \cot \theta) = \csc \theta$

6. $\sin \theta(\cot \theta + \tan \theta) = \sec \theta$

7. $\tan \theta \cot \theta - \cos^2 \theta = \sin^2 \theta$

8. $\sin \theta \csc \theta - \cos^2 \theta = \sin^2 \theta$

9. $(\sec \theta - 1)(\sec \theta + 1) = \tan^2 \theta$

10. $(\csc \theta - 1)(\csc \theta + 1) = \cot^2 \theta$

11. $(\sec \theta + \tan \theta)(\sec \theta - \tan \theta) = 1$

12. $(\csc \theta + \cot \theta)(\csc \theta - \cot \theta) = 1$

13. $\sin^2 \theta(1 + \cot^2 \theta) = 1$

14. $(1 - \sin^2 \theta)(1 + \tan^2 \theta) = 1$

15. $(\sin \theta + \cos \theta)^2 + (\sin \theta - \cos \theta)^2 = 2$

16. $\tan^2 \theta \cos^2 \theta + \cot^2 \theta \sin^2 \theta = 1$

17. $\sec^4 \theta - \sec^2 \theta = \tan^4 \theta + \tan^2 \theta$

18. $\csc^4 \theta - \csc^2 \theta = \cot^4 \theta + \cot^2 \theta$

19. $\sec \theta - \tan \theta = \dfrac{\cos \theta}{1 + \sin \theta}$

20. $\csc \theta - \cot \theta = \dfrac{\sin \theta}{1 + \cos \theta}$

21. $3 \sin^2 \theta + 4 \cos^2 \theta = 3 + \cos^2 \theta$

22. $9 \sec^2 \theta - 5 \tan^2 \theta = 5 + 4 \sec^2 \theta$

23. $1 - \dfrac{\cos^2 \theta}{1 + \sin \theta} = \sin \theta$

24. $1 - \dfrac{\sin^2 \theta}{1 - \cos \theta} = -\cos \theta$

25. $\dfrac{1 + \tan \theta}{1 - \tan \theta} = \dfrac{\cot \theta + 1}{\cot \theta - 1}$

26. $\dfrac{\csc \theta - 1}{\csc \theta + 1} = \dfrac{1 - \sin \theta}{1 + \sin \theta}$

27. $\dfrac{\sec \theta}{\csc \theta} + \dfrac{\sin \theta}{\cos \theta} = 2 \tan \theta$

28. $\dfrac{\csc \theta - 1}{\cot \theta} = \dfrac{\cot \theta}{\csc \theta + 1}$

29. $\dfrac{1 + \sin \theta}{1 - \sin \theta} = \dfrac{\csc \theta + 1}{\csc \theta - 1}$

30. $\dfrac{\cos \theta + 1}{\cos \theta - 1} = \dfrac{1 + \sec \theta}{1 - \sec \theta}$

31. $\dfrac{1 - \sin \theta}{\cos \theta} + \dfrac{\cos \theta}{1 - \sin \theta} = 2 \sec \theta$

32. $\dfrac{\cos \theta}{1 + \sin \theta} + \dfrac{1 + \sin \theta}{\cos \theta} = 2 \sec \theta$

33. $\dfrac{\sin \theta}{\sin \theta - \cos \theta} = \dfrac{1}{1 - \cot \theta}$

34. $1 - \dfrac{\sin^2 \theta}{1 + \cos \theta} = \cos \theta$

35. $\dfrac{1 - \sin \theta}{1 + \sin \theta} = (\sec \theta - \tan \theta)^2$

36. $\dfrac{1 - \cos \theta}{1 + \cos \theta} = (\csc \theta - \cot \theta)^2$

37. $\dfrac{\cos \theta}{1 - \tan \theta} + \dfrac{\sin \theta}{1 - \cot \theta} = \sin \theta + \cos \theta$

38. $\dfrac{\cot \theta}{1 - \tan \theta} + \dfrac{\tan \theta}{1 - \cot \theta} = 1 + \tan \theta + \cot \theta$

39. $\tan \theta + \dfrac{\cos \theta}{1 + \sin \theta} = \sec \theta$

40. $\dfrac{\sin \theta \cos \theta}{\cos^2 \theta - \sin^2 \theta} = \dfrac{\tan \theta}{1 - \tan^2 \theta}$

41. $\dfrac{\tan \theta + \sec \theta - 1}{\tan \theta - \sec \theta + 1} = \tan \theta + \sec \theta$

42. $\dfrac{\sin \theta - \cos \theta + 1}{\sin \theta + \cos \theta - 1} = \dfrac{\sin \theta + 1}{\cos \theta}$

43. $\dfrac{\tan \theta - \cot \theta}{\tan \theta + \cot \theta} = \sin^2 \theta - \cos^2 \theta$

44. $\dfrac{\sec \theta - \cos \theta}{\sec \theta + \cos \theta} = \dfrac{\sin^2 \theta}{1 + \cos^2 \theta}$

45. $\dfrac{\tan \theta - \cot \theta}{\tan \theta + \cot \theta} = 2 \sin^2 \theta - 1$

46. $\dfrac{\tan \theta - \cot \theta}{\tan \theta + \cot \theta} = 1 - 2 \cos^2 \theta$

47. $\dfrac{\sec\theta + \tan\theta}{\cot\theta + \cos\theta} = \tan\theta\sec\theta$

48. $\dfrac{\sec\theta}{1 + \sec\theta} = \dfrac{1 - \cos\theta}{\sin^2\theta}$

49. $\dfrac{1 - \tan^2\theta}{1 + \tan^2\theta} = 2\cos^2\theta - 1$

50. $\dfrac{1 - \cot^2\theta}{1 + \cot^2\theta} = 1 - 2\cos^2\theta$

51. $\dfrac{\sec\theta - \csc\theta}{\sec\theta\csc\theta} = \sin\theta - \cos\theta$

52. $\dfrac{\sin^2\theta - \tan\theta}{\cos^2\theta - \cot\theta} = \tan^2\theta$

53. $\sec\theta - \cos\theta = \sin\theta\tan\theta$

54. $\tan\theta + \cot\theta = \sec\theta\csc\theta$

55. $\dfrac{1}{1 - \sin\theta} + \dfrac{1}{1 + \sin\theta} = 2\sec^2\theta$

56. $\dfrac{1 + \sin\theta}{1 - \sin\theta} - \dfrac{1 - \sin\theta}{1 + \sin\theta} = 4\tan\theta\sec\theta$

57. $\dfrac{\sec\theta}{1 - \sin\theta} = \dfrac{1 + \sin\theta}{\cos^3\theta}$

58. $\dfrac{1 - \sin\theta}{1 + \sin\theta} = (\sec\theta - \tan\theta)^2$

59. $\dfrac{(\sec\theta - \tan\theta)^2 + 1}{\csc\theta(\sec\theta - \tan\theta)} = 2\tan\theta$

60. $\dfrac{\sec^2\theta - \tan^2\theta + \tan\theta}{\sec\theta} = \sin\theta + \cos\theta$

61. $\dfrac{\sin\theta + \cos\theta}{\cos\theta} - \dfrac{\sin\theta - \cos\theta}{\sin\theta} = \sec\theta\csc\theta$

62. $\dfrac{\sin\theta + \cos\theta}{\sin\theta} - \dfrac{\cos\theta - \sin\theta}{\cos\theta} = \sec\theta\csc\theta$

63. $\dfrac{\sin^3\theta + \cos^3\theta}{\sin\theta + \cos\theta} = 1 - \sin\theta\cos\theta$

64. $\dfrac{\sin^3\theta + \cos^3\theta}{1 - 2\cos^2\theta} = \dfrac{\sec\theta - \sin\theta}{\tan\theta - 1}$

65. $\dfrac{\cos^2\theta - \sin^2\theta}{1 - \tan^2\theta} = \cos^2\theta$

66. $\dfrac{\cos\theta + \sin\theta - \sin^3\theta}{\sin\theta} = \cot\theta + \cos^2\theta$

67. $\dfrac{(2\cos^2\theta - 1)^2}{\cos^4\theta - \sin^4\theta} = 1 - 2\sin^2\theta$

68. $\dfrac{1 - 2\cos^2\theta}{\sin\theta\cos\theta} = \tan\theta - \cot\theta$

69. $\dfrac{1 + \sin\theta + \cos\theta}{1 + \sin\theta - \cos\theta} = \dfrac{1 + \cos\theta}{\sin\theta}$

70. $\dfrac{1 + \cos\theta + \sin\theta}{1 + \cos\theta - \sin\theta} = \sec\theta + \tan\theta$

71. $(a\sin\theta + b\cos\theta)^2 + (a\cos\theta - b\sin\theta)^2 = a^2 + b^2$

72. $(2a\sin\theta\cos\theta)^2 + a^2(\cos^2\theta - \sin^2\theta)^2 = a^2$

73. $\dfrac{\tan\alpha + \tan\beta}{\cot\alpha + \cot\beta} = \tan\alpha\tan\beta$

74. $(\tan\alpha + \tan\beta)(1 - \cot\alpha\cot\beta) + (\cot\alpha + \cot\beta)(1 - \tan\alpha\tan\beta) = 0$

75. $(\sin\alpha + \cos\beta)^2 + (\cos\beta + \sin\alpha)(\cos\beta - \sin\alpha) = 2\cos\beta(\sin\alpha + \cos\beta)$

76. $(\sin\alpha - \cos\beta)^2 + (\cos\beta + \sin\alpha)(\cos\beta - \sin\alpha) = -2\cos\beta(\sin\alpha - \cos\beta)$

77. $\ln|\sec\theta| = -\ln|\cos\theta|$

78. $\ln|\tan\theta| = \ln|\sin\theta| - \ln|\cos\theta|$

79. $\ln|1 + \cos\theta| + \ln|1 - \cos\theta| = 2\ln|\sin\theta|$

80. $\ln|\sec\theta + \tan\theta| + \ln|\sec\theta - \tan\theta| = 0$

81. Write a few paragraphs outlining your strategy for establishing identities.

3.2

Sum and Difference Formulas

In this section, we continue our derivation of trigonometric identities by obtaining formulas that involve the sum or difference of two angles, such as $\cos(\alpha + \beta)$, $\cos(\alpha - \beta)$, or $\sin(\alpha + \beta)$. These formulas are referred to as the **sum and difference formulas.** We begin with the formulas for $\cos(\alpha + \beta)$ and $\cos(\alpha - \beta)$.

Theorem
Sum and Difference Formulas
for Cosines

$$\cos(\alpha + \beta) = \cos\alpha\cos\beta - \sin\alpha\sin\beta \qquad (1)$$

$$\cos(\alpha - \beta) = \cos\alpha\cos\beta + \sin\alpha\sin\beta \qquad (2)$$

In words, formula (1) states that the cosine of the sum of two angles equals the cosine of the first times the cosine of the second minus the sine of the first times the sine of the second.

FIGURE 1

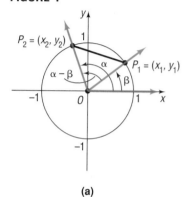

(a)

(b)

Proof   We shall prove formula (2) first. Although this formula is true for all numbers α and β, we shall assume in our proof that $0 < \beta < \alpha < 2\pi$. We begin with a circle with center at the origin $(0, 0)$ and radius of 1 unit (the unit circle), and we place the angles α and β in standard position, as shown in Figure 1(a). The point $P_1 = (x_1, y_1)$ lies on the terminal side of β, and the point $P_2 = (x_2, y_2)$ lies on the terminal side of α.

Now, place the angle $\alpha - \beta$ in standard position, as shown in Figure 1(b), where the point A has coordinates $(1, 0)$ and the point $P_3 = (x_3, y_3)$ is on the terminal side of the angle $\alpha - \beta$.

Looking at triangle OP_1P_2 in Figure 1(a) and triangle OAP_3 in Figure 1(b), we see that these triangles are congruent. (Do you see why? Two sides and the included angle, $\alpha - \beta$, are equal.) Hence, the unknown side of each triangle must be equal; that is,

$$d(A, P_3) = d(P_1, P_2)$$

Using the distance formula, we find that

$$\sqrt{(x_3 - 1)^2 + y_3^2} = \sqrt{(x_2 - x_1)^2 + (y_2 - y_1)^2}$$
$$(x_3 - 1)^2 + y_3^2 = (x_2 - x_1)^2 + (y_2 - y_1)^2 \quad \text{Square each side.}$$
$$x_3^2 - 2x_3 + 1 + y_3^2 = x_2^2 - 2x_1x_2 + x_1^2 + y_2^2 - 2y_1y_2 + y_1^2 \quad (3)$$

Since $P_1 = (x_1, y_1)$, $P_2 = (x_2, y_2)$, and $P_3 = (x_3, y_3)$ are points on the unit circle $x^2 + y^2 = 1$, it follows that

$$x_1^2 + y_1^2 = 1 \qquad x_2^2 + y_2^2 = 1 \qquad x_3^2 + y_3^2 = 1$$

Consequently, equation (3) simplifies to

$$x_3^2 + y_3^2 - 2x_3 + 1 = (x_2^2 + y_2^2) + (x_1^2 + y_1^2) - 2x_1x_2 - 2y_1y_2$$
$$2 - 2x_3 = 2 - 2x_1x_2 - 2y_1y_2$$
$$x_3 = x_1x_2 + y_1y_2 \quad (4)$$

But $P_1 = (x_1, y_1)$ is on the terminal side of angle β and is a distance of 1 unit from the origin. Thus,

$$\sin \beta = \frac{y_1}{1} = y_1 \qquad \cos \beta = \frac{x_1}{1} = x_1 \quad (5)$$

Similarly,

$$\sin \alpha = \frac{y_2}{1} = y_2 \qquad \cos \alpha = \frac{x_2}{1} = x_2 \qquad \cos(\alpha - \beta) = \frac{x_3}{1} = x_3 \quad (6)$$

Using equations (5) and (6) in equation (4), we get

$$\cos(\alpha - \beta) = \cos \alpha \cos \beta + \sin \alpha \sin \beta$$

which is formula (2).

The proof of formula (1) follows from formula (2). We use the fact that $\alpha + \beta = \alpha - (-\beta)$. Then

$$\cos(\alpha + \beta) = \cos[\alpha - (-\beta)]$$
$$= \cos \alpha \cos(-\beta) + \sin \alpha \sin(-\beta) \quad \text{Use formula (2).}$$
$$= \cos \alpha \cos \beta - \sin \alpha \sin \beta \quad \text{Even–odd identities} \quad \blacksquare$$

One use of formulas (1) and (2) is to obtain the exact value of the cosine of an angle that can be expressed as the sum or difference of angles whose sine and cosine are known exactly.

E X A M P L E   1   *Using the Sum Formula to Find Exact Values*

Find the exact value of $\cos 75°$.

Solution   Since $75° = 45° + 30°$, we use formula (2) to obtain

$$\cos 75° = \cos(45° + 30°) = \cos 45° \cos 30° - \sin 45° \sin 30°$$
$$\uparrow$$
$$\text{Formula (1)}$$

$$= \frac{\sqrt{2}}{2} \cdot \frac{\sqrt{3}}{2} - \frac{\sqrt{2}}{2} \cdot \frac{1}{2} = \frac{1}{4}(\sqrt{6} - \sqrt{2}) \qquad \blacksquare$$

E X A M P L E   2   *Using the Difference Formula to Find Exact Values*

Find the exact value of $\cos(\pi/12)$.

Solution

$$\cos \frac{\pi}{12} = \cos\left(\frac{3\pi}{12} - \frac{2\pi}{12}\right) = \cos\left(\frac{\pi}{4} - \frac{\pi}{6}\right)$$

$$= \cos \frac{\pi}{4} \cos \frac{\pi}{6} + \sin \frac{\pi}{4} \sin \frac{\pi}{6} \quad \text{Use formula (2).}$$

$$= \frac{\sqrt{2}}{2} \cdot \frac{\sqrt{3}}{2} + \frac{\sqrt{2}}{2} \cdot \frac{1}{2} = \frac{1}{4}(\sqrt{6} + \sqrt{2}) \qquad \blacksquare$$

$\blacksquare$ Now work Problem 3.

Another use of formulas (1) and (2) is to establish other identities. One important pair of identities is given next:

$$\cos\left(\frac{\pi}{2} - \theta\right) = \sin \theta \qquad\qquad (7a)$$

$$\sin\left(\frac{\pi}{2} - \theta\right) = \cos \theta \qquad\qquad (7b)$$

Proof   To prove formula (7a), we use the formula for $\cos(\alpha - \beta)$ with $\alpha = \pi/2$ and $\beta = \theta$:

$$\cos\left(\frac{\pi}{2} - \theta\right) = \cos \frac{\pi}{2} \cos \theta + \sin \frac{\pi}{2} \sin \theta$$

$$= 0 \cdot \cos \theta + 1 \cdot \sin \theta$$

$$= \sin \theta$$

To prove formula (7b), we make use of the identity (7a) just established:

$$\sin\left(\frac{\pi}{2} - \theta\right) = \cos\left[\frac{\pi}{2} - \left(\frac{\pi}{2} - \theta\right)\right] = \cos\theta$$

Use (7a).

Formulas (7a) and (7b) should look familiar. They are the basis for the theorem stated in Chapter 2: Cofunctions of complementary angles are equal.

Furthermore, since $\cos(\pi/2 - \theta) = \cos(\theta - \pi/2)$, it follows from formula (7a) that $\cos(\theta - \pi/2) = \sin\theta$. Thus, the graphs of $y = \sin x$ and $y = \cos(x - \pi/2)$ are identical, a fact we will use later in Section 4.5.

Now work Problem 29.

Formulas for $\sin(\alpha + \beta)$ and $\sin(\alpha - \beta)$

Having established the identities in formulas (7a) and (7b), we now can derive the sum and difference formulas for $\sin(\alpha + \beta)$ and $\sin(\alpha - \beta)$.

Proof
$$\sin(\alpha + \beta) = \cos\left[\frac{\pi}{2} - (\alpha + \beta)\right] \qquad \text{Formula (7a)}$$
$$= \cos\left[\left(\frac{\pi}{2} - \alpha\right) - \beta\right]$$
$$= \cos\left(\frac{\pi}{2} - \alpha\right)\cos\beta + \sin\left(\frac{\pi}{2} - \alpha\right)\sin\beta \quad \text{Formula (2)}$$
$$= \sin\alpha\cos\beta + \cos\alpha\sin\beta \qquad \text{Formulas (7a) and (7b)}$$

$$\sin(\alpha - \beta) = \sin[\alpha + (-\beta)]$$
$$= \sin\alpha\cos(-\beta) + \cos\alpha\sin(-\beta)$$
$$= \sin\alpha\cos\beta + \cos\alpha(-\sin\beta) \qquad \text{Even–odd identities}$$
$$= \sin\alpha\cos\beta - \cos\alpha\sin\beta$$

Thus,

Theorem
Sum and Difference Formulas
for Sines

$$\sin(\alpha + \beta) = \sin\alpha\cos\beta + \cos\alpha\sin\beta \qquad (8)$$
$$\sin(\alpha - \beta) = \sin\alpha\cos\beta - \cos\alpha\sin\beta \qquad (9)$$

In words, formula (8) states that the sine of the sum of two angles equals the sine of the first times the cosine of the second plus the cosine of the first times the sine of the second.

EXAMPLE 3   *Using the Sum Formula to Find Exact Values*
Find the exact value of $\sin(7\pi/12)$.

Solution
$$\sin\frac{7\pi}{12} = \sin\left(\frac{3\pi}{12} + \frac{4\pi}{12}\right) = \sin\left(\frac{\pi}{4} + \frac{\pi}{3}\right)$$

$$= \sin \frac{\pi}{4} \cos \frac{\pi}{3} + \cos \frac{\pi}{4} \sin \frac{\pi}{3} \quad \text{Formula (8)}$$

$$= \frac{\sqrt{2}}{2} \cdot \frac{1}{2} + \frac{\sqrt{2}}{2} \cdot \frac{\sqrt{3}}{2} = \frac{1}{4}(\sqrt{2} + \sqrt{6}) \qquad \blacksquare$$

E X A M P L E   4   *Using the Difference Formula to Find Exact Values*

Find the exact value of sin 165°.

Solution
$$\sin 165° = \sin(225° - 60°)$$
$$= \sin 225° \cos 60° - \cos 225° \sin 60° \quad \text{Formula (9)}$$
$$= \frac{-\sqrt{2}}{2} \cdot \frac{1}{2} - \frac{-\sqrt{2}}{2} \cdot \frac{\sqrt{3}}{2} = \frac{1}{4}(\sqrt{6} - \sqrt{2}) \qquad \blacksquare$$

■ Now work Problem 9.

E X A M P L E   5   *Using the Difference Formula to Find Exact Values*

Find the exact value of cos 80° cos 20° + sin 80° sin 20°.

Solution   The form of the expression cos 80° cos 20° + sin 80° sin 20° is that of the right side of the formula for $\cos(\alpha - \beta)$ with $\alpha = 80°$ and $\beta = 20°$. Thus,

$$\cos 80° \cos 20° + \sin 80° \sin 20° = \cos(80° - 20°) = \cos 60° = \frac{1}{2} \qquad \blacksquare$$

E X A M P L E   6   *Using the Sum Formula to Find Exact Values*

Find the exact value of $\sin \frac{\pi}{9} \cos \frac{\pi}{18} + \cos \frac{\pi}{9} \sin \frac{\pi}{18}$.

Solution   We observe that the form of the given expression is that of the right side of the formula for $\sin(\alpha + \beta)$ with $\alpha = \pi/9$ and $\beta = \pi/18$. Thus,

$$\sin \frac{\pi}{9} \cos \frac{\pi}{18} + \cos \frac{\pi}{9} \sin \frac{\pi}{18} = \sin\left(\frac{\pi}{9} + \frac{\pi}{18}\right) = \sin \frac{3\pi}{18} = \sin \frac{\pi}{6} = \frac{1}{2} \qquad \blacksquare$$

■ Now work Problem 19.

E X A M P L E   7   *Finding Exact Values*

If it is known that $\sin \alpha = \frac{4}{5}$, $\pi/2 < \alpha < \pi$, and that $\sin \beta = -2/\sqrt{5} = -2\sqrt{5}/5$, $\pi < \beta < 3\pi/2$, find the exact value of:

(a) cos α   (b) cos β   (c) cos(α + β)   (d) sin(α + β)

Solution   (a) See Figure 2. Since $(x, 4)$ is on the circle $x^2 + y^2 = 25$ and is in Quadrant II, we have

$$x^2 + 4^2 = 25, \quad x < 0$$
$$x^2 = 25 - 16 = 9$$
$$x = -3$$

Thus,

$$\cos \alpha = \frac{x}{r} = -\frac{3}{5}$$

FIGURE 2
Given $\sin \alpha = \frac{4}{5}$, $\pi/2 < \alpha < \pi$

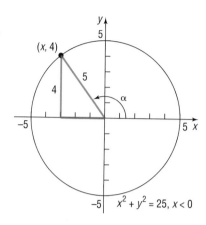

FIGURE 3
Given $\sin \beta = -2/\sqrt{5}$, $\pi < \beta < 3\pi/2$

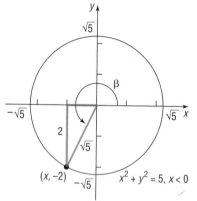

(b) See Figure 3. Since $(x, -2)$ is on the circle $x^2 + y^2 = 5$ and is in Quadrant III, we have

$$x^2 + (-2)^2 = 5, \qquad x < 0$$
$$x^2 = 5 - 4 = 1$$
$$x = -1$$

Thus,

$$\cos \beta = \frac{x}{r} = -\frac{1}{\sqrt{5}} = -\frac{\sqrt{5}}{5}$$

(c) Using the results found in parts (a) and (b) and formula (1), we have

$$\cos(\alpha + \beta) = \cos \alpha \cos \beta - \sin \alpha \sin \beta$$
$$= -\frac{3}{5}\left(-\frac{\sqrt{5}}{5}\right) - \frac{4}{5}\left(-\frac{2\sqrt{5}}{5}\right) = \frac{11\sqrt{5}}{25}$$

(d)
$$\sin(\alpha + \beta) = \sin \alpha \cos \beta + \cos \alpha \sin \beta$$
$$= \frac{4}{5}\left(-\frac{\sqrt{5}}{5}\right) + \left(-\frac{3}{5}\right)\left(-\frac{2\sqrt{5}}{5}\right) = \frac{2\sqrt{5}}{25}$$

∎

■ Now work Problem 23, parts (a), (b), and (c).

Formulas for $\tan(\alpha + \beta)$ and $\tan(\alpha - \beta)$

We use the identity $\tan \theta = (\sin \theta)/(\cos \theta)$ and the sum formulas for $\sin(\alpha + \beta)$ and $\cos(\alpha + \beta)$ to derive a formula for $\tan(\alpha + \beta)$.

Proof
$$\tan(\alpha + \beta) = \frac{\sin(\alpha + \beta)}{\cos(\alpha + \beta)} = \frac{\sin \alpha \cos \beta + \cos \alpha \sin \beta}{\cos \alpha \cos \beta - \sin \alpha \sin \beta}$$

Now we divide the numerator and denominator by $\cos \alpha \cos \beta$:

$$\tan(\alpha + \beta) = \frac{\dfrac{\sin \alpha \cos \beta + \cos \alpha \sin \beta}{\cos \alpha \cos \beta}}{\dfrac{\cos \alpha \cos \beta - \sin \alpha \sin \beta}{\cos \alpha \cos \beta}} = \frac{\dfrac{\sin \alpha \cos \beta}{\cos \alpha \cos \beta} + \dfrac{\cos \alpha \sin \beta}{\cos \alpha \cos \beta}}{\dfrac{\cos \alpha \cos \beta}{\cos \alpha \cos \beta} - \dfrac{\sin \alpha \sin \beta}{\cos \alpha \cos \beta}}$$

$$= \frac{\dfrac{\sin \alpha}{\cos \alpha} + \dfrac{\sin \beta}{\cos \beta}}{1 - \dfrac{\sin \alpha \sin \beta}{\cos \alpha \cos \beta}} = \frac{\tan \alpha + \tan \beta}{1 - \tan \alpha \tan \beta}$$

We use the sum formula for $\tan(\alpha + \beta)$ to get the difference formula:

$$\tan(\alpha - \beta) = \tan[\alpha + (-\beta)] = \frac{\tan \alpha + \tan(-\beta)}{1 - \tan \alpha \tan(-\beta)} = \frac{\tan \alpha - \tan \beta}{1 + \tan \alpha \tan \beta} \quad \blacksquare$$

Thus, we have proved the following results:

Theorem
Sum and Difference Formulas
for Tangents

$$\tan(\alpha + \beta) = \frac{\tan \alpha + \tan \beta}{1 - \tan \alpha \tan \beta} \quad (10)$$

$$\tan(\alpha - \beta) = \frac{\tan \alpha - \tan \beta}{1 + \tan \alpha \tan \beta} \quad (11)$$

In words, formula (10) states that the tangent of the sum of two angles equals the tangent of the first plus the tangent of the second divided by 1 minus their product.

E X A M P L E   8   *Establishing an Identity*

Prove the identity: $\tan(\theta + \pi) = \tan \theta$

Solution
$$\tan(\theta + \pi) = \frac{\tan \theta + \tan \pi}{1 - \tan \theta \tan \pi} = \frac{\tan \theta + 0}{1 - \tan \theta \cdot 0} = \tan \theta \quad \blacksquare$$

The result obtained in Example 8 verifies that the tangent function is periodic with period π, a fact we mentioned earlier.

Warning: Be careful when using formulas (10) and (11). These formulas can be used only for angles α and β for which $\tan \alpha$ and $\tan \beta$ are defined, that is, all angles except odd multiples of $\pi/2$.

E X A M P L E   9   *Establishing an Identity*

Prove the identity: $\tan\left(\theta + \dfrac{\pi}{2}\right) = -\cot \theta$

Solution   We cannot use formula (10), since $\tan(\pi/2)$ is not defined. Instead, we proceed as follows:

$$\tan\left(\theta + \frac{\pi}{2}\right) = \frac{\sin\left(\theta + \dfrac{\pi}{2}\right)}{\cos\left(\theta + \dfrac{\pi}{2}\right)} = \frac{\sin \theta \cos \dfrac{\pi}{2} + \cos \theta \sin \dfrac{\pi}{2}}{\cos \theta \cos \dfrac{\pi}{2} - \sin \theta \sin \dfrac{\pi}{2}}$$

$$= \frac{(\sin \theta)(0) + (\cos \theta)(1)}{(\cos \theta)(0) - (\sin \theta)(1)} = \frac{\cos \theta}{-\sin \theta} = -\cot \theta \quad \blacksquare$$

E X A M P L E   1 0   *Finding Exact Values*

Find the exact value of $\sin(\cos^{-1} \frac{1}{2} + \sin^{-1} \frac{3}{5})$.

Solution   Let $\alpha = \cos^{-1} \frac{1}{2}$ and $\beta = \sin^{-1} \frac{3}{5}$. Then

$$\cos \alpha = \tfrac{1}{2}, \quad 0 \le \alpha \le \pi, \quad \text{and} \quad \sin \beta = \tfrac{3}{5}, \quad -\frac{\pi}{2} \le \beta \le \frac{\pi}{2}$$

FIGURE 4

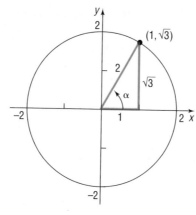

(a) $\cos \alpha = \frac{1}{2}, 0 \le \alpha \le \pi$

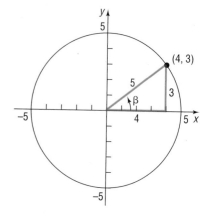

(b) $\sin \beta = \frac{3}{5}, -\pi/2 \le \beta \le \pi/2$

Based on Figure 4, we obtain $\sin \alpha = \sqrt{3}/2$ and $\cos \beta = \frac{4}{5}$. Thus,

$$\sin\left(\cos^{-1}\frac{1}{2} + \sin^{-1}\frac{3}{5}\right) = \sin(\alpha + \beta) = \sin \alpha \cos \beta + \cos \alpha \sin \beta$$

$$= \frac{\sqrt{3}}{2} \cdot \frac{4}{5} + \frac{1}{2} \cdot \frac{3}{5} = \frac{4\sqrt{3} + 3}{10}$$

■ Now work Problem 57.

E X A M P L E   1 1

Writing a Trigonometric Expression as an Algebraic Expression

Write $\sin(\sin^{-1} u + \cos^{-1} v)$ as an algebraic expression containing u and v (that is, without any trigonometric functions).

Solution   Let $\alpha = \sin^{-1} u$ and $\beta = \cos^{-1} v$. Then

$$\sin \alpha = u, \quad -\frac{\pi}{2} \le \alpha \le \frac{\pi}{2} \quad \text{and} \quad \cos \beta = v, \quad 0 \le \beta \le \pi$$

Since $-\pi/2 \le \alpha \le \pi/2$, we know that $\cos \alpha \ge 0$. As a result,

$$\cos \alpha = \sqrt{1 - \sin^2 \alpha} = \sqrt{1 - u^2}$$

Similarly, since $0 \le \beta \le \pi$, we know that $\sin \beta \ge 0$. Thus,

$$\sin \beta = \sqrt{1 - \cos^2 \beta} = \sqrt{1 - v^2}$$

Now,

$$\sin(\sin^{-1} u + \cos^{-1} v) = \sin(\alpha + \beta) = \sin \alpha \cos \beta + \cos \alpha \sin \beta$$
$$= uv + \sqrt{1 - u^2}\, \sqrt{1 - v^2}$$ ■

Summary

The following box summarizes the sum and difference formulas:

Sum and Difference Formulas

$$\cos(\alpha + \beta) = \cos \alpha \cos \beta - \sin \alpha \sin \beta$$
$$\cos(\alpha - \beta) = \cos \alpha \cos \beta + \sin \alpha \sin \beta$$

$$\sin(\alpha + \beta) = \sin \alpha \cos \beta + \cos \alpha \sin \beta$$
$$\sin(\alpha - \beta) = \sin \alpha \cos \beta - \cos \alpha \sin \beta$$
$$\tan(\alpha + \beta) = \frac{\tan \alpha + \tan \beta}{1 - \tan \alpha \tan \beta}$$
$$\tan(\alpha - \beta) = \frac{\tan \alpha - \tan \beta}{1 + \tan \alpha \tan \beta}$$

3.2

Exercise 3.2

In Problems 1–12, find the exact value of each trigonometric function.

1. $\sin \dfrac{5\pi}{12}$ 　　**2.** $\sin \dfrac{\pi}{12}$ 　　**3.** $\cos \dfrac{7\pi}{12}$ 　　**4.** $\tan \dfrac{7\pi}{12}$ 　　**5.** $\cos 165°$ 　　**6.** $\sin 105°$

7. $\tan 15°$ 　　**8.** $\tan 195°$ 　　**9.** $\sin \dfrac{17\pi}{12}$ 　　**10.** $\tan \dfrac{19\pi}{12}$ 　　**11.** $\sec\left(-\dfrac{\pi}{12}\right)$ 　　**12.** $\cot\left(-\dfrac{5\pi}{12}\right)$

In Problems 13–22, find the exact value of each expression.

13. $\sin 20° \cos 10° + \cos 20° \sin 10°$

14. $\sin 20° \cos 80° - \cos 20° \sin 80°$

15. $\cos 70° \cos 20° - \sin 70° \sin 20°$

16. $\cos 40° \cos 10° + \sin 40° \sin 10°$

17. $\dfrac{\tan 20° + \tan 25°}{1 - \tan 20° \tan 25°}$

18. $\dfrac{\tan 40° - \tan 10°}{1 + \tan 40° \tan 10°}$

19. $\sin \dfrac{\pi}{12} \cos \dfrac{7\pi}{12} - \cos \dfrac{\pi}{12} \sin \dfrac{7\pi}{12}$

20. $\cos \dfrac{5\pi}{12} \cos \dfrac{7\pi}{12} - \sin \dfrac{5\pi}{12} \sin \dfrac{7\pi}{12}$

21. $\sin \dfrac{\pi}{12} \cos \dfrac{5\pi}{12} - \sin \dfrac{5\pi}{12} \cos \dfrac{\pi}{12}$

22. $\sin \dfrac{\pi}{18} \cos \dfrac{5\pi}{18} + \cos \dfrac{\pi}{18} \sin \dfrac{5\pi}{18}$

In Problems 23–28, find the exact value of each of the following under the given conditions:

(a) $\sin(\alpha + \beta)$ 　　*(b)* $\cos(\alpha + \beta)$ 　　*(c)* $\sin(\alpha - \beta)$ 　　*(d)* $\tan(\alpha - \beta)$

23. $\sin \alpha = \frac{3}{5}, 0 < \alpha < \pi/2;$ 　 $\cos \beta = 2/\sqrt{5}, -\pi/2 < \beta < 0$

24. $\cos \alpha = 1/\sqrt{5}, 0 < \alpha < \pi/2;$ 　 $\sin \beta = -\frac{4}{5}, -\pi/2 < \beta < 0$

25. $\tan \alpha = -\frac{4}{3}, \pi/2 < \alpha < \pi;$ 　 $\cos \beta = \frac{1}{2}, 0 < \beta < \pi/2$

26. $\tan \alpha = \frac{5}{12}, \pi < \alpha < 3\pi/2;$ 　 $\sin \beta = -\frac{1}{2}, \pi < \beta < 3\pi/2$

27. $\sin \alpha = \frac{5}{13}, -3\pi/2 < \alpha < -\pi;$ 　 $\tan \beta = -\sqrt{3}, \pi/2 < \beta < \pi$

28. $\cos \alpha = \frac{1}{2}, -\pi/2 < \alpha < 0;$ 　 $\sin \beta = \frac{1}{3}, 0 < \beta < \pi/2$

In Problems 29–54, establish each identity.

29. $\sin\left(\dfrac{\pi}{2} + \theta\right) = \cos \theta$

30. $\cos\left(\dfrac{\pi}{2} + \theta\right) = -\sin \theta$

31. $\sin(\pi - \theta) = \sin \theta$

32. $\cos(\pi - \theta) = -\cos \theta$

33. $\sin(\pi + \theta) = -\sin \theta$

34. $\cos(\pi + \theta) = -\cos \theta$

35. $\tan(\pi - \theta) = -\tan \theta$

36. $\tan(2\pi - \theta) = -\tan \theta$

37. $\sin\left(\dfrac{3\pi}{2} + \theta\right) = -\cos \theta$

38. $\cos\left(\dfrac{3\pi}{2} + \theta\right) = \sin \theta$

39. $\sin(\alpha + \beta) + \sin(\alpha - \beta) = 2 \sin \alpha \cos \beta$

40. $\cos(\alpha + \beta) + \cos(\alpha - \beta) = 2 \cos \alpha \cos \beta$

41. $\dfrac{\sin(\alpha + \beta)}{\sin \alpha \cos \beta} = 1 + \cot \alpha \tan \beta$

42. $\dfrac{\sin(\alpha + \beta)}{\cos \alpha \cos \beta} = \tan \alpha + \tan \beta$

43. $\dfrac{\cos(\alpha + \beta)}{\cos \alpha \cos \beta} = 1 - \tan \alpha \tan \beta$

44. $\dfrac{\cos(\alpha - \beta)}{\sin \alpha \cos \beta} = \cot \alpha + \tan \beta$

45. $\dfrac{\sin(\alpha + \beta)}{\sin(\alpha - \beta)} = \dfrac{\tan \alpha + \tan \beta}{\tan \alpha - \tan \beta}$

46. $\dfrac{\cos(\alpha + \beta)}{\cos(\alpha - \beta)} = \dfrac{1 - \tan \alpha \tan \beta}{1 + \tan \alpha \tan \beta}$

47. $\cot(\alpha + \beta) = \dfrac{\cot \alpha \cot \beta - 1}{\cot \beta + \cot \alpha}$

48. $\cot(\alpha - \beta) = \dfrac{\cot \alpha \cot \beta + 1}{\cot \beta - \cot \alpha}$

49. $\sec(\alpha + \beta) = \dfrac{\csc \alpha \csc \beta}{\cot \alpha \cot \beta - 1}$

50. $\sec(\alpha - \beta) = \dfrac{\sec \alpha \sec \beta}{1 + \tan \alpha \tan \beta}$

51. $\sin(\alpha - \beta) \sin(\alpha + \beta) = \sin^2 \alpha - \sin^2 \beta$

52. $\cos(\alpha - \beta) \cos(\alpha + \beta) = \cos^2 \alpha - \sin^2 \beta$

53. $\sin(\theta + k\pi) = (-1)^k \cdot \sin \theta,\ k$ any integer

54. $\cos(\theta + k\pi) = (-1)^k \cdot \cos \theta,\ k$ any integer

In Problems 55–64, find the exact value for each expression.

55. $\sin(\sin^{-1} \tfrac{1}{2} + \cos^{-1} 0)$

56. $\sin\left(\sin^{-1} \dfrac{\sqrt{3}}{2} + \cos^{-1} 1\right)$

57. $\sin[\sin^{-1} \tfrac{3}{5} - \cos^{-1}(-\tfrac{4}{5})]$

58. $\sin[\sin^{-1}(-\tfrac{4}{5}) - \tan^{-1} \tfrac{3}{4}]$

59. $\cos(\tan^{-1} \tfrac{4}{3} + \cos^{-1} \tfrac{5}{13})$

60. $\sin[\tan^{-1} \tfrac{5}{12} - \sin^{-1}(-\tfrac{3}{5})]$

61. $\sec(\sin^{-1} \tfrac{5}{13} - \tan^{-1} \tfrac{3}{4})$

62. $\sec(\tan^{-1} \tfrac{4}{3} + \cot^{-1} \tfrac{5}{12})$

63. $\cot\left(\sec^{-1} \dfrac{5}{3} + \dfrac{\pi}{6}\right)$

64. $\cos\left(\dfrac{\pi}{4} - \csc^{-1} \dfrac{5}{3}\right)$

In Problems 65–70, write each trigonometric expression as an algebraic expression containing u and v.

65. $\cos(\cos^{-1} u + \sin^{-1} v)$

66. $\sin(\sin^{-1} u - \cos^{-1} v)$

67. $\sin(\tan^{-1} u - \sin^{-1} v)$

68. $\cos(\tan^{-1} u + \tan^{-1} v)$

69. $\tan(\sin^{-1} u - \cos^{-1} v)$

70. $\sec(\tan^{-1} u + \cos^{-1} v)$

71. *Calculus* Show that the difference quotient for $f(x) = \sin x$ is given by

$$\dfrac{f(x + h) - f(x)}{h} = \dfrac{\sin(x + h) - \sin x}{h} = \cos x \cdot \dfrac{\sin h}{h} - \sin x \cdot \dfrac{1 - \cos h}{h}$$

72. *Calculus* Show that the difference quotient for $f(x) = \cos x$ is given by

$$\dfrac{f(x + h) - f(x)}{h} = \dfrac{\cos(x + h) - \cos x}{h} = -\sin x \cdot \dfrac{\sin h}{h} - \cos x \cdot \dfrac{1 - \cos h}{h}$$

73. Show that $\sin(\sin^{-1} u + \cos^{-1} u) = 1$.

74. Show that $\cos(\sin^{-1} u + \cos^{-1} u) = 0$.

75. Explain why formula (11) cannot be used to show that

$$\tan\left(\dfrac{\pi}{2} - \theta\right) = \cot \theta$$

Establish this identity by using formulas (7a) and (7b).

76. If $\tan \alpha = x + 1$ and $\tan \beta = x - 1$, show that $2 \cot(\alpha - \beta) = x^2$.

77. *Geometry: Angle between Two Lines* Let L_1 and L_2 denote two nonvertical intersecting lines, and let θ denote the acute angle between L_1 and L_2 (see the figure). Show that

$$\tan \theta = \dfrac{m_2 - m_1}{1 + m_1 m_2}$$

where m_1 and m_2 are the slopes of L_1 and L_2, respectively. [*Hint:* Use the facts that $\tan \theta_1 = m_1$ and $\tan \theta_2 = m_2$.]

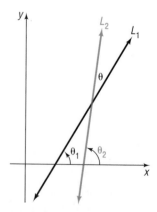

78. If $\alpha + \beta + \gamma = 180°$ and $\cot \theta = \cot \alpha + \cot \beta + \cot \gamma$, $0 < \theta < 90°$, show that

$$\sin^3 \theta = \sin(\alpha - \theta) \sin(\beta - \theta) \sin(\gamma - \theta)$$

79. Discuss the following derivation:

$$\tan(\theta + \frac{\pi}{2}) = \frac{\tan \theta + \tan(\pi/2)}{1 - \tan \theta \tan(\pi/2)} = \frac{\dfrac{\tan \theta}{\tan(\pi/2)} + 1}{\dfrac{1}{\tan(\pi/2)} - \tan \theta} = \frac{0 + 1}{0 - \tan \theta} = \frac{1}{-\tan \theta} = -\cot \theta$$

Can you justify each step?

3.3

Double-angle and Half-angle Formulas

In this section we derive formulas for $\sin 2\theta$, $\cos 2\theta$, $\sin \frac{1}{2} \theta$, and $\cos \frac{1}{2} \theta$ in terms of $\sin \theta$ and $\cos \theta$. They are easily derived using the sum formulas.

Double-angle Formulas

In the sum formulas for $\sin(\alpha + \beta)$ and $\cos(\alpha + \beta)$, let $\alpha = \beta = \theta$. Then

$$\sin(\alpha + \beta) = \sin \alpha \cos \beta + \cos \alpha \sin \beta$$
$$\sin(\theta + \theta) = \sin \theta \cos \theta + \cos \theta \sin \theta$$
$$\sin 2\theta = 2 \sin \theta \cos \theta \tag{1}$$

and

$$\cos(\alpha + \beta) = \cos \alpha \cos \beta - \sin \alpha \sin \beta$$
$$\cos(\theta + \theta) = \cos \theta \cos \theta - \sin \theta \sin \theta$$
$$\cos 2\theta = \cos^2 \theta - \sin^2 \theta \tag{2}$$

An application of the Pythagorean identity $\sin^2 \theta + \cos^2 \theta = 1$ results in two other ways to write formula (2) for $\cos 2\theta$:

$$\cos 2\theta = \cos^2 \theta - \sin^2 \theta = (1 - \sin^2 \theta) - \sin^2 \theta = 1 - 2 \sin^2 \theta$$

and

$$\cos 2\theta = \cos^2 \theta - \sin^2 \theta = \cos^2 \theta - (1 - \cos^2 \theta) = 2 \cos^2 \theta - 1$$

Thus, we have established the following **double-angle formulas:**

Theorem
Double-angle Formulas

$$\sin 2\theta = 2 \sin \theta \cos \theta \tag{3}$$
$$\cos 2\theta = \cos^2 \theta - \sin^2 \theta \tag{4a}$$
$$\cos 2\theta = 1 - 2 \sin^2 \theta \tag{4b}$$
$$\cos 2\theta = 2 \cos^2 \theta - 1 \tag{4c}$$

E X A M P L E   1

Finding Exact Values Using the Double-angle Formula

If $\sin \theta = \frac{3}{5}$, $\pi/2 < \theta < \pi$, find the exact value of:

(a) $\sin 2\theta$ (b) $\cos 2\theta$

Solution (a) Because $\sin 2\theta = 2 \sin \theta \cos \theta$ and we already know that $\sin \theta = \frac{3}{5}$, we only need to find $\cos \theta$ (see Figure 5). Since $\pi/2 < \theta < \pi$, it follows that $(x, 3)$ is on the circle $x^2 + y^2 = 25$ and is in Quadrant II. Thus,

FIGURE 5

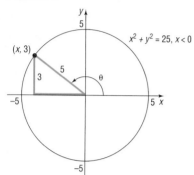

$$x^2 + 3^2 = 25, \quad x < 0$$
$$x^2 = 25 - 9 = 16$$
$$x = -4$$

Thus, $\cos\theta = -\frac{4}{5}$. Now we use formula (3) to obtain

$$\sin 2\theta = 2\sin\theta\cos\theta = 2(\tfrac{3}{5})(-\tfrac{4}{5}) = -\tfrac{24}{25}$$

(b) Because we are given $\sin\theta = \frac{3}{5}$, it is easiest to use formula (4b) to get $\cos 2\theta$:

$$\cos 2\theta = 1 - 2\sin^2\theta = 1 - 2(\tfrac{9}{25}) = 1 - \tfrac{18}{25} = \tfrac{7}{25} \qquad \blacksquare$$

Warning: In finding $\cos 2\theta$ in Example 1(b), we chose to use a version of the double-angle formula, formula (4b). Note that we are unable to use the Pythagorean identity $\cos 2\theta = \pm\sqrt{1 - \sin^2 2\theta}$, with $\sin 2\theta = -\frac{24}{25}$, because we have no way of knowing which sign to choose.

■ Now work Problems 1(a) and (b).

E X A M P L E   2 　　*Establishing Identities*

(a) Develop a formula for $\tan 2\theta$ in terms of $\tan\theta$.
(b) Develop a formula for $\sin 3\theta$ in terms of $\sin\theta$ and $\cos\theta$.

Solution 　(a) In the sum formula for $\tan(\alpha + \beta)$, let $\alpha = \beta = \theta$. Then

$$\tan(\alpha + \beta) = \frac{\tan\alpha + \tan\beta}{1 - \tan\alpha\tan\beta}$$

$$\tan(\theta + \theta) = \frac{\tan\theta + \tan\theta}{1 - \tan\theta\tan\theta}$$

$$\tan 2\theta = \frac{2\tan\theta}{1 - \tan^2\theta} \qquad (5)$$

(b) To get a formula for $\sin 3\theta$, we use the sum formula and write 3θ as $2\theta + \theta$.

$$\sin 3\theta = \sin(2\theta + \theta) = \sin 2\theta\cos\theta + \cos 2\theta\sin\theta$$

Now use the double-angle formulas to get

$$\sin 3\theta = (2\sin\theta\cos\theta)(\cos\theta) + (\cos^2\theta - \sin^2\theta)(\sin\theta)$$
$$= 2\sin\theta\cos^2\theta + \sin\theta\cos^2\theta - \sin^3\theta$$
$$= 3\sin\theta\cos^2\theta - \sin^3\theta \qquad \blacksquare$$

The formula obtained in Example 2(b) can also be written as

$$\sin 3\theta = 3\sin\theta\cos^2\theta - \sin^3\theta = 3\sin\theta(1 - \sin^2\theta) - \sin^3\theta$$
$$= 3\sin\theta - 4\sin^3\theta$$

That is, $\sin 3\theta$ is a third-degree polynomial in the variable $\sin\theta$. In fact, $\sin n\theta$, n a positive odd integer, can always be written as a polynomial of degree n in the variable $\sin\theta$.*

*Due to the work done by P. L. Chebyshëv, these polynomials are sometimes called *Chebyshëv polynomials.*

■ Now work Problem 47.

Other Variations of the Double-angle Formulas

By rearranging the double-angle formulas (4b) and (4c), we obtain other formulas that we will use a little later in this section.

If we solve formula (4b) for $\sin^2 \theta$, we get

$$\cos 2\theta = 1 - 2 \sin^2 \theta$$

$$2 \sin^2 \theta = 1 - \cos 2\theta$$

$$\sin^2 \theta = \frac{1 - \cos 2\theta}{2} \tag{6}$$

Similarly, we can solve for $\cos^2 \theta$ in formula (4c):

$$\cos 2\theta = 2 \cos^2 \theta - 1$$

$$2 \cos^2 \theta = 1 + \cos 2\theta$$

$$\cos^2 \theta = \frac{1 + \cos 2\theta}{2} \tag{7}$$

Formulas (6) and (7) can be used to develop a formula for $\tan^2 \theta$:

$$\tan^2 \theta = \frac{\sin^2 \theta}{\cos^2 \theta} = \frac{\dfrac{1 - \cos 2\theta}{2}}{\dfrac{1 + \cos 2\theta}{2}}$$

$$\tan^2 \theta = \frac{1 - \cos 2\theta}{1 + \cos 2\theta} \tag{8}$$

Formulas (6) through (8) do not have to be memorized since their derivations are so straightforward.

Formulas (6) and (7) are important in calculus. The next example illustrates a problem that arises in calculus requiring the use of formula (7).

E X A M P L E   3   *Establishing an Identity*

Write an equivalent expression for $\cos^4 \theta$ that does not involve any powers of sine or cosine greater than 1.

Solution   The idea here is to apply formula (7) twice:

$$\cos^4 \theta = (\cos^2 \theta)^2 = \left(\frac{1 + \cos 2\theta}{2} \right)^2 \qquad \text{Formula (7)}$$

$$= \frac{1}{4} (1 + 2 \cos 2\theta + \cos^2 2\theta)$$

$$= \frac{1}{4} + \frac{1}{2} \cos 2\theta + \frac{1}{4} \cos^2 2\theta$$

$$= \frac{1}{4} + \frac{1}{2} \cos 2\theta + \frac{1}{4} \left[\frac{1 + \cos 2(2\theta)}{2} \right] \quad \text{Formula (7)}$$

$$= \frac{1}{4} + \frac{1}{2} \cos 2\theta + \frac{1}{8} (1 + \cos 4\theta)$$

$$= \frac{3}{8} + \frac{1}{2} \cos 2\theta + \frac{1}{8} \cos 4\theta \qquad \blacksquare$$

■ Now work Problem 23.

Identities, such as the double-angle formulas, can sometimes be used to rewrite expressions in a more suitable form. Let's look at an example.

E X A M P L E   4   *Projectile Motion*

FIGURE 6

An object is propelled upward at an angle θ to the horizontal with an initial velocity of v_0 feet per second. See Figure 6. If air resistance is ignored, the horizontal distance R it travels, the **range,** is given by

$$R = \frac{1}{16} v_0^2 \sin \theta \cos \theta$$

(a) Show that $R = \frac{1}{32} v_0^2 \sin 2\theta$.

(b) Find the angle θ for which R is a maximum.

Solution   (a) We rewrite the given expression for the range using the double-angle formula $\sin 2\theta = 2 \sin \theta \cos \theta$. Then

$$R = \frac{1}{16} v_0^2 \sin \theta \cos \theta = \frac{1}{16} v_0^2 \frac{\sin 2\theta}{2} = \frac{1}{32} v_0^2 \sin 2\theta$$

(b) In this form, the largest value for the range R can be easily found. For a fixed initial speed v_0, the angle θ of inclination to the horizontal determines the value of R. Since the largest value of a sine function is 1, occurring when the argument is 90°, it follows that for maximum R we must have

$$2\theta = 90°$$
$$\theta = 45°$$

An inclination to the horizontal of 45° results in maximum range.   ■

Half-angle Formulas

Another important use of formulas (6) through (8) is to prove the **half-angle formulas.** In formulas (6) through (8), let $\theta = \alpha/2$. Then:

$$\sin^2 \frac{\alpha}{2} = \frac{1 - \cos \alpha}{2} \qquad \cos^2 \frac{\alpha}{2} = \frac{1 + \cos \alpha}{2}$$

$$\tan^2 \frac{\alpha}{2} = \frac{1 - \cos \alpha}{1 + \cos \alpha} \qquad (9)$$

In Problems 49–62, find the exact value of each expression.

49. $\sin(2 \sin^{-1} \frac{1}{2})$

50. $\sin[2 \sin^{-1}(\sqrt{3}/2)]$

51. $\cos(2 \sin^{-1} \frac{3}{5})$

52. $\cos(2 \cos^{-1} \frac{4}{5})$

53. $\tan[2 \cos^{-1}(-\frac{3}{5})]$

54. $\tan(2 \tan^{-1} \frac{3}{4})$

55. $\sin(2 \cos^{-1} \frac{4}{5})$

56. $\cos[2 \tan^{-1}(-\frac{4}{3})]$

57. $\sin^2(\frac{1}{2} \cos^{-1} \frac{3}{5})$

58. $\cos^2(\frac{1}{2} \sin^{-1} \frac{3}{5})$

59. $\sec(2 \tan^{-1} \frac{3}{4})$

60. $\csc[2 \sin^{-1} (-\frac{3}{5})]$

61. $\cot^2(\frac{1}{2} \tan^{-1} \frac{4}{3})$

62. $\cot^2(\frac{1}{2} \cos^{-1} \frac{5}{13})$

63. Find the value of the number C: $\frac{1}{2} \sin^2 x + C = -\frac{1}{4} \cos 2x$

64. Find the value of the number C: $\frac{1}{2} \cos^2 x + C = \frac{1}{4} \cos 2x$

65. *Area of an Isosceles Triangle* Show that the area A of an isosceles triangle whose equal sides are of length s and the angle between them is θ is

$$A = \frac{1}{2} s^2 \sin \theta$$

[*Hint:* See the illustration. The height h bisects the angle θ and is the perpendicular bisector of the base.]

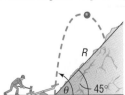

66. *Geometry* A rectangle is inscribed in a semicircle of radius 1. See the illustration.

(a) Express the area A of the rectangle as a function of the angle θ shown in the illustration.

(b) Show that $A = \sin 2\theta$.

(c) Find the angle θ that results in the largest area A.

(d) Find the dimensions of this largest rectangle.

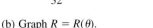

67. Graph $f(x) = \sin^2 x = (1 - \cos 2x)/2$ for $0 \le x \le 2\pi$ by using the ideas of shifts, compression, and the like.

68. Repeat Problem 67 for $g(x) = \cos^2 x$.

69. Use the fact that

$$\cos \frac{\pi}{12} = \frac{1}{4} (\sqrt{6} + \sqrt{2})$$

to find $\sin(\pi/24)$ and $\cos(\pi/24)$.

70. Show that

$$\sin \frac{\pi}{8} = \frac{\sqrt{2 - \sqrt{2}}}{2}$$

and use it to find $\sin(\pi/16)$ and $\cos(\pi/16)$.

71. Show that $\sin^3 \theta + \sin^3(\theta + 120°) + \sin^3(\theta + 240°) = -\frac{3}{4} \sin 3\theta$.

72. If $\tan \theta = a \tan(\theta/3)$, express $\tan(\theta/3)$ in terms of a.

In Problems 73 and 74, establish each identity.

73. $\ln|\sin \theta| = \frac{1}{2} (\ln|1 - \cos 2\theta| - \ln 2)$

74. $\ln|\cos \theta| = \frac{1}{2} (\ln|1 + \cos 2\theta| - \ln 2)$

75. *Projectile Motion* An object is propelled upward at an angle θ, $45° < \theta < 90°$, to the horizontal with an initial velocity of v_0 feet per second from the base of a plane that makes an angle of $45°$ with the horizontal. See the illustration. If air resistance is ignored, the distance R it travels up the inclined plane is given by

$$R = \frac{v_0^2 \sqrt{2}}{16} \cos \theta(\sin \theta - \cos \theta)$$

(a) Show that

$$R = \frac{v_0^2 \sqrt{2}}{32} (\sin 2\theta - \cos 2\theta - 1)$$

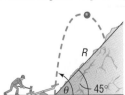

(b) Graph $R = R(\theta)$.

(c) What value of θ makes R the largest? (Use $v_0 = 32$ ft/sec).

76. *Sawtooth Curve* An oscilloscope often displays a sawtooth curve. This curve can be approximated by sinusoidal curves of varying periods and amplitudes. A first approximation to the sawtooth curve is given by

$$y = \frac{1}{2} \sin 2\pi x + \frac{1}{4} \sin 4\pi x.$$

Show that $y = \sin 2\pi x \cos^2 \pi x$.

77. Go to the library and research Chebyshëv polynomials. Write a report on your findings.

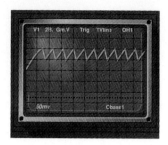

3.4

Product-to-Sum and Sum-to-Product Formulas

The sum and difference formulas can be used to derive formulas for writing the products of sines and/or cosines as sums or differences. These identities are usually called the **product-to-sum formulas.**

Theorem
Product-to-Sum Formulas

$$\sin \alpha \sin \beta = \tfrac{1}{2} [\cos(\alpha - \beta) - \cos(\alpha + \beta)] \tag{1}$$

$$\cos \alpha \cos \beta = \tfrac{1}{2} [\cos(\alpha - \beta) + \cos(\alpha + \beta)] \tag{2}$$

$$\sin \alpha \cos \beta = \tfrac{1}{2} [\sin(\alpha + \beta) + \sin(\alpha - \beta)] \tag{3}$$

■

These formulas do not have to be memorized. Instead, you should remember how they are derived. Then, when you want to use them, either look them up or derive them, as needed.

To derive formulas (1) and (2), write down the sum and difference formulas for the cosine:

$$\cos(\alpha - \beta) = \cos \alpha \cos \beta + \sin \alpha \sin \beta \tag{4}$$

$$\cos(\alpha + \beta) = \cos \alpha \cos \beta - \sin \alpha \sin \beta \tag{5}$$

Subtract equation (5) from equation (4) to get

$$\cos(\alpha - \beta) - \cos(\alpha + \beta) = 2 \sin \alpha \sin \beta$$

from which

$$\sin \alpha \sin \beta = \tfrac{1}{2} [\cos(\alpha - \beta) - \cos(\alpha + \beta)]$$

Now, add equations (4) and (5) to get

$$\cos(\alpha - \beta) + \cos(\alpha + \beta) = 2 \cos \alpha \cos \beta$$

from which

$$\cos \alpha \cos \beta = \tfrac{1}{2} [\cos(\alpha - \beta) + \cos(\alpha + \beta)]$$

To derive product-to-sum formula (3), use the sum and difference formulas for sine in a similar way. (You are asked to do this in Problem 41 at the end of this section.)

$\mathcal{M}$ISSION POSSIBLE

Chapter 3

$\mathbf{H}$OW FAR AND HOW HIGH DOES A BASEBALL NEED TO GO FOR AN OUT-OF-THE-PARK HOME RUN?

The sportscasters for the Cleveland Indians decided they had better be prepared for any eventuality when the new ballpark, Jacob's Field, opened in 1994. One eventuality they worried about was Albert Belle hitting a home run that went so high and so far that it left the ball park. What would they tell their listeners about the height and distance the ball went?

So they called in the Mission Possible team. The sportscasters wanted to know the distance from homeplate to the highest point of the stadium and the distance from ground level to the highest point of the stadium for every 5° from the 3rd base line around to the 1st base line. The problem was complicated by the fact that the distance from homeplate to the outfield fence varied from 325 feet to 410 feet. Furthermore, the height that would have to be cleared also varied, depending on where the ball was hit.

1. First, how many distances and heights do the sportscasters want?
2. To see one solution, use the distance from homeplate across 2nd base to the outfield fence, 410 feet. Using a transit, you find the angle of elevation from homeplate to the highest point of the stadium is 10° and the angle of elevation from the base of the outfield fence to the highest point of the stadium is 32.5°. Use this information to find the minimum distance from homeplate to the highest point of the stadium and the distance from ground level to the highest point of the stadium.
3. You tell the sportscasters that if they provide you the two angles of elevation for each 5° movement around the stadium, you can solve their problem. After receiving the two angles of elevation for each 5° movement around the stadium, you decide that it would be faster to develop a general formula, so that by inputting the two angles and the distance from homeplate to the fence, the distance from homeplate to the top of the roof and the height of the roof would be computed. Let α and β denote the angles of elevation of the top of the roof from homeplate and from the base of the outfield fence and let L be the distance from homeplate to the outfield fence. What are the correct formulas?
4. Compare your formulas with those of other groups. Are they all the same? Are they all equivalent? Which ones are simplest?
5. Now write each correct formula in the form shown below:

$$\text{height} = \frac{L \sin \alpha \sin \beta}{\sin(\beta - \alpha)} \qquad \text{distance} = \frac{L \sin \beta}{\sin(\beta - \alpha)}$$

6. What is the probable trajectory of a hit baseball on a windless day? How might the wind and other factors affect the path of the baseball?
7. If a player actually did hit an out-of-the-park home run, how would the distance the ball traveled compare to the figures you have been developing for the sportscasters? What other factors will add to the distance?
8. Could there be a longest home run? How would you measure it?

E X A M P L E   1   *Expressing Products as Sums*

Express each of the following products as a sum containing only sines or cosines:

(a) $\sin 6\theta \sin 4\theta$ (b) $\cos 3\theta \cos \theta$ (c) $\sin 3\theta \cos 5\theta$

Solution   (a) We use formula (1) to get

$$\sin 6\theta \sin 4\theta = \tfrac{1}{2}\left[\cos(6\theta - 4\theta) - \cos(6\theta + 4\theta)\right]$$
$$= \tfrac{1}{2}\left(\cos 2\theta - \cos 10\theta\right)$$

(b) We use formula (2) to get

$$\cos 3\theta \cos \theta = \tfrac{1}{2}\left[\cos(3\theta - \theta) + \cos(3\theta + \theta)\right]$$
$$= \tfrac{1}{2}\left(\cos 2\theta + \cos 4\theta\right)$$

(c) We use formula (3) to get

$$\sin 3\theta \cos 5\theta = \tfrac{1}{2}\left[\sin(3\theta + 5\theta) + \sin(3\theta - 5\theta)\right]$$
$$= \tfrac{1}{2}\left[\sin 8\theta + \sin(-2\theta)\right] = \tfrac{1}{2}\left(\sin 8\theta - \sin 2\theta\right) \qquad ■$$

■   Now work Problem 1.

The **sum-to-product formulas** are given next.

Theorem
Sum-to-Product Formulas

$$\sin \alpha + \sin \beta = 2 \sin \frac{\alpha + \beta}{2} \cos \frac{\alpha - \beta}{2} \qquad (6)$$

$$\sin \alpha - \sin \beta = 2 \sin \frac{\alpha - \beta}{2} \cos \frac{\alpha + \beta}{2} \qquad (7)$$

$$\cos \alpha + \cos \beta = 2 \cos \frac{\alpha + \beta}{2} \cos \frac{\alpha - \beta}{2} \qquad (8)$$

$$\cos \alpha - \cos \beta = -2 \sin \frac{\alpha + \beta}{2} \sin \frac{\alpha - \beta}{2} \qquad (9)$$

■

We shall derive formula (6) and leave the derivations of formulas (7) through (9) as exercises (see Problems 42 through 44).

Proof $2 \sin \dfrac{\alpha + \beta}{2} \cos \dfrac{\alpha - \beta}{2} = 2 \cdot \underset{\substack{\uparrow \\ \text{Product-to-sum formula (3)}}}{\dfrac{1}{2}} \left[\sin\left(\dfrac{\alpha + \beta}{2} + \dfrac{\alpha - \beta}{2}\right) + \sin\left(\dfrac{\alpha + \beta}{2} - \dfrac{\alpha - \beta}{2}\right) \right]$

$$= \sin \frac{2\alpha}{2} + \sin \frac{2\beta}{2} = \sin \alpha + \sin \beta \qquad ■$$

E X A M P L E   2   *Expressing Sums (or Differences) as a Product*

Express each sum or difference as a product of sines and/or cosines:

(a) $\sin 5\theta - \sin 3\theta$ (b) $\cos 3\theta + \cos 2\theta$

Solution (a) We use formula (7) to get

$$\sin 5\theta - \sin 3\theta = 2 \sin \frac{5\theta - 3\theta}{2} \cos \frac{5\theta + 3\theta}{2}$$

$$= 2 \sin \theta \cos 4\theta$$

(b) $\qquad \cos 3\theta + \cos 2\theta = 2 \cos \dfrac{3\theta + 2\theta}{2} \cos \dfrac{3\theta - 2\theta}{2}$ Formula (8)

$$= 2 \cos \frac{5\theta}{2} \cos \frac{\theta}{2}$$ ∎

■ Now work Problem 11.

3.4

Exercise 3.4

In Problems 1–10, express each product as a sum containing only sines or cosines

1. $\sin 4\theta \sin 2\theta$ **2.** $\cos 4\theta \cos 2\theta$ **3.** $\sin 4\theta \cos 2\theta$ **4.** $\sin 3\theta \sin 5\theta$

5. $\cos 3\theta \cos 5\theta$ **6.** $\sin 4\theta \cos 6\theta$ **7.** $\sin \theta \sin 2\theta$ **8.** $\cos 3\theta \cos 4\theta$

9. $\sin \dfrac{3\theta}{2} \cos \dfrac{\theta}{2}$ **10.** $\sin \dfrac{\theta}{2} \cos \dfrac{5\theta}{2}$

In Problems 11–18, express each sum or difference as a product of sines and/or cosines.

11. $\sin 4\theta - \sin 2\theta$ **12.** $\sin 4\theta + \sin 2\theta$ **13.** $\cos 2\theta + \cos 4\theta$ **14.** $\cos 5\theta - \cos 3\theta$

15. $\sin \theta + \sin 3\theta$ **16.** $\cos \theta + \cos 3\theta$ **17.** $\cos \dfrac{\theta}{2} - \cos \dfrac{3\theta}{2}$ **18.** $\sin \dfrac{\theta}{2} - \sin \dfrac{3\theta}{2}$

In Problems 19–36, establish each identity.

19. $\dfrac{\sin \theta + \sin 3\theta}{2 \sin 2\theta} = \cos \theta$ **20.** $\dfrac{\cos \theta + \cos 3\theta}{2 \cos 2\theta} = \cos \theta$ **21.** $\dfrac{\sin 4\theta + \sin 2\theta}{\cos 4\theta + \cos 2\theta} = \tan 3\theta$

22. $\dfrac{\cos \theta - \cos 3\theta}{\sin 3\theta - \sin \theta} = \tan 2\theta$ **23.** $\dfrac{\cos \theta - \cos 3\theta}{\sin \theta + \sin 3\theta} = \tan \theta$ **24.** $\dfrac{\cos \theta - \cos 5\theta}{\sin \theta + \sin 5\theta} = \tan 2\theta$

25. $\sin \theta(\sin \theta + \sin 3\theta) = \cos \theta(\cos \theta - \cos 3\theta)$ **26.** $\sin \theta(\sin 3\theta + \sin 5\theta) = \cos \theta(\cos 3\theta - \cos 5\theta)$

27. $\dfrac{\sin 4\theta + \sin 8\theta}{\cos 4\theta + \cos 8\theta} = \tan 6\theta$ **28.** $\dfrac{\sin 4\theta - \sin 8\theta}{\cos 4\theta - \cos 8\theta} = -\cot 6\theta$

29. $\dfrac{\sin 4\theta + \sin 8\theta}{\sin 4\theta - \sin 8\theta} = -\dfrac{\tan 6\theta}{\tan 2\theta}$ **30.** $\dfrac{\cos 4\theta - \cos 8\theta}{\cos 4\theta + \cos 8\theta} = \tan 2\theta \tan 6\theta$

31. $\dfrac{\sin \alpha + \sin \beta}{\sin \alpha - \sin \beta} = \tan \dfrac{\alpha + \beta}{2} \cot \dfrac{\alpha - \beta}{2}$ **32.** $\dfrac{\cos \alpha + \cos \beta}{\cos \alpha - \cos \beta} = -\cot \dfrac{\alpha + \beta}{2} \cot \dfrac{\alpha - \beta}{2}$

33. $\dfrac{\sin \alpha + \sin \beta}{\cos \alpha + \cos \beta} = \tan \dfrac{\alpha + \beta}{2}$ **34.** $\dfrac{\sin \alpha - \sin \beta}{\cos \alpha - \cos \beta} = -\cot \dfrac{\alpha + \beta}{2}$

35. $1 + \cos 2\theta + \cos 4\theta + \cos 6\theta = 4 \cos \theta \cos 2\theta \cos 3\theta$

36. $1 - \cos 2\theta + \cos 4\theta - \cos 6\theta = 4 \sin \theta \cos 2\theta \sin 3\theta$

37. *Touch-Tone Phones* On a Touch-Tone phone, each button produces a unique sound. The sound produced is the sum of two tones, given by

$$y = \sin 2\pi l t \quad \text{and} \quad y = \sin 2\pi h t$$

where l and h are the low and high frequencies (cycles per second) shown on the illustration. For example, if you touch 7, the low frequency is $l = 852$ cycles per second and the high frequency is $h = 1209$ cycles per second. The sound emitted by touching 7 is

$$y = \sin 2\pi(852)t + \sin 2\pi(1209)t$$

(a) Write this sound as a product of sines and/or cosines.
(b) Graph the sound emitted by touching 7.
(c) TRACE the graph to determine the maximum value of y.

38. *Touch-Tone Phones*
(a) Write the sound emitted by touching the # key as a product of sines and/or cosines.
(b) Graph the sound emitted by touching the # key.
(c) TRACE the graph to determine the maximum value of y.

Touch Tone Phone

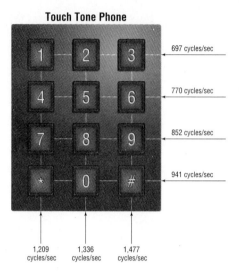

697 cycles/sec
770 cycles/sec
852 cycles/sec
941 cycles/sec

1,209 cycles/sec　1,336 cycles/sec　1,477 cycles/sec

39. If $\alpha + \beta + \gamma = \pi$, show that $\sin 2\alpha + \sin 2\beta + \sin 2\gamma = 4 \sin \alpha \sin \beta \sin \gamma$.

40. If $\alpha + \beta + \gamma = \pi$, show that $\tan \alpha + \tan \beta + \tan \gamma = \tan \alpha \tan \beta \tan \gamma$.

41. Derive formula (3).　　**42.** Derive formula (7).

43. Derive formula (8).　　**44.** Derive formula (9).

3.5

Trigonometric Equations

The previous sections of this chapter were devoted to trigonometric identities—that is, equations involving trigonometric functions that are satisfied by every value in the domain of the variable. In this section, we discuss **trigonometric equations**—that is, equations involving trigonometric functions that are satisfied only by some values of the variable (or, possibly, are not satisfied by any values of the variable). The values that satisfy the equation are called **solutions** of the equation.

E X A M P L E　1　　*Checking Whether a Given Number is a Solution of a Trigonometric Equation*

Determine whether $\theta = \pi/4$ is a solution of the equation $\sin \theta = \frac{1}{2}$. Is $\theta = \pi/6$ a solution?

Solution　Replace θ by $\pi/4$ in the given equation. The result is

$$\sin \frac{\pi}{4} = \frac{\sqrt{2}}{2} \neq \frac{1}{2}$$

We conclude that $\pi/4$ is not a solution.

Next, replace θ by $\pi/6$ in the equation. The result is

$$\sin \frac{\pi}{6} = \frac{1}{2}$$

Thus, $\pi/6$ is a solution of the given equation.　　■

FIGURE 7

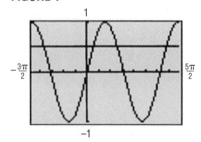

The equation given in Example 1 has other solutions besides $\theta = \pi/6$. For example, $\theta = 5\pi/6$ is also a solution, as is $\theta = 13\pi/6$. (You should check this for yourself.) In fact, the equation has an infinite number of solutions due to the periodicity of the sine function, as can be seen in Figure 7.

Solution   First, we observe that the given equation contains two cosine functions, but with different arguments, θ and 2θ. We use the double-angle formula $\cos 2\theta = 2\cos^2\theta - 1$ to obtain an equivalent equation containing only $\cos\theta$:

$$\cos 2\theta + 3 = 5\cos\theta$$
$$(2\cos^2\theta - 1) + 3 = 5\cos\theta$$
$$2\cos^2\theta - 5\cos\theta + 2 = 0$$
$$(\cos\theta - 2)(2\cos\theta - 1) = 0$$
$$\cos\theta = 2 \quad \text{or} \quad \cos\theta = \tfrac{1}{2}$$

For any angle θ, $-1 \le \cos\theta \le 1$; thus, the equation $\cos\theta = 2$ has no solution. The solutions of $\cos\theta = \tfrac{1}{2}$ are

$$\theta = \frac{\pi}{3} \qquad \theta = \frac{5\pi}{3}$$

Check: Graph $y = \cos 2x + 3 - 5\cos x$, $0 \le x \le 2\pi$, and find the x-intercepts. Compare your results with those of Example 9.   ∎

■   Now work Problem 33.

E X A M P L E   1 0   *Solving a Trigonometric Equation Using Identities*

Solve the equation: $\cos^2\theta + \sin\theta = 2, 0 \le \theta < 2\pi$

Solution   We use a form of the Pythagorean identity:

$$\cos^2\theta + \sin\theta = 2$$
$$(1 - \sin^2\theta) + \sin\theta = 2$$
$$\sin^2\theta - \sin\theta + 1 = 0$$

This is a quadratic equation in $\sin\theta$. The discriminant is $b^2 - 4ac = 1 - 4 = -3 < 0$. Therefore, the equation has no real solution.

Figure 13 shows the graph of $Y_1 = \cos^2 x + \sin x$ and $Y_2 = 2$. The graphs do not intersect anywhere so the equation $Y_1 = Y_2$ has no real solution.

FIGURE 13

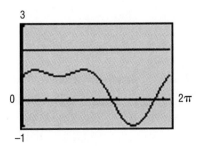

E X A M P L E   1 1   *Solving a Trigonometric Equation Using Identities*

Solve the equation: $\sin\theta\cos\theta = -\tfrac{1}{2}, 0 \le \theta < 2\pi$

Solution   The left side of the given equation is in the form of the double-angle formula $2\sin\theta\cos\theta = \sin 2\theta$, except for a factor of 2. Thus, we multiply each side by 2:

$$\sin \theta \cos \theta = -\tfrac{1}{2}$$
$$2 \sin \theta \cos \theta = -1$$
$$\sin 2\theta = -1$$

The argument here is 2θ. Thus, we need to write all the solutions of this equation and then list those that are in the interval $[0, 2\pi)$.

$$2\theta = \frac{3\pi}{2} + 2k\pi, \qquad k \text{ any integer}$$

$$\theta = \frac{3\pi}{4} + k\pi$$

The solutions in the interval $[0, 2\pi)$ are

$$\theta = \frac{3\pi}{4} \qquad \theta = \frac{7\pi}{4}$$ ■

Sometimes it is necessary to square both sides of an equation in order to obtain expressions that allow the use of identities. Remember, however, that when squaring both sides extraneous solutions may be introduced. As a result, apparent solutions must be checked.

E X A M P L E   1 2   *Other Methods for Solving a Trigonometric Equation*

Solve the equation: $\sin \theta + \cos \theta = 1, 0 \le \theta < 2\pi$

Solution A   Attempts to use available identities do not lead to equations that are easy to solve. (Try it yourself.) Given the form of this equation, we decide to square each side:

$$\sin \theta + \cos \theta = 1$$
$$(\sin \theta + \cos \theta)^2 = 1$$
$$\sin^2 \theta + 2 \sin \theta \cos \theta + \cos^2 \theta = 1$$
$$2 \sin \theta \cos \theta = 0 \quad \text{\small $\sin^2 \theta + \cos^2 \theta = 1$}$$
$$\sin \theta \cos \theta = 0$$

Thus,

$$\sin \theta = 0 \quad \text{or} \quad \cos \theta = 0$$

and the apparent solutions are

$$\theta = 0 \qquad \theta = \pi \qquad \theta = \frac{\pi}{2} \qquad \theta = \frac{3\pi}{2}$$

Because we squared both sides of the original equation, we must check these apparent solutions to see if any are extraneous:

$$\theta = 0: \quad \sin 0 + \cos 0 = 0 + 1 = 1 \qquad \text{\small A solution}$$
$$\theta = \pi: \quad \sin \pi + \cos \pi = 0 + (-1) = -1 \qquad \text{\small Not a solution}$$
$$\theta = \frac{\pi}{2}: \quad \sin \frac{\pi}{2} + \cos \frac{\pi}{2} = 1 + 0 = 1 \qquad \text{\small A solution}$$
$$\theta = \frac{3\pi}{2}: \quad \sin \frac{3\pi}{2} + \cos \frac{3\pi}{2} = -1 + 0 = -1 \qquad \text{\small Not a solution}$$

Thus, $\theta = 3\pi/2$ and $\theta = \pi$ are extraneous. The actual solutions are $\theta = 0$ and $\theta = \pi/2$. ■

74. *Projectile Motion* Refer to Problem 73.
   (a) If you can throw a baseball with an initial speed of 40 m/sec, at what angle of elevation θ should you direct the throw so that the ball travels a distance of 110 m before striking the ground.
   (b) Graph R, with $v_0 = 40$ m/sec.
   (c) TRACE R to verify the result obtained in part (a).
   (d) Determine the maximum distance you can throw the ball.

*The following discussion of **Snell's Law of Refraction** (named after Willebrod Snell, 1591–1626) is needed for Problems 75–81: Light, sound, and other waves travel at different speeds, depending on the media (air, water, wood, and so on) through which they pass. Suppose that light travels from a point A in one medium, where its speed is v_1, to a point B in another medium, where its speed is v_2. Refer to the figure, where the angle θ_1 is called the **angle of incidence** and the angle θ_2 is the **angle of refraction**. Snell's Law,* which can be proved using calculus, states that*

$$\frac{\sin \theta_1}{\sin \theta_2} = \frac{v_1}{v_2}$$

*The ratio v_1/v_2 is called the **index of refraction**. Some values are given in the following table.*

SOME INDEXES OF REFRACTION

MEDIUM	INDEX OF REFRACTION
Water	1.33
Ethyl alcohol	1.36
Carbon bisulfide	1.63
Air (1 atm and 20°C)	1.0003
Methylene iodide	1.74
Fused quartz	1.46
Glass, crown	1.52
Glass, dense flint	1.66
Sodium chloride	1.53

For light of wavelength 589 nanometers, measured with respect to a vacuum. The index with respect to air is negligibly different in most cases.

θ_1	θ_2
10°	7°45′
20°	15°30′
30°	22°30′
40°	29°0′
50°	35°0′
60°	40°30′
70°	45°30′
80°	50°0′

75. The index of refraction of light in passing from a vacuum into water is 1.33. If the angle of incidence is 40°, determine the angle of refraction.

76. The index of refraction of light in passing from a vacuum into dense glass is 1.66. If the angle of incidence is 50°, determine the angle of refraction.

77. Ptolemy, who lived in the city of Alexandria in Egypt during the second century AD, gave the measured values in the table in the margin for the angle of incidence θ_1 and the angle of refraction θ_2 for a light beam passing from air into water. Do these values agree with Snell's Law? If so, what index of refraction results? (These data are interesting as the oldest recorded physical measurements.)[†]

78. The speed of yellow sodium light (wavelength of 589 nanometers) in a certain liquid is measured to be 1.92×10^8 meters per second. What is the index of refraction of this liquid, with respect to air, for sodium light?[‡]

79. A beam of light with a wavelength of 589 nanometers traveling in air makes an angle of incidence of 40° upon a slab of transparent material, and the refracted beam makes an angle of refraction of 26°. Find the index of refraction of the material.[‡]

80. A light ray with a wavelength of 589 nanometers (produced by a sodium lamp) traveling through air makes an angle of incidence of 30° on a smooth, flat slab of crown glass. Find the angle of refraction.[‡]

81. A light beam passes from one medium to another through a thick slab of material whose index of refraction is n_2. Show that the emerging beam is parallel to the incident beam.[‡]

82. Explain in your own words how you would use your calculator to solve the equation $\sin x = 0.3$, $0 \le x < 2\pi$. How would you modify your approach in order to solve the equation $\cot x = 5$, $0 < x < 2\pi$?

---

*Because this law was also deduced by René Descartes, in France it is also known as Descartes' Law.
[†] Adapted from Halliday and Resnick, *Physics, Parts 1 & 2*, 3rd ed. New York: Wiley, 1978, p. 953.
[‡] Adapted from Serway, *Physics*, 3rd ed. Philadelphia: W.B. Saunders, p. 805.

Chapter Review

THINGS TO KNOW

Formulas

Sum and difference formulas

$$\cos(\alpha + \beta) = \cos \alpha \cos \beta - \sin \alpha \sin \beta$$
$$\cos(\alpha - \beta) = \cos \alpha \cos \beta + \sin \alpha \sin \beta$$
$$\sin(\alpha + \beta) = \sin \alpha \cos \beta + \cos \alpha \sin \beta$$
$$\sin(\alpha - \beta) = \sin \alpha \cos \beta - \cos \alpha \sin \beta$$
$$\tan(\alpha + \beta) = \frac{\tan \alpha + \tan \beta}{1 - \tan \alpha \tan \beta}$$
$$\tan(\alpha - \beta) = \frac{\tan \alpha - \tan \beta}{1 + \tan \alpha \tan \beta}$$

Double-angle formulas

$$\sin 2\theta = 2 \sin \theta \cos \theta$$
$$\cos 2\theta = \cos^2 \theta - \sin^2 \theta$$
$$\cos 2\theta = 1 - 2 \sin^2 \theta$$
$$\cos 2\theta = 2 \cos^2 \theta - 1$$
$$\tan 2\theta = \frac{2 \tan \theta}{1 - \tan^2 \theta}$$

Half-angle formulas

$$\sin^2 \frac{\alpha}{2} = \frac{1 - \cos \alpha}{2}$$

$$\cos^2 \frac{\alpha}{2} = \frac{1 + \cos \alpha}{2}$$

$$\tan^2 \frac{\alpha}{2} = \frac{1 - \cos \alpha}{1 + \cos \alpha}$$

$$\sin \frac{\alpha}{2} = \pm \sqrt{\frac{1 - \cos \alpha}{2}}$$

where the + or − sign is determined by the quadrant of the angle $\alpha/2$

$$\cos \frac{\alpha}{2} = \pm \sqrt{\frac{1 + \cos \alpha}{2}}$$

$$\tan \frac{\alpha}{2} = \pm \sqrt{\frac{1 - \cos \alpha}{1 + \cos \alpha}} = \frac{1 - \cos \alpha}{\sin \alpha} = \frac{\sin \alpha}{1 + \cos \alpha}$$

Product-to-sum formulas

$$\sin \alpha \sin \beta = \tfrac{1}{2} [\cos(\alpha - \beta) - \cos(\alpha + \beta)]$$

$$\cos \alpha \cos \beta = \tfrac{1}{2} [\cos(\alpha - \beta) + \cos(\alpha + \beta)]$$

$$\sin \alpha \cos \beta = \tfrac{1}{2} [\sin(\alpha + \beta) + \sin(\alpha - \beta)]$$

Sum-to-product formulas

$$\sin \alpha + \sin \beta = 2 \sin \frac{\alpha + \beta}{2} \cos \frac{\alpha - \beta}{2}$$

$$\sin \alpha - \sin \beta = 2 \sin \frac{\alpha - \beta}{2} \cos \frac{\alpha + \beta}{2}$$

$$\cos \alpha + \cos \beta = 2 \cos \frac{\alpha + \beta}{2} \cos \frac{\alpha - \beta}{2}$$

$$\cos \alpha - \cos \beta = -2 \sin \frac{\alpha + \beta}{2} \sin \frac{\alpha - \beta}{2}$$

HOW TO

Establish identities

Solve a trigonometric equation

EXAMPLE 6

Finding the Height of a Statue on a Building

Adorning the top of the Board of Trade building in Chicago is a statue of the Greek goddess Ceres, goddess of wheat. From street level, two observations are taken 400 feet from the center of the building. The angle of elevation to the base of the statue is found to be 45.0°; the angle of elevation to the top of the statue is 47.2°. See Figure 8(a). What is the height of the statue?

FIGURE 8

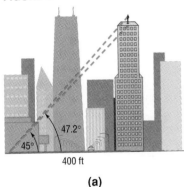

(a)

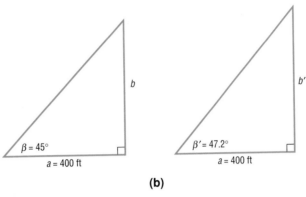

(b)

Solution   Figure 8(b) shows two triangles that replicate Figure 8(a). The height of the statue of Ceres will be $b' - b$. To find b and b', we refer to Figure 8(b):

$$\tan 45° = \frac{b}{400} \qquad\qquad \tan 47.2° = \frac{b'}{400}$$

$$b = 400 \tan 45° = 400 \qquad\qquad b' = 400 \tan 47.2° = 431.96$$

The height of the statue is approximately 32 feet.   ∎

When it is not possible to walk off a distance from the base of the object whose height we seek, a more imaginative solution is required.

EXAMPLE 7

Finding the Height of a Mountain

To measure the height of a mountain, a surveyor takes two sightings of the peak at a distance 900 meters apart on a direct line to the mountain.* See Figure 9(a). The first observation results in an angle of elevation of 47°, whereas the second results in an angle of elevation of 35°. If the transit is 2 meters high, what is the height h of the mountain?

FIGURE 9

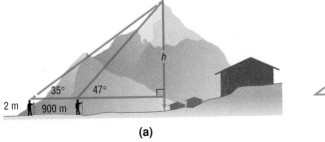

(a)

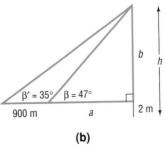

(b)

*For simplicity, we assume that these sightings are at the same level.

Solution   Figure 9(b) shows two triangles that replicate the illustration in Figure 9(a). From the two right triangles shown, we find

$$\tan \beta' = \frac{b}{a + 900} \qquad \tan \beta = \frac{b}{a}$$

$$\tan 35° = \frac{b}{a + 900} \qquad \tan 47° = \frac{b}{a}$$

This is a system of two equations involving two variables, a and b. Because we seek b, we choose to solve the right-hand equation for a and substitute the result, $a = b/\tan 47° = b \cot 47°$, in the left-hand equation. The result is

$$\tan 35° = \frac{b}{b \cot 47° + 900}$$

$$b = (b \cot 47° + 900) \tan 35°$$

$$b = b \cot 47° \tan 35° + 900 \tan 35°$$

$$b(1 - \cot 47° \tan 35°) = 900 \tan 35°$$

$$b = \frac{900 \tan 35°}{1 - \cot 47° \tan 35°} = \frac{900 \tan 35°}{1 - \dfrac{\tan 35°}{\tan 47°}} \approx 1816$$

The height of the peak from ground level is therefore approximately $1816 + 2 = 1818$ meters.   ∎

■ Now work Problem 29.

In navigation, the **direction** or **bearing** from O of an object at P means the positive angle measured clockwise* from the north (N) to the ray OP. See Figure 10. Based on Figure 10, we would say that the bearing of P from O is θ degrees.

FIGURE 10
The bearing of P from O is θ

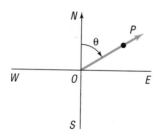

EXAMPLE 8

Finding the Bearing of an Airplane

A Boeing 777 aircraft takes off from O'Hare Airport on a runway having a bearing of 22°. After flying for 1 mile, the pilot of the aircraft requests permission to turn 90° and head toward the northwest. The request is granted.

(a) What is the new bearing?

(b) After the plane goes 2 miles in this direction, what bearing should the control tower use to locate the aircraft?

*Notice that this convention is just the opposite of what we are used to.

Solution (a) Figure 11 illustrates the situation. After flying 1 mile from the airport O (the control tower), the aircraft is at P. After turning $90°$ toward the northwest, we see that

$$\text{Angle } NOP = 22° \qquad \text{Angle } RQN = 90° - 22° = 68°$$

The bearing of this aircraft is therefore

$$360° - \text{Angle } RQN = 360° - 68° = 292°$$

(b) After flying 2 miles at a bearing of $292°$, the aircraft is at R. If $\theta =$ Angle ROP, then

$$\tan \theta = \frac{2}{1} = 2 \quad \text{so} \quad \theta = 63.4°$$

As a result,

$$\text{Angle } RON = \theta - 22° = 41.4°$$

The bearing of the plane from the central tower unit at O is

$$360° - \text{Angle } RON = 360° - 41.4° = 318.6°$$

FIGURE 11

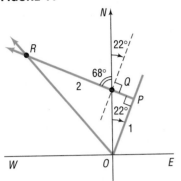

E X A M P L E   9

The Gibb's Hill Lighthouse, Southampton, Bermuda

In operation since 1846, it stands 117 feet high on a hill 245 feet high, so its beam of light is 362 feet above sea level. A brochure states that the light can be seen on the horizon about 26 miles distant. Verify the accuracy of this statement.

Solution Figure 12 illustrates the situation. The central angle θ obeys the equation

$$\cos \theta = \frac{3960}{3960 + \dfrac{362}{5280}} = 0.999982687$$

FIGURE 12

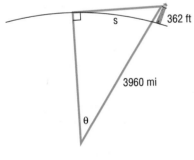

362 ft

3960 mi

so

$$\theta = 0.00588 \text{ radian} = 0.33715° = 20.23'$$

The brochure does not indicate whether the distance is measured in nautical miles or statute miles. The distance s in nautical miles is 20.23, the measurement of θ in minutes. (Refer to Problem 86, Section 2.1). The distance s in statute miles is

$$s = r\theta = 3960(0.00588) = 23.3 \text{ miles}$$

In either case, it would seem that the brochure overstated the distance somewhat. ■

4.1

Exercise 4.1

In Problems 1–14, use the right triangle shown in the margin. Then, using the given information, solve the triangle.

1. $b = 5$, $\beta = 20°$; find a, c, and α
2. $b = 4$, $\beta = 10°$; find a, c, and α
3. $a = 6$, $\beta = 40°$; find b, c, and α
4. $a = 7$, $\beta = 50°$; find b, c, and α
5. $b = 4$, $\alpha = 10°$; find a, c, and β
6. $b = 6$, $\alpha = 20°$; find a, c, and β

7. $a = 5$, $\alpha = 25°$; find b, c, and β

8. $a = 6$, $\alpha = 40°$; find b, c, and β

9. $c = 9$, $\beta = 20°$; find b, a, and α

10. $c = 10$, $\alpha = 40°$; find b, a, and β

11. $a = 5$, $b = 3$; find c, α, and β

12. $a = 2$, $b = 8$; find c, α, and β

13. $a = 2$, $c = 5$; find b, α, and β

14. $b = 4$, $c = 6$; find a, α, and β

15. A right triangle has a hypotenuse of length 3 inches. If one angle is 35°, find the length of each leg.

16. A right triangle has a hypotenuse of length 2 centimeters. If one angle is 40°, find the length of each leg.

17. A right triangle contains a 35° angle. If one leg is of length 5 inches, what is the length of the hypotenuse? [*Hint:* Two answers are possible.]

18. A right triangle contains an angle of $\pi/10$ radian. If one leg is of length 3 meters, what is the length of the hypotenuse? [*Hint:* Two answers are possible.]

19. The hypotenuse of a right triangle is 5 inches. If one leg is 2 inches, find the degree measure of each angle.

20. The hypotenuse of a right triangle is 3 feet. If one leg is 1 foot, find the degree measure of each angle.

21. *Finding the Width of a Gorge* Find the distance from A to C across the gorge illustrated in the figure.

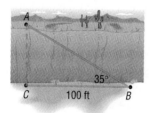

22. *Finding the Distance Across a Pond* Find the distance from A to C across the pond illustrated in the figure.

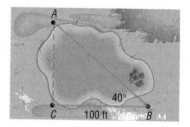

23. *The Eiffel Tower* The tallest tower built before the era of television masts, the Eiffel Tower was completed on March 31, 1889. Find the height of the Eiffel Tower (before a television mast was added to the top) using the information given in the illustration.

24. *Finding the Distance of a Ship from Shore* A ship, offshore from a vertical cliff known to be 100 feet in height, takes a sighting of the top of the cliff. If the angle of elevation is found to be 25°, how far offshore is the ship?

25. Suppose that you are headed toward a plateau 50 meters high. If the angle of elevation to the top of the plateau is 20°, how far are you from the base of the plateau?

26. *Statue of Liberty* A ship is just offshore of New York City. A sighting is taken of the Statue of Liberty, which is about 305 feet tall. If the angle of elevation to the top of the statue is 20°, how far is the ship from the base of the statue?

27. A 22 foot extension ladder leaning against a building makes a 70° angle with the ground. How far up the building does the ladder touch?

28. *Finding the Height of a Building* To measure the height of a building, two sightings are taken a distance of 50 feet apart. If the first angle of elevation is 40° and the second is 32°, what is the height of the building?

29. *Great Pyramid of Cheops* One of the original Seven Wonders of the World, the Great Pyramid of Cheops was built about 2580 BC. Its original height was 480 feet 11 inches, but due to the loss of its topmost stones, it is now shorter.* Find the current height of the Great Pyramid, using the information given in the illustration.

Source: Guinness Book of World Records.

47. *Surveillance Satellites* A surveillance satellite circles Earth at a height of h miles above the surface. Suppose d is the distance, in miles, on the surface of the Earth that can be observed from the satellite. See the illustration.

(a) Find an equation that relates the central angle θ to the height h.
(b) Find an equation that relates the observable distance d and θ.
(c) Find an equation that relates d and h.
(d) If d is to be 2500 miles, how high must the satellite orbit above Earth?
(e) If the satellite orbits at a height of 300 miles, what distance d on the surface can be observed?

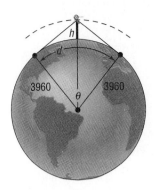

 48. Compare Problem 103 in Section 1 of the Appendix with Example 9 and Problem 40 of this section. One uses the Pythagorean Theorem in its solution, while the others use trigonometry. Write a short paper that contrasts the two methods of solution. Be sure to point out the benefits as well as the shortcomings of each method. Then decide which solution you prefer. Be sure to have reasons.

4.2

The Law of Sines

If none of the angles of a triangle is a right angle, the triangle is called **oblique.** Thus, an oblique triangle will have either three acute angles or two acute angles and one obtuse angle (an angle between $90°$ and $180°$). See Figure 13.

FIGURE 13
Oblique triangles

 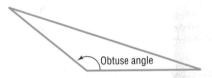

(a) All angles are acute   **(b)** Two acute angles and one obtuse angle

In the discussion that follows, we shall always label an oblique triangle so that side a is opposite angle α, side b is opposite angle β, and side c is opposite angle γ, as shown in Figure 14.

To **solve an oblique triangle** means to find the lengths of its sides and the measurements of its angles. To do this, we shall need to know the length of one side along with two other facts: either two angles, or the other two sides, or one angle and one other side.* Thus, there are four possibilities to consider:

FIGURE 14

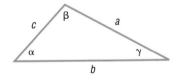

CASE 1:	One side and two angles are known (SAA or ASA).
CASE 2:	Two sides and the angle opposite one of them are known (SSA).
CASE 3:	Two sides and the included angle are known (SAS).
CASE 4:	Three sides are known (SSS).

Figure 15 illustrates the four cases.

FIGURE 15

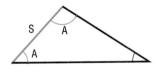 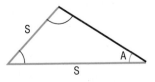 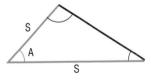 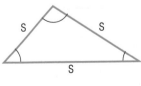

Case 1: SAA or ASA   Case 2: SSA   Case 3: SAS   Case 4: SSS

The **Law of Sines** is used to solve triangles for which Case 1 or 2 holds.

*Recall from plane geometry the fact that knowing three angles of a triangle determines a family of *similar triangles*—that is, triangles that have the same shape but different sizes.

Theorem
Law of Sines

For a triangle with sides a, b, c and opposite angles α, β, γ, respectively,

$$\frac{\sin \alpha}{a} = \frac{\sin \beta}{b} = \frac{\sin \gamma}{c} \tag{1}$$

Proof

FIGURE 16

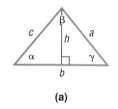

(a)

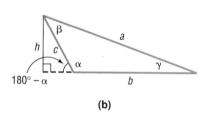

(b)

To prove the Law of Sines, we construct an altitude of length h from one of the vertices of such a triangle. Figure 16(a) shows h for a triangle with three acute angles, and Figure 16(b) shows h for a triangle with an obtuse angle. In each case, the altitude is drawn from the vertex at β. Using either illustration, we have

$$\sin \gamma = \frac{h}{a}$$

from which

$$h = a \sin \gamma \tag{2}$$

From Figure 16(a), it also follows that

$$\sin \alpha = \frac{h}{c}$$

from which

$$h = c \sin \alpha \tag{3}$$

From Figure 16(b), it follows that

$$\sin \alpha = \sin(180° - \alpha) = \frac{h}{c}$$

which again gives

$$h = c \sin \alpha$$

Thus, whether the triangle has three acute angles or has two acute angles and one obtuse angle, equations (2) and (3) hold. As a result, we may equate the expressions for h in equations (2) and (3) to get

$$a \sin \gamma = c \sin \alpha$$

from which

$$\frac{\sin \alpha}{a} = \frac{\sin \gamma}{c} \tag{4}$$

In a similar manner, by constructing the altitude h' from the vertex of angle α as shown in Figure 17, we can show that

$$\sin \beta = \frac{h'}{c} \quad \text{and} \quad \sin \gamma = \frac{h'}{b}$$

Thus,

$$h' = c \sin \beta = b \sin \gamma$$

and

$$\frac{\sin \beta}{b} = \frac{\sin \gamma}{c} \tag{5}$$

FIGURE 17

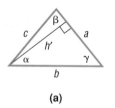

(a)

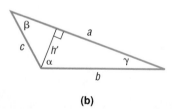

(b)

When equations (4) and (5) are combined, we have equation (1), the Law of Sines. ■

In applying the Law of Sines to solve triangles, we use the fact that the sum of the angles of any triangle equals 180°; that is,

$$\alpha + \beta + \gamma = 180° \qquad (6)$$

Our first two examples show how to solve a triangle when one side and two angles are known (Case 1: SAA or ASA).

E X A M P L E   1

Using the Law of Sines to Solve a SAA Triangle

Solve the triangle: $\alpha = 40°$, $\beta = 60°$, $a = 4$

Solution

Figure 18 shows the triangle that we want to solve. The third angle γ is easily found using equation (6):

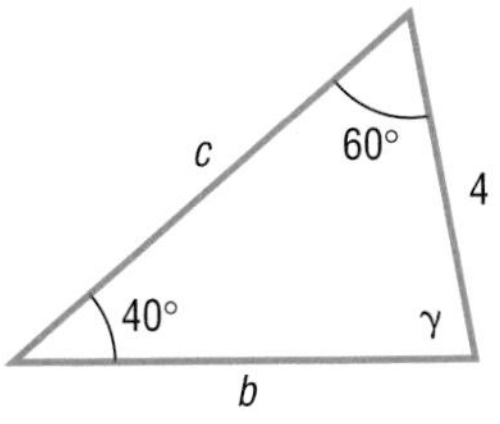

$$\alpha + \beta + \gamma = 180°$$
$$40° + 60° + \gamma = 180°$$
$$\gamma = 80°$$

FIGURE 18

Now we use the Law of Sines (twice) to find the unknown sides b and c:

$$\frac{\sin \alpha}{a} = \frac{\sin \beta}{b} \qquad \frac{\sin \alpha}{a} = \frac{\sin \gamma}{c}$$

Because $a = 4$, $\alpha = 40°$, $\beta = 60°$, and $\gamma = 80°$, we have

$$\frac{\sin 40°}{4} = \frac{\sin 60°}{b} \qquad \frac{\sin 40°}{4} = \frac{\sin 80°}{c}$$

Thus,

$$b = \frac{4 \sin 60°}{\sin 40°} \approx 5.39 \qquad c = \frac{4 \sin 80°}{\sin 40°} \approx 6.13$$

From a calculator   From a calculator   ■

Notice that in Example 1 we found b and c by working with the given side a. This is better than finding b first and working with a rounded value of b to find c.

E X A M P L E   2

Using the Law of Sines to Solve an ASA Triangle

Solve the triangle: $\alpha = 35°$, $\beta = 15°$, $c = 5$

Solution

Figure 19 illustrates the triangle that we want to solve. Because we know two angles ($\alpha = 35°$ and $\beta = 15°$), it is easy to find the third angle using equation (6):

$$\alpha + \beta + \gamma = 180°$$
$$35° + 15° + \gamma = 180°$$
$$\gamma = 130°$$

FIGURE 19

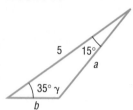

Now we know the three angles and one side ($c = 5$) of the triangle. To find the remaining two sides a and b, we use the Law of Sines (twice):

$$\frac{\sin \alpha}{a} = \frac{\sin \gamma}{a} \qquad\qquad \frac{\sin \beta}{b} = \frac{\sin \gamma}{c}$$

$$\frac{\sin 35°}{a} = \frac{\sin 130°}{5} \qquad\qquad \frac{\sin 15°}{b} = \frac{\sin 130°}{5}$$

$$a = \frac{5 \sin 35°}{\sin 130°} \approx 3.74 \qquad\qquad b = \frac{5 \sin 15°}{\sin 130°} \approx 1.69 \qquad ∎$$

∎   Now work Problem 1.

The Ambiguous Case

FIGURE 20

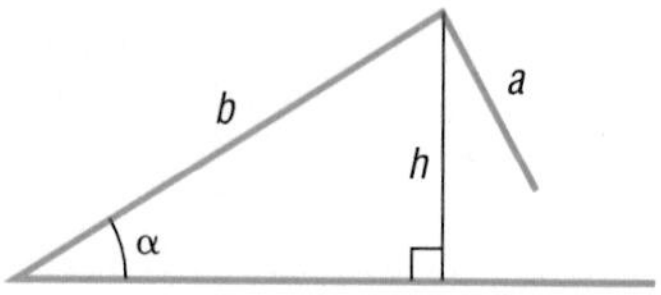

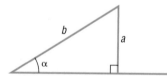

Case 2 (SSA), which applies to triangles where two sides and the angle opposite one of them are known, is referred to as the **ambiguous case,** because the known information may result in one triangle, two triangles, or no triangle at all. Suppose that we are given sides a and b and angle α, as illustrated in Figure 20. The key to determining the possible triangles, if any, that may be formed from the given information, lies primarily with the height h and the fact that $h = b \sin \alpha$.

No Triangle: If $a < b \sin \alpha = h$, then clearly side a is not sufficiently long to form a triangle. See Figure 21.

One Right Triangle: If $a = b \sin \alpha = h$, then side a is just long enough to form a right triangle. See Figure 22.

FIGURE 21
$a < b \sin \alpha$

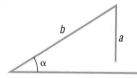

FIGURE 22
$a = b \sin \alpha$

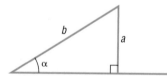

Two Triangles: If $a < b$ and $h = b \sin \alpha < a$, then two distinct triangles can be formed from the given information. See Figure 23.

One Triangle: If $a \geq b$, then only one triangle can be formed. See Figure 24.

FIGURE 23
$b \sin \alpha < a$ and $a < b$

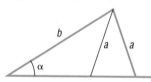

FIGURE 24
$a \geq b$

Fortunately, we do not have to rely on an illustration to draw the correct conclusion in the ambiguous case. The Law of Sines will lead us to the correct determination. Let's see how.

E X A M P L E   3

Using the Law of Sines to Solve a SSA Triangle (One Solution)

Solve the triangle: $a = 3, b = 2, \alpha = 40°$

Solution   (See Figure 25(a).) Because $a = 3$, $b = 2$, and $\alpha = 40°$ are known, we use the Law of Sines to find the angle β:

FIGURE 25(a)

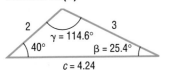

$$\frac{\sin \alpha}{a} = \frac{\sin \beta}{b}$$

Then

$$\frac{\sin 40°}{3} = \frac{\sin \beta}{2}$$

$$\sin \beta = \frac{2 \sin 40°}{3} \approx 0.43$$

There are two angles β, $0° < \beta < 180°$, for which $\sin \beta \approx 0.43$:

$$\beta \approx 25.4° \quad \text{and} \quad \beta \approx 154.6°$$

[**Note:** Here we computed β using the stored value of $\sin \beta$. If you use the rounded value, $\sin \beta \approx 0.43$, you will obtain slightly different results.]

The second possibility is ruled out, because $\alpha = 40°$, making $\alpha + \beta \approx$ 194.6° > 180°. Now, using $\beta \approx 25.4°$, we find

$$\gamma = 180° - \alpha - \beta \approx 180° - 40° - 25.4° = 114.6°$$

The third side c may now be determined using the Law of Sines:

$$\frac{\sin \gamma}{c} = \frac{\sin \alpha}{a}$$

$$\frac{\sin 114.6°}{c} = \frac{\sin 40°}{3}$$

FIGURE 25(b)

$$c = \frac{3 \sin 114.6°}{\sin 40°} \approx 4.24$$

Figure 25(b) illustrates the solved triangle.   ■

E X A M P L E   4   *Using the Law of Sines to Solve a SSA Triangle (Two Solutions)*
Solve the triangle: $a = 6$, $b = 8$, $\alpha = 35°$

Solution   Because $a = 6$, $b = 8$, and $\alpha = 35°$ are known, we use the Law of Sines to find the angle β:

$$\frac{\sin \alpha}{a} = \frac{\sin \beta}{b}$$

Then

$$\frac{\sin 35°}{6} = \frac{\sin \beta}{8}$$

$$\sin \beta = \frac{8 \sin 35°}{6} \approx 0.76$$

$$\beta_1 \approx 49.9° \quad \text{or} \quad \beta_2 \approx 130.1°$$

For both possibilities we have $\alpha + \beta < 180°$. Hence, there are two triangles—one containing the angle $\beta = \beta_1 \approx 49.9°$ and the other containing the angle $\beta = \beta_2 \approx$ 130.1°. The third angle γ is either

$$\gamma_1 = 180° - \alpha - \beta_1 \approx 95.1° \quad \text{or} \quad \gamma_2 = 180° - \alpha - \beta_2 \approx 14.9°$$

$$\underset{\substack{\alpha = 35° \\ \beta_1 = 49.9°}}{\uparrow} \qquad \qquad \underset{\substack{\alpha = 35° \\ \beta_2 = 130.1°}}{\uparrow}$$

The third side c obeys the Law of Sines, so we have

$$\frac{\sin \gamma}{c} = \frac{\sin \alpha}{a}$$

FIGURE 26

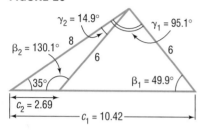

$$\frac{\sin 95.1°}{c_1} = \frac{\sin 35°}{6} \qquad \text{or} \qquad \frac{\sin 14.9°}{c_2} = \frac{\sin 35°}{6}$$

$$c_1 = \frac{6 \sin 95.1°}{\sin 35°} \approx 10.42 \qquad \qquad c_2 = \frac{6 \sin 14.9°}{\sin 35°} \approx 2.69$$

The two solved triangles are illustrated in Figure 26. ◼

E X A M P L E   5

Using the Law of Sines to Solve a SSA Triangle (No Solution)

Solve the triangle: $a = 2$, $c = 1$, $\gamma = 50°$

Solution

Because $a = 2$, $c = 1$, and $\gamma = 50°$ are known, we use the Law of Sines to find the angle α:

$$\frac{\sin \alpha}{a} = \frac{\sin \gamma}{c}$$

FIGURE 27

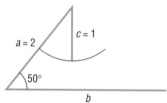

$$\frac{\sin \alpha}{2} = \frac{\sin 50°}{1}$$

$$\sin \alpha = 2 \sin 50° \approx 1.53$$

There is no angle α for which $\sin \alpha > 1$. Hence, there can be no triangle with the given measurements. Figure 27 illustrates the measurements given. Notice that, no matter how we attempt to position side c, it will never intersect side b to form a triangle. ◼

◼ Now work Problem 17.

Applied Problems

The Law of Sines is particularly useful for solving certain applied problems.

E X A M P L E   6

Rescue at Sea

Coast Guard Station Zulu is located 120 miles due west of Station X-ray. A ship at sea sends an SOS call that is received by each station. The call to Station Zulu indicates that the location of the ship is 40° east of north; the call to Station X-ray indicates that the location of the ship is 30° west of north.

(a) How far is each station from the ship?

(b) If a helicopter capable of flying 200 miles per hour is dispatched from the nearest station to the ship, how long will it take to reach the ship?

Solution

(a) Figure 28 illustrates the situation. The angle γ is found to be

$$\gamma = 180° - 50° - 60° = 70°$$

FIGURE 28

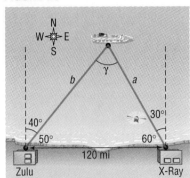

Zulu   X-Ray

The Law of Sines can now be used to find the two distances a and b that we seek:

$$\frac{\sin 50°}{a} = \frac{\sin 70°}{120}$$

$$a = \frac{120 \sin 50°}{\sin 70°} \approx 97.82 \text{ miles}$$

$$\frac{\sin 60°}{b} = \frac{\sin 70°}{120}$$

$$b = \frac{120 \sin 60°}{\sin 70°} \approx 110.59 \text{ miles}$$

Thus, Station Zulu is about 111 miles from the ship, and Station X-ray is about 98 miles from the ship.

(b) The time t needed for the helicopter to reach the ship from Station X-ray is found by using the formula

$$(\text{Velocity, } v)(\text{Time, } t) = \text{Distance, } a$$

Then

$$t = \frac{a}{v} = \frac{97.82}{200} \approx 0.49 \text{ hour} \approx 29 \text{ minutes}$$

It will take about 29 minutes for the helicopter to reach the ship. ■

■ Now work Problem 29.

4.2

Exercise 4.2

In Problems 1–8, solve each triangle.

1.
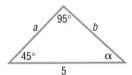

2.
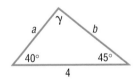

3.
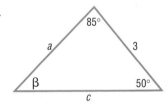

4.
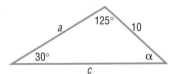

5.
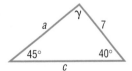

6.

7.
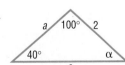

8.
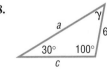

In Problems 9–16, solve each triangle.

9. $\alpha = 40°, \beta = 20°, a = 2$ **10.** $\alpha = 50°, \gamma = 20°, a = 3$ **11.** $\beta = 70°, \gamma = 10°, b = 5$

12. $\alpha = 70°, \beta = 60°, c = 4$ **13.** $\alpha = 110°, \gamma = 30°, c = 3$ **14.** $\beta = 10°, \gamma = 100°, b = 2$

15. $\alpha = 40°, \beta = 40°, c = 2$ **16.** $\beta = 20°, \gamma = 70°, a = 1$

In Problems 17–28, two sides and an angle are given. Determine whether the given information results in one triangle, two triangles, or no triangle at all. Solve any triangle(s) that results.

17. $a = 3, b = 2, \alpha = 50°$ **18.** $b = 4, c = 3, \beta = 40°$ **19.** $b = 5, c = 3, \beta = 100°$

20. $a = 2, c = 1, \alpha = 120°$ **21.** $a = 4, b = 5, \alpha = 60°$ **22.** $b = 2, c = 3, \beta = 40°$

23. $b = 4, c = 6, \beta = 20°$ **24.** $a = 3, b = 7, \alpha = 70°$ **25.** $a = 2, c = 1, \gamma = 100°$

26. $b = 4, c = 5, \beta = 95°$ **27.** $a = 2, c = 1, \gamma = 25°$ **28.** $b = 4, c = 5, \beta = 40°$

29. *Rescue at Sea* Coast Guard Station Able is located 150 miles due south of Station Baker. A ship at sea sends an SOS call that is received by each station. The call to Station Able indicates that the ship is located 35° north of east; the call to Station Baker indicates that the ship is located 30° south of east.
(a) How far is each station from the ship?
(b) If a helicopter capable of flying 200 miles per hour is dispatched from the nearest station to the ship, how long will it take to reach the ship?

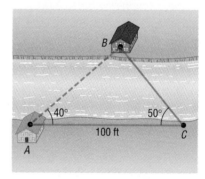

30. *Surveying* Consult the figure. To find the distance from the house at A to the house at B, a surveyor measures the angle BAC to be 40°, then walks off a distance of 100 feet to C, and measures the angle ACB to be 50°. What is the distance from A to B?

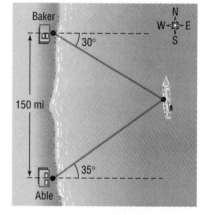

31. *Finding the Length of a Ski Lift* Consult the figure. To find the length of the span of a proposed ski lift from A to B, a surveyor measures the angle DAB to be 25°, then walks off a distance of 1000 feet to C, and measures the angle ACB to be 15°. What is the distance from A to B?

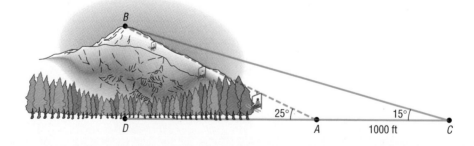

32. *Finding the Height of a Mountain* Use the illustration in Problem 31 to find the height *BD* of the mountain at *B*.

33. *Finding the Height of an Airplane* An aircraft is spotted by two observers who are 1000 feet apart. As the airplane passes over the line joining them, each observer takes a sighting of the angle of elevation to the plane, as indicated in the figure. How high is the airplane?

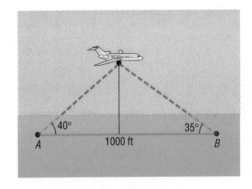

34. *Finding the Height of the Bridge Over the Royal Gorge* The highest bridge in the world is the bridge over the Royal Gorge of the Arkansas River in Colorado.* Sightings to the same point at water level directly under the bridge are taken from each side of the 880-foot-long bridge, as indicated in the figure. How high is the bridge?

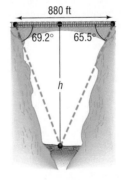

35. *Navigation* An airplane flies from city *A* to city *B*, a distance of 150 miles, then turns through an angle of 40° and heads toward city *C*, as shown in the figure.
 (a) If the distance between cities *A* and *C* is 300 miles, how far is it from city *B* to city *C*?
 (b) Through what angle should the pilot turn at city *C* to return to City *A*?

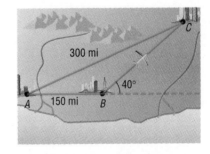

36. *Time Lost due to a Navigation Error* In attempting to fly from city *A* to city *B*, an aircraft followed a course that was 10° in error, as indicated in the figure. After flying a distance of 50 miles, the pilot corrected the course by turning at point *C* and flying 70 miles farther. If the constant speed of the aircraft was 250 miles per hour, how much time was lost due to the error?

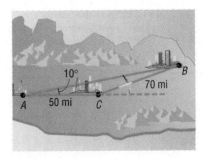

Source: Guinness Book of World Records.

37. *Finding the Lean of the Leaning Tower of Pisa* The famous Leaning Tower of Pisa was originally 184.5 feet high.* After walking 123 feet from the base of the tower, the angle of elevation to the top of the tower is found to be 60°. Find the angle *CAB* indicated in the figure. Also, find the perpendicular distance from *C* to *AB*.

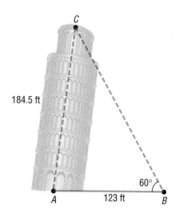

38. *Crankshafts on Cars* On a certain automobile, the crankshaft is 3 inches long and the connecting rod is 9 inches long (see the figure). At the time when the angle *OPA* is 15°, how far is the piston (*P*) from the center (*O*) of the crankshaft?

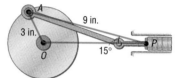

39. *Constructing a Highway* U.S. 41, a highway whose primary directions are north–south, is being constructed along the west coast of Florida. Near Naples, a bay obstructs the straight path of the road. Since the cost of a bridge is prohibitive, engineers decide to go around the bay. The illustration shows the path that they decide on and the measurements taken. What is the length of highway needed to go around the bay?

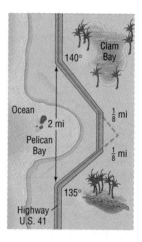

40. *Determining Distances at Sea* The navigator of a ship at sea spots two lighthouses that she knows to be 3 miles apart along a straight seashore. She determines that the angles formed between two line-of-sight observations of the lighthouses and the line from the ship directly to shore are 15° and 35°. See the illustration.
(a) How far is the ship from lighthouse *A*?
(b) How far is the ship from lighthouse *B*?
(c) How far is the ship from shore?

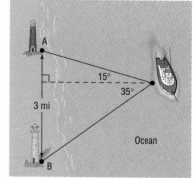

*In their 1986 report on the fragile 7-century-old bell tower, scientists in Pisa, Italy, said the Leaning Tower of Pisa had increased its famous lean by 1 millimeter, or 0.04 inch. That is about the annual average, although the tilting has slowed to about half that much in the previous 2 years. (*Source:* United Press International, June 29, 1986.)

Update PISA, ITALY. September, 1995. The Leaning Tower of Pisa has suddenly shifted, jeopardizing years of preservation work to stabilize it, Italian newspapers said Sunday. The tower, built on shifting subsoil between 1174 and 1350 as a belfry for the nearby cathedral, recently moved .07 inches in one night. The tower has been closed to tourists since 1990 but officials hope to partially reopen it next year.

41. *Calculating Distances at Sea* The navigator of a ship at sea has the harbor in sight at which the ship is to dock. She spots a lighthouse she knows is 1 mile down the coast from the mouth of the harbor, and she measures the angle between the line-of-sight observations of the harbor and lighthouse to be 20°. With the ship heading directly toward the harbor, she repeats this measurement after 5 minutes of traveling at 12 miles per hour. If the new angle is 30°, how far is the ship from the harbor?

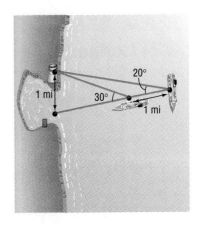

42. *Finding Distances* A forest ranger is walking on a path inclined at 5° to the horizontal directly toward a 100-foot-tall fire observation tower. The angle of elevation of the top of the tower is 40°. How far is the ranger from the tower at this time?

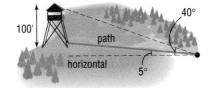

43. *Mollweide's Formula* For any triangle, **Mollweide's Formula** (named after Karl Mollweide, 1774–1825) states that

$$\frac{a + b}{c} = \frac{\cos \frac{1}{2}(\alpha - \beta)}{\sin \frac{1}{2}\gamma}$$

Derive it. [*Hint:* Use the Law of Sines and then a sum-to-product formula.] Notice that this formula involves all six parts of a triangle. As a result, it is sometimes used to check the solution of a triangle.

44. *Mollweide's Formula* Another form of Mollweide's Formula is

$$\frac{a - b}{c} = \frac{\sin \frac{1}{2}(\alpha - \beta)}{\cos \frac{1}{2}\gamma}$$

Derive it.

45. For any triangle, derive the formula

$$a = b \cos \gamma + c \cos \beta$$

[*Hint:* Use the fact that $\sin \alpha = \sin(180° - \beta - \gamma)$.]

46. *Law of Tangents* For any triangle, derive the **Law of Tangents:**

$$\frac{a - b}{a + b} = \frac{\tan \frac{1}{2}(\alpha - \beta)}{\tan \frac{1}{2}(\alpha + \beta)}$$

[*Hint:* Use Mollweide's Formula.]

47. *Circumscribing a Triangle* Show that

$$\frac{\sin \alpha}{a} = \frac{\sin \beta}{b} = \frac{\sin \gamma}{c} = \frac{1}{2r}$$

where r is the radius of the circle circumscribing the triangle ABC whose sides are a, b, c, as shown in the figure in the margin. [*Hint:* Draw the diameter AB'. Then $\beta = $ angle $ABC = $ angle $AB'C$ and angle $ACB' = 90°$.]

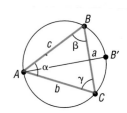

 48. Make up three problems involving oblique triangles. One should result in one triangle, the second in two triangles, and the third in no triangle.

4.3

The Law of Cosines

In Section 4.2, we used the Law of Sines to solve Case 1 (SAA or ASA) and Case 2 (SSA) of an oblique triangle. In this section, we derive the Law of Cosines and use it to solve the remaining cases, 3 and 4.

CASE 3: Two sides and the included angle are known (SAS).
CASE 4: Three sides are known (SSS).

Theorem
Law of Cosines

For a triangle with sides a, b, c and opposite angles α, β, γ, respectively.

$$c^2 = a^2 + b^2 - 2ab \cos \gamma \tag{1}$$
$$b^2 = a^2 + c^2 - 2ac \cos \beta \tag{2}$$
$$a^2 = b^2 + c^2 - 2bc \cos \alpha \tag{3}$$

Proof

We shall prove only formula (1) here. Formulas (2) and (3) may be proved using the same argument.

We begin by strategically placing a triangle on a rectangular coordinate system so that the vertex of angle γ is at the origin and side b lies along the positive x-axis. Regardless of whether γ is acute, as in Figure 29(a), or obtuse, as in Figure 29(b), the vertex B has coordinates $(a \cos \gamma, a \sin \gamma)$. Vertex A has coordinates $(b, 0)$.

FIGURE 29

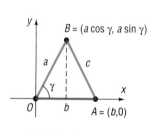

(a) Angle γ is acute

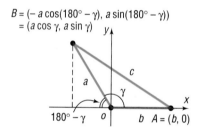

(b) Angle γ is obtuse

We can now use the distance formula to compute c^2:

$$
\begin{aligned}
c^2 &= (b - a \cos \gamma)^2 + (0 - a \sin \gamma)^2 \\
&= b^2 - 2ab \cos \gamma + a^2 \cos^2 \gamma + a^2 \sin^2 \gamma \\
&= a^2(\cos^2 \gamma + \sin^2 \gamma) + b^2 - 2ab \cos \gamma \\
&= a^2 + b^2 - 2ab \cos \gamma
\end{aligned}
$$

Each of formulas (1), (2), and (3) may be stated in words as follows:

Theorem
Law of Cosines

The square of one side of a triangle equals the sum of the squares of the other two sides minus twice their product times the cosine of their included angle. ∎

Observe that if the triangle is a right triangle (so that, say, $\gamma = 90°$), then formula (1) becomes the familiar Pythagorean Theorem: $c^2 = a^2 + b^2$. Thus, the Pythagorean Theorem is a special case of the Law of Cosines.

Let's see how to use the Law of Cosines to solve Case 3 (SAS), which applies to triangles for which two sides and the included angle are known.

E X A M P L E   1   *Using the Law of Cosines to Solve a SAS Triangle*

Solve the triangle: $a = 2$, $b = 3$, $\gamma = 60°$

Solution   See Figure 30. The Law of Cosines makes it easy to find the third side, c:

FIGURE 30

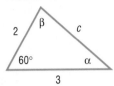

$$c^2 = a^2 + b^2 - 2ab \cos \gamma$$
$$= 4 + 9 - 2 \cdot 2 \cdot 3 \cdot \cos 60°$$
$$= 13 - (12 \cdot \tfrac{1}{2}) = 7$$
$$c = \sqrt{7}$$

Side c is of length $\sqrt{7}$. To find the angles α and β, we may use either the Law of Sines or the Law of Cosines. It is preferable to use the Law of Cosines, since it will lead to an equation with one solution. Using the Law of Sines would lead to an equation with two solutions that would need to be checked to determine which solution fits the given data. Thus, we choose to use formulas (2) and (3) of the Law of Cosines.

For α:

$$a^2 = b^2 + c^2 - 2bc \cos \alpha$$
$$2bc \cos \alpha = b^2 + c^2 - a^2$$
$$\cos \alpha = \frac{b^2 + c^2 - a^2}{2bc} = \frac{9 + 7 - 4}{2 \cdot 3\sqrt{7}} = \frac{12}{6\sqrt{7}} = \frac{2\sqrt{7}}{7}$$
$$\alpha \approx 40.9°$$

For β:

$$b^2 = a^2 + c^2 - 2ac \cos \beta$$
$$\cos \beta = \frac{a^2 + c^2 - b^2}{2ac} = \frac{4 + 7 - 9}{4\sqrt{7}} = \frac{1}{2\sqrt{7}} = \frac{\sqrt{7}}{14}$$
$$\beta \approx 79.1°$$

Notice that $\alpha + \beta + \gamma = 40.9° + 79.1° + 60° = 180°$, as required.   ■

■   Now work Problem 1.

The next example illustrates how the Law of Cosines is used when three sides of a triangle are known, Case 4 (SSS).

E X A M P L E   2   *Using the Law of Cosines to Solve a SSS Triangle*

Solve the triangle: $a = 4$, $b = 3$, $c = 6$

Solution   See Figure 31. To find the angles α, β, and γ, we proceed as we did in the latter part of the solution to Example 1.

For α:

$$\cos \alpha = \frac{b^2 + c^2 - a^2}{2bc} = \frac{9 + 36 - 16}{2 \cdot 3 \cdot 6} = \frac{29}{36}$$
$$\alpha \approx 36.3°$$

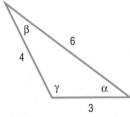

FIGURE 31

For β:

$$\cos \beta = \frac{a^2 + c^2 - b^2}{2ac} = \frac{16 + 36 - 9}{2 \cdot 4 \cdot 6} = \frac{43}{48}$$

$$\beta \approx 26.4°$$

Since we know α and β,

$$\gamma = 180° - \alpha - \beta \approx 180° - 36.3° - 26.4° = 117.3°$$ ■

■   Now work Problem 7.

E X A M P L E   3   *Correcting a Navigational Error*

A motorized sailboat leaves Naples, Florida, bound for Key West, 150 miles away. Maintaining a constant speed of 15 miles per hour, but encountering heavy cross-winds and strong currents, the crew finds, after 4 hours, that the sailboat is off course by 20°.

(a) How far is the sailboat from Key West at this time?

(b) Through what angle should the sailboat turn to correct its course?

(c) How much time has been added to the trip because of this? (Assume that the speed remains at 15 miles per hour.)

Solution   See Figure 32. With a speed of 15 miles per hour, the sailboat has gone 60 miles after 4 hours. We seek the distance x of the sailboat from Key West. We also seek the angle θ that the sailboat should turn through to correct its course.

FIGURE 32

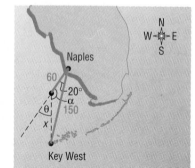

(a) To find x, we use the Law of Cosines, since we know two sides and the included angle.

$$x^2 = 150^2 + 60^2 - 2(150)(60) \cos 20° = 9186$$

$$x = 95.8$$

The sailboat is about 96 miles from Key West.

(b) We now know three sides of the triangle, so we can use the Law of Cosines again to find the angle α opposite the side of length 150 miles.

$$150^2 = 96^2 + 60^2 - 2(96)(60) \cos \alpha$$

$$9684 = -11{,}520 \cos \alpha$$

$$\cos \alpha \approx -0.8406$$

$$\alpha \approx 147.2°$$

The sailboat should turn through an angle of

$$\theta = 180° - \alpha \approx 180° - 147.2° = 32.8°$$

The sailboat should turn through an angle of about 33° to correct its course.

(c) The total length of the trip is now $60 + 96 = 156$ miles. The extra 6 miles will only require about 0.4 hours or 24 minutes more if the speed of 15 miles per hour is maintained.   ■

■   Now work Problem 27.

HISTORICAL FEATURE   ■ The Law of Sines was known vaguely long before it was explicitly stated by Nasîr ed-dîn (about AD 1250). Ptolemy (about AD 150) was aware of it in a form using a chord function instead of the sine function. But it was first clearly stated in Europe by Regiomontanus, writing in 1464.

The Law of Cosines appears first in Euclid's *Elements* (Book II), but in a well-disguised form in which squares built on the sides of triangles are added and a rectangle representing the cosine term is subtracted. It was thus known to all mathematicians because of their familiarity with Euclid's work. An early modern form of the Law of Cosines—that for finding the angle when the sides are known—was stated by François Viète (in 1593).

The Law of Tangents (see Problem 46 of Exercise 4.2) has become obsolete. In the past it was used in place of the Law of Cosines, because the Law of Cosines was very inconvenient for calculation with logarithms or slide rules. Mixing of addition and multiplication is now quite easy on a calculator, however, and the Law of Tangents has been shelved along with the slide rule.   ■

4.3

Exercise 4.3

In Problems 1–8, solve each triangle.

1.

2.

3.

4.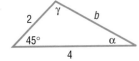

5.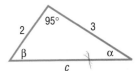

6.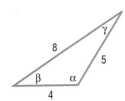

7.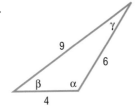

8.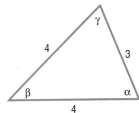

In Problems 9–24, solve each triangle.

9. $a = 3, b = 4, \gamma = 40°$

10. $a = 2, c = 1, \beta = 10°$

11. $b = 1, c = 3, \alpha = 80°$

12. $a = 6, b = 4, \gamma = 60°$

13. $a = 3, c = 2, \beta = 110°$

14. $b = 4, c = 1, \alpha = 120°$

15. $a = 2, b = 2, \gamma = 50°$

16. $a = 3, c = 2, \beta = 90°$

17. $a = 12, b = 13, c = 5$

18. $a = 4, b = 5, c = 3$

19. $a = 2, b = 2, c = 2$

20. $a = 3, b = 3, c = 2$

21. $a = 5, b = 8, c = 9$

22. $a = 4, b = 3, c = 6$

23. $a = 10, b = 8, c = 5$

24. $a = 9, b = 7, c = 10$

25. *Surveying* Consult the figure. To find the distance from the house at *A* to the house at *B*, a surveyor measures the angle *ACB*, which is found to be 70°, then walks off the distance to each house, 50 feet and 70 feet, respectively. How far apart are the houses?

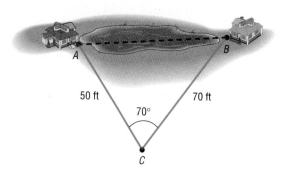

26. *Navigation* An airplane flies from city *A* to city *B*, a distance of 150 miles, then turns through an angle of 50° and flies to city *C*, a distance of 100 miles (see the figure).
(a) How far is it from city *A* to city *C*?
(b) Through what angle should the pilot turn at city *C* to return to city *A*?

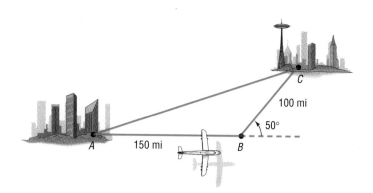

27. *Revising a Flight Plan* In attempting to fly from city *A* to city *B*, a distance of 330 miles, a pilot inadvertently took a course that was 10° in error, as indicated in the figure.
(a) If the aircraft maintains an average speed of 220 miles per hour and if the error in direction is discovered after 15 minutes, through what angle should the pilot turn to head toward city *B*?
(b) What new average speed should the pilot maintain so that the total time of the trip is 90 minutes?

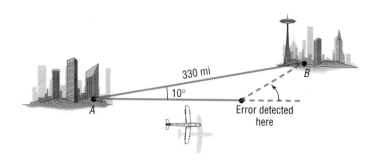

28. *Avoiding a Tropical Storm* A cruise ship maintains an average speed of 15 knots in going from San Juan, Puerto Rico, to Barbados, West Indies, a distance of 600 nautical miles. To avoid a tropical storm, the captain heads out of San Juan in a direction of 20° off a direct heading to Barbados. The captain maintains the 15 knot speed for 10 hours, after which time the path to Barbados becomes clear of storms.
(a) Through what angle should the captain turn to head directly to Barbados?
(b) How long will it be before the ship reaches Barbados if the same 15 knot speed is maintained?

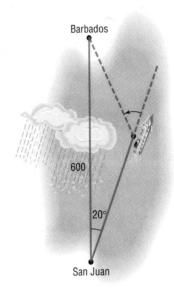

29. *Major League Baseball Field* A Major League baseball diamond is actually a square 90 feet on a side. The pitching rubber is located 60.5 feet from home plate on a line joining home plate and second base.
(a) How far is it from the pitching rubber to first base?
(b) How far is it from the pitching rubber to second base?
(c) If a pitcher faces home plate, through what angle does he need to turn to face first base?

30. *Little League Baseball Field* According to Little League baseball official regulations, a diamond is a square 60 feet on a side. The pitching rubber is located 46 feet from home plate on a line joining home plate and second base.
(a) How far is it from the pitching rubber to first base?
(b) How far is it from the pitching rubber to second base?
(c) If a pitcher faces home plate, through what angle does he need to turn to face first base?

31. *Finding the Length of a Guy Wire* The height of a radio tower is 500 feet, and the ground on one side of the tower slopes upward at an angle of 10° (see the figure).
(a) How long should a guy wire be if it is to connect to the top of the tower and be secured at a point on the slope 100 feet from the base of the tower?
(b) How long should a second guy wire be if it is to connect to the middle of the tower and be secured at a point 100 feet from the base?

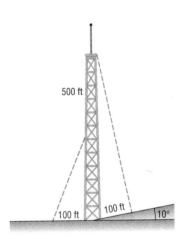

32. *Finding the Length of a Guy Wire* A radio tower 500 feet high is located on the side of a hill with an inclination to the horizontal of 5° (see the figure). How long should two guy wires be if they are to connect to the top of the tower and be secured at two points 100 feet directly above and directly below the base of the tower?

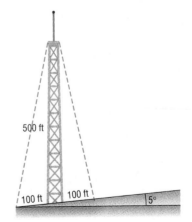

33. *Wrigley Field, Home of the Chicago Cubs* The distance from home plate to dead center in Wrigley Field is 400 feet (see the figure). How far is it from dead center to third base?

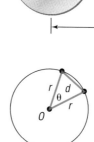

400 ft

90 ft   90 ft

34. *Little League Baseball* The distance from home plate to dead center at the Oak Lawn Little League field is 280 feet. How far is it from dead center to third base? [*Hint:* The distance between the bases in Little League is 60 feet.]

35. *Rods and Pistons* Rod *OA* (see the figure) rotates about the fixed point *O* so that point *A* travels on a circle of radius *r*. Connected to point *A* is another rod *AB* of length $L > r$, and point *B* is connected to a piston. Show that the distance *x* between point *O* and point *B* is given by

$$x = r \cos \theta + \sqrt{r^2 \cos^2 \theta + L^2 - r^2}$$

where θ is the angle of rotation of rod *OA*.

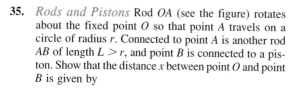

36. *Geometry* Show that the length *d* of a chord of a circle of radius *r* is given by the formula

$$d = 2r \sin \frac{\theta}{2}$$

where θ is the central angle formed by the radii to the ends of the chord (see the figure). Use this result to derive the fact that $\sin \theta < \theta$, where $\theta > 0$ is measured in radians.

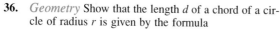

37. For any triangle, show that

$$\cos \frac{\gamma}{2} = \sqrt{\frac{s(s-c)}{ab}}$$

where $s = \frac{1}{2}(a + b + c)$. [*Hint:* Use a half-angle formula and the Law of Cosines.]

38. For any triangle, show that

$$\sin \frac{\gamma}{2} = \sqrt{\frac{(s-a)(s-b)}{ab}}$$

where $s = \frac{1}{2}(a + b + c)$.

39. Use the Law of Cosines to prove the identity

$$\frac{\cos \alpha}{a} + \frac{\cos \beta}{b} + \frac{\cos \gamma}{c} = \frac{a^2 + b^2 + c^2}{2abc}$$

40. Write down your strategy for solving an oblique triangle.

MISSION POSSIBLE

Chapter 4

LOCATING LOST TREASURE

While scuba diving off Wreck Hill in Bermuda, a group of 5 entrepreneurs discovered a treasure map in a small water-tight cask on a pirate schooner that had sunk in 1747. The map directed them to an area of Bermuda now known as The Flatts, but when they got there, they realized that the most important landmark on the map was gone. They called in the Mission Possible team to help them recreate the map. They promised you 25% of whatever treasure was found.

The directions on the map read as follows:

1. From the tallest palm tree, sight the highest hill. Drop your eyes vertically until you sight the base of the hill.
2. Turn 40° clockwise from that line and walk 70 paces to the big red rock.
3. From the red rock walk 50 paces back to the sight line between the palm tree and the hill. Dig there.

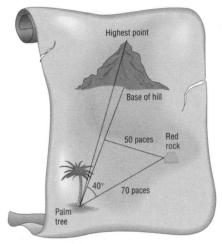

The 5 entrepreneurs said they believed they had found the red rock and the highest hill in the vicinity, but the "tallest palm tree" had long since fallen and disintegrated. It had occurred to them that the treasure must be located on a circle with radius 50 "paces" centered around the red rock, but they had decided against digging a trench 470 feet in circumference, especially since they had no assurance that the treasure was still there. (They had decided that a "pace" must be about a yard.)

1. Determine a plan to locate the position of the lost palm tree, and write out an explanation of your procedure for the entrepreneurs.
2. Unfortunately, it turns out that the entrepreneurs had more in common with the 18th century pirates than you had bargained for. Once you told them the location of the lost palm tree, they tied you all to the red rock, saying they could take it from there. From the location of the palm tree, they sighted 40° counter-clockwise from the rock to the hill, then ran about 30 yards to the circle they had traced about the rock and began to dig frantically. Nothing. After about an hour, they drove off shouting back at you, "25% of nothing is nothing!"
3. Fortunately, the entrepreneurs had left the shovels. After you managed to untie yourselves, you went to the correct location and found the treasure. Where was it? How far from the palm tree? Explain.
4. People who scuba dive for sunken treasure have certain legal obligations. What are they? Should you share the treasure with a lawyer, just to make sure you get to keep the rest?

4.4

The Area
of a Triangle

In this section, we shall derive several formulas for calculating the area A of a triangle. The most familiar of these is the following:

Theorem   The area A of a triangle is

$$A = \tfrac{1}{2}bh \qquad (1)$$

where b is the base and h is an altitude drawn to that base.   ■

Proof   The derivation of this formula is rather easy once a rectangle of base b and height h is constructed around the triangle. See Figures 33 and 34.

Triangles 1 and 2 in Figure 34 are equal in area, as are triangles 3 and 4. Consequently, the area of the triangle with base b and altitude h is exactly half the area of the rectangle, which is bh.   ■

If the base b and altitude h to that base are known, then we can easily find the area of such a triangle using formula (1). Usually, though, the information required to use formula (1) is not given. Suppose, for example, that we know two sides a and b and the included angle γ (see Figure 35). Then the altitude h can be found by noting that

$$\frac{h}{a} = \sin \gamma$$

so that

$$h = a \sin \gamma$$

Using this fact in formula (1) produces

$$A = \tfrac{1}{2}bh = \tfrac{1}{2}b(a \sin \gamma) = \tfrac{1}{2}ab \sin \gamma$$

Thus, we have the formula

$$A = \tfrac{1}{2}ab \sin \gamma \qquad (2)$$

By dropping altitudes from the other two vertices of the triangle, we obtain the following corresponding formulas:

$$A = \tfrac{1}{2}bc \sin \alpha \qquad (3)$$
$$A = \tfrac{1}{2}ac \sin \beta \qquad (4)$$

It is easiest to remember these formulas using the following wording:

Theorem   The area A of a triangle equals one-half the product of two of its sides times the sine of their included angle.   ■

FIGURE 33
$A = \tfrac{1}{2}bh$

FIGURE 34

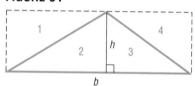

FIGURE 35
$h = a \sin \gamma$

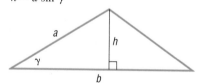

E X A M P L E   1   *Finding the Area of a SAS Triangle*

Find the area A of the triangle for which $a = 8$, $b = 6$, and $\gamma = 30°$.

Solution   We use formula (2) to get

$$A = \tfrac{1}{2}ab \sin \gamma = \tfrac{1}{2} \cdot 8 \cdot 6 \sin 30° = 12$$ ■

■ Now work Problem 1.

If the three sides of a triangle are known, another formula, called **Heron's Formula** (named after Heron of Alexandria), can be used to find the area of a triangle.

Theorem   The area A of a triangle with sides a, b, and c is
Heron's Formula

$$A = \sqrt{s(s - a)(s - b)(s - c)} \qquad (5)$$

where $s = \tfrac{1}{2}(a + b + c)$.   ■

Proof   The proof we shall give uses the Law of Cosines and is quite different from the proof given by Heron.

From the Law of Cosines,

$$c^2 = a^2 + b^2 - 2ab \cos \gamma$$

and the two half-angle formulas,

$$\cos^2 \frac{\gamma}{2} = \frac{1 + \cos \gamma}{2} \qquad \sin^2 \frac{\gamma}{2} = \frac{1 - \cos \gamma}{2}$$

we find

$$\cos^2 \frac{\gamma}{2} = \frac{1 + \cos \gamma}{2} = \frac{1 + \dfrac{a^2 + b^2 - c^2}{2ab}}{2}$$

$$= \frac{a^2 + 2ab + b^2 - c^2}{4ab} = \frac{(a + b)^2 - c^2}{4ab}$$

$$= \frac{(a + b - c)(a + b + c)}{4ab} = \frac{2(s - c) \cdot 2s}{4ab} = \frac{s(s - c)}{ab} \qquad (6)$$

$$\underset{\substack{\uparrow \\ a + b - c = a + b + c - 2c \\ = 2s - 2c}}{}$$

Similarly,

$$\sin^2 \frac{\gamma}{2} = \frac{(s - a)(s - b)}{ab} \qquad (7)$$

Now we use formula (2) for the area:

$$A = \frac{1}{2}ab\sin\gamma$$

$$= \frac{1}{2}ab \cdot 2\sin\frac{\gamma}{2}\cos\frac{\gamma}{2} \qquad \sin\gamma = \sin 2\left(\frac{\gamma}{2}\right) = 2\sin\frac{\gamma}{2}\cos\frac{\gamma}{2}$$

$$= ab\sqrt{\frac{(s-a)(s-b)}{ab}}\sqrt{\frac{s(s-c)}{ab}} \qquad \text{Use equations (6) and (7).}$$

$$= \sqrt{s(s-a)(s-b)(s-c)}$$ ∎

EXAMPLE 2 *Finding the Area of a SSS Triangle*

Find the area of a triangle whose sides are 4, 5, and 7.

Solution We let $a = 4$, $b = 5$, and $c = 7$. Then

$$s = \tfrac{1}{2}(a + b + c) = \tfrac{1}{2}(4 + 5 + 7) = 8$$

Heron's Formula then gives the area A as

$$A = \sqrt{s(s-a)(s-b)(s-c)} = \sqrt{8 \cdot 4 \cdot 3 \cdot 1} = \sqrt{96} = 4\sqrt{6}$$ ∎

■ Now work Problem 7.

HISTORICAL FEATURE ■ Heron's Formula (also known as *Hero's Formula*) is due to Heron of Alexandria (about AD 75), who had, besides his mathematical talents, a good deal of engineering skill. In various temples his mechanical devices produced effects that seemed supernatural, and visitors presumably were thus influenced to generosity. Heron's book *Metrica*, on making such devices, has survived and was discovered in 1896 in the city of Constantinople.

Heron's Formula for the area of a triangle caused some mild discomfort in Greek mathematics, because a product with two factors was an area and with three factors was a volume, but four factors seemed contradictory in Heron's time.

Karl Mollweide (1774–1875), a mathematician and astronomer, discovered the formulas named for him (see Problems 43 and 44 of Exercise 4.2). These formulas are not too important in themselves but often simplify the derivation of other formulas, as demonstrated in Historical Problems 1 and 2, which follow. ■

HISTORICAL PROBLEMS

■ 1. This derivation of Heron's formula uses Mollweide's Formula.
(a) Show that

$$\frac{s}{c} = \frac{\cos(\alpha/2)\cos(\beta/2)}{\sin(\gamma/2)}$$

where $s = \tfrac{1}{2}(a + b + c)$. *Hint:* Use Mollweide's Formula (see Problem 43 in Exercise 4.2) and add 1 to both sides. Then use the fact that

$$\sin\frac{\gamma}{2} = \sin\frac{180° - (\alpha + \beta)}{2} = \cos\frac{\alpha + \beta}{2}$$

(b) Similarly, show that

$$\frac{s}{a} = \frac{\cos(\beta/2)\cos(\gamma/2)}{\sin(\alpha/2)} \quad \text{and} \quad \frac{s}{b} = \frac{\cos(\alpha/2)\cos(\gamma/2)}{\sin(\beta/2)}$$

(c) Use the results of parts (a) and (b) to show that

$$\frac{s-a}{a} = \frac{\sin(\beta/2)\sin(\gamma/2)}{\sin(\alpha/2)} \qquad \frac{s-b}{b} = \frac{\sin(\alpha/2)\sin(\gamma/2)}{\sin(\beta/2)}$$

$$\frac{s-c}{c} = \frac{\sin(\alpha/2)\sin(\beta/2)}{\sin(\gamma/2)}$$

(d) Now form the product

$$\frac{s}{c} \cdot \frac{s-a}{a} \cdot \frac{s-b}{b} \cdot \frac{s-c}{c}$$

After cancellations, multiply each side by $cabc$, use the double-angle formulas, and use the fact that $A = \frac{1}{2}bc \sin \alpha = \frac{1}{2}ac \sin \beta$. Heron's Formula will then follow.

2. We again use Mollweide's Formula to derive some other interesting formulas. (These were derived in another way in Problems 37 and 38 of Exercise 4.3.)

   (a) Add 1 to both sides of the second form of Mollweide's Formula (see Problem 44 in Exercise 4.2) and simplify to obtain

   $$\frac{s-b}{c} = \frac{\sin(\alpha/2)\cos(\beta/2)}{\cos(\gamma/2)}$$

   (b) Similarly, show that

   $$\frac{s-c}{b} = \frac{\sin(\alpha/2)\cos(\gamma/2)}{\cos(\beta/2)}$$

   (c) Use the results of part (b) and Problem 1(b) and (c) to show that

   $$\cos^2 \frac{\gamma}{2} = \frac{s(s-c)}{ab} \quad \text{and} \quad \sin^2 \frac{\gamma}{2} = \frac{(s-a)(s-b)}{ab}$$

   (d) Show that

   $$\tan^2 \frac{\gamma}{2} = \frac{(s-a)(s-b)}{s(s-c)}$$

3. (a) If h_1, h_2, and h_3 are the altitudes dropped from A, B, and C, respectively, in a triangle (see the figure), show that

   $$\frac{1}{h_1} + \frac{1}{h_2} + \frac{1}{h_3} = \frac{s}{K}$$

   where K is the area of the triangle and $s = \frac{1}{2}(a + b + c)$. [*Hint*: $h_1 = 2K/a$.]

   (b) Show that a formula for the altitude h from a vertex to the opposite side a of a triangle is

   $$h = \frac{a \sin \beta \sin \gamma}{\sin \alpha}$$

4. *Inscribed Circle.* The lines that bisect each angle of a triangle meet in a single point O, and the perpendicular distance r from O to each side of the triangle is the same. The circle with center at O and radius r is called the *inscribed circle* of the triangle (see the figure).

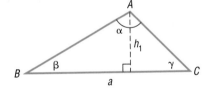

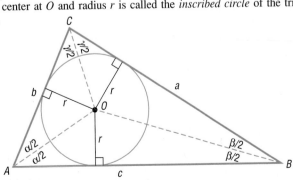

(a) Apply Problem 3(b) to triangle OAB to show that

$$r = \frac{c \sin(\alpha/2) \sin(\beta/2)}{\cos(\gamma/2)}$$

(b) Use the results of part (a) and Problem 1(c) to show that

$$r = (s - c) \tan \frac{\gamma}{2}$$

(c) Show that

$$\cot \frac{\alpha}{2} + \cot \frac{\beta}{2} + \cot \frac{\gamma}{2} = \frac{s}{r}$$

(d) Show that the area K of triangle ABC is $K = rs$. Then show that

$$r = \sqrt{\frac{(s - a)(s - b)(s - c)}{s}}$$

where $s = \frac{1}{2}(a + b + c)$.

5. Find the coordinates of the center O of the inscribed circle in terms of a, b, c, α, β, and γ. Then write the general equation of the inscribed circle. [*Hint:* Use a system of rectangular coordinates with the vertex A at the origin and side c along the positive x-axis.] ∎

4.4

Exercise 4.4

In Problems 1–8, find the area of each triangle. Round off answers to two decimal places.

1.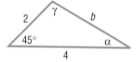

2.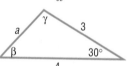

3.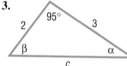

4.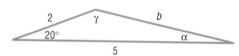

5.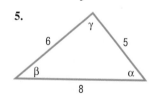

6.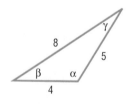

7.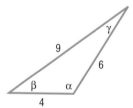

8.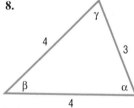

In Problems 9–24, find the area of each triangle. Round off answers to two decimal places.

9. $a = 3$, $b = 4$, $\gamma = 40°$

10. $a = 2$, $c = 1$, $\beta = 10°$

11. $b = 1$, $c = 3$, $\alpha = 80°$

12. $a = 6$, $b = 4$, $\gamma = 60°$

13. $a = 3$, $c = 2$, $\beta = 110°$

14. $b = 4$, $c = 1$, $\alpha = 120°$

15. $a = 2$, $b = 2$, $\gamma = 50°$

16. $a = 3$, $c = 2$, $\beta = 90°$

17. $a = 12$, $b = 13$, $c = 5$

18. $a = 4$, $b = 5$, $c = 3$

19. $a = 2$, $b = 2$, $c = 2$

20. $a = 3$, $b = 3$, $c = 2$

21. $a = 5$, $b = 8$, $c = 9$

22. $a = 4$, $b = 3$, $c = 6$

23. $a = 10$, $b = 8$, $c = 5$

24. $a = 9$, $b = 7$, $c = 10$

25. *Cost of a Triangular Lot* The dimensions of a triangular lot are 100 feet by 50 feet by 75 feet. If the price of such land is $3 per square foot, how much does the lot cost?

26. *Approximating the Area of a Lake* To approximate the area of a lake, a surveyor walks around the perimeter of the lake, taking the measurements shown in the illustration. Using this technique, what is the approximate area of the lake? [*Hint:* Use the Law of Cosines on the three triangles shown and then find the sum of their areas.]

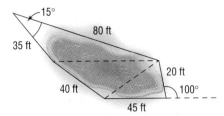

27. *Area of a Triangle* Prove that the area A of a triangle is given by the formula

$$A = \frac{a^2 \sin \beta \sin \gamma}{2 \sin \alpha}$$

28. *Area of a Triangle* Prove the two other forms of the formula given in Problem 27,

$$A = \frac{b^2 \sin \alpha \sin \gamma}{2 \sin \beta} \quad \text{and} \quad A = \frac{c^2 \sin \alpha \sin \beta}{2 \sin \gamma}$$

In Problems 29–36, use the results of Problem 27 or 28 to find the area of each triangle. Round off answers to two decimal places.

29. $\alpha = 40°, \beta = 20°, a = 2$

30. $\alpha = 50°, \gamma = 20°, a = 3$

31. $\beta = 70°, \gamma = 10°, b = 5$

32. $\alpha = 70°, \beta = 60°, c = 4$

33. $\alpha = 110°, \gamma = 30°, c = 3$

34. $\beta = 10°, \gamma = 100°, b = 2$

35. $\alpha = 40°, \beta = 40°, c = 2$

36. $\beta = 20°, \gamma = 70°, a = 1$

37. *Geometry* Consult the figure in the margin, which shows a circle of radius r with center at O. Find the area A of the shaded region as a function of the central angle θ.

4.5

Sinusoidal Graphs; Simple Harmonic Motion

Because $\sin x = \cos(x - \pi/2)$, it follows that the graph of $y = \sin x$ is the same as the graph of $y = \cos x$ after a horizontal shift of $\pi/2$ units to the right. See Figure 36.

FIGURE 36

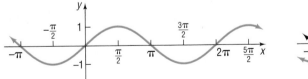

(a) $y = \sin x$

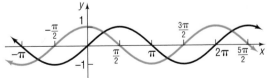

(b) $y = \cos x$ $y = \cos\left(x - \frac{\pi}{2}\right)$

Because of this, the graphs of sine functions and cosine functions are referred to as **sinusoidal graphs.**

Let's look at some general characteristics of sinusoidal graphs.

In Figure 37, we show the graph of $y = 2 \cos x$. Notice that the values of $y = 2 \cos x$ lie between -2 and 2, inclusive.

FIGURE 37
$y = 2 \cos x$

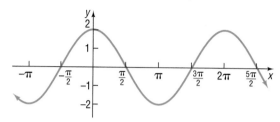

In general, the values of the functions $y = A \sin x$ and $y = A \cos x$, where $A \neq 0$, will always satisfy the inequalities.

$$-|A| \leq A \sin x \leq |A| \quad \text{and} \quad -|A| \leq A \cos x \leq |A|$$

respectively. The number $|A|$ is called the **amplitude** of $y = A \sin x$ or $y = A \cos x$. See Figure 38.

FIGURE 38

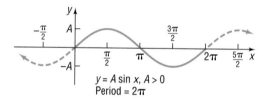

$y = A \sin x, A > 0$
Period $= 2\pi$

In Figure 39 we show the graph of $y = \cos 3x$. Notice that the period of this function is $2\pi/3$.

FIGURE 39
$y = \cos 3x$

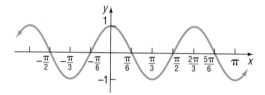

In general, if $\omega > 0$, the functions $y = \sin \omega x$ and $y = \cos \omega x$ will have period $T = 2\pi/\omega$. To see why, recall that the graph of $y = \sin \omega x$ is obtained from the graph of $y = \sin x$ by performing an appropriate horizontal compression or stretch. This horizontal compression replaces the interval $[0, 2\pi]$, which contains one period of the graph of $y = \sin x$, by the interval $[0, 2\pi/\omega]$, which contains one period of the graph of $y = \sin \omega x$. Thus, the period of the functions $y = \sin \omega x$ and $y = \cos \omega x$, $\omega > 0$, is $2\pi/\omega$. See Figure 40.

FIGURE 40

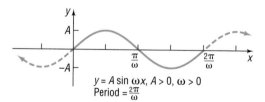

$y = A \sin \omega x, A > 0, \omega > 0$
Period $= \frac{2\pi}{\omega}$

If $\omega < 0$ in $y = \sin \omega x$ or $y = \cos \omega x$, we use the facts that

$$\sin \omega x = -\sin(-\omega x) \quad \text{and} \quad \cos \omega x = \cos(-\omega x)$$

to obtain an equivalent form in which the coefficient of x is positive.

Theorem   If $\omega > 0$, the amplitude and period of $y = A \sin \omega x$ and $y = A \cos \omega x$ are given by

$$\text{Amplitude} = |A| \qquad \text{Period} = T = \frac{2\pi}{\omega} \tag{1}$$

E X A M P L E   1   *Finding the Amplitude and Period of a Sinusoidal Function*

Determine the amplitude and period of $y = 3 \sin 4x$, and graph the function.

Solution   Comparing $y = 3 \sin 4x$ to $y = A \sin \omega x$, we find that $A = 3$ and $\omega = 4$. Thus, from equation (1), the amplitude is $|A| = 3$ and the period is $T = 2\pi/\omega = 2\pi/4 = \pi/2$. We can use this information to graph $y = 3 \sin 4x$ by beginning as shown in Figure 41.

FIGURE 41

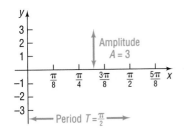

To graph $y = 3 \sin 4x$ using a graphing utility, we use the amplitude to set Ymin and Ymax and use the period to set Xmin and Xmax. It is best to show two periods and to set Xscl = period/4. See Figure 42(a). To graph $y = 3 \sin 4x$ by hand, we use the amplitude to scale the y-axis and the period to scale the x-axis. Then we fill in the graph of the sine function. See Figure 42(b).

FIGURE 42

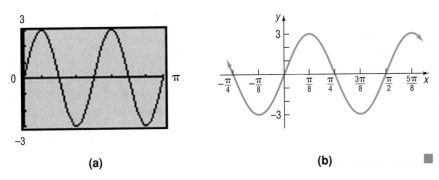

(a)   (b)

Now work Problem 5.

E X A M P L E   2   *Finding the Amplitude and Period of a Sinusoidal Function*

Determine the amplitude and period of $y = -4 \cos \pi x$, and graph the function.

Solution   Comparing $y = -4 \cos \pi x$ with $y = A \cos \omega x$, we find that $A = -4$ and $\omega = \pi$. Thus, the amplitude is $|A| = |-4| = 4$, and the period is $T = 2\pi/\omega = 2\pi/\pi = 2$.

Figure 43 shows the graph of $y = -4 \cos \pi x$ using a graphing utility.

FIGURE 43

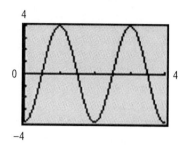

To graph $y = -4 \cos \pi x$ by hand, we use the amplitude to scale the y-axis, the period to scale the x-axis, and fill in the graph of the cosine function, thus obtaining the graph of $y = 4 \cos \pi x$ shown in Figure 44(a). Now, since we want the graph of $y = -4 \cos \pi x$, we reflect the graph of $y = 4 \cos \pi x$ about the x-axis, as shown in Figure 44(b).

FIGURE 44

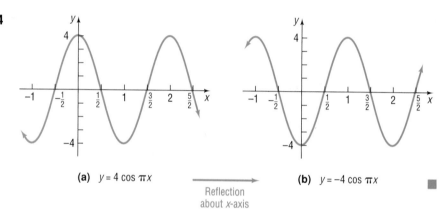

(a) $y = 4 \cos \pi x$

Reflection
about x-axis

(b) $y = -4 \cos \pi x$

$\blacksquare$

E X A M P L E   3   *Finding the Amplitude and Period of a Sinusoidal Function*
Determine the amplitude and period of $y = 2 \sin(-\pi x)$, and graph the function.

Solution   Since the sine function is odd, we use the equivalent form

$$y = -2 \sin \pi x$$

Comparing $y = -2 \sin \pi x$ to $y = A \sin \omega x$, we find that $A = -2$ and $\omega = \pi$. Thus, the amplitude is $|A| = 2$, and the period is $T = 2\pi/\omega = 2\pi/\pi = 2$.
Figure 45 shows the graph of $y = 2 \sin(-\pi x)$ using a graphing utility.

FIGURE 45

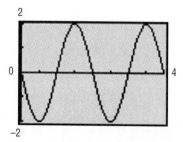

To graph $y = -2 \sin \pi x$ by hand, we use the amplitude to scale the y-axis, the period to scale the x-axis, and fill in the graph of the sine function. Figure 46(a) shows the resulting graph of $y = 2 \sin \pi x$. To obtain the graph of $y = -2 \sin \pi x$, we reflect the graph of $y = 2 \sin \pi x$ about the x-axis, as shown in Figure 46(b).

FIGURE 46

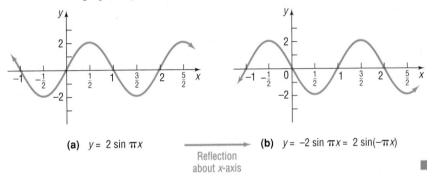

(a) $y = 2 \sin \pi x$ ⟶ Reflection about x-axis (b) $y = -2 \sin \pi x = 2 \sin(-\pi x)$

We also can use the ideas of amplitude and period to identify a sinusoidal function when its graph is given.

E X A M P L E   4 *Finding an Equation for a Sinusoidal Graph*

Find an equation for the graph shown in Figure 47.

FIGURE 47

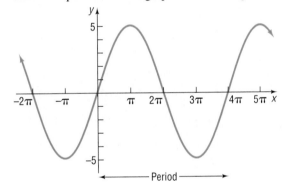

Period

Solution   This graph can be viewed as the graph of a sine function with amplitude $A = 5$.* The period T is observed to be 4π. Thus, by equation (1),

$$T = \frac{2\pi}{\omega}$$

$$4\pi = \frac{2\pi}{\omega}$$

$$\omega = \frac{2\pi}{4\pi} = \frac{1}{2}$$

A sine function whose graph is given in Figure 47 is

$$y = A \sin \omega x = 5 \sin \frac{x}{2}$$

Check: Graph $y = 5 \sin \dfrac{x}{2}$ and compare the result with Figure 47. ∎

*The equation could also be viewed as a cosine function with a horizontal shift, but viewing it as a sine function is easier.

E X A M P L E   5   *Finding an Equation for a Sinusoidal Graph*

Find an equation for the graph shown in Figure 48.

FIGURE 48

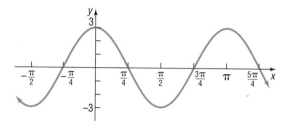

Solution   From the graph we conclude that it is easiest to view the equation as a cosine function with amplitude $A = 3$ and period $T = \pi$. Thus, $2\pi/\omega = \pi$, so $\omega = 2$. A cosine function whose graph is given in Figure 48 is

$$y = A \cos \omega x = 3 \cos 2x$$

Check: Graph $y = 3 \cos 2x$ and compare the result with Figure 48.   ■

E X A M P L E   6   *Finding an Equation for a Sinusoidal Graph*

Find an equation for the graph shown in Figure 49.

FIGURE 49

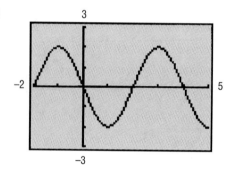

Solution   The graph is sinusoidal, with amplitude $A = 2$. The period is 4, so $2\pi/\omega = 4$ or $\omega = \pi/2$. Since the graph passes through the origin, it is easiest to view the equation as a sine function, but notice that the graph is actually the reflection of a sine function about the x-axis (since the graph is decreasing near the origin). Thus, we have

$$y = -A \sin \omega x = -2 \sin \frac{\pi}{2}x, \qquad A > 0$$

Check: Graph $y = -2 \sin \dfrac{\pi}{2}x$ and compare the result with Figure 49.   ■

FIGURE 50

One period of $y = A \sin \omega x$, $A > 0$, $\omega > 0$

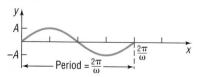

■   Now work Problem 37.

Phase Shift

We have seen that the graph of $y = A \sin \omega x$, $\omega > 0$, has amplitude $|A|$ and period $T = 2\pi/\omega$. Thus, one period can be drawn as x varies from 0 to $2\pi/\omega$ or, equivalently, as ωx varies from 0 to 2π. See Figure 50.

We now want to discuss the graph of

$$y = A \sin(\omega x - \phi)$$

where $\omega > 0$ and ϕ (the Greek letter phi) are real numbers. The graph will be a sine curve of amplitude $|A|$. As $\omega x - \phi$ varies from 0 to 2π, one period will be traced out. This period will begin when

$$\omega x - \phi = 0 \quad \text{or} \quad x = \frac{\phi}{\omega}$$

and will end when

$$\omega x - \phi = 2\pi \quad \text{or} \quad x = \frac{2\pi}{\omega} + \frac{\phi}{\omega}$$

FIGURE 51

One period of $y = A \sin(\omega x - \phi)$, $A > 0$, $\omega > 0$, $\phi > 0$

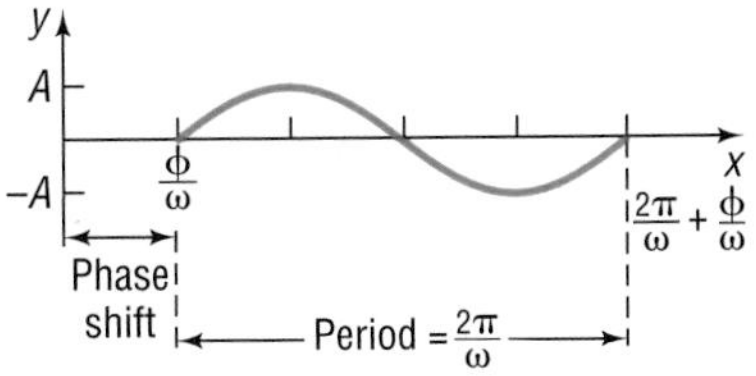

See Figure 51.

Thus, we see the graph of $y = A \sin(\omega x - \phi)$ is the same as the graph of $y = A \sin \omega x$, except that it has been shifted ϕ/ω units (to the right if $\phi > 0$ and to the left if $\phi < 0$). This number ϕ/ω is called the **phase shift** of the graph of $y = A \sin(\omega x - \phi)$.

For the graphs of $y = A \sin(\omega x - \phi)$ or $y = A \cos(\omega x - \phi)$, $\omega > 0$,

$$\text{Amplitude} = |A| \qquad \text{Period} = T = \frac{2\pi}{\omega} \qquad \text{Phase shift} = \frac{\phi}{\omega}$$

E X A M P L E   7

Finding the Amplitude, Period, and Phase Shift of a Sinusoidal Function

Find the amplitude, period, and phase shift of $y = 3 \sin(2x - \pi)$, and graph the function.

Solution   Comparing $y = 3 \sin(2x - \pi)$ to $y = A \sin(\omega x - \phi)$, we find that $A = 3$, $\omega = 2$, and $\phi = \pi$. The graph is a sine curve with amplitude $A = 3$ and period $T = 2\pi/\omega = 2\pi/2 = \pi$. One period of the sine curve begins at $2x - \pi = 0$ or $x = \pi/2$ (this is the phase shift) and ends at $2x - \pi = 2\pi$ or $x = 3\pi/2$. See Figure 52.

FIGURE 52

$y = 3 \sin(2x - \pi)$

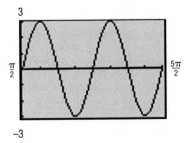

E X A M P L E   8

Finding the Amplitude, Period, and Phase Shift of a Sinusoidal Function

Find the amplitude, period, and phase shift of $y = 3 \cos(4x + 2\pi)$, and graph the function.

Solution   Comparing $y = 3 \cos(4x + 2\pi)$ to $y = A \cos(\omega x - \phi)$, we see that $A = 3$, $\omega = 4$, and $\phi = -2\pi$. The graph is a cosine curve with amplitude $A = 3$ and period

$T = 2\pi/\omega = 2\pi/4 = \pi/2$. One period of the cosine curve begins at $4x + 2\pi = 0$ or $x = -\pi/2$ (the phase shift) and ends at $4x + 2\pi = 2\pi$ or $x = 0$. See Figure 53.

FIGURE 53

$y = 3\cos(4x + 2\pi)$

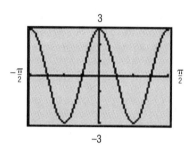

■

■ Now work Problem 51.

Simple Harmonic Motion

There are many physical phenomena that can be described as simple harmonic motion. Radio and television waves, light waves, sound waves, and water waves exhibit motion that is simple harmonic. Even yearly low and high temperatures at a given location can be modeled using an equation for simple harmonic motion.

The swinging of a pendulum, the vibrations of a tuning fork, and the bobbing of a weight attached to a coiled spring are also examples of vibrational motion. In this type of motion, an object swings back and forth over the same path. In each illustration in Figure 54, the point B is the **equilibrium (rest) position** of the vibrating object. The **amplitude** of vibration is the distance from the object's rest position to its point of greatest displacement (either point A or point C in Figure 54). The **period** of a vibrating object is the time required to complete one vibration—that is, the time it takes to go from, say, point A through B to C and back to A.

FIGURE 54

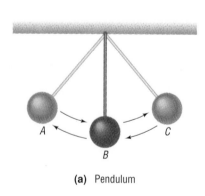

(a) Pendulum

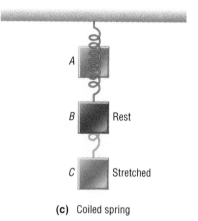

(b) Tuning fork

(c) Coiled spring

Simple Harmonic Motion

Simple harmonic motion is a special kind of vibrational motion in which the acceleration a of the object is directly proportional to the negative of its displacement d from its rest position. That is, $a = -kd, k > 0$.

For example, when the mass hanging from the spring in Figure 54(c) is pulled down from its rest position B to the point C, the force of the spring tries to restore the mass to its rest position. Assuming that there is no frictional force* to retard the motion, the amplitude will remain constant. The force increases in direct proportion to the distance the mass is pulled from its rest position. Since the force increases directly, the acceleration of the mass of the object must do likewise, because (by Newton's Second Law of Motion) force is directly proportional to acceleration. Thus, the acceleration of the object varies directly with its displacement, and the motion is an example of simple harmonic motion.

Simple harmonic motion is related to circular motion. To see this relationship, consider a circle of radius a, with center at $(0, 0)$. See Figure 55. Suppose that an object initially placed at $(a, 0)$ moves counterclockwise around the circle at constant angular speed ω. Suppose further that after time t has elapsed, the object is at the point $P = (x, y)$ on the circle. The angle θ, in radians, swept out by the ray $\overrightarrow{OP}$ in this time t is

FIGURE 55

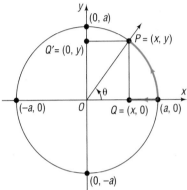

$$\theta = \omega t$$

The coordinates of the point P at time t are

$$x = a \cos \theta = a \cos \omega t$$
$$y = a \sin \theta = a \sin \omega t$$

Corresponding to each position $P = (x, y)$ of the object moving about the circle, there is the point $Q = (x, 0)$, called the **projection of P on the x-axis.** As P moves around the circle at a constant rate, the point Q moves back and forth between the points $(a, 0)$ and $(-a, 0)$ along the x-axis with a motion that is simple harmonic. Similarly, for each point P there is a point $Q' = (0, y)$, called the **projection of P on the y-axis.** As P moves around the circle, the point Q' moves back and forth between the points $(0, a)$ and $(0, -a)$ on the y-axis with a motion that is simple harmonic. Thus, simple harmonic motion can be described as the projection of constant circular motion on a coordinate axis.

Theorem
Simple Harmonic Motion

An object that moves on a coordinate axis so that its distance d from the origin at time t is given by either

$$d = a \cos \omega t \quad \text{or} \quad d = a \sin \omega t$$

where a and ω are constants, moves with simple harmonic motion. The motion has amplitude $|a|$ and period $2\pi/\omega$. ∎

The **frequency** f of an object in simple harmonic motion is the number of oscillations per unit time. Since the period is the time required for one oscillation, it follows that frequency is the reciprocal of the period; that is,

$$f = \frac{\omega}{2\pi}$$

*If friction is present, the amplitude will decrease with time to 0. This type of motion is an example of **damped motion**, which is discussed in the next section.

E X A M P L E   9   *Finding an Equation for an Object in Harmonic Motion*

Suppose that the object attached to the coiled spring in Figure 54(c) is pulled down a distance of 5 inches from its rest position and then released. If the time for one oscillation is 3 seconds, write an equation that relates the distance d of the object from its rest position after time t (in seconds). Assume no friction.

Solution   The motion of the object is simple harmonic. Since the object is released at time $t = 0$ when its distance d from the rest position is 5 inches, it is easiest to use the equation

$$d = a \cos \omega t$$

to describe the motion. (Do you see why? For this equation, when $t = 0$, then $d = a = 5$.) Now the amplitude is 5 and the period is 3. Thus,

$$a = 5 \quad \text{and} \quad \frac{2\pi}{\omega} = \text{Period} = 3 \quad \text{or} \quad \omega = \frac{2\pi}{3}$$

An equation of the motion of the object is

$$d = 5 \cos \frac{2\pi}{3}t \qquad \blacksquare$$

Note: In the solution to Example 9, we let $a = 5$, indicating that the positive direction of the motion is down. If we wanted the positive direction to be up, we would let $a = -5$.

■   Now work Problem 63.

E X A M P L E   1 0   *Analyzing the Motion of an Object*

Suppose that the distance x (in meters) an object travels in time t (in seconds) satisfies the equation

$$x = 10 \sin 5t$$

(a) Describe the motion of the object.

(b) What is the maximum displacement from its resting position?

(c) What is the time required for one oscillation?

(d) What is the frequency?

Solution   We observe that the given equation is of the form

$$d = a \sin \omega t \quad d = 10 \sin 5t$$

where $a = 10$ and $\omega = 5$.

(a) The motion is simple harmonic.

(b) The maximum displacement of the object from its resting position is the amplitude: $a = 10$ meters.

(c) The time required for one oscillation is the period:

$$\text{Period} = \frac{2\pi}{\omega} = \frac{2\pi}{5} \text{ seconds}$$

(d) The frequency is the reciprocal of the period. Thus,

$$\text{Frequency} = f = \frac{5}{2\pi} \text{ oscillation per second} \qquad \blacksquare$$

■   Now work Problem 71.

83. *Alternating Current (ac) Generators* The voltage V produced by an ac generator is

$$V = 220 \sin 120\pi t$$

(a) What is the amplitude? What is the period?
(b) Graph V over two periods, beginning at $t = 0$.
(c) If a resistance of $R = 10$ ohms is present, what is the current I? [*Hint:* Use Ohm's Law, $V = IR$].
(d) What is the amplitude and period of the current I?
(e) Graph I over two periods, beginning at $t = 0$.

84. *Alternating Current (ac) Generators* The voltage V produced by an ac generator is

$$V = 120 \sin 120\pi t$$

(a) What is the amplitude? What is the period?
(b) Graph V over two periods, beginning at $t = 0$.
(c) If a resistance of $R = 20$ ohms is present, what is the current I? [*Hint:* Use Ohm's Law, $V = IR$].
(d) What is the amplitude and period of the current I?
(e) Graph I over two periods, beginning at $t = 0$.

85. *Alternating Current (ac) Generators* The voltage V produced by an ac generator is sinusoidal. As a function of time, the voltage V is

$$V = V_0 \sin 2\pi ft$$

where f is the **frequency,** the number of complete oscillations (cycles) per second. [In the United States and Canada, f is 60 hertz (Hz)]. The **power** P delivered to a resistance R at any time t is defined as

$$P = \frac{V^2}{R}$$

(a) Show that $P = \dfrac{V_0{}^2}{R} \sin^2 2\pi ft$.
(b) The graph of P is shown in the figure. Express P as a sinusoidal function.
(c) Deduce that

$$\sin^2 2\pi ft = \frac{1}{2}(1 - \cos 4\pi ft)$$

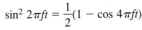

Power in an a-c generator

86. *Biorhythms* In the theory of biorhythms, a sine function of the form

$$P = 100 \sin \omega t$$

is used to measure the percent P of a person's potential at time t, where t is measured in days and $t = 0$ is the person's birthday. Three characteristics are commonly measured:

Physical potential: period of 23 days

Emotional potential: period of 28 days

Intellectual potential: period of 33 days

(a) Find ω for each characteristic.
(b) Graph all three functions.
(c) Is there a time t when all three characteristics have 100% potential? When is it?
(d) Suppose that you are 20 years old today ($t = 7305$ days). Describe your physical, emotional, and intellectual potential for the next 30 days.

87. Explain how the amplitude and period of a sinusoidal graph are used to establish the scale on each coordinate axis.

88. Find an application in your major field that leads to a sinusoidal graph. Write a paper about your findings.

89. *CBL Experiment* Part 1: A student puts his/her thumb over a light probe and points it at a light source. The student then begins lifting his/her thumb from the sensor and replacing it repeatedly. Light intensity is plotted over time and the period and frequency are determined. Part 2: A light probe is pointed toward a fluorescent light bulb and its light intensity is recorded. The frequency and period of the light bulb are recorded. (Activity 15, Real-World Math with the CBL System, 1994.)

90. *CBL Experiment* Pendulum motion is analyzed to estimate simple harmonic motion. A plot is generated with the position of the pendulum over time. The graph is used to find a sinusoidal curve of the form $y = A \cos B(x - C) + D$. Determine the amplitude, period and frequency. (Activity 16, Real-World Math with the CBL System.)

91. *CBL Experiment* The sound from a tuning fork is collected over time. The amplitude, frequency and period of the graph is determined. A model of the form $y = A \cos B(x - C)$ is fit to the data. (Activity 23, Real-World Math with the CBL System.)

92. *CBL Experiment* A weight is attached to a spring and allowed to "bob" up and down. The position of the weight is recorded over time. A model of the form $f(t) = a(t) \sin (bt + c) + d$ is developed from the data. (Experiment M3, CBL System Experiment Workbook, 1994.)

4.6

Damped Vibrations

Many physical and biological applications require the graphs of sums and products of functions, such as

$$f(x) = \sin x + \cos 2x \quad \text{and} \quad g(x) = e^x \sin x$$

For example, if two tones are emitted, the sound produced is the sum of the waves produced by the two tones. See Problem 29 for an explanation of Touch-Tone phones.

To analyze the graph of the sum or product of two or more functions, it is best to show the graphs of the component parts along with the graph of their sum or product. Let's look at an example.

E X A M P L E   1   *Analyzing the Graph of the Sum of Two Functions*

Analyze the graph of $f(x) = x + \sin x$.

Solution   First we graph the component functions:

$$Y_1 = x \quad \text{and} \quad Y_2 = \sin x$$

See Figure 56(a). Then we graph $Y_3 = x + \sin x$. See Figure 56(b). Notice that the graph of $Y_3 = x + \sin x$ is a sine curve drawn using $Y_1 = x$ as axis. Furthermore, notice that $Y_3 = x + \sin x$ will intersect $Y_1 = x$ whenever $\sin x = 0$. Finally, notice that Y_3 is not periodic.

FIGURE 56

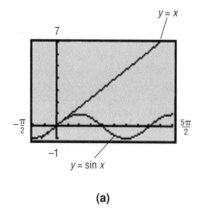

(a)

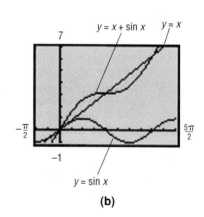

(b)

Check: Verify that the graphs of Y_3 and Y_1 intersect when $\sin x = 0$. ∎

The next example shows a periodic graph of the sum of two functions.

E X A M P L E 2   *Analyze the Graph of the Sum of Two Sinusoidal Functions*

Analyze the graph of $f(x) = \sin x + \cos 2x$

Solution   First we graph the component parts: $Y_1 = \sin x$ and $Y_2 = \cos 2x$. See Figure 57(a).
Then we graph the sum: $Y_3 = \sin x + \cos 2x$. See Figure 57(b).

FIGURE 57

$f(x) = \sin x + \cos 2x$

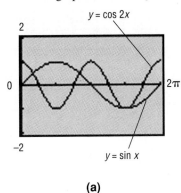

(a)   **(b)**

■   Now work Problem 1.

The amplitude of any real oscillating spring or swinging pendulum decreases
with time due to air resistance, friction, and so on. See Figure 58. The graph of
such phenomena is generally given by the product of two functions.

FIGURE 58

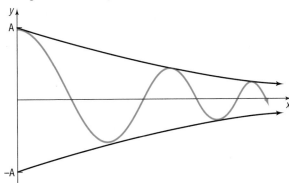

E X A M P L E 3   *Graphing the Product of Two Functions*

Graph $f(x) = x \sin x$.

Solution   Here, f is the product of $y = x$ and $y = \sin x$. Using properties of absolute value
and the fact that $|\sin x| \le 1$, we find that

$$|f(x)| = |x \sin x| = |x|\, |\sin x| \le |x|$$

If $x \ge 0$, this reduces to

$$|f(x)| \le x \quad \text{or} \quad -x \le f(x) \le x, \qquad x > 0$$

If $x < 0$ we have

$$|f(x)| \le -x \quad \text{or} \quad x \le f(x) \le -x, \qquad x < 0$$

Thus, in every case the graph of f will lie between the lines $y = x$ and $y = -x$.
Furthermore, we conclude that the graph of f will touch these lines when $\sin x = \pm 1$,

that is, when $x = -\pi/2, \pi/2, 3\pi/2$, and so on. Since $y = f(x) = x \sin x = 0$ when $x = -\pi, 0, \pi, 2\pi$, and so on, we know the location of the x-intercepts of the graph of f. Finally, the function f is an even function $[f(-x) = -x \sin(-x) = x \sin x = f(x)]$, so the graph is symmetric with respect to the y-axis. See Table 1. Figure 59 illustrates the hand drawn graph.

TABLE 1

x	**0**	**$\pi/2$**	**π**	**$3\pi/2$**	**2π**
$y = x$	0	$\pi/2$	π	$3\pi/2$	2π
$y = \sin x$	0	1	0	-1	0
$y = f(x) = x \sin x$	0	$\pi/2$	0	$-3\pi/2$	0
POINT ON GRAPH OF f	$(0, 0)$	$(\pi/2, \pi/2)$	$(\pi, 0)$	$(3\pi/2, -3\pi/2)$	$(2\pi, 0)$

FIGURE 59
$f(x) = x \sin x$

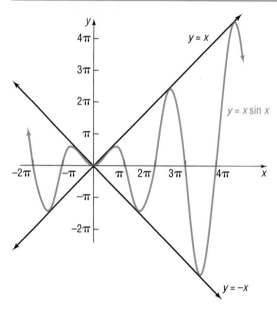

EXAMPLE 4 *Analyzing the Graph of $y = x \sin x$*

Graph $y = x \sin x$, along with $y = x$ and $y = -x$, for $0 \le x \le 4\pi$. Determine where $y = x \sin x$ touches $y = x$. Compare this to where $y = x \sin x$ has a local maximum. Express answers correct to two decimal places.

Solution Figure 60 shows the graphs of $y = x \sin x$, $y = x$, and $y = -x$. Using TRACE and BOX (or INTERSECT), we find that $y = x \sin x$ touches $y = x$ at $x = 1.57$ ($\pi/2$) and at $x = 7.85$ (2π). The local maxima occur just to the right of each of these values at $x = 2.02$ and at $x = 7.97$.

FIGURE 60

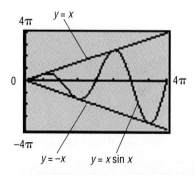

If $x \geq 0$, the graph of $f(x) = x \sin x$ is called an **amplified sine wave,** and x is called the **amplifying factor.** By using different factors, we can obtain other sinusoidal waves. The next example illustrates a **damping factor.**

E X A M P L E   5   *Graphing the Product of Two Functions*

The damped vibration curve

$$f(x) = e^{-x} \sin x, \qquad x \geq 0$$

is of importance in many applications, such as the motion of musical strings, the vibrations of a pendulum, and the current in an electrical circuit. Graph this function.

Solution   The function f is the product of $y = e^{-x}$ and $y = \sin x$. Using properties of absolute value and the fact that $|\sin x| \leq 1$, we find that

$$|f(x)| = |e^{-x} \sin x| = |e^{-x}| \, |\sin x| \leq |e^{-x}| = e^{-x}$$
$$\uparrow$$
$$e^{-x} > 0 \text{ for all } x$$

Thus,

$$-e^{-x} \leq f(x) \leq e^{-x}$$

and the graph of f will lie between the graphs of $y = e^{-x}$ and $y = -e^{-x}$. Also, the graph of f will touch these graphs when $|\sin x| = 1$, that is, when $x = -\pi/2$, $\pi/2$, $3\pi/2$, and so on. The x-intercepts of the graph of f occur at $x = -\pi$, 0, π, 2π, and so on. See Table 2. See Figure 61 for the graph.

TABLE 2

x	0	$\pi/2$	π	$3\pi/2$	2π
$y = e^{-x}$	1	$e^{-\pi/2}$	$e^{-\pi}$	$e^{-3\pi/2}$	$e^{-2\pi}$
$y = \sin x$	0	1	0	-1	0
$f(x) = e^{-x} \sin x$	0	$e^{-\pi/2}$	0	$-e^{-3\pi/2}$	0
POINT ON GRAPH OF f	$(0, 0)$	$(\pi/2, e^{-\pi/2})$	$(\pi, 0)$	$(3\pi/2, -e^{-3\pi/2})$	$(2\pi, 0)$

FIGURE 61
$f(x) = e^{-x} \sin x,$
$x \geq 0$

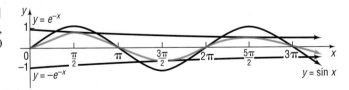

E X A M P L E   6   *Analyzing the Graph of $y = e^{-x} \sin x$*

Graph $y = e^{-x} \sin x$, along with $y = e^{-x}$ and $y = -e^{-x}$ for $0 \leq x \leq \pi$. Determine where $y = e^{-x} \sin x$ touches $y = e^{-x}$. Compare this to where $y = e^{-x} \sin x$ has a local maximum. Express answers correct to two decimal places.

Solution   Figure 62 shows the graphs of $y = e^{-x} \sin x$, $y = e^{-x}$, and $y = -e^{-x}$. Using TRACE and BOX (or INTERSECT), we find that $y = e^{-x} \sin x$ touches $y = e^{-x}$ at $x = \frac{\pi}{2} \approx 1.57$. The local maximum occurs at $x = 0.78$.

FIGURE 62

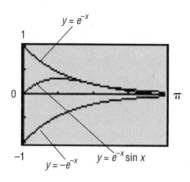

■ Now work Problem 21.

4.6

Exercise 4.6

In Problems 1–20, use a graphing utility to graph each function.

1. $f(x) = x + \cos x$

2. $f(x) = x + \cos 2x$

3. $f(x) = x - \sin x$

4. $f(x) = x - \cos x$

5. $f(x) = \sin x + \cos x$

6. $f(x) = \sin 2x + \cos x$

7. $g(x) = \sin x + \sin 2x$

8. $g(x) = \cos 2x + \cos x$

9. $h(x) = \sqrt{x} + \sin x$

10. $h(x) = \sqrt{x} + \cos x$

11. $F(x) = 2 \sin x - \cos 2x$

12. $F(x) = 2 \cos 2x - \sin x$

13. $f(x) = 2 \sin \pi x + \cos \pi x$

14. $f(x) = 2 \cos \dfrac{\pi}{2}x + \sin \dfrac{\pi}{2}x$

15. $f(x) = \dfrac{x^2}{\pi^2} + \sin 2x$

16. $f(x) = \dfrac{x^2}{\pi^2} - \cos 2x$

17. $f(x) = |x| + \sin \dfrac{\pi}{2}x$

18. $f(x) = |x| + \cos \pi x$

19. $f(x) = 3 \sin 2x + 2 \cos 3x$

20. $f(x) = 2 \sin 3x + 3 \cos 2x$

In Problems 21–28, graph each function over the indicated interval. Analyze the graph using a graphing utility.

21. $f(x) = x \cos x,\ [0, 3\pi]$

22. $f(x) = x \sin 2x,\ [0, 3\pi]$

23. $f(x) = x^2 \sin x,\ [0, 2\pi]$

24. $f(x) = x^2 \cos x,\ [0, 2\pi]$

25. $f(x) = |x| \cos x,\ [0, 2\pi]$

26. $f(x) = |x| \sin x,\ [0, 2\pi]$

27. $f(x) = e^{-x} \cos 2x,\ [0, \pi]$

28. $f(x) = e^{-x} \sin 2x,\ [0, \pi]$

29. *Touch-Tone Phones* On a Touch-Tone phone, each button produces a unique sound. The sound produced is the sum of two tones, given by

$$y = \sin 2\pi l t \quad \text{and} \quad y = \sin 2\pi h t$$

where l and h are the low and high frequencies (cycles per second) shown on the illustration. For example, if you touch 7, the low frequency is $l = 852$ cycles per second and the high frequency is $h = 1209$ cycles per second. The sound emitted by touching 7 is

$$y = \sin 2\pi(852)t + \sin 2\pi(1209)t$$

Graph the sound emitted by touching 7.

30. *Touch-Tone Phones* Graph the sound emitted by the star key (*) on a Touch-Tone phone. See Problem 29.

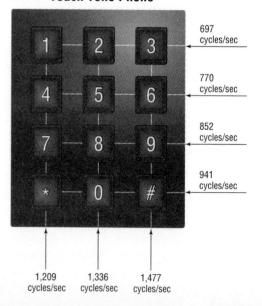

Touch Tone Phone

31. *The Sawtooth Curve* An oscilloscope often displays a *sawtooth curve*. This curve can be approximated by sinusoidal curves of varying periods and amplitudes.

(a) Graph the following function, which can be used to approximate the sawtooth curve:
$$f(x) = \tfrac{1}{2} \sin 2\pi x + \tfrac{1}{4} \sin 4\pi x \qquad 0 \le x \le 2$$

(b) A better approximation to the sawtooth curve is given by
$$f(x) = \tfrac{1}{2} \sin 2\pi x + \tfrac{1}{4} \sin 4\pi x + \tfrac{1}{8} \sin 8\pi x$$

Graph this function for $0 \le x \le 4$ and compare the result to the graph obtained in part (a).

(c) A third and even better approximation to the sawtooth curve is given by
$$f(x) = \tfrac{1}{2} \sin 2\pi x + \tfrac{1}{4} \sin 4\pi x + \tfrac{1}{8} \sin 8\pi x + \tfrac{1}{16} \sin 16\pi x$$

Graph this function for $0 \le x \le 4$ and compare the result to the graphs obtained in parts (a) and (b).

(d) What do you think the next approximation to the sawtooth curve is?

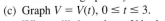

32. *Charging a Capacitor* If a charged capacitor is connected to a coil by closing a switch (see the figure), energy is transferred to the coil and then back to the capacitor in an oscillatory motion. The voltage V (in volts) across the capacitor will gradually diminish to 0 with time t.

(a) Graph the equation relating V and t:
$$V(t) = e^{-1.9t} \cos \pi t \qquad 0 \le t \le 3$$

(b) At what times t will the graph of V touch the graph of $y = e^{-1.9t}$? When does V touch the graph of $y = -e^{-1.9t}$?

(c) Graph $V = V(t)$, $0 \le t \le 3$.
When will the voltage V be between -0.1 and 0.1 volt?

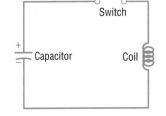

33. Graph the function $f(x) = (\sin x)/x$, $x > 0$. Based on the graph, what do you conjecture about the value of $(\sin x)/x$ for x close to 0?

34. Graph $y = x \sin x$, $y = x^2 \sin x$, and $y = x^3 \sin x$ for $x > 0$. What patterns do you observe?

35. Graph $y = \dfrac{1}{x} \sin x$, $y = \dfrac{1}{x^2} \sin x$, and $y = \dfrac{1}{x^3} \sin x$ for $x > 0$. What patterns do you observe?

36. How would you explain to a friend what simple harmonic motion is?

Chapter Review

THINGS TO KNOW

Formulas

Law of Sines	$\dfrac{\sin \alpha}{a} = \dfrac{\sin \beta}{b} = \dfrac{\sin \gamma}{c}$
Law of Cosines	$c^2 = a^2 + b^2 - 2ab \cos \gamma$
	$b^2 = a^2 + c^2 - 2ac \cos \beta$
	$a^2 = b^2 + c^2 - 2bc \cos \alpha$
Area of a triangle	$A = \tfrac{1}{2}bh$
	$A = \tfrac{1}{2}ab \sin \gamma$
	$A = \tfrac{1}{2}bc \sin \alpha$
	$A = \tfrac{1}{2}ac \sin \beta$
	$A = \sqrt{s(s-a)(s-b)(s-c)}$, where $s = \tfrac{1}{2}(a+b+c)$
Sinusoidal graphs	$y = A \sin \omega x$, $\omega > 0$ Period $= 2\pi/\omega$
	$y = A \cos \omega x$, $\omega > 0$ Amplitude $= |A|$
	$y = A \sin(\omega x - \phi)$ Phase shift $= \phi/\omega$
	$y = A \cos(\omega x - \phi)$

How To:

Solve right triangles

Use the Law of Sines to solve a SAA, ASA, or SSA triangle

Use the Law of Cosines to solve a SAS or SSS triangle

Find the area of a triangle

Find the period and amplitude of a sinusoidal function and use them to graph the function

Find a function whose sinusoidal graph is given

Graph certain sums of functions

Graph certain products of functions and analyze their graph

Fill-in-the-Blank Items

1. If two sides and the angle opposite one of them are known, the Law of _____ is used to determine whether the known information results in no triangle, one triangle, or two triangles.

2. If three sides of a triangle are given, the Law of _____ is used to solve the triangle.

3. If three sides of a triangle are given, _____ Formula is used to find the area of the triangle.

4. The following function has amplitude 3 and period 2:

$$y = \underline{\hspace{1.5cm}} \sin \underline{\hspace{1.5cm}} x$$

5. The function $y = 3 \sin 6x$ has amplitude _____ and period _____.

6. The amplifying factor of $f(x) = 5x \cos 3\pi x$ is _____.

7. The motion of an object obeys the equation $x = 4 \cos 6t$. Such motion is described as _____ _____ _____.

True/False Items

T F **1.** An oblique triangle in which two sides and an angle are given always results in at least one triangle.

T F **2.** Given three sides of a triangle, there is a formula for finding its area.

T F **3.** The graphs of $y = \sin x$ and $y = \cos x$ are identical except for a horizontal shift.

T F **4.** For $y = -2 \sin \pi x$, the amplitude is 2 and the period is $\pi/2$.

Review Exercises

In Problems 1–4, solve each triangle.

1. 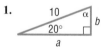   **2.** (triangle with sides c, b, 5 and angles 35°, β)   **3.** 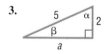   **4.** (triangle with sides c, 3, 1 and angles α, β)

In Problems 5–24, find the remaining angle(s) and side(s) of each triangle, if it exists. If no triangle exists, say "No triangle."

5. $\alpha = 50°$, $\beta = 30°$, $a = 1$

6. $\alpha = 10°$, $\gamma = 40°$, $c = 2$

7. $\alpha = 100°$, $a = 5$, $c = 2$

8. $a = 2$, $c = 5$, $\alpha = 60°$

9. $a = 3$, $c = 1$, $\gamma = 110°$

10. $a = 3$, $c = 1$, $\gamma = 20°$

11. $a = 3$, $c = 1$, $\beta = 100°$

12. $a = 3$, $b = 5$, $\beta = 80°$

13. $a = 2$, $b = 3$, $c = 1$

14. $a = 10$, $b = 7$, $c = 8$

15. $a = 1$, $b = 3$, $\gamma = 40°$

16. $a = 4$, $b = 1$, $\gamma = 100°$

17. $a = 5$, $b = 3$, $\alpha = 80°$

18. $a = 2$, $b = 3$, $\alpha = 20°$

19. $a = 1$, $b = \frac{1}{2}$, $c = \frac{4}{3}$

20. $a = 3$, $b = 2$, $c = 2$

21. $a = 3$, $\alpha = 10°$, $b = 4$

22. $a = 4$, $\alpha = 20°$, $\beta = 100°$

23. $c = 5$, $b = 4$, $\alpha = 70°$

24. $a = 1$, $b = 2$, $\gamma = 60°$

In Problems 25–34, find the area of each triangle.

25. $a = 2$, $b = 3$, $\gamma = 40°$

26. $b = 5$, $c = 4$, $\alpha = 20°$

27. $b = 4$, $c = 10$, $\alpha = 70°$

28. $a = 2$, $b = 1$, $\gamma = 100°$

29. $a = 4$, $b = 3$, $c = 5$

30. $a = 10$, $b = 7$, $c = 8$

31. $a = 4$, $b = 2$, $c = 5$

32. $a = 3$, $b = 2$, $c = 2$

33. $\alpha = 50°$, $\beta = 30°$, $a = 1$

34. $\alpha = 10°$, $\gamma = 40°$, $c = 3$

In Problems 35–38, determine the amplitude and period of each function without graphing.

35. $y = 4 \cos x$

36. $y = \sin 2x$

37. $y = -8 \sin \dfrac{\pi}{2}x$

38. $y = -2 \cos 3\pi x$

In Problems 39–46, find the amplitude, period, and phase shift of each function. Use a graphing utility to graph each function. Show at least one period.

39. $y = 4 \sin 3x$

40. $y = 2 \cos \frac{1}{3}x$

41. $y = -2 \sin\left(\dfrac{\pi}{2}x + \dfrac{1}{2}\right)$

42. $y = -6 \ \sin(2\pi x - 2)$

43. $y = \frac{1}{2} \sin(\frac{3}{2}x - \pi)$

44. $y = \frac{3}{2} \cos(6x + 3\pi)$

45. $y = -\frac{2}{3} \cos(\pi x - 6)$

46. $y = -7 \ \sin\left(\dfrac{\pi}{3}x + \dfrac{4}{3}\right)$

In Problems 47–50, find a function whose graph is given.

47.

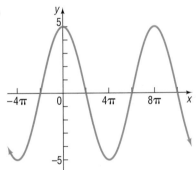

48.

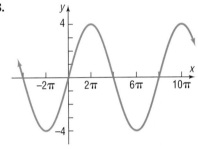

49.

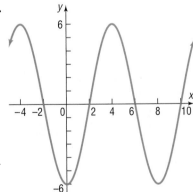

50.

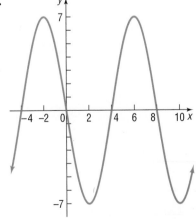

In Problems 51–56, use a graphing utility to graph each function.

51. $f(x) = 2x + \sin 2x$

52. $f(x) = 2x + \cos 2x$

53. $f(x) = \sin \pi x + \cos \dfrac{\pi}{2}x$

54. $f(x) = \sin \dfrac{\pi}{2}x + \cos \pi x$

55. $f(x) = 3 \sin \pi x + 2 \cos \pi x$

56. $f(x) = 3 \cos 2\pi x + 4 \sin 2\pi x$

In Problems 57–62, graph each function over the indicated interval. Analyze the graph using a graphing utility.

57. $f(x) = x \cos 2x$, $[0, \pi]$ **58.** $f(x) = x \sin \pi x$, $[0, 2]$ **59.** $f(x) = e^{-x} \sin \pi x$, $[0, 2]$

60. $f(x) = e^{-x} \cos \pi x$, $[0, 2]$ **61.** $f(x) = e^x \sin \pi x$, $[0, 2]$ **62.** $f(x) = e^x \cos \pi x$, $[0, 2]$

In Problems 63–66, the distance d (in feet) an object travels in time t (in seconds) is given.
(a) Describe the motion of the object.
(b) What is the maximum displacement from its resting position?
(c) What is the time required for one oscillation?
(d) What is the frequency?

63. $d = 6 \sin 2t$ **64.** $d = 2 \cos 4t$ **65.** $d = -2 \cos \pi t$ **66.** $d = -3 \sin \dfrac{\pi}{2} t$

67. *Alternating Voltage* The electromotive force E, in volts in a certain ac circuit obeys the equation

$$E = 120 \sin 120 \pi t, \ t \geq 0$$

where t is measured in seconds.
(a) What is the maximum value of E?
(b) What is the period?
(c) Graph this function over two periods.

68. *Alternating Current* The current I, in amperes, flowing through an ac (alternating current) circuit at time t is

$$I = 220 \sin\left(30\pi t + \frac{\pi}{6}\right), \qquad t \geq 0$$

(a) What is the period?   (b) What is the amplitude?
(c) What is the phase shift?   (d) Graph this function over two periods.

69. *Measuring the Length of a Lake* From a stationary hot air balloon 500 feet above the ground, two sightings of a lake are made (see the figure). How long is the lake?

70. *Finding the Speed of a Glider* From a glider 200 feet above the ground, two sightings of a stationary object directly in front are taken 1 minute apart (see the figure). What is the speed of the glider?

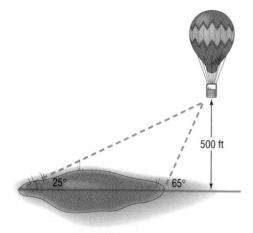

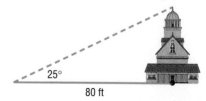

71. *Finding the Width of a River* Find the distance from A to C across the river illustrated in the figure.

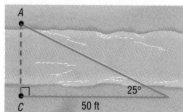

72. *Finding the Height of a Building* Find the height of the building shown in this figure.

73. *Finding the Distance to Shore* The Sears Tower in Chicago is 1454 feet tall and is situated about 1 mile inland from the shore of Lake Michigan, as indicated in the figure. An observer in a pleasure boat on the lake directly in front of the Sears Tower looks at the top of the tower and measures the angle of elevation as 5°. How far offshore is the boat?

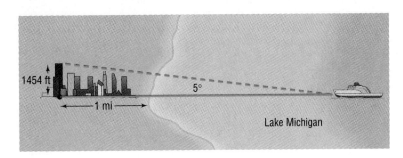

74. *Finding the Grade of a Mountain Trail* A straight trail with a uniform inclination leads from a hotel, elevation 5000 feet, to a lake in a valley, elevation 4100 feet. The length of the trail is 4100 feet. What is the inclination (grade) of the trail?

75. *Navigation* An airplane flies from city A to city B, a distance of 100 miles, then turns through an angle of 20° and heads toward city C, as indicated in the figure. If the distance from A to C is 300 miles, how far is it from city B to city C?

76. *Correcting a Navigation Error* Two cities A and B are 300 miles apart. In flying from city A to city B, a pilot inadvertently took a course that was 5° in error.
   (a) If the error was discovered after flying 10 minutes at a constant speed of 420 miles per hour, through what angle should the pilot turn to correct the course? (Consult the figure below.)
   (b) What new constant speed should be maintained so that no time is lost due to the error? (Assume that the speed would have been a constant 420 miles per hour if no error had occurred.)

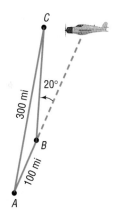

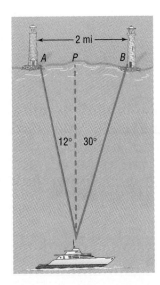

77. *Determining Distances at Sea* The navigator of a ship at sea spots two lighthouses that she knows to be 2 miles apart along a straight seashore. She determines that the angles formed between two line-of-sight observations of the lighthouses and the line from the ship directly to shore are 12° and 30°. See the illustration.
   (a) How far is the ship from lighthouse A?
   (b) How far is the ship from lighthouse B?
   (c) How far is the ship from shore?

78. *Constructing a Highway* A highway whose primary directions are north–south is being constructed along the west coast of Florida. Near Naples, a bay obstructs the straight path of the road. Since the cost of a bridge is prohibitive, engineers decide to go around the bay. The illustration shows the path they decide on and the measurements taken. What is the length of highway needed to go around the bay?

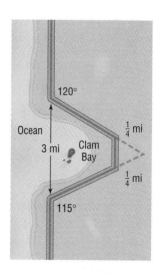

79. *Correcting a Navigational Error* A yacht leaves St. Thomas bound for an island in the British West Indies, 200 miles away. Maintaining a constant speed of 18 miles per hour, but encountering heavy crosswinds and strong currents, the crew finds after 4 hours that the sailboat is off course by 15°.

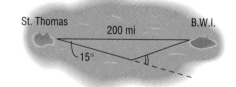

(a) How far is the sailboat from the island at this time?
(b) Through what angle should the sailboat turn to correct its course?
(c) How much time has been added to the trip because of this? (Assume that the speed remains at 18 miles per hour.)

80. *Surveying* Two homes are located on opposite sides of a small hill. See the illustration. To measure the distance between them, a surveyor walks a distance of 50 feet from house *A* to point *C*, uses a transit to measure the angle *ACB*, which is found to be 80°, and then walks to house *B*, a distance of 60 feet. How far apart are the houses?

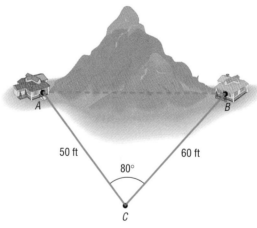

81. *Approximating the Area of a Lake* To approximate the area of a lake, a surveyor walks around the perimeter of the lake, taking the measurements shown in the illustration. Using this technique, what is the approximate area of the lake? [*Hint:* Use the Law of Cosines on the three triangles shown and then find the sum of their areas.]

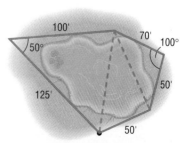

In using polar coordinates (r, θ), it is possible for the first entry r to be negative. When this happens, we follow the convention that the location of the point, instead of being on the terminal side of θ, is on the ray from the pole extending in the direction *opposite* the terminal side of θ at a distance $|r|$ from the pole. See Figure 5 for an illustration.

FIGURE 5

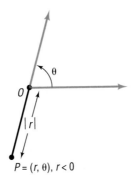

$P = (r, \theta), r < 0$

E X A M P L E   2

Polar Coordinates (r, θ), $r < 0$.

Consider again the point P with polar coordinates $(2, \pi/4)$, as shown in Figure 6(a). This same point P can be assigned the polar coordinates $(-2, 5\pi/4)$, as indicated in Figure 6(b). To locate the point $(-2, 5\pi/4)$, we use the ray in the opposite direction of $5\pi/4$ and go out 2 units along that ray to find the point P.

FIGURE 6

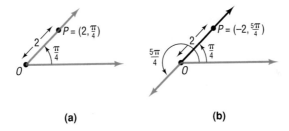

(a)　　　　　　　　　　(b)

These examples show a major difference between rectangular coordinates and polar coordinates. In the former, each point has exactly one pair of rectangular coordinates; in the latter, a point can have infinitely many pairs of polar coordinates.

Summary

A point with polar coordinates (r, θ) also can be represented by any of the following:

$$(r, \theta + 2k\pi) \quad \text{or} \quad (-r, \theta + \pi + 2k\pi), \quad k \text{ any integer}$$

The polar coordinates of the pole are $(0, \theta)$, where θ can be any angle.

E X A M P L E   3

Plotting Points Using Polar Coordinates

Plot the points with the following polar coordinates:

(a) $(3, 5\pi/3)$　　　(b) $(2, -\pi/4)$　　　(c) $(3, 0)$　　　(d) $(-2, \pi/4)$

Solution   Figure 7 shows the points.

FIGURE 7

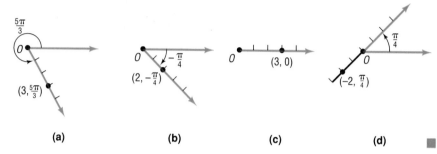

(a)   (b)   (c)   (d)   ■

■ Now work Problem 9.

E X A M P L E   4   *Finding Other Polar Coordinates of a Given Point*

Plot the point P with polar coordinates $(3, \pi/6)$, and find other polar coordinates (r, θ) of this same point for which:

(a) $r > 0$, $2\pi \le \theta < 4\pi$ (b) $r < 0$, $0 \le \theta < 2\pi$

(c) $r > 0$, $-2\pi \le \theta < 0$

Solution   The point $(3, \pi/6)$ is plotted in Figure 8.

(a) We add 1 revolution (2π radians) to the angle $\pi/6$ to get $P = (3, \pi/6 + 2\pi) = (3, 13\pi/6)$. See Figure 9.

FIGURE 8

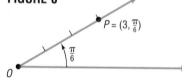

FIGURE 9

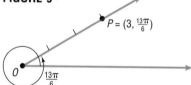

(b) We add $\frac{1}{2}$ revolution (π radians) to the angle and replace r by $-r$ to get $P = (-3, \pi/6 + \pi) = (-3, 7\pi/6)$. See Figure 10.

FIGURE 10

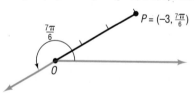

(c) We subtract 2π from the angle $\pi/6$ to get $P = (3, \pi/6 - 2\pi) = (3, -11\pi/6)$. See Figure 11.

FIGURE 11

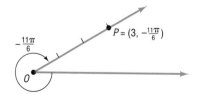

■

■ Now work Problem 13.

Conversion between Polar Coordinates and Rectangular Coordinates

It is sometimes convenient and, indeed, necessary to be able to convert coordinates or equations in rectangular form to polar form, and vice versa. To do this, we recall that the origin in rectangular coordinates is the pole in polar coordinates and that the positive x-axis in rectangular coordinates is the polar axis in polar coordinates.

Theorem
Conversion from Polar Coordinates to Rectangular Coordinates

If P is a point with polar coordinates (r, θ), the rectangular coordinates (x, y) of P are given by

$$x = r \cos \theta \qquad y = r \sin \theta \qquad (1)$$

Note: Most graphing calculators have the capability of converting from polar coordinates to rectangular coordinates. Consult your owner's manual to learn the proper key strokes.

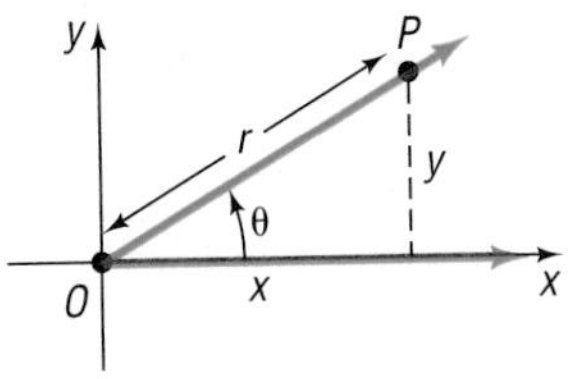

Proof Suppose that P has the polar coordinates (r, θ). We seek the rectangular coordinates (x, y) of P. Refer to Figure 12.

If $r = 0$, then, regardless of θ, the point P is the pole, for which the rectangular coordinates are $(0, 0)$. Thus, formula (1) is valid for $r = 0$.

If $r > 0$, the point P is on the terminal side of θ and $r = d(O, P)$. Thus,

$$\cos \theta = \frac{x}{r} \qquad \sin \theta = \frac{y}{r}$$

so

$$x = r \cos \theta \qquad y = r \sin \theta$$

If $r < 0$, then the point $P = (r, \theta)$ can be represented as $(-r, \pi + \theta)$, where $-r > 0$. Thus,

$$\cos(\pi + \theta) = -\cos \theta = \frac{x}{-r} \qquad \sin(\pi + \theta) = -\sin \theta = \frac{y}{-r}$$

so

$$x = r \cos \theta \qquad y = r \sin \theta \qquad ■$$

FIGURE 12

EXAMPLE 5

Converting from Polar Coordinates to Rectangular Coordinates

Find the rectangular coordinates of the points with the following polar coordinates:

(a) $(6, \pi/6)$ (b) $(-2, 5\pi/4)$ (c) $(-4, -\pi/4)$

Solution We use formula (1): $x = r \cos \theta$ and $y = r \sin \theta$.

(a) See Figure 13(a).

$$x = r \cos \theta = 6 \cos \frac{\pi}{6} = 6 \cdot \frac{\sqrt{3}}{2} = 3\sqrt{3}$$

$$y = r \sin \theta = 6 \sin \frac{\pi}{6} = 6 \cdot \frac{1}{2} = 3$$

The rectangular coordinates of the point $(6, \pi/6)$ are $(3\sqrt{3}, 3)$.

(b) See Figure 13(b).

$$x = r \cos \theta = -2 \cos \frac{5\pi}{4} = -2\left(-\frac{\sqrt{2}}{2}\right) = \sqrt{2}$$

$$y = r \sin \theta = -2 \sin \frac{5\pi}{4} = -2\left(-\frac{\sqrt{2}}{2}\right) = \sqrt{2}$$

The rectangular coordinates of the point $(-2, 5\pi/4)$ are $(\sqrt{2}, \sqrt{2})$.

(c) See Figure 13(c).

$$x = r \cos \theta = -4 \cos\left(-\frac{\pi}{4}\right) = -4 \cdot \frac{\sqrt{2}}{2} = -2\sqrt{2}$$

$$y = r \sin \theta = -4 \sin\left(-\frac{\pi}{4}\right) = -4\left(-\frac{\sqrt{2}}{2}\right) = 2\sqrt{2}$$

The rectangular coordinates of the given point $(-4, -\pi/4)$ are $(-2\sqrt{2}, 2\sqrt{2})$.

FIGURE 13

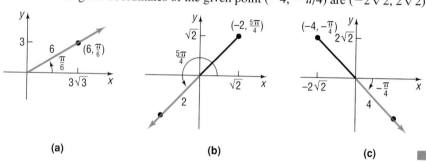

(a)　　　　　(b)　　　　　(c)

■ Now work Problems 21 and 33.

To convert from rectangular coordinates (x, y) to polar coordinates (r, θ) is a little more complicated. Let's look at some examples.

E X A M P L E　6

FIGURE 14

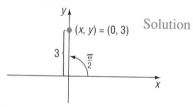

Converting from Rectangular Coordinates to Polar Coordinates

Find polar coordinates of a point whose rectangular coordinates are $(0, 3)$.

Solution　See Figure 14. The point $(0, 3)$ lies on the y-axis a distance of 3 units from the origin (pole). Polar coordinates for this point can be given by $(3, \pi/2)$.　■

Figure 15 shows polar coordinates of points that lie on either the x-axis or the y-axis.

FIGURE 15

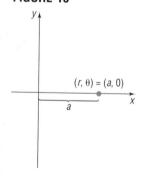

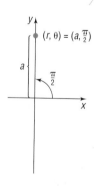

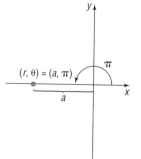

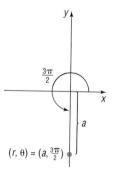

(a) $(x, y) = (a, 0), a > 0$　　**(b)** $(x, y) = (0, a), a > 0$　　**(c)** $(x, y) = (-a, 0), a > 0$　　**(d)** $(x, y) = (0, -a), a > 0$

■ Now work Problem 37.

EXAMPLE 7

Converting from Rectangular Coordinates to Polar Coordinates

Find polar coordinates of a point whose rectangular coordinates are $(2, -2)$.

Solution See Figure 16. The distance r from the origin to the point $(2, -2)$ is

$$r = \sqrt{x^2 + y^2} = \sqrt{(2)^2 + (-2)^2} = \sqrt{8} = 2\sqrt{2}$$

To find θ, we use the reference angle α. Then

$$\alpha = \tan^{-1}\left|\frac{y}{x}\right| = \tan^{-1}\left|\frac{-2}{2}\right| = \tan^{-1} 1 = \frac{\pi}{4}$$

Thus, $\theta = -\pi/4$, and a set of polar coordinates for this point is $(2\sqrt{2}, -\pi/4)$. Other possible representations include $(2\sqrt{2}, 7\pi/4)$ and $(-2\sqrt{2}, 3\pi/4)$. ■

FIGURE 16

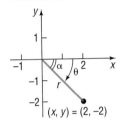

$(x, y) = (2, -2)$

EXAMPLE 8

Converting from Rectangular Coordinates to Polar Coordinates

Find polar coordinates of a point whose rectangular coordinates are $(-1, -\sqrt{3})$.

Solution See Figure 17. The distance r from the origin to the point $(-1, -\sqrt{3})$ is

$$r = \sqrt{x^2 + y^2} = \sqrt{(-1)^2 + (-\sqrt{3})^2} = \sqrt{4} = 2$$

To find θ, we use the reference angle α. Then

$$\alpha = \tan^{-1}\left|\frac{y}{x}\right| = \tan^{-1}\left|\frac{-\sqrt{3}}{-1}\right| = \tan^{-1}\sqrt{3} = \frac{\pi}{3}$$

Thus,

$$\theta = \pi + \alpha = \pi + \frac{\pi}{3} = \frac{4\pi}{3}$$

and a set of polar coordinates is $(2, 4\pi/3)$. Other possible representations include $(-2, \pi/3)$ and $(2, -2\pi/3)$. ■

FIGURE 17

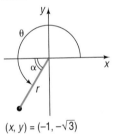

$(x, y) = (-1, -\sqrt{3})$

Figure 18 shows how to find polar coordinates of a point that lies in a quadrant when its rectangular coordinates (x, y) are given.

FIGURE 18

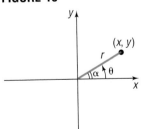

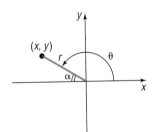

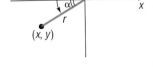

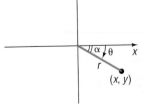

(a) $r = \sqrt{x^2 + y^2}$
$\theta = \alpha = \tan^{-1}\left|\frac{y}{x}\right|$

(b) $r = \sqrt{x^2 + y^2}$
$\theta = \pi - \alpha = \pi - \tan^{-1}\left|\frac{y}{x}\right|$

(c) $r = \sqrt{x^2 + y^2}$
$\theta = \pi + \alpha = \pi + \tan^{-1}\left|\frac{y}{x}\right|$

(d) $r = \sqrt{x^2 + y^2}$
$\theta = -\alpha = -\tan^{-1}\left|\frac{y}{x}\right|$

■ Now work Problem 41.

Based on the preceding discussion, we have the formulas

$$r^2 = x^2 + y^2 \qquad \tan \theta = \frac{y}{x} \qquad \text{if } x \neq 0 \tag{2}$$

Warning: Be careful when using formula (2). Always plot the point (x, y) first, as we did in Examples 7 and 8, so that r and θ are chosen to reflect the correct quadrant. Look again at Figure 18.

Note: Most graphing calculators have the capability of converting from rectangular coordinates to polar coordinates. Consult your owner's manual to learn the proper key strokes.

Formulas (1) and (2) may also be used to transform equations.

EXAMPLE 9

Transforming Equations from Polar to Rectangular Form

Transform the equation $r = 4 \sin \theta$ from polar coordinates to rectangular coordinates, and identify the graph.

Solution   If we multiply each side by r, it will be easier to apply formulas (1) and (2):

$$r = 4 \sin \theta$$
$$r^2 = 4r \sin \theta \quad \text{Multiply each side by } r.$$
$$x^2 + y^2 = 4y \qquad \text{Apply formulas (1) and (2).}$$

This is the equation of a circle:

$$x^2 + (y^2 - 4y) = 0$$
$$x^2 + (y^2 - 4y + 4) = 4$$
$$x^2 + (y - 2)^2 = 4$$

Its center is at $(0, 2)$, and its radius is 2.   ■

■ Now work Problem 57.

EXAMPLE 10

Transforming an Equation from Rectangular to Polar Form

Transform the equation $4xy = 9$ from rectangular coordinates to polar coordinates.

Solution   We use formula (1):

$$4xy = 9$$
$$4(r \cos \theta)(r \sin \theta) = 9 \quad \text{Formula (1)}$$
$$4r^2 \cos \theta \sin \theta = 9$$
$$2r^2 \sin 2\theta = 9 \quad \text{Double-angle formula}$$ ■

5.1

Exercise 5.1

In Problems 1–12, plot each point given in polar coordinates.

1. $(3, 90°)$	**2.** $(4, 270°)$	**3.** $(-2, 0)$	**4.** $(-3, \pi)$
5. $(6, \pi/6)$	**6.** $(5, 5\pi/3)$	**7.** $(-2, 135°)$	**8.** $(-3, 120°)$
9. $(-1, -\pi/3)$	**10.** $(-3, -3\pi/4)$	**11.** $(-2, -\pi)$	**12.** $(-3, -\pi/2)$

In Problems 13–20, plot each point given in polar coordinates, and find other polar coordinates (r, θ) of the point for which:

(a) $r > 0$, $-2\pi \leq \theta < 0$ *(b) $r < 0$, $0 \leq \theta < 2\pi$*   *(c) $r > 0$, $2\pi \leq \theta < 4\pi$*

13. $(5, 2\pi/3)$ **14.** $(4, 3\pi/4)$ **15.** $(-2, 3\pi)$ **16.** $(-3, 4\pi)$

17. $(1, \pi/2)$ **18.** $(2, \pi)$ **19.** $(-3, -\pi/4)$ **20.** $(-2, -2\pi/3)$

In Problems 21–36, polar coordinates of a point are given. Find the rectangular coordinates of each point.

21. $(3, \pi/2)$ **22.** $(4, 3\pi/2)$ **23.** $(-2, 0)$ **24.** $(-3, \pi)$

25. $(6, 150°)$ **26.** $(5, 300°)$ **27.** $(-2, 3\pi/4)$ **28.** $(-3, 2\pi/3)$

29. $(-1, -\pi/3)$ **30.** $(-3, -3\pi/4)$ **31.** $(-2, -180°)$ **32.** $(-3, -90°)$

33. $(7.5, 110°)$ **34.** $(-3.1, 182°)$ **35.** $(6.3, 3.8)$ **36.** $(8.1, 5.2)$

In Problems 37–48, the rectangular coordinates of a point are given. Find polar coordinates for each point.

37. $(3, 0)$ **38.** $(0, 2)$ **39.** $(-1, 0)$ **40.** $(0, -2)$

41. $(1, -1)$ **42.** $(-3, 3)$ **43.** $(\sqrt{3}, 1)$ **44.** $(-2, -2\sqrt{3})$

45. $(1.3, -2.1)$ **46.** $(-0.8, -2.1)$ **47.** $(8.3, 4.2)$ **48.** $(-2.3, 0.2)$

In Problems 49–56, the letters x and y represent rectangular coordinates. Write each equation using polar coordinates (r, θ).

49. $2x^2 + 2y^2 = 3$ **50.** $x^2 + y^2 = x$ **51.** $x^2 = 4y$ **52.** $y^2 = 2x$

53. $2xy = 1$ **54.** $4x^2y = 1$ **55.** $x = 4$ **56.** $y = -3$

In Problems 57–64, the letters r and θ represent polar coordinates. Write each equation using rectangular coordinates (x, y).

57. $r = \cos\theta$ **58.** $r = \sin\theta + 1$ **59.** $r^2 = \cos\theta$ **60.** $r = \sin\theta - \cos\theta$

61. $r = 2$ **62.** $r = 4$ **63.** $r = \dfrac{4}{1 - \cos\theta}$ **64.** $r = \dfrac{3}{3 - \cos\theta}$

65. Show that the formula for the distance d between two points $P_1 = (r_1, \theta_1)$ and $P_2 = (r_2, \theta_2)$ is

$$d = \sqrt{r_1^2 + r_2^2 - 2r_1r_2 \cos(\theta_2 - \theta_1)}$$

5.2

Polar Equations and Graphs

Polar Equation

An equation whose variables are polar coordinates is called a **polar equation.** The **graph of a polar equation** is the set of all points whose polar coordinates satisfy the equation.

Just as a rectangular grid may be used to plot points given by rectangular coordinates, as in Figure 19(a), we can use a grid consisting of concentric circles (with centers at the pole) and rays (with vertices at the pole) to plot points given by polar coordinates, as shown in Figure 19(b). We shall use such **polar grids** to graph polar equations by hand.

FIGURE 19

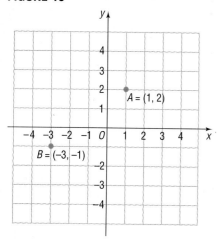

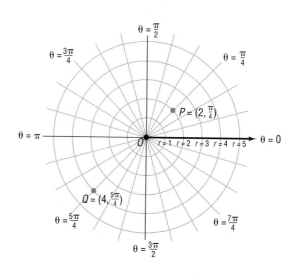

(a) Rectangular grid

(b) Polar grid

One method we can use to graph a polar equation by hand is to convert the equation to rectangular coordinates. In the discussion that follows, (x, y) represent the rectangular coordinates of a point P and (r, θ) represent polar coordinates of the point P.

E X A M P L E   1   *Identifying and Graphing a Polar Equation By Hand (Circle)*

Identify and graph the equation: $r = 3$

Solution   We convert the polar equation to a rectangular equation:

$$r = 3$$
$$r^2 = 9$$
$$x^2 + y^2 = 9$$

Thus, the graph of $r = 3$ is a circle, with center at the pole and radius 3. See Figure 20.

FIGURE 20

$r = 3$ or $x^2 + y^2 = 9$

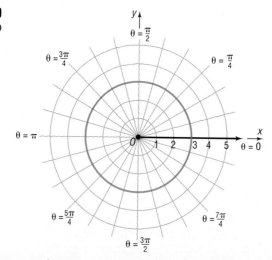

■ Now work Problem 1.

E X A M P L E   2

Identifying and Graphing a Polar Equation By Hand (Line)

Identify and graph the equation: $\theta = \pi/4$

Solution

We convert the polar equation to a rectangular equation:

$$\theta = \frac{\pi}{4}$$

$$\tan \theta = \tan \frac{\pi}{4} = 1$$

$$\frac{y}{x} = 1$$

$$y = x$$

The graph of $\theta = \pi/4$ is a line passing through the pole making an angle of $\pi/4$ with the polar axis. See Figure 21.

FIGURE 21

$\theta = \dfrac{\pi}{4}$ or $y = x$

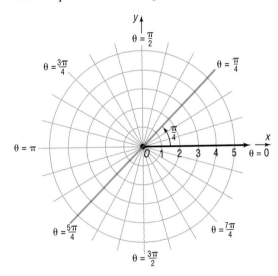

■

E X A M P L E   3

Identifying and Graphing a Polar Equation By Hand (Line)

Identify and graph the equation: $\theta = -\pi/6$

Solution

We convert the polar equation to a rectangular equation:

$$\theta = -\frac{\pi}{6}$$

$$\tan \theta = \tan\left(-\frac{\pi}{6}\right) = -\frac{\sqrt{3}}{3}$$

$$\frac{y}{x} = -\frac{\sqrt{3}}{3}$$

$$y = -\frac{\sqrt{3}}{3}x$$

The graph of $\theta = -\pi/6$ is a line passing through the pole making an angle of $-\pi/6$ with the polar axis. See Figure 22.

FIGURE 22

$\theta = -\dfrac{\pi}{6}$ or $y = -\dfrac{\sqrt{3}}{3}x$

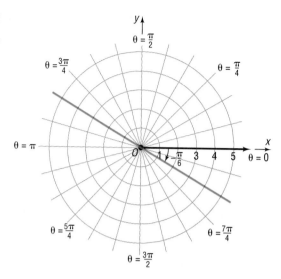

■   Now work Problem 3.

E X A M P L E   4

Identifying and Graphing a Polar Equation By Hand (Horizontal Line)

Identify and graph the equation: $r \sin \theta = 2$

Solution   Since $y = r \sin \theta$, we can simply write the equation as

$$y = 2$$

We conclude that the graph of $r \sin \theta = 2$ is a horizontal line 2 units above the pole. See Figure 23.

FIGURE 23

$r \sin \theta = 2$ or $y = 2$

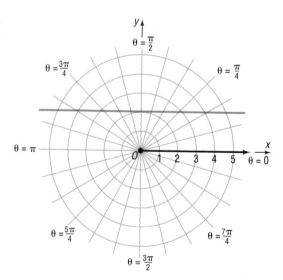

A second method we can use to graph a polar equation is to graph the equation using a graphing utility.

Most graphing utilities require the following steps in order to obtain the graph of an equation:

Graphing a Polar Equation Using a Graphing Utility

STEP 1: Solve the equation for r in terms of θ.

STEP 2: Select the viewing rectangle in polar mode. In addition to setting Xmin, Xmax, Xscl, etc., the viewing rectangle in polar mode requires setting minimum and maximum values for θ and an increment setting for θ (θ step). In addition, a square screen and radian measure should be used.

STEP 3: Enter the expression involving θ that you found in STEP 1. (Consult your manual for the correct way to enter the expression).

STEP 4: Execute

E X A M P L E 5

Graphing a Polar Equation Using a Graphing Utility

Use a graphing utility to graph the polar equation $r \sin \theta = 2$.

Solution

STEP 1: We solve the equation for r in terms of θ.

$$r \sin \theta = 2$$

$$r = \frac{2}{\sin \theta}$$

STEP 2: From the polar mode, select the viewing rectangle. We will use the one given next.

$$\theta\text{min} = 0$$
$$\theta\text{max} = 2\pi$$
$$\theta step = \pi/24$$
$$X\text{min} = -9$$
$$X\text{max} = 9$$
$$X scl = 1$$
$$Y\text{min} = -6$$
$$Y\text{max} = 6$$
$$Y scl = 1$$

$\theta step$ determines the number of points the graphing utility will plot. For example, if $\theta step$ is $\pi/24$, then the graphing utility will evaluate r at $\theta = 0(\theta\text{min})$, $\pi/24$, $2\pi/24$, $3\pi/24$, etc. up to $2\pi(\theta\text{max})$. The smaller $\theta step$, the more points the graphing utility will plot. The student is encouraged to experiment with different values for θmin, θmax, and $\theta step$ to see how the graph is affected.

STEP 3: Enter the expression $\dfrac{2}{\sin \theta}$ after the prompt $r =$.

STEP 4: Execute

The graph is shown in Figure 24.

FIGURE 24

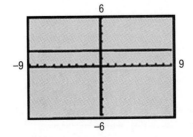

E X A M P L E   6   *Identifying and Graphing a Polar Equation (Vertical Line)*

Identify and graph the equation: $r \cos \theta = -3$

Solution   Since $x = r \cos \theta$, we can simply write the equation as

$$x = -3$$

We conclude that the graph of $r \cos \theta = -3$ is a vertical line 3 units to the left of the pole. Figure 25(a) shows the graph drawn by hand. Figure 25(b) shows the graph of $r \cos \theta = -3$ using a graphing utility with $\theta\min = 0$, $\theta\max = 2\pi$, and θstep $= \pi/24$.

FIGURE 25
$r \cos \theta = -3$ or $x = -3$

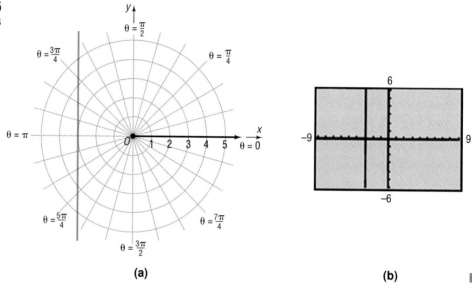

(a)   (b)

Based on Examples 4, 5, and 6, we are led to the following results. (The proofs are left as exercises.)

Theorem   If a is a nonzero real number:

The graph of the equation

$$r \sin \theta = a$$

is a horizontal line a units above the pole if $a > 0$ and $|a|$ units below the pole if $a < 0$.

The graph of the equation

$$r \cos \theta = a$$

is a vertical line a units to the right of the pole if $a > 0$ and $|a|$ units to the left of the pole if $a < 0$.

■   Now work Problem 7.

E X A M P L E   7   *Identifying and Graphing a Polar Equation (Circle)*

Identify and graph the equation: $r = 4 \sin \theta$

Solution   To transform the equation to rectangular coordinates, we multiply each side by r:

$$r^2 = 4r \sin \theta$$

Now we use the facts that $r^2 = x^2 + y^2$ and $y = r \sin \theta$. Then

$$x^2 + y^2 = 4y$$
$$x^2 + (y^2 - 4y) = 0$$
$$x^2 + (y - 2)^2 = 4$$

This is the equation of a circle with center at $(0, 2)$ in rectangular coordinates and radius 2. Figure 26(a) shows the graph drawn by hand. Figure 26(b) shows the graph of $r = 4 \sin \theta$ using a graphing utility with θmin $= 0$, θmax $= 2\pi$, and θstep $= \pi/24$.

FIGURE 26
$r = 4 \sin \theta$ or
$x^2 + (y - 2)^2 = 4$

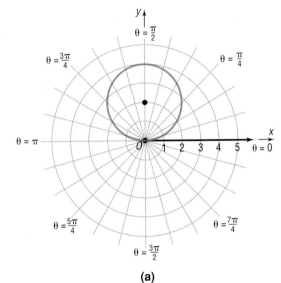
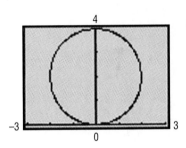

(a)   (b)

E X A M P L E   8   *Identifying and Graphing a Polar Equation (Circle)*
Identify and graph the equation: $r = -2 \cos \theta$

Solution   We proceed as in Example 7:

$$r^2 = -2r \cos \theta$$
$$x^2 + y^2 = -2x$$
$$x^2 + 2x + y^2 = 0$$
$$(x + 1)^2 + y^2 = 1$$

This is the equation of a circle with center at $(-1, 0)$ in rectangular coordinates and radius 1. Figure 27(a) shows the graph drawn by hand. Figure 27(b) shows the graph of $r = -2 \cos \theta$ using a graphing utility with θmin $= 0$, θmax $= 2\pi$, and θstep $= \pi/24$.

FIGURE 27
$r = -2 \cos \theta$ or
$(x + 1)^2 + y^2 = 1$

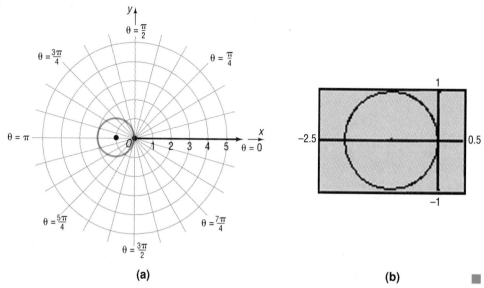

(a)   (b)

Exploration: Graph $r = \sin \theta$, $r = 2 \sin \theta$, and $r = 3 \sin \theta$. Do you see the pattern? Clear the screen and graph $r = -\sin \theta$, $r = -2 \sin \theta$, and $r = -3 \sin \theta$. Do you see the pattern? Clear the screen and graph $r = \cos \theta$, $r = 2 \cos \theta$, and $r = 3 \cos \theta$. Do you see the pattern? Clear the screen and graph $r = -\cos \theta$, $r = -2 \cos \theta$, and $r = -3 \cos \theta$. Do you see the pattern?

Based on Examples 7 and 8, we are led to the following results. (The proofs are left as exercises.)

Theorem   Let a be a positive real number. Then:

EQUATION	DESCRIPTION
(a) $r = 2a \sin \theta$	Circle: radius a; center at $(0, a)$ in rectangular coordinates
(b) $r = -2a \sin \theta$	Circle: radius a; center at $(0, -a)$ in rectangular coordinates
(c) $r = 2a \cos \theta$	Circle: radius a; center at $(a, 0)$ in rectangular coordinates
(d) $r = -2a \cos \theta$	Circle: radius a; center at $(-a, 0)$ in rectangular coordinates

The method of converting a polar equation to an identifiable rectangular equation in order to graph it is not always helpful, nor is it always necessary. Usually, we set up a table that lists several points on the graph. By checking for symmetry, it may be possible to reduce the number of points needed to draw the graph by hand.

Symmetry

In polar coordinates the points (r, θ) and $(r, -\theta)$ are symmetric with respect to the polar axis (and to the x-axis). See Figure 28(a). The points (r, θ) and $(r, \pi - \theta)$ are symmetric with respect to the line $\theta = \pi/2$ (y-axis). See Figure 28(b). The points (r, θ) and $(-r, \theta)$ are symmetric with respect to the pole (origin). See Figure 28(c).

FIGURE 28

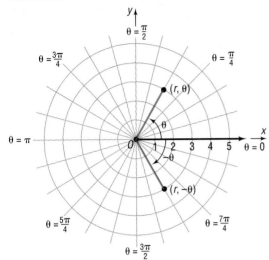

(a) Points symmetric with respect to the polar axis

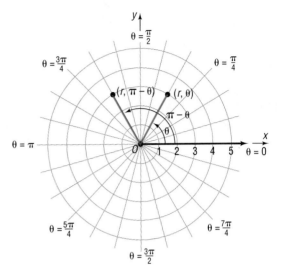

(b) Points symmetric with respect to the line $\theta = \frac{\pi}{2}$

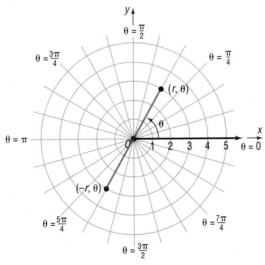

(c) Points symmetric with respect to the pole

The following tests are a consequence of these observations.

Theorem
Tests for Symmetry

Symmetry with Respect to the Polar Axis (x-Axis):
In a polar equation, replace θ by $-\theta$. If an equivalent equation results, the graph is symmetric with respect to the polar axis.

Symmetry with Respect to the Line $\theta = \pi/2$ (y-Axis):
In a polar equation, replace θ by $\pi - \theta$. If an equivalent equation results, the graph is symmetric with respect to the line $\theta = \pi/2$.

Symmetry with Respect to the Pole (Origin):
In a polar equation, replace r by $-r$. If an equivalent equation results, the graph is symmetric with respect to the pole. ■

The three tests for symmetry given here are *sufficient* conditions for symmetry, but they are not *necessary* conditions. That is, an equation may fail these tests and still have a graph that is symmetric with respect to the polar axis, the line $\theta = \pi/2$, or the pole. For example, the graph of $r = \sin 2\theta$ turns out to be symmetric with respect to the polar axis, the line $\theta = \pi/2$, and the pole, but all three tests given here fail. Refer to Problems 71, 72, and 73 in Exercise 5.2.

E X A M P L E   9

Graphing a Polar Equation (Cardioid)

Graph the equation: $r = 1 - \sin \theta$

Solution   We check for symmetry first:

Polar Axis: Replace θ by $-\theta$. The result is

$$r = 1 - \sin(-\theta) = 1 + \sin \theta$$

The test fails, so the graph may or may not be symmetric with respect to the polar axis.

The Line $\theta = \pi/2$: Replace θ by $\pi - \theta$. The result is

$$r = 1 - \sin(\pi - \theta) = 1 - (\sin \pi \cos \theta - \cos \pi \sin \theta)$$
$$= 1 - [0 \cdot \cos \theta - (-1) \sin \theta] = 1 - \sin \theta$$

Thus, the graph is symmetric with respect to the line $\theta = \pi/2$.

The Pole: Replace r by $-r$. Then the result is $-r = 1 - \sin \theta$, so $r = -1 + \sin \theta$. The test fails, so the graph may or may not be symmetric with respect to the pole.

Next, we identify points on the graph by assigning values to the angle θ and calculating the corresponding values of r. Due to the symmetry with respect to the line $\theta = \pi/2$, we only need to assign values to θ from $-\pi/2$ to $\pi/2$, as given in Table 1.

Now we plot the points (r, θ) from Table 1 and trace out the graph, beginning at the point $(2, -\pi/2)$ and ending at the point $(0, \pi/2)$. Then we reflect this portion of the graph about the line $\theta = \pi/2$ (y-axis) to obtain the complete graph.

Figure 29(a) shows the graph drawn by hand. Figure 29(b) shows the graph of $r = 1 - \sin \theta$ using a graphing utility with θmin $= 0$, θmax $= 2\pi$, and θstep $= \pi/24$.

TABLE 1

θ	$r = 1 - \sin \theta$
$-\pi/2$	$1 + 1 = 2$
$-\pi/3$	$1 + \sqrt{3}/2 \approx 1.87$
$-\pi/6$	$1 + \frac{1}{2} = \frac{3}{2}$
0	1
$\pi/6$	$1 - \frac{1}{2} = \frac{1}{2}$
$\pi/3$	$1 - \sqrt{3}/2 \approx 0.13$
$\pi/2$	0

FIGURE 29
$r = 1 - \sin \theta$

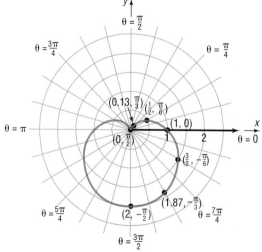

(a)

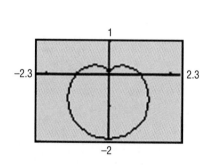

(b)

Exploration: Graph $r = 1 + \sin \theta$. Clear the screen and graph $r = 1 - \cos \theta$. Clear the screen and graph $r = 1 + \cos \theta$. Do you see a pattern?

The curve in Figure 29 is an example of a *cardioid* (a heart-shaped curve).

Cardioid

Cardioids are characterized by equations of the form

$$r = a(1 + \cos \theta) \qquad r = a(1 + \sin \theta)$$
$$r = a(1 - \cos \theta) \qquad r = a(1 - \sin \theta)$$

where $a > 0$. The graph of a cardioid contains the pole.

■ Now work Problem 31.

E X A M P L E   1 0

Graphing a Polar Equation (Limaçon)

Graph the equation: $r = 3 + 2 \cos \theta$

Solution

We check for symmetry first:

Polar Axis: Replace θ by $-\theta$. The result is

$$r = 3 + 2 \cos(-\theta) = 3 + 2 \cos \theta$$

Thus, the graph is symmetric with respect to the polar axis.

The Line $\theta = \pi/2$: Replace θ by $\pi - \theta$. The result is

$$r = 3 + 2 \cos(\pi - \theta) = 3 + 2(\cos \pi \cos \theta + \sin \pi \sin \theta)$$
$$= 3 - 2 \cos \theta$$

The test fails, so the graph may or may not be symmetric with respect to the line $\theta = \pi/2$.

The Pole: Replace r by $-r$. The test fails, so the graph may or may not be symmetric with respect to the pole.

Next, we identify points on the graph by assigning values to the angle θ and calculating the corresponding values of r. Due to the symmetry with respect to the polar axis, we only need to assign values to θ from 0 to π, as given in Table 2.

TABLE 2

θ	$r = 3 + 2 \cos \theta$
0	5
$\pi/6$	$3 + \sqrt{3} \approx 4.73$
$\pi/3$	4
$\pi/2$	3
$2\pi/3$	2
$5\pi/6$	$3 - \sqrt{3} \approx 1.27$
π	1

Now we plot the points (r, θ) from Table 2 and trace out the graph, beginning at the point $(5, 0)$ and ending at the point $(1, \pi)$. Then we reflect this portion of the graph about the pole (x-axis) to obtain the complete graph.

Figure 30(a) shows the graph drawn by hand. Figure 30(b) shows the graph of $r = 3 + 2 \cos \theta$ using a graphing utility with θmin $= 0$, θmax $= 2\pi$, and θstep $= \pi/24$.

FIGURE 30
$r = 3 + 2 \cos \theta$

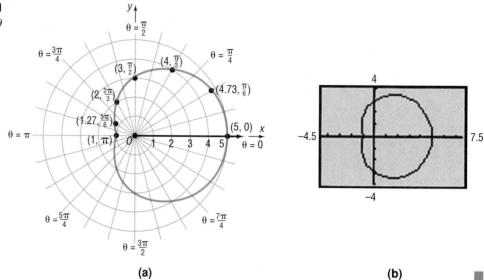

(a)   (b)

Exploration: Graph $r = 3 - 2 \cos \theta$. Clear the screen and graph $r = 3 + 2 \sin \theta$. Clear the screen and graph $r = 3 - 2 \sin \theta$. Do you see a pattern?

The curve in Figure 30 is an example of a *limaçon* (the French word for *snail*) without an inner loop.

Limaçon without an
Inner Loop

Limaçons without an inner loop are characterized by equations of the form

$$r = a + b \cos \theta \qquad r = a + b \sin \theta$$
$$r = a - b \cos \theta \qquad r = a - b \sin \theta$$

where $a > 0$, $b > 0$, and $a > b$. The graph of a limaçon without an inner loop does not contain the pole.

■   Now work Problem 37.

E X A M P L E   1 1   *Graphing a Polar Equation (Limaçon with Inner Loop)*

Graph the equation: $r = 1 + 2 \cos \theta$

Solution   First, we check for symmetry:

Polar Axis: Replace θ by $-\theta$. The result is

$$r = 1 + 2 \cos(-\theta) = 1 + 2 \cos \theta$$

Thus, the graph is symmetric with respect to the polar axis.

The Line $\theta = \pi/2$: Replace θ by $\pi - \theta$. The result is

$$r = 1 + 2 \cos(\pi - \theta) = 1 + 2(\cos \pi \cos \theta + \sin \pi \sin \theta)$$
$$= 1 - 2 \cos \theta$$

The test fails, so the graph may or may not be symmetric with respect to the line $\theta = \pi/2$.

The Pole: Replace r by $-r$. The test fails, so the graph may or may not be symmetric with respect to the pole.

FIGURE 34

$r = e^{\theta/5}$

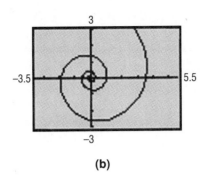

(b)

The curve in Figure 34 is called a **logarithmic spiral,** since its equation may be written as $\theta = 5 \ln r$ and it spirals infinitely both toward the pole and away from it.

Classification of Polar Equations

The equations of some lines and circles in polar coordinates and their corresponding equations in rectangular coordinates are given in Table 7. Also included are the names and the graphs of a few of the more frequently encountered polar equtions.

Calculus Comment

For those of you who are planning to study calculus, a comment about one important role of polar equations is in order.

In rectangular coordinates, the equation $x^2 + y^2 = 1$, whose graph is the unit circle, does not define a function. In fact, on the interval $[-1, 1]$ it defines two functions,

$$y = \sqrt{1 - x^2} \quad \text{Upper semicircle} \qquad y = -\sqrt{1 - x^2} \quad \text{Lower semicircle}$$

In polar coordinates, the equation $r = 1$, whose graph is also the unit circle, does define a function. That is, for each choice of θ there is only one corresponding value of r, that is, $r = 1$. Since many uses of calculus require that functions be used, the opportunity to express nonfunctions in rectangular coordinates as functions in polar coordinates becomes extremely useful.

Note also that the vertical line test for functions is valid only for equations in rectangular coordinates.

HISTORICAL FEATURE ■ Polar coordinates seem to have been invented by Jacob Bernoulli (1654–1705) about 1691, although, as with most such ideas, earlier traces of the notion exist. Early users of calculus remained committed to rectangular coordinates, and polar coordinates did not become widely used until the early 1800's. Even then, it was mostly geometers who used them for describing odd curves. Finally, about the mid-1800's, applied mathematicians realized the tremendous simplification polar coordinates make possible in the description of objects with circular or cylindrical symmetry. From then on their use became widespread. ■

TABLE 7

LINES

Description	Line passing through the pole making an angle α with the polar axis	Vertical line	Horizontal line
Rectangular equation	$y = (\tan \alpha)x$	$x = a$	$y = b$
Polar equation	$\theta = \alpha$	$r \cos \theta = a$	$r \sin \theta = b$
Typical graph	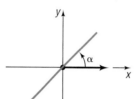	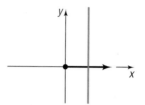	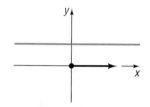

CIRCLES

Description	Center at the pole, radius a	Passing through the pole, tangent to the line $\theta = \pi/2$, center on the polar axis, radius a	Passing through the pole, tangent to the polar axis, center on the line $\theta = \pi/2$, radius a
Rectangular equation	$x^2 + y^2 = a^2, a > 0$	$x^2 + y^2 = \pm 2ax, a > 0$	$x^2 + y^2 = \pm 2ay, a > 0$
Polar equation	$r = a, a > 0$	$r = \pm 2a \cos \theta, a > 0$	$r = \pm 2a \sin \theta, a > 0$
Typical graph	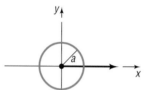	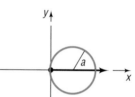	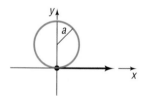

OTHER EQUATIONS

Name	Cardioid	Limaçon without inner loop	Limaçon with inner loop
Polar equations	$r = a \pm a \cos \theta, a > 0$ $r = a \pm a \sin \theta, a > 0$	$r = a \pm b \cos \theta, 0 < b < a$ $r = a \pm b \sin \theta, 0 < b < a$	$r = a \pm b \cos \theta, 0 < a < b$ $r = a \pm b \sin \theta, 0 < a < b$
Typical graph	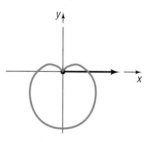	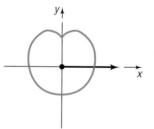	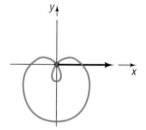

Name	Lemniscate	Rose with three petals	Rose with four petals
Polar equations	$r^2 = a^2 \cos 2\theta, a > 0$ $r^2 = a^2 \sin 2\theta, a > 0$	$r = a \sin 3\theta, a > 0$ $r = a \cos 3\theta, a > 0$	$r = a \sin 2\theta, a > 0$ $r = a \cos 2\theta, a > 0$
Typical graph	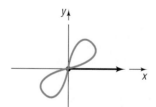	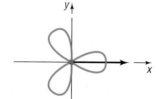	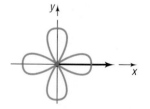

5.2

Exercise 5.2

In Problems 1–16, identify and graph each polar equation by hand. Verify your results using a graphing utility.

1. $r = 4$

2. $r = 2$

3. $\theta = \pi/3$

4. $\theta = -\pi/4$

5. $r \sin \theta = 4$

6. $r \cos \theta = 4$

7. $r \cos \theta = -2$

8. $r \sin \theta = -2$

9. $r = 2 \cos \theta$

10. $r = 2 \sin \theta$

11. $r = -4 \sin \theta$

12. $r = -4 \cos \theta$

13. $r \sec \theta = 4$

14. $r \csc \theta = 8$

15. $r \csc \theta = -2$

16. $r \sec \theta = -4$

In Problems 17–24, match each of the graphs (A) through (H) to one of the following polar equations.

17. $r = 2$

18. $\theta = \pi/4$

19. $r = 2 \cos \theta$

20. $r \cos \theta = 2$

21. $r = 1 + \cos \theta$

22. $r = 2 \sin \theta$

23. $\theta = 3\pi/4$

24. $r \sin \theta = 2$

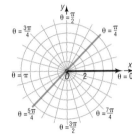

(A)

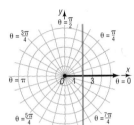

(B)

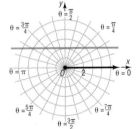

(C)

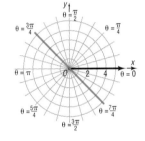

(D)

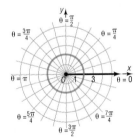

(E)

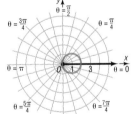

(F)

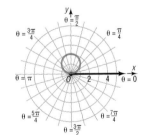

(G)

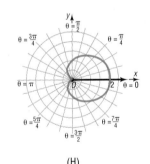

(H)

In Problems 25–30, match each of the graphs (A) through (F) to one of the following polar equations.

25. $r = 4$

26. $r = 3 \cos \theta$

27. $r = 3 \sin \theta$

28. $r \sin \theta = 3$

29. $r \cos \theta = 3$

30. $r = 2 + \sin \theta$

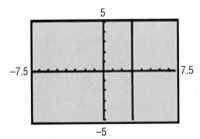

(A)

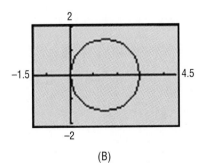

(B)

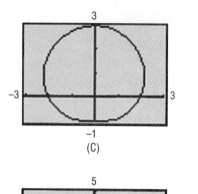

(C)

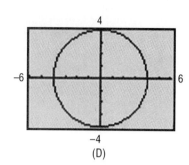

(D)

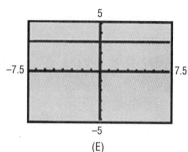

(E)

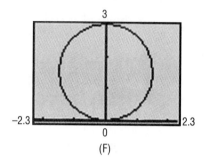

(F)

In Problems 31–54, identify and graph each polar equation by hand. Be sure to test for symmetry. Verify your results using a graphing utility.

31. $r = 2 + 2 \cos \theta$ **32.** $r = 1 + \sin \theta$ **33.** $r = 3 - 3 \sin \theta$ **34.** $r = 2 - 2 \cos \theta$

35. $r = 2 + \sin \theta$ **36.** $r = 2 - \cos \theta$ **37.** $r = 4 - 2 \cos \theta$ **38.** $r = 4 + 2 \sin \theta$

39. $r = 1 + 2 \sin \theta$ **40.** $r = 1 - 2 \sin \theta$ **41.** $r = 2 - 3 \cos \theta$ **42.** $r = 2 + 4 \cos \theta$

43. $r = 3 \cos 2\theta$ **44.** $r = 2 \sin 2\theta$ **45.** $r = 4 \sin 3\theta$ **46.** $r = 3 \cos 4\theta$

47. $r^2 = 9 \cos 2\theta$ **48.** $r^2 = \sin 2\theta$ **49.** $r = 2^\theta$ **50.** $r = 3^\theta$

51. $r = 1 - \cos \theta$ **52.** $r = 3 + \cos \theta$ **53.** $r = 1 - 3 \cos \theta$ **54.** $r = 4 \cos 3\theta$

In Problems 55–64, graph each polar equation by hand. Verify your results using a graphing utility.

55. $r = \dfrac{2}{1 - \cos \theta}$ (parabola)   **56.** $r = \dfrac{2}{1 - 2 \cos \theta}$ (hyperbola)

57. $r = \dfrac{1}{3 - 2 \cos \theta}$ (ellipse)   **58.** $r = \dfrac{1}{1 - \cos \theta}$ (parabola)

59. $r = \theta, \quad \theta \geq 0$ (spiral of Archimedes)   **60.** $r = \dfrac{3}{\theta}$ (reciprocal spiral)

61. $r = \csc \theta - 2, \quad 0 < \theta < \pi$ (conchoid)   **62.** $r = \sin \theta \tan \theta$ (cissoid)

63. $r = \tan \theta$ (kappa curve)   **64.** $r = \cos \dfrac{\theta}{2}$

65. Show that the graph of the equation $r \sin \theta = a$ is a horizontal line a units above the pole if $a > 0$ and $|a|$ units below the pole if $a < 0$.

66. Show that the graph of the equation $r \cos \theta = a$ is a vertical line a units to the right of the pole if $a > 0$ and $|a|$ units to the left of the pole if $a < 0$.

67. Show that the graph of the equation $r = 2a \sin \theta$, $a > 0$, is a circle of radius a with center at $(0, a)$ in rectangular coordinates.

68. Show that the graph of the equation $r = -2a \sin \theta$, $a > 0$, is a circle of radius a with center at $(0, -a)$ in rectangular coordinates.

ℳ ISSION POSSIBLE

Chapter 5

ℳ APPING INDIANAPOLIS

Supplies needed: Graph paper with squares 1/4 inch or 1 cm. or larger per side; compass, protractor, ruler; graphing calculator; map of Indianapolis (from any atlas)

Unlike many major American cities which grew up next to large bodies of water, and hence were limited in their ability to spread out in all directions, Indianapolis is surrounded by land on all sides. Major roads radiate out from the city center and intersect the ring road (Rt. 465). A map of the city looks somewhat like a rectangular grid overlaying a polar coordinate system.

The Mission Possible Team has been called to Indianapolis by a car rental company that wishes to create a computer system of interactive maps to be installed in their vehicles. The customers renting their cars would be able to indicate to the computer where they would like to go and the computer screen would then show them a map giving them the quickest route from their present location to their chosen destination. Today's project represents the beginning of a study to create such a computer system.

As a first step, your team needs to set up a polar coordinate system. Using a simplified model, we designate the intersection of Meridian St. and Tenth St. as the pole. The ring road is (roughly) a circle 7 miles out from the center.

1. On graph paper sketch a large polar model of the streets of Indianapolis, using 1 square of the graph paper to represent 1 mile. The ring road would be represented by a circle with radius 7 miles. Rt. 40 is the road that intersects the ring road due east of the pole, so we will designate Rt. 40 as 0°. The other roads and their approximate angle from Rt. 40 are as follows:

Rt. 36: 30°	Rt. 65: 135°	Mann Rd.: 240°
Rt. 37: 60°	Rt. 74: 165°	Rt. 31 (Meridian St.): 270°
Rt. 31 (Meridian St.): 90°	Tenth St.: 180°	Rt. 65: 300°
Rt. 421: 105°	Rt. 70: 210°	Rt. 421: 330°

2. All but 3 of these angles are multiples of 30°. What is the greatest common factor of ALL the angles?
3. In order to establish a mathematical structure for the computer map, your team decides to create a polar equation of the petalled rose variety that would pass through all these points of intersection with the ring road and connect each to the center. What would the equation be?
4. Sketch the graph of your rose on the polar map you created for question #1. By using your graphing calculator, you can determine the order in which the petals are drawn. Number the petals on your sketch to indicate that order.
5. Find the *xy* equation determined by taking your rose equation and replacing the *r* by *y* and the *θ* by *x*. (Example: If your polar equation was $r = 5 \cos 7\theta$, your corresponding rectangular equation would be $y = 5 \cos 7x$. Warning: These two equations are not equivalent in the sense that they will not give the same graph.) Sketch the resulting rectangular equation in the *xy*-plane, using the domain $0 \le x \le 2\pi/3$. Label the maximum and minimum points with the numbers that correspond to the rose petals of your polar graph.
6. If your original sketch for question #1 was done on graph paper, you can now see the lines of the graph paper as representing a rectangular grid overlaying the polar graph. If not, you will need to sketch vertical and horizontal lines on your polar graph to represent the major city streets. Again, for purposes of simplifying the problem, the major streets are assumed to be occurring at 1 mile intervals.
7. Suppose that travel on the Interstate (Rt. 465, the ring road) has an average speed of 55 mph. and travel on city streets has an average speed of 20 mph. Estimate the times it would take to go from the intersection of Rt. 74 and Rt. 465 to the intersection of Rt. 36 and Rt. 465 by each route. (Recall that travelling on city streets limits you to vertical and horizontal grid lines.)
8. What would be the longest time it would take to go from any intersection to another on Rt. 465, assuming you chose the shortest distance along the circle? How long would it take to go between those same two points using the city streets?

69. Show that the graph of the equation $r = 2a \cos \theta$, $a > 0$, is a circle of radius a with center at $(a, 0)$ in rectangular coordinates.

70. Show that the graph of the equation $r = -2a \cos \theta$, $a > 0$, is a circle of radius a with center at $(-a, 0)$ in rectangular coordinates.

 71. Explain why the following test for symmetry is valid: Replace r by $-r$ and θ by $-\theta$ in a polar equation. If an equivalent equation results, the graph is symmetric with respect to the line $\theta = \pi/2$ (y-axis).
   (a) Show that the test on page 327 fails for $r^2 = \cos \theta$, but this new test works.
   (b) Show that the test on page 327 works for $r^2 = \sin \theta$, yet this new test fails.

72. Develop a new test for symmetry with respect to the pole.
   (a) Find a polar equation for which this new test fails, yet the test on page 327 works.
   (b) Find a polar equation for which the test on page 327 fails, yet the new test works.

73. Write down two different tests for symmetry with respect to the polar axis. Find examples in which one works and the other fails. Which test do you prefer to use? Justify your position.

5.3

The Complex Plane; Demoivre's Theorem

When we first introduced complex numbers, we were not prepared to give a geometric interpretation of a complex number. Now we are ready. Although there are several such interpretations we could give, the one that follows is the easiest to understand.

A complex number $z = x + yi$ can be interpreted geometrically as the point (x, y) in the xy-plane. Thus, each point in the plane corresponds to a complex number and, conversely, each complex number corresponds to a point in the plane. We shall refer to the collection of such points as the **complex plane.** The x-axis will be referred to as the **real axis,** because any point that lies on the real axis is of the form $z = x + 0i = x$, a real number. The y-axis is called the **imaginary axis,** because any point that lies on it is of the form $z = 0 + yi = yi$, a pure imaginary number. See Figure 35.

FIGURE 35
Complex plane

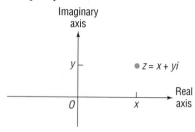

Magnitude

Let $z = x + yi$ be a complex number. The **magnitude** or **modulus** of z, denoted by $|z|$, is defined as the distance from the origin to the point (x, y). Thus,

$$|z| = \sqrt{x^2 + y^2} \qquad (1)$$

See Figure 36 for an illustration.

FIGURE 36

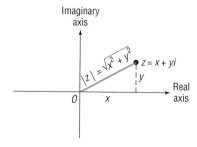

This definition for $|z|$ is consistent with the definition for the absolute value of a real number: If $z = x + yi$ is real, then $z = x + 0i$ and

$$|z| = \sqrt{x^2 + 0^2} = \sqrt{x^2} = |x|$$

Recall that if $z = x + yi$, then its **conjugate,** denoted by $\bar{z}$, is $\bar{z} = x - yi$. Because $z\bar{z} = x^2 + y^2$, it follows from equation (1) that the magnitude of z can be written as

$$|z| = \sqrt{z\bar{z}} \tag{2}$$

Polar Form of a Complex Number

When a complex number is written in the form $z = x + yi$, we say that it is in **rectangular,** or **Cartesian, form,** because (x, y) are the rectangular coordinates of the corresponding point in the complex plane. Suppose that (r, θ) are the polar coordinates of this point. Then

$$x = r \cos \theta \qquad y = r \sin \theta \tag{3}$$

If $r \geq 0$ and $0 \leq \theta < 2\pi$, the complex number $z = x + yi$ may be written in **polar form** as

Polar Form of z

$$z = x + yi = (r \cos \theta) + (r \sin \theta)i = r(\cos \theta + i \sin \theta) \tag{4}$$

FIGURE 37

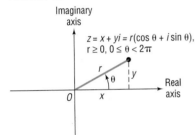

See Figure 37.

If $z = r(\cos \theta + i \sin \theta)$ is the polar form of a complex number, the angle θ, $0 \leq \theta < 2\pi$, is called the **argument of z.** Also, because $r \geq 0$, from equation (3) we have $r = \sqrt{x^2 + y^2}$. Thus, from equation (1) it follows that the magnitude of $z = r(\cos \theta + i \sin \theta)$ is

$$|z| = r$$

E X A M P L E 1

Plotting a Point in the Complex Plane and Writing a Complex Number in Polar Form

Plot the point corresponding to $z = \sqrt{3} - i$ in the complex plane, and write an expression for z in polar form.

Solution

FIGURE 38

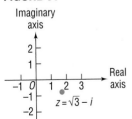

The point corresponding to $z = \sqrt{3} - i$ has the rectangular coordinates $(\sqrt{3}, -1)$. The point, located in quadrant IV, is plotted in Figure 38. Because $x = \sqrt{3}$ and $y = -1$, it follows that

$$r = \sqrt{x^2 + y^2} = \sqrt{(\sqrt{3})^2 + (-1)^2} = \sqrt{4} = 2$$

and

$$\sin \theta = \frac{y}{r} = \frac{-1}{2} \qquad \cos \theta = \frac{x}{r} = \frac{\sqrt{3}}{2}, \qquad 0 \leq \theta < 2\pi$$

Thus, $\theta = 11\pi/6$ and $r = 2$, so the polar form of $z = \sqrt{3} - i$ is

$$z = r(\cos \theta + i \sin \theta) = 2\left(\cos \frac{11\pi}{6} + i \sin \frac{11\pi}{6}\right)$$

■ Now work Problem 1.

E X A M P L E 2 *Plotting a Point in the Complex Plane and Converting from Polar to Rectangular Form*

Plot the point corresponding to $z = 2(\cos 30° + i \sin 30°)$ in the complex plane, and write an expression for z in rectangular form.

Solution To plot the complex number $z = 2(\cos 30° + i \sin 30°)$, we plot the point whose polar coordinates are $(r, \theta) = (2, 30°)$, as shown in Figure 39. In rectangular form,

FIGURE 39

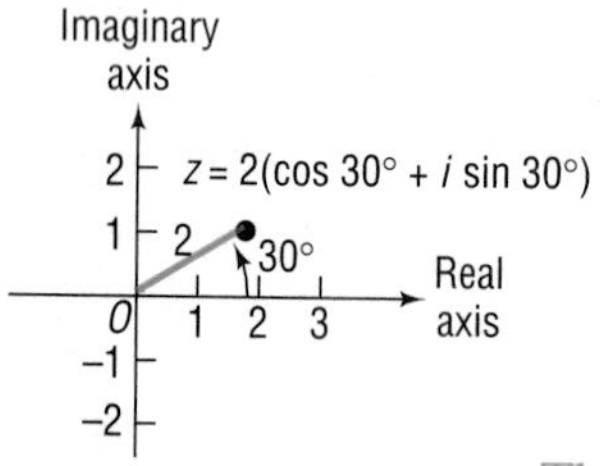

$$z = 2(\cos 30° + i \sin 30°) = 2\left(\frac{\sqrt{3}}{2} + \frac{1}{2}i\right) = \sqrt{3} + i \qquad ■$$

■ Now work Problem 13.

The polar form of a complex number provides an alternative for finding products and quotients of complex numbers.

Theorem Let $z_1 = r_1(\cos \theta_1 + i \sin \theta_1)$ and $z_2 = r_2(\cos \theta_2 + i \sin \theta_2)$ be two complex numbers. Then

$$z_1 z_2 = r_1 r_2 [\cos(\theta_1 + \theta_2) + i \sin(\theta_1 + \theta_2)] \qquad (5)$$

If $z_2 \ne 0$, then

$$\frac{z_1}{z_2} = \frac{r_1}{r_2}[\cos(\theta_1 - \theta_2) + i \sin(\theta_1 - \theta_2)] \qquad (6)$$

■

Proof We shall prove formula (5). The proof of formula (6) is left as an exercise (see Problem 56).

$$\begin{aligned}
z_1 z_2 &= [r_1(\cos \theta_1 + i \sin \theta_1)][r_2(\cos \theta_2 + i \sin \theta_2)] \\
&= r_1 r_2 [(\cos \theta_1 + i \sin \theta_1)(\cos \theta_2 + i \sin \theta_2)] \\
&= r_1 r_2 [(\cos \theta_1 \cos \theta_2 - \sin \theta_1 \sin \theta_2) + i(\sin \theta_1 \cos \theta_2 + \cos \theta_1 \sin \theta_2)] \\
&= r_1 r_2 [(\cos(\theta_1 + \theta_2) + i \sin(\theta_1 + \theta_2)] \qquad ■
\end{aligned}$$

Because the magnitude of a complex number z is r and its argument is θ, when $z = r(\cos \theta + i \sin \theta)$, we can restate this theorem as follows:

Theorem The magnitude of the product (quotient) of two complex numbers equals the product (quotient) of their magnitudes; the argument of the product (quotient) of two complex numbers equals the sum (difference) of their arguments. ■

Let's look at an example of how this theorem can be used.

E X A M P L E 3 *Finding Products and Quotients of Complex Numbers in Polar Form*

If $z = 3(\cos 20° + i \sin 20°)$ and $w = 5(\cos 100° + \sin 100°)$, find the following (leave your answers in polar form):

(a) zw (b) z/w

Solution   (a)

$$zw = [3(\cos 20° + i \sin 20°)][5(\cos 100° + i \sin 100°)]$$
$$= (3 \cdot 5)[\cos(20° + 100°) + i \sin(20° + 100°)]$$
$$= 15(\cos 120° + i \sin 120°)$$

(b)

$$\frac{z}{w} = \frac{3(\cos 20° + i \sin 20°)}{5(\cos 100° + i \sin 100°)}$$
$$= \tfrac{3}{5}[\cos(20° - 100°) + i \sin(20° - 100°)]$$
$$= \tfrac{3}{5}[\cos(-80°) + i \sin(-80°)]$$
$$= \tfrac{3}{5}(\cos 280° + i \sin 280°) \quad \text{Argument must lie between 0° and 360°}$$

■ Now work Problem 23.

Demoivre's Theorem

Demoivre's Theorem, stated by Abraham Demoivre (1667–1754) in 1730, but already known to many people by 1710, is important for the following reason: The fundamental processes of algebra are the four operations of addition, subtraction, multiplication, and division, together with powers and the extraction of roots. Demoivre's Theorem allows these latter fundamental algebraic operations to be applied to complex numbers.

    Demoivre's Theorem, in its most basic form, is a formula for raising a complex number z to the power n, where $n \geq 1$ is a positive integer. Let's see if we can guess the form of the result.

    Let $z = r(\cos \theta + i \sin \theta)$ be a complex number. Then, based on equation (5), we have

$$n = 2: \quad z^2 = r^2(\cos 2\theta + i \sin 2\theta)$$
$$n = 3: \quad z^3 = z^2 \cdot z$$
$$= [r^2(\cos 2\theta + i \sin 2\theta)][r(\cos \theta + i \sin \theta)]$$
$$= r^3(\cos 3\theta + i \sin 3\theta) \quad \text{Equation (5)}$$
$$n = 4: \quad z^4 = z^3 \cdot z$$
$$= [r^3(\cos 3\theta + i \sin 3\theta)][r(\cos \theta + i \sin \theta)]$$
$$= r^4(\cos 4\theta + i \sin 4\theta) \quad \text{Equation (5)}$$

The pattern should now be clear.

Theorem   If $z = r(\cos \theta + i \sin \theta)$ is a complex number, then
Demoivre's Theorem

$$z^n = r^n(\cos n\theta + i \sin n\theta) \tag{7}$$

where $n \geq 1$ is a positive integer. ■

    We shall not prove Demoivre's Theorem because it requires mathematical induction.
    Let's look at some examples.

E X A M P L E   4   *Using Demoivre's Theorem*

$$[2(\cos 20° + i \sin 20°)]^3 = 2^3[\cos(3 \cdot 20°) + i \sin(3 \cdot 20°)]$$
$$= 8(\cos 60° + i \sin 60°)$$
$$= 8\left(\frac{1}{2} + \frac{\sqrt{3}}{2}i\right) = 4 + 4\sqrt{3}\,i$$ ■

■ Now work Problem 31.

E X A M P L E   5   *Using Demoivre's Theorem*

Write $(1 + i)^5$ in the standard form $a + bi$.

Solution   To apply Demoivre's Theorem, we must first write the complex number in polar form. Thus, since the magnitude of $1 + i$ is $\sqrt{1^2 + 1^2} = \sqrt{2}$, we begin by writing

$$1 + i = \sqrt{2}\left(\frac{1}{\sqrt{2}} + \frac{1}{\sqrt{2}}i\right) = \sqrt{2}\left(\cos\frac{\pi}{4} + i \sin\frac{\pi}{4}\right)$$

Now

$$(1 + i)^5 = \left[\sqrt{2}\left(\cos\frac{\pi}{4} + i \sin\frac{\pi}{4}\right)\right]^5$$
$$= (\sqrt{2})^5\left[\cos\left(5 \cdot \frac{\pi}{4}\right) + i \sin\left(5 \cdot \frac{\pi}{4}\right)\right]$$
$$= 4\sqrt{2}\left(\cos\frac{5\pi}{4} + i \sin\frac{5\pi}{4}\right)$$
$$= 4\sqrt{2}\left[-\frac{1}{\sqrt{2}} + \left(-\frac{1}{\sqrt{2}}\right)i\right] = -4 - 4i$$ ■

E X A M P L E   6   *Using a Calculator with Demoivre's Theorem*

Write $(3 + 4i)^3$ in the standard form $a + bi$.

Solution   Again, we start by writing $3 + 4i$ in polar form. This time we will use degrees for the argument. The magnitude of $3 + 4i$ is $\sqrt{3^2 + 4^2} = \sqrt{25} = 5$, so we write

$$3 + 4i = 5\left(\frac{3}{5} + \frac{4}{5}i\right) \approx 5(\cos 53.1° + i \sin 53.1°)$$

Although we have written the angle rounded to one decimal place (53.1°), we keep the actual value of the angle in memory. Now

$$(3 + 4i)^3 \approx [5(\cos 53.1° + i \sin 53.1°)]^3$$
$$= 5^3[\cos(3 \cdot 53.1°) + i \sin(3 \cdot 53.1°)]$$
$$= 125(\cos 159.3° + i \sin 159.3°)$$
$$\approx 125[-0.935 + i(0.353)] = -117 + 44i$$

In this computation, we used the actual values stored in memory, not the rounded values shown. The final answer, $-117 + 44i$, is exact as you can verify by cubing $3 + 4i$. ■

Complex Roots

Let w be a given complex number, and let $n \geq 2$ denote a positive integer. Any complex number z that satisfies the equation

$$z^n = w$$

is called a **complex nth root** of w. In keeping with previous usage, if $n = 2$, the solutions of the equation $z^2 = w$ are called **complex square roots** of w, and if $n = 3$, the solutions of the equation $z^3 = w$ are called **complex cube roots** of w.

Theorem
Finding Complex Roots

Let $w = r(\cos \theta + i \sin \theta)$ be a complex number. If $w \neq 0$, there are n distinct complex nth roots of w, given by the formula

$$z_k = \sqrt[n]{r}\left[\cos\left(\frac{\theta}{n} + \frac{2k\pi}{n}\right) + i \sin\left(\frac{\theta}{n} + \frac{2k\pi}{n}\right)\right] \qquad (8)$$

where $k = 0, 1, 2, \ldots, n - 1$.

Proof (Outline)

We shall not prove this result in its entirety. Instead, we shall show only that each z_k in equation (8) obeys the equation $z_k^n = w$ and, hence, each z_k is a complex nth root of w.

$$z_k^n = \left\{\sqrt[n]{r}\left[\cos\left(\frac{\theta}{n} + \frac{2k\pi}{n}\right) + i \sin\left(\frac{\theta}{n} + \frac{2k\pi}{n}\right)\right]\right\}^n$$

$$= (\sqrt[n]{r})^n\left[\cos n\left(\frac{\theta}{n} + \frac{2k\pi}{n}\right) + i \sin n\left(\frac{\theta}{n} + \frac{2k\pi}{n}\right)\right] \quad \text{Demoivre's Theorem}$$

$$= r[\cos(\theta + 2k\pi) + i \sin(\theta + 2k\pi)]$$

$$= r(\cos \theta + i \sin \theta) = w$$

Thus, each z_k, $k = 0, 1, \ldots, n - 1$, is a complex nth root of w. To complete the proof, we would need to show that each z_k, $k = 0, 1, 2, \ldots, n - 1$, is, in fact, distinct and that there are no complex nth roots of w other than those given by equation (8). ∎

E X A M P L E   7

Finding Complex Cube Roots

Find the complex cube roots of $-1 + \sqrt{3}\, i$. Leave your answers in polar form, with θ in degrees.

Solution

First, we express $-1 + \sqrt{3}\, i$ in polar form using degrees:

$$-1 + \sqrt{3}\, i = 2\left(-\frac{1}{2} + \frac{\sqrt{3}}{2}i\right) = 2(\cos 120° + i \sin 120°)$$

The three complex cube roots of $-1 + \sqrt{3}\, i = 2(\cos 120° + i \sin 120°)$ are

$$z_k = \sqrt[3]{2}\left[\cos\left(\frac{120°}{3} + \frac{360°k}{3}\right) + i \sin\left(\frac{120°}{3} + \frac{360°k}{3}\right)\right], \qquad k = 0, 1, 2$$

Thus,

$$z_0 = \sqrt[3]{2}(\cos 40° + i \sin 40°)$$

$$z_1 = \sqrt[3]{2}(\cos 160° + i \sin 160°)$$

$$z_2 = \sqrt[3]{2}(\cos 280° + i \sin 280°)$$

Notice that each of the three complex cube roots of $-1 + \sqrt{3}\,i$ has the same magnitude, $\sqrt[3]{2}$. This means that the points corresponding to each cube root lie the same distance from the origin; hence, the three points lie on a circle with center at the origin and radius $\sqrt[3]{2}$. Furthermore, the arguments of these cube roots are 40°, 160°, and 280°, the difference of consecutive pairs being 120°. This means that the three points are equally spaced on the circle, as shown in Figure 40. These results are not coincidental. In fact, you are asked to show that these results hold for complex nth roots in Problems 53 through 55.

FIGURE 40

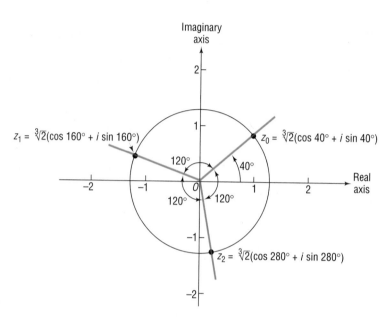

Now work Problem 43.

HISTORICAL PROBLEMS   ■ 1. The quadratic formula will work perfectly well if the coefficients are complex numbers. Solve the following, using Demoivre's Theorem where necessary. [*Hint:* The answers are "nice."]

(a) $z^2 - (2 + 5i)z - 3 + 5i = 0$ (b) $z^2 - (1 + i)\,z - 2 - i = 0$ ■

Solution   Figure 51 illustrates each graph.

FIGURE 51

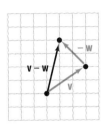

(a) v − w

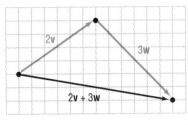

(b) 2v + 3w

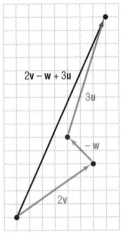

(c) 2v − w + 3u

■   Now work Problems 1 and 3.

Magnitudes of Vectors

If **v** is a vector, we use the symbol $\|\mathbf{v}\|$ to represent the **magnitude** of **v**. Since $\|\mathbf{v}\|$ equals the length of a directed line segment, it follows that $\|\mathbf{v}\|$ has the following properties:

Theorem   If **v** is a vector and if α is a scalar, then:
Properties of $\|\mathbf{v}\|$

(a) $\|\mathbf{v}\| \geq 0$ (b) $\|\mathbf{v}\| = 0$ if and only if $\mathbf{v} = \mathbf{0}$

(c) $\|-\mathbf{v}\| = \|\mathbf{v}\|$ (d) $\|\alpha\mathbf{v}\| = |\alpha|\|\mathbf{v}\|$ ■

Property (a) is a consequence of the fact that distance is a nonnegative number. Property (b) follows, because the length of the directed line segment $\overrightarrow{PQ}$ is positive unless P and Q are the same point, in which case the length is 0. Property (c) follows, because the length of the line segment $\overline{PQ}$ equals the length of the line segment $\overline{QP}$. Property (d) is a direct consequence of the definition of a scalar product.

Unit Vector   A vector **u** for which $\|\mathbf{u}\| = 1$ is called a **unit vector.**

To compute the magnitude and direction of a vector, we need an algebraic way of representing vectors.

Representing Vectors in the Plane

We use a rectangular coordinate system to represent vectors in the plane. Let **i** denote a unit vector whose direction is along the positive x-axis; let **j** denote a unit vector whose direction is along the positive y-axis. If **v** is a vector with initial point at the origin O and terminal point at $P = (a, b)$, then we can represent $\mathbf{v}$ in terms of the vectors $\mathbf{i}$ and $\mathbf{j}$ as

FIGURE 52

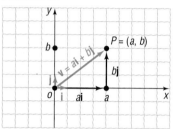

$$\mathbf{v} = a\mathbf{i} + b\mathbf{j}$$

See Figure 52. The scalars a and b are called the **components** of the vector $\mathbf{v} = a\mathbf{i} + b\mathbf{j}$, with a being the component in the direction $\mathbf{i}$ and b being the component in the direction $\mathbf{j}$.

A vector whose initial point is at the origin is called a **position vector.** The next result states that any vector whose initial point is not at the origin is equal to a unique position vector.

Theorem   Suppose that $\mathbf{v}$ is a vector with initial point $P_1 = (x_1, y_1)$, not necessarily the origin, and terminal point $P_2 = (x_2, y_2)$. If $\mathbf{v} = \overrightarrow{P_1P_2}$, then $\mathbf{v}$ is equal to the position vector

$$\mathbf{v} = (x_2 - x_1)\mathbf{i} + (y_2 - y_1)\mathbf{j}$$ ∎

To see why this is true, look at Figure 53. Triangle OPA and triangle P_1P_2Q are congruent. (Do you see why?) The line segments have the same magnitude, so $d(O, P) = d(P_1, P_2)$; and they have the same direction, so $\angle POA = \angle P_2P_1Q$. Since the triangles are right triangles, we have Angle–Side–Angle. Thus, it follows that corresponding sides are equal. As a result, $x_2 - x_1 = a$ and $y_2 - y_1 = b$, so $\mathbf{v}$ may be written as

$$\mathbf{v} = a\mathbf{i} + b\mathbf{j} = (x_2 - x_1)\mathbf{i} + (y_2 - y_1)\mathbf{j} \tag{1}$$

FIGURE 53

$$\mathbf{v} = a\mathbf{i} + b\mathbf{j} = (x_2 - x_1)\mathbf{i} + (y_2 - y_1)\mathbf{j}$$

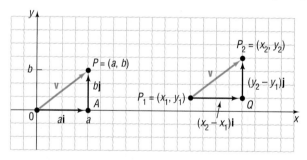

E X A M P L E   2   *Finding a Position Vector*

Find the position vector of the vector $\mathbf{v} = \overrightarrow{P_1P_2}$ if $P_1 = (-1, 2)$ and $P_2 = (4, 6)$.

Solution   By equation (1), the position vector equal to $\mathbf{v}$ is

$$\mathbf{v} = [4 - (-1)]\mathbf{i} + (6 - 2)\mathbf{j} = 5\mathbf{i} + 4\mathbf{j}$$

FIGURE 54

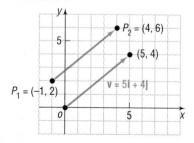

See Figure 54.   ∎

■ Now work Problem 21.

Two position vectors $\mathbf{v}$ and $\mathbf{w}$ are equal if and only if the terminal point of $\mathbf{v}$ is the same as the terminal point of $\mathbf{w}$. This leads to the following result:

Theorem   Two vectors **v** and **w** are equal if and only if their corresponding components are equal. That is:

Equality of Vectors

If $\mathbf{v} = a_1\mathbf{i} + b_1\mathbf{j}$ and $\mathbf{w} = a_2\mathbf{i} + b_2\mathbf{j}$,

then $\mathbf{v} = \mathbf{w}$ if and only if $a_1 = a_2$ and $b_1 = b_2$.

Because of the above result, we can replace any vector (directed line segment) by a unique position vector, and vice versa. This flexibility is one of the main reasons for the wide use of vectors. Unless otherwise specified, from now on the term *vector* will mean the unique position vector equal to it.

Next, we define addition, subtraction, scalar product, and magnitude in terms of the components of a vector.

Let $\mathbf{v} = a_1\mathbf{i} + b_1\mathbf{j}$ and $\mathbf{w} = a_2\mathbf{i} + b_2\mathbf{j}$ be two vectors, and let α be a scalar. Then:

$$\mathbf{v} + \mathbf{w} = (a_1 + a_2)\mathbf{i} + (b_1 + b_2)\mathbf{j} \qquad (2)$$

$$\mathbf{v} - \mathbf{w} = (a_1 - a_2)\mathbf{i} + (b_1 - b_2)\mathbf{j} \qquad (3)$$

$$\alpha\mathbf{v} = (\alpha a_1)\mathbf{i} + (\alpha b_1)\mathbf{j} \qquad (4)$$

$$\|\mathbf{v}\| = \sqrt{a_1^2 + b_1^2} \qquad (5)$$

These definitions are compatible with the geometric ones given earlier in this section. See Figure 55.

FIGURE 55

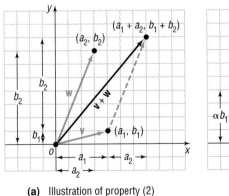

(a) Illustration of property (2)

(b) Illustration of property (4)

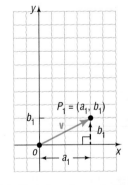

(c) Illustration of property (5):
$\| \mathbf{v} \|$ = Distance of O to P_1
$\| \mathbf{v} \| = \sqrt{a_1^2 + b_1^2}$

Thus, to add two vectors, simply add corresponding components. To subtract two vectors, subtract corresponding components.

E X A M P L E   3

Adding and Subtracting Vectors

If $\mathbf{v} = 2\mathbf{i} + 3\mathbf{j}$ and $\mathbf{w} = 3\mathbf{i} - 4\mathbf{j}$; find:

(a) $\mathbf{v} + \mathbf{w}$ (b) $\mathbf{v} - \mathbf{w}$

Solution
(a) $\mathbf{v} + \mathbf{w} = (2\mathbf{i} + 3\mathbf{j}) + (3\mathbf{i} - 4\mathbf{j}) = (2 + 3)\mathbf{i} + (3 - 4)\mathbf{j}$
$= 5\mathbf{i} - \mathbf{j}$
(b) $\mathbf{v} - \mathbf{w} = (2\mathbf{i} + 3\mathbf{j}) - (3\mathbf{i} - 4\mathbf{j}) = (2 - 3)\mathbf{i} + [3 - (-4)]\mathbf{j}$
$= -\mathbf{i} + 7\mathbf{j}$ ∎

E X A M P L E   4

Forming Scalar Products

If $\mathbf{v} = 2\mathbf{i} + 3\mathbf{j}$ and $\mathbf{w} = 3\mathbf{i} - 4\mathbf{j}$, find:

(a) $3\mathbf{v}$ (b) $2\mathbf{v} - 3\mathbf{w}$ (c) $\|\mathbf{v}\|$

Solution
(a) $3\mathbf{v} = 3(2\mathbf{i} + 3\mathbf{j}) = 6\mathbf{i} + 9\mathbf{j}$
(b) $2\mathbf{v} - 3\mathbf{w} = 2(2\mathbf{i} + 3\mathbf{j}) - 3(3\mathbf{i} - 4\mathbf{j}) = 4\mathbf{i} + 6\mathbf{j} - 9\mathbf{i} + 12\mathbf{j}$
$= -5\mathbf{i} + 18\mathbf{j}$
(c) $\|\mathbf{v}\| = \|2\mathbf{i} + 3\mathbf{j}\| = \sqrt{2^2 + 3^2} = \sqrt{13}$ ∎

■   Now work Problems 27 and 33.

Recall that a unit vector $\mathbf{u}$ is one for which $\|\mathbf{u}\| = 1$. In many applications, it is useful to be able to find a unit vector $\mathbf{u}$ that has the same direction as a given vector $\mathbf{v}$.

Theorem   For any nonzero vector $\mathbf{v}$, the vector

Unit Vector in Direction of $\mathbf{v}$

$$\mathbf{u} = \frac{\mathbf{v}}{\|\mathbf{v}\|}$$

is a unit vector that has the same direction as $\mathbf{v}$. ■

Proof   Let $\mathbf{v} = a\mathbf{i} + b\mathbf{j}$. Then $\|\mathbf{v}\| = \sqrt{a^2 + b^2}$ and

$$\mathbf{u} = \frac{\mathbf{v}}{\|\mathbf{v}\|} = \frac{a\mathbf{i} + b\mathbf{j}}{\sqrt{a^2 + b^2}} = \frac{a}{\sqrt{a^2 + b^2}}\mathbf{i} + \frac{b}{\sqrt{a^2 + b^2}}\mathbf{j}$$

The vector $\mathbf{u}$ is in the same direction as $\mathbf{v}$, since $\|\mathbf{v}\| > 0$, and

$$\|\mathbf{u}\| = \sqrt{\frac{a^2}{a^2 + b^2} + \frac{b^2}{a^2 + b^2}} = \sqrt{\frac{a^2 + b^2}{a^2 + b^2}} = 1$$

Thus, $\mathbf{u}$ is a unit vector in the direction of $\mathbf{v}$. ■

As a consequence of this theorem, if $\mathbf{u}$ is a unit vector in the same direction as a vector $\mathbf{v}$, then $\mathbf{v}$ may be expressed as

$$\mathbf{v} = \|\mathbf{v}\| \mathbf{u} \tag{6}$$

This way of expressing a vector is useful in many applications.

EXAMPLE 5   *Finding a Unit Vector*

Find a unit vector in the same direction as $\mathbf{v} = 4\mathbf{i} - 3\mathbf{j}$.

Solution   We find $\|\mathbf{v}\|$ first:

$$\|\mathbf{v}\| = \|4\mathbf{i} - 3\mathbf{j}\| = \sqrt{16 + 9} = 5$$

Now we multiply $\mathbf{v}$ by the scalar $1/\|\mathbf{v}\| = \frac{1}{5}$. The result is

$$\frac{\mathbf{v}}{\|\mathbf{v}\|} = \frac{4\mathbf{i} - 3\mathbf{j}}{5} = \frac{4}{5}\mathbf{i} - \frac{3}{5}\mathbf{j}$$

Check: This vector is, in fact, a unit vector because
$\left(\frac{4}{5}\right)^2 + \left(-\frac{3}{5}\right)^2 = \frac{16}{25} + \frac{9}{25} = \frac{25}{25} = 1$.   ■

■   Now work Problem 43.

Applications

FIGURE 56

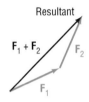
Resultant
$\mathbf{F}_1 + \mathbf{F}_2$
$\mathbf{F}_2$
$\mathbf{F}_1$

Forces provide an example of physical quantities that may be conveniently represented by vectors; two forces "combine" the way vectors "add." How do we know this? Laboratory experiments bear it out. Thus, if $\mathbf{F}_1$ and $\mathbf{F}_2$ are two forces simultaneously acting on an object, the vector sum $\mathbf{F}_1 + \mathbf{F}_2$ is the force that produces the same effect on the object as that obtained when the forces $\mathbf{F}_1$ and $\mathbf{F}_2$ act on the object. The force $\mathbf{F}_1 + \mathbf{F}_2$ is sometimes called the **resultant** of $\mathbf{F}_1$ and $\mathbf{F}_2$. See Figure 56.

Two important applications of the resultant of two vectors occur with aircraft flying in the presence of a wind and with boats cruising across a river with a current. For example, consider the velocity of wind acting on the velocity of an airplane (see Figure 57). Suppose that $\mathbf{w}$ is a vector describing the velocity of the wind; that is, $\mathbf{w}$ represents the direction and speed of the wind. If $\mathbf{v}$ is the velocity of the airplane in the absence of wind (called its **velocity relative to the air),** then $\mathbf{v} + \mathbf{w}$ is the vector equal to the actual velocity of the airplane (called its **velocity relative to the ground).**

Our next example illustrates this use of vectors in navigation.

FIGURE 57

(a) Velocity **w** of wind
relative to ground

(b) Velocity **v** of airplane
relative to air

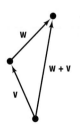

(c) Resultant **w + v** equals
velocity of airplane
relative to the ground

EXAMPLE 6

Finding the Actual Speed of an Aircraft

A Boeing 737 aircraft maintains a constant airspeed of 500 miles per hour in the direction due south. The velocity of the jet stream is 80 miles per hour in a north-easterly direction.

(a) Find a unit vector having northeast as direction.

(b) Find a vector 80 units in magnitude having the same direction as the unit vector found in part (a).

(c) Find the actual speed of the aircraft relative to the ground.

Solution We set up a coordinate system in which north (N) is along the positive y-axis. See Figure 58. Let

$$\mathbf{v}_a = \text{Velocity of aircraft relative to the air} = -500\mathbf{j}$$
$$\mathbf{v}_g = \text{Velocity of aircraft relative to ground}$$
$$\mathbf{v}_w = \text{Velocity of jet stream}$$

FIGURE 58

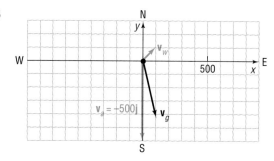

(a) A vector having northeast as direction is $\mathbf{i} + \mathbf{j}$. The unit vector in this direction is

$$\frac{\mathbf{i} + \mathbf{j}}{\|\mathbf{i} + \mathbf{j}\|} = \frac{\mathbf{i} + \mathbf{j}}{\sqrt{1 + 1}} = \frac{1}{\sqrt{2}}(\mathbf{i} + \mathbf{j})$$

(b) The velocity $\mathbf{v}_w$ of the jet stream is a vector with magnitude 80 in the direction of the unit vector $(1/\sqrt{2})(\mathbf{i} + \mathbf{j})$. Thus, from (6),

$$\mathbf{v}_w = 80\left[\frac{1}{\sqrt{2}}(\mathbf{i} + \mathbf{j})\right] = 40\sqrt{2}(\mathbf{i} + \mathbf{j})$$

(c) The velocity $\mathbf{v}_g$ of the aircraft relative to the ground is the resultant of the vectors $\mathbf{v}_a$ and $\mathbf{v}_w$. Thus,

$$\mathbf{v}_g = \mathbf{v}_a + \mathbf{v}_w = -500\mathbf{j} + 40\sqrt{2}(\mathbf{i} + \mathbf{j})$$
$$= 40\sqrt{2}\mathbf{i} + (40\sqrt{2} - 500)\mathbf{j}$$

The actual speed (speed relative to the ground) of the aircraft is

$$\|\mathbf{v}_g\| = \sqrt{(40\sqrt{2})^2 + (40\sqrt{2} - 500)^2} \approx 447 \text{ miles per hour} \qquad \blacksquare$$

We will find the actual direction of the aircraft in Example 4 of the next section.

Writing a Vector in Terms of Its Magnitude and Direction

FIGURE 61

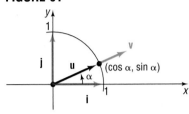

Many applications describe a vector in terms of its magnitude and direction, rather than in terms of its components. Suppose that we are given the magnitude $\|\mathbf{v}\|$ of a nonzero vector $\mathbf{v}$ and the angle α between $\mathbf{v}$ and $\mathbf{i}$. To express $\mathbf{v}$ in terms of $\|\mathbf{v}\|$ and α, we first find the unit vector $\mathbf{u}$ having the same direction as $\mathbf{v}$:

$$\mathbf{u} = \frac{\mathbf{v}}{\|\mathbf{v}\|} \quad \text{or} \quad \mathbf{v} = \|\mathbf{v}\|\mathbf{u} \qquad (9)$$

Look at Figure 61. The coordinates of the terminal point of $\mathbf{u}$ are $(\cos \alpha, \sin \alpha)$. Thus, $\mathbf{u} = \cos \alpha\mathbf{i} + \sin \alpha\mathbf{j}$ and, from (9),

$$\mathbf{v} = \|\mathbf{v}\|(\cos \alpha\mathbf{i} + \sin \alpha\mathbf{j}) \qquad (10)$$

EXAMPLE 3   *Writing a Vector When Its Magnitude and Direction Are Given*

A force $\mathbf{F}$ of 5 pounds is applied in a direction that makes an angle of $30°$ with the positive x-axis. Express $\mathbf{F}$ in terms of $\mathbf{i}$ and $\mathbf{j}$.

Solution   The magnitude of $\mathbf{F}$ is $\|\mathbf{F}\| = 5$ and the angle between the direction of $\mathbf{F}$ and $\mathbf{i}$, the positive x-axis, is $\alpha = 30°$. Thus, by (10),

$$\mathbf{F} = \|\mathbf{F}\|(\cos \alpha\mathbf{i} + \sin \alpha\mathbf{j}) = 5(\cos 30°\mathbf{i} + \sin 30°\mathbf{j})$$

$$= 5\left(\frac{\sqrt{3}}{2}\mathbf{i} + \frac{1}{2}\mathbf{j}\right) = \frac{5}{2}(\sqrt{3}\mathbf{i} + \mathbf{j}) \qquad \blacksquare$$

EXAMPLE 4   *Finding the Actual Direction of an Aircraft*

A Boeing 737 aircraft maintains a constant airspeed of 500 miles per hour in the direction due south. The velocity of the jet stream is 80 miles per hour in a north-easterly direction. Find the actual direction of the aircraft relative to the ground.

Solution   This is the same information given in Example 6 of Section 5.4. We repeat the figure from that example as Figure 62.

FIGURE 62

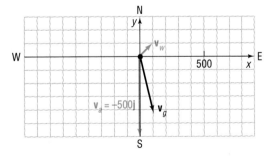

The velocity of the aircraft relative to the air is

$$\mathbf{v}_a = -500\mathbf{j}$$

The wind has magnitude 80 and direction $\alpha = 45°$. Thus, the velocity of the wind is

$$\mathbf{v}_w = 80(\cos 45°\mathbf{i} + \sin 45°\mathbf{j}) = 80\left(\frac{\sqrt{2}}{2}\mathbf{i} + \frac{\sqrt{2}}{2}\mathbf{j}\right)$$

$$= 40\sqrt{2}(\mathbf{i} + \mathbf{j})$$

The velocity of the aircraft relative to the ground is

$$\mathbf{v}_g = \mathbf{v}_a + \mathbf{v}_w = -500\mathbf{j} + 40\sqrt{2}(\mathbf{i} + \mathbf{j}) = 40\sqrt{2}\mathbf{i} + (40\sqrt{2} - 500)\mathbf{j}$$

The angle θ between $\mathbf{v}_g$ and the vector $\mathbf{v}_a = -500\mathbf{j}$ (the velocity of the aircraft relative to the air) is determined by the equation

$$\cos\theta = \frac{\mathbf{v}_g \cdot \mathbf{v}_a}{\|\mathbf{v}_g\| \|\mathbf{v}_a\|} = \frac{(40\sqrt{2} - 500)(-500)}{(447)(500)} \approx 0.9920$$

$$\theta \approx 7.2°$$

The direction of the aircraft relative to the ground is about 7.2° east of south.   ■

■   Now work Problem 19.

Parallel and Orthogonal Vectors

Two vectors $\mathbf{v}$ and $\mathbf{w}$ are said to be **parallel** if there is a nonzero scalar α so that $\mathbf{v} = \alpha\mathbf{w}$. In this case, the angle θ between $\mathbf{v}$ and $\mathbf{w}$ is 0 or π.

E X A M P L E   5   *Determining Whether Vectors are Parallel*

The vectors $\mathbf{v} = 3\mathbf{i} - \mathbf{j}$ and $\mathbf{w} = 6\mathbf{i} - 2\mathbf{j}$ are parallel, since $\mathbf{v} = \frac{1}{2}\mathbf{w}$. Furthermore, since

$$\cos\theta = \frac{\mathbf{v} \cdot \mathbf{w}}{\|\mathbf{v}\| \|\mathbf{w}\|} = \frac{18 + 2}{\sqrt{10}\sqrt{40}} = \frac{20}{\sqrt{400}} = 1$$

the angle θ between $\mathbf{v}$ and $\mathbf{w}$ is 0.   ■

FIGURE 63
$\mathbf{v} \cdot \mathbf{w} = 0$
$\mathbf{v}$ is orthogonal to $\mathbf{w}$

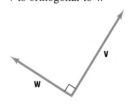

If the angle θ between two nonzero vectors $\mathbf{v}$ and $\mathbf{w}$ is $\pi/2$, the vectors $\mathbf{v}$ and $\mathbf{w}$ are called **orthogonal.***

It follows from formula (8) that if $\mathbf{v}$ and $\mathbf{w}$ are orthogonal then $\mathbf{v} \cdot \mathbf{w} = 0$, since $\cos(\pi/2) = 0$.

On the other hand, if $\mathbf{v} \cdot \mathbf{w} = 0$, then either $\mathbf{v} = \mathbf{0}$ or $\mathbf{w} = \mathbf{0}$ or $\cos\theta = 0$. In the latter case, $\theta = \pi/2$ and $\mathbf{v}$ and $\mathbf{w}$ are orthogonal. See Figure 63. If $\mathbf{v}$ or $\mathbf{w}$ is the zero vector, then, since the zero vector has no specific direction, we adopt the convention that the zero vector is orthogonal to every vector.

Theorem   Two vectors $\mathbf{v}$ and $\mathbf{w}$ are orthogonal if and only if

$$\mathbf{v} \cdot \mathbf{w} = 0$$

■

E X A M P L E   6   *Determining Whether Two Vectors Are Orthogonal*

The vectors

$$\mathbf{v} = 2\mathbf{i} - \mathbf{j} \quad \text{and} \quad \mathbf{w} = 3\mathbf{i} + 6\mathbf{j}$$

**Orthogonal, perpendicular,* and *normal* are all terms that mean "meet at a right angle." It is customary to refer to two vectors as being *orthogonal*, two lines as being *perpendicular*, and a line and a plane or a vector and a plane as being *normal*.

FIGURE 64

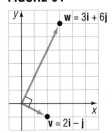

are orthogonal, since

$$\mathbf{v} \cdot \mathbf{w} = 6 - 6 = 0$$

See Figure 64.   ■

■   Now work Problem 11.

Projection of a Vector onto Another Vector

FIGURE 65

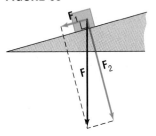

In many physical applications, it is necessary to find "how much" of a vector is applied in a given direction. Look at Figure 65. The force **F** due to gravity is pulling straight down (toward the center of Earth) on the block. To study the effect of gravity on the block, it is necessary to determine how much of **F** is actually pushing the block down the incline ($\mathbf{F}_1$) and how much is pressing the block against the incline ($\mathbf{F}_2$), at a right angle to the incline. Knowing the **decomposition** of **F** often will allow us to determine when friction is overcome and the block will slide down the incline.

Suppose that **v** and **w** are two nonzero vectors with the same initial point P. We seek to decompose **v** into two vectors: $\mathbf{v}_1$, which is parallel to **w**, and $\mathbf{v}_2$, which is orthogonal to **w**. See Figure 66(a) and 66(b). The vector $\mathbf{v}_1$ is called the **vector projection of v onto w** and is denoted by $\text{proj}_\mathbf{w} \mathbf{v}$.

FIGURE 66

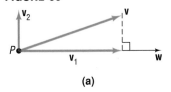

(a)

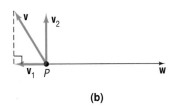

(b)

The vector $\mathbf{v}_1$ is obtained as follows: From the terminal point of **v**, drop a perpendicular to the line containing **w**. The vector $\mathbf{v}_1$ is the vector from P to the foot of this perpendicular. The vector $\mathbf{v}_2$ is given by $\mathbf{v}_2 = \mathbf{v} - \mathbf{v}_1$. Note that $\mathbf{v} = \mathbf{v}_1 + \mathbf{v}_2$, $\mathbf{v}_1$ is parallel to **w**, and $\mathbf{v}_2$ is orthogonal to **w**. This is the decomposition of **v** that we wanted.

Now we seek a formula for $\mathbf{v}_1$ that is based on a knowledge of the vectors **v** and **w**. Since $\mathbf{v} = \mathbf{v}_1 + \mathbf{v}_2$, we have

$$\mathbf{v} \cdot \mathbf{w} = (\mathbf{v}_1 + \mathbf{v}_2) \cdot \mathbf{w} = \mathbf{v}_1 \cdot \mathbf{w} + \mathbf{v}_2 \cdot \mathbf{w} \tag{11}$$

Since $\mathbf{v}_2$ is orthogonal to **w**, we have $\mathbf{v}_2 \cdot \mathbf{w} = 0$. Since $\mathbf{v}_1$ is parallel to **w**, we have $\mathbf{v}_1 = \alpha\mathbf{w}$ for some scalar α. Thus, equation (11) can be written as

$$\mathbf{v} \cdot \mathbf{w} = \alpha\mathbf{w} \cdot \mathbf{w} = \alpha\|\mathbf{w}\|^2$$

$$\alpha = \frac{\mathbf{v} \cdot \mathbf{w}}{\|\mathbf{w}\|^2}$$

Thus,

$$\mathbf{v}_1 = \alpha\mathbf{w} = \frac{\mathbf{v} \cdot \mathbf{w}}{\|\mathbf{w}\|^2}\mathbf{w}$$

Theorem   If **v** and **w** are two nonzero vectors, the vector projection of **v** onto **w** is

$$\text{proj}_\mathbf{w} \mathbf{v} = \frac{\mathbf{v} \cdot \mathbf{w}}{\|\mathbf{w}\|^2}\mathbf{w}$$

The decomposition of **v** into $\mathbf{v}_1$ and $\mathbf{v}_2$, where $\mathbf{v}_1$ is parallel to **w** and $\mathbf{v}_2$ is perpendicular to **w**, is

$$\mathbf{v}_1 = \text{proj}_\mathbf{w}\,\mathbf{v} = \frac{\mathbf{v}\cdot\mathbf{w}}{\|\mathbf{w}\|^2}\mathbf{w} \qquad \mathbf{v}_2 = \mathbf{v} - \mathbf{v}_1 \qquad (12)$$

E X A M P L E   7   *Decomposing a Vector into Two Orthogonal Vectors*

Find the vector projection of $\mathbf{v} = \mathbf{i} + 3\mathbf{j}$ onto $\mathbf{w} = \mathbf{i} + \mathbf{j}$. Decompose **v** into two vectors $\mathbf{v}_1$ and $\mathbf{v}_2$, where $\mathbf{v}_1$ is parallel to **w** and $\mathbf{v}_2$ is orthogonal to **w**.

Solution   We use formulas (12).

$$\mathbf{v}_1 = \text{proj}_\mathbf{w}\,\mathbf{v} = \frac{\mathbf{v}\cdot\mathbf{w}}{\|\mathbf{w}\|^2}\mathbf{w} = \frac{1+3}{(\sqrt{2})^2}\mathbf{w} = 2\mathbf{w} = 2(\mathbf{i}+\mathbf{j})$$

$$\mathbf{v}_2 = \mathbf{v} - \mathbf{v}_1 = (\mathbf{i}+3\mathbf{j}) - 2(\mathbf{i}+\mathbf{j}) = -\mathbf{i}+\mathbf{j}$$

See Figure 67.

FIGURE 67

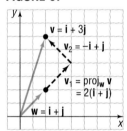

■ Now work Problem 13.

Work Done by a Constant Force

In elementary physics, the **work** W done by a constant force **F** in moving an object from a point A to a point B is defined as

$$W = (\text{Magnitude of force})(\text{Distance}) = \|\mathbf{F}\|\,\|\overrightarrow{AB}\|$$

FIGURE 68

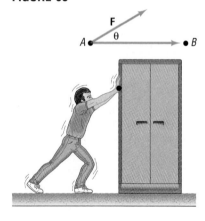

(Work is commonly measured in foot-pounds or in Newton-meters.)

In this definition, it is assumed that the force **F** is applied along the line of motion. If the constant force **F** is not along the line of motion, but, instead, is at an angle θ to the direction of motion, as illustrated in Figure 68, then the **work** W **done by F** in moving an object from A to B is defined as

$$W = \mathbf{F} \cdot \overrightarrow{AB} \qquad (13)$$

This definition is compatible with the force times distance definition given above, since

$$W = (\text{Amount of force in direction of } \overrightarrow{AB})(\text{Distance})$$

$$= \|\text{proj}_{\overrightarrow{AB}}\,\mathbf{F}\|\,\|\overrightarrow{AB}\| = \frac{\mathbf{F}\cdot\overrightarrow{AB}}{\|\overrightarrow{AB}\|^2}\|\overrightarrow{AB}\|\,\|\overrightarrow{AB}\| = \mathbf{F}\cdot\overrightarrow{AB}$$

E X A M P L E   8   *Computing Work*

Find the work done by a force of 5 pounds acting in the direction $\mathbf{i} + \mathbf{j}$ in moving an object 1 foot from $(0,0)$ to $(1,0)$.

Solution   First, we must express the force **F** as a vector. The force has magnitude 5 and direction $\mathbf{i} + \mathbf{j}$. The direction of **F** therefore makes an angle of $45°$ with **i**. Thus, the force **F** is

$$\mathbf{F} = 5(\cos 45°\mathbf{i} + \sin 45°\mathbf{j}) = 5\left(\frac{\sqrt{2}}{2}\mathbf{i} + \frac{\sqrt{2}}{2}\mathbf{j}\right) = \frac{5\sqrt{2}}{2}(\mathbf{i}+\mathbf{j})$$

The line of motion of the object is from $A = (0, 0)$ to $B = (1, 0)$, so $\overrightarrow{AB} = \mathbf{i}$. The work W is therefore

$$W = \mathbf{F} \cdot \overrightarrow{AB} = \frac{5\sqrt{2}}{2}(\mathbf{i} + \mathbf{j}) \cdot \mathbf{i} = \frac{5\sqrt{2}}{2} \text{ foot-pounds} \qquad \blacksquare$$

■ Now work Problem 25.

E X A M P L E   9

Computing Work

Figure 69(a) shows a girl pulling a wagon with a force of 50 pounds. How much work is done in moving the wagon 100 feet if the handle makes an angle of 30° with the ground?

FIGURE 69

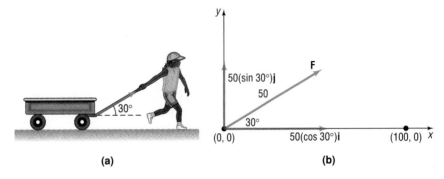

(a)   (b)

Solution   We position the vectors in a coordinate system in such a way that the wagon is moved from $(0, 0)$ to $(100, 0)$. Thus, the motion is from $A = (0, 0)$ to $B = (100, 0)$, so $\overrightarrow{AB} = 100\mathbf{i}$. The force vector $\mathbf{F}$, as shown in Figure 69(b), is

$$\mathbf{F} = 50(\cos 30°\mathbf{i} + \sin 30°\mathbf{j}) = 50\left(\frac{\sqrt{3}}{2}\mathbf{i} + \frac{1}{2}\mathbf{j}\right) = 25\sqrt{3}\mathbf{i} + 25\mathbf{j}$$

By formula (13), the work W done is

$$W = \mathbf{F} \cdot \overrightarrow{AB} = (25\sqrt{3}\mathbf{i} + 25\mathbf{j}) \cdot 100\mathbf{i} = 2500\sqrt{3} \text{ foot-pounds} \qquad \blacksquare$$

HISTORICAL PROBLEM   ■ 1. We stated in an earlier Historical Feature that complex numbers were used as vectors in the plane before the general notion of vector was clarified. Suppose that we make the correspondence

$$\text{Vector} \longleftrightarrow \text{Complex number}$$
$$a\mathbf{i} + b\mathbf{j} \longleftrightarrow a + bi$$
$$c\mathbf{i} + d\mathbf{j} \longleftrightarrow c + di$$

Show that

$$(a\mathbf{i} + b\mathbf{j}) \cdot (c\mathbf{i} + d\mathbf{j}) = \text{Real part}[\overline{(a + bi)}(c + di)]$$

This is how the dot product was found originally. The imaginary part is also interesting. It is a determinant and represents the area of the parallelogram whose edges are the vectors. This is close to some of Hermann Grassmann's ideas and is also connected with the scalar triple product of three-dimensional vectors.   ■

5.5

Exercise 5.5

In Problems 1–10, find the dot product $\mathbf{v} \cdot \mathbf{w}$ *and the cosine of the angle between* $\mathbf{v}$ *and* $\mathbf{w}$.

1. $\mathbf{v} = \mathbf{i} - \mathbf{j}, \quad \mathbf{w} = \mathbf{i} + \mathbf{j}$ **2.** $\mathbf{v} = \mathbf{i} + \mathbf{j}, \quad \mathbf{w} = -\mathbf{i} + \mathbf{j}$ **3.** $\mathbf{v} = 2\mathbf{i} + \mathbf{j}, \quad \mathbf{w} = \mathbf{i} + 2\mathbf{j}$

4. $\mathbf{v} = 2\mathbf{i} + 2\mathbf{j}, \quad \mathbf{w} = \mathbf{i} + 2\mathbf{j}$ **5.** $\mathbf{v} = \sqrt{3}\mathbf{i} - \mathbf{j}, \quad \mathbf{w} = \mathbf{i} + \mathbf{j}$ **6.** $\mathbf{v} = \mathbf{i} + \sqrt{3}\mathbf{j}, \quad \mathbf{w} = \mathbf{i} - \mathbf{j}$

7. $\mathbf{v} = 3\mathbf{i} + 4\mathbf{j}, \quad \mathbf{w} = 4\mathbf{i} + 3\mathbf{j}$ **8.** $\mathbf{v} = 3\mathbf{i} - 4\mathbf{j}, \quad \mathbf{w} = 4\mathbf{i} - 3\mathbf{j}$ **9.** $\mathbf{v} = 4\mathbf{i}, \quad \mathbf{w} = \mathbf{j}$

10. $\mathbf{v} = \mathbf{i}, \quad \mathbf{w} = -3\mathbf{j}$

11. Find a such that the angle between $\mathbf{v} = a\mathbf{i} - \mathbf{j}$ and $\mathbf{w} = 2\mathbf{i} + 3\mathbf{j}$ is $\pi/2$.

12. Find b such that the angle between $\mathbf{v} = \mathbf{i} + \mathbf{j}$ and $\mathbf{w} = \mathbf{i} + b\mathbf{j}$ is $\pi/2$.

In Problems 13–18, decompose $\mathbf{v}$ *into two vectors* $\mathbf{v}_1$ *and* $\mathbf{v}_2$, *where* $\mathbf{v}_1$ *is parallel to* $\mathbf{w}$ *and* $\mathbf{v}_2$ *is orthogonal to* $\mathbf{w}$.

13. $\mathbf{v} = 2\mathbf{i} - 3\mathbf{j}, \quad \mathbf{w} = \mathbf{i} - \mathbf{j}$ **14.** $\mathbf{v} = -3\mathbf{i} + 2\mathbf{j}, \quad \mathbf{w} = 2\mathbf{i} + \mathbf{j}$ **15.** $\mathbf{v} = \mathbf{i} - \mathbf{j}, \quad \mathbf{w} = \mathbf{i} + 2\mathbf{j}$

16. $\mathbf{v} = 2\mathbf{i} - \mathbf{j}, \quad \mathbf{w} = \mathbf{i} - 2\mathbf{j}$ **17.** $\mathbf{v} = 3\mathbf{i} + \mathbf{j}, \quad \mathbf{w} = -2\mathbf{i} - \mathbf{j}$ **18.** $\mathbf{v} = \mathbf{i} - 3\mathbf{j}, \quad \mathbf{w} = 4\mathbf{i} - \mathbf{j}$

19. *Finding the Actual Speed and Direction of an Aircraft*
A DC-10 jumbo jet maintains an airspeed of 550 miles per hour in a southwesterly direction. The velocity of the jet stream is a constant 80 miles per hour from the west. Find the actual speed and direction of the aircraft.

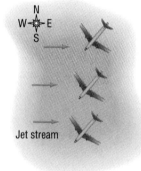

20. *Finding the Correct Compass Heading* The pilot of an aircraft wishes to head directly east, but is faced with a wind speed of 40 miles per hour from the northwest. If the pilot maintains an airspeed of 250 miles per hour, what compass heading should be maintained? What is the actual speed of the aircraft?

21. *Correct Direction for Crossing a Stream* A small stream has a constant current of 3 kilometers per hour. At what angle to a boat dock should a motorboat—capable of maintaining a constant speed of 20 kilometers per hour—be headed in order to reach a point directly opposite the dock? If the stream is $\frac{1}{2}$ kilometer wide, how long will it take to cross?

22. *Correct Direction for Crossing a Stream* Repeat Problem 21 if the current is 5 kilometers per hour.

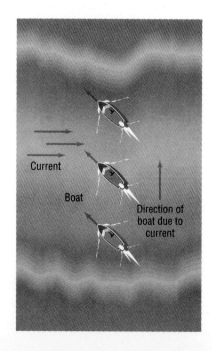

23. *Correct Direction for Crossing a Stream* A river is 500 meters wide and has a current of 1 kilometer per hour. If Sharon can swim at a rate of 2 kilometers per hour, at what angle to the shore should she swim if she wishes to cross the river to a point directly opposite? How long will it take to swim across the river?

24. An airplane travels 200 miles due west and then 150 miles 60° north of west. Determine the resultant displacement.

25. *Computing Work* Find the work done by a force of 3 pounds acting in the direction $2\mathbf{i} + \mathbf{j}$ in moving an object 2 feet from $(0, 0)$ to $(0, 2)$.

26. *Computing Work* Find the work done by a force of 1 pound acting in the direction $2\mathbf{i} + 2\mathbf{j}$ in moving an object 5 feet from $(0, 0)$ to $(3, 4)$.

27. *Computing Work* A wagon is pulled horizontally by exerting a force of 20 pounds on the handle at an angle of 30° with the horizontal. How much work is done in moving the wagon 100 feet?

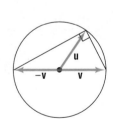

Current

Swimmer's direction

Direction of swimmer due to current

28. Find the acute angle that a constant unit force vector makes with the positive x-axis if the work done by the force in moving a particle from $(0, 0)$ to $(4, 0)$ equals 2.

29. Prove the distributive property, $\mathbf{u} \cdot (\mathbf{v} + \mathbf{w}) = \mathbf{u} \cdot \mathbf{v} + \mathbf{u} \cdot \mathbf{w}$.

30. Prove property (5), $\mathbf{0} \cdot \mathbf{v} = 0$.

31. If $\mathbf{v}$ is a unit vector and the angle between $\mathbf{v}$ and $\mathbf{i}$ is α, show that $\mathbf{v} = \cos \alpha \mathbf{i} + \sin \alpha \mathbf{j}$.

32. Suppose that $\mathbf{v}$ and $\mathbf{w}$ are unit vectors. If the angle between $\mathbf{v}$ and $\mathbf{i}$ is α and if the angle between $\mathbf{w}$ and $\mathbf{i}$ is β, use the idea of the dot product $\mathbf{v} \cdot \mathbf{w}$ to prove that

$$\cos(\alpha - \beta) = \cos \alpha \cos \beta + \sin \alpha \sin \beta$$

33. Show that the projection of $\mathbf{v}$ onto $\mathbf{i}$ is $(\mathbf{v} \cdot \mathbf{i})\mathbf{i}$. In fact, show that we can always write a vector $\mathbf{v}$ as

$$\mathbf{v} = (\mathbf{v} \cdot \mathbf{i})\mathbf{i} + (\mathbf{v} \cdot \mathbf{j})\mathbf{j}$$

34. (a) If $\mathbf{u}$ and $\mathbf{v}$ have the same magnitude, then show that $\mathbf{u} + \mathbf{v}$ and $\mathbf{u} - \mathbf{v}$ are orthogonal.
    (b) Use this to prove that an angle inscribed in a semicircle is a right angle (see the figure).

35. Let $\mathbf{v}$ and $\mathbf{w}$ denote two nonzero vectors. Show that the vector $\mathbf{v} - \alpha\mathbf{w}$ is orthogonal to $\mathbf{w}$ if $\alpha = (\mathbf{v} \cdot \mathbf{w})/\|\mathbf{w}\|^2$.

36. Let $\mathbf{v}$ and $\mathbf{w}$ denote two nonzero vectors. Show that the vectors $\|\mathbf{w}\|\mathbf{v} + \|\mathbf{v}\|\mathbf{w}$ and $\|\mathbf{w}\|\mathbf{v} - \|\mathbf{v}\|\mathbf{w}$ are orthogonal.

37. In the definition of work given in this section, what is the work done if $\mathbf{F}$ is orthogonal to $\overrightarrow{AB}$?

38. Prove the **polarization identity:** $\|\mathbf{u} + \mathbf{v}\|^2 - \|\mathbf{u} - \mathbf{v}\|^2 = 4(\mathbf{u} \cdot \mathbf{v})$

39. Make up an application different from any found in the text that requires the dot product.

Chapter Review

THINGS TO KNOW

Relationship between polar coordinates (r, θ) and rectangular coordinates (x, y)	$x = r \cos \theta, y = r \sin \theta$ $x^2 + y^2 = r^2, \tan \theta = \dfrac{y}{x},\ x \neq 0$
Polar form of a complex number	If $z = x + iy$, then $z = r(\cos \theta + i \sin \theta)$, where $r = |z| = \sqrt{x^2 + y^2}$, $\sin \theta = \dfrac{y}{r}$, $\cos \theta = \dfrac{x}{r}, 0 \leq \theta < 2\pi$
Demoivre's Theorem	If $z = r(\cos \theta + i \sin \theta)$, then $z^n = r^n(\cos n\theta + i \sin n\theta)$, where $n \geq 1$ is a positive integer

nth root of a complex number
$$\sqrt[n]{z} = \sqrt[n]{r}\left[\cos\left(\frac{\theta}{n} + \frac{2k\pi}{n}\right) + i \sin\left(\frac{\theta}{n} + \frac{2k\pi}{n}\right)\right], k = 0, \ldots, n - 1$$

Vector	Quantity having magnitude and direction; equivalent to a directed line segment $\overrightarrow{PQ}$
Position vector	Vector whose initial point is at the origin
Unit vector	Vector whose magnitude is 1
Dot product	If $\mathbf{v} = a_1\mathbf{i} + b_1\mathbf{j}$ and $\mathbf{w} = a_2\mathbf{i} + b_2\mathbf{j}$, then $\mathbf{v} \cdot \mathbf{w} = a_1a_2 + b_1b_2$.

Angle θ between two nonzero vectors $\mathbf{u}$ and $\mathbf{v}$ $\cos\theta = \dfrac{\mathbf{u} \cdot \mathbf{v}}{\|\mathbf{u}\|\,\|\mathbf{v}\|}$

How To:

Plot polar coordinates	Add and subtract vectors
Convert from polar to rectangular coordinates	Form scalar multiples of vectors
Convert from rectangular to polar coordinates	Find the magnitude of a vector
Graph polar equations by hand and using a graphing utility (see Table 7)	Solve problems involving vectors
	Find the dot product of two vectors
Write a complex number in polar form, $z = r(\cos\theta + i\sin\theta)$, $0° \le \theta < 360°$	Find the angle between two vectors
	Determine whether two vectors are parallel
Use Demoivre's Theorem to find powers of complex numbers	Determine whether two vectors are orthogonal
Use the complex roots theorem to find complex roots	Find the vector projection of $\mathbf{v}$ onto $\mathbf{w}$

Fill-in-the-Blank Items

1. In polar coordinates, the origin is called the _____, and the positive x-axis is referred to as the _____ _____.

2. Another representation in polar coordinates for the point $(2, \pi/3)$ is (_____, $4\pi/3$).

3. Using polar coordinates (r, θ), the circle $x^2 + y^2 = 2x$ takes the form _____.

4. In a polar equation, replace θ by $-\theta$. If an equivalent equation results, the graph is symmetric with respect to _____ _____.

5. When a complex number z is written in the polar form $z = r(\cos\theta + i\sin\theta)$, the nonnegative number r is the _____ _____ of z, and the angle θ, $0 \le \theta < 2\pi$, is the _____ of z.

6. A vector whose magnitude is 1 is called a(n) _____ vector.

7. If the angle between two vectors $\mathbf{v}$ and $\mathbf{w}$ is $\pi/2$, then the dot product $\mathbf{v} \cdot \mathbf{w}$ equals _____.

True/False Items

T   F   **1.** The polar coordinates of a point are unique.

T   F   **2.** The rectangular coordinates of a point are unique.

T   F   **3.** The tests for symmetry in polar coordinates are conclusive.

T   F   **4.** Demoivre's Theorem is useful for raising a complex number to a positive integer power.

T   F   **5.** Vectors are quantities that have magnitude and direction.

T   F   **6.** Force is a physical example of a vector.

T   F   **7.** If $\mathbf{u}$ and $\mathbf{v}$ are orthogonal vectors, then $\mathbf{u} \cdot \mathbf{v} = 0$.

REVIEW EXERCISES

In Problems 1–6, plot each point given in polar coordinates, and find its rectangular coordinates.

1. $(3, \pi/6)$ **2.** $(4, 2\pi/3)$ **3.** $(-2, 4\pi/3)$

4. $(-1, 5\pi/4)$ **5.** $(-3, -\pi/2)$ **6.** $(-4, -\pi/4)$

In Problems 7–12, the rectangular coordinates of a point are given. Find two pairs of polar coordinates (r, θ) for each point, one with $r > 0$ and the other with $r < 0$. Express θ in radians.

7. $(-3, 3)$ **8.** $(1, -1)$ **9.** $(0, -2)$

10. $(2, 0)$ **11.** $(3, 4)$ **12.** $(-5, 12)$

In Problems 13–18, the letters x and y represent rectangular coordinates. Write each equation using polar coordinates (r, θ).

13. $3x^2 + 3y^2 = 6y$ **14.** $2x^2 - 2y^2 = 5y$ **15.** $2x^2 - y^2 = \dfrac{y}{x}$

16. $x^2 + 2y^2 = \dfrac{y}{x}$ **17.** $x(x^2 + y^2) = 4$ **18.** $y(x^2 - y^2) = 3$

In Problems 19–24, write each polar equation as an equation in rectangular coordinates (x, y).

19. $r = 2 \sin \theta$ **20.** $3r = \sin \theta$ **21.** $r = 5$

22. $\theta = \pi/4$ **23.** $r \cos \theta + 3r \sin \theta = 6$ **24.** $r^2 \tan \theta = 1$

In Problems 25–30, sketch the graph of each polar equation by hand. Be sure to test for symmetry. Verify your results using a graphing utility.

25. $r = 4 \cos \theta$ **26.** $r = 3 \sin \theta$ **27.** $r = 3 - 3 \sin \theta$

28. $r = 2 + \cos \theta$ **29.** $r = 4 - \cos \theta$ **30.** $r = 1 - 2 \sin \theta$

In Problems 31–34, write each complex number in polar form. Express each argument in degrees.

31. $-1 - i$ **32.** $-\sqrt{3} + i$ **33.** $4 - 3i$ **34.** $3 - 2i$

In Problems 35–40, write each complex number in the standard form $a + bi$.

35. $2(\cos 150° + i \sin 150°)$ **36.** $3(\cos 60° + i \sin 60°)$ **37.** $3\left(\cos \dfrac{2\pi}{3} + i \sin \dfrac{2\pi}{3}\right)$

38. $4\left(\cos \dfrac{3\pi}{4} + i \sin \dfrac{3\pi}{4}\right)$ **39.** $0.1(\cos 350° + i \sin 350°)$ **40.** $0.5(\cos 160° + i \sin 160°)$

In Problems 41–46, find zw and z/w. Leave your answers in polar form.

41. $z = \cos 80° + i \sin 80°$
$w = \cos 50° + i \sin 50°$

42. $z = \cos 205° + i \sin 205°$
$w = \cos 85° + i \sin 85°$

43. $z = 3\left(\cos \dfrac{9\pi}{5} + i \sin \dfrac{9\pi}{5}\right)$
$w = 2\left(\cos \dfrac{\pi}{5} + i \sin \dfrac{\pi}{5}\right)$

44. $z = 2\left(\cos \dfrac{5\pi}{3} + i \sin \dfrac{5\pi}{3}\right)$
$w = 3\left(\cos \dfrac{\pi}{3} + i \sin \dfrac{\pi}{3}\right)$

45. $z = 5(\cos 10° + i \sin 10°)$
$w = \cos 355° + i \sin 355°$

46. $z = 4(\cos 50° + i \sin 50°)$
$w = \cos 340° + i \sin 340°$

In Problems 47–54, write each expression in the standard form $a + bi$.

47. $[3(\cos 20° + i \sin 20°)]^3$

48. $[2(\cos 50° + i \sin 50°)]^3$

49. $\left[\sqrt{2}\left(\cos \dfrac{5\pi}{8} + i \sin \dfrac{5\pi}{8}\right)\right]^4$

50. $\left[2\left(\cos \dfrac{5\pi}{16} + i \sin \dfrac{5\pi}{16}\right)\right]^4$

51. $(1 - \sqrt{3}\,i)^6$

52. $(2 - 2i)^8$

53. $(3 + 4i)^4$

54. $(1 - 2i)^4$

55. Find all the complex cube roots of 27.

56. Find all the complex fourth roots of -16.

In Problems 57–60, the vector $\mathbf{v}$ is represented by the directed line segment $\overrightarrow{PQ}$. Write $\mathbf{v}$ in the form $a\mathbf{i} + b\mathbf{j}$ and find $\|\mathbf{v}\|$.

57. $P = (1, -2);\quad Q = (3, -6)$

58. $P = (-3, 1);\quad Q = (4, -2)$

59. $P = (0, -2);\quad Q = (-1, 1)$

60. $P = (3, -4);\quad Q = (-2, 0)$

In Problems 61–68, use the vectors $\mathbf{v} = -2\mathbf{i} + \mathbf{j}$ and $\mathbf{w} = 4\mathbf{i} - 3\mathbf{j}$.

61. Find $4\mathbf{v} - 3\mathbf{w}$.

62. Find $-\mathbf{v} + 2\mathbf{w}$.

63. Find $\|\mathbf{v}\|$.

64. Find $\|\mathbf{v} + \mathbf{w}\|$.

65. Find $\|\mathbf{v}\| + \|\mathbf{w}\|$.

66. Find $\|2\mathbf{v}\| - 3\|\mathbf{w}\|$.

67. Find a unit vector having the same direction as $\mathbf{v}$.

68. Find a unit vector having the opposite direction of $\mathbf{w}$.

In Problems 69–72, find the dot product $\mathbf{v} \cdot \mathbf{w}$ and the cosine of the angle between $\mathbf{v}$ and $\mathbf{w}$.

69. $\mathbf{v} = -2\mathbf{i} + \mathbf{j},\quad \mathbf{w} = 4\mathbf{i} - 3\mathbf{j}$

70. $\mathbf{v} = 3\mathbf{i} - \mathbf{j},\quad \mathbf{w} = \mathbf{i} + \mathbf{j}$

71. $\mathbf{v} = \mathbf{i} - 3\mathbf{j},\quad \mathbf{w} = -\mathbf{i} + \mathbf{j}$

72. $\mathbf{v} = \mathbf{i} + 4\mathbf{j},\quad \mathbf{w} = 3\mathbf{i} - 2\mathbf{j}$

73. Find the vector projection of $\mathbf{v} = 2\mathbf{i} + 3\mathbf{j}$ onto $\mathbf{w} = 3\mathbf{i} + \mathbf{j}$.

74. Find the vector projection of $\mathbf{v} = -\mathbf{i} + 2\mathbf{j}$ onto $\mathbf{w} = 3\mathbf{i} - \mathbf{j}$.

75. Find the angle between the vectors $\mathbf{v} = 3\mathbf{i} - 4\mathbf{j}$ and $\mathbf{w} = 12\mathbf{i} - 5\mathbf{j}$.

76. Find the angle between the vectors $\mathbf{v} = \mathbf{i} - \mathbf{j}$ and $\mathbf{w} = 2\mathbf{i} + \mathbf{j}$.

77. *Actual Speed and Direction of a Swimmer* A swimmer can maintain a constant speed of 5 miles per hour. If the swimmer heads directly across a river that has a current moving at the rate of 2 miles per hour, what is the actual speed of the swimmer? (See the figure.) If the river is 1 mile wide, how far downstream will the swimmer end up from the point directly across the river?

78. *Actual Speed and Direction of a Motorboat* A small motorboat is moving at a true speed of 11 miles per hour in a southerly direction. The current is known to be from the northeast at 3 miles per hour. What is the speed of the motorboat relative to the water? In what direction does the compass indicate that the boat is headed?

79. *Correct Direction for Crossing a Stream* A stream 1 kilometer wide has a constant current of 5 kilometers per hour. At what angle to the shore should a person head a boat that is capable of maintaining a constant speed of 15 kilometers per hour in order to reach a point directly opposite?

80. *Actual Speed and Direction of an Airplane* An airplane has an airspeed of 500 kilometers per hour in a northerly direction. The wind velocity is 60 kilometers per hour in a southeasterly direction. Find the actual speed and direction of the plane relative to the ground.

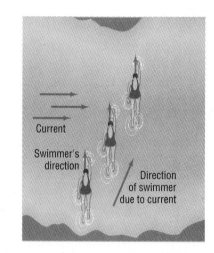
Current
Swimmer's direction
Direction of swimmer due to current

Chapter 6

ANALYTIC GEOMETRY

6.1 Conics
6.2 The Parabola
6.3 The Ellipse
6.4 The Hyperbola
6.5 Rotation of Axes; General Form of a Conic
6.6 Polar Equations of Conics
6.7 Plane Curves and Parametric Equations
    Chapter Review

PREPARING FOR THIS CHAPTER

Before getting started on this chapter, review the following concepts:

Distance formula (p. 5)
Completing the square (Appendix, Section 3)
Intercepts (pp. 21–23)
Symmetry (pp. 26–29)
Circles (pp. 47–51)
Double-angle and half-angle formulas (Section 3.3)
Polar coordinates (Section 5.1)
Amplitude and period of sinusoidal graphs (p. 285)

Preview   Satellite Dish

*A satellite dish is shaped like a **paraboloid of revolution,** a surface formed by rotating a parabola about its axis of symmetry. The signals that emanate from a satellite strike the surface of the dish and are reflected to a single point, where the receiver is located. If the dish is 8 feet across at its opening and is 3 feet deep at its center, at what position should the receiver be placed? [Example 10 in Section 6.2].* ■

istorically, Apollonius (200 BC) was among the first to study *conics* and discover some of their interesting properties. Today, conics are still studied because of their many uses. *Paraboloids of revolution* (parabolas rotated about their axes of symmetry) are used as signal collectors (the satellite dishes used with radar and cable TV, for example), as solar energy collectors, and as reflectors (telescopes, light projection, and so on). The planets circle the Sun in approximately *elliptical* orbits. Elliptical surfaces can be used to reflect signals such as light and sound from one place to another. And *hyperbolas* can be used to determine the positions of ships at sea.

The Greeks used the methods of Euclidean geometry to study conics. We shall use the more powerful methods of analytic geometry, bringing to bear both algebra and geometry, for our study of conics. Thus, we shall give a geometric description of each conic, and then, using rectangular coordinates and the distance formula, we shall find equations that represent conics. We used this same development, you may recall, when we first defined a circle in Section 1.3.

The chapter concludes with a section on equations of conics in polar coordinates, followed by a discussion of plane curves and parametric equations.

6.1

Conics

FIGURE 1

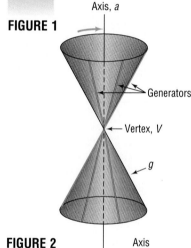

Axis, *a*

Generators

Vertex, *V*

g

The word *conic* derives from the word *cone*, which is a geometric figure that can be constructed in the following way: Let *a* and *g* be two distinct lines that intersect at a point *V*. Keep the line *a* fixed. Now rotate the line *g* about *a* while maintaining the same angle between *a* and *g*. The collection of points swept out (generated) by the line *g* is called a **(right circular) cone.** See Figure 1. The fixed line *a* is called the **axis** of the cone; the point *V* is called its **vertex;** the lines that pass through *V* and make the same angle with *a* as *g* are called **generators** of the cone. Thus, each generator is a line that lies entirely on the cone. The cone consists of two parts, called **nappes,** that intersect at the vertex.

Conics, an abbreviation for **conic sections,** are curves that result from the intersection of a (right circular) cone and a plane. The conics we shall study arise when the plane does not contain the vertex, as shown in Figure 2. These conics are **circles** when the plane is perpendicular to the axis of the cone and intersects each generator; **ellipses** when the plane is tilted slightly so that it intersects each generator, but intersects only one nappe of the cone; **parabolas** when the plane is

FIGURE 2

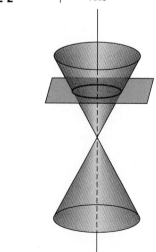

Axis

(a) Circle

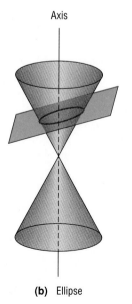

Axis

(b) Ellipse

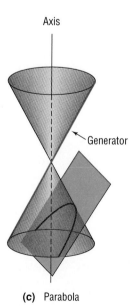

Axis

Generator

(c) Parabola

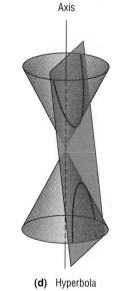

Axis

(d) Hyperbola

tilted further so that it is parallel to one (and only one) generator and intersects only one nappe of the cone; and **hyperbolas** when the plane intersects both nappes.

If the plane does contain the vertex, the intersection of the plane and the cone is a point, a line, or a pair of intersecting lines. These are usually called **degenerate conics.**

6.2

The Parabola

In algebra, we learn that the graph of a quadratic function is a parabola. In this section, we begin with a geometric definition of parabola and use it to obtain an equation.

Parabola

A **parabola** is defined as the collection of all points P in the plane that are the same distance from a fixed point F as they are from a fixed line D. The point F is called the **focus** of the parabola, and the line D is its **directrix.** As a result, a parabola is the set of points P for which

$$d(F, P) = d(P, D) \tag{1}$$

FIGURE 3

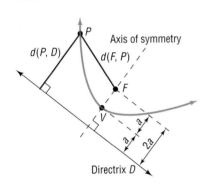

FIGURE 4

$y^2 = 4ax$

Figure 3 shows a parabola. The line through the focus F and perpendicular to the directrix D is called the **axis of symmetry** of the parabola. The point of intersection of the parabola with its axis of symmetry is called the **vertex** V.

Because the vertex V lies on the parabola, it must satisfy equation (1): $d(F, V) = d(V, D)$. Thus, the vertex is midway between the focus and the directrix. We shall let a equal the distance $d(F, V)$ from F to V. Now we are ready to derive an equation for a parabola. To do this, we use a rectangular system of coordinates, positioned so that the vertex V, focus F, and directrix D of the parabola are conveniently located. If we choose to locate the vertex V at the origin $(0, 0)$, then we can conveniently position the focus F on either the x-axis or the y-axis.

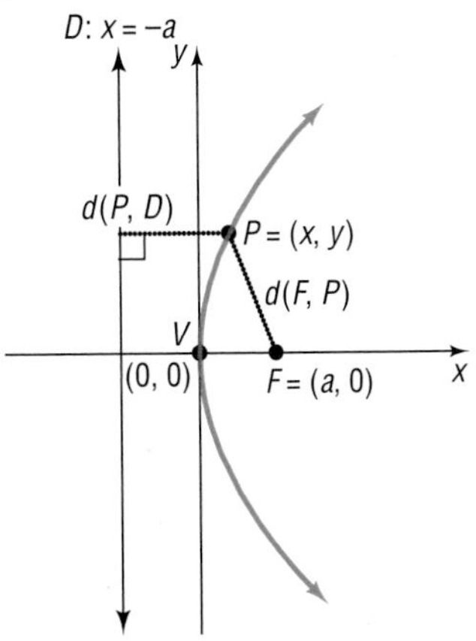

First, we consider the case where the focus F is on the positive x-axis, as shown in Figure 4. Because the distance from F to V is a, the coordinates of F will be $(a, 0)$ with $a > 0$. Similarly, because the distance from V to the directrix D is also a and because D must be perpendicular to the x-axis (since the x-axis is the axis of symmetry), the equation of the directrix D must be $x = -a$. Now, if $P = (x, y)$ is any point on the parabola, then P must obey equation (1):

$$d(F, P) = d(P, D)$$

So, we have

$$\sqrt{(x - a)^2 + y^2} = |x + a| \qquad \text{Use the distance formula.}$$
$$(x - a)^2 + y^2 = (x + a)^2 \qquad \text{Square both sides.}$$
$$x^2 - 2ax + a^2 + y^2 = x^2 + 2ax + a^2$$
$$y^2 = 4ax$$

Theorem   The equation of a parabola with vertex at $(0, 0)$, focus at $(a, 0)$, and directrix $x = -a, a > 0$, is

Equation of a Parabola;
Vertex at $(0, 0)$,
Focus at $(a, 0), a > 0$

$$y^2 = 4ax \tag{2}$$

■

E X A M P L E   1

Finding the Equation of a Parabola

Find an equation of the parabola with vertex at $(0, 0)$ and focus at $(3, 0)$. Graph the equation.

Solution   The distance from the vertex $(0, 0)$ to the focus $(3, 0)$ is $a = 3$. Based on equation (2), the equation of this parabola is

$$y^2 = 4ax$$
$$y^2 = 12x \qquad a = 3$$

To graph this parabola by hand, it is helpful to plot the two points on the graph above and below the focus. To locate them, we let $x = 3$. Then,

$$y^2 = 12x = 36$$
$$y = \pm 6$$

The points on the parabola above and below the focus are $(3, -6)$ and $(3, 6)$. See Figure 5.   ■

FIGURE 5
$y^2 = 12x$

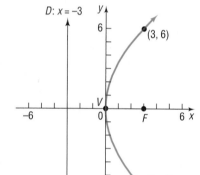

In general, the points on a parabola $y^2 = 4ax$ that lie above and below the focus $(a, 0)$ are each at a distance $2a$ from the focus. This follows from the fact that if $x = a$ then $y^2 = 4ax = 4a^2$, or $y = \pm 2a$. The line segment joining these two points is called the **latus rectum**; its length is $4a$.

E X A M P L E   2

Graphing a Parabola Using a Graphing Utility

Graph the parabola: $y^2 = 12x$

Solution   To graph the parabola $y^2 = 12x$, we need to graph the two functions $y_1 = \sqrt{12x}$ and $y_2 = -\sqrt{12x}$. Figure 6 shows the graph of $y^2 = 12x$. Notice that the graph fails the vertical line test, so $y^2 = 12x$ is not a function.   ■

■ Now work Problem 17.

FIGURE 6
$y^2 = 12x$

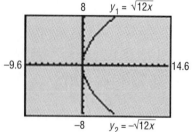

By reversing the steps we used to obtain equation (2), it follows that the graph of an equation of the form of equation (2) is a parabola; its vertex is at $(0, 0)$, its focus is at $(a, 0)$, its directrix is the line $x = -a$, and its axis of symmetry is the x-axis.

For the remainder of this section, the direction "Discuss the equation" will mean to find the vertex, focus, and directrix of the parabola and graph it.

E X A M P L E   3

Discussing the Equation of a Parabola

Discuss the equation: $y^2 = 8x$

Solution   Figure 7(a) shows the graph of $y^2 = 8x$ using a graphing utility. We now proceed to analyze the equation.

The equation $y^2 = 8x$ is of the form $y^2 = 4ax$, where $4a = 8$. Thus, $a = 2$.

Consequently, the graph of the equation is a parabola with vertex at (0, 0) and focus on the positive x-axis at (2, 0). The directrix is the vertical line $x = -2$. The two points defining the latus rectum are obtained by letting $x = 2$. Then $y^2 = 16$, or $y = \pm 4$. These points help in graphing the parabola by hand since they determine the "opening" of the graph. See Figure 7(b).

FIGURE 7
$y^2 = 8x$

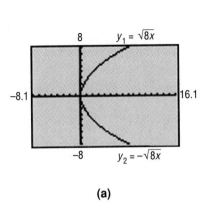

(a)

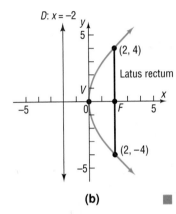

(b)

Recall that we arrived at equation (2) after placing the focus on the positive x-axis. If the focus is placed on the negative x-axis, positive y-axis, or negative y-axis, a different form of the equation for the parabola results. The four forms of the equation of a parabola with vertex at (0, 0) and focus on a coordinate axis a distance a from (0, 0) are given in Table 1, and their graphs are given in Figure 8. Notice that each graph is symmetric with respect to its axis of symmetry.

TABLE 1   EQUATIONS OF A PARABOLA: VERTEX AT (0, 0); FOCUS ON AXIS; $a > 0$

VERTEX	FOCUS	DIRECTRIX	EQUATION	DESCRIPTION
(0, 0)	(a, 0)	$x = -a$	$y^2 = 4ax$	Parabola, axis of symmetry is the x-axis, opens to right
(0, 0)	(-a, 0)	$x = a$	$y^2 = -4ax$	Parabola, axis of symmetry is the x-axis, opens to left
(0, 0)	(0, a)	$y = -a$	$x^2 = 4ay$	Parabola, axis of symmetry is the y-axis, opens up
(0, 0)	(0, -a)	$y = a$	$x^2 = -4ay$	Parabola, axis of symmetry is the y-axis, opens down

FIGURE 8

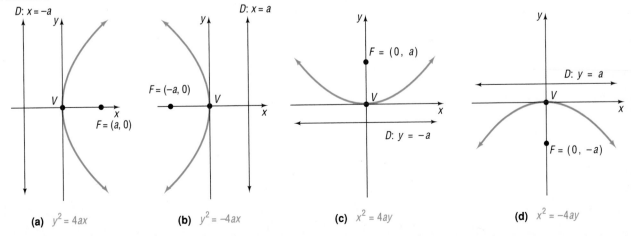

(a) $y^2 = 4ax$ (b) $y^2 = -4ax$ (c) $x^2 = 4ay$ (d) $x^2 = -4ay$

E X A M P L E   4   *Discussing the Equation of a Parabola*

Discuss the equation: $x^2 = -12y$

Solution   Figure 9(a) shows the graph of $x^2 = -12y$ using a graphing utility. We now proceed to analyze the equation.

The equation $x^2 = -12y$ is of the form $x^2 = -4ay$, with $a = 3$. Consequently, the graph of the equation is a parabola with vertex at $(0, 0)$, focus at $(0, -3)$, and directrix the line $y = 3$. The parabola opens down, and its axis of symmetry is the y-axis. To obtain the points defining the latus rectum, let $y = -3$. Then $x^2 = 36$, or $x = \pm 6$. See Figure 9(b).

FIGURE 9
$x^2 = -12y$

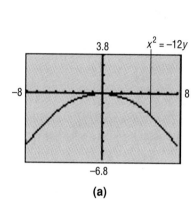

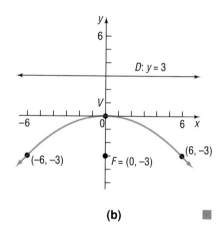

(a)   (b)

■   Now work Problem 35.

E X A M P L E   5   *Finding the Equation of a Parabola*

Find the equation of the parabola with focus at $(0, 4)$ and directrix the line $y = -4$. Graph the equation by hand.

Solution   A parabola whose focus is at $(0, 4)$ and whose directrix is the horizontal line $y = -4$ will have its vertex at $(0, 0)$. (Do you see why? The vertex is midway between the focus and the directrix.) Thus, the equation of this parabola is of the form $x^2 = 4ay$, with $a = 4$; that is,

$$x^2 = 16y$$

Figure 10 shows the graph.

FIGURE 10
$x^2 = 16y$

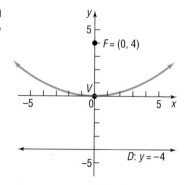

E X A M P L E   6   *Finding the Equation of a Parabola*

Find the equation of a parabola with vertex at $(0, 0)$ if its axis of symmetry is the *x*-axis and its graph contains the point $(-\frac{1}{2}, 2)$. Find its focus and directrix, and graph the equation by hand.

Solution   Because the vertex is at the origin and the axis of symmetry is the *x*-axis, we see from Table 1 that the form of the equation is

$$y^2 = kx$$

Because the point $(-\frac{1}{2}, 2)$ is on the parabola, the coordinates $x = -\frac{1}{2}$, $y = 2$ must satisfy the equation. Putting $x = -\frac{1}{2}$ and $y = 2$ into the equation, we find

$$4 = k(-\tfrac{1}{2})$$
$$k = -8$$

Thus, the equation of the parabola is

$$y^2 = -8x$$

Comparing this equation to $y^2 = -4ax$, we find that $a = 2$. The focus is therefore at $(-2, 0)$, and the directrix is the line $x = 2$. Letting $x = -2$, we find $y^2 = 16$ or $y = \pm 4$. The points $(-2, 4)$ and $(-2, -4)$ define the latus rectum. See Figure 11.

FIGURE 11
$y^2 = -8x$

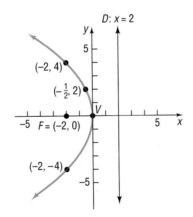

■   Now work Problem 27.

Vertex at (*h*, *k*)

If a parabola with vertex at the origin and axis of symmetry along a coordinate axis is shifted horizontally *h* units and then vertically *k* units, the result is a parabola with vertex at (h, k) and axis of symmetry parallel to a coordinate axis. The equations of such parabolas have the same forms as those in Table 1, but with *x* replaced by $x - h$ and *y* replaced by $y - k$. Table 2 gives the forms of the equations of such parabolas. Figure 12(a)–(d) illustrates the graphs for $h > 0$, $k > 0$.

TABLE 2   PARABOLAS WITH VERTEX AT (*h*,*k*), AXIS OF SYMMETRY PARALLEL TO A COORDINATE AXIS, *a* > 0

VERTEX	FOCUS	DIRECTRIX	EQUATION	DESCRIPTION
(h, k)	$(h + a, k)$	$x = -a + h$	$(y-k)^2 = 4a(x - h)$	Parabola, axis of symmetry parallel to *x*-axis, opens to right
(h, k)	$(h - a, k)$	$x = a + h$	$(y - k)^2 = -4a(x - h)$	Parabola, axis of symmetry parallel to *x*-axis, opens to left
(h, k)	$(h, k + a)$	$y = -a + k$	$(x - h)^2 = 4a(y - k)$	Parabola, axis of symmetry parallel to *y*-axis, opens up
(h, k)	$(h, k - a)$	$y = a + k$	$(x - h)^2 = -4a(y - k)$	Parabola, axis of symmetry parallel to *y*-axis, opens down

FIGURE 12

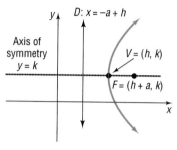

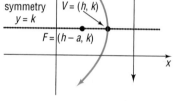

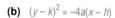

(a) $(y - k)^2 = 4a(x - h)$ **(b)** $(y - k)^2 = -4a(x - h)$

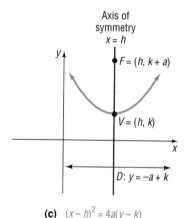

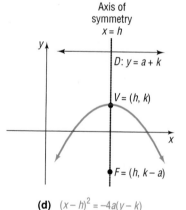

(c) $(x - h)^2 = 4a(y - k)$ **(d)** $(x - h)^2 = -4a(y - k)$

E X A M P L E   7   *Finding the Equation of a Parabola, Vertex Not at Origin*

Find an equation of the parabola with vertex at $(-2, 3)$ and focus at $(0, 3)$. Graph the equation by hand.

Solution   The vertex $(-2, 3)$ and focus $(0, 3)$ both lie on the horizontal line $y = 3$ (the axis of symmetry). The distance a from $(-2, 3)$ to $(0, 3)$, is $a = 2$. Also, because the focus lies to the right of the vertex, we know the parabola opens to the right. Consequently, the form of the equation is

$$(y - k)^2 = 4a(x - h)$$

where $(h, k) = (-2, 3)$ and $a = 2$. Therefore, the equation is

$$(y - 3)^2 = 4 \cdot 2[x - (-2)]$$
$$(y - 3)^2 = 8(x + 2)$$

If $x = 0$, then $(y - 3)^2 = 16$. Thus, $y - 3 = \pm 4$ and $y = -1$, $y = 7$. The points $(0, -1)$ and $(0, 7)$ define the latus rectum; the line $x = -4$ is the directrix. See Figure 13.

FIGURE 13
$(y - 3)^2 = 8(x + 2)$

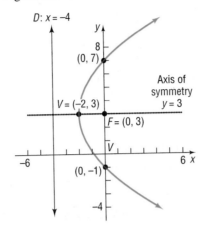

■   Now work Problem 25.

E X A M P L E   8   *Using a Graphing Utility to Graph a Parabola, Vertex Not at the Origin*

Using a graphing utility, graph the equation $(y - 3)^2 = 8(x + 2)$.

Solution   First, we must solve the equation for y.

$$(y - 3)^2 = 8(x + 2)$$
$$y - 3 = \pm\sqrt{8(x + 2)} \qquad \text{Take the square root of each side.}$$
$$y = 3 \pm\sqrt{8(x + 2)} \qquad \text{Solve for } y.$$

FIGURE 14

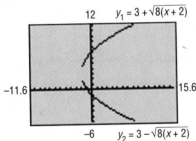

Figure 14 shows the graphs of the equations $y_1 = 3 + \sqrt{8(x + 2)}$ and $y_2 = 3 - \sqrt{8(x + 2)}$. ■

■   Now work Problem 41.

Polynomial equations define parabolas whenever they involve two variables that are quadratic in one variable and linear in the other. To discuss this type of equation, we first complete the square of the quadratic variable.

E X A M P L E   9   *Discussing the Equation of a Parabola*

Discuss the equation: $x^2 + 4x - 4y = 0$

Solution   Figure 15(a) shows the graph of $x^2 + 4x - 4y = 0$ using a graphing utility. We now proceed to analyze the equation.

To discuss the equation $x^2 + 4x - 4y = 0$, we complete the square involving the variable x. Thus,

$$x^2 + 4x - 4y = 0$$
$$x^2 + 4x = 4y \qquad \text{Isolate the terms involving } x \text{ on the left side.}$$
$$x^2 + 4x + 4 = 4y + 4 \qquad \text{Complete the square on the left side.}$$
$$(x + 2)^2 = 4(y + 1)$$

This equation is of the form $(x - h)^2 = 4a(y - k)$, with $h = -2$, $k = -1$, and $a = 1$. The graph is a parabola with vertex at $(h, k) = (-2, -1)$ that opens up. The focus is at $(-2, 0)$, and the directrix is the line $y = -2$. See Figure 15(b).

FIGURE 15
$x^2 + 4x - 4y = 0$

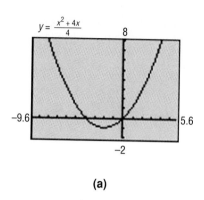

$y = \frac{x^2 + 4x}{4}$

(a)

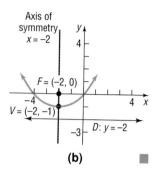

Axis of symmetry, $x = -2$

$F = (-2, 0)$

$V = (-2, -1)$

$D: y = -2$

(b)

Parabolas find their way into many applications. For example, suspension bridges have cables in the shape of a parabola. Another property of parabolas that is used in applications is their reflecting property.

FIGURE 16

Searchlight

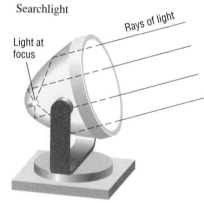

Light at focus

Rays of light

Reflecting Property

Suppose that a mirror is shaped like a **paraboloid of revolution,** a surface formed by rotating a parabola about its axis of symmetry. If a light (or any other emitting source) is placed at the focus of the parabola, all the rays emanating from the light will reflect off the mirror in lines parallel to the axis of symmetry. This principle is used in the design of searchlights, flashlights, certain automobile headlights, and other such devices. See Figure 16.

Conversely, suppose rays of light (or other signals) emanate from a distant source so that they are essentially parallel. When these rays strike the surface of a parabolic mirror whose axis of symmetry is parallel to these rays, they are reflected to a single point at the focus. This principle is used in the design of some solar energy devices, satellite dishes, and the mirrors used in some types of telescopes. See Figure 17.

FIGURE 17

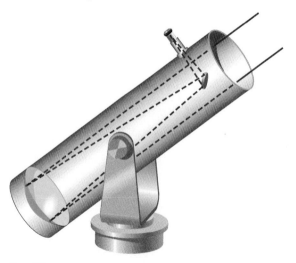

E X A M P L E   1 0   *Satellite Dish*

A satellite dish is shaped like a paraboloid of revolution. The signals that emanate from a satellite strike the surface of the dish and are reflected to a single point, where the receiver is located. If the dish is 8 feet across at its opening and is 3 feet deep at its center, at what position should the receiver be placed?

Solution   Figure 18(a) shows the satellite dish. We draw the parabola used to form the dish on a rectangular coordinate system so that the vertex of the parabola is at the origin and its focus is on the positive y-axis. See Figure 18(b). The form of the equation of the parabola is

$$x^2 = 4ay$$

and its focus is at $(0, a)$. Since $(4, 3)$ is a point on the graph, we have

$$4^2 = 4a(3)$$

$$a = \frac{4}{3}$$

The receiver should be located $1\frac{1}{3}$ feet from the base of the dish, along its axis of symmetry.

FIGURE 18

(a)

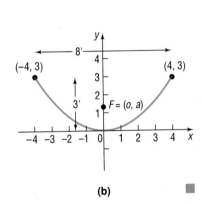

(b)

6.2

Exercise 6.2

In Problems 1–8, the graph of a parabola is given. Match each graph to its equation.

A. $y^2 = 4x$
D. $x^2 = -4y$
G. $(y - 1)^2 = -4(x - 1)$

B. $x^2 = 4y$
E. $(y - 1)^2 = 4(x - 1)$
H. $(x + 1)^2 = -4(y + 1)$

C. $y^2 = -4x$
F. $(x + 1)^2 = 4(y + 1)$

1.

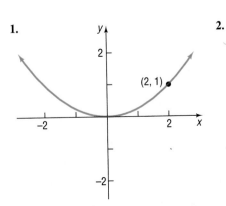

2.

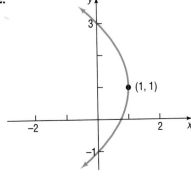

3.

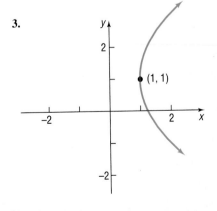

4.

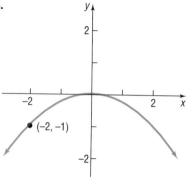

5.

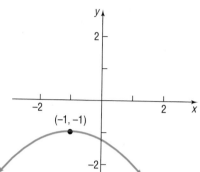

6.

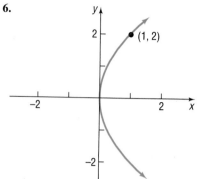

7.

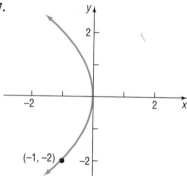

8.

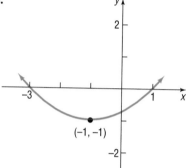

In Problems 9–16, the graph of a parabola is given. Match each graph to its equation.

A. $x^2 = 6y$

B. $x^2 = -6y$

C. $y^2 = 6x$

D. $y^2 = -6x$

E. $(y - 2)^2 = -6(x + 2)$

F. $(y - 2)^2 = 6(x + 2)$

G. $(x + 2)^2 = -6(y - 2)$

H. $(x + 2)^2 = 6(y - 2)$

9.

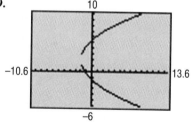

10.

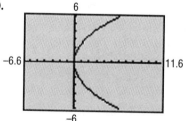

11.

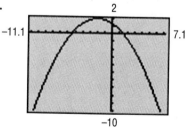

12.

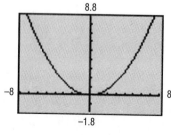

13.

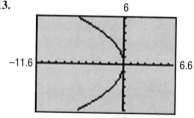

14.

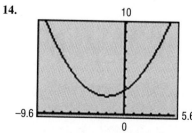

15.

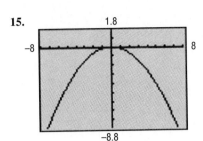

16.

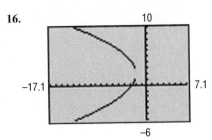

In Problems 17–32, find the equation of the parabola described. Find the two points that define the latus rectum, and graph the equation by hand.

17. Focus at (4, 0); vertex at (0, 0)

18. Focus at (0, 2); vertex at (0, 0)

19. Focus at (0, −3); vertex at (0, 0)

20. Focus at (−4, 0); vertex at (0, 0)

21. Focus at (−2, 0); directrix the line $x = 2$

22. Focus at (0, −1); directrix the line $y = 1$

23. Directrix the line $y = -\frac{1}{2}$; vertex at (0, 0)

24. Directrix the line $x = -\frac{1}{2}$; vertex at (0, 0)

25. Vertex at (2, −3); focus at (2, −5)

26. Vertex at (4, −2); focus at (6, −2)

27. Vertex at (0, 0); axis of symmetry the *y*-axis; containing the point (2, 3)

28. Vertex at (0, 0); axis of symmetry the *x*-axis; containing the point (2, 3)

29. Focus at (−3, 4); directrix the line $y = 2$

30. Focus at (2, 4); directrix the line $x = -4$

31. Focus at (−3, −2); directrix the line $x = 1$

32. Focus at (−4, 4); directrix the line $y = -2$

In Problems 33–50, find the vertex, focus, and directrix of each parabola. Graph the equation using a graphing utility.

33. $x^2 = 4y$

34. $y^2 = 8x$

35. $y^2 = -16x$

36. $x^2 = -4y$

37. $(y - 2)^2 = 8(x + 1)$

38. $(x + 4)^2 = 16(y + 2)$

39. $(x - 3)^2 = -(y + 1)$

40. $(y + 1)^2 = -4(x - 2)$

41. $(y + 3)^2 = 8(x - 2)$

42. $(x - 2)^2 = 4(y - 3)$

43. $y^2 - 4y + 4x + 4 = 0$

44. $x^2 + 6x - 4y + 1 = 0$

45. $x^2 + 8x = 4y - 8$

46. $y^2 - 2y = 8x - 1$

47. $y^2 + 2y - x = 0$

48. $x^2 - 4x = 2y$

49. $x^2 - 4x = y + 4$

50. $y^2 + 12y = -x + 1$

In Problems 51–58, write an equation for each parabola.

51.

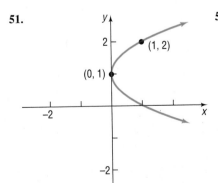

52.

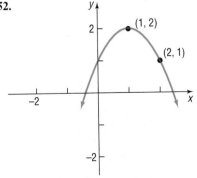

53.

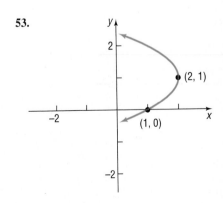

54.

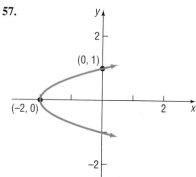

55.

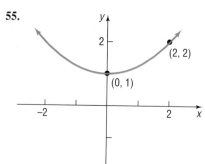

56.

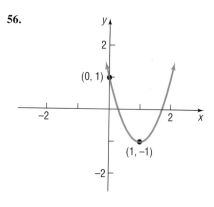

57.

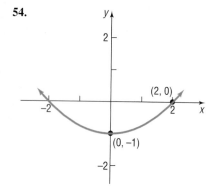

58.

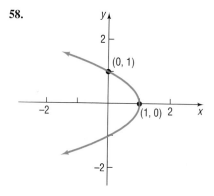

59. *Satellite Dish* A satellite dish is shaped like a paraboloid of revolution. The signals that emanate from a satellite strike the surface of the dish and are reflected to a single point, where the receiver is located. If the dish is 10 feet across at its opening and is 4 feet deep at its center, at what position should the receiver be placed?

60. *Constructing a TV Dish* A cable TV receiving dish is in the shape of a paraboloid of revolution. Find the location of the receiver, which is placed at the focus, if the dish is 6 feet across at its opening and 2 feet deep.

61. *Constructing a Flashlight* The reflector of a flashlight is in the shape of a paraboloid of revolution. Its diameter is 4 inches and its depth is 1 inch. How far from the vertex should the light bulb be placed so that the rays will be reflected parallel to the axis?

62. *Constructing a Headlight* A sealed-beam headlight is in the shape of a paraboloid of revolution. The bulb, which is placed at the focus, is 1 inch from the vertex. If the depth is to be 2 inches, what is the diameter of the headlight at its opening?

63. *Suspension Bridges* The cables of a suspension bridge are in the shape of a parabola, as shown in the figure. The towers supporting the cable are 600 feet apart and 80 feet high. If the cables touch the road surface midway between the towers, what is the height of the cable at a point 150 feet from a tower?

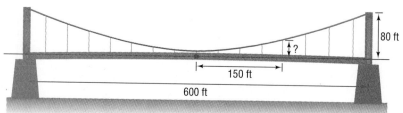

64. *Suspension Bridges* The cables of a suspension bridge are in the shape of a parabola. The towers supporting the cable are 400 feet apart and 100 feet high. If the cables are at a height of 10 feet midway between the towers, what is the height of the cable at a point 50 feet from a tower?

65. *Searchlights* A searchlight is shaped like a paraboloid of revolution. If the light source is located 2 feet from the base along the axis of symmetry and the opening is 5 feet across, how deep should the searchlight be?

66. *Searchlights* A searchlight is shaped like a paraboloid of revolution. If the light source is located 2 feet from the base along the axis of symmetry and the depth of the searchlight is 4 feet, what should the width of the opening be?

67. *Solar Heat* A mirror is shaped like a paraboloid of revolution and will be used to concentrate the rays of the sun at its focus, creating a heat source. If the mirror is 20 feet across at its opening and is 6 feet deep, where will the heat source be concentrated?

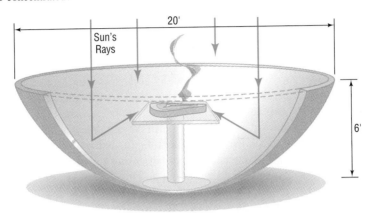

68. *Reflecting Telescopes* A reflecting telescope contains a mirror shaped like a paraboloid of revolution. If the mirror is 4 inches across at its opening and is 3 feet deep, where will the light collected be concentrated?

69. *Parabolic Arch Bridge* A bridge is built in the shape of a parabolic arch. The bridge has a span of 120 feet and a maximum height of 25 feet. See the illustration. Choose a suitable rectangular coordinate system and find the height of the arch at distances of 10, 30, and 50 feet from the center.

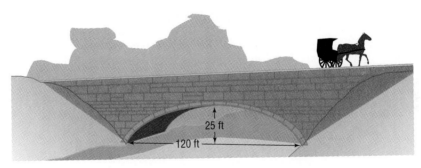

70. *Parabolic Arch Bridge* A bridge is built in the shape of a parabolic arch and is to have a span of 100 feet. The height of the arch at a distance of 40 feet from the center is to be 10 feet. Find the height of the arch at its center.

71. Show that an equation of the form

$$Ax^2 + Ey = 0 \qquad A \neq 0, E \neq 0$$

is the equation of a parabola with vertex at $(0, 0)$ and axis of symmetry the y-axis. Find its focus and directrix.

72. Show that an equation of the form

$$Cy^2 + Dx = 0 \qquad C \neq 0, D \neq 0$$

is the equation of a parabola with vertex at $(0, 0)$ and axis of symmetry the x-axis. Find its focus and directrix.

73. Show that the graph of an equation of the form

$$Ax^2 + Dx + Ey + F = 0 \qquad A \neq 0$$

(a) Is a parabola if $E \neq 0$.
(b) Is a vertical line if $E = 0$ and $D^2 - 4AF = 0$.
(c) Is two vertical lines if $E = 0$ and $D^2 - 4AF > 0$.
(d) Contains no points if $E = 0$ and $D^2 - 4AF < 0$.

74. Show that the graph of an equation of the form

$$Cy^2 + Dx + Ey + F = 0 \qquad C \neq 0$$

(a) Is a parabola if $D \neq 0$.
(b) Is a horizontal line if $D = 0$ and $E^2 - 4CF = 0$.
(c) Is two horizontal lines if $D = 0$ and $E^2 - 4CF > 0$.
(d) Contains no points if $D = 0$ and $E^2 - 4CF < 0$.

6.3

The Ellipse

Ellipse

FIGURE 19

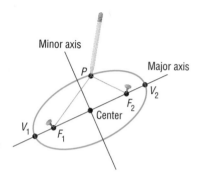

Minor axis

Major axis

P

V_2

F_2

Center

V_1

F_1

An ellipse is the collection of all points in the plane the sum of whose distances from two fixed points, called the **foci**, is a constant.

The definition actually contains within it a physical means for drawing an ellipse. Find a piece of string (the length of this string is the constant referred to in the definition). Then take two thumbtacks (the foci) and stick them on a piece of cardboard so that the distance between them is less than the length of the string. Now attach the ends of the string to the thumbtacks and, using the point of a pencil, pull the string taut. Keeping the string taut, rotate the pencil around the two thumbtacks. The pencil traces out an ellipse, as shown in Figure 19.

In Figure 19, the foci are labeled F_1 and F_2. The line containing the foci is called the **major axis**. The midpoint of the line segment joining the foci is called the **center** of the ellipse. The line through the center and perpendicular to the major axis is called the **minor axis.**

The two points of intersection of the ellipse and the major axis are the **vertices,** V_1 and V_2, of the ellipse. The distance from one vertex to the other is called the **length of the major axis.** The ellipse is symmetric with respect to its major axis and with respect to its minor axis.

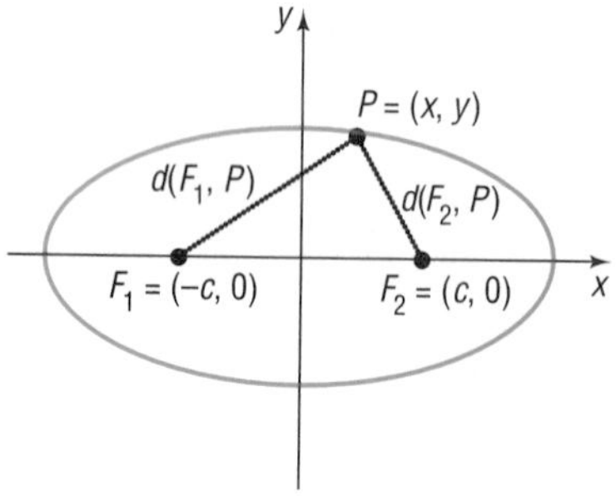

With these ideas in mind, we are now ready to find the equation of an ellipse in a rectangular coordinate system. First, we place the center of the ellipse at the origin. Second, we position the ellipse so that its major axis coincides with a coordinate axis. Suppose that the major axis coincides with the x-axis, as shown in Figure 20. If c is the distance from the center to a focus, then one focus will be at $F_1 = (-c, 0)$ and the other at $F_2 = (c, 0)$. As we shall see, it is convenient to let $2a$ denote the constant distance referred to in the definition. Thus, if $P = (x, y)$ is any point on the ellipse, we have

FIGURE 20

$d(F_1, P) + d(F_2, P) = 2a$

$P = (x, y)$

$d(F_1, P)$

$d(F_2, P)$

$F_1 = (-c, 0)$

$F_2 = (c, 0)$

x

y

$$d(F_1, P) + d(F_2, P) = 2a \qquad \text{Sum of the distances from } P \text{ to the foci equals a constant}$$

$$\sqrt{(x + c)^2 + y^2} + \sqrt{(x - c)^2 + y^2} = 2a \qquad \text{Use the distance formula.}$$

$$\sqrt{(x + c)^2 + y^2} = 2a - \sqrt{(x - c)^2 + y^2} \qquad \text{Isolate one radical.}$$

$$(x + c)^2 + y^2 = 4a^2 - 4a\sqrt{(x - c)^2 + y^2} \qquad \text{Square both sides.}$$
$$+ (x - c)^2 + y^2$$

$$x^2 + 2cx + c^2 + y^2 = 4a^2 - 4a\sqrt{(x - c)^2 + y^2} \qquad \text{Simplify}$$
$$+ x^2 - 2cx + c^2 + y^2$$

$$4cx - 4a^2 = -4a\sqrt{(x - c)^2 + y^2} \qquad \text{Isolate the radical.}$$

$$cx - a^2 = -a\sqrt{(x - c)^2 + y^2} \qquad \text{Divide each side by 4.}$$

$$c^2x^2 - 2a^2cx + a^4 = a^2[(x - c)^2 + y^2] \qquad \text{Square both sides again.}$$

$$c^2x^2 - 2a^2cx + a^4 = a^2(x^2 - 2cx + c^2 + y^2)$$

$$(c^2 - a^2)x^2 - a^2y^2 = a^2c^2 - a^4$$

$$(a^2 - c^2)x^2 + a^2y^2 = a^2(a^2 - c^2) \qquad \text{Multiply each side by } -1; \text{ factor } a^2 \text{ on the right side.} \quad (1)$$

MISSION POSSIBLE

Chapter 6

BUILDING A BRIDGE OVER THE EAST RIVER

Your team is working for the transportation authority in New York City. You have been asked to study the construction plans for a new bridge over the East River in New York City. The space between supports needs to be 1050 feet; the height at the center of the arch needs to be 350 feet. One company has suggested the support be in the shape of a parabola; another company suggests a semi-ellipse. The engineering team will determine the relative strengths of the two plans; your job is to find out if there are any differences in the channel widths.

An empty tanker needs a 280 foot clearance to pass beneath the bridge. You need to find the width of the channel for each of the two different plans.

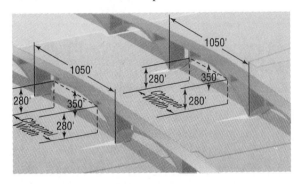

1. To determine the equation of a parabola with these characteristics, first place the parabola on coordinate axes in a convenient location and sketch it.
2. What is the equation of the parabola? (If using a decimal in the equation, you may want to carry 6 decimal places. If using a fraction, your answer will be more exact.)
3. How wide is the channel the tanker can pass through if the shape of the support is parabolic?
4. To determine the equation of a semi-ellipse with these characteristics, place the semi-ellipse on coordinate axes in a convenient location and sketch it.
5. What is the equation of the ellipse? How wide is the channel the tanker can pass through?
6. Now that you know which of the two provides the wider channel, consider some other factors. Your department is also in charge of channel depth, amount of traffic on the river, and various other factors. For example, if there were flooding in the river, and the water level rose by 10 feet, how would the clearances be affected? Make a decision about which plan you think would be the better one as far as your department is concerned, and explain why you think so.

To obtain points on the ellipse off the x-axis, it must be that $a > c$. To see why, look again at Figure 20:

$$d(F_1, P) + d(F_2, P) > d(F_1, F_2)$$

The sum of the lengths of two sides of a triangle is greater than the length of the third side.

$$2a > 2c$$

$d(F_1, P) + d(F_2, P) = 2a;$
$d(F_1, F_2) = 2c$

$$a > c$$

Since $a > c$, we also have $a^2 > c^2$, so $a^2 - c^2 > 0$. Let $b^2 = a^2 - c^2$, $b > 0$. Then $a > b$ and equation (1) can be written as

$$b^2x^2 + a^2y^2 = a^2b^2$$

$$\frac{x^2}{a^2} + \frac{y^2}{b^2} = 1 \quad \text{Divide each side by } a^2b^2.$$

Theorem   An equation of the ellipse with center at $(0, 0)$ and foci at $(-c, 0)$ and $(c, 0)$ is

**Equation of an Ellipse;
Center at $(0,0)$;
Foci at $(\pm c, 0)$;
Major Axis along
the x-Axis**

$$\frac{x^2}{a^2} + \frac{y^2}{b^2} = 1 \quad \text{where } a > b > 0 \text{ and } b^2 = a^2 - c^2 \qquad (2)$$

The major axis is the x-axis.   ■

As you can verify, the ellipse defined by equation (2) is symmetric with respect to the x-axis, y-axis, and origin.

To find the vertices of the ellipse defined by equation (2), let $y = 0$. The vertices satisfy the equation $x^2/a^2 = 1$, the solutions of which are $x = \pm a$. Consequently, the vertices of the ellipse given by equation (2) are $V_1 = (-a, 0)$ and $V_2 = (a, 0)$. The y-intercepts of the ellipse, found by letting $x = 0$, have co-ordinates $(0, -b)$ and $(0, b)$. These four intercepts, $(a, 0)$, $(-a, 0)$, $(0, b)$, and $(0,-b)$, are used to graph the ellipse by hand. See Figure 21.

Notice in Figure 21 the right triangle formed with the points $(0, 0)$, $(c, 0)$, and $(0, b)$. Because $b^2 = a^2 - c^2$ (or $b^2 + c^2 = a^2$), the distance from the focus at $(c, 0)$ to the point $(0, b)$ is a.

FIGURE 21

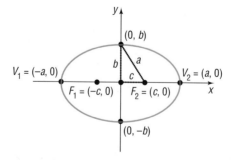

E X A M P L E   1   *Finding an Equation of an Ellipse*

Find an equation of the ellipse with center at the origin, one focus at $(3, 0)$, and a vertex at $(-4, 0)$. Graph the equation by hand.

Solution The ellipse has its center at the origin, and the major axis coincides with the *x*-axis. One focus is at $(c, 0) = (3, 0)$, so $c = 3$. One vertex is at $(-a, 0) = (-4, 0)$, so $a = 4$. From equation (2), it follows that

$$b^2 = a^2 - c^2 = 16 - 9 = 7$$

so an equation of the ellipse is

$$\frac{x^2}{16} + \frac{y^2}{7} = 1$$

Figure 22 shows the graph drawn by hand. ■

FIGURE 22

$$\frac{x^2}{16} + \frac{y^2}{7} = 1$$

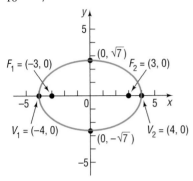

Notice in Figure 22 how we used the intercepts of the equation to graph the ellipse. Following this practice will make it easier for you to obtain an accurate graph of an ellipse when graphing by hand. It also tells you how to set the viewing rectangle when using a graphing utility.

E X A M P L E   2   *Graphing an Ellipse Using a Graphing Utility*

Use a graphing utility to graph the ellipse: $\dfrac{x^2}{16} + \dfrac{y^2}{7} = 1$.

Solution   First, we must solve $\dfrac{x^2}{16} + \dfrac{y^2}{7} = 1$ for *y*.

$$\frac{y^2}{7} = 1 - \frac{x^2}{16} \qquad \text{Subtract } \frac{x^2}{16} \text{ from each side.}$$

$$y^2 = 7\left(1 - \frac{x^2}{16}\right) \qquad \text{Multiply both sides by 7.}$$

$$y = \pm\sqrt{7\left(1 - \frac{x^2}{16}\right)} \qquad \text{Take the square root of each side.}$$

Figure 23 shows the graphs of $y_1 = \sqrt{7\left(1 - \dfrac{x^2}{16}\right)}$ and $y_2 = -\sqrt{7\left(1 - \dfrac{x^2}{16}\right)}$.

FIGURE 23

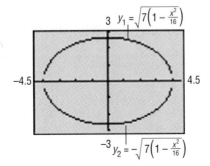

Notice in Figure 23 that we used a square screen. As with circles, this is done to avoid a distorted view of the graph.

An equation of the form of equation (2), with $a > b$, is the equation of an ellipse with center at the origin, foci on the x-axis at $(-c, 0)$ and $(c, 0)$, where $c^2 = a^2 - b^2$, and major axis along the x-axis.

For the remainder of this section, the direction "Discuss the equation" will mean to find the center, major axis, foci, and vertices of the ellipse and graph it.

E X A M P L E   3

Discussing the Equation of an Ellipse

Discuss the equation: $\dfrac{x^2}{25} + \dfrac{y^2}{9} = 1$

Solution   Figure 24(a) shows the graph of $\dfrac{x^2}{25} + \dfrac{y^2}{9} = 1$ using a graphing utility. We now proceed to analyze the equation. The given equation is of the form of equation (2), with $a^2 = 25$ and $b^2 = 9$. The equation is that of an ellipse with center $(0, 0)$ and major axis along the x-axis. The vertices are at $(\pm a, 0) = (\pm 5, 0)$. Because $b^2 = a^2 - c^2$, we find

$$c^2 = a^2 - b^2 = 25 - 9 = 16$$

The foci are at $(\pm c, 0) = (\pm 4, 0)$. Figure 24(b) shows the graph drawn by hand.

FIGURE 24

$\dfrac{x^2}{25} + \dfrac{y^2}{9} = 1$

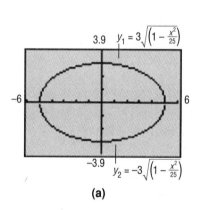

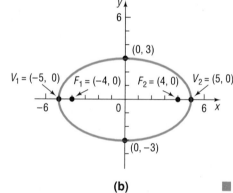

(a)   (b)

Now work Problems 9 and 19.

If the major axis of an ellipse with center at $(0, 0)$ coincides with the y-axis, then the foci are at $(0, -c)$ and $(0, c)$. Using the same steps as before, the definition of an ellipse leads to the following result:

Theorem   An equation of the ellipse with center at $(0, 0)$ and foci at $(0, -c)$ and $(0, c)$ is

Equation of an Ellipse;
Center at (0, 0);
Foci at (0, ± c);
Major Axis
along the y-Axis

$$\frac{x^2}{b^2} + \frac{y^2}{a^2} = 1 \qquad \text{where } a > b > 0 \text{ and } b^2 = a^2 - c^2 \qquad (3)$$

The major axis is the y-axis; the vertices are at $(0, -a)$ and $(0, a)$.

Figure 25 illustrates the graph of such an ellipse. Again, notice the right triangle with the points at $(0, 0)$, $(b, 0)$, and $(0, c)$.

FIGURE 25

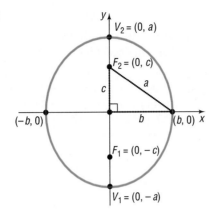

Look closely at equations (2) and (3). Although they may look alike, there is a difference! In equation (2), the larger number, a^2, is in the denominator of the x^2-term, so the major axis of the ellipse is along the x-axis. In equation (3), the larger number, a^2, is in the denominator of the y^2-term, so the major axis is along the y-axis.

E X A M P L E   4   *Discussing the Equation of an Ellipse*

Discuss the equation: $9x^2 + y^2 = 9$

Solution   Figure 26(a) shows the graph of $9x^2 + y^2 = 9$ using a graphing utility. We now proceed to analyze the equation. To put the equation in proper form, we divide each side by 9:

$$x^2 + \frac{y^2}{9} = 1$$

The larger number, 9, is in the denominator of the y^2-term so, based on equation (3), this is the equation of an ellipse with center at the origin and major axis along the y-axis. Also, we conclude that $a^2 = 9$, $b^2 = 1$, and $c^2 = a^2 - b^2 = 9 - 1 = 8$. The vertices are at $(0, \pm a) = (0, \pm 3)$, and the foci are at $(0, \pm c) = (0, \pm 2\sqrt{2})$. The graph, drawn by hand, is given in Figure 26(b).

FIGURE 26

$x^2 + \dfrac{y^2}{9} = 1$

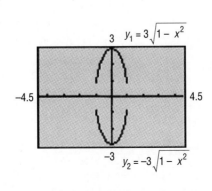

(a)

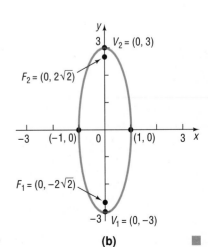

(b)

E X A M P L E   5   *Finding an Equation of an Ellipse*

Find an equation of the ellipse having one focus at $(0, 2)$ and vertices at $(0, -3)$ and $(0, 3)$. Graph the equation by hand.

Solution   Because the vertices are at $(0, -3)$ and $(0, 3)$, the center of this ellipse is at the origin. Also, its major axis coincides with the y-axis. The given information also reveals that $c = 2$ and $a = 3$, so $b^2 = a^2 - c^2 = 9 - 4 = 5$. The form of the equation of this ellipse is given by equation (3):

$$\frac{x^2}{b^2} + \frac{y^2}{a^2} = 1$$

$$\frac{x^2}{5} + \frac{y^2}{9} = 1$$

Figure 27 shows the graph.

FIGURE 27

$$\frac{x^2}{5} + \frac{y^2}{9} = 1$$

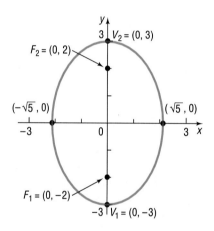

■   Now work Problems 13 and 21.

The circle may be considered a special kind of ellipse. To see why, let $a = b$ in equation (2) or in equation (3). Then

$$\frac{x^2}{a^2} + \frac{y^2}{a^2} = 1$$

$$x^2 + y^2 = a^2$$

This is the equation of a circle with center at the origin and radius a. The value of c is

$$c^2 = a^2 - b^2 = 0$$

We conclude that the closer the two foci of an ellipse are, the more the ellipse will look like a circle.

Center at *(h, k)*

If an ellipse with center at the origin and major axis coinciding with a coordinate axis is shifted horizontally h units and then vertically k units, the result is an ellipse with center at (h, k) and major axis parallel to a coordinate axis. Table 3 gives the forms of the equations of such ellipses, and Figure 28 shows their graphs.

TABLE 3   ELLIPSES WITH CENTER AT (h, k) AND MAJOR AXIS PARALLEL TO A COORDINATE AXIS

CENTER	MAJOR AXIS	FOCI	VERTICES	EQUATION
(h, k)	Parallel to x-axis	($h \pm c$, k)	($h \pm a$, k)	$\dfrac{(x-h)^2}{a^2} + \dfrac{(y-k)^2}{b^2} = 1$, $a > b$ and $b^2 = a^2 - c^2$
(h, k)	Parallel to y-axis	(h, $k \pm c$)	(h, $k \pm a$)	$\dfrac{(x-h)^2}{b^2} + \dfrac{(y-k)^2}{a^2} = 1$, $a > b$ and $b^2 = a^2 - c^2$

FIGURE 28

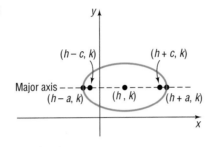
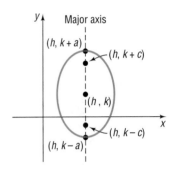

(a) $\dfrac{(x-h)^2}{a^2} + \dfrac{(y-k)^2}{b^2} = 1$ (b) $\dfrac{(x-h)^2}{b^2} + \dfrac{(y-k)^2}{a^2} = 1$

E X A M P L E   6   *Finding an Equation of an Ellipse, Center Not at the Origin*

Find an equation for the ellipse with center at (2, −3), one focus at (3, −3), and one vertex at (5, −3). Graph the equation by hand.

Solution   The center is at (h, k) = (2, −3), so $h = 2$ and $k = -3$. The major axis is parallel to the x-axis. The distance from the center (2, −3) to a focus (3, −3) is $c = 1$; the distance from the center (2, −3) to a vertex (5, −3) is $a = 3$. Thus, $b^2 = a^2 - c^2 = 9 - 1 = 8$. The form of the equation is

$$\frac{(x-h)^2}{a^2} + \frac{(y-k)^2}{b^2} = 1 \text{ where } h = 2, k = -3, a = 3, b = 2\sqrt{2}.$$

$$\frac{(x-2)^2}{9} + \frac{(y+3)^2}{8} = 1$$

Figure 29 shows the graph.

FIGURE 29

$$\frac{(x-2)^2}{9} + \frac{(y+3)^2}{8} = 1$$

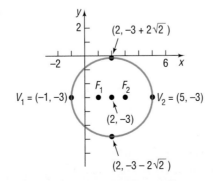

E X A M P L E   7   *Using a Graphing Utility to Graph an Ellipse, Center Not at the Origin*

Using a graphing utility, graph the ellipse: $\dfrac{(x-2)^2}{9} + \dfrac{(y+3)^2}{8} = 1$.

Solution   First, we must solve the equation $\dfrac{(x-2)^2}{9} + \dfrac{(y+3)^2}{8} = 1$ for y.

$$\frac{(y+3)^2}{8} = 1 - \frac{(x-2)^2}{9} \qquad \text{Subtract } \frac{(x-2)^2}{9} \text{ from each side.}$$

$$(y+3)^2 = 8\left[1 - \frac{(x-2)^2}{9}\right] \qquad \text{Multiply each side by 8.}$$

FIGURE 30

$$y + 3 = \pm\sqrt{8\left[1 - \frac{(x-2)^2}{9}\right]} \qquad \text{Take the square root of each side.}$$

$$y = -3 \pm \sqrt{8\left[1 - \frac{(x-2)^2}{9}\right]} \qquad \text{Subtract 3 from each side.}$$

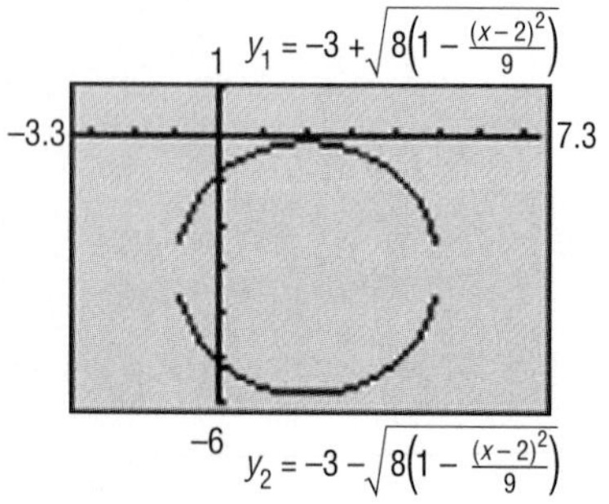

Figure 30 shows the graphs of $y_1 = -3 + \sqrt{8\left[1 - \dfrac{(x-2)^2}{9}\right]}$

and $y_2 = -3 - \sqrt{8\left[1 - \dfrac{(x-2)^2}{9}\right]}$. ■

■   Now work Problem 33.

E X A M P L E   8   *Discussing the Equation of an Ellipse*

Discuss the equation: $4x^2 + y^2 - 8x + 4y + 4 = 0$

Solution   We proceed to complete the square in x and in y:

$$4x^2 + y^2 - 8x + 4y + 4 = 0$$
$$4x^2 - 8x + y^2 + 4y = -4$$
$$4(x^2 - 2x) + (y^2 + 4y) = -4$$
$$4(x^2 - 2x + 1) + (y^2 + 4y + 4) = -4 + 4 + 4 \quad \text{Complete each square.}$$
$$4(x-1)^2 + (y+2)^2 = 4$$
$$(x-1)^2 + \frac{(y+2)^2}{4} = 1 \qquad \text{Divide each side by 4.}$$

Figure 31(a) shows the graph of $(x-1)^2 + \dfrac{(y+2)^2}{4} = 1$ using a graphing utility. We now proceed to analyze the equation.

This is the equation of an ellipse with center at $(1, -2)$ and major axis parallel to the y-axis. Since $a^2 = 4$ and $b^2 = 1$, we have $c^2 = a^2 - b^2 = 4 - 1 = 3$. The vertices are at $(h, k \pm a) = (1, -2 \pm 2)$ or $(1, 0)$ and $(1, -4)$. The foci are at $(h, k \pm c) = (1, -2 \pm \sqrt{3})$ or $(1, -2 - \sqrt{3})$ and $(1, -2 + \sqrt{3})$. Figure 31(b) shows the graph drawn by hand.

FIGURE 31

$$(x - 1)^2 + \frac{(y + 2)^2}{4} = 1$$

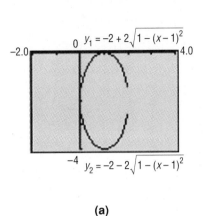

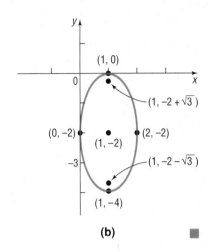

(a)

(b)

Applications

Ellipses are found in many applications in science and engineering. For example, the orbits of the planets around the Sun are elliptical, with the Sun's position at a focus. See Figure 32.

FIGURE 32

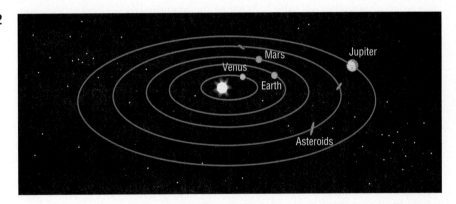

Stone and concrete bridges are often shaped as semielliptical arches. Elliptical gears are used in machinery when a variable rate of motion is required.

Ellipses also have an interesting reflection property. If a source of light (or sound) is placed at one focus, the waves transmitted by the source will reflect off the ellipse and concentrate at the other focus. This is the principal behind "whispering galleries," which are rooms designed with elliptical ceilings. A person standing at one focus of the ellipse can whisper and be heard by a person standing at the other focus, because all the sound waves that reach the ceiling are reflected to the other person.

E X A M P L E   9   *Whispering Galleries*

Figure 33 shows the specifications for an elliptical ceiling in a hall designed to be a whispering gallery. In a whispering gallery, a person standing at one focus of the ellipse can whisper and be heard by another person standing at the other focus, because all the sound waves that reach the ceiling from one focus are reflected to the other focus. Where in the hall are the foci located?

FIGURE 33

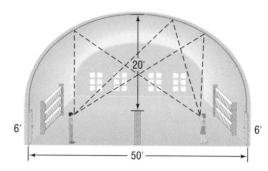

Solution   We set up a rectangular coordinate system so that the center of the ellipse is at the origin and the major axis is along the x-axis. See Figure 34. The equation of the ellipse is

$$\frac{x^2}{a^2} + \frac{y^2}{b^2} = 1$$

where $a = 25$ and $b = 20$. Since

$$c^2 = a^2 - b^2 = 25^2 - 20^2 = 625 - 400 = 225$$

we have $c = 15$. Thus, the foci are located 15 feet from the center of the ellipse along the major axis.

FIGURE 34

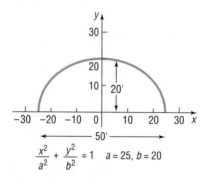

$\frac{x^2}{a^2} + \frac{y^2}{b^2} = 1$ $a = 25, b = 20$

6.3

Exercise 6.3

In Problems 1–4, the graph of an ellipse is given. Match each graph to its equation.

A. $\frac{x^2}{4} + y^2 = 1$ B. $x^2 + \frac{y^2}{4} = 1$ C. $\frac{x^2}{16} + \frac{y^2}{4} = 1$ D. $\frac{x^2}{4} + \frac{y^2}{16} = 1$

1.

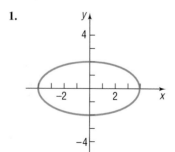

2.

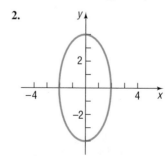

3.

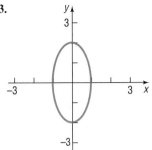

4.

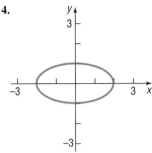

In Problems 5–8, the graph of an ellipse is given. Match each graph to its equation.

A. $\dfrac{(x + 1)^2}{4} + \dfrac{(y + 1)^2}{9} = 1$

B. $\dfrac{(x - 1)^2}{4} + \dfrac{(y + 1)^2}{9} = 1$

C. $\dfrac{(x - 1)^2}{9} + \dfrac{(y + 1)^2}{4} = 1$

D. $\dfrac{(x + 1)^2}{9} + \dfrac{(y - 1)^2}{4} = 1$

5.

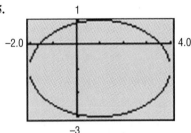

6.

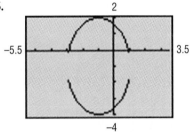

7.

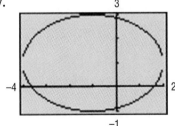

8.

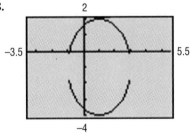

In Problems 9–18, find the vertices and foci of each ellipse. Graph each equation using a graphing utility.

9. $\dfrac{x^2}{25} + \dfrac{y^2}{4} = 1$

10. $\dfrac{x^2}{9} + \dfrac{y^2}{4} = 1$

11. $\dfrac{x^2}{9} + \dfrac{y^2}{25} = 1$

12. $\dfrac{x^2}{4} + \dfrac{y^2}{16} = 1$

13. $4x^2 + y^2 = 16$

14. $x^2 + 9y^2 = 18$

15. $4y^2 + x^2 = 8$

16. $4y^2 + 9x^2 = 36$

17. $x^2 + y^2 = 16$

18. $x^2 + y^2 = 4$

In Problems 19–28, find an equation for each ellipse. Graph the equation by hand.

19. Center at (0, 0); focus at (3, 0); vertex at (5, 0)

20. Center at (0, 0); focus at (−1, 0); vertex at (3, 0)

21. Center at (0, 0); focus at (0, −4); vertex at (0, 5)

22. Center at (0, 0); focus at (0, 1); vertex at (0, −2)

23. Foci at $(\pm 2, 0)$; length of the major axis is 6
24. Focus at $(0, -4)$; vertices at $(0, \pm 8)$
25. Foci at $(0, \pm 3)$; x-intercepts are ± 2
26. Foci at $(0, \pm 2)$; length of the major axis is 8
27. Center at $(0, 0)$; vertex at $(0, 4)$; $b = 1$
28. Vertices at $(\pm 5, 0)$; $c = 2$

In Problems 29–32, write an equation for each ellipse.

29.

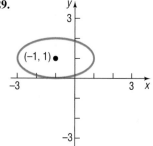

30.

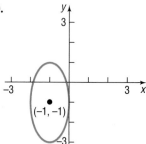

31.

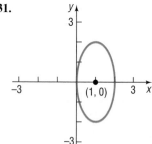

32.

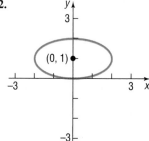

In Problems 33–44, find the center, foci, and vertices of each ellipse. Graph each equation using a graphing utility.

33. $\dfrac{(x-3)^2}{4} + \dfrac{(y+1)^2}{9} = 1$

34. $\dfrac{(x+4)^2}{9} + \dfrac{(y+2)^2}{4} = 1$

35. $(x+5)^2 + 4(y-4)^2 = 16$

36. $9(x-3)^2 + (y+2)^2 = 18$

37. $x^2 + 4x + 4y^2 - 8y + 4 = 0$

38. $x^2 + 3y^2 - 12y + 9 = 0$

39. $2x^2 + 3y^2 - 8x + 6y + 5 = 0$

40. $4x^2 + 3y^2 + 8x - 6y = 5$

41. $9x^2 + 4y^2 - 18x + 16y - 11 = 0$

42. $x^2 + 9y^2 + 6x - 18y + 9 = 0$

43. $4x^2 + y^2 + 4y = 0$

44. $9x^2 + y^2 - 18x = 0$

In Problems 45–54, find an equation for each ellipse. Graph the equation by hand.

45. Center at $(2, -2)$; vertex at $(7, -2)$; focus at $(4, -2)$
46. Center at $(-3, 1)$; vertex at $(-3, 3)$; focus at $(-3, 0)$
47. Vertices at $(4, 3)$ and $(4, 9)$; focus at $(4, 8)$
48. Foci at $(1, 2)$ and $(-3, 2)$; vertex at $(-4, 2)$
49. Foci at $(5, 1)$ and $(-1, 1)$; length of the major axis is 8

50. Vertices at $(2, 5)$ and $(2, -1)$; $c = 2$

51. Center at $(1, 2)$; focus at $(4, 2)$; contains the point $(1, 3)$

52. Center at $(1, 2)$; focus at $(1, 4)$; contains the point $(2, 2)$

53. Center at $(1, 2)$; vertex at $(4, 2)$; contains the point $(1, 3)$

54. Center at $(1, 2)$; vertex at $(1, 4)$; contains the point $(2, 2)$

In Problems 55–58, graph each function by hand. Use a graphing utility to verify your results. [Hint: Notice that each function is half an ellipse.]

55. $f(x) = \sqrt{16 - 4x^2}$

56. $f(x) = \sqrt{9 - 9x^2}$

57. $f(x) = -\sqrt{64 - 16x^2}$

58. $f(x) = -\sqrt{4 - 4x^2}$

59. *Semielliptical Arch Bridge* An arch in the shape of the upper half of an ellipse is used to support a bridge that is to span a river 20 meters wide. The center of the arch is 6 meters above the center of the river (see the figure). Write an equation for the ellipse in which the *x*-axis coincides with the water level and the *y*-axis passes through the center of the arch.

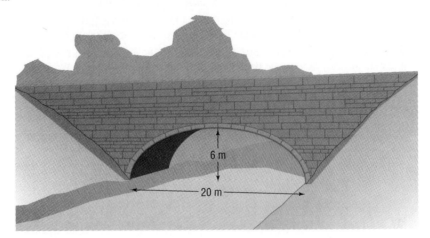

60. *Semielliptical Arch Bridge* The arch of a bridge is a semiellipse with a horizontal major axis. The span is 30 feet, and the top of the arch is 10 feet above the major axis. The roadway is horizontal and is 2 feet above the top of the arch. Find the vertical distance from the roadway to the arch at 5 foot intervals along the roadway.

61. *Whispering Galleries* A hall 100 feet in length is to be designed as a whispering gallery. If the foci are located 25 feet from the center, how high will the ceiling be at the center?

62. *Whispering Galleries* A person, standing at one focus of a whispering gallery, is 6 feet from the nearest wall. Her friend is standing at the other focus, 100 feet away. What is the length of this whispering gallery? How high is its elliptical ceiling at the center?

63. *Semielliptical Arch Bridge* A bridge is built in the shape of a semielliptical arch. The bridge has a span of 120 feet and a maximum height of 25 feet. Choose a suitable rectangular coordinate system and find the height of the arch at distances of 10, 30, and 50 feet from the center.

64. *Semielliptical Arch Bridge* A bridge is built in the shape of a semielliptical arch and is to have a span of 100 feet. The height of the arch, a distance of 40 feet from the center, is to be 10 feet. Find the height of the arch at its center.

65. *Semielliptical Arch* An arch in the form of half an ellipse is 40 feet wide and 15 feet high at the center. Find the height of the arch at intervals of 10 feet along its width.

66. *Semielliptical Arch Bridge* An arch for a bridge over a highway is in the form of half an ellipse. The top of the arch is 20 feet above the ground level (the major axis). The highway has four lanes, each 12 feet wide; a center safety strip 8 feet wide; and two side strips, each 4 feet wide. What should the span of the bridge be (the length of its major axis) if the height 28 feet from the center is to be 13 feet?

*In Problems 67–70, use the fact that the orbit of a planet about the Sun is an ellipse, with the Sun at one focus. The **aphelion** of a planet is its greatest distance from the Sun and the **perihelion** is its shortest distance. The **mean distance** of a planet from the Sun is the length of the semimajor axis of the elliptical orbit. See the illustration.*

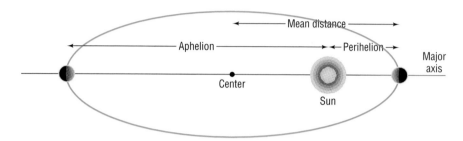

67. *Earth* The mean distance of Earth from the Sun is 93 million miles. If the aphelion of Earth is 94.5 million miles, what is the perihelion? Write an equation for the orbit of Earth around the Sun.

68. *Mars* The mean distance of Mars from the Sun is 142 million miles. If the perihelion of Mars is 128.5 million miles, what is the aphelion? Write an equation for the orbit of Mars about the Sun.

69. *Jupiter* The aphelion of Jupiter is 507 million miles. If the distance from the Sun to the center of its elliptical orbit is 23.2 million miles, what is the perihelion? What is the mean distance? Write an equation for the orbit of Jupiter around the Sun.

70. *Pluto* The perihelion of Pluto is 4551 million miles and the distance of the Sun from the center of its elliptical orbit is 897.5 million miles. Find the aphelion of Pluto. What is the mean distance of Pluto from the Sun? Write an equation for the orbit of Pluto about the Sun.

71. Consult the figure. A racetrack is in the shape of an ellipse, 100 feet long and 50 feet wide. What is the width 10 feet from the side?

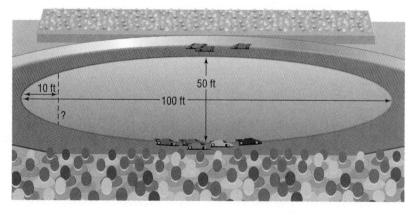

72. A racetrack is in the shape of an ellipse 80 feet long and 40 feet wide. What is the width 10 feet from the side?

73. Show that an equation of the form

$$Ax^2 + Cy^2 + F = 0 \qquad A \neq 0, C \neq 0, F \neq 0$$

where A and C are of the same sign and F is of opposite sign:

(a) Is the equation of an ellipse with center at $(0, 0)$ if $A \neq C$.

(b) Is the equation of a circle with center at $(0, 0)$ if $A = C$.

74. Show that the graph of an equation of the form

$$Ax^2 + Cy^2 + Dx + Ey + F = 0 \qquad A \neq 0, C \neq 0$$

where A and C are of the same sign:

(a) Is an ellipse if $(D^2/4A) + (E^2/4C) - F$ is the same sign as A.

(b) Is a point if $(D^2/4A) + (E^2/4C) - F = 0$.

(c) Contains no points if $(D^2/4A) + (E^2/4C) - F$ is of opposite sign to A.

75. The **eccentricity** e of an ellipse is defined as the number c/a, where a and c are the numbers given in equation (2). Because $a > c$, it follows that $e < 1$. Write a brief paragraph about the general shape of each of the following ellipses. Be sure to justify your conclusions.

(a) Eccentricity close to 0   (b) Eccentricity = 0.5   (c) Eccentricity close to 1

76. *CBL Experiment* The motion of a swinging pendulum is examined by plotting the velocity against the position of the pendulum. The result is the graph of an ellipse. (Activity 17, Real-World Math with the CBL System)

6.4

The Hyperbola

Hyperbola

A **hyperbola** is the collection of all points in the plane the difference of whose distances from two fixed points, called the **foci,** is a constant.

Figure 35 illustrates a hyperbola with foci F_1 and F_2. The line containing the foci is called the **transverse axis.** The midpoint of the line segment joining the foci is called the **center** of the hyperbola. The line through the center and perpendicular to the transverse axis is called the **conjugate axis.** The hyperbola consists of two separate curves, called **branches,** that are symmetric with respect to the transverse axis, conjugate axis, and center. The two points of intersection of the hyperbola and the transverse axis are the **vertices,** V_1 and V_2, of the hyperbola.

FIGURE 35

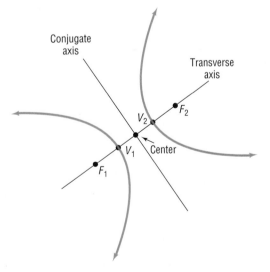

With these ideas in mind, we are now ready to find the equation of a hyperbola in a rectangular coordinate system. First, we place the center at the origin. Next, we position the hyperbola so that its transverse axis coincides with a coordinate axis. Suppose that the transverse axis coincides with the x-axis, as shown in Figure 36.

FIGURE 36

$d(F_1, P) - d(F_2, P) = \pm 2a$

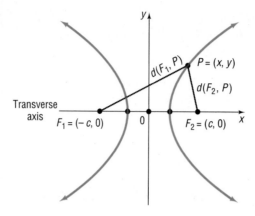

If c is the distance from the center to a focus, then one focus will be at $F_1 = (-c, 0)$ and the other at $F_2 = (c, 0)$. Now we let the constant difference of the distances from any point $P = (x, y)$ on the hyperbola to the foci F_1 and F_2 be denoted by $\pm 2a$. (If P is on the right branch, the $+$ sign is used; if P is on the left branch, the $-$ sign is used.) The coordinates of P must satisfy the equation

$$d(F_1, P) - d(F_2, P) = \pm 2a \qquad \text{Difference of the distances from } P \text{ to the foci equals } \pm 2a.$$

$$\sqrt{(x + c)^2 + y^2} - \sqrt{(x - c)^2 + y^2} = \pm 2a \qquad \text{Use the distance formula.}$$

$$\sqrt{(x + c)^2 + y^2} = \pm 2a + \sqrt{(x - c)^2 + y^2} \qquad \text{Isolate one radical.}$$

$$(x + c)^2 + y^2 = 4a^2 \pm 4a\sqrt{(x - c)^2 + y^2} \qquad \text{Square both sides.}$$
$$+ (x - c)^2 + y^2$$

Next, we remove the parentheses:

$$x^2 + 2cx + c^2 + y^2 = 4a^2 \pm 4a\sqrt{(x - c)^2 + y^2} + x^2 - 2cx + c^2 + y^2$$

$$4cx - 4a^2 = \pm 4a\sqrt{(x - c)^2 + y^2} \qquad \text{Isolate the radical.}$$

$$cx - a^2 = \pm a\sqrt{(x - c)^2 + y^2} \qquad \text{Divide each side by 4.}$$

$$(cx - a^2)^2 = a^2[(x - c)^2 + y^2] \qquad \text{Square both sides.}$$

$$c^2x^2 - 2ca^2x + a^4 = a^2(x^2 - 2cx + c^2 + y^2)$$

$$c^2x^2 + a^4 = a^2x^2 + a^2c^2 + a^2y^2$$

$$(c^2 - a^2)x^2 - a^2y^2 = a^2c^2 - a^4$$

$$(c^2 - a^2)x^2 - a^2y^2 = a^2(c^2 - a^2) \qquad (1)$$

To obtain points on the hyperbola off the x-axis, it must be that $a < c$. To see why, look again at Figure 36.

$$d(F_1, P) < d(F_2, P) + d(F_1, F_2) \qquad \text{Use triangle } F_1PF_2.$$

$$d(F_1, P) - d(F_2, P) < d(F_1, F_2) \qquad \begin{array}{l}P \text{ is on the right branch,} \\ \text{so } d(F_1, P) - d(F_2, P) = 2a.\end{array}$$

$$2a < 2c$$

$$a < c$$

Since $a < c$, we also have $a^2 < c^2$, so $c^2 - a^2 > 0$. Let $b^2 = c^2 - a^2$, $b > 0$. Then equation (1) can be written as

$$b^2x^2 - a^2y^2 = a^2b^2$$

$$\frac{x^2}{a^2} - \frac{y^2}{b^2} = 1$$

To find the vertices of the hyperbola defined by this equation, let $y = 0$. The vertices satisfy the equation $x^2/a^2 = 1$, the solutions of which are $x = \pm a$. Consequently, the vertices of the hyperbola are $V_1 = (-a, 0)$ and $V_2 = (a, 0)$.

Theorem   An equation of the hyperbola with center at $(0, 0)$, foci at $(-c, 0)$ and $(c, 0)$, and vertices at $(-a, 0)$ and $(a, 0)$ is

**Equation of a Hyperbola;
Center at $(0, 0)$; Foci
at $(\pm c, 0)$; Vertices at $(\pm a, 0)$;
Transverse Axis along x-Axis**

$$\frac{x^2}{a^2} - \frac{y^2}{b^2} = 1 \qquad \text{where } b^2 = c^2 - a^2 \tag{2}$$

The transverse axis is the x-axis.   ■

As you can verify, the hyperbola defined by equation (2) is symmetric with respect to the x-axis, y-axis, and origin. To find the y-intercepts, if any, let $x = 0$ in equation (2). This results in the equation $y^2/b^2 = -1$, which has no solution. We conclude that the hyperbola defined by equation (2) has no y-intercepts. In fact, since $x^2/a^2 - 1 = y^2/b^2 \geq 0$, it follows that $x^2/a^2 \geq 1$. Thus, there are no points on the graph for $-a < x < a$. See Figure 37.

FIGURE 37

$$\frac{x^2}{a^2} - \frac{y^2}{b^2} = 1,$$

$$b^2 = c^2 - a^2$$

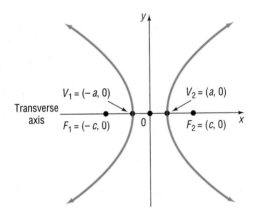

E X A M P L E   1   *Finding an Equation of a Hyperbola*

Find an equation of the hyperbola with center at the origin, one focus at $(3, 0)$, and one vertex at $(-2, 0)$. Graph the equation by hand.

Solution   The hyperbola has its center at the origin, and the transverse axis coincides with the x-axis. One focus is at $(c, 0) = (3, 0)$, so $c = 3$. One vertex is at $(-a, 0) = (-2, 0)$, so $a = 2$. From equation (2), it follows that $b^2 = c^2 - a^2 = 9 - 4 = 5$, so an equation of the hyperbola is

$$\frac{x^2}{4} - \frac{y^2}{5} = 1$$

See Figure 38.

FIGURE 38

$$\frac{x^2}{4} - \frac{y^2}{5} = 1$$

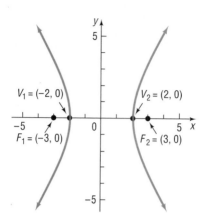

■ Now work Problem 9.

E X A M P L E   2

Using a Graphing Utility to Graph a Hyperbola

Use a graphing utility to graph the hyperbola: $\dfrac{x^2}{4} - \dfrac{y^2}{5} = 1$

Solution   To graph the hyperbola $\dfrac{x^2}{4} - \dfrac{y^2}{5} = 1$, we need to graph the two functions

$y_1 = \sqrt{5}\sqrt{\dfrac{x^2}{4} - 1}$ and $y_2 = -\sqrt{5}\sqrt{\dfrac{x^2}{4} - 1}$. As with graphing circles and ellipses

on a graphing utility, we use a square screen setting so that the graph is not distorted. Figure 39 shows the graph of the hyperbola.

FIGURE 39

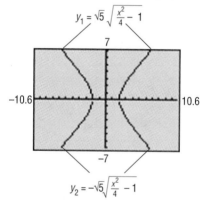

■

An equation of the form of equation (2) is the equation of a hyperbola with center at the origin, foci on the *x*-axis at $(-c, 0)$ and $(c, 0)$, where $c^2 = a^2 + b^2$, and transverse axis along the *x*-axis.

For the remainder of this section, the direction "Discuss the equation" will mean to find the center, transverse axis, vertices, and foci of the hyperbola and graph it.

E X A M P L E   3

Discussing the Equation of a Hyperbola

Discuss the equation: $\dfrac{x^2}{16} - \dfrac{y^2}{4} = 1$

Solution   Figure 40(a) shows the graph of $\dfrac{x^2}{16} - \dfrac{y^2}{4} = 1$ using a graphing utility. We now proceed to analyze the equation.

    The given equation is of the form of equation (2), with $a^2 = 16$ and $b^2 = 4$. Thus, the graph of the equation is a hyperbola with center at $(0, 0)$ and transverse axis along the x-axis. Also, we know that $c^2 = a^2 + b^2 = 16 + 4 = 20$. The vertices are at $(\pm a, 0) = (\pm 4, 0)$, and the foci are at $(\pm c, 0) = (\pm 2\sqrt{5}, 0)$. Figure 40(b) shows the graph drawn by hand.

FIGURE 40

$\dfrac{x^2}{16} - \dfrac{y^2}{4} = 1$

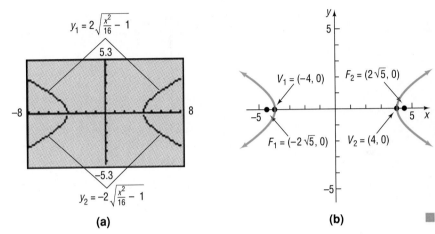

(a)   **(b)**

    Now work Problem 19.

    The next result gives the form of the equation of a hyperbola with center at the origin and transverse axis along the y-axis.

Theorem   An equation of the hyperbola with center at $(0, 0)$, foci at $(0, -c)$ and $(0, c)$, and vertices at $(0, -a)$ and $(0, a)$ is

---

Equation of a Hyperbola; Center at $(0, 0)$; Foci at $(0, \pm c)$; Vertices at $(0, \pm a)$; Transverse Axis along y-Axis

$$\dfrac{y^2}{a^2} - \dfrac{x^2}{b^2} = 1 \qquad \text{where } b^2 = c^2 - a^2 \qquad (3)$$

The transverse axis is the y-axis.

    Figure 41 shows the graph of a typical hyperbola defined by equation (3).

FIGURE 41

$\dfrac{y^2}{a^2} - \dfrac{x^2}{b^2} = 1,$

$b^2 = c^2 - a^2$

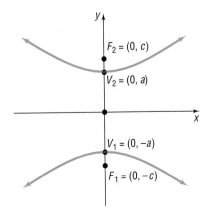

Notice the difference in the form of equations (2) and (3). When the y^2-term is subtracted from the x^2-term, the transverse axis is the x-axis. When the x^2-term is subtracted from the y^2-term, the transverse axis is the y-axis.

E X A M P L E   4   *Discussing the Equation of a Hyperbola*

Discuss the equation: $y^2 - 4x^2 = 4$

Solution   Figure 42(a) shows the graph of $y^2 - 4x^2 = 4$ using a graphing utility. We now proceed to analyze the equation.

To put the equation in proper form, we divide each side by 4:

$$\frac{y^2}{4} - x^2 = 1$$

Since the x^2-term is subtracted from the y^2-term, the equation is that of a hyperbola with center at the origin and transverse axis along the y-axis. Also, comparing the above equation to equation (3), we find $a^2 = 4$, $b^2 = 1$, and $c^2 = a^2 + b^2 = 5$. The vertices are at $(0, \pm a) = (0, \pm 2)$, and the foci are at $(0, \pm c) = (0, \pm\sqrt{5})$. The graph, drawn by hand, is given in Figure 42(b).

FIGURE 42

$$\frac{y^2}{4} - x^2 = 1$$

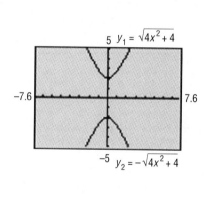

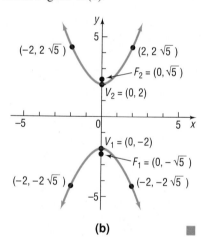

(a)   (b)

E X A M P L E   5   *Finding an Equation of a Hyperbola*

Find an equation of the hyperbola having one vertex at $(0, 2)$ and foci at $(0, -3)$ and $(0, 3)$. Graph the equation by hand.

Solution   Since the foci are at $(0, -3)$ and $(0, 3)$, the center of the hyperbola is at the origin. Also, the transverse axis is along the y-axis. The given information also reveals that $c = 3$, $a = 2$, and $b^2 = c^2 - a^2 = 9 - 4 = 5$. The form of the equation of the hyperbola is given by equation (3):

$$\frac{y^2}{a^2} - \frac{x^2}{b^2} = 1$$

$$\frac{y^2}{4} - \frac{x^2}{5} = 1$$

See Figure 43.

FIGURE 43

$$\frac{y^2}{4} - \frac{x^2}{5} = 1$$

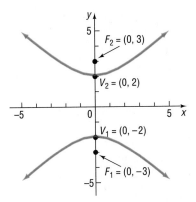

■ Now work Problem 11.

Look at the equations of the hyperbolas in Examples 4 and 5. For the hyperbola in Example 4, $a^2 = 4$ and $b^2 = 1$, so $a > b$; for the hyperbola in Example 5, $a^2 = 4$ and $b^2 = 5$, so $a < b$. We conclude that, for hyperbolas, there are no requirements involving the relative size of a and b. Contrast this situation to the case of an ellipse, in which the relative sizes of a and b dictate which axis is the major axis. Hyperbolas have another feature to distinguish them from ellipses and parabolas: Hyperbolas have asymptotes.

Asymptotes

A **horizontal** or **oblique (slant) asymptote** of a graph is a line with the property that the distance from the line to points on the graph approaches 0 as $x \to -\infty$ or as $x \to \infty$.

Theorem   The hyperbola $\dfrac{x^2}{a^2} - \dfrac{y^2}{b^2} = 1$

has the two oblique asymptotes

Asymptotes of a Hyperbola

$$y = \frac{b}{a}x \quad \text{and} \quad y = -\frac{b}{a}x$$

Proof   We begin by solving for y in the equation of the hyperbola:

$$\frac{x^2}{a^2} - \frac{y^2}{b^2} = 1$$

$$\frac{y^2}{b^2} = \frac{x^2}{a^2} - 1$$

$$y^2 = b^2\left(\frac{x^2}{a^2} - 1\right)$$

Since $x \neq 0$, we can rearrange the right side in the form

$$y^2 = \frac{b^2 x^2}{a^2}\left(1 - \frac{a^2}{x^2}\right)$$

$$y = \pm\frac{bx}{a}\sqrt{1 - \frac{a^2}{x^2}}$$

Now, as $x \to -\infty$ or as $x \to \infty$, the term a^2/x^2 approaches 0, so the expression under the radical approaches 1. Thus, as $x \to -\infty$ or as $x \to \infty$, the value of y approaches $\pm bx/a$; that is, the graph of the hyperbola approaches the lines

$$y = -\frac{b}{a}x \quad \text{and} \quad y = \frac{b}{a}x$$

Thus, these lines are oblique asymptotes of the hyperbola. ■

The asymptotes of a hyperbola are not part of the hyperbola, but they do serve as a guide for graphing a hyperbola by hand. For example, suppose that we want to graph the equation

$$\frac{x^2}{a^2} - \frac{y^2}{b^2} = 1$$

We begin by plotting the vertices $(-a, 0)$ and $(a, 0)$. Then we plot the points $(0, -b)$ and $(0, b)$ and use these four points to construct a rectangle, as shown in Figure 44. The diagonals of this rectangle have slopes b/a and $-b/a$, and their extensions are the asymptotes $y = (b/a)x$ and $y = -(b/a)x$ of the hyperbola.

FIGURE 44

$$\frac{x^2}{a^2} - \frac{y^2}{b^2} = 1$$

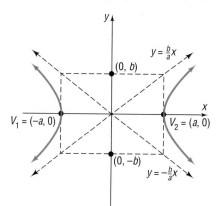

Theorem   The hyperbola $\dfrac{y^2}{a^2} - \dfrac{x^2}{b^2} = 1$

has the two oblique asymptotes

Asymptotes of a Hyperbola

$$y = \frac{a}{b}x \quad \text{and} \quad y = -\frac{a}{b}x$$

■

You are asked to prove this result in Problem 64.

E X A M P L E   6   *Discussing the Equation of a Hyperbola*

Discuss the equation: $9x^2 - 4y^2 = 36$

Solution   First, we divide each side by 36 to put the equation in proper form:

$$\frac{x^2}{4} - \frac{y^2}{9} = 1$$

Figure 45(a) shows the graph of $\dfrac{x^2}{4} - \dfrac{y^2}{9} = 1$ and its asymptotes, $y = \dfrac{3}{2}x$ and $y = -\dfrac{3}{2}x$, using a graphing utility. We now proceed to analyze the equation.

The equation is that of a hyperbola with center at the origin and transverse axis along the x-axis. Using $a^2 = 4$ and $b^2 = 9$, we find $c^2 = a^2 + b^2 = 13$. The vertices are at $(\pm a, 0) = (\pm 2, 0)$, the foci are at $(\pm c, 0) = (\pm\sqrt{13}, 0)$, and the asymptotes have the equations

$$y = \frac{3}{2}x \quad \text{and} \quad y = -\frac{3}{2}x$$

Now form the rectangle containing the points $(\pm a, 0)$ and $(0, \pm b)$, that is, $(-2, 0)$ $(2, 0)$, $(0, -3)$, and $(0, 3)$. The extensions of the diagonals of this rectangle are the asymptotes. See Figure 45(b) for the graph drawn by hand.

FIGURE 45

$$\frac{x^2}{4} - \frac{y^2}{9} = 1$$

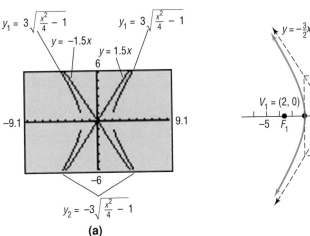

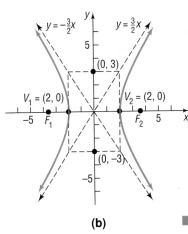

(a)   (b)

Exploration: Use TRACE to see what happens as x becomes unbounded in the positive direction for both the upper and the lower portions of the hyperbola.

■   Now work Problem 21.

Center at (h, k)

If a hyperbola with center at the origin and transverse axis coinciding with a coordinate axis is shifted horizontally h units and then vertically k units, the result is a hyperbola with center at (h, k) and transverse axis parallel to a coordinate axis. Table 4 gives the forms of the equations of such hyperbolas. See Figure 46 for the graphs.

TABLE 4   HYPERBOLAS WITH CENTER AT (h,k) AND TRANSVERSE AXIS PARALLEL TO A COORDINATE AXIS

CENTER	TRANSVERSE AXIS	FOCI	VERTICES	EQUATION	ASYMPTOTES
(h, k)	Parallel to x-axis	$(h \pm c, k)$	$(h \pm a, k)$	$\dfrac{(x-h)^2}{a^2} - \dfrac{(y-k)^2}{b^2} = 1,$ $b^2 = c^2 - a^2$	$y - k = \pm\dfrac{b}{a}(x - h)$
(h, k)	Parallel to y-axis	$(h, k \pm c)$	$(h, k \pm a)$	$\dfrac{(y-k)^2}{a^2} - \dfrac{(x-h)^2}{b^2} = 1,$ $b^2 = c^2 - a^2$	$y - k = \pm\dfrac{a}{b}(x - h)$

FIGURE 46

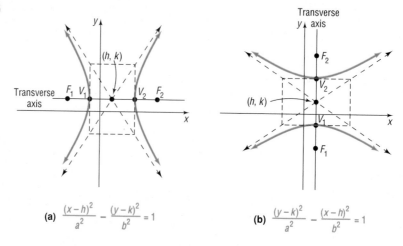

(a) $\dfrac{(x-h)^2}{a^2} - \dfrac{(y-k)^2}{b^2} = 1$

(b) $\dfrac{(y-k)^2}{a^2} - \dfrac{(x-h)^2}{b^2} = 1$

E X A M P L E   7

Finding an Equation of a Hyperbola, Center Not at the Origin

Find an equation for the hyperbola with center at $(1, -2)$, one focus at $(4, -2)$, and one vertex at $(3, -2)$. Graph the equation by hand.

Solution   The center is at $(h, k) = (1, -2)$, so $h = 1$ and $k = -2$. The transverse axis is parallel to the x-axis. The distance from the center $(1, -2)$ to the focus $(4, -2)$ is $c = 3$; the distance from the center $(1, -2)$ to the vertex $(3, -2)$ is $a = 2$. Thus, $b^2 = c^2 - a^2 = 9 - 4 = 5$. The equation is

$$\frac{(x-h)^2}{a^2} - \frac{(y-k)^2}{b^2} = 1$$

$$\frac{(x-1)^2}{4} - \frac{(y+2)^2}{5} = 1$$

See Figure 47.

FIGURE 47

$\dfrac{(x-1)^2}{4} - \dfrac{(y+2)^2}{5} = 1$

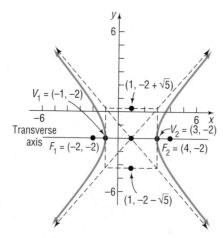

■   Now work Problem 31.

E X A M P L E   8

Using a Graphing Utility to Graph a Hyperbola, Center Not at the Origin

Use a graphing utility to graph the hyperbola $\dfrac{(x-1)^2}{4} - \dfrac{(y+2)^2}{5} = 1$

Solution   First, we must solve the equation for y:

$$\frac{(x-1)^2}{4} - \frac{(y+2)^2}{5} = 1$$

$$\frac{(y+2)^2}{5} = \frac{(x-1)^2}{4} - 1$$

$$(y+2)^2 = 5\left[\frac{(x-1)^2}{4} - 1\right]$$

$$y + 2 = \pm\sqrt{5\left[\frac{(x-1)^2}{4} - 1\right]}$$

$$y = -2 \pm\sqrt{5\left[\frac{(x-1)^2}{4} - 1\right]}$$

FIGURE 48

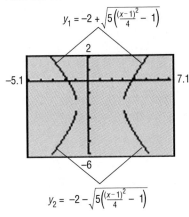

$$y_1 = -2 + \sqrt{5\left(\frac{(x-1)^2}{4} - 1\right)}$$

$$y_2 = -2 - \sqrt{5\left(\frac{(x-1)^2}{4} - 1\right)}$$

Figure 48 shows the graph of $y_1 = -2 + \sqrt{5\left[\frac{(x-1)^2}{4} - 1\right]}$ and $y_2 = -2 - \sqrt{5\left[\frac{(x-1)^2}{4} - 1\right]}$.   ◼

E X A M P L E   9   *Discussing the Equation of a Hyperbola*

Discuss the equation: $-x^2 + 4y^2 - 2x - 16y + 11 = 0$

Solution   We complete the squares in x and in y:

$$-x^2 + 4y^2 - 2x - 16y + 11 = 0$$

$$-(x^2 + 2x) + 4(y^2 - 4y) = -11 \qquad \text{Group terms.}$$

$$-(x^2 + 2x + 1) + 4(y^2 - 4y + 4) = -1 + 16 - 11 \quad \text{Complete each square.}$$

$$-(x+1)^2 + 4(y-2)^2 = 4$$

$$(y-2)^2 - \frac{(x+1)^2}{4} = 1 \qquad \text{Divide by 4.}$$

Figure 49(a) shows the graph of $(y-2)^2 - \frac{(x+1)^2}{4} = 1$ using a graphing utility. We now proceed to analyze the equation.

It is the equation of a hyperbola with center at $(-1, 2)$ and transverse axis parallel to the y-axis. Also, $a^2 = 1$ and $b^2 = 4$, so $c^2 = a^2 + b^2 = 5$. The vertices are at $(h, \ k \pm a) = (-1, \ 2 \pm 1)$, or $(-1, 1)$ and $(-1, 3)$. The foci are at $(h, \ k \pm c) = (-1, \ 2 \pm \sqrt{5})$. The asymptotes are $y - 2 = \frac{1}{2}(x + 1)$ and $y - 2 = -\frac{1}{2}(x + 1)$. Figure 49(b) shows the graph drawn by hand.

FIGURE 49

$$(y-2)^2 - \frac{(x+1)^2}{4} = 1$$

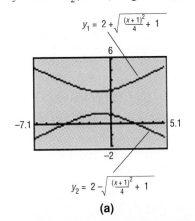

$$y_1 = 2 + \sqrt{\frac{(x+1)^2}{4} + 1}$$

$$y_2 = 2 - \sqrt{\frac{(x+1)^2}{4} + 1}$$

(a)

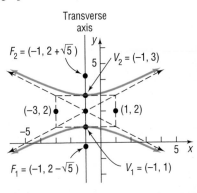

(b)

◼

FIGURE 50

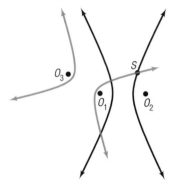

Applications

Suppose that a gun is fired from an unknown source S. An observer at O_1 hears the report (sound of gun shot) 1 second after another observer at O_2. Because sound travels at about 1100 feet per second, it follows that the point S must be 1100 feet closer to O_2 than to O_1. Thus, S lies on one branch of a hyperbola with foci at O_1 and O_2. (Do you see why? The difference of the distances from S to O_1 and from S to O_2 is the constant 1100.) If a third observer at O_3 hears the same report 2 seconds after O_1 hears it, then S will lie on a branch of a second hyperbola with foci at O_1 and O_3. The intersection of the two hyperbolas will pinpoint the location of S. See Figure 50 for an illustration.

LORAN

In the LOng RAnge Navigation system (LORAN), a master radio sending station and a secondary sending station emit signals that can be received by a ship at sea. (See Figure 51.) Because a ship monitoring the two signals will usually be nearer to one of the two stations, there will be a difference in the distance the two signals travel, which will register as a slight time difference between the signals. As long as the time difference remains constant, the difference of the two distances will also be constant. If the ship follows a path corresponding to the fixed time difference, it will follow the path of a hyperbola whose foci are located at the positions of the two sending stations. So for each time difference a different hyperbolic path results, each bringing the ship to a different shore location. Navigation charts show the various hyperbolic paths corresponding to different time differences.

FIGURE 51

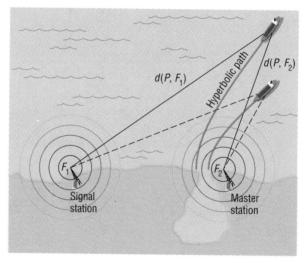

$d(P, F_1) - d(P, F_2) = $ constant

E X A M P L E   1 0   *LORAN*

Two LORAN stations are positioned 250 miles apart along a straight shore.

(a) A ship records a time difference of 0.00086 second between the LORAN signals. Set up an appropriate rectangular coordinate system to determine where the ship would reach shore if it were to follow the hyperbola corresponding to this time difference.

(b) If the ship wants to enter a harbor located between the two stations 25 miles from the master station, what time difference should it be looking for?

(c) If the ship is 80 miles off shore when the desired time difference is obtained, what is the exact location of the ship? [*Note:* The speed of each radio signal is 186,000 miles per second.]

Solution (a) We set up a rectangular coordinate system so that the two stations lie on the *x*-axis and the origin is midway between them. See Figure 52.

FIGURE 52

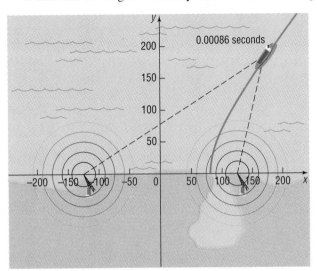

The ship lies on a hyperbola whose foci are the locations of the two stations. The reason for this is that the constant time difference of the signals from each station results in a constant difference in the distance of the ship from each station. Since the time difference is 0.00086 second and the speed of the signal is 186,000 miles per second, the difference of the distances from the ship to each station (foci) is

$$\text{Distance} = \text{Speed} \times \text{Time} = 186{,}000 \times 0.00086 = 160 \text{ miles}$$

The difference of the distances from the ship to each station, 160, equals $2a$, so $a = 80$ and the vertex of the corresponding hyperbola is at $(80, 0)$. Since the focus is at $(125, 0)$, following this hyperbola the ship would reach shore 45 miles from the master station.

(b) To reach shore 25 miles from the master station, the ship should follow a hyperbola with vertex at $(100, 0)$. For this hyperbola, $a = 100$, so the constant difference of the distances from the ship to each station is 200. The time difference the ship should look for is

$$\text{Time} = \frac{\text{Distance}}{\text{Speed}} = \frac{200}{186{,}000} = 0.001075 \text{ second}$$

(c) To find the exact location of the ship, we need to find the equation of the hyperbola with vertex at $(100, 0)$ and a focus at $(125, 0)$. The form of the equation of this hyperbola is

$$\frac{x^2}{a^2} - \frac{y^2}{b^2} = 1$$

where $a = 100$. Since $c = 125$, we have

$$b^2 = c^2 - a^2 = 125^2 - 100^2 = 5625$$

The equation of the hyperbola is

$$\frac{x^2}{100^2} - \frac{y^2}{5625} = 1$$

Since the ship is 80 miles from shore, we use $y = 80$ in the equation and solve for x.

$$\frac{x^2}{100^2} - \frac{80^2}{5625} = 1$$

$$\frac{x^2}{100^2} = 1 + \frac{80^2}{5625} = 2.14$$

$$x^2 = 100^2(2.14)$$

$$x = 146$$

The ship is at the position (146, 80). See Figure 53.

FIGURE 53

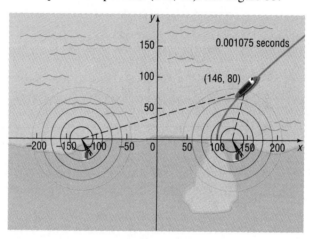

6.4

Exercise 6.4

In Problems 1–4, the graph of a hyperbola is given. Match each graph to its equation.

A. $\dfrac{x^2}{4} - y^2 = 1$ B. $x^2 - \dfrac{y^2}{4} = 1$ C. $\dfrac{y^2}{4} - x^2 = 1$ D. $y^2 - \dfrac{x^2}{4} = 1$

1.

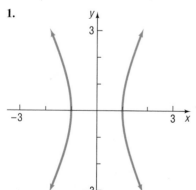

2.

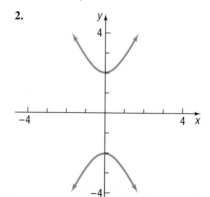

3.

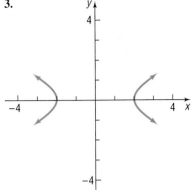

4.

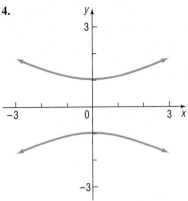

In Problems 5–8, the graph of a hyperbola is given. Match each graph to its equation.

A. $\dfrac{x^2}{16} - \dfrac{y^2}{9} = 1$ B. $\dfrac{x^2}{9} - \dfrac{y^2}{16} = 1$ C. $\dfrac{y^2}{16} - \dfrac{x^2}{9} = 1$ D. $\dfrac{y^2}{9} - \dfrac{x^2}{16} = 1$

5.

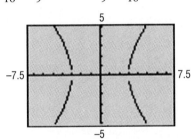

6.

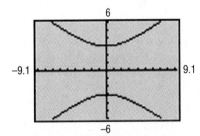

7.

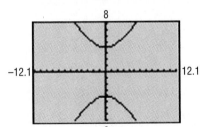

8.

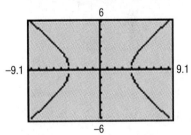

In Problems 9–18, find an equation for the hyperbola described. Graph the equation by hand.

9. Center at $(0, 0)$; focus at $(3, 0)$; vertex at $(1, 0)$

10. Center at $(0, 0)$; focus at $(0, 5)$; vertex at $(0, 3)$

11. Center at $(0, 0)$; focus at $(0, -6)$; vertex at $(0, 4)$

12. Center at $(0, 0)$; focus at $(-3, 0)$; vertex at $(2, 0)$

13. Foci at $(-5, 0)$ and $(5, 0)$; vertex at $(3, 0)$

14. Focus at $(0, 6)$; vertices at $(0, -2)$ and $(0, 2)$

15. Vertices at $(0, -6)$ and $(0, 6)$; asymptote the line $y = 2x$

16. Vertices at $(-4, 0)$ and $(4, 0)$; asymptote the line $y = 2x$

17. Foci at $(-4, 0)$ and $(4, 0)$; asymptote the line $y = -x$

18. Foci at $(0, -2)$ and $(0, 2)$; asymptote the line $y = -x$

In Problems 19–26, find the center, transverse axis, vertices, foci, and asymptotes. Graph each equation using a graphing utility.

19. $\dfrac{x^2}{16} - \dfrac{y^2}{4} = 1$ **20.** $\dfrac{y^2}{16} - \dfrac{x^2}{4} = 1$ **21.** $4x^2 - y^2 = 16$ **22.** $y^2 - 4x^2 = 16$

23. $y^2 - 9x^2 = 9$ **24.** $x^2 - y^2 = 4$ **25.** $y^2 - x^2 = 25$ **26.** $2x^2 - y^2 = 4$

In Problems 27–30, write an equation for each hyperbola.

27.
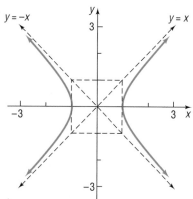

28.
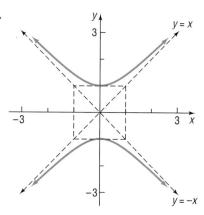

29.
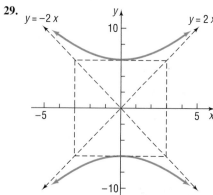

30.
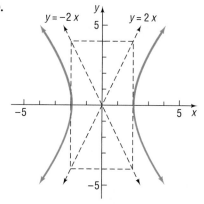

In Problems 31–38, find an equation for the hyperbola described. Graph the equation by hand.

31. Center at $(4, -1)$; focus at $(7, -1)$; vertex at $(6, -1)$

32. Center at $(-3, 1)$; focus at $(-3, 6)$; vertex at $(-3, 4)$

33. Center at $(-3, -4)$; focus at $(-3, -8)$; vertex at $(-3, -2)$

34. Center at $(1, 4)$; focus at $(-2, 4)$; vertex at $(0, 4)$

35. Foci at $(3, 7)$ and $(7, 7)$; vertex at $(6, 7)$

36. Focus at $(-4, 0)$; vertices at $(-4, 4)$ and $(-4, 2)$

37. Vertices at $(-1, -1)$ and $(3, -1)$; asymptote the line $(x - 1)/2 = (y + 1)/3$

38. Vertices at $(1, -3)$ and $(1, 1)$; asymptote the line $(x - 1)/2 = (y + 1)/3$

In Problems 39–52, find the center, transverse axis, vertices, foci, and asymptotes. Graph each equation using a graphing utility.

39. $\dfrac{(x - 2)^2}{4} - \dfrac{(y + 3)^2}{9} = 1$ **40.** $\dfrac{(y + 3)^2}{4} - \dfrac{(x - 2)^2}{9} = 1$

Using the results of Problems 73 and 74 in Exercise 6.2, it follows that, except for the degenerate cases, the equation is a parabola.

(b) If $AC > 0$, then A and C are of the same sign. Using the results of Problems 73 and 74 in Exercise 6.3, except for the degenerate cases, the equation is an ellipse if $A \neq C$ or a circle if $A = C$.

(c) If $AC < 0$, then A and C are of opposite sign. Using the results of Problems 65 and 66 in Exercise 6.4, except for the degenerate cases, the equation is a hyperbola. ■

We shall not be concerned with the degenerate cases of equation (2). However, in practice, you should be alert to the possibility of degeneracy.

E X A M P L E   1   *Identifying a Conic without Completing the Squares*

Identify each equation without completing the squares.

(a) $3x^2 + 6y^2 + 6x - 12y = 0$ (b) $2x^2 - 3y^2 + 6y + 4 = 0$

(c) $y^2 - 2x + 4 = 0$

Solution   (a) We compare the given equation to equation (2) and conclude that $A = 3$ and $C = 6$. Since $AC = 18 > 0$, the equation is an ellipse.

(b) Here, $A = 2$ and $C = -3$, so $AC = -6 < 0$. The equation is a hyperbola.

(c) Here, $A = 0$ and $C = 1$, so $AC = 0$. The equation is a parabola. ■

■   Now work Problem 1.

Although we can now identify the type of conic represented by any equation of the form of equation (2) without completing the squares, we will still need to complete the squares if we desire additional information about a conic.

Now we turn our attention to equations of the form of equation (1), where $B \neq 0$. To discuss this case, we first need to investigate a new procedure: *rotation of axes*.

Rotation of Axes

FIGURE 54

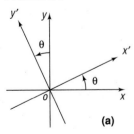

(a)

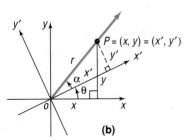

(b)

In a **rotation of axes,** the origin remains fixed while the x-axis and y-axis are rotated through an angle θ to a new position; the new positions of the x- and y-axes are denoted by x' and y', respectively, as shown in Figure 54(a).

Now look at Figure 54(b). There the point P has the coordinates (x, y) relative to the xy-plane, while the same point P has coordinates (x', y') relative to the $x'y'$-plane. We seek relationships that will enable us to express x and y in terms of x', y', and θ.

As Figure 54(b) shows, r denotes the distance from the origin O to the point P, and α denotes the angle between the positive x'-axis and the ray from O through P. Then, using the definitions of sine and cosine, we have

$$x' = r \cos \alpha \qquad y' = r \sin \alpha \qquad (3)$$

$$x = r \cos(\theta + \alpha) \qquad y = r \sin(\theta + \alpha) \qquad (4)$$

Now

$$
\begin{aligned}
x &= r \cos(\theta + \alpha) \\
&= r(\cos \theta \cos \alpha - \sin \theta \sin \alpha) \qquad \text{Sum formula}\\
&= (r \cos \alpha)(\cos \theta) - (r \sin \alpha)(\sin \theta) \\
&= x' \cos \theta - y' \sin \theta \qquad \text{By equation (3)}
\end{aligned}
$$

Similarly,

$$y = r \sin(\theta + \alpha)$$
$$= r(\sin \theta \cos \alpha + \cos \theta \sin \alpha)$$
$$= x' \sin \theta + y' \cos \theta$$

Theorem
Rotation Formulas

If the x- and y-axes are rotated through an angle θ, the coordinates (x, y) of a point P relative to the xy-plane and the coordinates (x', y') of the same point relative to the new x'- and y'-axes are related by the formulas

$$x = x' \cos \theta - y' \sin \theta \qquad y = x' \sin \theta + y' \cos \theta \qquad (5)$$

∎

E X A M P L E   2

Rotating Axes

Express the equation $xy = 1$ in terms of new $x'y'$-coordinates by rotating the axes through a 45° angle. Discuss the new equation.

Solution

Figure 55(a) shows the graph of $xy = 1$ using a graphing utility. We now proceed to express the equation in terms of new $x'y'$-coordinates and discuss the new equation.
Let $\theta = 45°$ in equation (5). Then

$$x = x' \cos 45° - y' \sin 45° = x' \frac{\sqrt{2}}{2} - y' \frac{\sqrt{2}}{2} = \frac{\sqrt{2}}{2}(x' - y')$$

$$y = x' \sin 45° + y' \cos 45° = x' \frac{\sqrt{2}}{2} + y' \frac{\sqrt{2}}{2} = \frac{\sqrt{2}}{2}(x' + y')$$

FIGURE 55

$xy = 1$ or $\dfrac{x'^2}{2} - \dfrac{y'^2}{2} = 1$

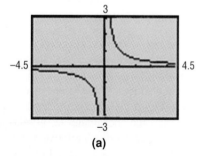

(a)

Substituting these expressions for x and y in $xy = 1$ gives

$$\left[\frac{\sqrt{2}}{2}(x' - y') \right]\left[\frac{\sqrt{2}}{2}(x' + y') \right] = 1$$

$$\frac{1}{2}(x'^2 - y'^2) = 1$$

$$\frac{x'^2}{2} - \frac{y'^2}{2} = 1$$

This is the equation of a hyperbola with center at $(0, 0)$ and transverse axis along the x'-axis. The vertices are at $(\pm\sqrt{2}, 0)$ on the x'-axis; the asymptotes are $y' = x'$ and $y' = -x'$ (which correspond to the original x- and y-axes). See Figure 55(b) for the graph drawn by hand.

∎

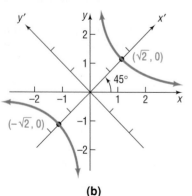

(b)

As Example 2 illustrates, a rotation of axes through an appropriate angle can transform a second-degree equation in x and y containing an xy-term into one in x' and y' in which no $x'y'$-term appears. In fact, we will show that a rotation of axes through an appropriate angle will transform any equation of the form of equation (1) into an equation in x' and y' without an $x'y'$-term.

To find the formula for choosing an appropriate angle θ through which to rotate the axes, we begin with equation (1),

$$Ax^2 + Bxy + Cy^2 + Dx + Ey + F = 0, \qquad B \neq 0$$

Next we rotate through an angle θ using rotation formulas (5):

$$A(x' \cos \theta - y' \sin \theta)^2 + B(x' \cos \theta - y' \sin \theta)(x' \sin \theta + y' \cos \theta)$$
$$+ C(x' \sin \theta + y' \cos \theta)^2 + D(x' \cos \theta - y' \sin \theta)$$
$$+ E(x' \sin \theta + y' \cos \theta) + F = 0$$

By expanding and collecting like terms, we obtain

$$(A \cos^2 \theta + B \sin \theta \cos \theta + C \sin^2 \theta)x'^2 + [B(\cos^2 \theta - \sin^2 \theta) + 2(C - A)(\sin \theta \cos \theta)]x'y'$$
$$+ (A \sin^2 \theta - B \sin \theta \cos \theta + C \cos^2 \theta)y'^2 \qquad (6)$$
$$+ (D \cos \theta + E \sin \theta)x'$$
$$+ (-D \sin \theta + E \cos \theta)y' + F = 0$$

In equation (6), the coefficient of $x'y'$ is

$$B' = 2(C - A)(\sin \theta \cos \theta) + B(\cos^2 \theta - \sin^2 \theta)$$

Since we want to eliminate the $x'y'$-term, we select an angle θ so that $B' = 0$. Thus,

$$2(C - A)(\sin \theta \cos \theta) + B(\cos^2 \theta - \sin^2 \theta) = 0$$
$$(C - A)(\sin 2\theta) + B \cos 2\theta = 0 \quad \text{Double-angle formulas}$$
$$B \cos 2\theta = (A - C)(\sin 2\theta)$$
$$\cot 2\theta = \frac{A - C}{B}, \qquad B \neq 0$$

Theorem   To transform the equation

$$Ax^2 + Bxy + Cy^2 + Dx + Ey + F = 0, \qquad B \neq 0$$

into an equation in x' and y' without an $x'y'$-term, rotate the axes through an angle θ that satisfies the equation

$$\cot 2\theta = \frac{A - C}{B} \qquad (7)$$

Equation (7) has an infinite number of solutions for θ. We shall adopt the convention of choosing the acute angle θ that satisfies (7). Then we have the following two possibilities:

If $\cot 2\theta > 0$, then $0 < 2\theta \leq \pi/2$ so that $0 < \theta \leq \pi/4$.

If $\cot 2\theta < 0$, then $\pi/2 < 2\theta < \pi$ so that $\pi/4 < \theta < \pi/2$.

Each of these results in a counterclockwise rotation of the axes through an acute angle θ.*

---

*Any rotation (clockwise or counterclockwise) through an angle θ that satisfies $\cot 2\theta = (A - C)/B$ will eliminate the $x'y'$-term. However, the final form of the transformed equation may be different (but be equivalent), depending on the angle chosen.

Warning: Be careful if you use a calculator to solve equation (7).

1. If $\cot 2\theta = 0$, then $2\theta = \pi/2$ and $\theta = \pi/4$.
2. If $\cot 2\theta \neq 0$, first find $\cos 2\theta$. Then use the inverse cosine function key(s) to obtain 2θ, $0 < 2\theta < \pi$. Finally, divide by 2 to obtain the correct acute angle θ.

E X A M P L E 3

Discussing an Equation Using a Rotation of Axes

Discuss the equation: $x^2 + \sqrt{3}xy + 2y^2 - 10 = 0$

Solution

We need to solve the equation for y in order to graph the equation using a graphing utility. Rearranging the terms we observe the equation is quadratic in the variable y: $2y^2 + \sqrt{3}xy + (x^2 - 10) = 0$. We can solve the equation for y using the quadratic formula with $a = 2$, $b = \sqrt{3}x$, and $c = x^2 - 10$.

$$y_1 = \frac{-\sqrt{3}x + \sqrt{(\sqrt{3}x)^2 - 4(2)(x^2 - 10)}}{2(2)} = \frac{-\sqrt{3}x + \sqrt{-5x^2 + 80}}{4}$$

and

$$y_2 = \frac{-\sqrt{3}x - \sqrt{(\sqrt{3}x)^2 - 4(2)(x^2 - 10)}}{2(2)} = \frac{-\sqrt{3}x - \sqrt{-5x^2 + 80}}{4}$$

Figure 56(a) shows the graph of y_1 and y_2.

We now proceed to analyze the equation. Since an xy-term is present, we must rotate the axes. Using $A = 1$, $B = \sqrt{3}$, and $C = 2$ in equation (7), the appropriate acute angle θ through which to rotate the axes satisfies the equation

$$\cot 2\theta = \frac{A - C}{B} = \frac{-1}{\sqrt{3}} = \frac{-\sqrt{3}}{3}, \qquad 0° < 2\theta < 180°$$

Since $\cot 2\theta = -\sqrt{3}/3$, we find $2\theta = 120°$, so $\theta = 60°$. Using $\theta = 60°$ in rotation formulas (5), we find

$$x = \frac{1}{2}x' - \frac{\sqrt{3}}{2}y' = \frac{1}{2}(x' - \sqrt{3}y')$$

$$y = \frac{\sqrt{3}}{2}x' + \frac{1}{2}y' = \frac{1}{2}(\sqrt{3}x' + y')$$

Substituting these values into the original equation and simplifying, we have

$$x^2 + \sqrt{3}xy + 2y^2 - 10 = 0$$

$$\frac{1}{4}(x' - \sqrt{3}y')^2 + \sqrt{3}\left[\frac{1}{2}(x' - \sqrt{3}y')\right]\left[\frac{1}{2}(\sqrt{3}x' + y')\right] + 2\left[\frac{1}{4}(\sqrt{3}x' + y')^2\right] = 10$$

Multiply both sides by 4 and expand to obtain

$$x'^2 - 2\sqrt{3}x'y' + 3y'^2 + \sqrt{3}(\sqrt{3}x'^2 - 2x'y' - \sqrt{3}y'^2) + 2(3x'^2 + 2\sqrt{3}x'y' + y'^2) = 40$$

$$10x'^2 + 2y'^2 = 40$$

$$\frac{x'^2}{4} + \frac{y'^2}{20} = 1$$

This is the equation of an ellipse with center at $(0, 0)$ and major axis along the y'-axis. The vertices are at $(0, \pm 2\sqrt{5})$ on the y'-axis. See Figure 56(b) for the graph drawn by hand. ∎

FIGURE 56

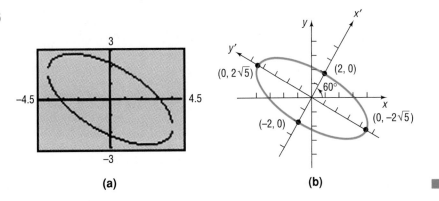

(a)   (b)

■   Now work Problem 21.

In Example 3, the acute angle θ through which to rotate the axes was easy to find because of the numbers we used in the given equation. In general, the equation $\cot 2\theta = (A - C)/B$ will not have such a "nice" solution. As the next example shows, we can still find the appropriate rotation formulas without using a calculator approximation by applying half-angle formulas.

E X A M P L E   4

Discussing an Equation Using a Rotation of Axes

Discuss the equation: $4x^2 - 4xy + y^2 + 5\sqrt{5}x + 5 = 0$

Solution

We need to solve the equation for y in order to graph the equation using a graphing utility. Rearranging the terms we observe the equation is quadratic in the variable y: $y^2 - 4xy + (4x^2 + 5\sqrt{5}x + 5) = 0$. We can solve the equation for y using the quadratic formula with $a = 1$, $b = -4x$, and $c = 4x^2 + 5\sqrt{5}x + 5$.

$$y_1 = \frac{-(-4x) + \sqrt{(-4x)^2 - 4(1)(4x^2 + 5\sqrt{5}x + 5)}}{2(1)} = 2x + \sqrt{-5(\sqrt{5}x + 1)}$$

$$y_2 = \frac{-(-4x) - \sqrt{(-4x)^2 - 4(1)(4x^2 + 5\sqrt{5}x + 5)}}{2(1)} = 2x - \sqrt{-5(\sqrt{5}x + 1)}$$

Figure 57(a) shows the graph of y_1 and y_2.

We now proceed to analyze the equation. Letting $A = 4$, $B = -4$, and $C = 1$ in equation (7), the appropriate angle θ through which to rotate the axes satisfies

$$\cot 2\theta = \frac{A - C}{B} = \frac{3}{-4}$$

In order to use rotation formulas (5), we need to know the values of $\sin \theta$ and $\cos \theta$. Since we seek an acute angle θ, we know that $\sin \theta > 0$ and $\cos \theta > 0$. Thus, we use the half-angle formulas in the form

$$\sin \theta = \sqrt{\frac{1 - \cos 2\theta}{2}} \qquad \cos \theta = \sqrt{\frac{1 + \cos 2\theta}{2}}$$

Now we need to find the value of $\cos 2\theta$. Since $\cot 2\theta = -\frac{3}{4}$ and $\pi/2 < 2\theta < \pi$, it follows that $\cos 2\theta = -\frac{3}{5}$. Thus,

$$\sin \theta = \sqrt{\frac{1 - \cos 2\theta}{2}} = \sqrt{\frac{1 - (-\frac{3}{5})}{2}} = \sqrt{\frac{4}{5}} = \frac{2}{\sqrt{5}} = \frac{2\sqrt{5}}{5}$$

$$\cos \theta = \sqrt{\frac{1 + \cos 2\theta}{2}} = \sqrt{\frac{1 + (-\frac{3}{5})}{2}} = \sqrt{\frac{1}{5}} = \frac{1}{\sqrt{5}} = \frac{\sqrt{5}}{5}$$

With these values, rotation formulas (5) give us

$$x = \frac{\sqrt{5}}{5}x' - \frac{2\sqrt{5}}{5}y' = \frac{\sqrt{5}}{5}(x' - 2y')$$

$$y = \frac{2\sqrt{5}}{5}x' + \frac{\sqrt{5}}{5}y' = \frac{\sqrt{5}}{5}(2x' + y')$$

FIGURE 57

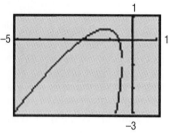

(a)

(b)

Substituting these values in the original equation and simplifying, we obtain

$$4x^2 - 4xy + y^2 + 5\sqrt{5}x + 5 = 0$$

$$4\left[\frac{\sqrt{5}}{5}(x' - 2y')\right]^2 - 4\left[\frac{\sqrt{5}}{5}(x' - 2y')\right]\left[\frac{\sqrt{5}}{5}(2x' + y')\right]$$

$$+ \left[\frac{\sqrt{5}}{5}(2x' + y')\right]^2 + 5\sqrt{5}\left[\frac{\sqrt{5}}{5}(x' - 2y')\right] = -5$$

Multiply both sides by 5 and expand to obtain

$$4(x'^2 - 4x'y' + 4y'^2) - 4(2x'^2 - 3x'y' - 2y'^2)$$
$$+ 4x'^2 + 4x'y' + y'^2 + 25(x' - 2y') = -25$$
$$25y'^2 - 50y' + 25x' = -25$$
$$y'^2 - 2y' + x' = -1$$
$$y'^2 - 2y' + 1 = -x' \quad \text{Complete the}$$
$$(y' - 1)^2 = -x' \quad \text{square in } y'.$$

This is the equation of a parabola with vertex at $(0, 1)$ in the $x'y'$-plane. The axis of symmetry is parallel to the x'-axis. Using a calculator to solve $\sin \theta = 2\sqrt{5}/5$, we find that $\theta \approx 63.4°$. See Figure 57(b) for the graph drawn by hand. ■

■ Now work Problem 27.

Identifying Conics without a Rotation of Axes

Suppose that we are required only to identify (rather than discuss) an equation of the form

$$Ax^2 + Bxy + Cy^2 + Dx + Ey + F = 0, \quad B \neq 0 \tag{8}$$

If we apply rotation formulas (5) to this equation, we obtain an equation of the form

$$A'x'^2 + B'x'y' + C'y'^2 + D'x' + E'y' + F' = 0 \tag{9}$$

where A', B', C', D', E', and F' can be expressed in terms of A, B, C, D, E, F, and the angle θ of rotation (see Problem 43 at the end of this section). It can be shown that the value of $B^2 - 4AC$ in equation (8) and the value of $B'^2 - 4A'C'$ in equation (9) are equal no matter what angle θ of rotation is chosen (see Problem 45). In particular, if the angle θ of rotation satisfies equation (7), then $B' = 0$ in equation (9), and $B^2 - 4AC = -4A'C'$. Since equation (9) then has the form of equation (2),

$$A'x'^2 + C'y'^2 + D'x' + E'y' + F' = 0$$

we can identify it without completing the squares, as we did in the beginning of this section. In fact, now we can identify the conic described by any equation of the form of equation (8) without a rotation of axes.

Theorem
Identifying Conics without a Rotation of Axes

Except for degenerate cases, the equation

$$Ax^2 + Bxy + Cy^2 + Dx + Ey + F = 0$$

(a) Defines a parabola if $B^2 - 4AC = 0$.
(b) Defines an ellipse (or a circle) if $B^2 - 4AC < 0$.
(c) Defines a hyperbola if $B^2 - 4AC > 0$.   ■

You are asked to prove this theorem in Problem 46.

E X A M P L E   5

Identifying a Conic without a Rotation of Axes
Identify the equation: $8x^2 - 12xy + 17y^2 - 4\sqrt{5}x - 2\sqrt{5}y - 15 = 0$

Solution   Here, $A = 8$, $B = -12$, and $C = 17$, so that $B^2 - 4AC = -400$. Since $B^2 - 4AC < 0$, the equation defines an ellipse.   ■

■   Now work Problem 33.

6.5

Exercise 6.5

In Problems 1–10, identify each equation without completing the squares.

1. $x^2 + 4x + y + 3 = 0$
2. $2y^2 - 3y + 3x = 0$
3. $6x^2 + 3y^2 - 12x + 6y = 0$
4. $2x^2 + y^2 - 8x + 4y + 2 = 0$
5. $3x^2 - 2y^2 + 6x + 4 = 0$
6. $4x^2 - 3y^2 - 8x + 6y + 1 = 0$
7. $2y^2 - x^2 - y + x = 0$
8. $y^2 - 8x^2 - 2x - y = 0$
9. $x^2 + y^2 - 8x + 4y = 0$
10. $2x^2 + 2y^2 - 8x + 8y = 0$

In Problems 11–20, determine the appropriate rotation formulas to use so that the new equation contains no xy-term.

11. $x^2 + 4xy + y^2 - 3 = 0$
12. $x^2 - 4xy + y^2 - 3 = 0$
13. $5x^2 + 6xy + 5y^2 - 8 = 0$
14. $3x^2 - 10xy + 3y^2 - 32 = 0$
15. $13x^2 - 6\sqrt{3}xy + 7y^2 - 16 = 0$
16. $11x^2 + 10\sqrt{3}xy + y^2 - 4 = 0$
17. $4x^2 - 4xy + y^2 - 8\sqrt{5}x - 16\sqrt{5}y = 0$
18. $x^2 + 4xy + 4y^2 + 5\sqrt{5}y + 5 = 0$
19. $25x^2 - 36xy + 40y^2 - 12\sqrt{13}x - 8\sqrt{13}y = 0$
20. $34x^2 - 24xy + 41y^2 - 25 = 0$

In Problems 21–32, graph the equation using a graphing utility. Rotate the axes so that the new equation contains no xy-term. Discuss and, by hand, graph the new equation. (Refer to Problems 11–20 for Problems 21–30.)

21. $x^2 + 4xy + y^2 - 3 = 0$

22. $x^2 - 4xy + y^2 - 3 = 0$

23. $5x^2 + 6xy + 5y^2 - 8 = 0$

24. $3x^2 - 10xy + 3y^2 - 32 = 0$

25. $13x^2 - 6\sqrt{3}xy + 7y^2 - 16 = 0$

26. $11x^2 + 10\sqrt{3}xy + y^2 - 4 = 0$

27. $4x^2 - 4xy + y^2 - 8\sqrt{5}x - 16\sqrt{5}y = 0$

28. $x^2 + 4xy + 4y^2 + 5\sqrt{5}y + 5 = 0$

29. $25x^2 - 36xy + 40y^2 - 12\sqrt{13}x - 8\sqrt{13}y = 0$

30. $34x^2 - 24xy + 41y^2 - 25 = 0$

31. $16x^2 + 24xy + 9y^2 - 130x + 90y = 0$

32. $16x^2 + 24xy + 9y^2 - 60x + 80y = 0$

In Problems 33–42, identify each equation without applying a rotation of axes.

33. $x^2 + 3xy - 2y^2 + 3x + 2y + 5 = 0$

34. $2x^2 - 3xy + 4y^2 + 2x + 3y - 5 = 0$

35. $x^2 - 7xy + 3y^2 - y - 10 = 0$

36. $2x^2 - 3xy + 2y^2 - 4x - 2 = 0$

37. $9x^2 + 12xy + 4y^2 - x - y = 0$

38. $10x^2 + 12xy + 4y^2 - x - y + 10 = 0$

39. $10x^2 - 12xy + 4y^2 - x - y - 10 = 0$

40. $4x^2 + 12xy + 9y^2 - x - y = 0$

41. $3x^2 - 2xy + y^2 + 4x + 2y - 1 = 0$

42. $3x^2 + 2xy + y^2 + 4x - 2y + 10 = 0$

In Problems 43–46, apply rotation formulas (5) to

$$Ax^2 + Bxy + Cy^2 + Dx + Ey + F = 0$$

to obtain the equation

$$A'x'^2 + B'x'y + C'y'^2 + D'x' + E'y' + F' = 0$$

43. Express A', B', C', D', E', and F' in terms of A, B, C, D, E, F, and the angle θ of rotation.

44. Show that $A + C = A' + C'$, and thus show that $A + C$ is **invariant;** that is, its value does not change under a rotation of axes.

45. Refer to Problem 44. Show that $B^2 - 4AC$ is invariant.

46. Prove that, except for degenerate cases, the equation

$$Ax^2 + Bxy + Cy^2 + Dx + Ey + F = 0$$

(a) Defines a parabola if $B^2 - 4AC = 0$.

(b) Defines an ellipse (or a circle) if $B^2 - 4AC < 0$.

(c) Defines a hyperbola if $B^2 - 4AC > 0$.

47. Use rotation formulas (5) to show that distance is invariant under a rotation of axes. That is, show that the distance from $P_1 = (x_1, y_1)$ to $P_2 = (x_2, y_2)$ in the xy-plane equals the distance from $P_1 = (x'_1, y'_1)$ to $P_2 = (x'_2, y'_2)$ in the $x'y'$-plane.

48. Show that the graph of the equation $x^{1/2} + y^{1/2} = a^{1/2}$ is part of the graph of a parabola.

49. Formulate a strategy for discussing and graphing an equation of the form $Ax^2 + Cy^2 + Dx + Ey + F = 0$. How does your strategy change if the equation is of the form $Ax^2 + Bxy + Cy^2 + Dx + Ey + F = 0$?

6.6

Polar Equations of Conics

In Sections 6.2, 6.3, and 6.4, we gave separate definitions for the parabola, ellipse, and hyperbola based on geometric properties and the distance formula. In this section, we present an alternative definition that simultaneously defines all these conics. As we shall see, this approach is well suited to polar coordinate representation. (Refer to Section 5.1.)

Conic

Let D denote a fixed line called the **directrix;** let F denote a fixed point called the **focus,** which is not on D; and let e be a fixed positive number called the **eccentricity.** A **conic** is the set of points P in the plane such that the ratio of the distance from F to P to the distance from P to D equals e. Thus, a conic is the collection of points P for which

$$\frac{d(F, P)}{d(P, D)} = e \qquad (1)$$

If $e = 1$, the conic is a **parabola.**

If $e < 1$, the conic is an **ellipse.**

If $e > 1$, the conic is a **hyperbola.**

Observe that if $e = 1$ the definition of a parabola in equation (1) is exactly the same as the definition used earlier in Section 6.2.

In the case of an ellipse, the **major axis** is a line through the focus perpendicular to the directrix. In the case of a hyperbola, the **transverse axis** is a line through the focus perpendicular to the directrix. For both an ellipse and a hyperbola, the eccentricity e satisfies

$$e = \frac{c}{a} \qquad (2)$$

where c is the distance from the center to the focus and a is the distance from the center to a vertex.

Just as we did earlier using rectangular coordinates, we derive equations for the conics in polar coordinates by choosing a convenient position for the focus F and the directrix D. The focus F is positioned at the pole, and the directrix D is either parallel to the polar axis or perpendicular to it.

Suppose that we start with the directrix D perpendicular to the polar axis at a distance p units to the left of the pole (the focus F). See Figure 58.

If $P = (r, \theta)$ is any point on the conic, then, by equation (1),

$$\frac{d(F, P)}{d(D, P)} = e \quad \text{or} \quad d(F, P) = e \cdot d(D, P) \qquad (3)$$

Now we use the point Q obtained by dropping the perpendicular from P to the polar axis to calculate $d(D, P)$:

$$d(D, P) = p + d(O, Q) = p + r \cos \theta$$

Using this expression and the fact that $d(F, P) = d(O, P) = r$ in equation (3), we get

$$d(F, P) = e \cdot d(D, P)$$
$$r = e(p + r \cos \theta)$$
$$r = ep + er \cos \theta$$
$$r - er \cos \theta = ep$$
$$r(1 - e \cos \theta) = ep$$
$$r = \frac{ep}{1 - e \cos \theta}$$

FIGURE 58

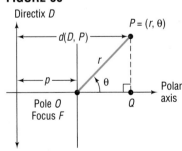

Directix D

$P = (r, \theta)$

$d(D, P)$

r

p

θ

Pole O
Focus F

Q

Polar axis

Theorem

Polar Equation of a Conic; Focus at Pole; Directrix Perpendicular to Polar Axis a Distance p to the Left of the Pole

The polar equation of a conic with focus at the pole and directrix perpendicular to the polar axis at a distance p to the left of the pole is

$$r = \frac{ep}{1 - e \cos \theta} \qquad (4)$$

where e is the eccentricity of the conic. ∎

E X A M P L E   1

Discussing and Graphing the Polar Equation of a Conic

Discuss and graph the equation: $r = \dfrac{4}{2 - \cos \theta}$

Solution   Figure 59(a) shows the graph of the equation using a graphing utility in POLar mode with $\theta_{min} = 0$, $\theta_{max} = 2\pi$, and $\theta_{step} = \pi/24$. We now proceed to discuss the equation.

The given equation is not quite in the form of equation (4), since the first term in the denominator is 2 instead of 1. Thus, we divide the numerator and denominator by 2 to obtain

$$r = \frac{2}{1 - \frac{1}{2} \cos \theta}$$

This equation is in the form of equation (4), with

$$e = \tfrac{1}{2} \quad \text{and} \quad ep = \tfrac{1}{2}p = 2$$

Thus, $e = \frac{1}{2}$ and $p = 4$. We conclude that the conic is an ellipse, since $e = \frac{1}{2} < 1$. One focus is at the pole, and the directrix is perpendicular to the polar axis, a distance of 4 units to the left of the pole. It follows that the major axis is along the polar axis. To find the vertices, we let $\theta = 0$ and $\theta = \pi$. Thus, the vertices of the ellipse are $(4, 0)$ and $(\frac{4}{3}, \pi)$. At the midpoint of the vertices, we locate the center of the ellipse at $(\frac{4}{3}, 0)$. [Do you see why? The vertices $(4, 0)$ and $(\frac{4}{3}, \pi)$ in polar coordinates are $(4, 0)$ and $(-\frac{4}{3}, 0)$ in rectangular coordinates. The midpoint in rectangular coordinates is $(\frac{4}{3}, 0)$, which is also $(\frac{4}{3}, 0)$ in polar coordinates.] Thus, $a =$ distance from the center to a vertex $= \frac{8}{3}$. Using $a = \frac{8}{3}$ and $e = \frac{1}{2}$ in equation (2), $e = c/a$, we find $c = \frac{4}{3}$. Finally, using $a = \frac{8}{3}$ and $c = \frac{4}{3}$ in $b^2 = a^2 - c^2$, we have

$$b^2 = a^2 - c^2 = \frac{64}{9} - \frac{16}{9} = \frac{48}{9}$$

$$b = \frac{4\sqrt{3}}{3}$$

Figure 59(b) shows the graph drawn by hand.

FIGURE 59

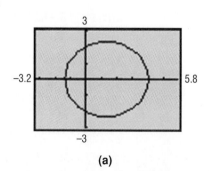

(a)

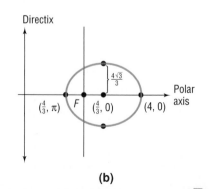

(b)

Exploration: Graph $r = 4/(2 + \cos \theta)$ and compare the result with Figure 59(a). What do you conclude? Clear the screen and graph $r = 4/(2 - \sin \theta)$ and then $r = 4/(2 + \sin \theta)$. Compare each of these graphs with Figure 59(a). What do you conclude? ◼

◼ Now work Problem 5.

Equation (4) was obtained under the assumption that the directrix was perpendicular to the polar axis at a distance p units to the left of the pole. A similar derivation (see Problem 37), in which the directrix is perpendicular to the polar axis at a distance p units to the right of the pole, results in the equation

$$r = \frac{ep}{1 + e \cos \theta}$$

In Problems 38 and 39 you are asked to derive the polar equations of conics with focus at the pole and directrix parallel to the polar axis. Table 5 summarizes the polar equations of conics.

TABLE 5   POLAR EQUATIONS OF CONICS (FOCUS AT THE POLE, ECCENTRICITY e)

EQUATION	DESCRIPTION
(a) $r = \dfrac{ep}{1 - e \cos \theta}$	Directrix is perpendicular to the polar axis at a distance p units to the left of the pole.
(b) $r = \dfrac{ep}{1 + e \cos \theta}$	Directrix is perpendicular to the polar axis at a distance p units to the right of the pole.
(c) $r = \dfrac{ep}{1 + e \sin \theta}$	Directrix is parallel to the polar axis at a distance p units above the pole.
(d) $r = \dfrac{ep}{1 - e \sin \theta}$	Directrix is parallel to the polar axis at a distance p units below the pole.

ECCENTRICITY

If $e = 1$, the conic is a parabola; the axis of symmetry is perpendicular to the directrix.
If $e < 1$, the conic is an ellipse; the major axis is perpendicular to the directrix.
If $e > 1$, the conic is a hyperbola; the transverse axis is perpendicular to the directrix.

E X A M P L E   2

Discussing and Graphing the Polar Equation of a Conic

Discuss and graph the equation: $\quad r = \dfrac{6}{3 + 3 \sin \theta}$

Solution

Figure 60(a) shows the graph of the equation using a graphing utility in POLar mode with $\theta_{\min} = 0$, $\theta_{\max} = 2\pi$, and $\theta_{step} = \pi/24$. We now proceed to discuss the equation. To place the equation in proper form, we divide the numerator and denominator by 3 to get

$$r = \frac{2}{1 + \sin \theta}$$

Referring to Table 5, we conclude that this equation is in the form of equation (c) with

$$e = 1 \quad \text{and} \quad ep = 2$$

Thus, $e = 1$ and $p = 2$. The conic is a parabola with focus at the pole. The directrix is parallel to the polar axis at a distance 2 units above the pole; the axis

FIGURE 60

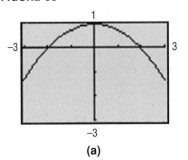

(a)

of symmetry is perpendicular to the polar axis. The vertex of the parabola is at $(1, \pi/2)$. (Do you see why?) See Figure 60(b) for the graph drawn by hand. Notice that we plotted two additional points, $(2, 0)$ and $(2, \pi)$, to assist in graphing.

FIGURE 60

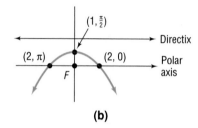

(b)

■   Now work Problem 7.

E X A M P L E   3   *Discussing and Graphing the Polar Equation of a Conic*

Discuss and graph the equation: $r = \dfrac{3}{1 + 3 \cos \theta}$

Solution   Figure 61(a) and (b) show the graph of the equation using a graphing utility in POLar mode with $\theta_{\min} = 0$, $\theta_{\max} = 2\pi$, and $\theta_{step} = \pi/24$, using both dot mode and connected mode. Notice the extraneous asymptotes in the connected mode. We now proceed to discuss the equation.

This equation is in the form of equation (b) in Table 5. We conclude that

FIGURE 61

$$e = 3 \quad \text{and} \quad ep = 3p = 3$$

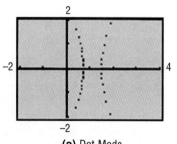

(a) Dot Mode

Thus, $e = 3$ and $p = 1$. This is the equation of a hyperbola with a focus at the pole. The directrix is perpendicular to the polar axis, 1 unit to the right of the pole. The transverse axis is along the polar axis. To find the vertices, we let $\theta = 0$ and $\theta = \pi$. Thus, the vertices are $(\frac{3}{4}, 0)$ and $(-\frac{3}{2}, \pi)$. The center is at the midpoint of $(\frac{3}{4}, 0)$ and $(-\frac{3}{2}, \pi)$, which is $(\frac{9}{8}, 0)$. Thus, $c = $ distance from the center to a focus $= \frac{9}{8}$. Since $e = 3$, it follows from equation (2), $e = c/a$, that $a = \frac{3}{8}$. Finally, using $a = \frac{3}{8}$ and $c = \frac{9}{8}$ in $b^2 = c^2 - a^2$, we find

$$b^2 = c^2 - a^2 = \frac{81}{64} - \frac{9}{64} = \frac{72}{64} = \frac{9}{8}$$

$$b = \frac{3}{2\sqrt{2}} = \frac{3\sqrt{2}}{4}$$

Figure 61(c) shows the graph drawn by hand. Notice that we plotted two additional points, $(3, \pi/2)$ and $(3, 3\pi/2)$, on the left branch and used symmetry to obtain the right branch. The asymptotes of this hyperbola were found in the usual way by constructing the rectangle shown.

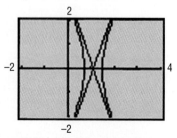

(b) Connected Mode

■   Now work Problem 11.

FIGURE 61

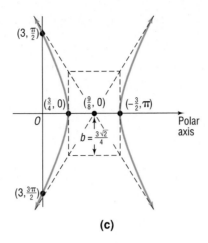

$(3, \frac{\pi}{2})$

$(\frac{3}{4}, 0)$ $(\frac{9}{8}, 0)$ $(-\frac{3}{2}, \pi)$

0

Polar axis

$b = \frac{3\sqrt{2}}{4}$

$(3, \frac{3\pi}{2})$

(c)

E X A M P L E   4   *Converting a Polar Equation to a Rectangular Equation*

Convert the polar equation

$$r = \frac{1}{3 - 3 \cos \theta}$$

to a rectangular equation.

Solution   The strategy here is first to rearrange the equation and square each side, before using the transformation equations:

$$r = \frac{1}{3 - 3 \cos \theta}$$

$$3r - 3r \cos \theta = 1$$

$$3r = 1 + 3r \cos \theta \qquad \text{Rearrange the equation.}$$

$$9r^2 = (1 + 3r \cos \theta)^2 \qquad \text{Square each side.}$$

$$9(x^2 + y^2) = (1 + 3x)^2 \qquad \text{Use the transformation equations.}$$

$$9x^2 + 9y^2 = 9x^2 + 6x + 1$$

$$9y^2 = 6x + 1$$

This is the equation of a parabola in rectangular coordinates.

■   Now work Problem 19.

6.6

Exercise 6.6

In Problems 1–6, identify the conic that each polar equation represents. Also, give the position of the directrix.

1. $r = \dfrac{1}{1 + \cos \theta}$

2. $r = \dfrac{3}{1 - \sin \theta}$

3. $r = \dfrac{4}{2 - 3 \sin \theta}$

4. $r = \dfrac{2}{1 + 2 \cos \theta}$

5. $r = \dfrac{3}{4 - 2 \cos \theta}$

6. $r = \dfrac{6}{8 + 2 \sin \theta}$

In Problems 7–18, graph each equation using a graphing utility. Discuss each equation and graph it by hand.

7. $r = \dfrac{1}{1 + \cos\theta}$

8. $r = \dfrac{3}{1 - \sin\theta}$

9. $r = \dfrac{8}{4 + 3\sin\theta}$

10. $r = \dfrac{10}{5 + 4\cos\theta}$

11. $r = \dfrac{9}{3 - 6\cos\theta}$

12. $r = \dfrac{12}{4 + 8\sin\theta}$

13. $r = \dfrac{8}{2 - \sin\theta}$

14. $r = \dfrac{8}{2 + 4\cos\theta}$

15. $r(3 - 2\sin\theta) = 6$

16. $r(2 - \cos\theta) = 2$

17. $r = \dfrac{6\sec\theta}{2\sec\theta - 1}$

18. $r = \dfrac{3\csc\theta}{\csc\theta - 1}$

In Problems 19–30, convert each polar equation to a rectangular equation.

19. $r = \dfrac{1}{1 + \cos\theta}$

20. $r = \dfrac{3}{1 - \sin\theta}$

21. $r = \dfrac{8}{4 + 3\sin\theta}$

22. $r = \dfrac{10}{5 + 4\cos\theta}$

23. $r = \dfrac{9}{3 - 6\cos\theta}$

24. $r = \dfrac{12}{4 + 8\sin\theta}$

25. $r = \dfrac{8}{2 - \sin\theta}$

26. $r = \dfrac{8}{2 + 4\cos\theta}$

27. $r(3 - 2\sin\theta) = 6$

28. $r(2 - \cos\theta) = 2$

29. $r = \dfrac{6\sec\theta}{2\sec\theta - 1}$

30. $r = \dfrac{3\csc\theta}{\csc\theta - 1}$

In Problems 31–36, find a polar equation for each conic. For each, a focus is at the pole.

31. $e = 1$; directrix is parallel to the polar axis 1 unit above the pole

32. $e = 1$; directrix is parallel to the polar axis 2 units below the pole

33. $e = \frac{4}{5}$; directrix is perpendicular to the polar axis 3 units to the left of the pole

34. $e = \frac{2}{3}$; directrix is parallel to the polar axis 3 units above the pole

35. $e = 6$; directrix is parallel to the polar axis 2 units below the pole

36. $e = 5$; directrix is perpendicular to the polar axis 5 units to the right of the pole

37. Derive equation (b) in Table 5: $\quad r = \dfrac{ep}{1 + e\cos\theta}$

38. Derive equation (c) in Table 5: $\quad r = \dfrac{ep}{1 + e\sin\theta}$

39. Derive equation (d) in Table 5: $\quad r = \dfrac{ep}{1 - e\sin\theta}$

40. The planet Mercury travels around the Sun in an elliptical orbit given approximately by

$$r = \frac{(3.442)10^7}{1 - 0.206\cos\theta}$$

where r is measured in miles and the Sun is at the pole. Find the distance from Mercury to the Sun at *aphelion* (greatest distance from the Sun) and at *perihelion* (shortest distance from the Sun). See the figure on the right. Use the aphelion and perihelion to graph the orbit of Mercury using a graphing utility.

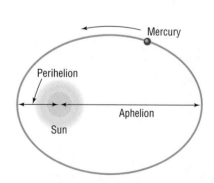

6.7

Plane Curves and Parametric Equations

Equations of the form $y = f(x)$, where f is a function, have graphs that are intersected no more than once by any vertical line. The graphs of many of the conics and certain other, more complicated graphs do not have this characteristic. Yet each graph, like the graph of a function, is a collection of points (x, y) in the xy-plane; that is, each is a *plane curve*. In this section, we discuss another way of representing such graphs.

Plane Curve

Let $x = f(t)$ and $y = g(t)$, where f and g are two functions whose common domain is some interval I. The collection of points defined by

$$(x, y) = (f(t), g(t))$$

is called a **plane curve.** The equations

$$x = f(t) \qquad y = g(t)$$

where t is in I, are called **parametric equations** of the curve. The variable t is called a **parameter.**

Parametric equations are particularly useful in describing movement along a curve. Suppose that a curve is defined by the parametric equations

$$x = f(t) \qquad y = g(t)$$

where f and g are each defined over some interval I. For a given value of t in I, we can find the value of $x = f(t)$ and $y = g(t)$, thus obtaining a point (x, y) on the curve. In fact, as t varies over the interval I in some order, say from left to right, successive values of t give rise to a directed movement along the curve. That is, as t varies over I in some order, the curve is traced out in a certain direction by the corresponding succession of points (x, y). Let's look at an example.

E X A M P L E   1

Discussing a Curve Defined by Parametric Equations

Discuss the curve defined by the parametric equations

$$x = 3t^2 \qquad y = 2t, \qquad -2 \le t \le 2 \tag{1}$$

Solution

For each number t, $-2 \le t \le 2$, there corresponds a number x and a number y. For example, when $t = -2$, then $x = 12$ and $y = -4$. When $t = 0$, then $x = 0$ and $y = 0$. Indeed, we can set up a table listing various choices of the parameter t and the corresponding values for x and y, as shown in Table 6. Plotting these points and connecting them with a smooth curve leads to Figure 62.

TABLE 6

t	x	y	(x, y)
−2	12	−4	(12, −4)
−1	3	−2	(3, −2)
0	0	0	(0, 0)
1	3	2	(3, 2)
2	12	4	(12, 4)

FIGURE 62

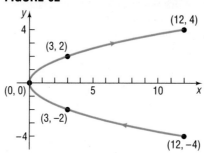

Notice the arrows on the curve in Figure 62. They indicate the direction, or **orientation,** of the curve for increasing values of the parameter t.

Most graphing utilities have the capability of graphing parametric equations. The following steps are usually required in order to obtain the graph of parametric equations. Check your owner's manual to see how yours works.

Graphing Parametric Equations Using a Graphing Utility

STEP 1: Set the mode to PARametric. Enter $x(t)$ and $y(t)$.
STEP 2: Select the viewing rectangle. In addition to setting Xmin, Xmax, Xscl, etc., the viewing rectangle in parametric mode requires setting minimum and maximum values for the parameter t and an increment setting for t (Tstep).
STEP 3: Execute

E X A M P L E   2   *Graphing a Curve Defined by Parametric Equations Using a Graphing Utility*

Graph the curve defined by the parametric equations

$$x = 3t^2 \qquad y = 2t \qquad -2 \le t \le 2$$

Solution   STEP 1: Enter the equations $x(t) = 3t^2$, $y(t) = 2t$ with the graphing utility in PARametric mode.
STEP 2: Select the viewing rectangle. The interval I is $-2 \le t \le 2$, so we select the following viewing rectangle:

$$\text{Tmin} = -2$$
$$\text{Tmax} = 2$$
$$\text{Tstep} = 0.1$$
$$\text{Xmin} = 0$$
$$\text{Xmax} = 12$$
$$\text{Xscl} = 1$$
$$\text{Ymin} = -4$$
$$\text{Ymax} = 4$$
$$\text{Yscl} = 1$$

We choose Xmin = 0 and Xmax = 12 because $x = 3t^2$ and $-2 \le t \le 2$. We choose Ymin = −4 and Ymax = 4 because $y = 2t$ and $-2 \le t \le 2$. We choose Tmin = −2 and Tmax = 2 because $-2 \le t \le 2$. Finally, the choice for Tstep will determine the number of points the graphing utility will plot. For example, with Tstep at 0.1, the graphing utility will evaluate x and y at −2, −1.9, −1.8, etc. The smaller the Tstep, the more points the graphing utility will plot. The reader is encouraged to experiment with different values of Tstep to see how the graph is affected.
STEP 3: Execute. Notice the direction the graph is drawn. This direction shows the orientation of the curve.

FIGURE 63

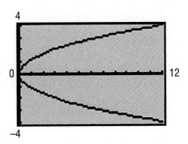

The graph shown in Figure 63 is complete.   ▪

▪ Now work Problem 1.

Parametric equations defining a curve are not unique. You can verify that the curve given in Examples 1 and 2 may also be defined by any of the following parametric equations:

$$x = \frac{3t^2}{4} \qquad y = t, \qquad -4 \leq t \leq 4$$

or

$$x = 3t^2 + 12t + 12 \qquad y = 2t + 4, \qquad -4 \leq t \leq 0$$

or

$$x = 3t^{2/3} \qquad y = 2\sqrt[3]{t}, \qquad -8 \leq t \leq 8$$

The curve given in Examples 1 and 2 should be familiar. To identify it accurately, we find the corresponding rectangular equation by eliminating the parameter t from the parametric equations (1) given in Example 1,

$$x = 3t^2 \qquad y = 2t, \qquad -2 \leq t \leq 2$$

Noting that we can readily solve for t in $y = 2t$, obtaining $t = y/2$, we substitute this expression in the other equation:

$$x = 3t^2 = 3\left(\frac{y}{2}\right)^2 = \frac{3y^2}{4}$$
$$\underset{t = \frac{y}{2}}{\uparrow}$$

This equation, $x = 3y^2/4$, is the equation of a parabola with vertex at $(0, 0)$ and axis along the x-axis.

Note that the parameterized curve defined by equation (1) and shown in Figure 62 (or 63) is only a part of the parabola $x = 3y^2/4$. Thus, the graph of the rectangular equation obtained by eliminating the parameter will, in general, contain more points than the original parameterized curve. Care must therefore be taken when a parameterized curve is sketched by hand after eliminating the parameter. Even so, the process of eliminating the parameter t of a parameterized curve in order to identify it accurately is sometimes a better approach than merely plotting points. However, the elimination process sometimes requires a little ingenuity.

E X A M P L E 3

Finding the Rectangular Equation of a Curve Defined Parametrically

Find the rectangular equation of the curve whose parametric equations are

$$x = a \cos t \qquad y = a \sin t$$

where $a > 0$ is a constant. Graph this curve, indicating its orientation.

Solution   The presence of sines and cosines in the parametric equations suggests that we use a Pythagorean identity. In fact, since

$$\cos t = \frac{x}{a} \qquad \sin t = \frac{y}{a}$$

we find that

$$\cos^2 t + \sin^2 t = 1$$

$$\left(\frac{x}{a}\right)^2 + \left(\frac{y}{a}\right)^2 = 1$$

$$x^2 + y^2 = a^2$$

FIGURE 64

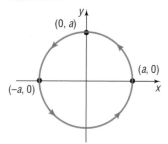

Thus, the curve is a circle with center at $(0, 0)$ and radius a. As the parameter t increases, say from $t = 0$ [the point $(a, 0)$] to $t = \pi/2$ [the point $(0, a)$] to $t = \pi$ [the point $(-a, 0)$], we see that the corresponding points are traced in a counterclockwise direction around the circle. Hence, the orientation is as indicated in Figure 64. ∎

▪ Now work Problem 13.

Let's discuss the curve in Example 3 further. The domain of each parametric equation is $-\infty < t < \infty$. Thus, the graph in Figure 64 is actually being repeated each time t increases by 2π. If we wanted the curve to consist of exactly 1 revolution in the counterclockwise direction, we could write

$$x = a \cos t \qquad y = a \sin t, \qquad 0 \le t \le 2\pi$$

This curve starts at $t = 0$ [the point $(a, 0)$] and, proceeding counterclockwise around the circle, ends at $t = 2\pi$ [also the point $(a, 0)$].

If we wanted the curve to consist of exactly three revolutions in the counterclockwise direction, we could write

$$x = a \cos t \qquad y = a \sin t, \qquad -2\pi \le t \le 4\pi$$

FIGURE 65

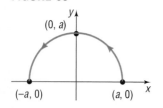

or

$$x = a \cos t \qquad y = a \sin t, \qquad 0 \le t \le 6\pi$$

or

$$x = a \cos t \qquad y = a \sin t, \qquad 2\pi \le t \le 8\pi$$

If we wanted the curve to consist of the upper semicircle of radius a with a counterclockwise orientation, we could write

$$x = a \cos t \qquad y = a \sin t, \qquad 0 \le t \le \pi$$

See Figure 65.

If we wanted the curve to consist of the left semicircle of radius a with a clockwise orientation, we could write

FIGURE 66

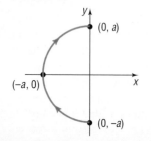

$$x = -a \sin t \qquad y = -a \cos t, \qquad 0 \le t \le \pi$$

See Figure 66.

Conic in polar coordinates	$\dfrac{d(F, P)}{d(P, D)} = e$	Parabola if $e = 1$ Ellipse if $e < 1$ Hyperbola if $e > 1$
Plane curve	$(x, y) = (f(t), g(t))$, t a parameter	
Polar equations of a conic	See Table 5.	

Definitions

Parabola	Set of points P in the plane for which $d(F, P) = d(P, D)$, where F is the focus and D is the directrix
Ellipse	Set of points P in the plane, the sum of whose distances from two fixed points (the foci) is a constant
Hyperbola	Set of points P in the plane, the difference of whose distances from two fixed points (the foci) is a constant

Formulas

Rotation formulas	$x = x' \cos\theta - y' \sin\theta$ $y = x' \sin\theta + y' \cos\theta$
Angle θ of rotation that eliminates the $x'y'$-term	$\cot 2\theta = \dfrac{A - C}{B}$

How To:

Find the vertex, focus, and directrix of a parabola given its equation

Graph a parabola given its equation

Find an equation of a parabola given certain information about the parabola

Find the center, foci, and vertices of an ellipse given its equation

Graph an ellipse given its equation

Find an equation of an ellipse given certain information about the ellipse

Find the center, foci, vertices, and asymptotes of a hyperbola given its equation

Graph a hyperbola given its equation

Find an equation of a hyperbola given certain information about the hyperbola

Identify conics without completing the square

Identify conics without a rotation of axes

Use rotation formulas to transform second-degree equations so that no xy-term is present

Identify and graph (by hand and using a graphing utility) conics given by a polar equation

Graph parametric equations by hand and using a graphing utility

Find the rectangular equation given the parametric equations

Fill-in-the-Blank Items

1. A(n) _____ is the collection of all points in the plane such that the distance from each point to a fixed point equals its distance to a fixed line.

2. A(n) _____ is the collection of all points in the plane the sum of whose distances from two fixed points is a constant.

3. A(n) _____ is the collection of all points in the plane the difference of whose distances from two fixed points is a constant.

4. For an ellipse, the foci lie on the _____ axis; for a hyperbola, the foci lie on the _____ axis.

5. For the ellipse $(x^2/9) + (y^2/16) = 1$, the major axis is along the _____.

6. The equations of the asymptotes of the hyperbola $(y^2/9) - (x^2/4) = 1$ are _____ and _____.

7. To transform the equation

$$Ax^2 + Bxy + Cy^2 + Dx + Ey + F = 0, \qquad B \neq 0$$

into one in x' and y' without an $x'y'$-term, rotate the axes through an acute angle θ that satisfies the equation _____.

8. The polar equation

$$r = \frac{8}{4 - 2\sin\theta}$$

is a conic whose eccentricity is _____. It is a(n) _____ whose directrix is _____ to the polar axis at a distance _____ units _____ the pole.

9. The parametric equations $x = 2\sin t$ and $y = 3\cos t$ represent a(n) _____.

True/False Items

T F **1.** On a parabola, the distance from any point to the focus equals the distance from that point to the directrix.
T F **2.** The foci of an ellipse lie on its minor axis.
T F **3.** The foci of a hyperbola lie on its transverse axis.
T F **4.** Hyperbolas always have asymptotes, and ellipses never have asymptotes.
T F **5.** A hyperbola never intersects its conjugate axis.
T F **6.** A hyperbola always intersects its transverse axis.
T F **7.** The equation $ax^2 + 6y^2 - 12y = 0$ defines an ellipse if $a > 0$.
T F **8.** The equation $3x^2 + bxy + 12y^2 = 10$ defines a parabola if $b = -12$.
T F **9.** If (r, θ) are polar coordinates, the equation $r = 2/(2 + 3\sin\theta)$ defines a hyperbola.
T F **10.** Parametric equations defining a curve are unique.

Review Exercises

In Problems 1–20, identify each equation. If it is a parabola, give its vertex, focus, and directrix; if it is an ellipse, give its center, vertices, and foci; if it is a hyperbola, give its center, vertices, foci, and asymptotes.

1. $y^2 = -16x$

2. $16x^2 = y$

3. $\dfrac{x^2}{25} - y^2 = 1$

4. $\dfrac{y^2}{25} - x^2 = 1$

5. $\dfrac{y^2}{25} + \dfrac{x^2}{16} = 1$

6. $\dfrac{x^2}{9} + \dfrac{y^2}{16} = 1$

7. $x^2 + 4y = 4$

8. $3y^2 - x^2 = 9$

9. $4x^2 - y^2 = 8$

10. $9x^2 + 4y^2 = 36$

11. $x^2 - 4x = 2y$

12. $2y^2 - 4y = x - 2$

13. $y^2 - 4y - 4x^2 + 8x = 4$

14. $4x^2 + y^2 + 8x - 4y + 4 = 0$

15. $4x^2 + 9y^2 - 16x - 18y = 11$

16. $4x^2 + 9y^2 - 16x + 18y = 11$

17. $4x^2 - 16x + 16y + 32 = 0$

18. $4y^2 + 3x - 16y + 19 = 0$

19. $9x^2 + 4y^2 - 18x + 8y = 23$

20. $x^2 - y^2 - 2x - 2y = 1$

In Problems 21–36, obtain an equation of the conic described. Graph the equation by hand.

21. Parabola; focus at $(-2, 0)$; directrix the line $x = 2$
22. Ellipse; center at $(0, 0)$; focus at $(0, 3)$; vertex at $(0, 5)$
23. Hyperbola; center at $(0, 0)$; focus at $(0, 4)$; vertex at $(0, -2)$
24. Parabola; vertex at $(0, 0)$; directrix the line $y = -3$
25. Ellipse; foci at $(-3, 0)$ and $(3, 0)$; vertex at $(4, 0)$

26. Hyperbola; vertices at $(-2, 0)$ and $(2, 0)$; focus at $(4, 0)$

27. Parabola; vertex at $(2, -3)$; focus at $(2, -4)$

28. Ellipse; center at $(-1, 2)$; focus at $(0, 2)$; vertex at $(2, 2)$

29. Hyperbola; center at $(-2, -3)$; focus at $(-4, -3)$; vertex at $(-3, -3)$

30. Parabola; focus at $(3, 6)$; directrix the line $y = 8$

31. Ellipse; foci at $(-4, 2)$ and $(-4, 8)$; vertex at $(-4, 10)$

32. Hyperbola; vertices at $(-3, 3)$ and $(5, 3)$; focus at $(7, 3)$

33. Center at $(-1, 2)$; $a = 3$; $c = 4$; transverse axis parallel to the x-axis

34. Center at $(4, -2)$; $a = 1$; $c = 4$; transverse axis parallel to the y-axis

35. Vertices at $(0, 1)$ and $(6, 1)$; asymptote the line $3y + 2x - 9 = 0$

36. Vertices at $(4, 0)$ and $(4, 4)$; asymptote the line $y + 2x - 10 = 0$

In Problems 37–46, identify each conic without completing the squares and without applying a rotation of axes.

37. $y^2 + 4x + 3y - 8 = 0$

38. $2x^2 - y + 8x = 0$

39. $x^2 + 2y^2 + 4x - 8y + 2 = 0$

40. $x^2 - 8y^2 - x - 2y = 0$

41. $9x^2 - 12xy + 4y^2 + 8x + 12y = 0$

42. $4x^2 + 4xy + y^2 - 8\sqrt{5}x + 16\sqrt{5}y = 0$

43. $4x^2 + 10xy + 4y^2 - 9 = 0$

44. $4x^2 - 10xy + 4y^2 - 9 = 0$

45. $x^2 - 2xy + 3y^2 + 2x + 4y - 1 = 0$

46. $4x^2 + 12xy - 10y^2 + x + y - 10 = 0$

In Problems 47–52, rotate the axes so that the new equation contains no xy-term. Discuss and graph the new equation by hand. Verify your result using a graphing utility.

47. $2x^2 + 5xy + 2y^2 - \frac{9}{2} = 0$

48. $2x^2 - 5xy + 2y^2 - \frac{9}{2} = 0$

49. $6x^2 + 4xy + 9y^2 - 20 = 0$

50. $x^2 + 4xy + 4y^2 + 16\sqrt{5}x - 8\sqrt{5}y = 0$

51. $4x^2 - 12xy + 9y^2 + 12x + 8y = 0$

52. $9x^2 - 24xy + 16y^2 + 80x + 60y = 0$

In Problems 53–58, identify the conic that each polar equation represents, and graph it by hand. Verify your result using a graphing utility.

53. $r = \dfrac{4}{1 - \cos\theta}$

54. $r = \dfrac{6}{1 + \sin\theta}$

55. $r = \dfrac{6}{2 - \sin\theta}$

56. $r = \dfrac{2}{3 + 2\cos\theta}$

57. $r = \dfrac{8}{4 + 8\cos\theta}$

58. $r = \dfrac{10}{5 + 20\sin\theta}$

In Problems 59–62, convert each polar equation to a rectangular equation.

59. $r = \dfrac{4}{1 - \cos\theta}$

60. $r = \dfrac{6}{2 - \sin\theta}$

61. $r = \dfrac{8}{4 + 8\cos\theta}$

62. $r = \dfrac{2}{3 + 2\cos\theta}$

In Problems 63–68, graph the curve whose parametric equations are given using a graphing utility. Find the rectangular equation of each curve. Graph the curve by hand and show its orientation.

63. $x = 4t - 2$, $y = 1 - t$; $-\infty < t < \infty$

64. $x = 2t^2 + 6$, $y = 5 - t$; $-\infty < t < \infty$

65. $x = 3\sin t$, $y = 4\cos t + 2$; $0 \le t \le 2\pi$

66. $x = \ln t$, $y = t^3$; $t > 0$

67. $x = \sec^2 t$, $y = \tan^2 t$; $0 \le t \le \pi/4$

68. $x = t^{3/2}$, $y = 2t + 4$; $t \ge 0$

69. Find an equation of the hyperbola whose foci are the vertices of the ellipse $4x^2 + 9y^2 = 36$ and whose vertices are the foci of this ellipse.

70. Find an equation of the ellipse whose foci are the vertices of the hyperbola $x^2 - 4y^2 = 16$ and whose vertices are the foci of this hyperbola.

71. Describe the collection of points in a plane so that the distance from each point to the point $(3, 0)$ is three-fourths of its distance from the line $x = \frac{16}{3}$.

72. Describe the collection of points in a plane so that the distance from each point to the point $(5, 0)$ is five-fourths of its distance from the line $x = \frac{16}{5}$.

73. *Mirrors* A mirror is shaped like a paraboloid of revolution. If a light source is located 1 foot from the base along the axis of symmetry and the opening is 2 feet across, how deep should the mirror be?

74. *Parabolic Arch Bridge* A bridge is built in the shape of a parabolic arch. The bridge has a span of 60 feet and a maximum height of 20 feet. Find the height of the arch at distances of 5, 10, and 20 feet from the center.

75. *Semielliptical Arch Bridge* A bridge is built in the shape of a semielliptical arch. The bridge has a span of 60 feet and a maximum height of 20 feet. Find the height of the arch at distances of 5, 10, and 20 feet from the center.

76. *Whispering Galleries* The figure shows the specifications for an elliptical ceiling in a hall designed to be a whispering gallery. Where in the hall are the foci located?

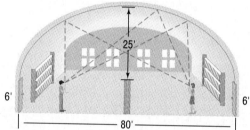

77. *LORAN* Two LORAN stations are positioned 150 miles apart along a straight shore.
  (a) A ship records a time difference of 0.00032 second between the LORAN signals. Set up an appropriate rectangular coordinate system to determine where the ship would reach shore if it were to follow the hyperbola corresponding to this time difference.
  (b) If the ship wants to enter a harbor located between the two stations 15 miles from the master station, what time difference should it be looking for?
  (c) If the ship is 20 miles offshore when the desired time difference is obtained, what is the exact location of the ship? [*Note:* The speed of each radio signal is 186,000 miles per second.]

78. Formulate a strategy for discussing and graphing an equation of the form $Ax^2 + Bxy + Cy^2 + Dx + Ey + F = 0$.

Chapter 7

PREPARING FOR THIS CHAPTER

Before getting started on this chapter, review the following concepts:

Radicals; Rational Exponents (Appendix, Section 3)
Functions (Sections 1.4 and 1.5)
Inverse functions (pp. 109–115)
Linear Curve Fitting (Appendix, Section 5)

EXPONENTIAL AND LOGARITHMIC FUNCTIONS

7.1 Exponential Functions
7.2 Logarithmic Functions
7.3 Properties of Logarithms
7.4 Logarithmic and Exponential Equations
7.5 Compound Interest
7.6 Growth and Decay
7.7 Nonlinear Curve Fitting
7.8 Logarithmic Scales
    Chapter Review

Preview   Alcohol and Driving

The concentration of alcohol in a person's blood is measurable. Recent medical research suggests that the risk R (given as a percent) of having an accident while driving a car can be modeled by the equation

$$R = 6e^{kx}$$

where x is the variable concentration of alcohol in the blood and k is a constant.

(a) *Suppose that a concentration of alcohol in the blood of 0.04 results in a 10% risk (R = 10) of an accident. Find the constant k in the equation. Graph $R = 6e^{kx}$.*

(b) *Using this value of k, what is the risk if the concentration is 0.17?*

(c) *Using the same value of k, what concentration of alcohol corresponds to a risk of 100%?*

(d) *If the law asserts that anyone with a risk of having an accident of 20% or more should not have driving privileges, at what concentration of alcohol in the blood should a driver be arrested and charged with DUI (Driving Under the Influence)?*
[Example 9 in Section 7.2.]

*F*unctions that can be expressed in terms of sums, differences, products, quotients, powers, or roots of polynomials are called **algebraic functions.** Functions that are not algebraic are termed **transcendental** (they transcend, or go beyond, algebraic functions). The trigonomet- ric functions are examples of transcendental functions.

In this chapter, we study two other transcendental functions: the *exponential* and *logarithmic functions.* These functions occur frequently in a wide variety of applications.

Exponential Functions

In the Appendix, Section 1, we gave a definition for raising a real number a to a rational power. Based on that discussion, we gave meaning to expressions of the form

$$a^r$$

where the base a is a positive real number and the exponent r is a rational number.

But what is the meaning of a^x, where the base a is a positive real number and the exponent x is an irrational number? Although a rigorous definition requires methods discussed in calculus, the basis for the definition is easy to follow: Select a rational number r that is formed by truncating (removing) all but a finite number of digits from the irrational number x. Then it is reasonable to expect that

$$a^x \approx a^r$$

For example, take the irrational number $\pi = 3.14159\ldots$. Then, an approximation to a^π is

$$a^\pi \approx a^{3.14}$$

where the digits after the hundredths position have been removed from the value for π. A better approximation would be

$$a^\pi \approx a^{3.14159}$$

where the digits after the hundred-thousandths position have been removed. Continuing in this way, we can obtain approximations to a^π to any desired degree of accuracy.

Graphing calculators can easily evaluate expressions of the form a^r as follows. Enter the base a, then press the caret key (^), enter the exponent y, and press enter.

E X A M P L E   1

Using a Graphing Calculator to Evaluate Powers of 2

Using a graphing calculator, evaluate:

(a) $2^{1.4}$ (b) $2^{1.41}$ (c) $2^{1.414}$ (d) $2^{1.4142}$ (e) $2^{\sqrt{2}}$

Solution   Figure 1 shows the solution to (a) using a TI-82 graphing calculator.

FIGURE 1

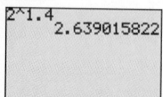

(a) $2^{1.4} \approx 2.639015822$ (b) $2^{1.41} \approx 2.657371628$
(c) $2^{1.414} \approx 2.664749765$ (d) $2^{1.4142} \approx 2.665119089$
(e) $2^{\sqrt{2}} \approx 2.665144143$

■

■   Now work Problem 1.

It can be shown that the familiar laws of rational exponents hold for real exponents.

Solution   Figure 10 shows the various steps.

FIGURE 10

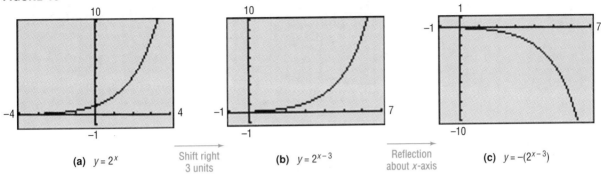

(a) $y = 2^x$ → Shift right 3 units   (b) $y = 2^{x-3}$ → Reflection about x-axis   (c) $y = -(2^{x-3})$

■   Now work Problems 11 and 13.

The Base e

As we shall see shortly, many problems that occur in nature require the use of an exponential function whose base is a certain irrational number, symbolized by the letter e.

Let's look now at one way of arriving at this important number e.

The Number e

> The **number e** is defined as the number that the expression
> $$\left(1 + \frac{1}{n}\right)^n \tag{2}$$
> approaches as $n \to \infty$. In calculus, this is expressed using limit notation as
> $$e = \lim_{n \to \infty} \left(1 + \frac{1}{n}\right)^n$$

TABLE 3

n	$\dfrac{1}{n}$	$1 + \dfrac{1}{n}$	$\left(1 + \dfrac{1}{n}\right)^n$
1	1	2	2
2	0.5	1.5	2.25
5	0.2	1.2	2.48832
10	0.1	1.1	2.59374246
100	0.01	1.01	2.704813829
1,000	0.001	1.001	2.716923932
10,000	0.0001	1.0001	2.718145927
100,000	0.00001	1.00001	2.718268237
1,000,000	0.000001	1.000001	2.718280469
1,000,000,000	10^{-9}	$1 + 10^{-9}$	2.718281827

Table 3 illustrates what happens to the defining expression (2) as n takes on increasingly large values. The last number in the last column in the table is correct to nine decimal places and is the same as the entry given for e on your calculator (if expressed correct to nine decimal places).

The exponential function $f(x) = e^x$, whose base is the number e, occurs with such frequency in applications that it is usually referred to as *the* exponential function. Indeed, graphing calculators have the key $\boxed{e^x}$ or $\boxed{exp(x),}$ which may be used to evaluate the exponential function for a given value of x. Now use your calculator to find e^x for $x = -2$, $x = -1$, $x = 0$, $x = 1$, and $x = 2$, as we have done to create Table 4. The graph of the exponential function $f(x) = e^x$ is given in Figure 11. Since $2 < e < 3$, the graph of $y = e^x$ lies between the graphs of $y = 2^x$ and $y = 3^x$. (Refer to Figures 2 and 4.)

TABLE 4   **FIGURE 11**

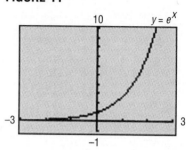

■   Now work Problem 25.

There are many applications involving the exponential function. Let's look at one.

E X A M P L E   6   *Response to Advertising*

Suppose that the percent R of people who respond to a newspaper advertisement for a new product and purchase the item advertised after t days is found using the formula

$$R = 50 - 100e^{-0.3t}$$

(a) What percent has responded and purchased after 5 days?

(b) What percent has responded and purchased after 10 days?

(c) What is the highest percent of people expected to respond and purchase?

(d) Graph $R = 50 - 100e^{-0.3t}$, $t > 0$. TRACE and compare the values of R for $t = 5$ and $t = 10$ to the ones obtained in (a) and (b). How many days are required for R to exceed 40%?

Solution   (a) After 5 days, we have $t = 5$. The corresponding percent R of people responding and purchasing is

$$R = 50 - 100e^{(-0.3)(5)} = 50 - 100e^{-1.5}$$

We evaluate this expression in Figure 12.

FIGURE 12

Thus, about 28% will have responded after 5 days.

(b) After 10 days, we have $t = 10$. The corresponding percent R of people responding and purchasing is

$$R = 50 - 100e^{(-0.3)(10)} = 50 - 100e^{-3} \approx 45.021$$

About 45% will have responded and purchased after 10 days.

(c) As time passes, more people are expected to respond and purchase. The highest percent expected is therefore found for the value of R as $t \to \infty$. Since $e^{-0.3t} = 1/e^{0.3t}$, it follows that $e^{-0.3t} \to 0$ as $t \to \infty$. Thus, the highest percent expected is 50%.

(d) See Figure 13.

FIGURE 13

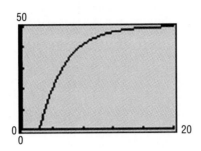

It will require 8 days to exceed 40%.

■ Now work Problem 43.

Summary

PROPERTIES OF THE EXPONENTIAL FUNCTION

$f(x) = a^x, a > 1$ Domain: $(-\infty, \infty)$; Range: $(0, \infty)$; x-intercepts: none; y-intercept: 1; horizontal asymptote: x-axis as $x \to -\infty$; increasing; one-to-one See Figure 2 for a typical graph.

$f(x) = a^x, 0 < a < 1$ Domain: $(-\infty, \infty)$; Range: $(0, \infty)$; x-intercepts: none, y-intercept: 1; horizontal asymptote: x-axis as $x \to \infty$; decreasing; one-to-one See Figure 6 for a typical graph.

7.1

Exercise 7.1

In Problems 1–10, approximate each number using a calculator. Express your answer rounded to three decimal places.

1. (a) $3^{2.2}$ (b) $3^{2.23}$ (c) $3^{2.236}$ (d) $3^{\sqrt{5}}$

2. (a) $5^{1.7}$ (b) $5^{1.73}$ (c) $5^{1.732}$ (d) $5^{\sqrt{3}}$

3. (a) $2^{3.14}$ (b) $2^{3.141}$ (c) $2^{3.1415}$ (d) 2^{π}

4. (a) $2^{2.7}$ (b) $2^{2.71}$ (c) $2^{2.718}$ (d) 2^{e}

5. (a) $3.1^{2.7}$ (b) $3.14^{2.71}$ (c) $3.141^{2.718}$ (d) π^{e}

6. (a) $2.7^{3.1}$ (b) $2.71^{3.14}$ (c) $2.718^{3.141}$ (d) e^{π}

7. $e^{1.2}$ **8.** $e^{-1.3}$ **9.** $e^{-0.85}$ **10.** $e^{2.1}$

In Problems 11–18, the graph of an exponential function is given. Match each graph to one of the following functions:

A. $y = 3^x$ *B.* $y = 3^{-x}$ *C.* $y = -3^x$ *D.* $y = -3^{-x}$

E. $y = 3^x - 1$ *F.* $y = 3^{x-1}$ *G.* $y = 3^{1-x}$ *H.* $y = 1 - 3^x$

11.
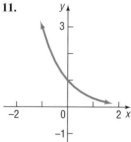

12.
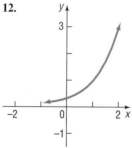

13.
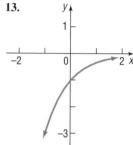

14.
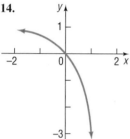

15.
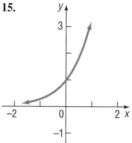

16.
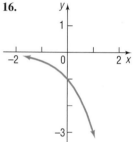

17.
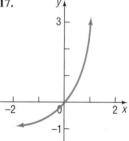

18.
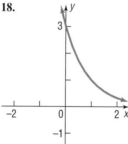

In Problems 19–24, the graph of an exponential function is given. Match each graph to one of the following functions:

A. $y = 4^x$ *B.* $y = 4^{-x}$ *C.* $y = 4^{x-1}$ *D.* $y = 4^x - 1$ *E.* $y = -4^x$ *F.* $y = 1 - 4^x$

19.
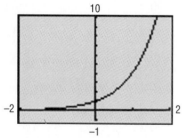

20.
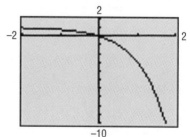

21.

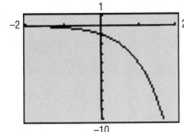

22.

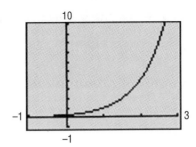

23.

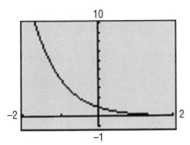

24.

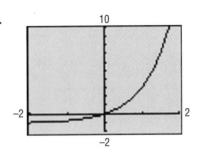

In Problems 25–32, using a graphing utility, show the stages required to graph each function.

25. $y = e^{-x}$ **26.** $y = -e^x$ **27.** $y = e^{x+2}$ **28.** $y = e^x - 1$

29. $y = 5 - e^{-x}$ **30.** $y = 9 - 3e^{-x}$ **31.** $y = 2 - e^{-x/2}$ **32.** $y = 7 - 3e^{-2x}$

33. If $4^x = 7$, what does 4^{-2x} equal?   **34.** If $2^x = 3$, what does 4^{-x} equal?

35. If $3^{-x} = 2$, what does 3^{2x} equal?   **36.** If $5^{-x} = 3$, what does 5^{3x} equal?

37. *Optics* If a single pane of glass obliterates 3% of the light passing through it, then the percent p of light that passes through n successive panes is given approximately by the equation

$$p = 100e^{-0.03n}$$

(a) What percent of light will pass through 10 panes?

(b) What percent of light will pass through 25 panes?

38. *Atmospheric Pressure* The atmospheric pressure p on a balloon or plane decreases with increasing height. This pressure, measured in millimeters of mercury, is related to the number of kilometers h above sea level by the formula

$$p = 760e^{-0.145h}$$

(a) Find the atmospheric pressure at a height of 2 kilometers (over a mile).

(b) What is it at a height of 10 kilometers (over 30,000 feet)?

39. *Space Satellites* The number of watts w provided by a space satellite's power supply over a period of d days is given by the formula

$$w = 50e^{-0.004d}$$

(a) How much power will be available after 30 days?

(b) How much power will be available after 1 year (365 days)?

40. *Healing of Wounds* The normal healing of wounds can be modeled by an exponential function. If A_0 represents the original area of the wound and if A equals the area of the wound after n days, then the formula

$$A = A_0 e^{-0.35n}$$

describes the area of a wound on the nth day following an injury when no infection is present to retard the healing. Suppose a wound initially had an area of 100 square centimeters.

(a) If healing is taking place, how large should the area of the wound be after 3 days?

(b) How large should it be after 10 days?

41. *Drug Medication* The formula

$$D = 5e^{-0.4h}$$

can be used to find the number of milligrams D of a certain drug that is in a patient's bloodstream h hours after the drug has been administered. How many milligrams will be present after 1 hour? After 6 hours?

42. *Spreading of Rumors* A model for the number of people N in a college community who have heard a certain rumor is

$$N = P(1 - e^{-0.15d})$$

where P is the total population of the community and d is the number of days that have elapsed since the rumor began. In a community of 1000 students, how many students will have heard the rumor after 3 days?

43. *Response to TV Advertising* The percent R of viewers who respond to a television commercial for a new product after t days is found by using the formula

$$R = 70 - 100e^{-0.2t}$$

(a) What percent is expected to respond after 10 days?

(b) What percent has responded after 20 days?

(c) What is the highest percent of people expected to respond?

(d) Graph $R = 70 - 100e^{-0.2t}$, $t > 0$. TRACE and compare the values of R for $t = 10$ and $t = 20$ to the ones obtained in parts (a) and (b). How many days are required for R to exceed 40%?

44. *Profit* The annual profit P of a company due to the sales of a particular item after it has been on the market x years is determined to be

$$P = \$100,000 - \$60,000\left(\tfrac{1}{2}\right)^x$$

(a) What is the profit after 5 years?

(b) What is the profit after 10 years?

(c) What is the most profit the company can expect from this product?

(d) Graph the profit function. TRACE and compare the values of P for $x = 5$ and $x = 10$ to the one obtained in parts (a) and (b). How many years does it take before a profit of \$65,000 is obtained?

45. *Alternating Current in a RL Circuit* The equation governing the amount of current I (in amperes) after time t (in seconds) in a single RL circuit consisting of a resistance R (in ohms), an inductance L (in henrys), and an electromotive force E (in volts) is

$$I = \frac{E}{R}[1 - e^{-(R/L)t}]$$

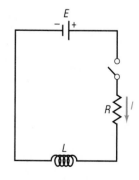

(a) If $E = 120$ volts, $R = 10$ ohms, and $L = 5$ henrys, how much current I_1 is available after 0.3 second? After 0.5 second? After 1 second?

(b) What is the maximum current?

(c) Graph this function $I = I_1(t)$, measuring I along the y-axis and t along the x-axis.

(d) If $E = 120$ volts, $R = 5$ ohms, and $L = 10$ henrys, how much current I_2 is available after 0.3 second? After 0.5 second? After 1 second?

(e) What is the maximum current?

(f) Graph this function $I = I_2(t)$ on the same viewing window as $I_1(t)$.

46. *Alternating Current in a RC Circuit* The equation governing the amount of current *I* (in milliamperes) after time *t* (in milliseconds) in a single *RC* circuit consisting of a resistance *R* (in ohms), a capacitance *C* (in microfarads), and an electromotive force *E* (in volts) is

$$I = \frac{E}{R}e^{-t/(RC)}$$

(a) If $E = 120$ volts, $R = 2000$ ohms, and $C = 1.0$ microfarad, how much current I_1 is available initially ($t = 0$)? After 1000 milliseconds? After 3000 milliseconds?

(b) What is the maximum current?

(c) Graph this function $I = I_1(t)$, measuring *I* along the *y*-axis and *t* along the *x*-axis.

(d) If $E = 120$ volts, $R = 1000$ ohms, and $C = 2.0$ microfarads, how much current I_2 is available initially? After 1000 milliseconds? After 3000 milliseconds?

(e) What is the maximum current?

(f) Graph this function $I = I_2(t)$ on the same viewing window as $I_1(t)$.

47. *The Challenger Disaster** After the *Challenger* disaster in 1986, a study of the 23 launches that preceded the fatal flight was made. A mathematical model was developed involving the relationship between the Fahrenheit temperature *x* around the O-rings and the number *y* of eroded or leaky primary O-rings. The model stated that

$$y = 6\left[1 + e^{-(5.085-0.1156x)}\right]^{-1}$$

where the number 6 indicates the 6 primary O-rings on the spacecraft.

(a) What is the predicted number of eroded or leaky primary O-rings at a temperature of 100°F?

(b) What is the predicted number of eroded or leaky primary O-rings at a temperature of 60°F?

(c) What is the predicted number of eroded or leaky primary O-rings at a temperature of 30°F?

(d) Graph the equation and TRACE. At what temperature is the predicted number of eroded or leaky O-rings 1? 3? 5?

48. *Postage Stamps†* The cumulative number *y* of different postage stamps (regular and commemorative only) issued by the U.S. Post Office can be approximated (modeled) by the exponential function

$$y = 78e^{0.025x}$$

where *x* is the number of years since 1848.

(a) What is the predicted cumulative number of stamps that will have been issued by the year 1995? Check with the Postal Service and comment on the accuracy of using the function.

(b) What is the predicted cumulative number of stamps that will have been issued by the year 1998?

(c) The cumulative number of stamps actually issued by the United States was 2 in 1848, 88 in 1868, 218 in 1888, and 341 in 1908. What conclusion can you draw about using the given function as a model over the first few decades in which stamps were issued?

49. *Another Formula for e* Use a calculator to compute the value of

$$2 + \frac{1}{2!} + \frac{1}{3!} + \ldots + \frac{1}{n!}$$

for $n = 4, 6, 8$, and 10. Compare each result with *e*.

[*Hint:* $1! = 1, 2! = 2 \cdot 1, 3! = 3 \cdot 2 \cdot 1, n! = n(n-1) \cdot \ldots \cdot (3)(2)(1)$].

---

*Linda Tappin, "Analyzing Data Relating to the *Challenger* Disaster," *Mathematics Teacher,* Vol. 87, No. 6, September 1994, pp. 423–426.
†David Kullman, "Patterns of Postage-stamp Production," *Mathematics Teacher,* Vol. 85, No. 3, March 1992, pp. 188–189.

50. *Another Formula for e* Use a calculator to compute the various values of the expression

$$2 + \cfrac{1}{1 + \cfrac{1}{2 + \cfrac{2}{3 + \cfrac{3}{4 + \cfrac{4}{\text{etc.}}}}}}$$

Compare the values to e.

51. If $f(x) = a^x$, show that: $\dfrac{f(x + h) - f(x)}{h} = a^x \left(\dfrac{a^h - 1}{h} \right)$

52. If $f(x) = a^x$, show that: $f(A + B) = f(A) \cdot f(B)$

53. If $f(x) = a^x$, show that: $f(-x) = \dfrac{1}{f(x)}$

54. If $f(x) = a^x$, show that: $f(\alpha x) = [f(x)]^\alpha$

Problems 55 and 56 provide definitions for two other transcendental functions.

55. The **hyperbolic sine function,** designated by sinh x, is defined as

$$\sinh x = \frac{1}{2}(e^x - e^{-x})$$

(a) Show that $f(x) = \sinh x$ is an odd function.
(b) Graph $y = e^x$, $y = -e^{-x}$, and $f(x) = \sinh x$ on the same set of coordinate axes.
(c) How do you explain the relationship of the three graphs?

56. The **hyperbolic cosine function,** designated by cosh x, is defined as

$$\cosh x = \frac{1}{2}(e^x + e^{-x})$$

(a) Show that $f(x) = \cosh x$ is an even function.
(b) Graph $y = e^x$, $y = e^{-x}$, and $f(x) = \cosh x$ on the same set of coordinate axes.
(c) Refer to Problem 55. Show that, for every x,

$$(\cosh x)^2 - (\sinh x)^2 = 1$$

57. *Historical problem* Pierre de Fermat (1601–1665) conjectured that the function

$$f(x) = 2^{(2^x)} + 1$$

for $x = 1, 2, 3, \ldots$, would always have a value equal to a prime number. But Leonhard Euler (1707–1783) showed that this formula fails for $x = 5$. Use a calculator to determine the prime numbers produced by f for $x = 1, 2, 3, 4$. Then show that $f(5) = 641 \times 6{,}700{,}417$, which is not prime.

58. The bacteria in a 4 liter container double every minute. After 60 minutes the container is full. How long did it take to fill half the container?

59. Explain in your own words what the number e is. Provide at least two applications that require the use of this number.

60. Do you think there is a power function that increases more rapidly than an exponential function whose base is greater than 1? Explain.

7.2

Logarithmic Functions

Recall that a one-to-one function $y = f(x)$ has an inverse that is defined (implicitly) by the equation $x = f(y)$. In particular, the exponential function $y = f(x) = a^x$, $a > 0$, $a \neq 1$, is one-to-one and, hence, has an inverse that is defined implicitly by the equation

$$x = a^y \qquad a > 0, a \neq 1$$

This inverse is so important that it is given a name, the *logarithmic function*.

Logarithmic Function

> The **logarithmic function to the base a,** where $a > 0$ and $a \neq 1$, is denoted by $y = \log_a x$ (read as "y is the logarithm to the base a of x") and is defined by
>
> $$y = \log_a x \quad \text{if and only if} \quad x = a^y$$

E X A M P L E   1   *Relating Logarithms to Exponents*

(a) If $y = \log_3 x$, then $x = 3^y$. Thus, if $x = 9$, then $y = 2$, so $9 = 3^2$ is equivalent to $2 = \log_3 9$.

(b) If $y = \log_5 x$, then $x = 5^y$. Thus, if $x = \frac{1}{5} = 5^{-1}$, then $y = -1$, so $\frac{1}{5} = 5^{-1}$ is equivalent to $-1 = \log_5 \left(\frac{1}{5}\right)$. ■

E X A M P L E   2   *Changing Exponential Expressions to Logarithmic Expressions*

Change each exponential expression to an equivalent expression involving a logarithm.

(a) $1.2^3 = m$ (b) $e^b = 9$ (c) $a^4 = 24$

Solution   We use the fact that $y = \log_a x$ and $x = a^y$, $a > 0$, $a \neq 1$, are equivalent.

(a) If $1.2^3 = m$, then $3 = \log_{1.2} m$.

(b) If $e^b = 9$, then $b = \log_e 9$.

(c) If $a^4 = 24$, then $4 = \log_a 24$. ■

■ Now work Problem 1.

E X A M P L E   3   *Changing Logarithmic Expressions to Exponential Expressions*

Change each logarithmic expression to an equivalent expression involving an exponent.

(a) $\log_a 4 = 5$ (b) $\log_e b = -3$ (c) $\log_3 5 = c$

Solution   (a) If $\log_a 4 = 5$, then $a^5 = 4$.

(b) If $\log_e b = -3$, then $e^{-3} = b$.

(c) If $\log_3 5 = c$, then $3^c = 5$. ■

■ Now work Problem 13.

To find the exact value of a logarithm, we write the logarithm in exponential notation and use the following fact:

$$\text{If } a^u = a^v, \quad \text{then} \quad u = v. \tag{1}$$

The result (1) is a consequence of the fact that exponential functions are one-to-one.

EXAMPLE 4

Finding the Exact Value of a Logarithmic Function

Find the exact value of:

(a) $\log_2 8$ (b) $\log_3 \frac{1}{3}$ (c) $\log_5 25$

Solution   (a) For $y = \log_2 8$, we have the equivalent exponential equation $2^y = 8 = 2^3$, so, by (1), $y = 3$. Thus, $\log_2 8 = 3$.

(b) For $y = \log_3 \frac{1}{3}$, we have $3^y = \frac{1}{3} = 3^{-1}$, so $y = -1$. Thus, $\log_3 \frac{1}{3} = -1$.

(c) For $y = \log_5 25$, we have $5^y = 25 = 5^2$, so $y = 2$. Thus, $\log_5 25 = 2$.   ■

■   Now work Problem 25.

Domain of a Logarithmic Function

The logarithmic function $y = \log_a x$ has been defined as the inverse of the exponential function $y = a^x$. That is, if $f(x) = a^x$, then $f^{-1}(x) = \log_a x$. Based on the discussion given in Section 1.7 on inverse functions, we know that for a function f and its inverse f^{-1}

$$\text{Domain } f^{-1} = \text{Range } f \quad \text{and} \quad \text{Range } f^{-1} = \text{Domain } f$$

Consequently, it follows that:

Domain of logarithmic function = Range of exponential function   = $(0, \infty)$
Range of logarithmic function   = Domain of exponential function = $(-\infty, \infty)$

In the next box, we summarize some properties of the logarithmic function:

$$y = \log_a x \quad \text{(defining equation: } x = a^y\text{)}$$
$$\text{Domain:} \quad 0 < x < \infty \quad \text{Range:} \quad -\infty < y < \infty$$

Notice that the domain of a logarithmic function consists of the *positive* real numbers.

EXAMPLE 5

Finding the Domain of a Logarithmic Function

Find the domain of each logarithmic function:

(a) $F(x) = \log_2 (1 - x)$ (b) $g(x) = \log_5 \left(\dfrac{1 + x}{1 - x} \right)$ (c) $h(x) = \log_{1/2} |x|$

Solution (a) The domain of F consists of all x for which $(1 - x) > 0$; that is, all $x < 1$, or $(-\infty, 1)$.

(b) The domain of g is restricted to

$$\frac{1 + x}{1 - x} > 0$$

Solving this inequality, we find that the domain of g consists of all x between -1 and 1, that is, $-1 < x < 1$, or $(-1, 1)$.

(c) Since $|x| > 0$ provided $x \neq 0$, the domain of h consists of all nonzero real numbers. ■

■ Now work Problem 39.

Graphs of Logarithmic Functions

Since exponential functions and logarithmic functions are inverses of each other, the graph of a logarithmic function $y = \log_a x$ is the reflection about the line $y = x$ of the graph of the exponential function $y = a^x$, as shown in Figure 14.

FIGURE 14

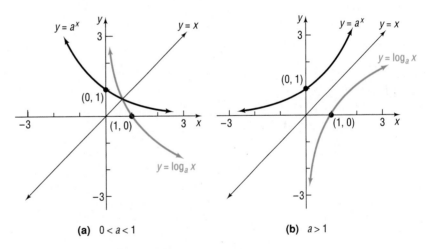

(a) $0 < a < 1$ **(b)** $a > 1$

Facts about the Graph of a Logarithmic Function $f(x) = \log_a x$

1. The x-intercept of the graph is 1. There is no y-intercept.
2. The y-axis is a vertical asymptote of the graph.
3. A logarithmic function is decreasing if $0 < a < 1$ and increasing if $a > 1$.
4. The graph is smooth and continuous, with no corners or gaps.

If the base of a logarithmic function is the number e, then we have the **natural logarithm function.** This function occurs so frequently in applications that it is given a special symbol, **ln** (from the Latin, *logarithmus naturalis*). Thus,

$$y = \ln x \quad \text{if and only if} \quad x = e^y$$

Since $y = \ln x$ and the exponential function $y = e^x$ are inverse functions, we can obtain the graph of $y = \ln x$ by reflecting the graph of $y = e^x$ about the line $y = x$. See Figure 15.

Table 5 displays other points on the graph of $f(x) = \ln x$. Notice for $x < 0$ we obtain an error message. Do you recall why?

FIGURE 15   **TABLE 5**

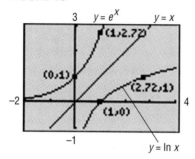

E X A M P L E   6   *Graphing Functions That Are Basically Logarithmic Using Shifting, Reflection, and the Like*

Graph $y = -\ln x$ by starting with the graph of $y = \ln x$.

Solution   The graph of $y = -\ln x$ is obtained by a reflection about the x-axis of the graph of $y = \ln x$. See Figure 16.

FIGURE 16

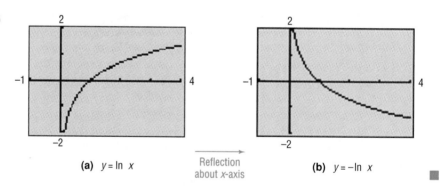

(a) $y = \ln x$ Reflection about x-axis   **(b)** $y = -\ln x$ ∎

E X A M P L E   7   *Graphing Functions That Are Basically Logarithmic Using Shifting, Reflection, and the Like*

Graph: $y = \ln(x + 2)$

Solution   The domain consists of all x for which

$$x + 2 > 0 \quad \text{or} \quad x > -2$$

The graph is obtained by applying a horizontal shift to the left 2 units, as shown in Figure 17. Notice that the line $x = -2$ is a vertical asymptote.

FIGURE 17

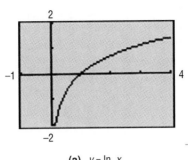

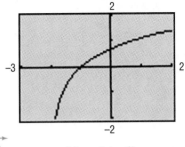

(a) $y = \ln x$ Shift left 2 units   (b) $y = \ln(x + 2)$

EXAMPLE 8   *Graphing Functions That Are Basically Logarithmic Using Shifting, Reflection, and the Like*

Graph: $y = \ln(1 - x)$

Solution   The domain consists of all x for which

$$1 - x > 0 \quad \text{or} \quad x < 1$$

To obtain the graph of $y = \ln(1 - x)$, we use the steps illustrated in Figure 18. Note that the vertical asymptote of $y = \ln(1 - x)$ is $x = 1$.

FIGURE 18

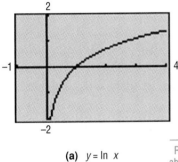

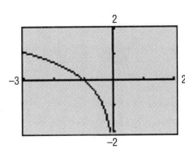

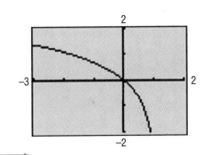

(a) $y = \ln x$ Reflection about y-axis   (b) $y = \ln(-x)$ Shift right 1 unit   (c) $y = \ln(-x + 1)$ $= \ln(1 - x)$

▣   Now work Problem 67.

EXAMPLE 9   *Alcohol and Driving*

The concentration of alcohol in a person's blood is measurable. Recent medical research suggests that the risk R (given as a percent) of having an accident while driving a car can be modeled by the equation

$$R = 6e^{kx}$$

where x is the variable concentration of alcohol in the blood and k is a constant.

(a) Suppose that a concentration of alcohol in the blood of 0.04 results in a 10% risk ($R = 10$) of an accident. Find the constant k in the equation. Graph $R = 6e^{kx}$.

(b) Using this value of k, what is the risk if the concentration is 0.17?

(c) Using the same value of k, what concentration of alcohol corresponds to a risk of 100%?

(d) If the law asserts that anyone with a risk of having an accident of 20% or more should not have driving privileges, at what concentration of alcohol in the blood should a driver be arrested and charged with a DUI—Driving Under the Influence?

Solution   (a) For a concentration of alcohol in the blood of 0.04 and a risk of 10%, we let $x = 0.04$ and $R = 10$ in the equation and solve for k.

$$R = 6e^{kx}$$

$$10 = 6e^{k(0.04)}$$

$$\frac{10}{6} = e^{0.04k} \qquad \text{Change to a logarithmic expression.}$$

$$0.04k = \ln\frac{10}{6} = 0.5108256$$

$$k = 12.77$$

See Figure 19 for the graph of $R = 6e^{12.77x}$.

FIGURE 19

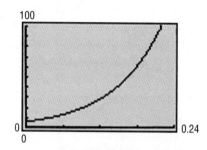

(b) Using $k = 12.77$ and $x = 0.17$ in the equation, we find the risk R to be

$$R = 6e^{kx} = 6e^{(12.77)(0.17)} = 52.6$$

For a concentration of alcohol in the blood of 0.17, the risk of an accident is about 52.6%. Verify this answer by TRACEing the graph.

(c) Using $k = 12.77$ and $R = 100$ in the equation, we find the concentration x of alcohol in the blood to be

$$R = 6e^{kx}$$

$$100 = 6e^{12.77x}$$

$$\frac{100}{6} = e^{12.77x} \qquad \text{Change to a logarithmic expression.}$$

$$12.77x = \ln\frac{100}{6} = 2.8134$$

$$x = 0.22$$

For a concentration of alcohol in the blood of 0.22, the risk of an accident is 100%. Verify this answer by TRACEing the graph.

(d) Using $k = 12.77$ and $R = 20$ in the equation, we find the concentration x of alcohol in the blood to be

$$R = 6e^{kx}$$

$$20 = 6e^{12.77x}$$

$$\frac{20}{6} = e^{12.77x}$$

$$12.77x = \ln \frac{20}{6} = 1.204$$

$$x = 0.094$$

A driver with a concentration of alcohol in the blood of 0.094 or more should be arrested and charged with DUI. Verify by TRACEing the graph.　■

Note: Most states use 0.10 as the blood alcohol content at which a DUI citation is given. A few states use 0.08.

Summary

PROPERTIES OF THE LOGARITHMIC FUNCTION

$f(x) = \log_a x, \quad a > 1$

($y = \log_a x$ means $x = a^y$)

Domain: $(0, \infty)$; Range: $(-\infty, \infty)$; x-intercept: 1; y-intercept: none; vertical asymptote: y-axis; increasing; one-to-one See Figure 14(b) for a typical graph.

$f(x) = \log_a x, \quad 0 < a < 1$

($y = \log_a x$ means $x = a^y$)

Domain: $(0, \infty)$; Range: $(-\infty, \infty)$; x-intercept: 1; y-intercept: none; vertical asymptote: y-axis; decreasing; one-to-one See Figure 14(a) for a typical graph.

7.2

Exercise 7.2

In Problems 1–12, change each exponential expression to an equivalent expression involving a logarithm.

1. $9 = 3^2$　　**2.** $16 = 4^2$　　**3.** $a^2 = 1.6$　　**4.** $a^3 = 2.1$　　**5.** $1.1^2 = M$　　**6.** $2.2^3 = N$

7. $2^x = 7.2$　　**8.** $3^x = 4.6$　　**9.** $x^{\sqrt{2}} = \pi$　　**10.** $x^\pi = e$　　**11.** $e^x = 8$　　**12.** $e^{2.2} = M$

In Problems 13–24, change each logarithmic expression to an equivalent expression involving an exponent.

13. $\log_2 8 = 3$　　　　**14.** $\log_3(\frac{1}{9}) = -2$　　　　**15.** $\log_a 3 = 6$　　　　**16.** $\log_b 4 = 2$

17. $\log_3 2 = x$　　　　**18.** $\log_2 6 = x$　　　　**19.** $\log_2 M = 1.3$　　　　**20.** $\log_3 N = 2.1$

21. $\log_{\sqrt{2}} \pi = x$　　　**22.** $\log_\pi x = \frac{1}{2}$　　　　**23.** $\ln 4 = x$　　　　**24.** $\ln x = 4$

In Problems 25–36, find the exact value of each logarithm without using a calculator.

25. $\log_2 1$　　　**26.** $\log_8 8$　　　**27.** $\log_5 25$　　　**28.** $\log_3(\frac{1}{9})$　　　**29.** $\log_{1/2} 16$　　　**30.** $\log_{1/3} 9$

31. $\log_{10} \sqrt{10}$　　**32.** $\log_5 \sqrt[3]{25}$　　**33.** $\log_{\sqrt{2}} 4$　　**34.** $\log_{\sqrt{3}} 9$　　**35.** $\ln \sqrt{e}$　　**36.** $\ln e^3$

In Problems 37–46, find the domain of each function.

37. $f(x) = \ln(3 - x)$　　**38.** $g(x) = \ln(x^2 - 1)$　　　**39.** $F(x) = \log_2 x^2$

40. $H(x) = \log_5 x^3$　　**41.** $h(x) = \log_{1/2}(x^2 - x - 6)$　　**42.** $G(x) = \log_{1/2}\left(\frac{1}{x}\right)$

43. $f(x) = \dfrac{1}{\ln x}$　　　**44.** $g(x) = \ln(x - 5)$　　　**45.** $g(x) = \log_5\left(\dfrac{x + 1}{x}\right)$　　**46.** $h(x) = \log_3\left(\dfrac{x^2}{x - 1}\right)$

In Problems 47–50, use a calculator to evaluate each expression. Round your answer to three decimal places.

47. $\ln \dfrac{5}{3}$ **48.** $\dfrac{\ln 5}{3}$ **49.** $\dfrac{\ln 10/3}{0.04}$ **50.** $\dfrac{\ln 2/3}{-0.1}$

51. Find a such that the graph of $f(x) = \log_a x$ contains the point $(2, 2)$.

52. Find a such that the graph of $f(x) = \log_a x$ contains the point $(\frac{1}{2}, -4)$.

In Problems 53–60, the graph of a logarithmic function is given. Match each graph to one of the following functions:

A. $y = \log_3 x$ B. $y = \log_3(-x)$ C. $y = -\log_3 x$ D. $y = -\log_3(-x)$

E. $y = \log_3 x - 1$ F. $y = \log_3(x - 1)$ G. $y = \log_3(1 - x)$ H. $y = 1 - \log_3 x$

53. 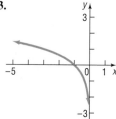   **54.** 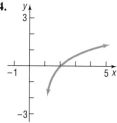   **55.** 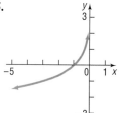   **56.**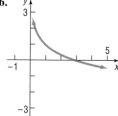

57. 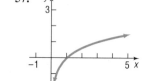   **58.** 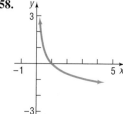   **59.** 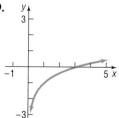   **60.**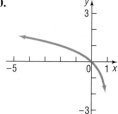

In Problems 61–67, the graph of a logarithmic function is given. Match each graph to one of the following functions:

A. $y = \log_4 x$ B. $y = \log_4(-x)$ C. $y = \log_4(x - 1)$

D. $y = -\log_4 x$ E. $y = 1 - \log_4 x$ F. $y = -\log_4(-x)$

61. 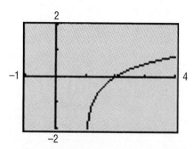   **62.**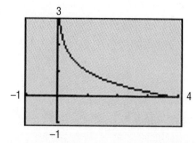

63.

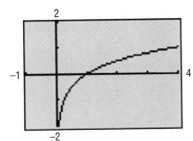

64.

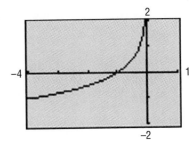

65.

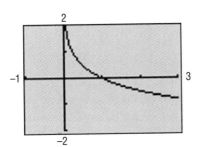

66.

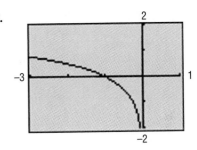

In Problems 67–76, using a graphing utility, show the stages required to graph each function.

67. $f(x) = \ln(x + 4)$

68. $f(x) = \ln(x - 3)$

69. $f(x) = \ln(-x)$

70. $f(x) = -\ln(-x)$

71. $g(x) = \ln 2x$

72. $h(x) = \ln \frac{1}{2}x$

73. $f(x) = 3 \ln x$

74. $f(x) = -2 \ln x$

75. $g(x) = \ln(3 - x)$

76. $h(x) = \ln(4 - x)$

77. *Optics* If a single pane of glass obliterates 10% of the light passing through it, then the percent P of light that passes through n successive panes is given approximately by the equation

$$P = 100e^{-0.1n}$$

(a) How many panes are necessary to block at least 50% of the light?
(b) How many panes are necessary to block at least 75% of the light?

78. *Chemistry* The pH of a chemical solution is given by the formula

$$pH = -\log_{10} [H^+]$$

where $[H^+]$ is the concentration of hydrogen ions in moles per liter. Values of pH range from 0 (acidic) to 14 (alkaline).

(a) Find the pH of a 1 liter container of water with 0.0000001 mole of hydrogen ion.
(b) Find the hydrogen ion concentration of a mildly acidic solution with a pH of 4.2.

79. *Space Satellites* The number of watts w provided by a space satellite's power supply after d days is given by the formula

$$w = 50e^{-0.004d}$$

(a) How long will it take for the available power to drop to 30 watts?
(b) How long will it take for the available power to drop to only 5 watts?

80. *Healing of Wounds* The normal healing of wounds can be modeled by an exponential function. If A_0 represents the original area of the wound and if A equals the area of the wound after n days, then the formula

$$A = A_0 e^{-0.35n}$$

describes the area of a wound on the nth day following an injury when no infection is present to retard the healing. Suppose a wound initially had an area of 100 square centimeters.
(a) If healing is taking place, how many days should pass before the wound is one-half its original size?
(b) How long before the wound is 10% of its original size?

81. *Drug Medication* The formula

$$D = 5e^{-0.4h}$$

can be used to find the number of milligrams D of a certain drug that is in a patient's bloodstream h hours after the drug has been administered. When the number of milligrams reaches 2, the drug is to be administered again. What is the time between injections?

82. *Spreading of Rumors* A model for the number of people N in a college community who have heard a certain rumor is

$$N = P(1 - e^{-0.15d})$$

where P is the total population of the community and d is the number of days that have elapsed since the rumor began. In a community of 1000 students, how many days will elapse before 450 students have heard the rumor?

83. *Current in a RL Circuit* The equation governing the amount of current I (in amperes) after time t (in seconds) in a simple RL circuit consisting of a resistance R (in ohms), an inductance L (in henrys), and an electromotive force E (in volts) is

$$I = \frac{E}{R}[1 - e^{-(R/L)t}]$$

If $E = 12$ volts, $R = 10$ ohms, and $L = 5$ henrys, how long does it take to obtain a current of 0.5 ampere? Of 1.0 ampere? Graph the equation.

84. *Learning Curve* Psychologists sometimes use the function

$$L(t) = A(1 - e^{-kt})$$

to measure the amount L learned at time t. The number A represents the amount to be learned, and the number k measures the rate of learning. Suppose that a student has an amount A of 200 vocabulary words to learn. A psychologist determines that the student learned 20 vocabulary words after 5 minutes.
(a) Determine the rate of learning k.
(b) Approximately how many words will the student have learned after 10 minutes?
(c) After 15 minutes?
(d) How long does it take for the student to learn 180 words?

85. *Alcohol and Driving* The concentration of alcohol in a person's blood is measurable. Suppose the risk R (given as a percent) of having an accident while driving a car can be modeled by the equation

$$R = 3e^{kx}$$

where x is the variable concentration of alcohol in the blood and k is a constant.
(a) Suppose a concentration of alcohol in the blood of 0.06 results in a 10% risk ($R = 10$) of an accident. Find the constant k in the equation.
(b) Using this value of k, what is the risk if the concentration is 0.17?
(c) Using the same value of k, what concentration of alcohol corresponds to a risk of 100%?
(d) If the law asserts that anyone with a risk of having an accident of 15% or more should not have driving privileges, at what concentration of alcohol in the blood should a driver be arrested and charged with a DUI?

(e) Compare this situation with that of Example 9. If you were a lawmaker, which situation would you support? Give your reasons.

86. Is there any function of the form $y = x^\alpha$, $0 < \alpha < 1$, that increases more slowly than a logarithmic function whose base is greater than 1? Explain.

87. *Constructing a Function* Look back at Figure 3 in Section 7.1. Assuming that the points (1950, 20) and (1990, 50) are on the graph, find an exponential equation $y = Ae^{bt}$, that fits the data. [*Hint:* Let $t = 0$ correspond to the year 1950. Then show that $A = 20$. Now find b.] Is the projection of 102 million in 2015 confirmed by your model? Try to obtain similar data about the United States birthrate and construct a function to fit those data.

88. *Critical Thinking* In buying a new car, one consideration might be how well the price of the car holds up over time. Different makes of cars have different depreciation rates. One way to compute a depreciation rate for a car is given here. Suppose the current prices of a certain Mercedes automobile are as follows:

NEW	1 YEAR OLD	2 YEARS OLD	3 YEARS OLD	4 YEARS OLD	5 YEARS OLD
$38,000	$36,600	$32,400	$28,750	$25,400	$21,200

Use the formula New = Old(e^{Rt}) to find R, the annual depreciation rate, for a specific time t. When might be the best time to trade the car in? Consult the NADA ("blue") book and compare two like models that you are interested in. Which has the better depreciation rate?

7.3

Properties of Logarithms

Logarithms have some very useful properties that can be derived directly from the definition and the laws of exponents.

EXAMPLE 1 *Establishing Properties of Logarithms*
  (a) Show that $\log_a 1 = 0$.   (b) Show that $\log_a a = 1$.

Solution (a) This fact was established when we graphed $y = \log_a x$ (see Figure 14). Algebraically, for $y = \log_a 1$, we have $a^y = 1 = a^0$, so $y = 0$.
  (b) For $y = \log_a a$, we have $a^y = a = a^1$, so $y = 1$.

$$\log_a 1 = 0 \qquad \log_a a = 1$$

Theorem
Properties of Logarithms
In the properties given next, M and a are positive real numbers, with $a \neq 1$, and r is any real number.

The number $\log_a M$ is the exponent to which a must be raised to obtain M. That is,

$$a^{\log_a M} = M \tag{1}$$

The logarithm to the base a of a raised to a power equals that power. That is,

$$\log_a a^r = r \tag{2}$$

Proof of Property (1) Let $x = \log_a M$. Change this logarithmic expression to the equivalent exponential expression:

$$a^x = M$$

But $x = \log_a M$, so

$$a^{\log_a M} = M$$

Proof of Property (2)   Let $x = a^r$. Change this exponential expression to the equivalent logarithmic expression:

$$\log_a x = r$$

But $x = a^r$, so

$$\log_a a^r = r$$ ■

E X A M P L E   2   *Using Properties (1) and (2)*

(a) $2^{\log_2 \pi} = \pi$ (b) $\log_{0.2} 0.2^{-\sqrt{2}} = -\sqrt{2}$ (c) $\ln e^{kt} = kt$ ■

Other useful properties of logarithms are given now.

Theorem   In the following properties, M, N, and a are positive real numbers, with $a \neq 1$, and r is any real number.

The Log of a Product Equals the Sum of the Logs

$$\log_a MN = \log_a M + \log_a N \tag{3}$$

The Log of a Quotient Equals the Difference of the Logs

$$\log_a\left(\frac{M}{N}\right) = \log_a M - \log_a N \tag{4}$$

$$\log_a\left(\frac{1}{N}\right) = -\log_a N \tag{5}$$

$$\log_a M^r = r \log_a M \tag{6}$$

■

We shall derive properties (3) and (6) and leave the derivations of properties (4) and (5) as exercises (see Problems 69 and 70).

Proof of Property (3)   Let $A = \log_a M$ and let $B = \log_a N$. These expressions are equivalent to the exponential expressions

$$a^A = M \quad \text{and} \quad a^B = N$$

Now

$$\log_a MN = \log_a a^A a^B = \log_a a^{A+B} \qquad \text{Law of exponents}$$
$$= A + B \qquad \text{Property (2) of logarithms}$$
$$= \log_a M + \log_a N \qquad ■$$

Proof of Property (6)   Let $A = \log_a M$. This expression is equivalent to

$$a^A = M$$

Now

$$\log_a M^r = \log_a (a^A)^r = \log_a a^{rA} \qquad \text{Law of exponents}$$
$$= rA \qquad \text{Property (2) of logarithms}$$
$$= r \log_a M \qquad ■$$

Logarithms can be used to transform products into sums, quotients into differences, and powers into factors. Such transformations prove useful in certain types of calculus problems.

E X A M P L E   3   *Writing a Logarithmic Expression as a Sum of Logarithms*

Write $\log_a(x\sqrt{x^2 + 1})$ as a sum of logarithms. Express all powers as factors.

Solution
$$\log_a(x\sqrt{x^2 + 1}) = \log_a x + \log_a \sqrt{x^2 + 1} \quad \text{Property (3)}$$
$$= \log_a x + \log_a(x^2 + 1)^{1/2}$$
$$= \log_a x + \tfrac{1}{2}\log_a(x^2 + 1) \quad \text{Property (6)}$$

E X A M P L E   4   *Writing a Logarithmic Expression as a Difference of Logarithms*

Write
$$\log_a \frac{x^2}{(x-1)^3}$$
as a difference of logarithms. Express all powers as factors.

Solution
$$\log_a \frac{x^2}{(x-1)^3} \underset{\substack{\uparrow \\ \text{Property (4)}}}{=} \log_a x^2 - \log_a(x-1)^3 \underset{\substack{\uparrow \\ \text{Property (6)}}}{=} 2\log_a x - 3\log_a(x-1)$$

■ Now work Problem 13.

E X A M P L E   5   *Writing a Logarithmic Expression as a Sum and Difference of Logarithms*

Write
$$\log_a \frac{x^3\sqrt{x^2 + 1}}{(x + 1)^4}$$
as a sum and difference of logarithms. Express all powers as factors.

Solution
$$\log_a \frac{x^3\sqrt{x^2 + 1}}{(x + 1)^4} = \log_a(x^3\sqrt{x^2 + 1}) - \log_a(x + 1)^4$$
$$= \log_a x^3 + \log_a \sqrt{x^2 + 1} - \log_a(x + 1)^4$$
$$= \log_a x^3 + \log_a(x^2 + 1)^{1/2} - \log_a(x + 1)^4$$
$$= 3\log_a x + \tfrac{1}{2}\log_a(x^2 + 1) - 4\log_a(x + 1)$$

Another use of properties (3) through (6) is to write sums and/or differences of logarithms with the same base as a single logarithm.

E X A M P L E   6   *Writing Expressions as a Single Logarithm*

Write each of the following as a single logarithm:

(a) $\log_a 7 + 4\log_a 3$

(b) $\tfrac{2}{3}\log_a 8 - \log_a(3^4 - 8)$

(c) $\log_a x + \log_a 9 + \log_a(x^2 + 1) - \log_a 5$

Solution (a) $\log_a 7 + 4 \log_a 3 = \log_a 7 + \log_a 3^4$ Property (6)

$$= \log_a 7 + \log_a 81$$
$$= \log_a(7 \cdot 81) \qquad \text{Property (3)}$$
$$= \log_a 567$$

(b) $\frac{2}{3} \log_a 8 - \log_a(3^4 - 8) = \log_a 8^{2/3} - \log_a(81 - 8)$ Property (6)

$$= \log_a 4 - \log_a 73$$
$$= \log_a\left(\tfrac{4}{73}\right) \qquad\qquad \text{Property (4)}$$

(c) $\log_a x + \log_a 9 + \log_a(x^2 + 1) - \log_a 5 = \log_a 9x + \log_a(x^2 + 1) - \log_a 5$

$$= \log_a[9x(x^2 + 1)] - \log_a 5$$
$$= \log_a\left[\frac{9x(x^2 + 1)}{5}\right] \qquad\qquad\blacksquare$$

Warning: A common error made by some students is to express the logarithm of a sum as the sum of logarithms:

$$\log_a(M + N) \quad \text{is not equal to} \quad \log_a M + \log_a N$$

Correct Statement $\log_a MN = \log_a M + \log_a N$ Property (3)

Another common error is to express the difference of logarithms as the quotient of logarithms:

$$\log_a M - \log_a N \quad \text{is not equal to} \quad \frac{\log_a M}{\log_a N}$$

Correct Statement $\log_a M - \log_a N = \log_a\left(\dfrac{M}{N}\right)$ Property (4)

■   Now work Problem 23.

There remain two other properties of logarithms we need to know. They are a consequence of the fact that the logarithmic function $y = \log_a x$ is one-to-one.

Theorem   In the following properties, M, N, and a are positive real numbers, with $a \neq 1$:

$$\text{If} \quad M = N, \quad \text{then} \quad \log_a M = \log_a N. \tag{7}$$
$$\text{If} \quad \log_a M = \log_a N, \quad \text{then} \quad M = N. \tag{8}$$

■

Properties (7) and (8) are useful for solving *logarithmic equations,* a topic discussed in the next section.

Using a Calculator to Evaluate and Graph Logarithms with Bases Other Than e or 10

Logarithms to the base 10, called **common logarithms,** were used to facilitate arithmetic computations before the widespread use of calculators. (See the Historic Feature at the end of this section.) Natural logarithms, that is, logarithms whose base is the number e, remain very important because they arise frequently in the study of natural phenomena.

Common logarithms are usually abbreviated by writing **log,** with the base understood to be 10, just as natural logarithms are abbreviated by **ln,** with the base understood to be e.

Graphing calculators have both $\boxed{\log}$ and $\boxed{\ln}$ keys to calculate the common logarithm and natural logarithm of a number. Let's look at an example to see how to calculate logarithms having a base other than 10 or e.

E X A M P L E   7   *Evaluating Logarithms Whose Base Is Neither 10 nor e*

Evaluate: $\log_2 7$

Solution   Let $y = \log_2 7$. Then $2^y = 7$, so

$$2^y = 7$$
$$\ln 2^y = \ln 7 \qquad \text{Property (7)}$$
$$y \ln 2 = \ln 7 \qquad \text{Property (6)}$$
$$y = \frac{\ln 7}{\ln 2} \qquad \text{Solve for } y.$$
$$= 2.8074 \qquad \text{Use calculator } (\boxed{\ln}\ \text{key}).$$ ■

Example 7 shows how to change the base from 2 to e. In general, to change from the base b to the base a, we use the **Change-of-Base Formula.**

Theorem   If $a \neq 1$, $b \neq 1$, and M are positive real numbers, then
Change-of-Base Formula

$$\log_a M = \frac{\log_b M}{\log_b a} \qquad (9)$$

■

Proof   We derive this formula as follows: Let $y = \log_a M$. Then $a^y = M$, so

$$\log_b a^y = \log_b M \qquad \text{Property (7)}$$
$$y \log_b a = \log_b M \qquad \text{Property (6)}$$
$$y = \frac{\log_b M}{\log_b a} \qquad \text{Solve for } y.$$
$$\log_a M = \frac{\log_b M}{\log_b a} \qquad y = \log_a M$$ ■

Since calculators have only keys for $\boxed{\log}$ and $\boxed{\ln}$, in practice, the Change-of-Base Formula uses either $b = 10$ or $b = e$. Thus,

$$\log_a M = \frac{\log M}{\log a} \quad \text{and} \quad \log_a M = \frac{\ln M}{\ln a} \qquad (10)$$

E X A M P L E   8   *Using the Change-of-Base Formula*

Calculate:

(a) $\log_5 89$ (b) $\log_{\sqrt{2}} \sqrt{5}$

Solution (a) $\log_5 89 = \dfrac{\log 89}{\log 5} \approx \dfrac{1.94939}{0.69897} = 2.7889$

or

$\log_5 89 = \dfrac{\ln 89}{\ln 5} \approx \dfrac{4.4886}{1.6094} = 2.7889$

(b) $\log_{\sqrt{2}} \sqrt{5} = \dfrac{\log \sqrt{5}}{\log \sqrt{2}} = \dfrac{\frac{1}{2} \log 5}{\frac{1}{2} \log 2} \approx \dfrac{0.69897}{0.30103} = 2.3219$

or

$\log_{\sqrt{2}} \sqrt{5} = \dfrac{\ln \sqrt{5}}{\ln \sqrt{2}} = \dfrac{\frac{1}{2} \ln 5}{\frac{1}{2} \ln 2} \approx \dfrac{1.6094}{0.6931} = 2.3219$ ■

■ Now work Problem 33.

We also use the Change-of-Base-Formula to graph logarithmic functions whose base is neither 10 nor e.

E X A M P L E 9

Graphing a Logarithmic Function Whose Base Is Neither 10 Nor e

Use a graphing utility to graph $y = \log_2 x$.

Solution Since graphing utilities only have logarithms with the base 10 or the base e, we need to use the Change-of-Base Formula to express $y = \log_2 x$ in terms of logarithms with base 10 or base e. We can graph either $y = \ln x/\ln 2$ or $y = \log x/\log 2$ to obtain the graph of $y = \log_2 x$. See Figure 20. ■

FIGURE 20

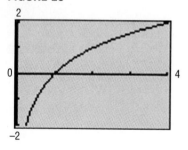

Check: Verify that $y = \ln x/\ln 2$ and $y = \log x/\log 2$ result in the same graph by graphing each one on the same screen.

■ Now work Problem 41.

Summary of Properties of Logarithms

In the list that follows, $a > 0$, $a \neq 1$, and $b > 0$, $b \neq 1$; also, $M > 0$ and $N > 0$.

Definition	$y = \log_a x$ means $x = a^y$
Properties of logarithms	$\log_a 1 = 0; \quad \log_a a = 1$
	$a^{\log_a M} = M; \quad \log_a a^r = r$
	$\log_a MN = \log_a M + \log_a N$
	$\log_a\left(\dfrac{M}{N}\right) = \log_a M - \log_a N$
	$\log_a\left(\dfrac{1}{N}\right) = -\log_a N$
	$\log_a M^r = r \log_a M$
Change-of-Base Formula	$\log_a M = \dfrac{\log_b M}{\log_b a}$

HISTORICAL FEATURE ■ Logarithms were invented about 1590 by John Napier (1550–1617) and Jobst Bürgi (1552–1632), working independently. Napier, whose work had the greater influence, was a Scottish lord, a secretive man whose neighbors were inclined to believe him to be in league with the devil. His approach to logarithms was quite different from ours; it was based on the relationship between arithmetic and geometric sequences, and not on the inverse function relationship of logarithms to exponential functions (described in Section 7.2). Napier's tables, published in 1614, listed what would now be called *natural logarithms* of sines and were rather difficult to use. A London professor, Henry Briggs, became interested in the tables and visited Napier. In their conversations, they developed the idea of common logarithms, and Briggs then converted Napier's tables into tables of common logarithms, which were published in 1617. Their importance for calculation was immediately recognized, and by 1650 they were being printed as far away as China. They remained an important calculation tool until the advent of the inexpensive handheld calculator about 1972, which has decreased their calculational, but not their theoretical, importance.

A side effect of the invention of logarithms was the popularization of the decimal system of notation for real numbers. ■

7.3

Exercise 7.3

In Problems 1–12, suppose $\ln 2 = a$ *and* $\ln 3 = b$. *Use properties of logarithms to write each logarithm in terms of* a *and* b.

1. $\ln 6$

2. $\ln \frac{2}{3}$

3. $\ln 1.5$

4. $\ln 0.5$

5. $\ln 2e$

6. $\ln\left(\dfrac{3}{e}\right)$

7. $\ln 12$

8. $\ln 24$

9. $\ln \sqrt[5]{18}$

10. $\ln \sqrt[4]{48}$

11. $\log_2 3$

12. $\log_3 2$

In Problems 13–22, write each expression as a sum and/or difference of logarithms. Express powers as factors.

13. $\ln(x^2\sqrt{1-x})$

14. $\ln(x\sqrt{1+x^2})$

15. $\log_2\left(\dfrac{x^3}{x-3}\right)$

16. $\log_5\left(\dfrac{\sqrt[3]{x^2+1}}{x^2-1}\right)$

17. $\log\left[\dfrac{x(x+2)}{(x+3)^2}\right]$

18. $\log\dfrac{x^3\sqrt{x+1}}{(x-2)^2}$

19. $\ln\left[\dfrac{x^2-x-2}{(x+4)^2}\right]^{1/3}$

20. $\ln\left[\dfrac{(x-4)^2}{x^2-1}\right]^{2/3}$

21. $\ln\dfrac{5x\sqrt{1-3x}}{(x-4)^3}$

22. $\ln\left[\dfrac{5x^2\sqrt[3]{1-x}}{4(x+1)^2}\right]$

In Problems 23–32, write each expression as a single logarithm.

23. $3\log_5 u + 4\log_5 v$

24. $\log_3 u^2 - \log_3 v$

25. $\log_{1/2}\sqrt{x} - \log_{1/2} x^3$

26. $\log_2\left(\dfrac{1}{x}\right) + \log_2\left(\dfrac{1}{x^2}\right)$

27. $\ln\left(\dfrac{x}{x-1}\right) + \ln\left(\dfrac{x+1}{x}\right) - \ln(x^2-1)$

28. $\log\left(\dfrac{x^2+2x-3}{x^2-4}\right) - \log\left(\dfrac{x^2+7x+6}{x+2}\right)$

29. $8\log_2\sqrt{3x-2} - \log_2\left(\dfrac{4}{x}\right) + \log_2 4$

30. $21\log_3\sqrt[3]{x} + \log_3 9x^2 - \log_5 25$

31. $2\log_a 5x^3 - \frac{1}{2}\log_a(2x+3)$

32. $\frac{1}{3}\log(x^3+1) + \frac{1}{2}\log(x^2+1)$

In Problems 33–40, use the change of base formula and a calculator to evaluate each logarithm. Round your answer to three decimal places.

33. $\log_3 21$

34. $\log_5 18$

35. $\log_{1/3} 71$

36. $\log_{1/2} 15$

37. $\log_{\sqrt{2}} 7$

38. $\log_{\sqrt{5}} 8$

39. $\log_{\pi} e$

40. $\log_{\pi} \sqrt{2}$

For Problems 41–46, graph each function using a graphing utility and the Change-of-Base Formula.

41. $y = \log_4 x$

42. $y = \log_5 x$

43. $y = \log_2(x + 2)$

44. $y = \log_4(x - 3)$

45. $y = \log_{x-1}(x + 1)$

46. $y = \log_{x+2}(x - 2)$

47. Show that: $\log_a(x + \sqrt{x^2 - 1}) + \log_a(x - \sqrt{x^2 - 1}) = 0$

48. Show that: $\log_a(\sqrt{x} + \sqrt{x - 1}) + \log_a(\sqrt{x} - \sqrt{x - 1}) = 0$

49. Show that: $\ln(1 + e^{2x}) = 2x + \ln(1 + e^{-2x})$

50. If $f(x) = \log_a x$, show that: $\dfrac{f(x + h) - f(x)}{h} = \log_a\left(1 + \dfrac{h}{x}\right)^{1/h}$, $h \neq 0$

51. If $f(x) = \log_a x$, show that: $-f(x) = \log_{1/a} x$

52. If $f(x) = \log_a x$, show that: $f(1/x) = -f(x)$

53. If $f(x) = \log_a x$, show that: $f(AB) = f(A) + f(B)$

54. If $f(x) = \log_a x$, show that: $f(x^\alpha) = \alpha f(x)$

In Problems 55–64, express y as a function of x. The constant C is a positive number.

55. $\ln y = \ln x + \ln C$

56. $\ln y = \ln(x + C)$

57. $\ln y = \ln x + \ln(x + 1) + \ln C$

58. $\ln y = 2 \ln x - \ln(x + 1) + \ln C$

59. $\ln y = 3x + \ln C$

60. $\ln y = -2x + \ln C$

61. $\ln(y - 3) = -4x + \ln C$

62. $\ln(y + 4) = 5x + \ln C$

63. $3 \ln y = \frac{1}{2} \ln(2x + 1) - \frac{1}{3} \ln(x + 4) + \ln C$

64. $2 \ln y = -\frac{1}{2} \ln x + \frac{1}{3} \ln(x^2 + 1) + \ln C$

65. Find the value of $\log_2 3 \cdot \log_3 4 \cdot \log_4 5 \cdot \log_5 6 \cdot \log_6 7 \cdot \log_7 8$.

66. Find the value of $\log_2 4 \cdot \log_4 6 \cdot \log_6 8$.

67. Find the value of $\log_2 3 \cdot \log_3 4 \cdot \ldots \cdot \log_n(n + 1) \cdot \log_{n+1} 2$.

68. Find the value of $\log_2 2 \cdot \log_2 4 \cdot \ldots \cdot \log_2 2^n$.

69. Show that $\log_a(M/N) = \log_a M - \log_a N$, where a, M, and N are positive real numbers, with $a \neq 1$.

70. Show that $\log_a (1/N) = -\log_a N$, where a and N are positive real numbers, with $a \neq 1$.

71. Find the domain of $f(x) = \log_a x^2$ and the domain of $g(x) = 2 \log_a x$. Since $\log_a x^2 = 2 \log_a x$, how do you reconcile the fact that the domains are not equal? Write a brief explanation.

7.4

Logarithmic and Exponential Equations

Logarithmic Equations

Equations that contain terms of the form $\log_a x$, where a is a positive real number, with $a \neq 1$, are often called **logarithmic equations.**

As before, our practice will be to solve equations, whenever possible, by finding exact solutions using algebraic methods. In such cases, we shall also verify the solution obtained by using a graphing utility. When algebraic methods cannot be used, approximate solutions will be obtained using a graphing utility. The reader is encouraged to pay particular attention to the form of equations for which exact solutions are possible.

$\mathcal{M}$ISSION POSSIBLE

Chapter 7

McNEWTON'S COFFEE

Your team has been called in to solve a problem encountered by a fast food restaurant. They believe that their coffee should be brewed at 170° Fahrenheit; however, at that temperature it is too hot to drink, and a customer who accidentally spills the coffee on himself might receive third degree burns.

What they need is a special container that will heat the water to 170°, brew the coffee at that temperature, then cool it quickly to a drinkable temperature, say 140°F, and hold it there, or at least keep it at or above 120°F for a reasonable period of time without further cooking. To cool down the coffee, three companies have submitted proposals with these specifications:

(a) The CentiKeeper Company has a container that will reduce the temperature of a liquid from 200°F to 100°F in 90 minutes by maintaining a constant temperature of 70°F.

(b) The TempControl Company has a container that will reduce the temperature of a liquid from 200°F to 110°F in 60 minutes by maintaining a constant temperature of 60°F.

(c) The Hot'n'Cold, Inc., has a container that will reduce the temperature of a liquid from 210°F to 90°F in 30 minutes by maintaining a constant temperature of 50°F.

Your job is to make a recommendation as to which container to purchase. For this you will need Newton's Law of Cooling which follows:

$$u = T + (u_0 - T)e^{kt}, \quad k < 0$$

In this formula, T represents the temperature of the surrounding medium, u_0 is the initial temperature of the heated object, t is the length of time in minutes, k is a negative constant, and u represents the temperature at time t.

1. Use Newton's Law of Cooling to find the constant k of the formula for each container.
2. Use a graphing utility to graph each relation.
3. How long does it take each container to lower the coffee temperature from 170°F to 140°F?
4. How long will the coffee temperature remain between 120°F and 140°F?
5. On the basis of this information, which company should get the contract with McNewton's? What are your reasons?
6. What are "capital cost" and "operating cost"? How might they affect your choice?

E X A M P L E   1   *Solving a Logarithmic Equation*

Solve: $\log_3(4x - 7) = 2$

Solution   We can obtain an exact solution by changing the logarithm to exponential form.

$$\log_3(4x - 7) = 2$$
$$4x - 7 = 3^2$$
$$4x - 7 = 9$$
$$4x = 16$$
$$x = 4 \qquad \blacksquare$$

We verify the solution by graphing the function $f(x) = \log_3(4x - 7) - 2 = \dfrac{\ln(4x - 7)}{\ln 3} - 2$ to determine where the graph crosses the x-axis. See Figure 21. The graph of f crosses the x-axis at $x = 4$.

FIGURE 21

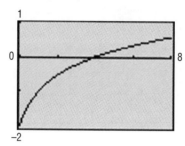

E X A M P L E   2   *Solving a Logarithmic Equation*

Solve: $2 \log_5 x = \log_5 9$

Solution   Because each logarithm is to the same base, 5, we can obtain an exact solution as follows:

$$2 \log_5 x = \log_5 9$$
$$\log_5 x^2 = \log_5 9 \qquad \text{Property (6), Section 7.3}$$
$$x^2 = 9 \qquad \text{Property (8), Section 7.3}$$
$$x = 3 \quad \text{or} \quad \cancel{x = -3} \qquad \text{Recall that logarithms of negative numbers are not defined, so, in the expression } 2 \log_5 x, x \text{ must be positive. Therefore, } -3 \text{ is extraneous and we discard it.}$$

The equation has only one solution, 3. $\qquad \blacksquare$

You should verify for yourself that 3 is the only solution using a graphing utility.

■   Now work Problem 1.

E X A M P L E   3   *Solving a Logarithmic Equation*

Solve: $\log_4(x + 3) + \log_4(2 - x) = 1$

Solution   To obtain an exact solution, we need to express the left side as a single logarithm. Then we will change the expression to exponential form.

$$\log_4(x + 3) + \log_4(2 - x) = 1$$
$$\log_4[(x + 3)(2 - x)] = 1 \qquad \text{Property (3), Section 7.3}$$
$$(x + 3)(2 - x) = 4^1 = 4$$
$$-x^2 - x + 6 = 4$$
$$x^2 + x - 2 = 0$$
$$(x + 2)(x - 1) = 0$$
$$x = -2 \quad \text{or} \quad x = 1$$

You should verify that both of these are solutions using a graphing utility.   ■

■   Now work Problem 11.

Care must be taken when solving logarithmic equations. Be sure to check each apparent solution in the original equation and discard any that are extraneous. In the expression $\log_a M$, remember that a and M are positive and $a \neq 1$.

Exponential Equations

Equations that involve terms of the form a^x, $a > 0$, $a \neq 1$, are often referred to as **exponential equations.** Such equations sometimes can be solved by appropriately applying the laws of exponents and equation (1), namely,

$$\text{If } a^u = a^v, \quad \text{then } u = v. \tag{1}$$

To use equation (1), each side of the equality must be written with the same base.

E X A M P L E   4

Solving an Exponential Equation

Solve the equation: $3^{x+1} = 81$

Solution   Since $81 = 3^4$, we can write the equation as

$$3^{x+1} = 81 = 3^4$$

Now we have the same base, 3, on each side, so we can apply (1) to obtain

$$x + 1 = 4$$
$$x = 3$$

We verify the solution by graphing $f(x) = 3^{x+1} - 81$ to determine where the graph crosses the x-axis. Figure 22 shows that the graph of f crosses the x-axis at $x = 3$.   ■

FIGURE 22

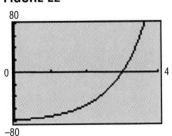

■   Now work Problem 19.

E X A M P L E   5   *Solving an Exponential Equation*

Solve the equation: $e^{-x^2} = (e^x)^2 \cdot \dfrac{1}{e^3}$

Solution   We use some laws of exponents first to get the same base e on each side:

$$e^{-x^2} = (e^x)^2 \cdot \frac{1}{e^3} = e^{2x} \cdot e^{-3} = e^{2x-3}$$

Now apply (1) to get

$$-x^2 = 2x - 3$$
$$x^2 + 2x - 3 = 0$$
$$(x + 3)(x - 1) = 0$$
$$x = -3 \quad \text{or} \quad x = 1 \qquad \blacksquare$$

You should verify these solutions using a graphing utility.

E X A M P L E   6   *Solving an Exponential Equation*

Solve the equation: $4^x - 2^x - 12 = 0$

Solution   We note that $4^x = (2^2)^x = 2^{2x} = (2^x)^2$, so the equation is actually quadratic in form, and we can rewrite it as

$$(2^x)^2 - 2^x - 12 = 0$$

Now we can factor as usual:

$$(2^x - 4)(2^x + 3) = 0$$
$$2^x - 4 = 0 \quad \text{or} \quad 2^x + 3 = 0$$
$$2^x = 4 \qquad\qquad 2^x = -3$$

The equation on the left has the solution $x = 2$, since $2^x = 4 = 2^2$; the equation on the right has no solution, since $2^x > 0$ for all x. $\blacksquare$

In each of the preceding three examples, we were able to write each exponential expression using the same base, obtaining exact solutions to the equation. When this is not possible, logarithms can sometimes be used to obtain the solution.

E X A M P L E   7   *Solving an Exponential Equation*

Solve for x: $2^x = 5$

Solution   We write the exponential equation as the equivalent logarithmic equation:

$$2^x = 5$$

$$x = \log_2 5 = \frac{\ln 5}{\ln 2}$$
$$\uparrow$$
$$\text{Change-of-Base Formula (10)}$$

Alternatively, we can solve the equation $2^x = 5$ by taking the natural logarithm (or common logarithm) of each side. Taking the natural logarithm,

$$2^x = 5$$
$$\ln 2^x = \ln 5$$
$$x \ln 2 = \ln 5$$
$$x = \frac{\ln 5}{\ln 2}$$

Using a calculator, the solution, rounded to three decimal places, is:

$$x = \frac{\ln 5}{\ln 2} = 2.322$$ ∎

■ Now work Problem 33.

E X A M P L E   8   *Solving an Exponential Equation*

Solve for x: $8 \cdot 3^x = 5$

Solution

$$8 \cdot 3^x = 5 \qquad \text{Isolate } 3^x \text{ on the left side.}$$
$$3^x = \tfrac{5}{8} \qquad \text{Proceed as in Example 7.}$$
$$x = \log_3\!\left(\tfrac{5}{8}\right) = \frac{\ln \tfrac{5}{8}}{\ln 3}$$

Using a calculator, the solution, rounded to three decimal places, is:

$$x = \frac{\ln\left(\tfrac{5}{8}\right)}{\ln 3} = -0.428$$ ∎

E X A M P L E   9   *Solving an Exponential Equation*

Solve for x: $5^{x-2} = 3^{3x+2}$

Solution

Because the bases are different, we take the natural logarithm of each side and apply appropriate properties of logarithms. The result is an equation in x that we can solve.

$$5^{x-2} = 3^{3x+2}$$
$$\ln 5^{x-2} = \ln 3^{3x+2} \qquad \text{Property (7)}$$
$$(x-2)\ln 5 = (3x+2)\ln 3 \qquad \text{Property (6)}$$
$$(\ln 5)x - 2 \ln 5 = (3 \ln 3)x + 2 \ln 3$$
$$(\ln 5 - 3 \ln 3)x = 2 \ln 3 + 2 \ln 5$$
$$x = \frac{2(\ln 3 + \ln 5)}{\ln 5 - 3 \ln 3} = -3.212$$ ∎

■ Now work Problem 41.

Graphing Utility Solutions

The techniques introduced in this section only apply to certain types of logarithmic and exponential equations. Solutions for other types are usually studied in calculus, using numerical methods. For example, the logarithmic equation $\log_3 x + \log_4 x = 4$ cannot be solved by previous methods. (Notice the bases are different.) We can use a graphing utility to approximate the solution.

E X A M P L E   1 0   *Solving Equations Using a Graphing Utility*

Solve: $\log_3 x + \log_4 x = 4$

Express the solution(s) correct to two decimal places.

Solution   This type of logarithmic equation cannot be solved by previous methods. However, a graphing utility can be used here. We begin by noting that the equation to be solved is equivalent to

$$\log_3 x + \log_4 x = 4$$
$$\log_3 x + \log_4 x - 4 = 0$$

The solution(s) of this equation is the same as the x-intercepts of the graph of the function $f(x) = \log_3 x + \log_4 x - 4$. (Remember to use the Change-of-Base Formula to graph f.) This function f is increasing (do you see why?) and so has at most one x-intercept. Since $f(10) \approx -0.24 < 0$ and $f(15) \approx 0.41 > 0$, it follows that there is one x-intercept and it is between 10 and 15. So we graph the function with $10 \le x \le 15$. See Figure 23.

FIGURE 23

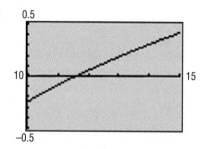

By using TRACE, ZOOM-IN, and/or BOX, we obtain the solution $x = 11.60$ correct to two decimal places.   ■

E X A M P L E   1 1   *Solving Equations Using a Graphing Utility*

Solve: $x + e^x = 2$

Express the solution(s) correct to two decimal places.

Solution   This type of exponential equation cannot be solved by previous methods. A graphing utility, though, can be used here. We begin by noting that the equation to be solved is equivalent to

$$x + e^x = 2$$
$$x + e^x - 2 = 0$$

The solution(s) of this equation is the same as the x-intercept(s) of the graph of the function $f(x) = x + e^x - 2$. This function f is increasing (do you see why?) and so has at most one x-intercept. Since $f(0) = -1 < 0$ and $f(1) = 1 + e - 2 > 0$, it follows that there is one x-intercept and it is between 0 and 1. So we graph the function, with $0 \le x \le 1$. See Figure 24.

FIGURE 24

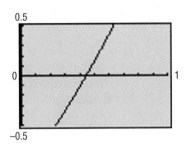

By using TRACE, ZOOM-IN, and/or BOX, we obtain the solution $x = 0.44$, correct to two decimal places. ∎

7.4

Exercise 7.4

In Problems 1–58, solve each equation. Verify your solution using a graphing utility.

1. $\log_2(2x + 1) = 3$

2. $\log_3(3x - 2) = 2$

3. $\log_3(x^2 + 1) = 2$

4. $\log_5(x^2 + x + 4) = 2$

5. $\frac{1}{2}\log_3 x = 2\log_3 2$

6. $-2\log_4 x = \log_4 9$

7. $2\log_5 x = 3\log_5 4$

8. $3\log_2 x = -\log_2 27$

9. $3\log_2(x - 1) + \log_2 4 = 5$

10. $2\log_3(x + 4) - \log_3 9 = 2$

11. $\log_{10} x + \log_{10}(x + 15) = 2$

12. $\log_4 x + \log_4(x - 3) = 1$

13. $\log_x 4 = 2$

14. $\log_x\left(\frac{1}{8}\right) = 3$

15. $\log_3(x - 1)^2 = 2$

16. $\log_2(x + 4)^3 = 6$

17. $\log_{1/2}(3x + 1)^{1/3} = -2$

18. $\log_{1/3}(1 - 2x)^{1/2} = -1$

19. $2^{2x+1} = 4$

20. $5^{1-2x} = \frac{1}{5}$

21. $3^{x^3} = 9^x$

22. $4^{x^2} = 2^x$

23. $8^{x^2-2x} = \frac{1}{2}$

24. $9^{-x} = \frac{1}{3}$

25. $2^x \cdot 8^{-x} = 4^x$

26. $\left(\frac{1}{2}\right)^{1-x} = 4$

27. $2^{2x} - 2^x - 12 = 0$

28. $3^{2x} + 3^x - 2 = 0$

29. $3^{2x} + 3^{x+1} - 4 = 0$

30. $4^x - 2^x = 0$

31. $4^x = 8$

32. $9^{2x} = 27$

33. $2^x = 10$

34. $3^x = 14$

35. $8^{-x} = 1.2$

36. $2^{-x} = 1.5$

37. $3^{1-2x} = 4^x$

38. $2^{x+1} = 5^{1-2x}$

39. $\left(\frac{3}{5}\right)^x = 7^{1-x}$

40. $\left(\frac{4}{3}\right)^{1-x} = 5^x$

41. $1.2^x = (0.5)^{-x}$

42. $(0.3)^{1+x} = 1.7^{2x-1}$

43. $\pi^{1-x} = e^x$

44. $e^{x+3} = \pi^x$

45. $5(2^{3x}) = 8$

46. $0.3(4^{0.2x}) = 0.2$

47. $400e^{0.2x} = 600$

48. $500e^{0.3x} = 600$

49. $\log_a(x - 1) - \log_a(x + 6) = \log_a(x - 2) - \log_a(x + 3)$

50. $\log_a x + \log_a(x - 2) = \log_a(x + 4)$

51. $\log_{1/3}(x^2 + x) - \log_{1/3}(x^2 - x) = -1$

52. $\log_4(x^2 - 9) - \log_4(x + 3) = 3$

53. $\log_2 8^x = -3$

54. $\log_3 3^x = -1$

55. $\log_2(x^2 + 1) - \log_4 x^2 = 1$
   [*Hint:*   Change $\log_4 x^2$ to base 2.]

56. $\log_2(3x + 2) - \log_4 x = 3$

57. $\log_{16} x + \log_4 x + \log_2 x = 7$

58. $\log_9 x + 3\log_3 x = 14$

In Problems 59–74, use a graphing utility to solve each equation. Express your answer correct to two decimal places.

59. $\log_5 x + \log_3 x = 1$

60. $\log_2 x + \log_6 x = 3$

61. $\log_5(x + 1) - \log_4(x - 2) = 1$

62. $\log_2(x - 1) - \log_6(x + 2) = 2$

63. $e^x = -x$

64. $e^{2x} = x + 2$

65. $e^x = x^2$

66. $e^x = x^3$

67. $\ln x = -x$

68. $\ln 2x = -x + 2$

69. $\ln x = x^3 - 1$

70. $\ln x = -x^2$

71. $e^x + \ln x = 4$

72. $e^x - \ln x = 4$

73. $e^{-x} = \ln x$

74. $e^{-x} = -\ln x$

7.5

Compound Interest

Interest is money paid for the use of money. The total amount borrowed (whether by an individual from a bank in the form of a loan or by a bank from an individual in the form of a savings account) is called the **principal.** The **rate of interest,** expressed as a percent, is the amount charged for the use of the principal for a given period of time, usually on a yearly (that is, per annum) basis.

If a principal of P dollars is borrowed for a period of t years at a per annum interest rate r, expressed as a decimal, the interest I charged is

Simple Interest Formula

$$I = Prt \tag{1}$$

Interest charged according to formula (1) is called **simple interest.**

In working with problems involving interest we use the term **payment period** as follows:

Annually	Once per year
Semiannually	Twice per year
Quarterly	4 times per year
Monthly	12 times per year
Daily	365 times per year*

When the interest due at the end of a payment period is added to the principal so that the interest computed at the end of the next payment period is based on this new principal amount (old principal + interest), the interest is said to have been **compounded.** Thus, **compound interest** is interest paid on previously earned interest.

E X A M P L E   1

Computing Compound Interest

A credit union pays interest of 8% per annum compounded quarterly on a certain savings plan. If $1000 is deposited in such a plan and the interest is left to accumulate, how much is in the account after 1 year?

Solution

We use the simple interest formula, $I = Prt$. The principal P is $1000 and the rate of interest is 8% $= 0.08$. After the first quarter of a year, the time t is $\frac{1}{4}$ year, so the interest earned is

$$I = Prt = (\$1000)(0.08)\left(\tfrac{1}{4}\right) = \$20$$

The new principal is $P + I = \$1000 + \$20 = \$1020$. At the end of the second quarter, the interest on this principal is

$$I = (\$1020)(0.08)\left(\tfrac{1}{4}\right) = \$20.40$$

At the end of the third quarter, the interest on the new principal of $1020 + $20.40 = $1040.40 is

$$I = (\$1040.40)(0.08)\left(\tfrac{1}{4}\right) = \$20.81$$

*Some banks use a 360 day "year."

Finally, after the fourth quarter, the interest is

$$I = (\$1061.21)(0.08)\left(\tfrac{1}{4}\right) = \$21.22$$

Thus, after 1 year the account contains $1082.43. ∎

The pattern of the calculations performed in Example 1 leads to a general formula for compound interest. To fix our ideas, let P represent the principal to be invested at a per annum interest rate r, which is compounded n times per year. (For computing purposes, r is expressed as a decimal.) The interest earned after each compounding period is the principal times r/n. Thus, the amount A after one compounding period is

$$A = P + P\left(\frac{r}{n}\right) = P\left(1 + \frac{r}{n}\right)$$

After two compounding periods, the amount A, based on the new principal $P(1 + r/n)$, is

$$A = \underbrace{P\left(1 + \frac{r}{n}\right)}_{\substack{\text{New} \\ \text{principal}}} + \underbrace{P\left(1 + \frac{r}{n}\right)\left(\frac{r}{n}\right)}_{\substack{\text{Interest on} \\ \text{new principal}}} = P\left(1 + \frac{r}{n}\right)\left(1 + \frac{r}{n}\right) = P\left(1 + \frac{r}{n}\right)^2$$

After three compounding periods,

$$A = P\left(1 + \frac{r}{n}\right)^2 + P\left(1 + \frac{r}{n}\right)^2\left(\frac{r}{n}\right) = P\left(1 + \frac{r}{n}\right)^2\left(1 + \frac{r}{n}\right) = P\left(1 + \frac{r}{n}\right)^3$$

Continuing in this way, after n compounding periods (1 year),

$$A = P\left(1 + \frac{r}{n}\right)^n$$

Because t years will contain $n \cdot t$ compounding periods, after t years we have

$$A = P\left(1 + \frac{r}{n}\right)^{nt}$$

Theorem The amount A after t years due to a principal P invested at an annual interest rate r compounded n times per year is

Compound Interest Formula

$$A = P\left(1 + \frac{r}{n}\right)^{nt} \tag{2}$$

∎

E X A M P L E   2 *Comparing Investments Using Different Compounding Periods*

Investing $1000 at an annual rate of 10% compounded annually, quarterly, monthly, and daily will yield the following amounts after 1 year:

Annual compounding: $A = P(1 + r)$
$$= (\$1000)(1 + 0.10) = \$1100.00$$

Quarterly compounding: $A = P\left(1 + \dfrac{r}{4}\right)^4$

$\qquad\qquad\qquad\qquad\qquad = (\$1000)(1 + 0.025)^4 = \$1103.81$

Monthly compounding: $A = P\left(1 + \dfrac{r}{12}\right)^{12}$

$\qquad\qquad\qquad\qquad\qquad = (\$1000)(1 + 0.00833)^{12} = \$1104.71$

Daily compounding: $A = P\left(1 + \dfrac{r}{365}\right)^{365}$

$\qquad\qquad\qquad\qquad\qquad = (\$1000)(1 + 0.000274)^{365} = \$1105.16$ ■

■   Now work Problem 1.

From Example 2, we can see that the effect of compounding more frequently is that the account after 1 year is higher: $\$1000$ compounded 4 times a year at 10% results in $\$1103.81$; $\$1000$ compounded 12 times a year at 10% results in $\$1104.71$; and $\$1000$ compounded 365 times a year at 10% results in $\$1105.16$. This leads to the following question: What would happen to the amount after 1 year if the number of times the interest is compounded were increased without bound?

Let's find the answer. Suppose P is the principal, r is the per annum interest rate, and n is the number of times the interest is compounded each year. The amount after 1 year is

$$A = P\left(1 + \frac{r}{n}\right)^n$$

Now suppose that the number n of times the interest is compounded per year gets larger and larger; that is, suppose that $n \to \infty$. Then,

$$A = P\left(1 + \frac{r}{n}\right)^n = P\left[1 + \frac{1}{n/r}\right]^n = P\left(\left[1 + \frac{1}{n/r}\right]^{n/r}\right)^r = P\left[\left(1 + \frac{1}{h}\right)^h\right]^r \quad (3)$$

$$\underset{\uparrow}{}$$
$$h = \frac{n}{r}$$

Thus, in (3), as $n \to \infty$, then $h = n/r \to \infty$, and the expression in brackets equals e, [Refer to (2) on p. 453] so $A \to Pe^r$. Table 6 compares $(1 + r/n)^n$, for large values of n, to e^r for $r = 0.05$, $r = 0.10$, $r = 0.15$, and $r = 1$. The larger n gets, the closer $(1 + r/n)^n$ gets to e^r. Thus, no matter how frequent the compounding, the amount after 1 year has the definite ceiling Pe^r.

TABLE 6

$$\left(1 + \frac{r}{n}\right)^n$$

	$n = 100$	$n = 1000$	$n = 10{,}000$	e^r
$r = 0.05$	1.0512579	1.05127	1.051271	1.0512711
$r = 0.10$	1.1051157	1.1051654	1.1051703	1.1051709
$r = 0.15$	1.1617037	1.1618212	1.1618329	1.1618342
$r = 1$	2.7048138	2.7169239	2.7181459	2.7182818

When interest is compounded so that the amount after 1 year is Pe^r, we say the interest is **compounded continuously.**

Theorem

The amount A after t years due to a principal P invested at an annual interest rate r compounded continuously is

Continuous Compounding

$$A = Pe^{rt} \qquad (4)$$

EXAMPLE 3

Using Continuous Compounding

The amount A that results from investing a principal P of \$1000 at an annual rate r of 10% compounded continuously for a time t of 1 year is

$$A = \$1000e^{0.10} = (\$1000)(1.10517) = \$1105.17$$

■ Now work Problem 9.

The **effective rate of interest** is the equivalent annual simple rate of interest that would yield the same amount as compounding after 1 year. For example, based on Example 3, a principal of \$1000 will result in \$1105.17 at a rate of 10% compounded continuously. To get this same amount using a simple rate of interest would require that interest of \$1105.17 − \$1000.00 = \$105.17 be earned on the principal. Since \$105.17 is 10.517% of \$1000, a simple rate of interest of 10.517% is needed to equal 10% compounded continuously. Thus, the effective rate of interest of 10% compounded continuously is 10.517%.

Based on the results of Examples 2 and 3, we find the following comparisons:

	ANNUAL RATE	EFFECTIVE RATE
Annual compounding	10%	10%
Quarterly compounding	10%	10.381%
Monthly compounding	10%	10.471%
Daily compounding	10%	10.516%
Continuous compounding	10%	10.517%

■ Now work Problem 21.

EXAMPLE 4

Computing the Value of an IRA

On January 2, 1996, \$2000 is placed in an Individual Retirement Account (IRA) that will pay interest of 10% per annum compounded continuously. What will the IRA be worth on January 1, 2016?

Solution

The amount A after 20 years is

$$A = Pe^{rt} = \$2000e^{(0.10)(20)} = \$14{,}778.11$$

Check: Graph $y = 2000e^{0.1x}$ and use TRACE to verify that when $x = 20$ then $y = \$14{,}778.11$. ■

Exploration: How long will it be until $y = \$40{,}000$? ■

When people engaged in finance speak of the "time value of money," they are usually referring to the **present value** of money. The present value of A dollars to be received at a future date is the principal you would need to invest now so that

FIGURE 25

Time is Money

it would grow to A dollars in the specified time period. Thus, the present value of money to be received at a future date is always less than the amount to be received, since the amount to be received will equal the present value (money invested now) *plus* the interest accrued over the time period.

We use the compound interest formula (2) to get a formula for present value. If P is the present value of A dollars to be received after t years at a per annum interest rate r compounded n times per year, then, by formula (2),

$$A = P\left(1 + \frac{r}{n}\right)^{nt}$$

To solve for P, we divide both sides by $(1 + r/n)^{nt}$, and the result is

$$\frac{A}{(1 + r/n)^{nt}} = P \quad \text{or} \quad P = A\left(1 + \frac{r}{n}\right)^{-nt}$$

Theorem   The present value P of A dollars to be received after t years, assuming a per annum interest rate r compounded n times per year, is

Present Value Formulas

$$P = A\left(1 + \frac{r}{n}\right)^{-nt} \tag{5}$$

If the interest is compounded continuously, then

$$P = Ae^{-rt} \tag{6}$$

∎

To prove (6), solve formula (4) for P.

E X A M P L E   5   *Computing the Value of a Zero-Coupon Bond*

A zero-coupon (noninterest-bearing) bond can be redeemed in 10 years for $1000. How much should you be willing to pay for it now if you want a return of:

(a)   8% compounded monthly?

(b)   7% compounded continuously?

Solution   (a)   We are seeking the present value of $1000. Thus, we use formula (5) with $A = \$1000$, $n = 12$, $r = 0.08$, and $t = 10$:

$$P = A\left(1 + \frac{r}{n}\right)^{-nt}$$

$$= \$1000\left(1 + \frac{0.08}{12}\right)^{-12(10)}$$

$$= \$450.52$$

For a return of 8% compounded monthly, you should pay $450.52 for the bond.

(b) Here, we use formula (6) with $A = \$1000$, $r = 0.07$, and $t = 10$:

$$P = Ae^{-rt}$$
$$= \$1000e^{-(0.07)(10)}$$
$$= \$496.59$$

For a return of 7% compounded continuously, you should pay \$496.59 for the bond. ∎

■ Now work Problem 11.

EXAMPLE 6   *Rate of Interest Required to Double an Investment*

What annual rate of interest compounded annually should you seek if you want to double your investment in 5 years?

Algebraic Solution   If P is the principal and we want P to double, the amount A will be $2P$. We use the compound interest formula with $n = 1$ and $t = 5$ to find r:

$$2P = P(1 + r)^5$$
$$2 = (1 + r)^5$$
$$1 + r = \sqrt[5]{2}$$
$$r = \sqrt[5]{2} - 1 = 1.148698 - 1 = 0.148698$$

The annual rate of interest needed to double the principal in 5 years is 14.87%.

Graphing Solution   We solve the equation

$$2 = (1 + r)^5$$

for r by graphing the two functions $y_1 = 2$ and $y_2 = (1 + x)^5$. The x-coordinate of their point of intersection is the rate r we seek. See Figure 26. [We could also graph $y = (1 + x)^5 - 2$. Then the x-intercept is the rate r we seek.] Using the INTERSECT command*, we find the point of intersection of y_1 and y_2 is (0.14869835, 2).

FIGURE 26

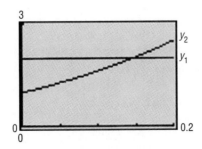

■ Now work Problem 23.

EXAMPLE 7   *Doubling and Tripling Time for an Investment*

(a) How long will it take for an investment to double in value if it earns 5% compounded continuously?

(b) How long will it take to triple at this rate?

*Note: If your utility does not have an INTERSECT command, you will have to use BOX or ZOOM-IN with TRACE to find the point of intersection.

Algebraic Solution   (a)   If P is the initial investment and we want P to double, the amount A will be $2P$. We use formula (4) for continuously compounded interest with $r = 0.05$. Then

$$A = Pe^{rt}$$
$$2P = Pe^{0.05t}$$
$$2 = e^{0.05t}$$
$$0.05t = \ln 2$$
$$t = \frac{\ln 2}{0.05} = 13.86$$

It will take about 14 years to double the investment.

Graphing Solution   We solve the equation

$$2 = e^{0.05t}$$

for t by graphing the two functions $y_1 = 2$ and $y_2 = e^{0.05x}$. Their point of intersection is $(13.86, 2)$. See Figure 27.

FIGURE 27

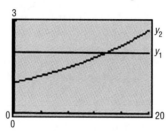

Algebraic Solution   (b)   To triple the investment, we set $A = 3P$ in formula (4).

$$A = Pe^{rt}$$
$$3P = Pe^{0.05t}$$
$$3 = e^{0.05t}$$
$$0.05t = \ln 3$$
$$t = \frac{\ln 3}{0.05} = 21.97$$

It will take about 22 years to triple the investment.

Graphing Solution   We solve the equation

$$3 = e^{0.05t}$$

for t by graphing the two functions $y_1 = 3$ and $y_2 = e^{0.05x}$. Their point of intersection is $(21.97, 3)$. See Figure 28.

FIGURE 28

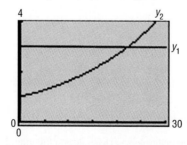

■   Now work Problem 29.

7.5

Exercise 7.5

In Problems 1–10, find the amount that results from each investment.

1. $100 invested at 4% compounded quarterly after a period of 2 years
2. $50 invested at 6% compounded monthly after a period of 3 years
3. $500 invested at 8% compounded quarterly after a period of $2\frac{1}{2}$ years
4. $300 invested at 12% compounded monthly after a period of $1\frac{1}{2}$ years
5. $600 invested at 5% compounded daily after a period of 3 years
6. $700 invested at 6% compounded daily after a period of 2 years
7. $10 invested at 11% compounded continuously after a period of 2 years
8. $40 invested at 7% compounded continuously after a period of 3 years
9. $100 invested at 10% compounded continuously after a period of $2\frac{1}{4}$ years
10. $100 invested at 12% compounded continuously after a period of $3\frac{3}{4}$ years

In Problems 11–20, find the principal needed now to get each amount; that is, find the present value.

11. To get $100 after 2 years at 6% compounded monthly
12. To get $75 after 3 years at 8% compounded quarterly
13. To get $1000 after $2\frac{1}{2}$ years at 6% compounded daily
14. To get $800 after $3\frac{1}{2}$ years at 7% compounded monthly
15. To get $600 after 2 years at 4% compounded quarterly
16. To get $300 after 4 years at 3% compounded daily
17. To get $80 after $3\frac{1}{4}$ years at 9% compounded continuously
18. To get $800 after $2\frac{1}{2}$ years at 8% compounded continuously
19. To get $400 after 1 year at 10% compounded continuously
20. To get $1000 after 1 year at 12% compounded continuously

21. Find the effective rate of interest for $5\frac{1}{4}$% compounded quarterly.
22. What interest rate compounded quarterly will give an effective interest rate of 7%?
23. What annual rate of interest is required to double an investment in 3 years? Verify your answer using a graphing utility.
24. What annual rate of interest is required to double an investment in 10 years? Verify your answer using a graphing utility.

In Problems 25–28, which of the two rates would yield the larger amount in 1 year? [Hint: Start with a principal of $10,000 in each instance.]

25. 6% compounded quarterly or $6\frac{1}{4}$% compounded annually?
26. 9% compounded quarterly or $9\frac{1}{4}$% compounded annually?
27. 9% compounded monthly or 8.8% compounded daily?
28. 8% compounded semiannually or 7.9% compounded daily?

29. How long does it take for an investment to double in value if it is invested at 8% per annum compounded monthly? Compounded continuously? Verify your answer using a graphing utility.
30. How long does it take for an investment to double in value if it is invested at 10% per annum compounded monthly? Compounded continuously? Verify your answer using a graphing utility.
31. If you have $100 to invest at 8% per annum compounded monthly, how long will it be before the amount is $150? If the compounding is continuous, how long will it be?

32. If you have $100 to invest at 10% per annum compounded monthly, how long will it be before the amount is $175? If the compounding is continuous, how long will it be?

33. How many years will it take for an initial investment of $10,000 to grow to $25,000? Assume a rate of interest of 6% compounded continuously.

34. How many years will it take for an initial investment of $25,000 to grow to $80,000? Assume a rate of interest of 7% compounded continuously.

35. What will a $90,000 house cost 5 years from now if the inflation rate over that period averages 3% compounded annually?

36. A department store charges 1.25% per month on the unpaid balance for customers with charge accounts (interest is compounded monthly). A customer charges $200 and does not pay her bill for 6 months. What is the bill at that time?

37. You will be buying a new car for $15,000 in 3 years. How much money should you ask your parents for now so that, if you invest it at 5% compounded continuously, you will have enough to buy the car?

38. You will require $3000 in 6 months to pay off a loan that has no prepayment privileges. If you have the $3000 now, how much of it should you save in an account paying 3% compounded monthly so that in 6 months you will have exactly $3000?

39. You are contemplating the purchase of 100 shares of a stock selling for $15 per share. The stock pays no dividends. The history of the stock indicates that it should grow at an annual rate of 15% per year. How much will the 100 shares of stock be worth in 5 years?

40. You are contemplating the purchase of 100 shares of a stock selling for $15 per share. The stock pays no dividends. Your broker says the stock will be worth $20 per share in 2 years. What is the annual rate of return on this investment?

41. A business purchased for $650,000 in 1994 is sold in 1997 for $850,000. What is the annual rate of return for this investment?

42. You have just inherited a diamond ring appraised at $5000. If diamonds have appreciated in value at an annual rate of 8%, what was the value of the ring 10 years ago when the ring was purchased?

43. You place $1000 in a bank account that pays 5.6% compounded continuously. After 1 year, will you have enough money to buy a computer system that costs $1060? If another bank will pay you 5.9% compounded monthly, is this a better deal?

44. On January 1, you place $1000 in a Certificate of Deposit that pays 6.8% compounded continuously and matures in 3 months. Then you place the $1000 and the interest in a passbook account that pays 5.25% compounded monthly. How much do you have in the passbook account on May 1?

45. You invest $2000 in a bond trust that pays 9% interest compounded semiannually. Your friend invests $2000 in a Certificate of Deposit (CD) that pays $8\frac{1}{2}$% compounded continuously. Who has more money after 20 years, you or your friend?

46. Suppose that you have access to an investment that will pay 10% interest compounded continuously. Which is better: To be given $1000 now so that you can take advantage of this investment opportunity or to be given $1325 after 3 years?

47. You have just purchased a house for $150,000, with the seller holding a second mortgage of $50,000. You promise to pay the seller $50,000 plus all accrued interest 5 years from now. The seller offers you three interest options on the second mortgage:
(a) Simple interest at 12% per annum   (b) $11\frac{1}{2}$% interest compounded monthly
(c) $11\frac{1}{4}$% interest compounded continuously

Which option is best; that is, which results in the least interest on the loan?

48. A bank advertises that it pays interest on saving accounts at the rate of 4.25% compounded daily. Find the effective rate if the bank uses (a) 360 days or (b) 365 days in determining the daily rate.

Problems 49–52 involve zero-coupon bonds. A zero-coupon bond *is a bond that is sold now at a discount and will pay its face value at some time in the future when it matures; no interest payments are made.*

49. A zero-coupon bond can be redeemed in 20 years for $10,000. How much should you be willing to pay for it now if you want a return of:
(a) 10% compounded monthly?   (b) 10% compounded continuously?

50. A child's grandparents are considering buying a $40,000 face value zero-coupon bond at birth so that she will have enough money for her college education 17 years later. If money is worth 8% compounded annually, what should they pay for the bond?

51. How much should a $10,000 face value zero-coupon bond, maturing in 10 years, be sold for now if its rate of return is to be 8% compounded annually?

52. If you pay $12,485.52 for a $25,000 face value zero-coupon bond that matures in 8 years, what is your annual rate of return?

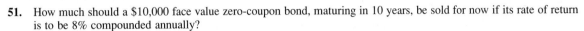

53. Explain in your own words what the term *compound interest* means. What does *continuous compounding* mean?

54. Explain in your own words the meaning of present value.

55. Write a program that will calculate the amount after n years if a principal P is invested at r% per annum compounded quarterly. Use it to verify your answers to Problem 1 and 3.

56. Write a program that will calculate the principal needed now to get the amount A in n years at r% per annum compounded daily. Use it to verify your answer to Problem 13.

57. Write a program that will calculate the annual rate of interest required to double an investment in n years. Use it to verify your answers to Problems 23 and 24.

58. Write a program that will calculate the number of months required for an initial investment of x dollars to grow to y dollars at r% per annum compounded continuously. Use it to verify your answer to Problems 33 and 34.

59. *Time to Double or Triple an Investment* The formula

$$y = \frac{\ln m}{n \ln\left(1 + \dfrac{r}{n}\right)}$$

can be used to find the number of years y required to multiply an investment m times when r is the per annum interest rate compounded n times a year.
(a) How many years will it take to double the value of an IRA that compounds annually at the rate of 12%?
(b) How many years will it take to triple the value of a savings account that compounds quarterly at an annual rate of 6%?
(c) Give a derivation of this formula.

60. *Time to Reach an Investment Goal* The formula

$$y = \frac{\ln A - \ln P}{r}$$

can be used to find the number of years y required for an investment P to grow to a value A when compounded continuously at an annual rate r.
(a) How long will it take to increase an initial investment of $1000 to $8000 at an annual rate of 10%?
(b) What annual rate is required to increase the value of a $2000 IRA to $30,000 in 35 years?
(c) Give a derivation of this formula.

61. *Critical Thinking* You have just contracted to buy a house and will seek financing in the amount of $100,000. You go to several banks. Bank 1 will lend you $100,000 at the rate of 8.75% amortized over 30 years with a loan origination fee of 1.75%. Bank 2 will lend you $100,000 at the rate of 8.375% amortized over 15 years with a loan origination fee of 1.5%. Bank 3 will lend you $100,000 at the rate of 9.125% amortized over 30 years with no loan origination fee. Bank 4 will lend you $100,000 at the rate of 8.625% amortized over 15 years with no loan origination fee. Which loan would you take? Why? Be sure to have sound reasons for your choice. If the amount of the monthly payment does not matter to you, which loan would you take? Again, have sound reasons for your choice. Use the information in the table to assist you. Compare your final decision with others in the class. Discuss.

	MONTHLY PAYMENT	LOAN ORIGINATION FEE
Bank 1	$786.70	$1750.00
Bank 2	$977.42	$1500.00
Bank 3	$813.63	$0.00
Bank 4	$990.68	$0.00

7.6

Growth and Decay

Many natural phenomena have been found to follow the law that an amount A varies with time t according to

$$A = A_0 e^{kt} \tag{1}$$

where A_0 is the original amount ($t = 0$) and $k \neq 0$ is a constant.

If $k > 0$, then equation (1) states that the amount A is increasing over time; if $k < 0$, the amount A is decreasing over time. In either case, when an amount A varies over time according to equation (1), it is said to follow the **exponential law** or the **law of uninhibited growth** ($k > 0$) **or decay** ($k < 0$). See Figure 29.

FIGURE 29

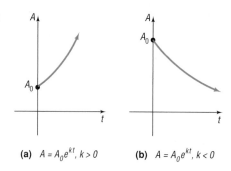

(a) $A = A_0 e^{kt}, k > 0$ **(b)** $A = A_0 e^{kt}, k < 0$

For example, we saw in Section 7.5 that continuously compounded interest follows the law of uninhibited growth. In this section we shall look at three additional phenomena that follow the exponential law.

Biology

Mitosis, or division of cells, is a universal process in the growth of living organisms such as amebas, plants, human skin cells, and many others. Based on an ideal situation in which no cells die and no by-products are produced, the number of cells present at a given time follows the law of uninhibited growth. Actually, however, after enough time has passed, growth at an exponential rate will cease due to the influence of factors such as lack of living space, dwindling food supply, and so on. The law of uninhibited growth accurately reflects only the early stages of the mitosis process.

The mitosis process begins with a culture containing N_0 cells. Each cell in the culture grows for a certain period of time and then divides into two identical cells. We assume that the time needed for each cell to divide in two is constant and does not change as the number of cells increases. These cells then grow, and eventually each divides in two; and so on.

A formula that gives the number N of cells in the culture after a time t has passed (in the early stages of growth) is

Uninhibited Growth of Cells

$$N(t) = N_0 e^{kt} \qquad k > 0 \tag{2}$$

where k is a positive constant.

In using equation (2) to model the growth of cells, we are using a function that yields positive real numbers, even though we are counting the number of cells, which must be an integer. This is a common practice in many applications.

E X A M P L E   1

Bacterial Growth

A colony of bacteria increases according to the law of uninhibited growth.

(a) If the number of bacteria doubles in 3 hours, how long will it take for the size of the colony to triple?

(b) Using a graphing utility, verify that the population doubles in 3 hours.

(c) Using a graphing utility, BOX and TRACE to approximate the time it takes for the population to double a second time (i.e., increase four times). Does this answer seem reasonable?

Solution (a) Using formula (2), the number N of cells at a time t is

$$N(t) = N_0 e^{kt}$$

where N_0 is the initial number of bacteria present and k is a positive number. We first seek the number k. The number of cells doubles in 3 hours; thus, we have

$$N(3) = 2N_0$$

But $N(3) = N_0 e^{k(3)}$, so

$$N_0 e^{k(3)} = 2N_0$$
$$e^{3k} = 2$$
$$3k = \ln 2 \quad \text{Write the exponential equation as a logarithm.}$$
$$k = \tfrac{1}{3} \ln 2 \approx \tfrac{1}{3}(0.6931) = 0.2310$$

Formula (2) for this growth process is therefore

$$N(t) = N_0 e^{0.2310t}$$

The time t needed for the size of the colony to triple requires that $N = 3N_0$. Thus, we substitute $3N_0$ for N to get

$$3N_0 = N_0 e^{0.2310t}$$
$$3 = e^{0.2310t}$$
$$0.2310t = \ln 3$$
$$t = \frac{1}{0.2310} \ln 3 \approx \frac{1.0986}{0.2310} = 4.756 \text{ hours}$$

It will take about 4.756 hours for the size of the colony to triple.

(b) Figure 30 shows the graph of $y = e^{0.2310x}$ where x represents the time in hours and y represents the multiple of the original population. Using BOX and TRACE, it takes 3 hours for the population to double.

(c) Using BOX and TRACE, it takes 6 hours for the population to double again.

∎

FIGURE 30

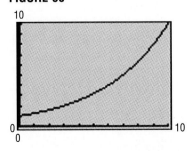

■ Now work Problem 1.

Radioactive Decay

Radioactive materials follow the law of uninhibited decay. Thus, the amount A of a radioactive material present at time t is given by the formula

---
Uninhibited Radioactive
Decay

$$A = A_0 e^{kt} \qquad k < 0 \qquad\qquad (3)$$

where A_0 is the original amount of radioactive material and k is a negative number.

All radioactive substances have a specific **half-life,** which is the time required for half of the radioactive substance to decay. In **carbon dating,** we use the fact that all living organisms contain two kinds of carbon, carbon-12 (a stable carbon) and carbon-14 (a radioactive carbon, with a half-life of 5600 years). While an organism is living, the ratio of carbon-12 to carbon-14 is constant. But when an organism dies, the original amount of carbon-12 present remains unchanged, whereas the amount of carbon-14 begins to decrease. This change in the amount of carbon-14 present relative to the amount of carbon-12 present makes it possible to calculate when an organism died.

E X A M P L E   2

Estimating the Age of Ancient Tools

Traces of burned wood found along with ancient stone tools in an archaeological dig in Chile were found to contain approximately 1.67% of the original amount of carbon-14.

(a) If the half-life of carbon-14 is 5600 years, approximately when was the tree cut and burned?

(b) Using a graphing utility, graph the relation between the percentage of carbon-14 remaining and time.

(c) Using BOX and TRACE, determine the time that elapses until half of the carbon-14 remains. This answer should equal the half-life of carbon-14.

(d) Verify the answer found in (a).

Solution   (a) Using equation (3), the amount A of carbon-14 present at time t is

$$A = A_0 e^{kt}$$

where A_0 is the original amount of carbon-14 present and k is a negative number. We first seek the number k. To find it, we use the fact that after 5600 years half of the original amount of carbon-14 remains. Thus,

$$\frac{1}{2} A_0 = A_0 e^{k(5600)}$$

$$\frac{1}{2} = e^{5600k}$$

$$5600k = \ln \frac{1}{2}$$

$$k = \frac{1}{5600} \ln \frac{1}{2} \approx -0.000124$$

Formula (3) therefore becomes

$$A = A_0 e^{-0.000124t}$$

If the amount A of carbon-14 now present is 1.67% of the original amount, it follows that

$$0.0167A_0 = A_0e^{-0.000124t}$$

$$0.0167 = e^{-0.000124t}$$

$$-0.000124t = \ln 0.0167$$

$$t = \frac{1}{-0.000124} \ln 0.0167 \approx 33{,}000 \text{ years}$$

The tree was cut and burned about 33,000 years ago. Some archaeologists use this conclusion to argue that humans lived in the Americas 33,000 years ago, much earlier than is generally accepted.

(b) Figure 31 shows the graph of $y = e^{-0.000124x}$, where y in the fraction of carbon-14 present and x is the time.

FIGURE 31

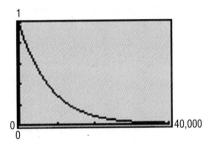

(c) Using BOX and TRACE, it takes 5600 years until half the carbon-14 remains.

(d) Using BOX and TRACE, it takes 33,000 years until 1.67% of the carbon-14 remains. ■

■ Now work Problem 3.

Newton's Law of Cooling

Newton's Law of Cooling* states that the temperature of a heated object decreases exponentially over time toward the temperature of the surrounding medium. That is, the temperature u of a heated object at a given time t satisfies the equation

Newton's Law of Cooling

$$u = T + (u_0 - T)e^{kt} \qquad k < 0 \qquad (4)$$

where T is the constant temperature of the surrounding medium, u_0 is the initial temperature of the heated object, and k is a negative number.

E X A M P L E 3   *Using Newton's Law of Cooling*

An object is heated to 100°C and is then allowed to cool in a room whose air temperature is 30°C.

(a) If the temperature of the object is 80°C after 5 minutes, when will its temperature be 50°C?

(b) Using a graphing utility, graph the relation found between the temperature and time.

*Named after Sir Isaac Newton (1642–1727), one of the cofounders of calculus.

(c) Using BOX and TRACE, verify that when $x = 18.6$, then $y = 50$.

(d) Using BOX and TRACE, determine the elapsed time before the object is 35°C.

(e) TRACE the function until $x = 100$. What do you notice about y, the temperature?

Solution (a) Using equation (4) with $T = 30$ and $u_0 = 100$, the temperature (in degrees Celsius) of the object at time t (in minutes) is

$$u = 30 + (100 - 30)e^{kt} = 30 + 70e^{kt} \qquad (5)$$

where k is a negative number. To find k, we use the fact that $u = 80$ when $t = 5$. Then

$$80 = 30 + 70e^{k(5)}$$
$$50 = 70e^{5k}$$
$$e^{5k} = \frac{50}{70}$$
$$5k = \ln \frac{5}{7}$$
$$k = \frac{1}{5} \ln \frac{5}{7} \approx -0.0673$$

Formula (5) therefore becomes

$$u = 30 + 70e^{-0.0673t}$$

Now, we want to find t when $u = 50$°C, so

$$50 = 30 + 70e^{-0.0673t}$$
$$20 = 70e^{-0.0673t}$$
$$e^{-0.0673t} = \frac{20}{70}$$
$$-0.0673t = \ln \frac{2}{7}$$
$$t = \frac{1}{-0.0673} \ln \frac{2}{7} \approx 18.6 \text{ minutes}$$

Thus, the temperature of the object will be 50°C after about 18.6 minutes.

(b) Figure 32 shows the graph of $y = 30 + 70e^{-0.0673x}$, where y is the temperature and x is the time.

FIGURE 32

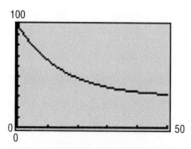

(c) Using BOX and TRACE, we find that $y = 50$ when $x = 18.6$.

(d) Using BOX and TRACE, $x = 39.21$ when $y = 35$°C.

(e) As x increases, y gets closer to 30, but never actually reaches 30. ∎

7.6

Exercise 7.6

1. *Growth of an Insect Population* The size P of a certain insect population at time t (in days) obeys the equation $P = 500e^{0.02t}$.
   (a) After how many days will the population reach 1000? 2000?
   (b) Using a graphing utility, graph the relation between P and t. Verify your answers in (a) using BOX and TRACE.

2. *Growth of Bacteria* The number N of bacteria present in a culture at time t (in hours) obeys the equation $N = 1000e^{0.01t}$.
   (a) After how many hours will the population equal 1500? 2000?
   (b) Using a graphing utility, graph the relation between N and t. Verify your answers in (a) using BOX and TRACE.

3. *Radioactive Decay* Strontium-90 is a radioactive material that decays according to the law $A = A_0 e^{-0.0244t}$, where A_0 is the initial amount present and A is the amount present at time t (in years). What is the half-life of strontium-90?

4. *Radioactive Decay* Iodine-131 is a radioactive material that decays according to the law $A = A_0 e^{-0.087t}$, where A_0 is the initial amount present and A is the amount present at time t (in days). What is the half-life of iodine-131?

5. (a) Use the information in Problem 3 to determine how long it takes for 100 grams of strontium-90 to decay to 10 grams.
   (b) Graph the relation $A = 100e^{-0.0244t}$ and verify your answer found in (a).

6. (a) Use the information in Problem 4 to determine how long it takes for 100 grams of iodine-131 to decay to 10 grams.
   (b) Graph the relation $A = 100e^{-0.087t}$ and verify your answer found in (a).

7. *Growth of a Colony of Mosquitoes* The population of a colony of mosquitoes obeys the law of uninhibited growth. If there are 1000 mosquitoes initially, and there are 1800 after 1 day, what is the size of the colony after 3 days? How long is it until there are 10,000 mosquitoes?

8. *Bacterial Growth* A culture of bacteria obeys the law of uninhibited growth. If 500 bacteria are present initially, and there are 800 after 1 hour, how many will be present in the culture after 5 hours? How long is it until there are 20,000 bacteria?

9. *Population Growth* The population of a southern city follows the exponential law. If the population doubled in size over an 18 month period and the current population is 10,000, what will the population be 2 years from now?

10. *Population Growth* The population of a midwestern city follows the exponential law. If the population decreased from 900,000 to 800,000 from 1993 to 1995, what will the population be in 1997?

11. *Radioactive Decay* The half-life of radium is 1690 years. If 10 grams are present now, how much will be present in 50 years?

12. *Radioactive Decay* The half-life of radioactive potassium is 1.3 billion years. If 10 grams are present now, how much will be present in 100 years? In 1000 years?

13. *Estimating the Age of A Tree* A piece of charcoal is found to contain 30% of the carbon-14 it originally had.
    (a) When did the tree from which the charcoal came die? Use 5600 years as the half-life of carbon-14.
    (b) Using a graphing utility, graph the relation between the percentage of carbon-14 remaining and time.
    (c) Using BOX and TRACE, determine the time that elapses until half of the carbon-14 remains.
    (d) Verify the answer found in (a).

14. *Estimating the Age of a Fossil* A fossilized leaf contains 70% of its normal amount of carbon-14.
    (a) How old is the fossil?
    (b) Using a graphing utility, graph the relation between the percentage of carbon-14 remaining and time.
    (c) Using BOX and TRACE, determine the time that elapses until half of the carbon-14 remains.
    (d) Verify the answer found in (a).

15. *Cooling Time of a Pizza* A pizza baked at 450°F is removed from the oven at 5:00 PM into a room that is a constant 70°F. After 5 minutes, the pizza is at 300°F.
    (a) At what time can you begin eating the pizza if you want its temperature to be 135°F?
    (b) Using a graphing utility, graph the relation between temperature and time.
    (c) Using BOX and TRACE, determine the time needed to elapse before the pizza is 160°F.
    (d) TRACE the function for large values of time. What do you notice about y, the temperature?

16. A thermometer reading 72°F is placed in a refrigerator where the temperature is a constant 38°F. If the thermometer reads 60°F after 2 minutes, what will it read after 7 minutes?
    (a) How long will it take before the thermometer reads 39°F?
    (b) Using a graphing utility, graph the relation between temperature and time.
    (c) Using BOX and TRACE, determine the time needed to elapse before the thermometer reads 45°F.
    (d) TRACE the function for large values of time. What do you notice about y, the temperature?

17. A thermometer reading 8°C is brought into a room with a constant temperature of 35°C. If the thermometer reads 15°C after 3 minutes, what will it read after being in the room for 5 minutes? For 10 minutes? [*Hint:* You need to construct a formula similar to equation (4).] Graph the relation between temperature and time. TRACE to verify your answer is correct.

18. *Thawing Time of a Steak* A frozen steak has a temperature of 28°F. It is placed in a room with a constant temperature of 70°F. After 10 minutes, the temperature of the steak has risen to 35°F. What will the temperature of the steak be after 30 minutes? How long will it take the steak to thaw to a temperature of 45°F? [See the hint given for Problem 17.] Graph the relation between temperature and time. TRACE to verify your answer is correct.

19. *Decomposition of Salt in Water* Salt (NaCl) decomposes in water into sodium (NA^+) and chloride (Cl^-) ions according to the law of uninhibited decay. If the initial amount of salt is 25 kilograms and, after 10 hours, 15 kilograms of salt are left, how much salt is left after 1 day? How long does it take until $\frac{1}{2}$ kilogram of salt is left?

20. *Voltage of a Condenser* The voltage of a certain condenser decreases over time according to the law of uninhibited decay. If the initial voltage is 40 volts, and 2 seconds later it is 10 volts, what is the voltage after 5 seconds?

21. *Radioactivity from Chernobyl* After the release of radioactive material into the atmosphere from a nuclear power plant at Chernobyl (Ukraine) in 1986, the hay in Austria was contaminated by iodine-131 (see Problem 4). If it is all right to feed the hay to cows when 10% of the iodine-131 remains, how long do the farmers need to wait to use this hay?

22. *Pig Roasts* The hotel Bora-Bora is having a pig roast. At noon, the chef put the pig in a large earthen oven. The pig's original temperature was 75°F. At 2:00 PM, the chef checked the pig's temperature and was upset because it had reached only 100°F. If the oven's temperature remains a constant 325°F, at what time may the hotel serve its guests, assuming that pork is done when it reaches 175°F?

7.7

Nonlinear Curve Fitting

In the Appendix, Section 5, we discussed how to find an equation that describes the relation that exists between two variables using data. We assumed there was a linear relationship of the form $y = ax + b$, where x is the independent variable and y is the dependent variable.

Many real-world phenomena, however, do not follow a linear relationship. For example, we saw in Section 7.6 that the uninhibited growth of bacteria follows the model $N(t) = N_o e^{kt}$, $k > 0$. Also radioactive decay follows the model $A(t) = A_o e^{kt}$, $k < 0$. In this section we will discuss how to use a graphing utility to find equations that describe the relation between two variables when the relation is nonlinear.

Three types of nonlinear models are popular:

1. Exponential $y = ab^x$
2. Power $y = ax^b$
3. Logarithmic $y = a + b \ln x$

Figure 33 shows scatter diagrams that typically will be observed for the three nonlinear models for different values of the parameters, a and b. Below each scatter diagram are any restrictions on the values of the independent or dependent variable.

Determining the appropriate model to use requires statistical techniques that are beyond the scope of this text. However, there are some rudimentary techniques

FIGURE 33

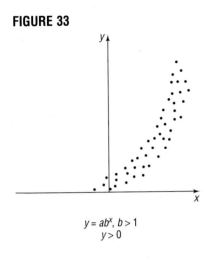

$y = ab^x, b > 1$
$y > 0$

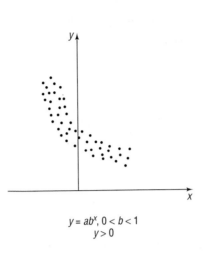

$y = ab^x, 0 < b < 1$
$y > 0$

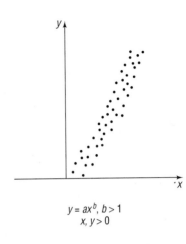

$y = ax^b, b > 1$
$x, y > 0$

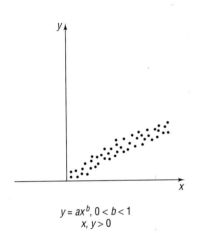

$y = ax^b, 0 < b < 1$
$x, y > 0$

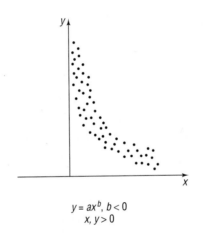

$y = ax^b, b < 0$
$x, y > 0$

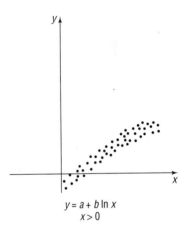

$y = a + b \ln x$
$x > 0$

we can use that will allow us to determine which model (exponential, power, or logarithmic) is most appropriate. Before we do this, however, we need to discuss how graphing utilities are used to fit curves to data.

Most graphing utilities have REGression options that fit data to a specific type of curve. Once the data have been entered and a scatter diagram obtained, the type of curve you want to fit to the data is selected. Then that REGression option is used to obtain the curve of "best fit" of the type selected. The number r that appears measures how good the fit is. The closer $|r|$ is to 1, the better the fit.

Let's look at some examples.

Exponential Curve Fitting

As we saw in Section 7.6, many growth and decay models are exponential.

E X A M P L E   1

Fitting a Curve to an Exponential Model

A researcher is interested in finding a model that explains the relation between the amount of an unknown radioactive material and time. She obtains 100 grams of the radioactive material and every day for 9 days measures the amount of radioactive material in the sample and obtains the following data:

DAY	WEIGHT (IN GRAMS)
0	100
1	91.6
2	84.1
3	74.2
4	68.3
5	63.7
6	58.5
7	53.8
8	50.2
9	47.3

(a) Draw a scatter diagram.

(b) Fit an exponential curve to the data.

(c) Express the curve in the form $A = A_o e^{kt}$.

(d) Using the solution to (c), estimate the half-life of the unknown radioactive material.

(e) Predict how much of the material will remain after 30 days.

Solution (a) After entering the data into the graphing utility, we obtain the scatter diagram shown in Figure 34.

FIGURE 34

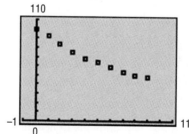

(b) A graphing utility fits the data in Figure 34 to an exponential equation of the form $y = ab^x$ by using the EXPonential REGression option.* See Figure 35.

FIGURE 35

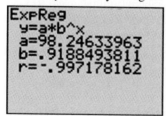

*If your utility does not have such an option but does have a LINear REGressin option, you can transform the exponential model to a linear model by the following technique:

$$
\begin{array}{ll}
y = ab^x & \text{exponential form} \\
\ln y = \ln(ab^x) & \text{take natural logs of each side} \\
\ln y = \ln a + \ln b^x & \text{property of logarithms} \\
\ln y = \ln a + (\ln b)x & \text{property of logarithms}
\end{array}
$$

Now apply the LINear REGression techniques discussed in the Appendix, Section 5. Be careful though. The dependent variable is now $\ln y$, while the independent variable remains x. Once the line of best fit is obtained, you can find a and b and obtain the exponential curve $y = ab^x$.

The exponential equation of best fit to the data is

$$y = 98.24633963(0.9188493811)^x$$

Notice that $r = -0.997178162$, so $|r|$ is close to 1, indicating a good fit.

(c) To express $y = ab^x$ in the form $A = A_o e^{kt}$, we proceed as follows.

$$ab^x = A_o e^{kt}$$

$$a = A_o \qquad b^x = e^{kt} = (e^k)^t$$

$$b = e^k$$

Thus,

$$A_o = 98.2463393 \text{ and } 0.9188493811 = e^k$$

$$k = -0.0846331$$

As a result, $A = A_o e^{kt} = 98.24633963 e^{-0.0846331t}$

(d) We set $A = \frac{1}{2}A_o$ and solve the equation $\frac{1}{2}A_o = A_o e^{-0.0846331t}$ for t. The result is $t = 8.2$ days.

(e) After $t = 30$ days, the amount A of the radioactive material left is

$$A = 98.24633963 e^{-0.0846331(30)} = 7.8 \text{ grams.} \qquad \blacksquare$$

Exploration: Graph the equation found in (b) on the same screen as the scatter diagram. Does the 'fit' seem good? ∎

Logarithmic Curve Fitting

Many models do not follow uninhibited exponential growth but instead, as the independent variable increases, the growth of the dependent variable diminishes. Under these circumstances, a logarithmic model may be appropriate.

E X A M P L E   2

Fitting a Curve to a Logarithmic Model

The following data, obtained from the U.S. Census Bureau, represent the population of New York State over time. An urban economist is interested in finding a model that describes the population of New York over time.

YEAR	POPULATION
1900	7,268,894
1910	9,113,614
1920	10,385,227
1930	12,588,066
1940	13,479,142
1950	14,830,192
1960	16,782,304
1970	18,241,391
1980	17,558,165
1990	17,990,455

(a) Using a graphing utility, draw a scatter diagram of the data with time as the independent variable.

(b) Using a graphing utility, fit a logarithmic growth model to the data.

(c) Use the model found in (b) to predict the population of New York State in 1995.

Solution   (a) After entering the data into the graphing utility, we obtain the scatter diagram shown in Figure 36. Notice the data appear to follow a logarithmic growth model.

FIGURE 36

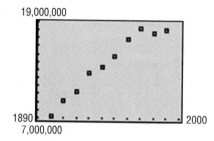

(b) A graphing utility fits the data in Figure 36 to a logarithmic equation of the form $y = a + b \ln x$ by using the Logarithm REGression option. See Figure 37.

FIGURE 37

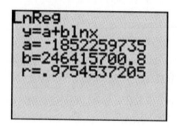

The logarithm equation of best fit to the data is

$$y = -1,852,259,735 + 246,415,700.8 \ln x$$

where y is the state's population and x is the year.

Notice that the value of r is close to 1, indicating a good fit.

(c) Using the equation found in (b), we predict the population of New York State in 1995 to be

$$y = -1,852,259,735 + 246,415,700.8 \ln(1995)$$
$$\approx 20,105,161$$

Power Curve Fitting

If the dependent variable is proportional to a power of the independent variable, then a power curve of the form

$$y = ax^b$$

where y is the dependent variable and x is the independent variable may be the appropriate model.

E X A M P L E   3   *Fitting a Curve to a Power Function*

The *period* of a pendulum is the time required for one oscillation; the pendulum is usually referred to as *simple* when the angle made to the vertical is less than 5°. An experiment is conducted in which simple pendulums are constructed with different lengths, l, and the corresponding period T is recorded. The following data are collected:

LENGTH l (IN FEET)	PERIOD T (IN SECONDS)
1	1.1
2	1.5
3	1.8
4	2.1
5	2.5
6	2.8
7	3.0

(a) Using a graphing utility, draw a scatter diagram of the data with length as the independent variable and the period as the dependent variable.

(b) Using a graphing utility, fit a power model to the data.

(c) If the period of a simple pendulum is known to be 2.3 seconds, what is the length of the pendulum?

Solution   (a) After entering the data into the graphing utility, we obtain the scatter diagram shown in Figure 38.

FIGURE 38

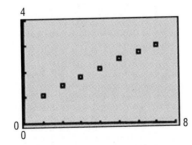

(b) A graphing utility fits the data in Figure 38 to a power curve of the form $y = ax^b$ by using the PoWeR REGression option. See Figure 39.

FIGURE 39

```
PwrReg
y=a*x^b
a=1.059518033
b=.5243681281
r=.9953260793
```

The period T of a simple pendulum is related to its length l by the power function

$$T = 1.0595 \; l^{0.524}$$

Notice that the value of r is close to l, indicating a good fit.*

(c) The length l of string corresponding to a period of 2.3 seconds obeys

$$2.3 = 1.0595 \, l^{0.524}$$
$$l = 4.39 \text{ feet}$$

■

Choosing the Best Model

We have just discussed three different nonlinear models that can be used to explain the relation between two variables. However, how can we be certain the model we used is the "best" model to explain this relation? For example, why did we use an exponential model in Example 1 and a power model in Example 3? Unfortunately, there is no such thing as *the* correct model. Modeling is not only a science but also an art form. Selecting an appropriate model requires experience and skill in the field in which you are modeling. For example, knowledge of economics is imperative when trying to determine a model to predict unemployment. The main reason for this is that there are theories in the field that can help the modeler select appropriate relations and variables. Nonetheless, there are two important tools that can be used to help determine the "best" model: scatter diagrams and the number, r. For our purposes, we will say the "best" model is the one which yields the value r for which $|r|$ is closest to 1.

When there is no theoretical basis for choosing a particular model, the best approach to follow is to first draw a scatter diagram and obtain a general idea of the shape of the data. Then fit several models to the data and determine which seems to explain the relation between the two variables best. Again, we will define "best" as the model that yields the value of $|r|$ closest to 1.

E X A M P L E   4

Selecting the Best Model

Suppose an astronomer claims to have discovered a tenth planet that is 4.2 billion miles from the sun. How long would it take for this planet to completely orbit the sun?

Solution   In order to solve this problem, we need to develop a model that relates the distance a planet is from the sun and the time it takes to make a complete orbit. This time is called the *sidereal year.* The data in the following table show the distances the planets are from the sun and their sidereal years.

PLANET	DISTANCE FROM SUN (MILLIONS OF MILES)	SIDEREAL YEAR
Mercury	36	0.24
Venus	67	0.62
Earth	93	1.00
Mars	142	1.88
Jupiter	483	11.9
Saturn	887	29.5
Uranus	1,785	84.0
Neptune	2,797	165.0
Pluto	3,675	248.0

*In physics, it is proven that $T = \frac{2\pi}{\sqrt{32}} \sqrt{l}$, provided friction is ignored. Since $\sqrt{l} = l^{1/2} = l^{0.5}$ and

$\frac{2\pi}{\sqrt{32}} = 1.1107$, the power function obtained is reasonably close to what is expected.

The first step in determining the best model is to draw a scatter diagram. Figure 40 shows a scatter diagram for the planet data.

FIGURE 40

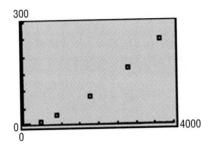

From the scatter diagram in Figure 40 it appears the data follow either a linear model, an exponential model ($k > 0$), or a power model ($b > 1$). Therefore, we shall find equations for all three models and use the value of r to determine which is best. Figure 41(a) shows the results obtained from the graphing utility for an exponential model, Figure 41(b) shows the results for a power model, and Figure 41(c) shows the results for a linear model.

FIGURE 41

(a)   (b)   (c)

By looking at the value of r, we conclude that the power model is superior to both the exponential model and the linear model. Thus, we conclude the relation between the distance a planet is from the sun (x) and its sidereal year (y) is
$y = 0.00112x^{1.499938}$.*

We can predict the sidereal year of the newly discovered planet by substituting $x = 4200$ into the equation just found:

$$y = 0.00112(4200)^{1.499938}$$

$$\approx 304.7 \text{ sidereal years}$$

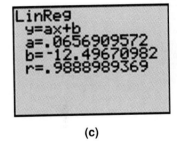

*If we square both sides of this equation we obtain
$$y^2 \approx 0.0000012x^3$$

which implies the square of the sidereal year is proportional to the cube of the distance from the sun to the planet. This is Kepler's Third Law, named after Johann Kepler. Kepler discovered this relation in 1618 through experimentation without the aid of data analysis.

7.7

Exercise 7.7

1. *Biology* A certain bacteria increases according to the law of uninhibited growth. A biologist collects the following data for this bacteria:

TIME (HOURS)	POPULATION
0	1000
1	1415
2	2000
3	2828
4	4000
5	5656
6	8000

(a) Draw a scatter diagram.

(b) Fit an exponential curve to the data.

(c) Express the curve in the form $N = N_o e^{kt}$.

(d) Using the solution to (b), predict the population at $t = 7$ hours.

2. *Biology* A colony of bacteria increases according to the law of uninhibited growth. A biologist collects the following data for this bacteria:

TIME (DAYS)	POPULATION
0	50
1	153
2	234
3	357
4	547
5	839
6	1280

(a) Draw a scatter diagram.

(b) Fit an exponential curve to the data.

(c) Express the curve in the form $N = N_o e^{kt}$.

(d) Using the solution to (b), predict the population at $t = 7$ days.

3. *Economics/Marketing* A store manager collected the following data regarding price and quantity sold of a product:

PRICE ($/UNIT)	QUANTITY DEMANDED
79	10
67	20
54	30
46	40
38	50
31	60

(a) Draw a scatter diagram with quantity demanded on the x-axis and price on the y-axis.

(b) Fit an exponential curve to the data.

(c) Predict the quantity demanded if the price is $60.

4. *Economics/Marketing* A store manager collected the following data regarding price and quantity supplied of a product:

PRICE ($/UNIT)	QUANTITY SUPPLIED
25	10
32	20
40	30
46	40
60	50
74	60

(a) Draw a scatter diagram with quantity supplied on the x-axis and price on the y-axis.

(b) Fit an exponential curve to the data.

(c) Predict the quantity demanded if the price is $45.

5. *Chemistry* A chemist has a 100-gram sample of a radioactive material. He records the amount of radioactive material every week for 6 weeks and obtains the following data:

WEEK	WEIGHT (IN GRAMS)
0	100.0
1	88.3
2	75.9
3	69.4
4	59.1
5	51.8
6	45.5

(a) Draw a scatter diagram with week as the independent variable.

(b) Fit an exponential curve to the data.

(c) Express the curve in the form $A = A_o e^{kt}$.

(d) From the result found in (b), determine the half-life of the radioactive material.

(e) How much radioactive material will be left after 50 weeks?

6. *Chemistry* A chemist has a 1000-gram sample of a radioactive material. She records the amount of radioactive material remaining in the sample every day for a week and obtains the following data:

DAY	WEIGHT (IN GRAMS)
0	1000.0
1	897.1
2	802.5
3	719.8
4	651.1
5	583.4
6	521.7
7	468.3

(a) Draw a scatter diagram with day as the independent variable.

(b) Fit an exponential curve to the data.

(c) Express the curve in the form $A = A_o e^{kt}$.

(d) From the result found in (b), find the half-life of the radioactive material.

(e) How much radioactive material will be left after 20 days?

7. *Finance* The following data represent the amount of money an investor has in an investment account each year for 10 years. She wishes to determine the average annual rate of return on her investment.

YEAR	VALUE OF ACCOUNT
1985	$10,000
1986	$10,573
1987	$11,260
1988	$11,733
1989	$12,424
1990	$13,269
1991	$13,968
1992	$14,823
1993	$15,297
1994	$16,539

(a) Draw a scatter diagram with time as the independent variable and the value of the account as the dependent variable.

(b) Fit an exponential curve to the data.

(c) Based on the answer to (b), what was the average annual rate of return from this account over the past 10 years?

(d) If the investor plans on retiring in 2020, what will the predicted value of this account be?

8. *Finance* The following data show the amount of money an investor has in an investment account each year for 7 years. He wishes to determine the average annual rate of return on his investment.

YEAR	VALUE OF ACCOUNT
1988	$20,000
1989	$21,516
1990	$23,355
1991	$24,885
1992	$27,434
1993	$30,053
1994	$32,622

(a) Draw a scatter diagram with time as the independent variable and the value of the account as the dependent variable.

(b) Fit an exponential curve to the data.

(c) Based on the answer to (b), what was the average annual rate of return from this account over the past 7 years?

(d) If the investor plans on retiring in 2020, what will the predicted value of this account be?

9. *CBL Experiment* The following data were collected by placing a probe in a cup of hot water, removing the probe, and then recording temperature over time:

TIME	TEMPERATURE (°F)	TIME	TEMPERATURE (°F)
0	175.69	13	148.50
1	173.52	14	146.84
2	171.21	15	145.33
3	169.07	16	143.83
4	166.59	17	142.38
5	164.21	18	141.22
6	161.89	19	140.09
7	159.66	20	138.69
8	157.86	21	137.59
9	155.75	22	136.78
10	153.70	23	135.70
11	151.93	24	134.91
12	150.08	25	133.86

According to Newton's Law of Cooling, these data should follow an exponential model.

(a) Using a graphing utility, draw a scatter diagram for the data.

(b) Fit an exponential model to the data.

(c) Predict how long it will take for the water to reach a temperature of 110°F.

10. *CBL Experiment* The following data were collected by placing a probe in a portable heater, removing the probe, and then recording temperature over time:

TIME	TEMPERATURE (°F)	TIME	TEMPERATURE (°F)
0	165.07	8	159.35
1	164.77	9	158.61
2	163.99	10	157.89
3	163.22	11	156.83
4	162.82	12	156.11
5	161.96	13	155.08
6	161.20	14	154.40
7	160.45	15	153.72

According to Newton's Law of Cooling, these data should follow an exponential model.

(a) Using a graphing utility, draw a scatter diagram for the data.

(b) Fit an exponential model to the data.

(c) Predict how long it will take for the air blown out of the heater to reach a temperature of 110°F.

11. *Population Model* The following data obtained from the U.S. Census Bureau represent the population of Illinois. An urban economist is interested in finding a model that describes the population of Illinois.

YEAR	POPULATION
1900	4,821,550
1910	5,638,591
1920	6,485,280
1930	7,630,654
1940	7,897,241
1950	8,712,176
1960	10,081,158
1970	11,110,285
1980	11,427,409
1990	11,430,602

(a) Using a graphing utility, draw a scatter diagram of the data with time as the independent variable.

(b) Using a graphing utility, fit a logarithmic growth model to the data.

(c) Use the model found in (b) to predict the population of Illinois in 1995.

12. *Population Model* The following data obtained from the U.S. Census Bureau represent the population of Pennsylvania. An urban economist is interested in finding a model that describes the population of Pennsylvania.

YEAR	POPULATION
1900	6,302,115
1910	7,665,111
1920	8,720,017
1930	9,631,350
1940	9,900,180
1950	10,498,012
1960	11,319,366
1970	11,800,766
1980	11,864,720
1990	11,881,643

(a) Using a graphing utility, draw a scatter diagram of the data with time as the independent variable.

(b) Using a graphing utility, fit a logarithmic growth model to the data.

(c) Use the model found in (b) to predict the population of Pennsylvania in 1995.

13. *CBL Experiment* Yolanda is conducting an experiment to measure the relation between a light bulb's intensity and the distance from the light source. She measures a 100-watt light bulb's intensity 1 meter from the bulb and at 0.1-meter intervals up to 2 meters from the bulb and obtains the following data.

DISTANCE (METERS)	INTENSITY
1.0	0.29645
1.1	0.25215
1.2	0.20547
1.3	0.17462
1.4	0.15342
1.5	0.13521
1.6	0.11450
1.7	0.10243
1.8	0.09231
1.9	0.08321
2.0	0.07342

(a) Using your graphing utility, draw a scatter diagram with distance as the independent variable and intensity as the dependent variable.
(b) Using your graphing utility, fit a power model to the data.
(c) It is known that intensity I is inversely proportional to the square of the distance x, so that $I = \dfrac{a}{x^2}$. How close is the estimate of the exponent?
(d) What will the intensity of a 100-watt light bulb be if you stand 2.3 meters away?

14. *CBL Experiment* Yolanda repeats the experiment from Problem 13, but this time uses a 40-watt light bulb and obtains the following data.

DISTANCE (METERS)	INTENSITY
1.0	0.0972
1.1	0.0804
1.2	0.0674
1.3	0.0572
1.4	0.0495
1.5	0.0433
1.6	0.0384
1.7	0.0339
1.8	0.0294
1.9	0.0268
2.0	0.0224

(a) Using your graphing utility, draw a scatter diagram with distance as the independent variable and intensity as the dependent variable.
(b) Using your graphing utility, fit a power model to the data.
(c) What will the intensity of a 40-watt light bulb be if you stand 2.3 meters away?

15. *CBL Experiment* Jim is conducting an experiment in order to estimate the acceleration of an object due to gravity. Jim takes a ball and drops it from different heights and records the time it takes for the ball to hit the ground. Using an optic laser connected to a stop watch in order to determine the time, he collects the following data:

TIME (SECONDS)	DISTANCE (FEET)
1.003	16
1.365	30
1.769	50
2.093	70
2.238	80

(a) Draw a scatter diagram using time as the independent variable and distance as the dependent variable.
(b) Fit a power model to the data.
(c) Physics theory states the distance an object falls is directly proportional to the time squared. Rewrite the model found in (b) so it is of the form $s = \frac{1}{2}gt^2$, where s is distance, g is the acceleration due to gravity, and t is time. What is Jim's estimate of g? (It is known that the acceleration due to gravity is approximately 32 feet/sec^2.)
(d) Predict how long it will take an object to fall 100 feet.

16. *CBL Experiment* Bill, Jim's friend, doesn't believe Jim's estimate of acceleration due to gravity is correct. Bill repeats the experiment described in Problem 15 and obtains the following data:

TIME (SECONDS)	DISTANCE (FEET)
0.7907	10
1.1160	20
1.4760	35
1.6780	45
1.9380	60

   (a) Draw a scatter diagram using time as the dependent variable and distance as the independent variable.
   (b) Fit a power model to the data.
   (c) Physics theory states the distance an object falls is directly proportional to the time squared. Rewrite the model found in (b) so it is of the form $s = \frac{1}{2}gt^2$, where s is distance, g is the acceleration due to gravity, and t is time. What is Bill's estimate of g? [It is known that the acceleration due to gravity is approximately 32 feet/sec^2.]
   (d) Predict how long it will take an object to fall 100 feet. Why do you think this estimate is different than the one from Problem 15?

For Problems 17–22 the independent variable is x and the dependent variable is y. (a) For each set of data draw a scatter diagram of the data. (b) Use an exponential model, logarithmic model, power model, and linear model to find the "best" model that explains the relationship between the variables.

17.

x	1	2	3	4	5
y	164	269	450	740	1220

18.

x	1	2	3	4	5
y	110	245	553	1228	2728

19.

x	10	20	30	40	50
y	6.94	13.78	20.45	21.23	22.95

20.

x	10	20	30	40	50
y	4.95	6.46	7.23	7.78	8.23

21.

x	1	1.3	1.7	2.1	2.4	2.7
y	9.92	4.56	1.42	0.37	0.12	0.032

22.

x	1	1.3	1.7	2.1	2.4	2.7
y	1.72	1.94	2.48	3.28	4.01	5.07

23. *Economics* The following data are levels of the Consumer Price Index for food items (Source: Bureau of Labor Statistics).

YEAR	CPI
1984	103.2
1985	105.6
1986	109.0
1987	113.5
1988	118.2
1989	125.1
1990	132.4
1991	136.3
1992	137.9
1993	140.9

   (a) Using a graphing utility, draw a scatter diagram of the data using time as the independent variable and the CPI as the dependent variable.
   (b) Use an exponential model, logarithmic model, power model, and linear model to find the "best" model to describe the relation between time and the CPI.
   (c) Use this model to predict the CPI for food items for 1994.

24. *Economics* The following data are levels of the Consumer Price Index for electricity (Source: Bureau of Labor Statistics).

YEAR	CPI
1984	105.3
1985	108.9
1986	110.4
1987	110.0
1988	111.5
1989	114.7
1990	117.4
1991	121.8
1992	124.2
1993	126.7

(a) Using a graphing utility, draw a scatter diagram of the data using time as the independent variable and the CPI as the dependent variable.

(b) Use an exponential model, logarithmic model, power model, and linear model to find the "best" model to describe the relation between time and the CPI.

(c) Use this model to predict the CPI for electricity for 1994.

25. *Economics* The following data represent the per capita gross domestic product (GDP) in constant 1987 dollars (this removes the effects of inflation). (Source: U.S. Bureau of Economic Analysis)

YEAR	PER CAPITA GROSS DOMESTIC PRODUCT (1987 DOLLARS)
1988	19,252
1989	19,556
1990	19,593
1991	19,238
1992	19,518
1993	19,888

(a) Using a graphing utility, draw a scatter diagram of the data using time as the independent variable and the per capita GDP as the dependent variable.

(b) Use an exponential model, logarithmic model, power model, and linear model to find the "best" model to describe the relation between time and the per capita GDP.

(c) Use this model to predict per capita GDP for 1994.

26. *Economics* The following data represent the per capita gross national product (GNP) in constant 1987 dollars (this removes the effects of inflation). (Source: U.S. Bureau of Economic Analysis)

YEAR	PER CAPITA GROSS NATIONAL PRODUCT (1987 DOLLARS)
1988	19,284
1989	19,615
1990	19,670
1991	19,290
1992	19,548
1993	19,897

(a) Using a graphing utility, draw a scatter diagram of the data using time as the independent variable and the GNP as the dependent variable.

(b) Use an exponential model, logarithmic model, power model, and linear model to find the "best" model to describe the relation between time and the per capita GNP.

(c) Use this model to predict the per capita GNP for 1994.

27. *Population Model* The following data represent the population of Florida. (Source: US Census Bureau)

YEAR	POPULATION
1900	528,542
1910	752,619
1920	968,470
1930	1,468,211
1940	1,897,414
1950	2,771,305
1960	4,951,560
1970	6,791,418
1980	9,746,961
1990	12,937,926

(a) Using a graphing utility, draw a scatter diagram using time as the independent variable and population as the dependent variable.

(b) Use an exponential model, logarithmic model, power model, and linear model to find the "best" model that describes the relation between time and population.

(c) Use the model found in (b) to predict the population of Florida in 1995.

28. *Population Model* The following data represent the population of Arizona. (Source: US Census Bureau)

YEAR	POPULATION
1900	122,931
1910	204,354
1920	334,162
1930	435,573
1940	499,261
1950	749,587
1960	1,302,161
1970	1,775,399
1980	2,716,546
1990	3,665,228

(a) Using a graphing utility, draw a scatter diagram using time as the independent variable and population as the dependent variable.

(b) Use an exponential model, logarithmic model, power model, and linear model to find the "best" model that describes the relation between time and population.

(c) Use the model found in (b) to predict the population of Arizona in 1995.

7.8

Logarithmic Scales

Common logarithms often appear in the measurement of quantities, because they provide a way to scale down positive numbers that vary from very small to very large. For example, if a certain quantity can take on values from $0.0000000001 = 10^{-10}$ to $10,000,000,000 = 10^{10}$, the common logarithms of such numbers would be between -10 and 10.

Loudness of Sound

Our first application utilizes a logarithmic scale to measure the loudness of a sound. Physicists define the **intensity of a sound wave** as the amount of energy the wave transmits through a given area. For example, the least intense sound that a human ear can detect at a frequency of 100 hertz is about 10^{-12} watt per square meter. The **loudness** $L(x)$, measured in **decibels** (named in honor of Alexander Graham Bell), of a sound of intensity x (measured in watts per square meter) is defined as

Loudness

$$L(x) = 10 \log \frac{x}{I_0} \qquad (1)$$

where $I_0 = 10^{-12}$ watt per square meter is the least intense sound that a human ear can detect. If we let $x = I_0$ in equation (1), we get

$$L(I_0) = 10 \log \frac{I_0}{I_0} = 10 \log 1 = 0$$

Thus, at the threshold of human hearing, the loudness is 0 decibels. Figure 42 gives the loudness of some common sounds.

FIGURE 42

Loudness of common sounds (in decibels)

DECIBELS

140	Shotgun blast, jet 100 feet away at takeoff	Pain
130	Motor test chamber	Human ear pain threshold
120	Firecrackers, severe thunder, pneumatic jackhammer, hockey crowd	Uncomfortably loud
110	Amplified rock music	
100	Textile loom, subway train, elevated train, farm tractor, power lawn mower, newspaper press	Loud
90	Heavy city traffic, noisy factory	
80	Diesel truck going 40 mi/hr 50 feet away, crowded restaurant, garbage disposal, average factory, vacuum cleaner	Moderately loud
70	Passenger car going 50 mi/hr 50 feet away	
60	Quiet typewriter, singing birds, window air conditioner, quiet automobile	Quiet
50	Normal conversation, average office	
40	Household refrigerator, quiet office	Very quiet
30	Average home, dripping faucet, whisper 5 feet away	
20	Light rainfall, rustle of leaves	Average person's threshold of hearing
10	Whisper across room	Just audible
0		Threshold for acute hearing

Note that a decibel is not a linear unit like the meter. For example, a noise level of 10 decibels is 10 times as great as a noise level of 0 decibels. [If $L(x) = 10$, then $x = 10I_0$.] A noise level of 20 decibels is 100 times as great as a noise level of 0 decibels. [If $L(x) = 20$, then $x = 100I_0$.] A noise level of 30 decibels is 1000 times as great as a noise level of 0 decibels, and so on.

EXAMPLE 1

Finding the Intensity of a Sound

Use Figure 42 to find the intensity of the sound of a dripping faucet.

Solution   From Figure 42, we see that the loudness of the sound of dripping water is 30 decibels. Thus, by equation (1), its intensity x may be found as follows:

$$30 = 10 \log\left(\frac{x}{I_0}\right)$$

$$3 = \log\left(\frac{x}{I_0}\right) \qquad \text{Divide by 10.}$$

$$\frac{x}{I_0} = 10^3 \qquad \text{Write in exponential form.}$$

$$x = 1000I_0$$

where $I_0 = 10^{-12}$ watt per square meter. Thus, the intensity of the sound of a dripping faucet is 1000 times as great as a noise level of 0 decibels; that is, such a sound has an intensity of $1000 \cdot 10^{-12} = 10^{-9}$ watt per square meter.   ■

■   Now work Problem 5.

E X A M P L E   2   *Finding the Loudness of a Sound*

Use Figure 42 to determine the loudness of a subway train if it is known that this sound is 10 times as intense as the sound due to heavy city traffic.

Solution   The sound due to heavy city traffic has a loudness of 90 decibels. Its intensity, therefore, is the value of x in the equation

$$90 = 10 \log\left(\frac{x}{I_0}\right)$$

A sound 10 times as intense as x has loudness $L(10x)$. Thus, the loudness of the subway train is

$$L(10x) = 10 \log\left(\frac{10x}{I_0}\right) \qquad \text{Replace } x \text{ by } 10x.$$

$$= 10 \log\left(10 \cdot \frac{x}{I_0}\right)$$

$$= 10 \left[\log 10 + \log\left(\frac{x}{I_0}\right)\right] \qquad \text{Log of product = Sum of logs}$$

$$= 10 \log 10 + 10 \log\left(\frac{x}{I_0}\right) \qquad \log 10 = 1$$

$$= 10 + 90 = 100 \text{ decibels} \qquad\qquad ■$$

Magnitude of an Earthquake

Our second application uses a logarithmic scale to measure the magnitude of an earthquake.

The **Richter scale*** is one way of converting seismographic readings into numbers that provide an easy reference for measuring the magnitude M of an earthquake. All earthquakes are compared to a **zero-level earthquake** whose seismographic reading measures 0.001 millimeter at a distance of 100 kilometers from the epicenter. An earthquake whose seismographic reading measures x millimeters has **magnitude $M(x)$** given by

Magnitude of an Earthquake

$$M(x) = \log\left(\frac{x}{x_0}\right) \qquad\qquad (2)$$

where $x_0 = 10^{-3}$ is the reading of a zero-level earthquake the same distance from its epicenter.

E X A M P L E   3   *Finding the Magnitude of an Earthquake*

What is the magnitude of an earthquake whose seismographic reading is 0.1 millimeter at a distance of 100 kilometers from its epicenter?

*Named after the American scientist, C.F. Richter, who devised it in 1935.

Solution   If $x = 0.1$, the magnitude $M(x)$ of this earthquake is

$$M(0.1) = \log\left(\frac{x}{x_0}\right) = \log\left(\frac{0.1}{0.001}\right) = \log\left(\frac{10^{-1}}{10^{-3}}\right) = \log 10^2 = 2$$

This earthquake thus measures 2.0 on the Richter scale.   ■

■ Now work Problem 7.

Based on formula (2), we define the **intensity of an earthquake** as the ratio of x to x_0. For example, the intensity of the earthquake described in Example 3 is $\frac{0.1}{0.001} = 10^2 = 100$. That is, it is 100 times as intense as a zero-level earthquake.

E X A M P L E   4   *Comparing the Intensity of Two Earthquakes*

The devastating San Francisco earthquake of 1906 measured 8.9 on the Richter scale. How did the intensity of that earthquake compare to the Papua New Guinea earthquake of 1988, which measured 6.7 on the Richter scale?

FIGURE 43

Solution   Let x_1 and x_2 denote the seismographic readings, respectively, of the 1906 San Francisco earthquake and the Papua New Guinea earthquake. Then, based on formula (2),

$$8.9 = \log\left(\frac{x_1}{x_0}\right) \qquad 6.7 = \log\left(\frac{x_2}{x_0}\right)$$

Consequently,

$$\frac{x_1}{x_0} = 10^{8.9} \qquad \frac{x_2}{x_0} = 10^{6.7}$$

The 1906 San Francisco earthquake was $10^{8.9}$ times as intense as a zero-level earthquake. The Papua New Guinea earthquake was $10^{6.7}$ times as intense as a zero-level earthquake. Thus,

$$\frac{x_1}{x_2} = \frac{10^{8.9}x_0}{10^{6.7}x_0} = 10^{2.2} \approx 158$$

$$x_1 \approx 158x_2$$

Hence, the San Francisco earthquake was 158 times as intense as the Papua New Guinea earthquake.   ■

Example 4 demonstrates that the relative intensity of two earthquakes can be found by raising 10 to a power equal to the difference of their readings on the Richter scale.

7.8

Exercise 7.8

1. *Loudness of a Dishwasher* Find the loudness of a dishwasher that operates at an intensity of 10^{-5} watt per square meter. Express your answer in decibels.

2. *Loudness of a Diesel Engine* Find the loudness of a diesel engine that operates at an intensity of 10^{-3} watt per square meter. Express your answer in decibels.

3. *Loudness of a Jet Engine* With engines at full throttle, a Boeing 727 jetliner produces noise at an intensity of 0.15 watt per square meter. Find the loudness of the engines in decibels.

4. *Loudness of a Whisper* A whisper produces noise at an intensity of $10^{-9.8}$ watt per square meter. What is the loudness of a whisper in decibels?

5. *Intensity of a Sound at Threshold of Pain* For humans, the threshold of pain due to sound averages 130 decibels. What is the intensity of such a sound in watts per square meter?

6. If one sound is 50 times as intense as another, what is the difference in the loudness of the two sounds? Express your answer in decibels.

7. *Magnitude of an Earthquake* Find the magnitude of an earthquake whose seismographic reading is 10.0 millimeters at a distance of 100 kilometers from its epicenter.

8. *Magnitude of an Earthquake* Find the magnitude of an earthquake whose seismographic reading is 1210 millimeters at a distance of 100 kilometers from its epicenter.

9. *Comparing Earthquakes* The Mexico City earthquake of 1978 registered 7.85 on the Richter scale. What would a seismograph 100 kilometers from the epicenter have measured for this earthquake? How does this earthquake compare in intensity to the 1906 San Francisco earthquake, which registered 8.9 on the Richter scale?

10. *Comparing Earthquakes* Two earthquakes differ by 1.0 when measured on the Richter scale. How would the seismographic readings differ at a distance of 100 kilometers from the epicenter? How do their intensities compare?

Chapter Review

THINGS TO KNOW

Properties of the exponential function

$f(x) = a^x, a > 1$
Domain: $(-\infty, \infty)$; Range: $(0, \infty)$; x-intercepts: none; y-intercept: 1; horizontal asymptote: x-axis as $x \to -\infty$; increasing; one-to-one
See Figure 2 for a typical graph.

$f(x) = a^x, 0 < a < 1$
Domain: $(-\infty, \infty)$; Range: $(0, \infty)$; x-intercepts: none, y-intercept: 1; horizontal asymptote: x-axis as $x \to \infty$; decreasing; one-to-one
See Figure 6 for a typical graph.

Properties of the logarithmic function

$f(x) = \log_a x, a > 1$
$(y = \log_a x$ means $x = a^y)$
Domain: $(0, \infty)$; Range: $(-\infty, \infty)$; x-intercept: 1; y-intercept: none; vertical asymptote: y-axis; increasing; one-to-one
See Figure 14(b) for a typical graph.

$f(x) = \log_a x, 0 < a < 1$
$(y = \log_a x$ means $x = a^y)$
Domain: $(0, \infty)$; Range: $(-\infty, \infty)$; x-intercept: 1; y-intercept: none; vertical asymptote: y-axis; decreasing; one-to-one
See Figure 14(a) for a typical graph.

Number e

Value approached by the expression $\left(1 + \dfrac{1}{n}\right)^n$ as $n \to \infty$; that is, $\displaystyle\lim_{n \to \infty} \left(1 + \dfrac{1}{n}\right)^n = e$

Natural logarithm

$y = \ln x$ means $x = e^y$

Properties of logarithms

$\log_a 1 = 0 \qquad \log_a a = 1 \qquad a^{\log_a M} = M \qquad \log_a a^r = r$

$\log_a MN = \log_a M + \log_a N \qquad \log_a\!\left(\dfrac{M}{N}\right) = \log_a M - \log_a N \qquad \log_a\!\left(\dfrac{1}{N}\right) = -\log_a N \qquad \log_a M^r = r \log_a M$

FORMULAS

Change-of-Base Formula	$\log_a M = \dfrac{\log_b M}{\log_b a}$
Compound interest	$A = P\left(1 + \dfrac{r}{n}\right)^{nt}$
Continuous compounding	$A = Pe^{rt}$
Present value	$P = A\left(1 + \dfrac{r}{n}\right)^{-nt}$ or $P = Ae^{-rt}$
Growth and decay	$A = A_0 e^{kt}$

HOW TO:

Graph exponential and logarithmic functions

Solve certain exponential equations

Solve certain logarithmic equations

Solve problems involving compound interest

Solve problems involving growth and decay

Solve problems involving intensity of sound and intensity of earthquakes

Fit nonlinear models to data

FILL-IN-THE-BLANK ITEMS

1. The graph of every exponential function $f(x) = a^x$, $a > 0$, $a \neq 1$, passes through the two points _____.
2. If the graph of an exponential function $f(x) = a^x$, $a > 0$, $a \neq 1$, is decreasing, then its base must be less than _____.
3. If $3^x = 3^4$, then $x =$ _____.
4. The logarithm of a product equals the _____ of the logarithms.
5. For every base, the logarithm of _____ equals 0.
6. If $\log_8 M = \log_5 7 / \log_5 8$, then $M =$ _____.
7. The domain of the logarithmic function $f(x) = \log_a x$ consists of _____.
8. The graph of every logarithmic function $f(x) = \log_a x$, $a > 0$, $a \neq 1$, passes through the two points _____.
9. If the graph of a logarithmic function $f(x) = \log_a x$, $a > 0$, $a \neq 1$, is increasing, then its base must be larger than _____.
10. If $\log_3 x = \log_3 7$, then $x =$ _____.

TRUE/FALSE ITEMS

T   F   **1.** The graph of every exponential function $f(x) = a^x$, $a > 0$, $a \neq 1$, will contain the points $(0, 1)$ and $(1, a)$.

T   F   **2.** The graphs of $y = 3^{-x}$ and $y = (\frac{1}{3})^x$ are identical.

T   F   **3.** The present value of $1000 to be received after 2 years at 10% per annum compounded continuously is approximately $1205.

T   F   **4.** If $y = \log_a x$, then $y = a^x$.

T   F   **5.** The graph of every logarithmic function $f(x) = \log_a x$, $a > 0$, $a \neq 1$, will contain the points $(1, 0)$ and $(a, 1)$.

T   F   **6.** $a^{\log_M a} = M$, where $a > 0$, $a \neq 1$, $M > 0$

T   F   **7.** $\log_a(M + N) = \log_a M + \log_a N$, where $a > 0$, $a \neq 1$, $M > 0$, $N > 0$

T   F   **8.** $\log_a M - \log_a N = \log_a(M/N)$, where $a > 0$, $a \neq 1$, $M > 0$, $N > 0$

REVIEW EXERCISES

In Problems 1–6, evaluate each expression.

1. $\log_2(\frac{1}{8})$ **2.** $\log_3 81$ **3.** $\ln e^{\sqrt{2}}$ **4.** $e^{\ln 0.1}$ **5.** $2^{\log_2 0.4}$ **6.** $\log_2 2^{\sqrt{3}}$

In Problems 7–12, write each expression as a single logarithm.

7. $3 \log_4 x^2 + \frac{1}{2} \log_4 \sqrt{x}$

8. $-2 \log_3\left(\frac{1}{x}\right) + \frac{1}{3} \log_3 \sqrt{x}$

9. $\ln\left(\frac{x - 1}{x}\right) + \ln\left(\frac{x}{x + 1}\right) - \ln(x^2 - 1)$

10. $\log(x^2 - 9) - \log(x^2 + 7x + 12)$

11. $2 \log 2 + 3 \log x - \frac{1}{2}[\log(x + 3) + \log(x - 2)]$

12. $\frac{1}{2} \ln(x^2 + 1) - 4 \ln\frac{1}{2} - \frac{1}{2}[\ln(x - 4) + \ln x]$

In Problems 13–20, find y as a function of x. The constant C is a positive number.

13. $\ln y = 2x^2 + \ln C$

14. $\ln(y - 3) = \ln 2x^2 + \ln C$

15. $\frac{1}{2} \ln y = 3x^2 + \ln C$

16. $\ln 2y = \ln(x + 1) + \ln(x + 2) + \ln C$

17. $\ln(y - 3) + \ln(y + 3) = x + C$

18. $\ln(y - 1) + \ln(y + 1) = -x + C$

19. $e^{y+C} = x^2 + 4$

20. $e^{3y-C} = (x + 4)^2$

In Problems 21–30, using a graphing utility, show the stages required to graph each function.

21. $f(x) = e^{-x}$ **22.** $f(x) = \ln(-x)$ **23.** $f(x) = 1 - e^x$ **24.** $f(x) = 3 + \ln x$

25. $f(x) = 3e^x$ **26.** $f(x) = \frac{1}{2} \ln x$ **27.** $f(x) = e^{|x|}$ **28.** $f(x) = \ln|x|$

29. $f(x) = 3 - e^{-x}$ **30.** $f(x) = 4 - \ln(-x)$

In Problems 31–50, solve each equation. Verify your result using a graphing utility.

31. $4^{1-2x} = 2$ **32.** $8^{6+3x} = 4$ **33.** $3^{x^2+x} = \sqrt{3}$

34. $4^{x-x^2} = \frac{1}{2}$ **35.** $\log_x 64 = -3$ **36.** $\log_{\sqrt{2}} x = -6$

37. $5^x = 3^{x+2}$ **38.** $5^{x+2} = 7^{x-2}$ **39.** $9^{2x} = 27^{3x-4}$

40. $25^{2x} = 5^{x^2-12}$ **41.** $\log_3 \sqrt{x - 2} = 2$ **42.** $2^{x+1} \cdot 8^{-x} = 4$

43. $8 = 4^{x^2} \cdot 2^{5x}$ **44.** $2^x \cdot 5 = 10^x$ **45.** $\log_6(x + 3) + \log_6(x + 4) = 1$

46. $\log_{10}(7x - 12) = 2 \log_{10} x$ **47.** $e^{1-x} = 5$ **48.** $e^{1-2x} = 4$

49. $2^{3x} = 3^{2x+1}$ **50.** $2^{x^3} = 3^{x^2}$

In Problems 51–54, use the following result: If x is the atmospheric pressure (measured in millimeters of mercury), then the formula for the altitude $h(x)$ (measured in meters above sea level) is

$$h(x) = (30T + 8000)\log\left(\frac{P_0}{x}\right)$$

where T is the temperature (in degrees Celsius) and P_0 is the atmospheric pressure at sea level, which is approximately 760 millimeters of mercury.

51. *Finding the Altitude of an Airplane* At what height is an aircraft whose instruments record an outside temperature of 0°C and a barometric pressure of 300 millimeters of mercury?

52. *Finding the Height of a Mountain* How high is a mountain if instruments placed on its peak record a temperature of 5°C and a barometric pressure of 500 millimeters of mercury?

53. *Atmospheric Pressure Outside an Airplane* What is the atmospheric pressure outside an aircraft flying at an altitude of 10,000 meters if the outside air temperature is −100°C?

54. *Atmospheric Pressure at High Altitudes* What is the atmospheric pressure (in millimeters of mercury) on Mt. Everest, which has an altitude of approximately 8900 meters, if the air temperature is 5°C?

55. *Amplifying Sound* An amplifier's power output P (in watts) is related to its decibel voltage gain d by the formula $P = 25e^{0.1d}$.

  (a) Find the power output for a decibel voltage gain of 4 decibels.
  (b) For a power output of 50 watts, what is the decibel voltage gain?

56. *Limiting Magnitude of a Telescope* A telescope is limited in its usefulness by the brightness of the star it is aimed at and by the diameter of its lens. One measure of a star's brightness is its *magnitude:* the dimmer the star, the larger its magnitude. A formula for the limiting magnitude L of a telescope, that is, the magnitude of the dimmest star it can be used to view, is given by

$$L = 9 + 5.1 \log d$$

where d is the diameter (in inches) of the lens.

  (a) What is the limiting magnitude of a 3.5 inch telescope?
  (b) What diameter is required to view a star of magnitude 14?

57. *Product Demand* The demand for a new product increases rapidly at first and then levels off. The percent P of actual purchases of this product after it has been on the market t months is

$$P = 90 - 80\left(\frac{3}{4}\right)^t$$

  (a) What is the percent of purchases of the product after 5 months?
  (b) What is the percent of purchases of the product after 10 months?
  (c) What is the maximum percent of purchases of the product?
  (d) How many months does it take before 40% of purchases occurs?
  (e) How many months before 70% of purchases occurs?

58. *Disseminating Information* A survey of a certain community of 10,000 residents shows that the number of residents N who have heard a piece of information after m months is given by the formula

$$m = 55.3 - 6 \ln(10,000 - N)$$

How many months will it take for half of the citizens to learn about a community program of free blood pressure readings?

59. *Salvage Value* The number of years n for a piece of machinery to depreciate to a known salvage value can be found using the formula

$$n = \frac{\log_{10} s - \log_{10} i}{\log_{10}(1 - d)}$$

where s is the salvage value of the machinery, i is its initial value, and d is the annual rate of depreciation.

  (a) How many years will it take for a piece of machinery to decline in value from \$90,000 to \$10,000 if the annual rate of depreciation is 0.20 (20%)?
  (b) How many years will it take for a piece of machinery to lose half of its value if the annual rate of depreciation is 15%?

60. *Funding a College Education* A child's grandparents purchase a $10,000 bond fund that matures in 18 years to be used for her college education. The bond fund pays 4% interest compounded semiannually. How much will the bond fund be worth at maturity?

61. *Funding a College Education* A child's grandparents wish to purchase a bond fund that matures in 18 years to be used for her college education. The bond fund pays 4% interest compounded semiannually. How much should they purchase so that the bond fund will be worth $85,000 at maturity?

62. *Funding an IRA* First Colonial Bankshares Corporation advertised the following IRA investment plans.
(a) Assuming continuous compounding, what was the annual rate of interest they offered?
(b) First Colonial Bankshares claims that $4000 invested today will have a value of over $32,000 in 20 years. Use the answer found in part (a) to find the actual value of $4000 in 20 years. Assume continuous compounding.

TARGET IRA PLANS

FOR EACH $5000 MATURITY VALUE DESIRED	
DEPOSIT:	AT A TERM OF:
$620.17	20 years
$1045.02	15 years
$1760.92	10 years
$2967.26	5 years

63. *Loudness of a Garbage Disposal* Find the loudness of a garbage disposal unit that operates at an intensity of 10^{-4} watt per square meter. Express your answer in decibels.

64. *Comparing Earthquakes* On September 9, 1985, the western suburbs of Chicago experienced a mild earthquake that registered 3.0 on the Richter scale. How did this earthquake compare in intensity to the great San Francisco earthquake of 1906, which registered 8.9 on the Richter scale?

65. *Estimating the Date a Prehistoric Man Died* The bones of a prehistoric man found in the desert of New Mexico contain approximately 5% of the original amount of carbon-14. If the half-life of carbon-14 is 5600 years, approximately how long ago did the man die?

66. *Temperature of a Skillet* A skillet is removed from an oven whose temperature is 450°F and placed in a room whose temperature is 70°F. After 5 minutes, the temperature of the skillet is 400°F. How long will it be until its temperature is 150°F?

In Problems 67–68, the independent variable is x and the dependent variable is y. For each set of data:
(a) draw a scatter diagram of the data; (b) use an exponential model, logarithmic model, power model, and linear model to find the "best" model that explains the relationship between the variable.

67.

x	1	4	5	7	10	12
y	1.3084	0.8932	0.8123	0.7737	0.6954	0.6659

68.

x	0.8	1.2	1.7	2.4	2.9	3.4
y	1.1843	2.8634	4.3104	5.6459	6.4683	7.2081

69. *Population Model* The following data represent the population of the United States from 1900 to 1990.

YEAR	POPULATION
1900	76,212,168
1910	92,228,496
1920	106,021,537
1930	123,202,624
1940	132,164,569
1950	151,325,798
1960	179,323,175
1970	203,302,031
1980	226,542,203
1990	248,709,873

(a) Using a graphing utility, draw a scatter diagram for the data using time as the independent variable and population as the dependent variable.
(b) Use an exponential model, logarithmic model, power model, and linear model to find the best model to describe the relationship between the variables.
(c) Using the model found in (b), predict the population of the United States for 1995.

70. *Economics* The following data represent personal consumption expenditures and per capita gross national product figures in the United States from 1984 to 1993 (Source: U.S. Bureau of Economic Analysis).

YEAR	PER CAPITA PERSONAL CONSUMPTION	PER CAPITA GROSS NATIONAL PRODUCT
1984	11,617	17,659
1985	12,015	18,007
1986	12,336	18,337
1987	12,568	18,712
1988	12,903	19,284
1989	13,029	19,615
1990	13,093	19,670
1991	12,895	19,290
1992	13,081	19,548
1993	13,372	19,897

(a) Using a graphing utility, draw a scatter diagram of the data using gross national product as the independent variable and personal consumption as the dependent variable.

(b) Use an exponential model, logarithmic model, power model, and linear model to find the best model to describe the relationship between the variables.

(c) Using the model found in (b), predict personal consumption expenditures if per capita gross national product is $20,000.

71. In a room whose constant temperature is 70°F, will a pizza heated to 450°F ever reach exactly 70°F? Newton's Law of Cooling seems to say no! What really happens? What assumptions are made in using Newton's law? Write a brief paragraph explaining your conclusions.

Appendix

1. Topics from Algebra and Geometry
2. Solving Equations
3. Completing the Square
4. Complex Numbers
5. Linear Curve Fitting

1

Topics from Algebra and Geometry

Sets

When we want to treat a collection of similar but distinct objects as a whole, we use the idea of a **set.** For example, the set of *digits* consists of the collection of numbers 0, 1, 2, 3, 4, 5, 6, 7, 8, and 9. If we use the symbol D to denote the set of digits, then we can write

$$D = \{0, 1, 2, 3, 4, 5, 6, 7, 8, 9\}$$

In this notation, the braces { } are used to enclose the objects, or **elements,** in the set. This method of denoting a set is called the **roster method.** A second way to denote a set is to use **set-builder notation,** where the set D of digits is written as

$$D = \{ \quad x \quad | \quad x \text{ is a digit}\}$$

Read as "D is the set of all x such that x is a digit."

E X A M P L E   1

Using Set-builder Notation and the Roster Method

(a) $E = \{x | x \text{ is an even digit}\} = \{0, 2, 4, 6, 8\}$
(b) $O = \{x | x \text{ is an odd digit}\} = \{1, 3, 5, 7, 9\}$ ∎

In listing the elements of a set, we do not list an element more than once because the elements of a set are distinct. Also, the order in which the elements are listed is not relevant. Thus, for example, $\{2, 3\}$ and $\{3, 2\}$ both represent the same set.

If every element of a set A is also an element of a set B, then we say that A is a **subset** of B. If two sets A and B have the same elements, then we say that A **equals** B. For example, $\{1, 2, 3\}$ is a subset of $\{1, 2, 3, 4, 5\}$; and $\{1, 2, 3\}$ equals $\{2, 3, 1\}$.

Real Numbers

Real numbers are represented by symbols such as

$$25, \quad 0, \quad -3, \quad \tfrac{1}{2}, \quad -\tfrac{5}{4}, \quad 0.125, \quad \sqrt{2}, \quad \pi, \quad \sqrt[3]{-2}, \quad 0.666\ldots$$

527

The set of **counting numbers,** or **natural numbers,** is the set {1, 2, 3, 4, ... }. (The three dots, called an **ellipsis,** indicate that the pattern continues indefinitely.) The set of **integers** is the set { ... , −3, −2, −1, 0, 1, 2, 3, ... }. A **rational number** is a number that can be expressed as a quotient a/b of two integers, where the integer b cannot be 0. Examples of rational numbers are $\frac{3}{4}$, $\frac{5}{2}$, $\frac{0}{4}$, and $-\frac{2}{3}$. Since $a/1 = a$ for any integer a, every integer is also a rational number. Real numbers that are not rational are called **irrational.** Examples of irrational numbers are $\sqrt{2}$ and π (the Greek letter pi), which equals the constant ratio of the circumference to the diameter of a circle. See Figure 1.

FIGURE 1

$$\pi = \frac{C}{d}$$

Real numbers can be represented as **decimals.** Rational real numbers have decimal representations that either **terminate** or are nonterminating with **repeating** blocks of digits. For example, $\frac{3}{4} = 0.75$, which terminates; and $\frac{2}{3} = 0.666...$, in which the digit 6 repeats indefinitely. Irrational real numbers have decimal representations that neither repeat nor terminate. For example, $\sqrt{2} = 1.414213...$ and $\pi = 3.14159...$. In practice, irrational numbers are generally represented by approximations. We use the symbol $\approx$ (read as "approximately equal to") to write $\sqrt{2} \approx 1.4142$ and $\pi \approx 3.1416$.

Often, letters are used to represent numbers. If the letter used is to represent *any* number from a given set of numbers, it is referred to as a **variable.** A **constant** is either a fixed number, such as 5, $\sqrt{2}$, and so on, or a letter that represents a fixed (possibly unspecified) number. In general, we will follow the practice of using letters near the beginning of the alphabet, such as a, b, and c, for constants and using those near the end, such as x, y, and z, as variables.

In working with expressions or formulas involving variables, the variables may only be allowed to take on values from a certain set of numbers, called the **domain of the variable.** For example, in the expression $1/x$, the variable x cannot take on the value 0, since division by 0 is not allowed.

It can be shown that there is a one-to-one correspondence between real numbers and points on a line. That is, every real number corresponds to a point on the line and, conversely, each point on the line has a unique real number associated with it. We establish this correspondence of real numbers with points on a line in the following manner.

We start with a line that is, for convenience, drawn horizontally. Pick a point on the line and label it O, for **origin.** Then pick another point some fixed distance to the right of O and label it U, for **unit.** The fixed distance, which may be 1 inch, 1 centimeter, 1 light-year, or any unit distance, determines the **scale.** We associate the real number 0 with the origin O and the number 1 with the point U. Refer to Figure 2. The point to the right of U that is twice as far from O as U is associated with the number 2. The point to the right of U that is three times as far from O as U is associated with the number 3. The point midway between O and U is assigned the number 0.5, or $\frac{1}{2}$. Corresponding points to the left of the origin O are assigned the numbers $-\frac{1}{2}$, -1, -2, -3, and so on. The real number x associated with a point P is called the **coordinate** of P, and the line whose points have been assigned coordinates is called the **real number line.** Notice in Figure 2 that we placed an arrowhead on the right end of the line to indicate the direction in which the assigned numbers increase. Figure 2 also shows the points associated with the irrational numbers $\sqrt{2}$ and π.

FIGURE 2

Real number line

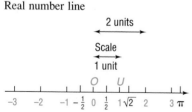

The real number line divides the real numbers into three classes: the **negative real numbers** are the coordinates of points to the left of the origin O; the real number **zero** is the coordinate of the origin O; the **positive real numbers** are the coordinates of points to the right of the origin O.

Let a and b be two real numbers. If the difference $a - b$ is positive, then we say that a is **greater than** b and write $a > b$. Alternatively, if $a - b$ is positive, we can also say that b is **less than** a and write $b < a$. Thus, $a > b$ and $b < a$ are equivalent statements.

On the real number line, if $a > b$, the point with coordinate a is to the right of the point with coordinate b. For example, $0 > -1$, $\pi > 3$, and $\sqrt{2} < 2$. Furthermore,

> $a > 0$ is equivalent to a is positive
>
> $a < 0$ is equivalent to a is negative

If the difference $a - b$ of two real numbers is positive or 0, that is, if $a > b$ or $a = b$, then we say that a is **greater than or equal to** b and write $a \geq b$. Alternatively, if $a \geq b$, we can also say that b is **less than or equal to** a and write $b \leq a$.

Statements of the form $a < b$ or $b > a$ are called **strict inequalities;** statements of the form $a \leq b$ or $b \geq a$ are called **nonstrict inequalities.** The symbols $>$, $<$, $\geq$, and $\leq$ are called **inequality signs.**

If x is a real number and $x \geq 0$, then x is either positive or 0. As a result, we describe the inequality $x \geq 0$ by saying that x is nonnegative.

Inequalities are useful in representing certain subsets of real numbers. In so doing, though, other variations of the inequality notation may be used.

E X A M P L E   2   *Graphing Inequalities*

(a) In the inequality $x > 4$, x is any number greater than 4. In Figure 3, we use a left parenthesis to indicate that the number 4 is not part of the graph.

FIGURE 3
$x > 4$

(b) In the inequality $4 < x \leq 6$, x is any number between 4 and 6, including 6 but excluding 4. In Figure 4, we use a right bracket to indicate that 6 is part of the graph.

FIGURE 4
$x > 4$ and $x \leq 6$

■ Now work Problem 15 (in the exercise set at the end of this section).

Let a and b represent two real numbers with $a < b$: A **closed interval,** denoted by $[a, b]$, consists of all real numbers x for which $a \leq x \leq b$. An **open interval,** denoted by (a, b), consists of all real numbers x for which $a < x < b$. The **half-open,** or **half-closed, intervals** are $(a, b]$, consisting of all real numbers x for which $a < x \leq b$, and $[a, b)$, consisting of all real numbers x for which $a \leq x < b$. In each of these definitions, a is called the **left end point** and b the **right end point** of the interval. Figure 5 illustrates each type of interval.

FIGURE 5

(a) Closed interval [a,b]; $a \le x \le b$

(b) Open interval (a,b); $a < x < b$

(c) Half-open (half-closed) intervals [a,b); $a \le x < b$ (a,b]; $a < x \le b$

The symbol ∞ (read as "infinity") is not a real number, but a notational device used to indicate unboundedness in the positive direction. The symbol $-\infty$ (read as "minus infinity") also is not a real number, but a notational device used to indicate unboundedness in the negative direction. Using the symbols ∞ and $-\infty$, we can define five other kinds of intervals:

$[a, \infty)$ consists of all real numbers x for which $a \le x < \infty$ $(x \ge a)$

(a, ∞) consists of all real numbers x for which $a < x < \infty$ $(x > a)$

$(-\infty, a]$ consists of all real numbers x for which $-\infty < x \le a$ $(x \le a)$

$(-\infty, a)$ consists of all real numbers x for which $-\infty < x < a$ $(x < a)$

$(-\infty, \infty)$ consists of all real numbers x for which $-\infty < x < \infty$ (all real numbers)

Figure 6 illustrates these types of intervals.

FIGURE 6

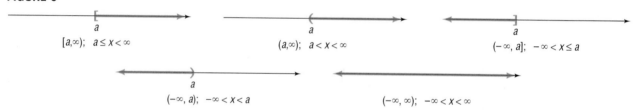

$[a,\infty)$; $a \le x < \infty$

(a,∞); $a < x < \infty$

$(-\infty, a]$; $-\infty < x \le a$

$(-\infty, a)$; $-\infty < x < a$

$(-\infty, \infty)$; $-\infty < x < \infty$

■ Now work Problems 21 and 25.

The *absolute value* of a number a is the distance from the point whose coordinate is a to the origin. For example, the point whose coordinate is -4 is 4 units from the origin. The point whose coordinate is 3 is 3 units from the origin. See Figure 7. Thus, the absolute value of -4 is 4, and the absolute value of 3 is 3.

FIGURE 7

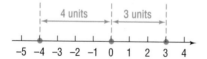

A more formal definition of absolute value is given next.

Absolute Value

The **absolute value** of a real number a, denoted by the symbol $|a|$, is defined by the rules

$$|a| = a \quad \text{if } a \ge 0 \quad \text{and} \quad |a| = -a \quad \text{if } a < 0$$

For example, since $-4 < 0$, then the second rule must be used to get $|-4| = -(-4) = 4$.

E X A M P L E   3

Computing Absolute Value

(a) $|8| = 8$ (b) $|0| = 0$ (c) $|-15| = 15$ ■

■ Now work Problem 29.

Look again at Figure 7. The distance from the point whose coordinate is -4 to the point whose coordinate is 3 is 7 units. This distance is the difference $3 - (-4)$, obtained by subtracting the smaller coordinate from the larger. However, since $|3 - (-4)| = |7| = 7$ and $|-4 - 3| = |-7| = 7$, we can use the absolute value to calculate the distance between two points without being concerned about which coordinate is smaller.

Distance between P and Q

If P and Q are two points on a real number line with coordinates a and b, respectively, the **distance between P and Q,** denoted by $d(P, Q)$, is

$$d(P, Q) = |b - a|$$

Since $|b - a| = |a - b|$, it follows that $d(P, Q) = d(Q, P)$.

E X A M P L E   4

Finding Distance on a Number Line

Let P, Q, and R be points on the real number line with coordinates $-5, 7$, and -3, respectively. Find the distance:

(a) Between P and Q (b) Between Q and R

Solution   (a) $d(P, Q) = |7 - (-5)| = |12| = 12$ (See Figure 8.)

(b) $d(Q, R) = |-3 - 7| = |-10| = 10$

FIGURE 8

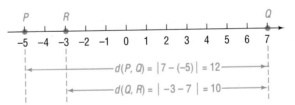

■

Exponents

Integer exponents provide a shorthand device for representing repeated multiplications of a real number.

If a is a real number and n is a positive integer, then the symbol a^n represents the product of n factors of a. That is,

$$a^n = \underbrace{a \cdot a \cdot \ldots \cdot a}_{n \text{ factors}}$$

where it is understood that $a^1 = a$. Thus, $a^2 = a \cdot a$, $a^3 = a \cdot a \cdot a$, and so on. In the expression a^n, a is called the **base** and n is called the **exponent,** or **power.** We read a^n as "a raised to the power n" or as "a to the nth power." We usually read a^2 as "a squared" and a^3 as "a cubed."

Care must be taken when parentheses are used in conjunction with exponents. For example, $-2^4 = -(2 \cdot 2 \cdot 2 \cdot 2) = -16$, whereas $(-2)^4 = (-2) \cdot (-2) \cdot (-2) \cdot (-2) = 16$. Notice the difference: The exponent applies only to the number or parenthetical expression immediately preceding it.

If $a \neq 0$, we define

$$a^0 = 1 \qquad \text{if } a \neq 0$$

If $a \neq 0$ and if n is a positive integer, then we define

$$a^{-n} = \frac{1}{a^n} \qquad \text{if } a \neq 0$$

With these definitions, the symbol a^n is defined for any integer n.

The following properties, called the **laws of exponents,** can be proved using the preceding definitions. In the list, a and b are real numbers, and m and n are integers.

Laws of Exponents

$$a^m a^n = a^{m+n} \qquad (a^m)^n = a^{mn} \qquad (ab)^n = a^n b^n$$

$$\frac{a^m}{a^n} = a^{m-n} = \frac{1}{a^{n-m}}, \quad \text{if } a \neq 0 \qquad \left(\frac{a}{b}\right)^n = \frac{a^n}{b^n}, \qquad \text{if } b \neq 0$$

EXAMPLE 5

Using the Laws of Exponents

Write each expression so that all exponents are positive.

(a) $\dfrac{x^5 y^{-2}}{x^3 y}$, $\quad x \neq 0, y \neq 0$ $\qquad$ (b) $\dfrac{xy}{x^{-1} - y^{-1}}$, $\quad x \neq 0, y \neq 0$

Solution

(a) $\dfrac{x^5 y^{-2}}{x^3 y} = \dfrac{x^5}{x^3} \cdot \dfrac{y^{-2}}{y} = x^{5-3} \cdot y^{-2-1} = x^2 y^{-3} = x^2 \cdot \dfrac{1}{y^3} = \dfrac{x^2}{y^3}$

(b) $\dfrac{xy}{x^{-1} - y^{-1}} = \dfrac{xy}{\dfrac{1}{x} - \dfrac{1}{y}} = \dfrac{xy}{\dfrac{y - x}{xy}} = \dfrac{(xy)(xy)}{y - x} = \dfrac{x^2 y^2}{y - x}$ ∎

Now work Problem 57.

The **principal nth root of a number** a, symbolized by $\sqrt[n]{a}$, is defined as follows:

$$\sqrt[n]{a} = b \quad \text{means} \quad a = b^n \qquad \text{where } a \geq 0 \text{ and } b \geq 0 \text{ if } n \text{ is even}$$
$$\text{and } a, b \text{ are any real numbers if } n \text{ is odd}$$

Notice that if a is negative and n is even, then $\sqrt[n]{a}$ is not defined. When it is defined, the principal nth root of a number is unique.

The symbol $\sqrt[n]{a}$ for the principal nth root of a is sometimes called a **radical;** the integer n is called the **index,** and a is called the **radicand.** If the index of a radical is 2, we call $\sqrt[2]{a}$ the **square root** of a and omit the index 2 by simply writing $\sqrt{a}$. If the index is 3, we call $\sqrt[3]{a}$ the **cube root** of a.

E X A M P L E   6

Simplifying Principal nth Roots

(a) $\sqrt[3]{8} = 2$ because $8 = 2^3$

(b) $\sqrt{64} = 8$ because $64 = 8^2$

(c) $\sqrt[3]{-64} = -4$ because $-64 = (-4)^3$

(d) $\sqrt[4]{\frac{1}{16}} = \frac{1}{2}$ because $\frac{1}{16} = \left(\frac{1}{2}\right)^4$

(e) $\sqrt{0} = 0$ because $0 = 0^2$ ◼

These are examples of **perfect roots.** Thus, 8 and -64 are perfect cubes, since $8 = 2^3$ and $-64 = (-4)^3$; 64 and 0 are perfect squares, since $64 = 8^2$ and $0 = 0^2$; and $\frac{1}{2}$ is a perfect 4th root of $\frac{1}{16}$, since $\frac{1}{16} = \left(\frac{1}{2}\right)^4$.

In general, if $n \geq 2$ is a positive integer and a is a real number,

$$\sqrt[n]{a^n} = a \qquad \text{if } n \text{ is odd} \qquad (1a)$$
$$\sqrt[n]{a^n} = |a| \qquad \text{if } n \text{ is even} \qquad (1b)$$

Notice the need for the absolute value in equation (1b). If n is even, then a^n is positive whether $a > 0$ or $a < 0$. But if n is even, the principal nth root must be nonnegative. Hence, the reason for using the absolute value—it gives a nonnegative result.

E X A M P L E   7

Simplifying Radicals

(a) $\sqrt{8} = \sqrt{4 \cdot 2} = \sqrt{4} \cdot \sqrt{2} = 2\sqrt{2}$

(b) $\sqrt[3]{-16} = \sqrt[3]{-8 \cdot 2} = \sqrt[3]{-8} \cdot \sqrt[3]{2} = -2\sqrt[3]{2}$

(c) $\sqrt{x^2} = |x|$ ◼

◼ Now work Problem 49.

Radicals are used to define **rational exponents.** If a is a real number and $n \geq 2$ is an integer, then

$$a^{1/n} = \sqrt[n]{a}$$

provided $\sqrt[n]{a}$ exists.

If a is a real number and m and n are integers containing no common factors with $n \geq 2$, then

$$a^{m/n} = \sqrt[n]{a^m} = (\sqrt[n]{a})^m \qquad (2)$$

provided $\sqrt[n]{a}$ exists.

In simplifying $a^{m/n}$, either $\sqrt[n]{a^m}$ or $(\sqrt[n]{a})^m$ may be used. Generally, taking the root first, as in $(\sqrt[n]{a})^m$, is preferred.

EXAMPLE 8   *Using Equation (2)*

(a) $8^{2/3} = (\sqrt[3]{8})^2 = 2^2 = 4$ (b) $16^{3/2} = (\sqrt{16})^3 = 4^3 = 64$

(c) $(-8x^5)^{1/3} = \sqrt[3]{-8x^3 \cdot x^2} = \sqrt[3]{(-2x)^3 \cdot x^2} = \sqrt[3]{(-2x)^3}\sqrt[3]{x^2}$

$$= -2x\sqrt[3]{x^2}$$

■ Now work Problem 47.

Polynomials

Monomial

A **monomial** in one variable is the product of a constant times a variable raised to a nonnegative integer power. Thus, a monomial is of the form

$$ax^k$$

where a is a constant, x is a variable, and $k \geq 0$ is an integer.

Two monomials ax^k and bx^k, when added or subtracted, can be combined into a single monomial by using the distributive property. For example,

$$2x^2 + 5x^2 = (2 + 5)x^2 = 7x^2 \quad \text{and} \quad 8x^3 - 5x^3 = (8 - 5)x^3 = 3x^3$$

Polynomial

A **polynomial** in one variable is an algebraic expression of the form

$$a_nx^n + a_{n-1}x^{n-1} + \cdots + a_1x + a_0$$

where $a_n, a_{n-1}, \ldots, a_1, a_0$ are constants,* called the **coefficients** of the polynomial, $n \geq 0$ is an integer, and x is a variable. If $a_n \neq 0$, it is called the **leading coefficient,** and n is called the **degree** of the polynomial.

The monomials that make up a polynomial are called its **terms.** If all the coefficients are 0, the polynomial is called the **zero polynomial,** which has no degree.

---

*The notation a_n is read as "a sub n." The number n is called a **subscript** and should not be confused with an exponent. We use subscripts to distinguish one constant from another when a large or undetermined number of constants is required.

Polynomials are usually written in **standard form,** beginning with the nonzero term of highest degree and continuing with terms in descending order according to degree. Examples of polynomials are

POLYNOMIAL	COEFFICIENTS	DEGREE
$3x^2 - 5 = 3x^2 + 0 \cdot x + (-5)$	$3, 0, -5$	2
$8 - 2x + x^2 = 1 \cdot x^2 - 2x + 8$	$1, -2, 8$	2
$5x + \sqrt{2} = 5x^1 + \sqrt{2}$	$5, \sqrt{2}$	1
$3 = 3 \cdot 1 = 3 \cdot x^0$	3	0
0	0	No degree

Although we have been using x to represent the variable, letters such as y or z are also commonly used. Thus,

$3x^4 - x^2 + 2$ is a polynomial (in x) of degree 4.

$9y^3 - 2y^2 + y - 3$ is a polynomial (in y) of degree 3.

$z^5 + \pi$ is a polynomial (in z) of degree 5.

Pythagorean Theorem

FIGURE 9

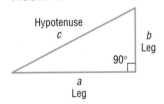

The *Pythagorean Theorem* is a statement about *right triangles*. A **right triangle** is one that contains a **right angle,** that is, an angle of 90°. The side of the triangle opposite the 90° angle is called the **hypotenuse;** the remaining two sides are called **legs.** In Figure 9 we have used c to represent the length of the hypotenuse and a and b to represent the lengths of the legs. Notice the use of the symbol ⌐ to show the 90° angle. We now state the Pythagorean Theorem.

Pythagorean Theorem

In a right triangle, the square of the length of the hypotenuse is equal to the sum of the squares of the lengths of the legs. That is, in the right triangle shown in Figure 9,

$$c^2 = a^2 + b^2 \qquad (3)$$

■

E X A M P L E   9

Finding the Hypotenuse of a Right Triangle

In a right triangle, one leg is of length 4 and the other is of length 3. What is the length of the hypotenuse?

Solution

Since the triangle is a right triangle, we use the Pythagorean Theorem with $a = 4$ and $b = 3$ to find the length c of the hypotenuse. Thus, from equation (3), we have

$$c^2 = a^2 + b^2$$
$$c^2 = 4^2 + 3^2 = 16 + 9 = 25$$
$$c = 5$$

■

■ Now work Problem 69.

The converse of the Pythagorean Theorem is also true.

Converse of the Pythagorean Theorem

In a triangle, if the square of the length of one side equals the sum of the squares of the lengths of the other two sides, then the triangle is a right triangle. The 90° angle is opposite the longest side. ■

E X A M P L E 1 0   *Verifying That a Triangle Is a Right Triangle*

Show that a triangle whose sides are of lengths 5, 12, and 13 is a right triangle. Identify the hypotenuse.

FIGURE 10

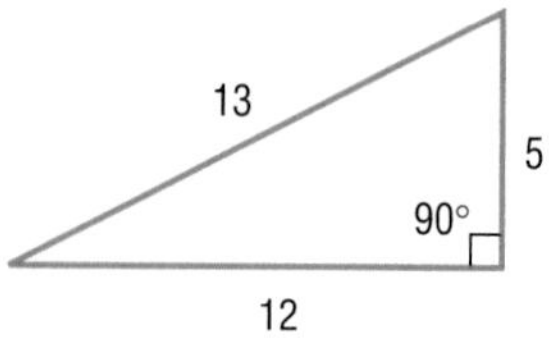

Solution   We square the lengths of the sides:

$$25, \quad 144, \quad 169$$

Notice that the sum of the first two squares (25 and 144) equals the third square (169). Hence, the triangle is a right triangle. The longest side, 13, is the hypotenuse. See Figure 10. ■

■ Now work Problem 79.

Geometry Formulas

Certain formulas from geometry are useful in solving trigonometry problems. We list some of these formulas next.

For a rectangle of length l and width w,

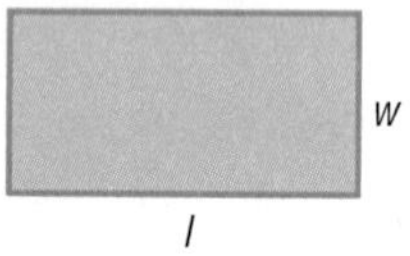

$$\text{Area} = lw \qquad \text{Perimeter} = 2l + 2w$$

For a triangle with base b and altitude h,

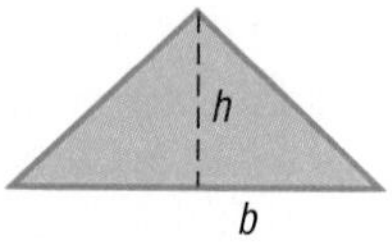

$$\text{Area} = \frac{1}{2}bh$$

For a circle of radius r (diameter $d = 2r$),

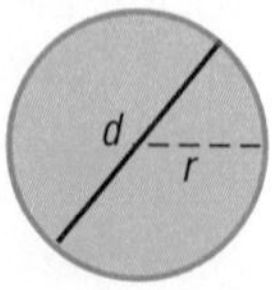

$$\text{Area} = \pi r^2 \qquad \text{Circumference} = 2\pi r = \pi d$$

For a rectangular box of length l, width w, and height h,

$$\text{Volume} = lwh$$

Exercise 1

In Problems 1–10, replace the question mark by $<$, $>$, or $=$, whichever is correct.

1. $\frac{1}{2}$? 0   **2.** 5 ? 6   **3.** -1 ? -2 **4.** -3 ? $-\frac{5}{2}$ **5.** π ? 3.14

6. $\sqrt{2}$? 1.41   **7.** $\frac{1}{2}$? 0.5   **8.** $\frac{1}{3}$? 0.33   **9.** $\frac{2}{3}$? 0.67   **10.** $\frac{1}{4}$? 0.25

11. On the real number line, label the points with coordinates 0, 1, -1, $\frac{5}{2}$, -2.5, $\frac{3}{4}$, and 0.25.

12. Repeat Problem 11 for the coordinates 0, -2, 2, -1.5, $\frac{3}{2}$, $\frac{1}{3}$, and $\frac{2}{3}$.

In Problems 13–20, write each statement as an inequality.

13. x is positive   **14.** z is negative

15. x is less than 2   **16.** y is greater than -5

17. x is less than or equal to 1   **18.** x is greater than or equal to 2

19. x is less than 5 and x is greater than 2

20. y is less than or equal to 2 and y is greater than 0

In Problems 21–24, write each inequality using interval notation, and illustrate each inequality using the real number line.

21. $0 \le x \le 4$ **22.** $-1 < x < 5$ **23.** $4 \le x < 6$ **24.** $-2 < x \le 0$

In Problems 25–28, write each interval as an inequality involving x, and illustrate each inequality using the real number line.

25. $[2, 5]$ **26.** $(1, 2)$ **27.** $[4, \infty)$ **28.** $(-\infty, 2]$

In Problems 29–32, find the value of each expression if $x = 2$ and $y = -3$.

29. $|x + y|$ **30.** $|x - y|$ **31.** $|x| + |y|$ **32.** $|x| - |y|$

In Problems 33–52, simplify each expression.

33. 3^0 **34.** 3^2 **35.** 4^{-2} **36.** $(-3)^2$

37. $\left(\frac{2}{3}\right)^2$ **38.** $\left(\frac{-4}{5}\right)^3$ **39.** $3^{-6} \cdot 3^4$ **40.** $4^{-2} \cdot 4^3$

41. $\left(\frac{2}{3}\right)^{-2}$ **42.** $\left(\frac{3}{2}\right)^{-3}$ **43.** $\dfrac{2^3 \cdot 3^2}{2 \cdot 3^{-2}}$ **44.** $\dfrac{3^{-2} \cdot 5^3}{3 \cdot 5}$

45. $9^{3/2}$ **46.** $16^{3/4}$ **47.** $(-8)^{4/3}$ **48.** $(-27)^{2/3}$

49. $\sqrt{32}$ **50.** $\sqrt[3]{24}$ **51.** $\sqrt[3]{-\frac{8}{27}}$ **52.** $\sqrt{\frac{4}{9}}$

In Problems 53–68, simplify each expression so that all exponents are positive. Whenever an exponent is negative or 0, we assume that the base does not equal 0.

53. $x^0 y^2$ **54.** $x^{-1} y$ **55.** $x^{-2} y$ **56.** $x^4 y^0$

57. $\dfrac{x^{-2} y^3}{xy^4}$ **58.** $\dfrac{x^{-2} y}{xy^2}$ **59.** $\left(\dfrac{4x}{5y}\right)^{-2}$ **60.** $(xy)^{-2}$

61. $\dfrac{x^{-1} y^{-2} z}{x^2 y z^3}$ **62.** $\dfrac{3x^{-2} y z^2}{x^4 y^{-3} z}$ **63.** $\dfrac{(-2)^3 x^4 (yz)^2}{3^2 xy^3 z^4}$ **64.** $\dfrac{4x^{-2}(yz)^{-1}}{(-5)^2 x^4 y^2 z^{-2}}$

65. $\dfrac{x^{-2}}{x^{-2} + y^{-2}}$ **66.** $\dfrac{x^{-1} + y^{-1}}{x^{-1} - y^{-1}}$ **67.** $\left(\dfrac{3x^{-1}}{4y^{-1}}\right)^{-2}$ **68.** $\left(\dfrac{5x^{-2}}{6y^{-2}}\right)^{-3}$

In Problems 69–78, a and b are the lengths of the legs of a right triangle and c is the length of the hypotenuse. Find the missing length.

69. $a = 5, b = 12, c = ?$ **70.** $a = 6, b = 8, c = ?$ **71.** $a = 10, b = 24, c = ?$

72. $a = 4, b = 3, c = ?$ **73.** $a = 7, b = 24, c = ?$ **74.** $a = 14, b = 48, c = ?$

75. $a = 3, c = 5, b = ?$ **76.** $b = 6, c = 10, a = ?$ **77.** $b = 7, c = 25, a = ?$

78. $a = 10, c = 13, b = ?$

In Problems 79–84, the lengths of the sides of a triangle are given. Determine which are right triangles. For those that are right triangles, identify the hypotenuse.

79. 3, 4, 5   **80.** 6, 8, 10   **81.** 4, 5, 6

82. 2, 2, 3   **83.** 7, 24, 25   **84.** 10, 24, 26

85. *Geometry* Find the diagonal of a rectangle whose length is 8 inches and whose width is 5 inches.

86. *Geometry* Find the length of a rectangle of width 3 inches if its diagonal is 20 inches long.

87. *Finding the Length of a Guy Wire* A radio transmission tower is 100 feet high. How long does a guy wire need to be if it is to connect a point halfway up the tower to a point 30 feet from the base?

88. Answer Problem 87 if the guy wire is attached to the top of the tower.

89. *How Far Can You See?* The tallest inhabited building in North America is the Sears Tower in Chicago.* If the observation tower is 1454 feet above ground level, use the figure to determine how far a person standing in the observation tower can see (with the aid of a telescope). Use 3960 miles for the radius of Earth. [*Note:* 1 mile = 5280 feet]

 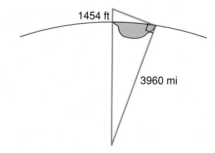

In Problems 90–92, use the fact that the radius of Earth is 3960 miles.

90. *How Far Can You See?* The conning tower of the USS *Silversides,* a World War II submarine now permanently stationed in Muskegon, Michigan, is approximately 20 feet above sea level. How far can one see from the conning tower?

91. *How Far Can You See?* A person who is 6 feet tall is standing on the beach in Fort Lauderdale, Florida and looks out onto the Atlantic Ocean. Suddenly, a ship appears on the horizon. How far is the ship from shore?

92. *How Far Can You See?* The deck of a destroyer is 100 feet above sea level. How far can a person see from the deck? How far can a person see from the bridge, which is 150 feet above sea level?

93. If $a \leq b$ and $c > 0$, show that $ac \leq bc$. [*Hint:* Since $a \leq b$, it follows that $a - b \leq 0$. Now multiply each side by c.]

94. If $a \leq b$ and $c < 0$, show that $ac \geq bc$.

95. If $a < b$, show that $a < (a + b)/2 < b$. The number $(a + b)/2$ is called the **arithmetic mean** of a and b.

Source: Guinness Book of World Records.

96. Refer to Problem 95. Show that the arithmetic mean of a and b is equidistant from a and b.

97. Are there any real numbers that are both rational and irrational? Are there any real numbers that are neither? Explain your reasoning.

98. Explain why the sum of a rational number and an irrational number must be irrational.

99. What rational number does the repeating decimal $0.9999 \ldots$ equal?

100. Is there a positive real number "closest" to 0?

101. I'm thinking of a number! It lies between 1 and 10; its square is rational and lies between 1 and 10. The number is larger than π. Correct to two decimal places, name the number. Now think of your own number, describe it, and challenge a fellow student to name it.

102. Write a brief paragraph that illustrates the similarities and differences between "less than" ($<$) and "less than or equal" ($\leq$).

103. *The Gibb's Hill Lighthouse, Southampton, Bermuda,* in operation since 1846, stands 117 feet high on a hill 245 feet high, so its beam of light is 362 feet above sea level. A brochure states that the light itself can be seen on the horizon about 26 miles distant. Verify the correctness of this information. The brochure further states that ships 40 miles away can see the light and planes flying at 10,000 feet can see it 120 miles away. Verify the accuracy of these statements. What assumption did the brochure make about the height of the ship?

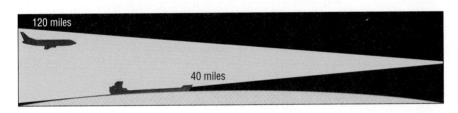

104. You have 1000 feet of flexible pool siding and wish to construct a swimming pool. Experiment with rectangular-shaped pools with perimeters of 1000 feet. How do their areas vary? What is the shape of the rectangle with the largest area? Now compute the area enclosed by a circular pool with a perimeter (circumference) of 1000 feet. What would be your choice of shape for the pool? If rectangular, what is your preference for dimensions? Justify your choice. If your only consideration is to have a pool that encloses the most area, what shape should you use?

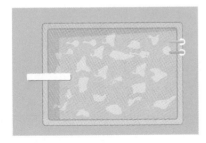 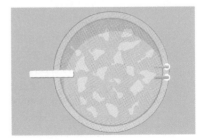

2

Solving Equations

An **equation in one variable** is a statement in which two expressions, at least one containing the variable, are equal. The expressions are called the **sides** of the equation. Since an equation is a statement, it may or may not be true, depending on the value of the variable. Unless otherwise restricted, the admissible values of the variable are those in the domain of the variable. Those admissible values of the variable, if any, that result in a true statement are called **solutions, or roots,** of the equation. To **solve an equation** means to find all the solutions of the equation.

For example, the following are all equations in one variable, x:

$$x + 5 = 9 \qquad x^2 + 5x = 2x - 2 \qquad \frac{x^2 - 4}{x + 1} = 0 \qquad x^2 + 9 = 5$$

The first of these statements, $x + 5 = 9$, is true when $x = 4$ and false for any other choice of x. Thus, 4 is a solution of the equation $x + 5 = 9$. We also say that 4 **satisfies** the equation $x + 5 = 9$, because, when x is replaced by 4, a true statement results.

Sometimes an equation will have more than one solution. For example, the equation

$$\frac{x^2 - 4}{x + 1} = 0$$

has either $x = -2$ or $x = 2$ as a solution.

Sometimes we will write the solutions of an equation in set notation. This set is called the **solution set** of the equation. For example, the solution set of the equation $x^2 - 9 = 0$ is $\{-3, 3\}$.

Unless indicated otherwise, we will limit ourselves to real solutions. Some equations have no real solution. For example, $x^2 + 9 = 5$ has no real solution, because there is no real number whose square when added to 9 equals 5.

An equation that is satisfied for every choice of the variable for which both sides are defined is called an **identity.** For example, the equation

$$3x + 5 = x + 3 + 2x + 2$$

is an identity, because this statement is true for any real number x.

Two or more equations that have precisely the same solutions are called **equivalent equations.** For example, all the following equations are equivalent, because each has only the solution $x = 5$:

$$2x + 3 = 13$$
$$2x = 10$$
$$x = 5$$

These three equations illustrate one method for solving many types of equations: Replace the original equation by an equivalent equation, and continue until an equation with an obvious solution, such as $x = 5$, is reached. The question, though, is: "How do I obtain an equivalent equation?" In general, there are five ways to do so.

Procedures That Result in Equivalent Equations

1. Interchange the two sides of the equation:

$$\text{Replace} \quad 3 = x \quad \text{by} \quad x = 3$$

2. Simplify the sides of the equation by combining like terms, eliminating parentheses, and so on:

$$\text{Replace} \quad (x + 2) + 6 = 2x + (x + 1)$$
$$\text{by} \quad x + 8 = 3x + 1$$

3. Add or subtract the same expression on both sides of the equation:

$$\text{Replace} \qquad 3x - 5 = 4$$
$$\text{by} \qquad (3x - 5) + 5 = 4 + 5$$

4. Multiply or divide both sides of the equation by the same nonzero expression:

$$\text{Replace} \qquad \frac{3x}{x - 1} = \frac{6}{x - 1} \qquad x \neq 1$$

$$\text{by} \qquad \frac{3x}{x - 1} \cdot (x - 1) = \frac{6}{x - 1} \cdot (x - 1)$$

5. If one side of the equation is 0 and the other side can be factored, then we may use the product law* and set each factor equal to 0:

$$\text{Replace} \qquad x(x - 3) = 0$$
$$\text{by} \qquad x = 0 \quad \text{or} \quad x - 3 = 0$$

Whenever it is possible to solve an equation in your head, do so. For example:

The solution of $2x = 8$ is $x = 4$.

The solution of $3x - 15 = 0$ is $x = 5$.

Often, though, some rearrangement is necessary.

E X A M P L E   1   *Solving an Equation*

Solve the equation: $(x + 1)(2x) = (x + 1)(2)$

Solution   We begin by collecting all terms on the left side:

$$(x + 1)(2x) = (x + 1)(2)$$
$$(x + 1)(2x) - (x + 1)(2) = 0$$
$$(x + 1)(2x - 2) = 0 \qquad \text{Factor.}$$
$$x + 1 = 0 \quad \text{or} \quad 2x - 2 = 0 \qquad \text{Apply the product law.}$$
$$x = -1 \qquad\qquad 2x = 2$$
$$x = 1$$

The solution set is $\{-1, 1\}$.   ■

E X A M P L E   2   *Solving an Equation*

Solve the equation: $\dfrac{3x}{x - 1} + 2 = \dfrac{3}{x - 1}$

*The product law states that if $ab = 0$ then $a = 0$ or $b = 0$ or both equal 0.

Solution   First, we note that the domain of the variable is $\{x | x \neq 1\}$. Since the two quotients in the equation have the same denominator, $x - 1$, we can simplify by multiplying both sides by $x - 1$. The resulting equation is equivalent to the original equation, since we are multiplying by $x - 1$, which is not 0 (remember, $x \neq 1$).

$$\frac{3x}{x-1} + 2 = \frac{3}{x-1}$$

$$\left(\frac{3x}{x-1} + 2\right) \cdot (x - 1) = \frac{3}{x-1} \cdot (x - 1) \qquad \text{Multiply both sides by } x - 1; \\ \text{cancel on the right.}$$

$$\frac{3x}{x-1} \cdot (x - 1) + 2 \cdot (x - 1) = 3 \qquad \text{Use the distributive property on} \\ \text{the left side; cancel on the left.}$$

$$3x + (2x - 2) = 3 \qquad \text{Simplify.}$$

$$5x - 2 = 3$$

$$5x = 5 \qquad \text{Add 2 to each side.}$$

$$x = 1 \qquad \text{Divide both sides by 5.}$$

The solution appears to be 1. But recall that $x = 1$ is not in the domain of the variable. Thus, the equation has no solution. ▪

■   Now work Problem 19.

Steps for Solving Equations

STEP 1: List any restrictions on the domain of the variable.
STEP 2: Simplify the equation by replacing the original equation by a succession of equivalent equations following the procedures listed earlier.
STEP 3: If the result of Step 2 is a product of factors equal to 0, use the product law and set each factor equal to 0 (procedure 5).
STEP 4: Check your solution(s).

E X A M P L E   3

Solving an Equation

Solve the equation: $x^3 = 25x$

Solution   We first rearrange the equation to get 0 on the right side:

$$x^3 = 25x$$

$$x^3 - 25x = 0$$

We notice that x is a factor of each term on the left:

$$x(x^2 - 25) = 0$$

$$x(x + 5)(x - 5) = 0 \qquad \text{Difference of two squares}$$

$$x = 0 \quad \text{or} \quad x + 5 = 0 \quad \text{or} \quad x - 5 = 0 \qquad \text{Set each factor equal to 0.}$$

$$x = -5 \quad \text{or} \qquad x = 5 \qquad \text{Solve.}$$

The solution set is $\{-5, 0, 5\}$. ▪

■   Now work Problem 21.

2

Exercise 2

In Problems 1–36, solve each equation.

1. $6 - x = 2x + 9$

2. $3 - 2x = 2 - x$

3. $2(3 + 2x) = 3(x - 4)$

4. $3(2 - x) = 2x - 1$

5. $8x - (2x + 1) = 3x - 10$

6. $5 - (2x - 1) = 10$

7. $\frac{1}{2}x - 4 = \frac{3}{4}x$

8. $1 - \frac{1}{2}x = 5$

9. $0.9t = 0.4 + 0.1t$

10. $0.9t = 1 + t$

11. $\frac{2}{y} + \frac{4}{y} = 3$

12. $\frac{4}{y} - 5 = \frac{5}{2y}$

13. $(x + 7)(x - 1) = (x + 1)^2$

14. $(x + 2)(x - 3) = (x - 3)^2$

15. $x(2x - 3) = (2x + 1)(x - 4)$

16. $x(1 + 2x) = (2x - 1)(x - 2)$

17. $z(z^2 + 1) = 3 + z^3$

18. $w(4 - w^2) = 8 - w^3$

19. $\frac{x}{x - 3} + 3 = \frac{3}{x - 3}$

20. $\frac{3x}{x + 2} = \frac{-6}{x + 2} - 2$

21. $x^2 = 9x$

22. $x^3 = x^2$

23. $t^3 - 9t^2 = 0$

24. $4z^3 - 8z^2 = 0$

25. $\frac{2x}{x^2 - 4} = \frac{4}{x^2 - 4} - \frac{1}{x + 2}$

26. $\frac{x}{x^2 - 9} + \frac{1}{x + 3} = \frac{3}{x^2 - 9}$

27. $\frac{x}{x + 2} = \frac{1}{2}$

28. $\frac{3x}{x - 1} = 2$

29. $\frac{3}{2x - 3} = \frac{2}{x + 5}$

30. $\frac{-2}{x + 4} = \frac{-3}{x + 1}$

31. $(x + 2)(3x) = (x + 2)(6)$

32. $(x - 5)(2x) = (x - 5)(4)$

33. $\frac{6t + 7}{4t - 1} = \frac{3t + 8}{2t - 4}$

34. $\frac{8w + 5}{10w - 7} = \frac{4w - 3}{5w + 7}$

35. $\frac{2}{x - 2} = \frac{3}{x + 5} + \frac{10}{(x + 5)(x - 2)}$

36. $\frac{1}{2x + 3} + \frac{1}{x - 1} = \frac{1}{(2x + 3)(x - 1)}$

3

Completing the Square

The idea behind the method of **completing the square** is to "adjust" the left side of a second degree polynomial, $ax^2 + bx + c$, so that it becomes a perfect square—the square of a first-degree polynomial. For example, $x^2 + 6x + 9$ and $x^2 - 4x + 4$ are perfect squares because

$$x^2 + 6x + 9 = (x + 3)^2 \quad \text{and} \quad x^2 - 4x + 4 = (x - 2)^2$$

How do we adjust the second degree polynomial? We do it by adding the appropriate number to create a perfect square. For example, to make $x^2 + 6x$ a perfect square, we add 9.

Let's look at several examples of completing the square when the coefficient of x^2 is 1:

START	ADD	RESULT
$x^2 + 4x$	4	$x^2 + 4x + 4 = (x + 2)^2$
$x^2 + 12x$	36	$x^2 + 12x + 36 = (x + 6)^2$
$x^2 - 6x$	9	$x^2 - 6x + 9 = (x - 3)^2$
$x^2 + x$	$\frac{1}{4}$	$x^2 + x + \frac{1}{4} = \left(x + \frac{1}{2}\right)^2$

Do you see the pattern? Provided the coefficient of x^2 is 1, we complete the square by adding the square of one-half the coefficient of x:

START	ADD	RESULT
$x^2 + mx$	$\left(\dfrac{m}{2}\right)^2$	$x^2 + mx + \left(\dfrac{m}{2}\right)^2 = \left(x + \dfrac{m}{2}\right)^2$

■   Now work Problem 1.

E X A M P L E   1

Completing the Square of an Equation Containing Two Variables

Complete the squares of x and y in the equation

$$x^2 + y^2 - 2x + 4y - 4 = 0$$

Solution   We rearrange the equation, grouping the terms involving the variable x and the variable y.

$$(x^2 - 2x) + (y^2 + 4y) = 4$$

Next, we complete the square of each parenthetical expression. Of course, since we want an equivalent equation, whatever we add to the left side, we also add to the right side.

$$(x^2 - 2x + 1) + (y^2 + 4y + 4) = 4 + 1 + 4$$
$$(x - 1)^2 + (y + 2)^2 = 9$$

The terms involving the variables x and y now appear as perfect squares. ■

■   Now work Problem 7.

The next example illustrates how the procedure of completing the square can be used to solve a quadratic equation.

E X A M P L E   2

Solving a Quadratic Equation by Completing the Square

Solve by completing the square: $x^2 + 5x + 4 = 0$

Solution   We always begin this procedure by rearranging the equation so that the constant is on the right side:

$$x^2 + 5x + 4 = 0$$
$$x^2 + 5x = -4$$

Since the coefficient of x^2 is 1, we can complete the square on the left side by adding $\left(\frac{1}{2} \cdot 5\right)^2 = \frac{25}{4}$. Of course, in an equation, whatever we add to the left side must also be added to the right side. Thus, we add $\frac{25}{4}$ to *both* sides:

$$x^2 + 5x + \tfrac{25}{4} = -4 + \tfrac{25}{4}$$
$$\left(x + \tfrac{5}{2}\right)^2 = \tfrac{9}{4}$$
$$x + \tfrac{5}{2} = \pm\sqrt{\tfrac{9}{4}}$$
$$x + \tfrac{5}{2} = \pm\tfrac{3}{2}$$
$$x = -\tfrac{5}{2} \pm \tfrac{3}{2}$$
$$x = -\tfrac{5}{2} + \tfrac{3}{2} = -1 \quad \text{or} \quad x = -\tfrac{5}{2} - \tfrac{3}{2} = -4$$

The solution set is $\{-4, -1\}$. ∎

■ Now work Problem 13.

E X A M P L E   3 *Solving a Quadratic Equation by Completing the Square*

Solve by completing the square: $2x^2 - 8x - 5 = 0$

Solution First, we rewrite the equation:

$$2x^2 - 8x - 5 = 0$$
$$2x^2 - 8x = 5$$

Next, we divide by 2 so that the coefficient of x^2 is 1. (This enables us to complete the square at the next step.)

$$x^2 - 4x = \tfrac{5}{2}$$

Finally, we complete the square by adding 4 to each side:

$$x^2 - 4x + 4 = \tfrac{5}{2} + 4$$
$$(x - 2)^2 = \tfrac{13}{2}$$
$$x - 2 = \pm\sqrt{\tfrac{13}{2}} = \pm\tfrac{\sqrt{26}}{2}$$
$$x = 2 \pm \tfrac{\sqrt{26}}{2}$$

We choose to leave our answer in this compact form. Thus, the solution set is $\{2 - \sqrt{26}/2,\ 2 + \sqrt{26}/2\}$. ∎

Note: If we wanted an approximation, say, to two decimal places, of these solutions, we would use a calculator to get $\{-0.55, 4.55\}$.

■ Now work Problem 17.

3

Exercise 3

In Problems 1–6, tell what number should be added to complete the square of each expression.

1. $x^2 - 4x$ **2.** $x^2 - 2x$ **3.** $x^2 + \tfrac{1}{2}x$

4. $x^2 - \tfrac{1}{3}x$ **5.** $x^2 - \tfrac{2}{3}x$ **6.** $x^2 - \tfrac{2}{5}x$

In Problems 7–12, complete the squares of x and y in each equation.

7. $x^2 + y^2 - 4x + 4y - 1 = 0$ **8.** $x^2 + y^2 + 4x + 4y - 8 = 0$ **9.** $x^2 + y^2 + 6x - 2y + 1 = 0$

10. $x^2 + y^2 - 8x + 2y + 1 = 0$ **11.** $x^2 + y^2 + x - y - \tfrac{1}{2} = 0$ **12.** $x^2 + y^2 - x + y - \tfrac{3}{2} = 0$

In Problems 13–18, solve each equation by completing the square.

13. $x^2 + 4x - 21 = 0$ **14.** $x^2 - 6x = 13$ **15.** $x^2 - \frac{1}{2}x = \frac{3}{16}$

16. $x^2 + \frac{2}{3}x = \frac{1}{3}$ **17.** $3x^2 + x - \frac{1}{2} = 0$ **18.** $2x^2 - 3x = 1$

4

Complex Numbers

One property of a real number is that its square is nonnegative. For example, there is no real number x for which

$$x^2 = -1$$

To remedy this situation, we introduce a number called the **imaginary unit,** which we denote by i and whose square is -1. Thus,

$$i^2 = -1$$

This should not surprise you. If our universe were to consist only of integers, there would be no number x for which $2x = 1$. This unfortunate circumstance was remedied by introducing numbers such as $\frac{1}{2}$ and $\frac{2}{3}$, the *rational numbers*. If our universe were to consist only of rational numbers, there would be no number x whose square equals 2. That is, there would be no number x for which $x^2 = 2$. To remedy this, we introduced numbers such as $\sqrt{2}$ and $\sqrt[3]{5}$, the *irrational numbers*. The *real numbers*, you will recall, consist of the rational numbers and the irrational numbers. Now, if our universe were to consist only of real numbers, then there would be no number x whose square is -1. To remedy this, we introduce a number i, whose square is -1.

In the progression outlined, each time we encountered a situation that was unsuitable, we introduced a new number system to remedy this situation. And each new number system contained the earlier number system as a subset. The number system that results from introducing the number i is called the **complex number system.**

Complex Numbers   **Complex numbers** are numbers of the form $a + bi$, where a and b are real numbers. The real number a is called the **real part** of the number $a + bi$; the real number b is called the **imaginary part** of $a + bi$.

For example, the complex number $-5 + 6i$ has the real part -5 and the imaginary part 6.

When a complex number is written in the form $a + bi$, where a and b are real numbers, we say it is in **standard form.** However, if the imaginary part of a complex number is negative, such as in the complex number $3 + (-2)i$, we agree to write it instead in the form $3 - 2i$.

Also, the complex number $a + 0i$ is usually written merely as a. This serves to remind us that the real numbers are a subset of the complex numbers. The complex number $0 + bi$ is usually written as bi. Sometimes the complex number bi is called a **pure imaginary number.**

Equality, addition, subtraction, and multiplication of complex numbers are defined so as to preserve the familiar rules of algebra for real numbers. Thus, two complex numbers are equal if and only if their real parts are equal and their imaginary parts are equal. That is,

Equality of Complex Numbers

$$a + bi = c + di \quad \text{if and only if} \quad a = c \text{ and } b = d \qquad (1)$$

Two complex numbers are added by forming the complex number whose real part is the sum of the real parts and whose imaginary part is the sum of the imaginary parts. That is,

Sum of Complex Numbers

$$(a + bi) + (c + di) = (a + c) + (b + d)i \qquad (2)$$

To subtract two complex numbers, we follow the rule

Difference of Complex Numbers

$$(a + bi) - (c + di) = (a - c) + (b - d)i \qquad (3)$$

E X A M P L E   1   *Adding and Subtracting Complex Numbers*

(a) $(3 + 5i) + (-2 + 3i) = [3 + (-2)] + (5 + 3)i = 1 + 8i$

(b) $(6 + 4i) - (3 + 6i) = (6 - 3) + (4 - 6)i = 3 + (-2)i = 3 - 2i$ ■

■   Now work Problem 5.

Products of complex numbers are calculated as illustrated in Example 2.

E X A M P L E   2   *Multiplying Complex Numbers*

$$(5 + 3i) \cdot (2 + 7i) = 5 \cdot (2 + 7i) + 3i(2 + 7i) = 10 + 35i + 6i + 21i^2$$

$\uparrow$ Distributive property $\uparrow$ Distributive property

$$= 10 + 41i + 21(-1)$$

$\uparrow$ $i^2 = -1$

$$= -11 + 41i \qquad ■$$

Based on the procedure of Example 2, we define the **product** of two complex numbers by the formula

Product of Complex Numbers

$$(a + bi) \cdot (c + di) = (ac - bd) + (ad + bc)i \qquad (4)$$

Do not bother to memorize formula (4). Instead, whenever it is necessary to multiply two complex numbers, follow the usual rules for multiplying two binomials, as in Example 2, remembering that $i^2 = -1$. For example,

$$(2i)(2i) = 4i^2 = -4$$

$$(2 + i)(1 - i) = 2 - 2i + i - i^2 = 3 - i$$

■   Now work Problem 11.

Algebraic properties for addition and multiplication, such as the commutative, associative, and distributive properties, hold for complex numbers. Of these, the property that every nonzero complex number has a multiplicative inverse, or reciprocal, requires a closer look.

Conjugates

Conjugate | If $z = a + bi$ is a complex number, then its **conjugate,** denoted by $\bar{z}$, is defined as

$$\bar{z} = \overline{a + bi} = a - bi$$

For example, $\overline{2 + 3i} = 2 - 3i$ and $\overline{-6 - 2i} = -6 + 2i$.

E X A M P L E   3

Multiplying a Complex Number by Its Conjugate

Find the product of the complex number $z = 3 + 4i$ and its conjugate $\bar{z}$.

Solution | Since $\bar{z} = 3 - 4i$, we have

$$z\bar{z} = (3 + 4i)(3 - 4i) = 9 + 12i - 12i - 16i^2 = 9 + 16 = 25 \qquad \blacksquare$$

The result obtained in Example 3 has an important generalization:

Theorem | The product of a complex number and its conjugate is a nonnegative real number. Thus, if $z = a + bi$, then

$$z\bar{z} = a^2 + b^2 \tag{5}$$

Proof: | If $z = a + bi$, then

$$z\bar{z} = (a + bi)(a - bi) = a^2 - (bi)^2 = a^2 - b^2i^2 = a^2 + b^2 \qquad \blacksquare$$

To express the reciprocal of a nonzero complex number z in standard form, multiply the numerator and denominator by its conjugate $\bar{z}$. Thus, if $z = a + bi$ is a nonzero complex number, then

$$\frac{1}{a + bi} = \frac{1}{z} = \frac{1}{z} \cdot \frac{\bar{z}}{\bar{z}} = \frac{\bar{z}}{z\bar{z}} = \frac{a - bi}{(a + bi)(a - bi)}$$

$$= \frac{a - bi}{a^2 + b^2}$$

$\uparrow$
Use (5).

$$= \frac{a}{a^2 + b^2} - \frac{b}{a^2 + b^2}i$$

E X A M P L E   4   *Writing the Reciprocal of a Complex Number in Standard Form*

Write $\dfrac{1}{3 + 4i}$ in standard form $a + bi$; that is, find the reciprocal of $3 + 4i$.

Solution   The idea is to multiply the numerator and denominator by the conjugate of $3 + 4i$, that is, the complex number $3 - 4i$. The result is

$$\frac{1}{3 + 4i} = \frac{1}{3 + 4i} \cdot \frac{3 - 4i}{3 - 4i} = \frac{3 - 4i}{9 + 16} = \frac{3}{25} - \frac{4}{25}i \qquad \blacksquare$$

To express the quotient of two complex numbers in standard form, we multiply the numerator and denominator of the quotient by the conjugate of the denominator.

E X A M P L E   5   *Writing the Quotient of Complex Numbers in Standard Form*

Write each of the following in standard form:

(a) $\dfrac{1 + 4i}{5 - 12i}$ (b) $\dfrac{2 - 3i}{4 - 3i}$

Solution   (a) $\dfrac{1 + 4i}{5 - 12i} = \dfrac{1 + 4i}{5 - 12i} \cdot \dfrac{5 + 12i}{5 + 12i} = \dfrac{5 + 20i + 12i + 48i^2}{25 + 144}$

$$= \frac{-43 + 32i}{169} = \frac{-43}{169} + \frac{32}{169}i$$

(b) $\dfrac{2 - 3i}{4 - 3i} = \dfrac{2 - 3i}{4 - 3i} \cdot \dfrac{4 + 3i}{4 + 3i} = \dfrac{8 - 12i + 6i - 9i^2}{16 + 9} = \dfrac{17 - 6i}{25} = \dfrac{17}{25} - \dfrac{6}{25}i$

$\blacksquare$

■ Now work Problem 19.

E X A M P L E   6   *Writing Other Expressions in Standard Form*

If $z = 2 - 3i$ and $w = 5 + 2i$, write each of the following expressions in standard form:

(a) $\dfrac{z}{w}$ (b) $\overline{z + w}$ (c) $z + \bar{z}$

Solution   (a) $\dfrac{z}{w} = \dfrac{z \cdot \overline{w}}{w \cdot \overline{w}} = \dfrac{(2 - 3i)(5 - 2i)}{(5 + 2i)(5 - 2i)} = \dfrac{10 - 15i - 4i + 6i^2}{25 + 4}$

$$= \frac{4 - 19i}{29} = \frac{4}{29} - \frac{19}{29}i$$

(b) $\overline{z + w} = \overline{(2 - 3i) + (5 + 2i)} = \overline{7 - i} = 7 + i$

(c) $z + \bar{z} = (2 - 3i) + (2 + 3i) = 4$

$\blacksquare$

The conjugate of a complex number has certain general properties that we shall find useful later.

For a real number $a = a + 0i$, the conjugate is $\bar{a} = \overline{a + 0i} = a - 0i = a$. That is,

Theorem   The conjugate of a real number is the real number itself.   ∎

Other properties of the conjugate that are direct consequences of the definition are given next. In each statement, z and w represent complex numbers.

Theorem   The conjugate of the conjugate of a complex number is the complex number itself:

$$\overline{(\bar{z})} = z \qquad (6)$$

The conjugate of the sum of two complex numbers equals the sum of their conjugates:

$$\overline{z + w} = \bar{z} + \bar{w} \qquad (7)$$

The conjugate of the product of two complex numbers equals the product of their conjugates:

$$\overline{z \cdot w} = \bar{z} \cdot \bar{w} \qquad (8)$$

    ∎

We leave the proofs of equations (6), (7), and (8) as exercises.

Powers of i

The **powers of i** follow a pattern that is useful to know:

$$
\begin{aligned}
i^1 &= i & i^5 &= i^4 \cdot i = 1 \cdot i = i \\
i^2 &= -1 & i^6 &= i^4 \cdot i^2 = -1 \\
i^3 &= i^2 \cdot i = -i & i^7 &= i^4 \cdot i^3 = -i \\
i^4 &= i^2 \cdot i^2 = (-1)(-1) = 1 & i^8 &= i^4 \cdot i^4 = 1
\end{aligned}
$$

And so on. Thus, the powers of i repeat with every fourth power.

E X A M P L E   7   *Evaluating Powers of i*

(a) $i^{27} = i^{24} \cdot i^3 = (i^4)^6 \cdot i^3 = 1^6 \cdot i^3 = -i$

(b) $i^{101} = i^{100} \cdot i^1 = (i^4)^{25} \cdot i = 1^{25} \cdot i = i$ ∎

E X A M P L E   8   *Writing the Power of a Complex Number in Standard Form*

Write $(2 + i)^3$ in standard form.

Solution   We use the special product formula for $(x + a)^3$:

$$(x + a)^3 = x^3 + 3ax^2 + 3a^2x + a^3$$

Thus,

$$(2 + i)^3 = 2^3 + 3 \cdot i \cdot 2^2 + 3 \cdot i^2 \cdot 2 + i^3$$
$$= 8 + 12i + 6(-1) + (-i)$$
$$= 2 + 11i$$

■   Now work Problem 33.

4

Exercise 4

In Problems 1–38, write each expression in the standard form $a + bi$.

1. $(2 - 3i) + (6 + 8i)$ **2.** $(4 + 5i) + (-8 + 2i)$ **3.** $(-3 + 2i) - (4 - 4i)$ **4.** $(3 - 4i) - (-3 - 4i)$

5. $(2 - 5i) - (8 + 6i)$ **6.** $(-8 + 4i) - (2 - 2i)$ **7.** $3(2 - 6i)$ **8.** $-4(2 + 8i)$

9. $2i(2 - 3i)$ **10.** $3i(-3 + 4i)$ **11.** $(3 - 4i)(2 + i)$ **12.** $(5 + 3i)(2 - i)$

13. $(-6 + i)(-6 - i)$ **14.** $(-3 + i)(3 + i)$ **15.** $\dfrac{10}{3 - 4i}$ **16.** $\dfrac{13}{5 - 12i}$

17. $\dfrac{2 + i}{i}$ **18.** $\dfrac{2 - i}{-2i}$ **19.** $\dfrac{6 - i}{1 + i}$ **20.** $\dfrac{2 + 3i}{1 - i}$

21. $\left(\dfrac{1}{2} + \dfrac{\sqrt{3}}{2}i\right)^2$ **22.** $\left(\dfrac{\sqrt{3}}{2} - \dfrac{1}{2}i\right)^2$ **23.** $(1 + i)^2$ **24.** $(1 - i)^2$

25. i^{23} **26.** i^{14} **27.** i^{-15} **28.** i^{-23}

29. $i^6 - 5$ **30.** $4 + i^3$ **31.** $6i^3 - 4i^5$ **32.** $4i^3 - 2i^2 + 1$

33. $(1 + i)^3$ **34.** $(3i)^4 + 1$ **35.** $i^7(1 + i^2)$ **36.** $2i^4(1 + i^2)$

37. $i^6 + i^4 + i^2 + 1$ **38.** $i^7 + i^5 + i^3 + i$

In Problems 39–44, perform the indicated operations and express your answer in the form $a + bi$.

39. $\sqrt{-4}$ **40.** $\sqrt{-9}$ **41.** $\sqrt{-25}$

42. $\sqrt{-64}$ **43.** $\sqrt{(3 + 4i)(4i - 3)}$ **44.** $\sqrt{(4 + 3i)(3i - 4)}$

In Problems 45–48, $z = 3 - 4i$ and $w = 8 + 3i$. Write each expression in the standard form $a + bi$.

45. $z + \bar{z}$ **46.** $w - \bar{w}$ **47.** $z\bar{z}$ **48.** $\overline{z - w}$

49. Use $z = a + bi$ to show that $z + \bar{z} = 2a$ and that $z - \bar{z} = 2bi$.

50. Use $z = a + bi$ to show that $(\bar{\bar{z}}) = z$.

51. Use $z = a + bi$ and $w = c + di$ to show that $\overline{z + w} = \bar{z} + \bar{w}$.

52. Use $z = a + bi$ and $w = c + di$ to show that $\overline{z \cdot w} = \bar{z} \cdot \bar{w}$.

53. Explain to a friend how you would add two complex numbers and how you would multiply two complex numbers. Explain any differences in the two explanations.

54. Write a brief paragraph that compares the method used to rationalize the denominator of a rational expression and the method used to write a complex number in standard form.

5

Linear Curve Fitting

Curve fitting is an area of statistics in which a relation between two or more variables is explained through an equation. For example, the equation Sales = $100000 + 12Advertising implies that if advertising expenditures were $0, sales would be $100,000 + 12(0) = $100,000 and if advertising expenditures were $10,000, sales would be $100,000 + 12($10,000) = $220,000. In this model, the variable advertising is called the predictor (independent) variable and sales is called the response (dependent) variable because if the level of advertising is known, it can be used to predict sales. Curve fitting is used to find an equation that relates two or more variables, using observed or experimental data.

There are three steps to follow to determine whether a relation exists between two variables.

STEP 1: Ask whether the variables are logically related to each other.
STEP 2: Obtain data and verify a relation exists. Then plot the points. The graph obtained is called a **scatter diagram.**
STEP 3: Find an equation which describes this relation.

Scatter Diagrams

Scatter diagrams are used to help us determine whether a relation exists between two variables. They are useful because of the ease with which they are constructed and also because they allow us to see the type of relation that might exist between the two variables.

To construct a scatter diagram simply plot the data with the independent variable as the *x*-coordinate and the corresponding dependent variable as the *y*-coordinate.

E X A M P L E   1

Drawing a Scatter Diagram

Use the following data to construct a scatter diagram. Comment on the type of relation you think exists between the two variables.

X(INDEPENDENT)	Y(DEPENDENT)
3	6
5	10
5	8
8	14
9	15
9	18
10	20
11	23

Solution   To draw a scatter diagram by hand, we plot the points (3, 6), (5, 10), (5, 8), and so on, to obtain a scatter diagram. See Figure 11.

FIGURE 11

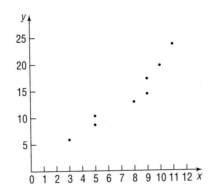

Scatter diagrams can also be constructed using a graphing utility (consult your owners manual). We enter the data into our utility and obtain the scatter diagram shown in Figure 12.

FIGURE 12

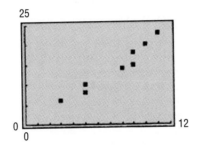

The scatter diagram appears to follow a line with positive slope. ■

You may observe scatter plots which show other patterns. For example, the data in Figure 13 are not linear, but a pattern is evident. Fitting an equation to this type of data is discussed in Chapter 7.

FIGURE 13

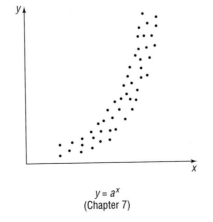

$y = a^x$
(Chapter 7)

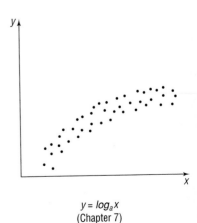

$y = \log_a x$
(Chapter 7)

Lines of Best Fit

E X A M P L E   2   *Finding the Equation of a Line From a Scatter Diagram*

A farmer collected the following data, which shows crop yields for various amounts of fertilizer used.

(a) Use a graphing utility to draw a scatter diagram.

(b) Use a graphing utility to fit a straight line to the data.

PLOT	1	2	3	4	5	5	6	7	8	9	10	11
FERTILIZER, X (POUNDS/100 FT²)	0	0	5	5	10	10	15	15	20	20	25	25
YIELD, Y (POUNDS)	4	6	10	7	12	10	15	17	18	21	23	22

Solution   (a) The data collected indicates a relation exists between the amount of fertilizer used and crop yield. To draw a scatter diagram, we plot points, using fertilizer as the *x*-coordinate and yield as the *y*-coordinate. See Figure 14. From the scatter diagram, it appears a linear relation exists between the amount of fertilizer used and yield.

FIGURE 14

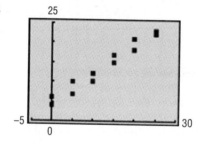

(b) Graphing utilities contain built-in programs that find the linear equation of "best fit" for a collection of points in a scatter diagram.* (Look in your owner's manual under Linear Regression or Line of Best Fit for details on how to execute the program.) Upon executing the LINear REGression program, we obtain the results shown in Figure 15. The output the utility provides shows us the equation, $y = ax + b$, where a is the slope of the line and b is the y-intercept. The line of best fit which relates fertilizer and yield is:

$$\text{Yield} = 0.7171428571(\text{fertilizer}) + 4.785714286$$

FIGURE 15

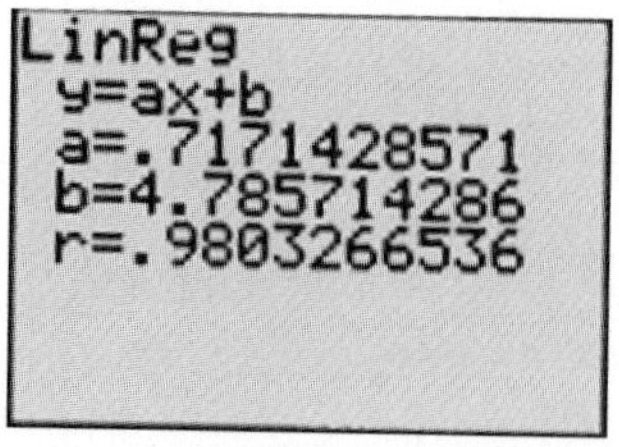

```
LinReg
y=ax+b
a=.7171428571
b=4.785714286
r=.9803266536
```

*We shall not discuss in this book the underlying mathematics of lines of best fit. Most books in statistics and many in linear algebra discuss this topic.

FIGURE 16

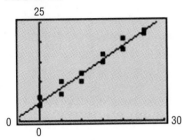

Figure 16 shows the graph of this line, along with the scatter diagram. ■

Does the line of best fit appear to be a good "fit"? In other words, does the line appear to accurately describe the relation between yield and fertilizer?

And just how 'good' is this line of 'best fit'? How accurately does this line describe the relationship between yield and fertilizer? The answers are given by what is called the *correlation coefficient*.

■ Now work Problem 7.

Correlation Coefficients

Look again at Figure 15. The last line of output is 0.98. This number, called the **correlation coefficient, *r*,** $0 \leq |r| \leq 1$, is a measure of the strength of the linear relation that exists between two variables. The closer $|r|$ is to 1, the more perfect the linear relationship is. If r is close to 0, there is little or no linear relationship between the variables. A negative value of r, $r < 0$, indicates that as x increases, then y decreases; a positive value of r, $r > 0$, indicates that as x increases, then y does also. Thus, the data given in Example 2, having a correlation coefficient of 0.98, are strongly indicative of a linear relationship.

Figure 17 illustrates a variety of scatter diagrams and the relations they suggest.

FIGURE 17

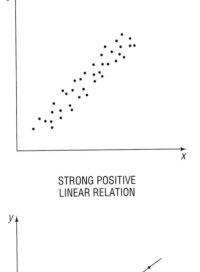

STRONG POSITIVE
LINEAR RELATION

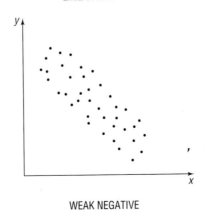

STRONG NEGATIVE
LINEAR RELATION

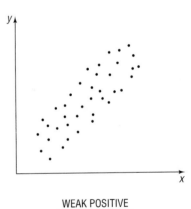

WEAK POSITIVE
LINEAR RELATION

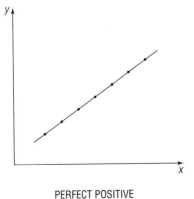

PERFECT POSITIVE
LINEAR RELATION

WEAK NEGATIVE
LINEAR RELATION

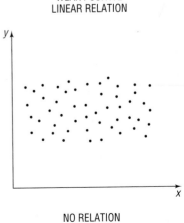

NO RELATION

Prediction and Accuracy

Once an equation with a correlation coefficient close to one has been obtained, then the equation can be used to predict values for the dependent variable for a given independent variable.

EXAMPLE 3   *Prediction*

Use the results of Example 2 to estimate the yield if the farmer uses 17 pounds of fertilizer/100 square feet.

Solution   To determine the yield, we substitute the value $x = 17$ pounds into the equation found in Example 2.

$$(\text{Yield}) = 0.7171428571(17) + 4.785714286 \approx 17 \text{ pounds}$$

So, 17 pounds of fertilizer per 100 square feet will, on average, produce a yield of 17 pounds. Notice that we rounded our prediction to the nearest whole number. This is because our data is measured to the nearest whole number as well. Our predictions cannot be more precise than our data.   ■

Warning: It is important not to round the estimates of the slope or the y-intercept of the line of best fit, since this will decrease the accuracy of predictions.

EXAMPLE 4   *Interpreting the Slope*

Use the equation of the line found in Example 2 and interpret the slope.

Solution   The slope of the line is 0.7171428571. This can be interpreted as follows: For every 1 pound per square foot increase in fertilizer, the yield is expected to increase by 0.7171428571 pounds.

It is important that predictions are made within the scope of the model. That is, we can only make predictions regarding this model for $0 \leq$ fertilizer ≤ 25, since this is the range for which we have observable data. Unless we collect additional data for $x > 25$, we can't make predictions for $x > 25$. The reason should be clear. It is not clear whether adding more fertilizer will continue to increase crop yield. In fact, it is conceivable that adding more fertilizer may actually reduce crop yield.   ■

5

Exercise 5

In Problems 1–6, examine the scatter diagram and determine if there is a strong or weak linear relation, a nonlinear relation, or no relation at all.

1.

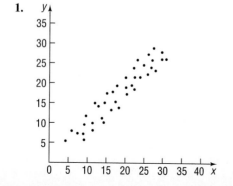

2.

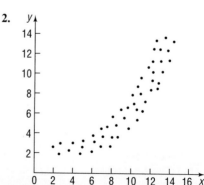

3.

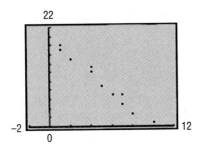

4.

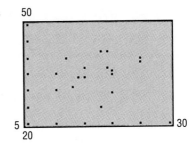

5.

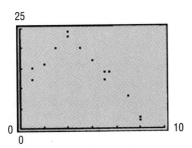

6.

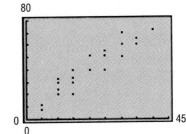

In Problems 7–14, (a) use a graphing utility to draw a scatter diagram; (b) use a graphing utility to fit a straight line to the data.

7.

x	3	4	5	6	7	8	9
y	4	6	7	10	12	14	16

8.

x	3	5	7	9	11	13
y	0	2	3	6	9	11

9.

x	−2	−1	0	1	2
y	−4	0	1	4	5

10.

x	−2	−1	0	1	2
y	7	6	3	2	0

11.

x	20	30	40	50	60
y	100	95	91	83	70

12.

x	5	10	15	20	25
y	2	4	7	11	18

13.

x	−20	−17	−15	−14	−10
y	100	120	118	130	140

14.

x	−30	−27	−25	−20	−14
y	10	12	13	13	18

15. *Consumption and Disposable Income* An economist wishes to estimate a line which relates personal consumption expenditures (C) and disposable income (I). Both C and I are in thousands of dollars. She interviews 8 heads of households for families of size 4 and obtains the following data:

C (000)	I (000)
16	20
18	20
13	18
21	27
27	36
26	37
36	45
39	50

Let I represent the independent variable and C the dependent variable.
(a) Use a graphing utility to draw a scatter diagram.
(b) Use a graphing utility to fit a straight line to the data.
(c) Interpret the slope. The slope of this line is called the **marginal propensity to consume.**
(d) Predict the consumption of a family whose disposable income is $42,000.

16. *Marginal Propensity to Save* The same economist as the one in Problem 15 wants to estimate a line which relates savings (S) and disposable income (I). Let $S = I - C$ be the dependent variable and I the independent variable. The slope of this line is called the **marginal propensity to save.**
(a) Use a graphing utility to draw a scatter diagram.
(b) Use a graphing utility to fit a straight line to the data.
(c) Interpret the slope.
(d) Predict the savings of a family whose income is $42,000.

17. *Average Speed of a Car* An individual wanted to determine the relation that might exist between speed and miles per gallon of an automobile. Let X be the average speed of a car on the highway measured in miles per hour and let Y represent the miles per gallon of the automobile. The following data is collected:

X	50	55	55	60	60	62	65	65
Y	28	26	25	22	20	20	17	15

(a) Use a graphing utility to draw a scatter diagram.
(b) Use a graphing utility to fit a straight line to the data.
(c) Interpret the slope.
(d) Predict the miles per gallon of a car traveling 61 miles per hour.

18. *Height versus Weight* A doctor wished to determine whether a relation exists between the height of a female and her weight. She obtained the heights and weights of 10 females aged 18–24. Let height be the independent variable, X, measured in inches, and weight be the dependent variable, Y; measured in pounds.

X (height)	60	61	62	62	64	65	65	67	68	68
Y (weight)	105	110	115	120	120	125	130	135	135	145

(a) Use a graphing utility to draw a scatter diagram.
(b) Use a graphing utility to fit a straight line to the data.
(c) Interpret the slope.
(d) Predict the weight of a female aged 18–24 whose height is 66 inches.

19. *Sales Data versus Income* The following data represent sales and net income before taxes (both are in billions of dollars) for all manufacturing firms within the United States for 1980–1989. Treat sales as the independent variable and net income before taxes as the dependent variable.

YEAR	SALES	NET INCOME BEFORE TAXES
1980	1912.8	92.6
1981	2144.7	101.3
1982	2039.4	70.9
1983	2114.3	85.8
1984	2335.0	107.6
1985	2331.4	87.6
1986	2220.9	83.1
1987	2378.2	115.6
1988	2596.2	154.6
1989	2745.1	136.3

(a) Use a graphing utility to draw a scatter diagram.
(b) Use a graphing utility to fit a straight line to the data.
(c) Interpret the slope.
(d) Predict the net income before taxes of manufacturing firms in 1990, if sales are $2456.4 billion.

Source: Economic Report of the President, February, 1995.

20. *Employment and the Labor Force* The following data represent the civilian labor force (people aged 16 years and older, excluding those serving in the military) and the number of employed people in the United States for the years 1981–1991. Treat the size of the labor force as the independent variable and the number employed as the dependent variable. Both the size of the labor force and the number employed are measured in thousands of people.

YEAR	CIVILIAN LABOR FORCE	NUMBER EMPLOYED
1981	108,670	100,397
1982	110,204	99,526
1983	111,550	100,834
1984	113,544	105,005
1985	115,461	107,150
1986	117,834	109,597
1987	119,865	112,440
1988	121,669	114,968
1989	123,869	117,342
1990	124,787	117,914
1991	125,303	116,877

(a) Use a graphing utility to draw a scatter diagram.
(b) Use a graphing utility to fit a straight line to the data.
(c) Interpret the slope.
(d) Predict the number of employed if the civilian labor force is 122,340,000 people.

Source: Business Statistics, 1963–1991, U.S. Department of Commerce, Economics and Statistics Administration, Bureau of Economic Analysis, June 1992.

ANSWERS

CHAPTER 1 *Exercise 1.1*

1. (a) Quadrant II
(b) Positive x-axis
(c) Quadrant III
(d) Quadrant I
(e) Negative y-axis
(f) Quadrant IV

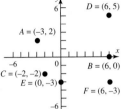

3. The points will be on a vertical line that is 2 units to the right of the y-axis.

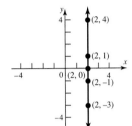

5. $(-1, 4)$ **7.** $(3, 1)$

9. $x\min = -11$
$x\max = 5$
$x\text{scl} = 1$
$y\min = -3$
$y\max = 6$
$y\text{scl} = 1$

11. $x\min = -30$
$x\max = 50$
$x\text{scl} = 10$
$y\min = -90$
$y\max = 50$
$y\text{scl} = 10$

13. $x\min = -10$
$x\max = 110$
$x\text{scl} = 10$
$y\min = -10$
$y\max = 160$
$y\text{scl} = 10$

15. $x\min = -6$
$x\max = 6$
$x\text{scl} = 2$
$y\min = -4$
$y\max = 4$
$y\text{scl} = 2$

17. $x\min = -9$
$x\max = 9$
$x\text{scl} = 3$
$y\min = -4$
$y\max = 4$
$y\text{scl} = 2$

19. $x\min = -6$
$x\max = 6$
$x\text{scl} = 1$
$y\min = -8$
$y\max = 8$
$y\text{scl} = 2$

21. $x\min = -6$
$x\max = 6$
$x\text{scl} = 2$
$y\min = -1$
$y\max = 3$
$y\text{scl} = 1$

23. $x\min = 3$
$x\max = 9$
$x\text{scl} = 1$
$y\min = 2$
$y\max = 10$
$y\text{scl} = 2$

25. $\sqrt{5}$ **27.** $2\sqrt{2}$ **29.** $2\sqrt{17}$ **31.** $\sqrt{85}$ **33.** $\sqrt{53}$ **35.** 2.625   **37.** $\sqrt{a^2 + b^2}$ **39.** $4\sqrt{10}$ **41.** $2\sqrt{17}$

43. $d(A, B) = \sqrt{13}$
$d(B, C) = \sqrt{13}$
$d(A, C) = \sqrt{26}$
$(\sqrt{13})^2 + (\sqrt{13})^2 = (\sqrt{26})^2$
Area $= \frac{13}{2}$ square units

45. $d(A, B) = \sqrt{130}$
$d(B, C) = \sqrt{26}$
$d(A, C) = \sqrt{104}$
$(\sqrt{26})^2 + (\sqrt{104})^2 = (\sqrt{130})^2$
Area = 26 square units

47. $d(A, B) = 4$
$d(A, C) = 5$
$d(B, C) = \sqrt{41}$
$4^2 + 5^2 = 16 + 25 = (\sqrt{41})^2$
Area = 10 square units

49. $(2, 2)$; $(2, -4)$ **51.** $(0, 0)$; $(8, 0)$ **53.** $(4, -1)$ **55.** $(\frac{3}{2}, 1)$ **57.** $(5, -1)$ **59.** $(1.05, 0.7)$ **61.** $(a/2, b/2)$ **63.** $\sqrt{73}$; $2\sqrt{13}$; 5

65. $d(P_1, P_2) = 6$; $d(P_2, P_3) = 4$; $d(P_1, P_3) = 2\sqrt{13}$; right triangle

67. $d(P_1, P_2) = \sqrt{68}$; $d(P_2, P_3) = \sqrt{34}$; $d(P_1, P_3) = \sqrt{34}$; isosceles right triangle

69. $\dfrac{x - x_1}{x_2 - x_1} = r$ and $\dfrac{y - y_1}{y_2 - y_1} = r$; thus, $x = x_1 + r(x_2 - x_1)$ and $y = y_1 + r(y_2 - y_1)$ **71.** P_2 **73.** $(8, 8)$ **75.** $90\sqrt{2} \approx 127.28$ ft

77. (a) $(90, 0), (90, 90), (0, 90)$ **(b)** 232.4 ft   **(c)** 366.2 ft   **79.** $M_1 = (s/2, s/2)$; $M_2 = (s/2, s/2)$ **81.** $d = 50t$

Exercise 1.2

1.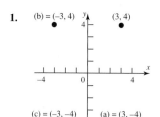

3.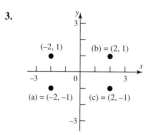

5.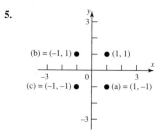

7.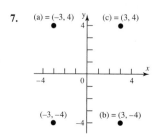

9.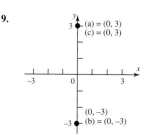

11. (a) 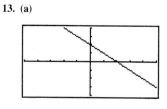 **(b)** 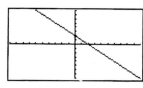 **(c)** 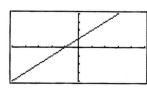 **(d)**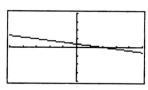

13. (a) 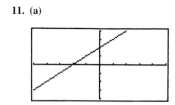 **(b)** 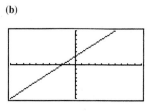 **(c)** 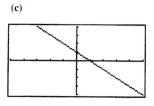 **(d)**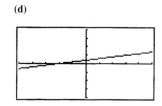

15. (a) 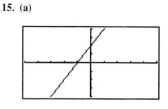 **(b)** 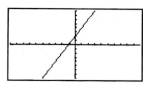 **(c)** 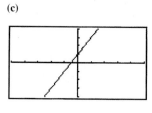 **(d)**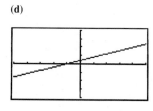

17. (a) 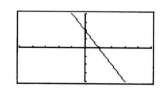 **(b)** 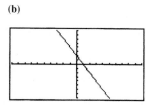 **(c)** 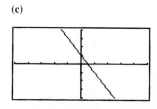 **(d)**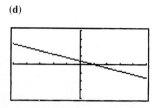

19. (a)　　　(b)　　　(c)　　　(d)

21. (a)　　　(b)　　　(c)　　　(d)

23. (a)　　　(b)　　　(c)　　　(d)

25. (a)　　　(b)　　　(c)　　　(d)

27. (a)　　　(b)　　　(c)　　　(d)

29. (a)　　　(b)　　　(c)　　　(d)

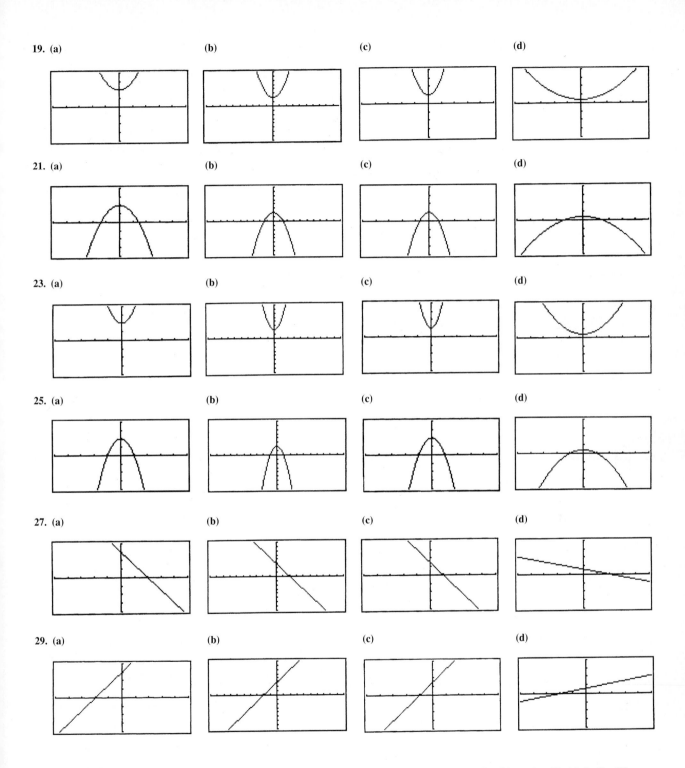

31. (a) $(-1, 0)$, $(1, 0)$ **(b)** x-axis, y-axis, origin **33. (a)** $(-\frac{\pi}{2}, 0)$, $(\frac{\pi}{2}, 0)$, $(0, 1)$ **(b)** y-axis **35. (a)** $(0, 0)$ **(b)** x-axis **37. (a)** $(1, 0)$ **(b)** none
39. (a) $(-3, 0)$, $(0, 2)$, $(3, 0)$ **(b)** y-axis **41. (a)** $(x, 0)$, $0 \le x < 2$ **(b)** none **43. (a)** $(-1.5, 0)$, $(0, -2)$, $(1.5, 0)$ **(b)** y-axis
45. (a) none **(b)** origin **47.** $(0, 0)$ is on the graph **49.** $(0, 2)$ is on the graph **51.** $(0, 2)$ and $(\sqrt{2}, \sqrt{2})$ are on the graph **53.** $-\frac{2}{5}$

55. $2a + 3b = 6$

57.

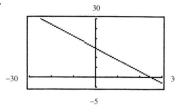

59.

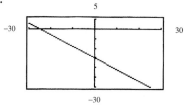

61.

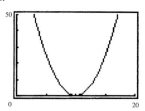

63.

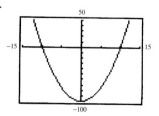

65.

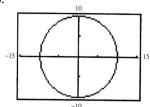

67.

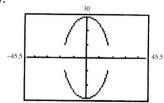

69.

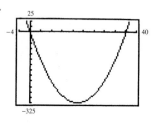

71.

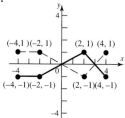

73.

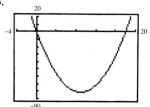

75.

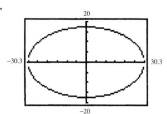

77.

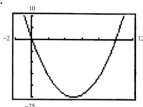

79.

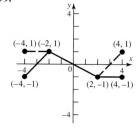

81.

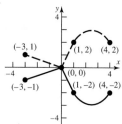

83.

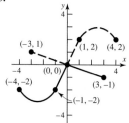

85. (0, 0); symmetric with respect to the y-axis

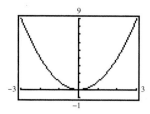

87. (0, 0); symmetric with respect to the origin

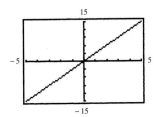

89. (0, 9), (3, 0), (−3, 0); symmetric with respect to the y-axis

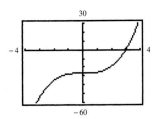

91. (−2, 0), (2, 0), (0, −3), (0, 3); symmetric with respect to the x-axis, y-axis, and origin

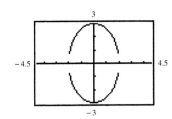

93. (0, −27), (3, 0); no symmetry

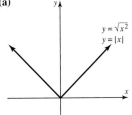

95. (0, −4), (4, 0), (−1, 0); no symmetry

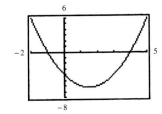

97. (0, 0); symmetric with respect to the origin

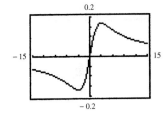

99. (a)

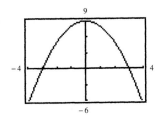 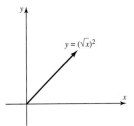 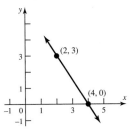

(b) Since $\sqrt{x^2} = |x|$, for all x, the graphs of $y = \sqrt{x^2}$ and $y = |x|$ are the same. **(c)** For $y = (\sqrt{x})^2$, the domain of the variable x is $x \geq 0$; for $y = x$, the domain of the variable x is all real numbers. Thus, $(\sqrt{x})^2 = x$ only for $x \geq 0$. **(d)** For $y = \sqrt{x^2}$, the range of the variable y is $y \geq 0$; for $y = x$, the range of the variable y is all real numbers. Also, $\sqrt{x^2} = |x|$, which equals x only if $x \geq 0$.

Exercise 1.3

1. $\frac{1}{2}$ **3.** -1

5. Slope $= -\frac{3}{2}$

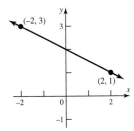

7. Slope $= -\frac{1}{2}$

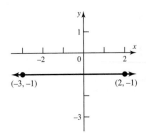

9. Slope $= 0$

11. Slope undefined

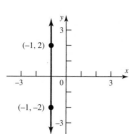

13. Slope $= \dfrac{\sqrt{3} - 3}{1 - \sqrt{2}} \approx 3.06$

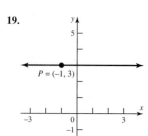

15.

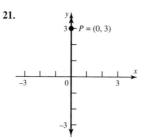

17.

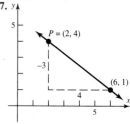

19.

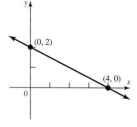

21.

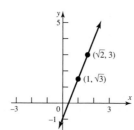

23. $x - 2y = 0$ or $y = \frac{1}{2}x$ **25.** $x + y - 2 = 0$ or $y = -x + 2$ **27.** $3x - y + 9 = 0$ or $y = 3x + 9$

29. $2x + 3y + 1 = 0$ or $y = -\frac{2}{3}x - \frac{1}{3}$ **31.** $x - 2y + 5 = 0$ or $y = \frac{1}{2}x + \frac{5}{2}$ **33.** $3x + y - 3 = 0$ or $y = -3x + 3$

35. $x - 2y - 2 = 0$ or $y = \frac{1}{2}x - 1$ **37.** $x - 2 = 0$; no slope-intercept form

39. Slope $= 2$; y-intercept $= 3$

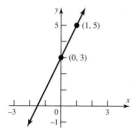

41. $y = 2x - 2$; Slope $= 2$; y-intercept $= -2$

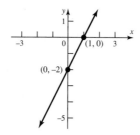

43. Slope $= \frac{1}{2}$; y-intercept $= 2$

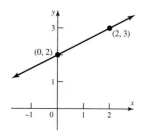

45. $y = -\frac{1}{2}x + 2$; Slope $= -\frac{1}{2}$; y-intercept $= 2$

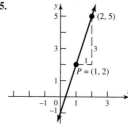

47. $y = \frac{2}{3}x - 2$; Slope $= \frac{2}{3}$; y-intercept $= -2$

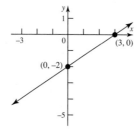

49. $y = -x + 1$; Slope $= -1$; y-intercept $= 1$

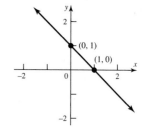

51. Slope undefined; no y-intercept

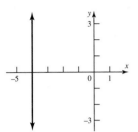

53. Slope = 0; y-intercept = 5

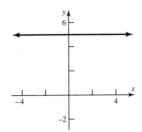

55. $y = x$; Slope = 1; y-intercept = 0

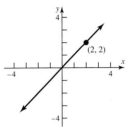

57. $y = \frac{3}{2}x$; Slope = $\frac{3}{2}$; y-intercept = 0

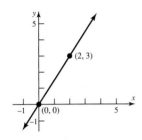

59. Center $(2, 1)$; Radius 2; $(x - 2)^2 + (y - 1)^2 = 4$ **61.** Center $(\frac{5}{2}, 2)$; Radius $\frac{3}{2}$; $(x - \frac{5}{2})^2 + (y - 2)^2 = \frac{9}{4}$

63. $(x - 1)^2 + (y + 1)^2 = 1$;
$x^2 + y^2 - 2x + 2y + 1 = 0$

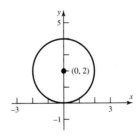

65. $x^2 + (y - 2)^2 = 4$;
$x^2 + y^2 - 4y = 0$

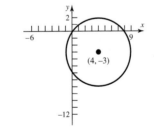

67. $(x - 4)^2 + (y + 3)^2 = 25$;
$x^2 + y^2 - 8x + 6y = 0$

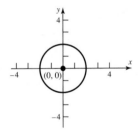

69. $x^2 + y^2 = 4$;
$x^2 + y^2 - 4 = 0$

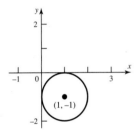

71. $(x - \frac{1}{2})^2 + y^2 = \frac{1}{4}$;
$x^2 + y^2 - x = 0$

73. $r = 2$;
$(h, k) = (0, 0)$

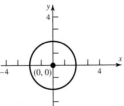

75. $r = 2$;
$(h, k) = (3, 0)$

77. $r = 3$;
$(h, k) = (-2, 2)$

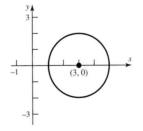

79. $r = \frac{1}{2}$;
$(h, k) = (\frac{1}{2}, -1)$

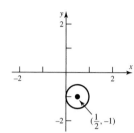

81. $r = 5$;
$(h, k) = (3, -2)$

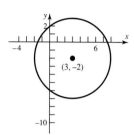

83. $x^2 + y^2 - 13 = 0$ **85.** $x^2 + y^2 - 4x - 6y + 4 = 0$ **87.** $x^2 + y^2 + 2x - 6y + 5 = 0$ **89.** (b) **91.** (d) **93.** $x - y + 2 = 0$ or $y = x + 2$
95. $x + 3y - 3 = 0$ or $y = -\frac{1}{3}x + 1$ **97.** $2x + 3y = 0$ or $y = -\frac{2}{3}x$ **99.** (c) **101.** (b) **103.** $(x + 3)^2 + (y - 1)^2 = 16$
105. $(x - 2)^2 + (y - 2)^2 = 9$ **107.** $y = 0$ **109.** $°C = \frac{5}{9}(°F - 32)$; approx. $21°C$
111. (a) $P = 0.5x - 100$ (b) \$400 (c) \$2400 **113.** $C = 0.10819x + 9.06$; for 300 kWhr, $C = \$41.52$; for 900 kWhr, $C = \$106.43$

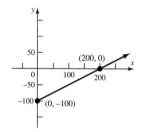

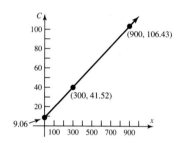

115. (a)

$$x^2 + (mx + b)^2 = r^2$$
$$(1 + m^2)x^2 + 2mbx + b^2 - r^2 = 0$$
One solution if and only if discriminant $= 0$
$$(2mb)^2 - 4(1 + m^2)(b^2 - r^2) = 0$$
$$-4b^2 + 4r^2 + 4m^2r^2 = 0$$
$$r^2(1 + m^2) = b^2$$

(b) $x = \dfrac{-2mb}{2(1 + m^2)} = \dfrac{-2mb}{2b^2/r^2} = \dfrac{-r^2m}{b}$

$$y = m\left(\dfrac{-r^2m}{b}\right) + b = \dfrac{-r^2m^2}{b} + b$$
$$= \dfrac{-r^2m^2 + b^2}{b} = \dfrac{r^2}{b}$$

(c) Slope of tangent line $= m$
Slope of line joining center to point of
tangency $= \dfrac{r^2/b}{-r^2m/b} = -\dfrac{1}{m}$

117. $\sqrt{2}x + 4y - 11\sqrt{2} + 12 = 0$ **119.** $x + 5y + 13 = 0$ **121.** All have the same slope, 2; the lines are parallel.

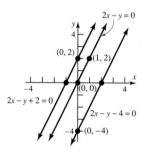

127. Its slope is -1 **129.** Yes, if the y-intercept $= 0$

Exercise 1.4

1. (a) -4 (b) -5 (c) -9 (d) -25 **3.** (a) 0 (b) $\frac{1}{2}$ (c) $-\frac{1}{2}$ (d) $\frac{3}{10}$ **5.** (a) 4 (b) 5 (c) 5 (d) 7
7. (a) $-\frac{1}{5}$ (b) $-\frac{3}{2}$ (c) $\frac{1}{8}$ (d) $\frac{7}{4}$ **9.** $f(0) = 3; f(-6) = -3$ **11.** Positive **13.** $-3, 6$, and 10 **15.** $\{x| -6 \le x \le 11\}$
17. $(-3, 0), (6, 0) (10, 0)$ **19.** 3 times **21.** (a) No (b) -3 (c) 14 (d) $\{x| x \ne 6\}$ **23.** (a) Yes (b) $\frac{8}{17}$ (c) $-1, 1$ (d) All real numbers
25. Not a function
27. Function (a) Domain: $\{x| -\pi \le x \le \pi\}$; Range: $\{y| -1 \le y \le 1\}$ (b) Intercepts: $(-\pi/2, 0), (\pi/2, 0), (0, 1)$ (c) y-axis
29. Not a function **31.** Function (a) Domain: $\{x| x > 0\}$; Range: all real numbers (b) Intercept: $(1, 0)$ (c) None
33. Function (a) Domain: all real numbers; Range: $\{y|y \le 2\}$ (b) Intercepts: $(-3, 0), (3, 0), (0, 2)$ (c) y-axis

35. Function   **(a)** Domain: $\{x|x \neq 2\}$; Range: $\{y|y \neq 1\}$ **(b)** Intercept: (0, 0)   **(c)** None   **37.** All real numbers   **39.** All real numbers
41. $\{x|x \neq -1, x \neq 1\}$ **43.** $\{x|x \neq 0\}$ **45.** $\{x|x \geq 4\}$ **47.** $(-\infty, -3]$ or $[3, \infty)$ **49.** $(-\infty, 1)$ or $[2, \infty)$ **51.** $A = -\frac{7}{2}$ **53.** $A = -4$
55. $A = 8$; undefined at $x = 3$
57. (a)

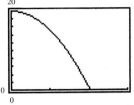

59. $A(x) = \frac{1}{2}x^2$ **61.** $G(x) = 5x$
63. (a) $C(x) = 10x + 14\sqrt{x^2 - 10x + 29}$, $0 < x < 5$
(b) $C(1) = \$72.61$
(c) $C(3) = \$69.60$
(d)

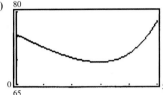

(e) least cost: $x = 2.95$ miles

(b) 15.1 m, 14.07 m, 12.94 m, 11.72 m
(c) 1.01 seconds, 1.42 seconds, 1.74 seconds
(d) 2.02 sec

65. (a) $A(x) = (8.5 - 2x)(11 - 2x)$,
(b) $0 \leq x \leq 4.25$, $0 \leq A \leq 93.5$
(c) $A(1) = 58.5$ in.^{2}, $A(1.2) = 52.46$ in.^{2}, $A(1.5) = 44$ in.^{2}
(d)

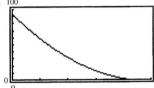

(e) $A(x) = 70$ when $x = 0.64$ in; $A(x) = 50$ when $x = 1.28$ in.

67. (a)

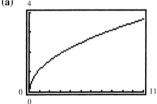

(b) The period T varies from 1.107 seconds to 3.502 seconds.
(c) 81.56 feet

69. Function   **71.** Function   **73.** Not a function   **75.** Function   **77.** Only $h(x) = 2x$

Exercise 1.5

1. C   **3.** E   **5.** B   **7.** F
9. (a) Domain: $\{x| -3 \leq x \leq 4\}$; Range: $\{y|0 \leq y \leq 3\}$ **(b)** Increasing on $[-3, 0]$ and on $[2, 4]$; decreasing on $[0, 2]$ **(c)** Neither
(d) $(-3, 0)$, $(0, 3)$, $(2, 0)$
11. (a) Domain: all real numbers; Range: $\{y|0 < y < \infty\}$ **(b)** Increasing on $(-\infty, \infty)$ **(c)** Neither   **(d)** $(0, 1)$
13. (a) Domain: $\{x| -\pi \leq x \leq \pi\}$; Range: $\{y| -1 \leq y \leq 1\}$
(b) Increasing on $[-\pi/2, \pi/2]$; decreasing on $[-\pi, -\pi/2]$ and on $[\pi/2, \pi]$ **(c)** Odd (symmetric with respect to the origin)
(d) $(-\pi, 0)$, $(0, 0)$, $(\pi, 0)$
15. (a) Domain: $\{x|x \neq 2\}$; Range: $\{y|y \neq 1\}$ **(b)** Decreasing on $(-\infty, 2)$ and on $(2, \infty)$ **(c)** Neither   **(d)** $(0, 0)$
17. (a) Domain: $\{x|x \neq 0\}$; Range: all real numbers   **(b)** Increasing on $(-\infty, 0)$ and on $(0, \infty)$ **(c)** Odd   **(d)** $(-1, 0)$, $(1, 0)$
19. (a) Domain: $\{x|x \neq -2, x \neq 2\}$; Range: $\{y|-\infty < y \leq 0 \text{ and } 1 < y < \infty\}$
(b) Increasing on $(-\infty, -2)$ and on $(-2, 0]$; decreasing on $[0, 2)$ and on $(2, \infty)$ **(c)** Even   **(d)** $(0, 0)$
21. (a) Domain: $\{x|-4 \leq x \leq 4\}$; Range: $\{y|0 \leq y \leq 2\}$ **(b)** Increasing on $[-2, 0]$ and $[2, 4]$; Decreasing on $[-4, -2]$ and $[0, 2]$ **(c)** Even
(d) $(-2, 0)$, $(0, 2)$, $(2, 0)$
23. (a) Domain: $\{x|-4 \leq x \leq 4\}$; Range: $\{y|0 < y \leq 4\}$ **(b)** Increasing on $[-4, 0)$; Decreasing on $(0, 4]$ **(c)** Even   **(d)** None
25. (a) 2   **(b)** 3   **(c)** -4 **27. (a)** 4   **(b)** 2   **(c)** 5
29. (a) $-2x + 5$ **(b)** $-2x - 5$ **(c)** $4x + 5$ **(d)** $2x - 1$ **(e)** $\dfrac{2}{x} + 5 = \dfrac{5x + 2}{x}$ **(f)** $\dfrac{1}{2x + 5}$
31. (a) $2x^2 - 4$ **(b)** $-2x^2 + 4$ **(c)** $8x^2 - 4$ **(d)** $2x^2 - 12x + 14$ **(e)** $\dfrac{2 - 4x^2}{x^2}$ **(f)** $\dfrac{1}{2x^2 - 4}$
33. (a) $-x^3 + 3x$ **(b)** $-x^3 + 3x$ **(c)** $8x^3 - 6x$ **(d)** $x^3 - 9x^2 + 24x - 18$ **(e)** $\dfrac{1}{x^3} - \dfrac{3}{x}$ **(f)** $\dfrac{1}{x^3 - 3x}$
35. (a) $-\dfrac{x}{x^2 + 1}$ **(b)** $-\dfrac{x}{x^2 + 1}$ **(c)** $\dfrac{2x}{4x^2 + 1}$ **(d)** $\dfrac{x - 3}{x^2 - 6x + 10}$ **(e)** $\dfrac{x}{x^2 + 1}$ **(f)** $\dfrac{x^2 + 1}{x}$

37. (a) $|x|$ (b) $-|x|$ (c) $2|x|$ (d) $|x-3|$ (e) $\dfrac{1}{|x|}$ (f) $\dfrac{1}{|x|}$

39. (a) $1-\dfrac{1}{x}$ (b) $-1-\dfrac{1}{x}$ (c) $1+\dfrac{1}{2x}$ (d) $1+\dfrac{1}{x-3}$ (e) $1+x$ (f) $\dfrac{x}{x+1}$

41. 3 **43.** -3 **45.** $3x+1$ **47.** $x(x+1)$ **49.** $\dfrac{-1}{x+1}$ **51.** $\dfrac{1}{\sqrt{x}+1}$

53. Odd **55.** Even **57.** Odd **59.** Neither **61.** Even **63.** Odd **65.** At most one

67. (a) All real numbers (c)
(b) $(0,-3)$, $(1,0)$
(d) All real numbers

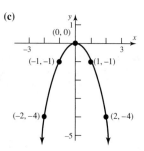

69. (a) All real numbers (c)
(b) $(-2,0)$, $(2,0)$, $(0,-4)$
(d) $\{y\,|\,-4\le y<\infty\}$

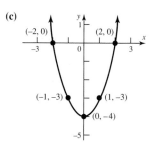

71. (a) All real numbers (c)
(b) $(0,0)$
(d) $\{y\,|\,-\infty<y\le 0\}$

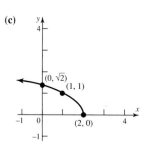

73. (a) $\{x\,|\,2\le x<\infty\}$ (c)
(b) $(2,0)$
(d) $\{y\,|\,0\le y<\infty\}$

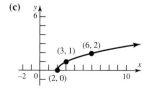

75. (a) $\{x\,|\,-\infty<x\le 2\}$ (c)
(b) $(2,0)$, $(0,\sqrt{2})$
(d) $\{y\,|\,0\le y<\infty\}$

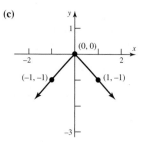

77. (a) All real numbers (c)
(b) $(0,3)$
(d) $\{y\,|\,3\le y<\infty\}$

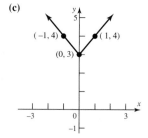

79. (a) All real numbers (c)
(b) $(0,0)$
(d) $\{y\,|\,-\infty<y\le 0\}$

81. (a) All real numbers (c)
(b) $(0,0)$
(d) All real numbers

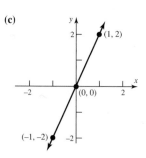

83. (a) All real numbers
(b) (0, 0), (−1, 0)
(d) All real numbers
(c)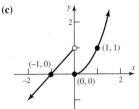

85. (a) $\{x| -2 \le x < \infty\}$
(b) (0, 1)
(d) $\{y|0 < y < \infty\}$
(c)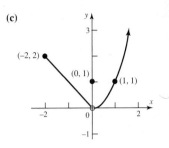

87. (a) All real numbers
(b) (0, 1)
(d) $\{-1, 1\}$
(c)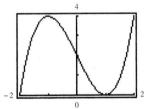

89. (a) All real numbers
(b) (x, 0) for $0 \le x < 1$
(d) Set of even integers
(c)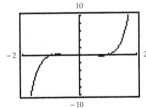

91. (a) All real numbers
(b) (−2, 0), (0, 4), (2, 0)
(d) $\{y|y \ge 0\}$
(c)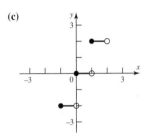

93. 2 **95.** $2x + h + 2$ **97.** $f(x) = \begin{cases} -x & \text{if } -1 \le x \le 0 \\ \frac{1}{2}x & \text{if } 0 < x \le 2 \end{cases}$ (Other answers are possible.)

99. $f(x) = \begin{cases} -x & \text{if } x \le 0 \\ -x + 2 & \text{if } 0 < x \le 2 \end{cases}$ (Other answers are possible.) **101.** No; $f(-2) = 8$ and $f(2) = 6$

103.

Increasing: $[-2, -1]$, $[1, 2]$
Decreasing: $[-1, 1]$
Local maxima: $(-1, 4)$, local minima: $(1, 0)$

105.

Increasing: $[-2, -0.77]$, $[0.77, 2]$
Decreasing: $[-0.77, 0.77]$
Local maxima: $(-0.77, 0.18)$, local minima: $(0.77, -0.18)$

107. Each graph is that of $y = x^2$, but shifted vertically. If $y = x^2 + k$, $k > 0$, the shift is up k units; if $y = x^2 + k$, $k < 0$, the shift is down $|k|$ units.
109. Each graph is that of $y = |x|$, but either compressed or stretched. If $y = k|x|$ and $k > 1$, the graph is stretched; if $y = k|x|$, $0 < k < 1$, the graph is compressed.
111. The graph of $y = f(-x)$ is the reflection about the y-axis of the graph of $y = f(x)$.

113.

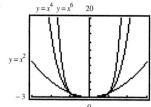

They are all U-shaped and open up. All three go through the point (0, 0). As the exponent increases, the steepness of the curve increases.

115. (a) $30.70
   (b) $203.75

   (c) $C = \begin{cases} 7 + 0.47395x & \text{if } 0 \le x \le 90 \\ 15.83 + 0.37583x & \text{if } x > 90 \end{cases}$

   (d)
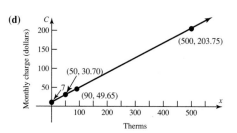

117. (a) $E(-x) = \frac{1}{2}[f(-x) + f(x)] = E(x)$ **(b)** $O(-x) = \frac{1}{2}[f(-x) - f(x)] = -\frac{1}{2}[f(x) - f(-x)] = -O(x)$
   (c) $E(x) + O(x) = \frac{1}{2}[f(x) + f(-x)] + \frac{1}{2}[f(x) - f(-x)] = f(x)$ **(d)** Combine the results of parts **(a)**, **(b)**, and **(c)**.

Exercise 1.6

1. B **3.** H **5.** I **7.** L **9.** F **11.** G **13.** C **15.** B **17.** $y = (x - 4)^3$ **19.** $y = x^3 + 4$ **21.** $y = -x^3$ **23.** $y = 4x^3$

25.
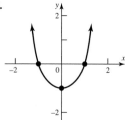

27.
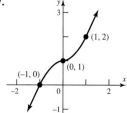

29.
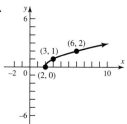

31.
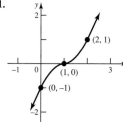

33.
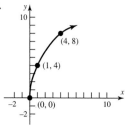

35.
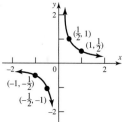

37.
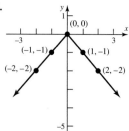

39.
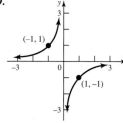

41.
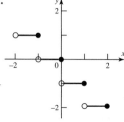

43.
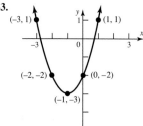

45.
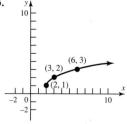

47.
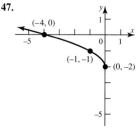

49.

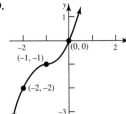

51.

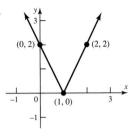

53.

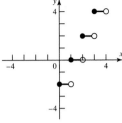

55. (a) $F(x) = f(x) + 3$

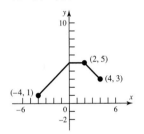

(b) $G(x) = f(x + 2)$

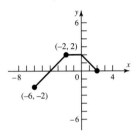

(c) $P(x) = -f(x)$

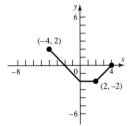

(d) $Q(x) = \frac{1}{2}f(x)$

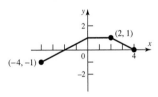

(e) $g(x) = f(-x)$

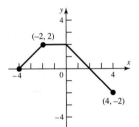

(f) $h(x) = 3f(x)$

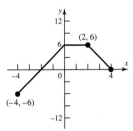

57. (a) $F(x) = f(x) + 3$

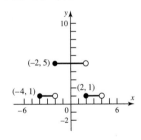

(b) $G(x) = f(x + 2)$

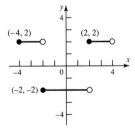

(c) $P(x) = -f(x)$

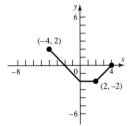

(d) $Q(x) = \frac{1}{2}f(x)$

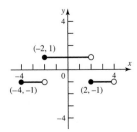

(e) $g(x) = f(-x)$

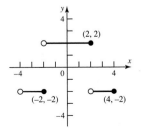

(f) $h(x) = 3f(x)$

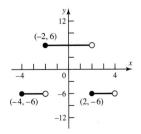

59. (a) $F(x) = f(x) + 3$

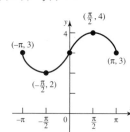

(b) $G(x) = f(x + 2)$

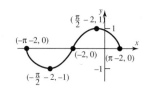

(c) $P(x) = -f(x)$

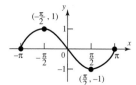

(d) $Q(x) = \frac{1}{2}f(x)$

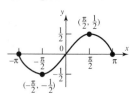

(e) $g(x) = f(-x)$

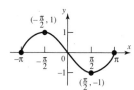

(f) $h(x) = 3f(x)$

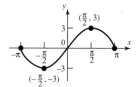

61. (a)

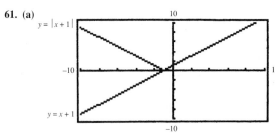

(b)

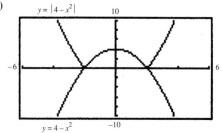

(c)

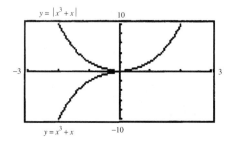

(d) Any part of the graph of $y = f(x)$ that lies below the x-axis is reflected about the x-axis to obtain the graph of $y = |f(x)|$

63. (a)

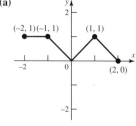

(b)

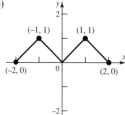

65. $f(x) = (x + 1)^2 - 1$

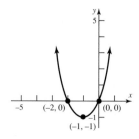

67. $f(x) = (x - 4)^2 - 15$

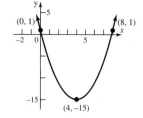

69. $f(x) = (x + \frac{1}{2})^2 + \frac{3}{4}$

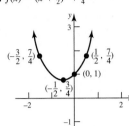

71.

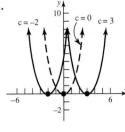

73.

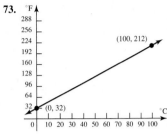

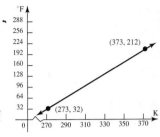

Exercise 1.7

1. One-to-one
3. Not one-to-one
5. One-to-one

7.

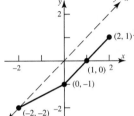

9.

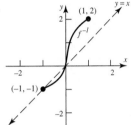

11.

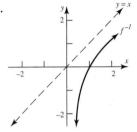

13. $f(g(x)) = f(\frac{1}{3}(x - 4)) = 3[\frac{1}{3}(x - 4)] + 4 = x;\ g(f(x)) = g(3x + 4) = \frac{1}{3}[(3x + 4) - 4] = x$

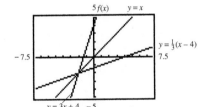

15. $f(g(x)) = 4\left[\dfrac{x}{4} + 2\right] - 8 = x;\ g(f(x)) = \dfrac{4x - 8}{4} + 2 = x$ **17.** $f(g(x)) = (\sqrt[3]{x + 8})^3 - 8 = x;\ g(f(x)) = \sqrt[3]{(x^3 - 8) + 8} = x$

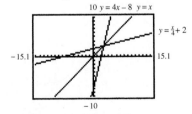

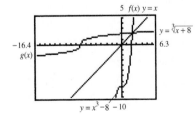

19. $f(g(x)) = \dfrac{1}{1/x} = x; \ g(f(x)) = \dfrac{1}{1/x} = x$

21. $f(g(x)) = \dfrac{2\left(\dfrac{4x-3}{2-x}\right) + 3}{\dfrac{4x-3}{2-x} + 4} = \dfrac{8x - 3x}{-3 + 8} = x; \ g(f(x)) = \dfrac{4\left(\dfrac{2x+3}{x+4}\right) - 3}{2 - \dfrac{2x+3}{x+4}} = x$

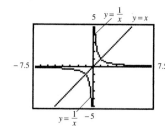

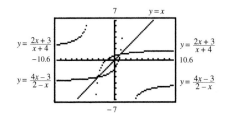

23. $f^{-1}(x) = \tfrac{1}{3}x$
$f(f^{-1}(x)) = 3(\tfrac{1}{3}x) = x$
$f^{-1}(f(x)) = \tfrac{1}{3}(3x) = x$
Domain f = Range $f^{-1} = (-\infty, \infty)$
Range f = Domain $f^{-1} = (-\infty, \infty)$

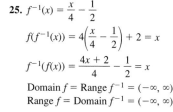

25. $f^{-1}(x) = \dfrac{x}{4} - \dfrac{1}{2}$

$f(f^{-1}(x)) = 4\left(\dfrac{x}{4} - \dfrac{1}{2}\right) + 2 = x$

$f^{-1}(f(x)) = \dfrac{4x + 2}{4} - \dfrac{1}{2} = x$

Domain f = Range $f^{-1} = (-\infty, \infty)$
Range f = Domain $f^{-1} = (-\infty, \infty)$

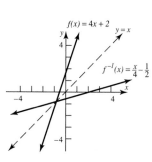

27. $f^{-1}(x) = \sqrt[3]{x + 1}$
$f(f^{-1}(x)) = (\sqrt[3]{x+1})^3 - 1 = x$
$f^{-1}(f(x)) = \sqrt[3]{x^3 - 1 + 1} = x$
Domain f = Range $f^{-1} = (-\infty, \infty)$
Range f = Domain $f^{-1} = (-\infty, \infty)$

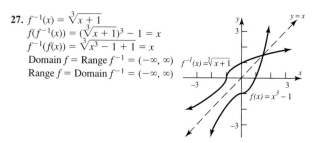

29. $f^{-1}(x) = \sqrt{x - 4}$
$f(f^{-1}(x)) = (\sqrt{x-4})^2 + 4 = x$
$f^{-1}(f(x)) = \sqrt{(x^2 + 4) - 4} = \sqrt{x^2} = |x| = x$
Domain f = Range $f^{-1} = [0, \infty)$
Range f = Domain $f^{-1} = [4, \infty)$

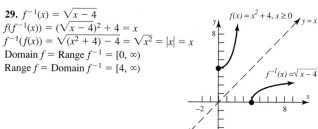

31. $f^{-1}(x) = \dfrac{4}{x}$

$f(f^{-1}(x)) = \dfrac{4}{4/x} = x$

$f^{-1}(f(x)) = \dfrac{4}{4/x} = x$

Domain f = Range f^{-1} = All real numbers except 0
Range f = Domain f^{-1} = All real numbers except 0

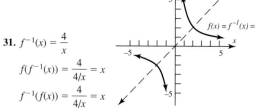

33. $f^{-1}(x) = \dfrac{2x + 1}{x}$

$f(f^{-1}(x)) = \dfrac{1}{\dfrac{2x+1}{x} - 2} = x$

$f^{-1}(f(x)) = \dfrac{2\left(\dfrac{1}{x-2}\right) + 1}{\dfrac{1}{x-2}} = x$

Domain f = Range f^{-1} = All real numbers except 2
Range f = Domain f^{-1} = All real numbers except 0

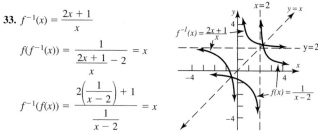

35. $f^{-1}(x) = \dfrac{2 - 3x}{x}$

$$f(f^{-1}(x)) = \dfrac{2}{3 + \dfrac{2 - 3x}{x}} = x$$

$$f^{-1}(f(x)) = \dfrac{2 - 3\left(\dfrac{2}{3 + x}\right)}{\dfrac{2}{3 + x}} = x$$

Domain f = Range f^{-1} = All real numbers except -3
Range f = Domain f^{-1} = All real numbers except 0

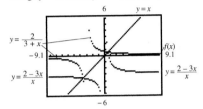

37. $f^{-1}(x) = \sqrt{x} - 2$

$$f(f^{-1}(x)) = (\sqrt{x} - 2 + 2)^2 = x$$
$$f^{-1}(f(x)) = \sqrt{(x + 2)^2} - 2 = |x + 2| - 2 = x$$

Domain f = Range f^{-1} = $[-2, \infty)$
Range f = Domain f^{-1} = $[0, \infty)$

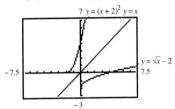

39. $f^{-1}(x) = \dfrac{x}{x - 2}$

$$f(f^{-1}(x)) = \dfrac{2\left(\dfrac{x}{x - 2}\right)}{\dfrac{x}{x - 2} - 1} = x$$

$$f^{-1}(f(x)) = \dfrac{\dfrac{2x}{x - 1}}{\dfrac{2x}{x - 1} - 2} = x$$

Domain f = Range f^{-1} = All real numbers except 1
Range f = Domain f^{-1} = All real numbers except 2

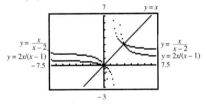

41. $f^{-1}(x) = \dfrac{3x + 4}{2x - 3}$

$$f(f^{-1}(x)) = \dfrac{3\left(\dfrac{3x + 4}{2x - 3}\right) + 4}{2\left(\dfrac{3x + 4}{2x - 3}\right) - 3} = x$$

$$f^{-1}(f(x)) = \dfrac{3\left(\dfrac{3x + 4}{2x - 3}\right) + 4}{2\left(\dfrac{3x + 4}{2x - 3}\right) - 3} = x$$

Domain f = Range f^{-1} = All real numbers except $\frac{3}{2}$
Range f = Domain f^{-1} = All real numbers except $\frac{3}{2}$

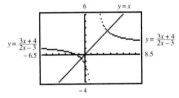

43. $f^{-1}(x) = \dfrac{-2x + 3}{x - 2}$

$$f(f^{-1}(x)) = \dfrac{2\left(\dfrac{-2x + 3}{x - 2}\right) + 3}{\dfrac{-2x + 3}{x - 2} + 2} = x$$

$$f^{-1}(f(x)) = \dfrac{-2\left(\dfrac{2x + 3}{x + 2}\right) + 3}{\dfrac{2x + 3}{x + 2} - 2} = x$$

Domain f = Range f^{-1} = All real numbers except -2
Range f = Domain f^{-1} = All real numbers except 2

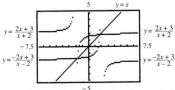

45. $f^{-1}(x) = \dfrac{x^3}{8}$

$$f(f^{-1}(x)) = 2\sqrt[3]{\dfrac{x^3}{8}} = x$$

$$f^{-1}(f(x)) = \dfrac{(2\sqrt[3]{x})^3}{8} = x$$

Domain f = Range f^{-1} = $(-\infty, \infty)$
Range f = Domain f^{-1} = $(-\infty, \infty)$

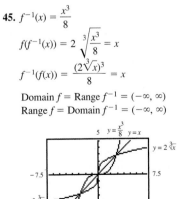

47. $f^{-1}(x) = \dfrac{1}{m}(x - b)$, $m \neq 0$ **49.** No; whenever x and $-x$ are in the domain of f, two equal y values, $f(x)$ and $f(-x)$, are present.

51. Quadrant I   **53.** $f(x) = |x|$, $x \geq 0$ is one-to-one; this may be written as $f(x) = x$; $f^{-1}(x) = x$

55. $f(g(x)) = \frac{9}{5}\left[\frac{5}{9}(x - 32)\right] + 32 = x$; $g(f(x)) = \frac{5}{9}\left[\left(\frac{9}{5}x + 32\right) - 32\right] = x$ **57.** $l(T) = gT^2/4\pi^2$ **59.** $f^{-1}(x) = \dfrac{-dx + b}{cx - a}$; $f = f^{-1}$ if $a = -d$

Fill-in-the-Blank Items

1. x-coordinate; y-coordinate   **2.** midpoint   **3.** y-axis   **4.** circle; radius; center   **5.** undefined; 0   **6.** independent; dependent   **7.** vertical
8. even; odd   **9.** horizontal; right   **10.** one-to-one   **11.** $y = x$

True/False Items

1. F   **2.** T   **3.** T   **4.** F   **5.** T   **6.** T   **7.** T   **8.** F   **9.** T   **10.** F   **11.** T

Review Exercises

1. $2x + y - 5 = 0$ or $y = -2x + 5$ **3.** $x + 3 = 0$; no y-intercept form   **5.** $x + 5y + 10 = 0$ or $y = -\frac{1}{5}x - 2$
7. $2x - 3y + 19 = 0$ or $y = \frac{2}{3}x + \frac{19}{3}$ **9.** $-x + y + 7 = 0$ or $y = x - 7$

11. 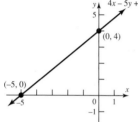   **13.** 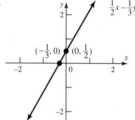   **15.** $\sqrt{2}x + \sqrt{3}y = \sqrt{6}$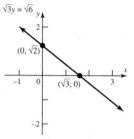

17. Center $(1, -2)$; Radius $= 3$ **19.** Center $(1, -2)$; Radius $= \sqrt{5}$ **21.** Slope $= \frac{1}{5}$; distance $= 2\sqrt{26}$; midpoint $= (2, 3)$
23. Intercept: $(0, 0)$; symmetric with respect to the x-axis
25. Intercepts: $(0, -2)$, $(0, 2)$, $(4, 0)$, $(-4, 0)$; symmetric with respect to the x-axis, y-axis, and origin
27. Intercept: $(0, 1)$; symmetric with respect to the y-axis   **29.** Intercepts: $(0, 0)$, $(0, -2)$, $(-1, 0)$; no symmetry

31. $f(x) = -2x + 3$ **33.** $A = 11$ **35.** (a) B, C, D   (b) D   **37.** (a) $f(-x) = \dfrac{-3x}{x^2 - 4}$ (b) $-f(x) = \dfrac{-3x}{x^2 - 4}$ (c) $f(x + 2) = \dfrac{3x + 6}{x^2 + 4x}$

(d) $f(x - 2) = \dfrac{3x - 6}{x^2 - 4x}$ **39.** (a) $f(-x) = \sqrt{x^2 - 4}$ (b) $-f(x) = -\sqrt{x^2 - 4}$ (c) $f(x + 2) = \sqrt{x^2 + 4x}$ (d) $f(x - 2) = \sqrt{x^2 - 4x}$

41. (a) $f(x) = \dfrac{x^2 - 4}{x^2}$ (b) $-f(x) = -\dfrac{x^2 - 4}{x^2}$ (c) $f(x + 2) = \dfrac{x^2 + 4x}{x^2 + 4x + 4}$ (d) $f(x - 2) = \dfrac{x^2 - 4x}{x^2 - 4x + 4}$ **43.** Odd   **45.** Even

47. Neither   **49.** $\{x \,|\, x \neq -3, x \neq 3\}$ **51.** $(-\infty, 2]$ **53.** $(0, \infty)$ **55.** $\{x \,|\, x \neq -3, x \neq 1\}$ **57.** $[-1, \infty)$ **59.** $[0, \infty)$
61. (a) All real numbers   (c)
(b) $(0, -4)$, $(-4, 0)$, $(4, 0)$
(d) $\{y \,|\, -4 \leq y < \infty\}$

63. (a) All real numbers   (c)
(b) $(0, 0)$
(d) $\{y \,|\, -\infty < y \leq 0\}$

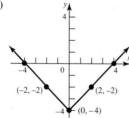

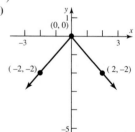

65. (a) $\{x\,|\,1 \le x < \infty\}$
(b) $(1, 0)$
(d) $\{y\,|\,0 \le y < \infty\}$

(c)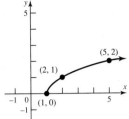

67. (a) $\{x\,|\,-\infty < x \le 1\}$
(b) $(1, 0), (0, 1)$
(d) $\{y\,|\,0 \le y < \infty\}$

(c)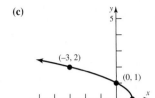

69. (a) All real numbers
(b) $(0, 4), (2, 0)$
(d) All real numbers

(c)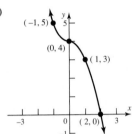

71. (a) All real numbers
(b) $(0, 3)$
(d) $\{y\,|\,2 \le y < \infty\}$

(c)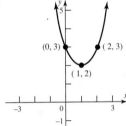

73. (a) All real numbers
(b) $(0, 0)$
(d) All real numbers

(c)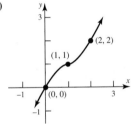

75. (a) $\{x\,|\,0 < x < \infty\}$
(b) None
(d) $\{y\,|\,0 < y < \infty\}$

(c)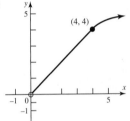

77. (a) $\{x\,|\,x \ne 1\}$
(b) $(0, 0)$
(d) $\{y\,|\,y \ne 1\}$

(c)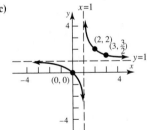

79. (a) All real numbers
(b) $-1 < x \le 0$ are x-intercepts
0 is the y-intercept
(d) Set of integers

(c)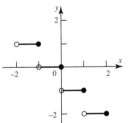

81. $f^{-1}(x) = \dfrac{2x + 3}{5x - 2}$

$$f(f^{-1}(x)) = \dfrac{2\left(\dfrac{2x + 3}{5x - 2}\right) + 3}{5\dfrac{2x + 3}{5x - 2} - 2} = x$$

$$f^{-1}(f(x)) = \dfrac{2\left(\dfrac{2x + 3}{5x - 2}\right) + 3}{5\dfrac{2x + 3}{5x - 2} - 2} = x$$

Domain f = Range f^{-1} = All real numbers except 2/5
Range f = Domain f^{-1} = All real numbers except 2/5

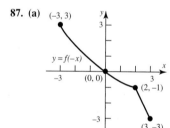

83. $f^{-1}(x) = \dfrac{x + 1}{x}$

$$f(f^{-1}(x)) = \dfrac{1}{\dfrac{x + 1}{x} - 1} = x$$

$$f^{-1}(f(x)) = \dfrac{\dfrac{1}{x - 1} + 1}{\dfrac{1}{x - 1}} = x$$

Domain f = Range f^{-1} = All real numbers except 1
Range f = Domain f^{-1} = All real numbers except 0

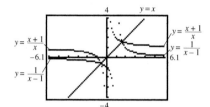

85. $f^{-1}(x) = \dfrac{27}{x^3}$

$$f(f^{-1}(x)) = \dfrac{3}{(27/x^3)^{1/3}} = x$$

$$f^{-1}(f(x)) = \dfrac{27}{(3/x^{1/3})^3} = x$$

Domain f = Range f^{-1} = All real numbers except 0
Range f = Domain f^{-1} = All real numbers except 0

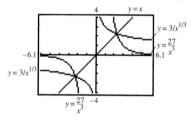

87. (a) 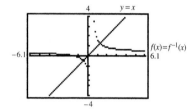   (b) 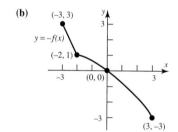   (c)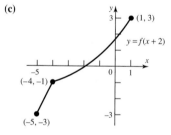

(d) 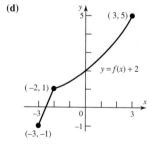   (e) 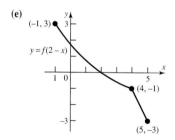   (f)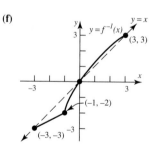

89. $T(h) = -0.0025h + 30$ **91.** $d(A, B) = \sqrt{13}$; $d(B, C) = \sqrt{13}$ **93.** Slope using A and B is -1; slope using B and C is -1
95. Center $(1, -2)$; radius $= 4\sqrt{2}$; $x^2 + y^2 - 2x + 4y - 27 = 0$

CHAPTER 2 *Exercise 2.1*

1.   **3.**  **5.**

7. 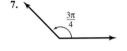 **9.** 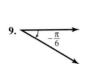 **11.**

13. $\frac{\pi}{6}$ **15.** $\frac{4\pi}{3}$ **17.** $\frac{-\pi}{3}$ **19.** π **21.** $\frac{3\pi}{4}$ **23.** 60° **25.** −225° **27.** 90° **29.** 15° **31.** 120° **33.** 5 m **35.** 6 ft **37.** 0.6 radian

39. $\frac{\pi}{3} \approx 1.047$ in. **41.** 0.30 **43.** −0.70 **45.** 2.18 **47.** 5.93 **49.** 179.91° **51.** 587.28° **53.** 114.59° **55.** 362.11° **57.** 40.17° **59.** 1.03°

61. 9.15° **63.** 40°19′12″ **65.** 18°15′18″ **67.** 19°59′24″ **69.** $3\pi \approx 9.4248$ in.; $5\pi \approx 15.7080$ in. **71.** $\omega = \frac{1}{60}$ radian/sec; $v = \frac{1}{12}$ cm/sec
73. 452.5 rpm **75.** 37.7 in. **77.** 2292 mph **79.** $\frac{3}{4}$ rpm **81.** 2.86 mph **83.** 31.47 revolutions/second **85.** 1.152 mi

Exercise 2.2

1. $\sin\theta = \frac{4}{5}$, $\cos\theta = -\frac{3}{5}$, $\tan\theta = -\frac{4}{3}$, $\csc\theta = \frac{5}{4}$, $\sec\theta = -\frac{5}{3}$, $\cot\theta = -\frac{3}{4}$
3. $\sin\theta = -3\sqrt{13}/13$, $\cos\theta = 2\sqrt{13}/13$, $\tan\theta = -\frac{3}{2}$, $\csc\theta = -\sqrt{13}/3$, $\sec\theta = \sqrt{13}/2$, $\cot\theta = -\frac{2}{3}$
5. $\sin\theta = -\sqrt{2}/2$, $\cos\theta = -\sqrt{2}/2$, $\tan\theta = 1$, $\csc\theta = -\sqrt{2}$, $\sec\theta = -\sqrt{2}$, $\cot\theta = 1$
7. $\sin\theta = -2\sqrt{13}/13$, $\cos\theta = -3\sqrt{13}/13$, $\tan\theta = \frac{2}{3}$, $\csc\theta = -\sqrt{13}/2$, $\sec\theta = -\sqrt{13}/3$, $\cot\theta = \frac{3}{2}$
9. $\sin\theta = -\frac{3}{5}$, $\cos\theta = \frac{4}{5}$, $\tan\theta = -\frac{3}{4}$, $\csc\theta = -\frac{5}{3}$, $\sec\theta = \frac{5}{4}$, $\cot\theta = -\frac{4}{3}$
11. $\frac{1}{2}(\sqrt{2}+1)$ **13.** 2 **15.** $\frac{1}{2}$ **17.** $\sqrt{6}$ **19.** 4 **21.** 0 **23.** 0 **25.** $2\sqrt{2}+4\sqrt{3}/3$ **27.** −1 **29.** 1
31. $\sin(2\pi/3) = \sqrt{3}/2$, $\cos(2\pi/3) = -\frac{1}{2}$, $\tan(2\pi/3) = -\sqrt{3}$, $\csc(2\pi/3) = 2\sqrt{3}/3$, $\sec(2\pi/3) = -2$, $\cot(2\pi/3) = -\sqrt{3}/3$
33. $\sin 150° = \frac{1}{2}$, $\cos 150° = -\sqrt{3}/2$, $\tan 150° = -\sqrt{3}/3$, $\csc 150° = 2$, $\sec 150° = -2\sqrt{3}/3$, $\cot 150° = -\sqrt{3}$
35. $\sin(-\pi/6) = -\frac{1}{2}$, $\cos(-\pi/6) = \sqrt{3}/2$, $\tan(-\pi/6) = -\sqrt{3}/3$, $\csc(-\pi/6) = -2$, $\sec(-\pi/6) = 2\sqrt{3}/3$, $\cot(-\pi/6) = -\sqrt{3}$
37. $\sin 225° = -\sqrt{2}/2$, $\cos 225° = -\sqrt{2}/2$, $\tan 225° = 1$, $\csc 225° = -\sqrt{2}$, $\sec 225° = -\sqrt{2}$, $\cot 225° = 1$
39. $\sin(5\pi/2) = 1$, $\cos(5\pi/2) = 0$, $\tan(5\pi/2)$ is not defined, $\csc(5\pi/2) = 1$, $\sec(5\pi/2)$ is not defined, $\cot(5\pi/2) = 0$
41. $\sin(-180°) = 0$, $\cos(-180°) = -1$, $\tan(-180°) = 0$, $\csc(-180°)$ is not defined, $\sec(-180°) = -1$, $\cot(-180°)$ is not defined
43. $\sin(3\pi/2) = -1$, $\cos(3\pi/2) = 0$, $\tan(3\pi/2)$ is not defined, $\csc(3\pi/2) = -1$, $\sec(3\pi/2)$ is not defined, $\cot(3\pi/2) = 0$
45. $\sin 450° = 1$, $\cos 450° = 0$, $\tan 450°$ is not defined, $\csc 450° = 1$, $\sec 450°$ is not defined, $\cot 450° = 0$ **47.** 0.47 **49.** 0.38 **51.** 1.33
53. 0.36 **55.** 0.31 **57.** 3.73 **59.** 1.04 **61.** 5.67 **63.** 0.84 **65.** 0.02 **67.** 0.93 **69.** 0.31 **71.** $\sqrt{3}/2$ **73.** $\frac{1}{2}$ **75.** $\frac{3}{4}$ **77.** $\sqrt{3}/2$
79. $\sqrt{3}$ **81.** $\sqrt{3}/4$ **83.** 0 **85.** −0.1 **87.** 3 **89.** 5 **91.** $R \approx 310.56$ ft, $H \approx 77.64$ ft **93.** $R \approx 19{,}542$ m, $H \approx 2278$ m
95. (a) 1.2 sec (b) 1.12 sec (c) 1.2 sec **97.** (a) 1.9 hr (b) 1.69 hr (c) 1.63 hr (d) 1.67 hr
99. (a) 16.6 ft (b) (c) $\theta = 67.50°$

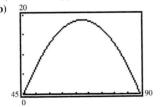

Exercise 2.3

1. $\sqrt{2}/2$ **3.** 1 **5.** 1 **7.** $\sqrt{3}$ **9.** $\sqrt{2}/2$ **11.** 0 **13.** $\sqrt{2}$ **15.** $\sqrt{3}/3$ **17.** II **19.** IV **21.** IV **23.** II
25. $\tan\theta = 2$, $\cot\theta = \frac{1}{2}$, $\sec\theta = \sqrt{5}$, $\csc\theta = \sqrt{5}/2$ **27.** $\tan\theta = \sqrt{3}/3$, $\cot\theta = \sqrt{3}$, $\sec\theta = 2\sqrt{3}/3$, $\csc\theta = 2$
29. $\tan\theta = -\sqrt{2}/4$, $\cot\theta = -2\sqrt{2}$, $\sec\theta = 3\sqrt{2}/4$, $\csc\theta = -3$
31. $\tan\theta = 0.2679$, $\cot\theta = 3.7322$, $\sec\theta = 1.0353$, $\csc\theta = 3.8640$
33. $\cos\theta = -\frac{5}{13}$, $\tan\theta = -\frac{12}{5}$, $\csc\theta = \frac{13}{12}$, $\sec\theta = -\frac{13}{5}$, $\cot\theta = -\frac{5}{12}$
35. $\sin\theta = -\frac{3}{5}$, $\tan\theta = \frac{3}{4}$, $\csc\theta = -\frac{5}{3}$, $\sec\theta = -\frac{5}{4}$, $\cot\theta = \frac{4}{3}$

37. $\cos\theta = -\frac{12}{13}$, $\tan\theta = -\frac{5}{12}$, $\cot\theta = -\frac{12}{5}$, $\sec\theta = -\frac{13}{12}$, $\csc\theta = \frac{13}{5}$

39. $\sin\theta = 2\sqrt{2}/3$, $\tan\theta = -2\sqrt{2}$, $\cot\theta = -\sqrt{2}/4$, $\sec\theta = -3$, $\csc\theta = 3\sqrt{2}/4$

41. $\cos\theta = -\sqrt{5}/3$, $\tan\theta = -2\sqrt{5}/5$, $\cot\theta = -\sqrt{5}/2$, $\sec\theta = -3\sqrt{5}/5$, $\csc\theta = \frac{3}{2}$

43. $\sin\theta = -\sqrt{3}/2$, $\cos\theta = \frac{1}{2}$, $\tan\theta = -\sqrt{3}$, $\cot\theta = -\sqrt{3}/3$, $\csc\theta = -2\sqrt{3}/3$

45. $\sin\theta = -\frac{3}{5}$, $\cos\theta = -\frac{4}{5}$, $\cot\theta = \frac{4}{3}$, $\sec\theta = -\frac{5}{4}$, $\csc\theta = -\frac{5}{3}$

47. $\sin\theta = -\sqrt{10}/10$, $\cos\theta = -3\sqrt{10}/10$, $\cot\theta = -3$, $\sec\theta = -\sqrt{10}/3$, $\csc\theta = \sqrt{10}$ **49.** $-\sqrt{3}/2$ **51.** $-\sqrt{3}/3$ **53.** 2

55. -1 **57.** -1 **59.** $\sqrt{2}/2$ **61.** 0 **63.** $-\sqrt{2}$ **65.** $2\sqrt{3}/3$ **67.** -1 **69.** -2 **71.** $1 - \sqrt{2}/2$ **73.** 1 **75.** 1 **77.** 0 **79.** 0.9 **81.** 9

83. 15.8 min   **85.** At odd multiples of $\pi/2$ **87.** At odd multiples of $\pi/2$ **89.** 0

91. Let $P = (x, y)$ be the point on the unit circle that corresponds to θ. Consider the equation $\tan\theta = y/x = a$. Then $y = ax$. But $x^2 + y^2 = 1$ so that $x^2 + a^2x^2 = 1$. Thus, $x = \pm 1/\sqrt{1 + a^2}$ and $y = \pm a/\sqrt{1 + a^2}$; that is, for any real number a, there is a point $P = (x, y)$ on the unit circle for which $\tan\theta = a$. In other words, $-\infty < \tan\theta < \infty$, and the range of the tangent function is the set of all real numbers.

93. Suppose there is a number p, $0 < p < 2\pi$, for which $\sin(\theta + p) = \sin\theta$ for all θ. If $\theta = 0$, then $\sin(0 + p) = \sin p = \sin 0 = 0$; so that $p = \pi$. If $\theta = \pi/2$, then $\sin(\pi/2 + p) = \sin(\pi/2)$. But $p = \pi$. Thus, $\sin(3\pi/2) = -1 = \sin(\pi/2) = 1$. This is impossible. The smallest positive number p for which $\sin(\theta + p) = \sin\theta$ for all θ is therefore $p = 2\pi$.

95. $\sec\theta = 1/(\cos\theta)$; since $\cos\theta$ has period 2π, so does $\sec\theta$

97. If $P = (a, b)$ is the point on the unit circle corresponding to θ, then $Q = (-a, -b)$ is the point on the unit circle corresponding to $\theta + \pi$. Thus, $\tan(\theta + \pi) = (-b)/(-a) = b/a = \tan\theta$; that is, the period of the tangent function is π.

99. Let $P = (a, b)$ be the point on the unit circle corresponding to θ. Then $\csc\theta = 1/b = 1/(\sin\theta)$; $\sec\theta = 1/a = 1/(\cos\theta)$; $\cot\theta = a/b = 1/(b/a) = 1/(\tan\theta)$.

101. $(\sin\theta\cos\phi)^2 + (\sin\theta\sin\phi)^2 + \cos^2\theta = \sin^2\theta\cos^2\phi + \sin^2\theta\sin^2\phi + \cos^2\theta = \sin^2\theta(\cos^2\phi + \sin^2\phi) + \cos^2\theta = \sin^2\theta + \cos^2\theta = 1$

Exercise 2.4

1. $\sin\theta = \frac{5}{13}$, $\cos\theta = \frac{12}{13}$, $\tan\theta = \frac{5}{12}$, $\csc\theta = \frac{13}{5}$, $\sec\theta = \frac{13}{12}$, $\cot\theta = \frac{12}{5}$

3. $\sin\theta = 2\sqrt{13}/13$, $\cos\theta = 3\sqrt{13}/13$, $\tan\theta = \frac{2}{3}$, $\csc\theta = \sqrt{13}/2$, $\sec\theta = \sqrt{13}/3$, $\cot\theta = \frac{3}{2}$

5. $\sin\theta = \sqrt{3}/2$, $\cos\theta = \frac{1}{2}$, $\tan\theta = \sqrt{3}$, $\csc\theta = 2\sqrt{3}/3$, $\sec\theta = 2$, $\cot\theta = \sqrt{3}/3$

7. $\sin\theta = \sqrt{6}/3$, $\cos\theta = \sqrt{3}/3$, $\tan\theta = \sqrt{2}$, $\csc\theta = \sqrt{6}/2$, $\sec\theta = \sqrt{3}$, $\cot\theta = \sqrt{2}/2$

9. $\sin\theta = \sqrt{5}/5$, $\cos\theta = 2\sqrt{5}/5$, $\tan\theta = \frac{1}{2}$, $\csc\theta = \sqrt{5}$, $\sec\theta = \sqrt{5}/2$, $\cot\theta = 2$ **11.** $30°$ **13.** $60°$ **15.** $30°$ **17.** $\pi/4$ **19.** $\pi/3$

21. $45°$ **23.** $\pi/3$ **25.** $60°$ **27.** $\frac{1}{2}$ **29.** $\sqrt{2}/2$ **31.** -2 **33.** $-\sqrt{3}$ **35.** $\sqrt{2}/2$ **37.** $\sqrt{3}$ **39.** $\frac{1}{2}$ **41.** $-\sqrt{3}/2$ **43.** $-\sqrt{3}$ **45.** $\sqrt{2}$

47. 0 **49.** 1 **51.** 0 **53.** 0 **55.** 1 **57.** **(a)** $\frac{1}{3}$ **(b)** $\frac{8}{9}$ **(c)** 3 **(d)** 3 **59.** **(a)** 17 **(b)** $\frac{1}{4}$ **(c)** 4 **(d)** $\frac{17}{16}$

61. **(a)** $\frac{1}{4}$ **(b)** 15 **(c)** 4 **(d)** $\frac{16}{15}$ **63.** 0.6 **65.** 0 **67.** $20°$

69. **(a)** $T(\theta) = 1 + \dfrac{2}{3\sin\theta} - \dfrac{1}{4\tan\theta}$ **(b)** $\theta = 67.97°$ for least time; the least time is $T = 1.62$ hrs. Sally is on the road for 0.9 hr.

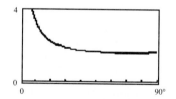

71. **(a)** 10 min **(b)** 20 min **(c)** $T(\theta) = 5\left(1 - \dfrac{1}{3\tan\theta} + \dfrac{1}{\sin\theta}\right)$ **(d)** 10.4 min

(e) T is least for $\theta = 70.5°$; the least time is 9.7 min; $x = 177$ ft

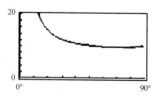

75.

θ	0.5	0.4	0.2	0.1	0.01	0.001	0.0001	0.00001
$\sin\theta$	0.4794	0.3894	0.1987	0.0998	0.0100	0.0010	0.0001	0.00001
$\dfrac{\sin\theta}{\theta}$	0.9589	0.9735	0.9933	0.9983	1.0000	1.0000	1.0000	1.0000

$\dfrac{\sin\theta}{\theta}$ approaches 1 as θ approaches 0

77. (a) $|OA| = |OC| = 1$; Angle OAC + Angle OAC + $180° - \theta = 180°$; Angle $OAC = \theta/2$

(b) $\sin \theta = \dfrac{|CD|}{|OC|} = |CD|$; $\cos \theta = \dfrac{|OD|}{|OC|} = |OD|$ **(c)** $\tan \dfrac{\theta}{2} = \dfrac{|CD|}{|AD|} = \dfrac{\sin \theta}{1 + |OD|} = \dfrac{\sin \theta}{1 + \cos \theta}$

79. $h = x \tan \theta$ and $h = (1 - x) \tan n\theta$; thus, $x \tan \theta = (1 - x) \tan n\theta$ and $x = \dfrac{\tan n\theta}{\tan \theta + \tan n\theta}$

81. (a) Area $\triangle OAC = \frac{1}{2}|OC\,||AC| = \frac{1}{2} \cdot \dfrac{|OC|}{1} \cdot \dfrac{|AC|}{1} = \frac{1}{2} \sin \alpha \cos \alpha$

(b) Area $\triangle OCB = \frac{1}{2}|BC\,||OC| - \frac{1}{2}|OB|^2 \dfrac{|BC|}{|OB|} \cdot \dfrac{|OC|}{|OB|} = \frac{1}{2}|OB|^2 \sin \beta \cos \beta$

(c) Area $\triangle OAB = \frac{1}{2}|BD||OA| = \frac{1}{2}|OB|\dfrac{|BD|}{|OB|} = \frac{1}{2}|OB| \sin(\alpha + \beta)$ **(d)** $\dfrac{\cos \alpha}{\cos \beta} = \dfrac{|OC|/1}{|OC|/|OB|} = |OB|$

(e) Use the hint and results from parts (a)–(d).

83. $\sin \alpha = \tan \alpha \cos \alpha = \cos \beta \cos \alpha = \cos \beta \tan \beta = \sin \beta$; $\sin^2 \alpha + \cos^2 \alpha = 1$, thus,

$$\sin^2 \alpha + \tan^2 \beta = 1$$
$$\sin^2 \alpha + \dfrac{\sin^2 \beta}{\cos^2 \beta} = 1$$
$$\sin^2 \alpha + \dfrac{\sin^2 \alpha}{1 - \sin^2 \alpha} = 1$$
$$\sin^2 \alpha - \sin^4 \alpha + \sin^2 \alpha = 1 - \sin^2 \alpha$$
$$\sin^4 \alpha - 3 \sin^2 \alpha + 1 = 0$$
$$\sin^2 \alpha = \dfrac{3 \pm \sqrt{5}}{2}$$
$$\sin^2 \alpha = \dfrac{3 - \sqrt{5}}{2}$$
$$\sin \alpha = \sqrt{\dfrac{3 - \sqrt{5}}{2}}$$

Exercise 2.5

1. 0 **3.** $-\pi/2 \le x \le \pi/2$ **5.** 1 **7.** $0, \pi, 2\pi$ **9.** $\sin x = 1$ for $x = -3\pi/2, \pi/2$; $\sin x = -1$ for $x = -\pi/2, 3\pi/2$ **11.** 0 **13.** 1
15. $\sec x = 1$ for $x = -2\pi, 0, 2\pi$; $\sec x = -1$ for $x = -\pi, \pi$ **17.** $-3\pi/2, -\pi/2, \pi/2, 3\pi/2$ **19.** $-3\pi/2, -\pi/2, \pi/2, 3\pi/2$ **21.** B, C, F
23. C **25.** D **27.** B **29.** A

31.

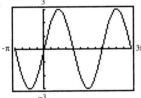

33.

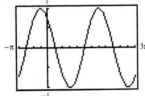

35.

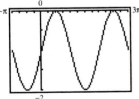

37.

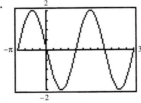

39.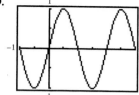

41.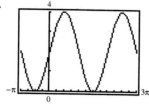

43.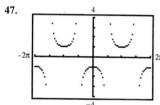

45.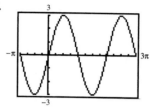

47.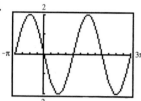

49.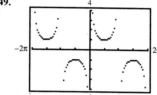

51.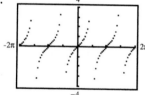

53.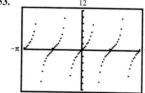

55.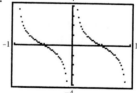

57.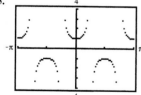

59.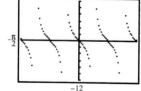

61.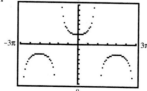

63. (a) $L = 3/\cos\theta + 4/\sin\theta = 3\sec\theta + 4\csc\theta$ **(c)** L is smallest when $\theta = 0.83$ **(d)** $L(0.83) = 9.86$ ft.

(b)

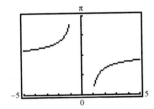

65. $y = \sin\omega x$ has period $2\pi/\omega$ **67.** The graphs are the same; yes

Exercise 2.6

1. 0   **3.** $-\pi/2$ **5.** 0   **7.** $\pi/4$ **9.** $\pi/3$ **11.** $5\pi/6$ **13.** 0.10   **15.** 1.37   **17.** 0.51   **19.** -0.38 **21.** -0.12 **23.** 1.08   **25.** $\sqrt{2}/2$ **27.** $-\sqrt{3}/3$
29. 2   **31.** $\sqrt{2}$ **33.** $-\sqrt{2}/2$ **35.** $2\sqrt{3}/3$ **37.** $\sqrt{2}/4$ **39.** $\sqrt{5}/2$ **41.** $-\sqrt{14}/2$ **43.** $-3\sqrt{10}/10$ **45.** $\sqrt{5}$ **47.** 0.58   **49.** 0.10   **51.** 0.57
53. 0.43   **55.** 0.37
57. Let $\theta = \tan^{-1} v$. Then $\tan\theta = v$, $-\pi/2 < \theta < \pi/2$. Now, $\sec\theta > 0$ and $\tan^2\theta + 1 = \sec^2\theta$. Thus, $\sec\theta = \sec(\tan^{-1} v) = \sqrt{1 + v^2}$.

59. Let $\theta = \cos^{-1} v$. Then $\cos\theta = v$, $0 \le \theta \le \pi$, and $\tan(\cos^{-1} v) = \tan\theta = \dfrac{\sin\theta}{\cos\theta} = \dfrac{\sqrt{1 - \cos^2\theta}}{\cos\theta} = \dfrac{\sqrt{1 - v^2}}{v}$.

61. Let $\theta = \sin^{-1} v$. Then $\sin\theta = v$, $-\pi/2 \le \theta \le \pi/2$, and $\cos(\sin^{-1} v) = \cos\theta = \sqrt{1 - \sin^2\theta} = \sqrt{1 - v^2}$.
63. Let $\alpha = \sin^{-1} v$ and $\beta = \cos^{-1} v$. Then $\sin\alpha = v = \cos\beta$, so α and β are complementary angles. Thus, $\alpha + \beta = \pi/2$.

65. Let $\alpha = \tan^{-1}\dfrac{1}{v}$. Then $\dfrac{1}{v} = \tan\alpha$, $\dfrac{-\pi}{2} < \alpha < \dfrac{\pi}{2}$, $\alpha \ne 0$. Let $\beta = \tan^{-1} v$. Then $v = \tan\beta$, $\dfrac{-\pi}{2} < \beta < \dfrac{\pi}{2}$, $\beta \ne 0$. Thus, $\tan\alpha \tan\beta = 1$ so

that $\tan\alpha = \cot\beta$. Thus, $\alpha + \beta = \dfrac{\pi}{2}$ and $\tan^{-1}\left(\dfrac{1}{v}\right) = \dfrac{\pi}{2} - \tan^{-1} v$.

67. 1.32   **69.** 0.46   **71.** -0.34 **73.** 2.72   **75.** -0.73 **77.** 2.55   **79.** $t = 2.77$ minutes   **81.** $-1 \le x \le 1$ **83.** $0 \le x \le \pi$
85. $y = \sec^{-1}(x) = \cos^{-1}(1/x)$

Fill-in-the-Blanks

1. angle; initial side; terminal side   **2.** radians   **3.** π **4.** complementary   **5.** cosine   **6.** standard position   **7.** $45°$ **8.** 2π, π
9. $y = \cos x$, $y = \sec x$ **10.** $y = \sin x$, $y = \tan x$, $y = \csc x$, $y = \cot x$ **11.** $-1 \le x \le 1$; $-\pi/2 \le y \le \pi/2$ **12.** 0

True/False Items

1. F   **2.** T   **3.** F   **4.** T   **5.** F   **6.** F   **7.** T   **8.** F   **9.** T

Review Exercises

1. $3\pi/4$ **3.** $\pi/10$ **5.** $135°$ **7.** $-450°$ **9.** $\frac{1}{2}$ **11.** $3\sqrt{2}/2 - 4\sqrt{3}/3$ **13.** $-3\sqrt{2} - 2\sqrt{3}$ **15.** 3   **17.** 0   **19.** 0   **21.** 1   **23.** 1   **25.** 1
27. -1 **29.** 1   **31.** $\cos\theta = \frac{3}{5}$, $\tan\theta = -\frac{4}{3}$, $\csc\theta = -\frac{5}{4}$, $\sec\theta = \frac{5}{3}$, $\cot\theta = -\frac{3}{4}$ **33.** $\sin\theta = -\frac{12}{13}$, $\cos\theta = -\frac{5}{13}$, $\csc\theta = -\frac{13}{12}$, $\sec\theta = -\frac{13}{5}$, $\cot\theta = \frac{5}{12}$
35. $\sin\theta = \frac{3}{5}$, $\cos\theta = -\frac{4}{5}$, $\tan\theta = -\frac{3}{4}$, $\csc\theta = \frac{5}{3}$, $\cot\theta = -\frac{4}{3}$ **37.** $\cos\theta = -\frac{5}{13}$, $\tan\theta = -\frac{12}{5}$, $\csc\theta = \frac{13}{12}$, $\sec\theta = -\frac{13}{5}$, $\cot\theta = -\frac{5}{12}$
39. $\cos\theta = \frac{12}{13}$, $\tan\theta = -\frac{5}{12}$, $\csc\theta = -\frac{13}{5}$, $\sec\theta = \frac{13}{12}$, $\cot\theta = -\frac{12}{5}$
41. $\sin\theta = -\sqrt{10}/10$, $\cos\theta = -3\sqrt{10}/10$, $\csc\theta = -\sqrt{10}$, $\sec\theta = -\sqrt{10}/3$, $\cot\theta = 3$
43. $\sin\theta = -2\sqrt{2}/3$, $\cos\theta = \frac{1}{3}$, $\tan\theta = -2\sqrt{2}$, $\csc\theta = -3\sqrt{2}/4$, $\cot\theta = -\sqrt{2}/4$
45. $\sin\theta = \sqrt{5}/5$, $\cos\theta = -2\sqrt{5}/5$, $\tan\theta = -\frac{1}{2}$, $\csc\theta = \sqrt{5}$, $\sec\theta = -\sqrt{5}/2$

47.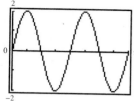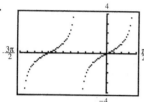

49.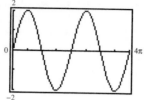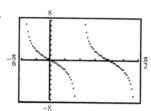

51.

53.

55. $\pi/2$ **57.** $\pi/4$ **59.** $5\pi/6$ **61.** $\sqrt{2}/2$ **63.** $-\sqrt{3}$ **65.** $2\sqrt{3}/3$ **67.** $\frac{3}{5}$ **69.** $-\frac{4}{3}$ **71.** $\pi/3$ ft **73.** 114.59 revolutions per hour

CHAPTER 3 *Exercise 3.1*

1. $\csc\theta \cdot \cos\theta = \dfrac{1}{\sin\theta} \cdot \cos\theta = \dfrac{\cos\theta}{\sin\theta} = \cot\theta$ **3.** $1 + \tan^2(-\theta) = 1 + (-\tan\theta)^2 = 1 + \tan^2\theta = \sec^2\theta$

5. $\cos\theta(\tan\theta + \cot\theta) = \cos\theta\left(\dfrac{\sin\theta}{\cos\theta} + \dfrac{\cos\theta}{\sin\theta}\right) = \cos\theta\left(\dfrac{\sin^2\theta + \cos^2\theta}{\cos\theta\sin\theta}\right) = \dfrac{1}{\sin\theta} = \csc\theta$

7. $\tan\theta\cot\theta - \cos^2\theta = \dfrac{\sin\theta}{\cos\theta} \cdot \dfrac{\cos\theta}{\sin\theta} - \cos^2\theta = 1 - \cos^2\theta = \sin^2\theta$ **9.** $(\sec\theta - 1)(\sec\theta + 1) = \sec^2\theta - 1 = \tan^2\theta$

11. $(\sec\theta + \tan\theta)(\sec\theta - \tan\theta) = \sec^2\theta - \tan^2\theta = 1$ **13.** $\sin^2\theta(1 + \cot^2\theta) = \sin^2\theta\csc^2\theta = \sin^2\theta\left(\dfrac{1}{\sin^2\theta}\right) = 1$

15. $(\sin\theta + \cos\theta)^2 + (\sin\theta - \cos\theta)^2 = \sin^2\theta + 2\sin\theta\cos\theta + \cos^2\theta + \sin^2\theta - 2\sin\theta\cos\theta + \cos^2\theta$
$$= \sin^2\theta + \cos^2\theta + \sin^2\theta + \cos^2\theta = 1 + 1 = 2$$

17. $\sec^4\theta - \sec^2\theta = \sec^2\theta(\sec^2\theta - 1) = (1 + \tan^2\theta)\tan^2\theta = \tan^4\theta + \tan^2\theta$

19. $\sec\theta - \tan\theta = \dfrac{1}{\cos\theta} - \dfrac{\sin\theta}{\cos\theta} = \dfrac{1 - \sin\theta}{\cos\theta} \cdot \dfrac{1 + \sin\theta}{1 + \sin\theta} = \dfrac{1 - \sin^2\theta}{\cos\theta(1 + \sin\theta)} = \dfrac{\cos^2\theta}{\cos\theta(1 + \sin\theta)} = \dfrac{\cos\theta}{1 + \sin\theta}$

21. $3\sin^2\theta + 4\cos^2\theta = 3\sin^2\theta + 3\cos^2\theta + \cos^2\theta = 3(\sin^2\theta + \cos^2\theta) + \cos^2\theta = 3 + \cos^2\theta$

23. $1 - \dfrac{\cos^2\theta}{1 + \sin\theta} = 1 - \dfrac{1 - \sin^2\theta}{1 + \sin\theta} = 1 - (1 - \sin\theta) = \sin\theta$ **25.** $\dfrac{1 + \tan\theta}{1 - \tan\theta} = \dfrac{1 + \dfrac{1}{\cot\theta}}{1 - \dfrac{1}{\cot\theta}} = \dfrac{\dfrac{\cot\theta + 1}{\cot\theta}}{\dfrac{\cot\theta - 1}{\cot\theta}} = \dfrac{\cot\theta + 1}{\cot\theta - 1}$

27. $\dfrac{\sec\theta}{\csc\theta} + \dfrac{\sin\theta}{\cos\theta} = \dfrac{1/\cos\theta}{1/\sin\theta} + \tan\theta = \dfrac{\sin\theta}{\cos\theta} + \tan\theta = \tan\theta + \tan\theta = 2\tan\theta$ **29.** $\dfrac{1 + \sin\theta}{1 - \sin\theta} = \dfrac{1 + \dfrac{1}{\csc\theta}}{1 - \dfrac{1}{\csc\theta}} = \dfrac{\dfrac{\csc\theta + 1}{\csc\theta}}{\dfrac{\csc\theta - 1}{\csc\theta}} = \dfrac{\csc\theta + 1}{\csc\theta - 1}$

31. $\dfrac{1 - \sin\theta}{\cos\theta} + \dfrac{\cos\theta}{1 - \sin\theta} = \dfrac{(1 - \sin\theta)^2 + \cos^2\theta}{\cos\theta(1 - \sin\theta)} = \dfrac{1 - 2\sin\theta + \sin^2\theta + \cos^2\theta}{\cos\theta(1 - \sin\theta)} = \dfrac{2 - 2\sin\theta}{\cos\theta(1 - \sin\theta)} = \dfrac{2(1 - \sin\theta)}{\cos\theta(1 - \sin\theta)} = \dfrac{2}{\cos\theta} = 2\sec\theta$

33. $\dfrac{\sin\theta}{\sin\theta - \cos\theta} = \dfrac{1}{\dfrac{\sin\theta - \cos\theta}{\sin\theta}} = \dfrac{1}{1 - \dfrac{\cos\theta}{\sin\theta}} = \dfrac{1}{1 - \cot\theta}$

35. $(\sec\theta - \tan\theta)^2 = \sec^2\theta - 2\sec\theta\tan\theta + \tan^2\theta = \dfrac{1}{\cos^2\theta} - \dfrac{2\sin\theta}{\cos^2\theta} + \dfrac{\sin^2\theta}{\cos^2\theta} = \dfrac{1 - 2\sin\theta + \sin^2\theta}{\cos^2\theta} = \dfrac{(1 - \sin\theta)^2}{1 - \sin^2\theta}$
$$= \dfrac{(1 - \sin\theta)^2}{(1 - \sin\theta)(1 + \sin\theta)} = \dfrac{1 - \sin\theta}{1 + \sin\theta}$$

37. $\dfrac{\cos\theta}{1 - \tan\theta} + \dfrac{\sin\theta}{1 - \cot\theta} = \dfrac{\cos\theta}{1 - \dfrac{\sin\theta}{\cos\theta}} + \dfrac{\sin\theta}{1 - \dfrac{\cos\theta}{\sin\theta}} = \dfrac{\cos\theta}{\dfrac{\cos\theta - \sin\theta}{\cos\theta}} + \dfrac{\sin\theta}{\dfrac{\sin\theta - \cos\theta}{\sin\theta}} = \dfrac{\cos^2\theta}{\cos\theta - \sin\theta} + \dfrac{\sin^2\theta}{\sin\theta - \cos\theta} = \dfrac{\cos^2\theta - \sin^2\theta}{\cos\theta - \sin\theta}$
$$= \dfrac{(\cos\theta - \sin\theta)(\cos\theta + \sin\theta)}{\cos\theta - \sin\theta} = \sin\theta + \cos\theta$$

39. $\tan\theta + \dfrac{\cos\theta}{1+\sin\theta} = \dfrac{\sin\theta}{\cos\theta} + \dfrac{\cos\theta}{1+\sin\theta} = \dfrac{\sin\theta(1+\sin\theta)+\cos^2\theta}{\cos\theta(1+\sin\theta)} = \dfrac{\sin\theta+\sin^2\theta+\cos^2\theta}{\cos\theta(1+\sin\theta)} = \dfrac{\sin\theta+1}{\cos\theta(1+\sin\theta)} = \dfrac{1}{\cos\theta} = \sec\theta$

41. $\dfrac{\tan\theta+\sec\theta-1}{\tan\theta-\sec\theta+1} = \dfrac{\tan\theta+(\sec\theta-1)}{\tan\theta-(\sec\theta-1)} \cdot \dfrac{\tan\theta+(\sec\theta-1)}{\tan\theta+(\sec\theta-1)} = \dfrac{\tan^2\theta+2\tan\theta(\sec\theta-1)+\sec^2\theta-2\sec\theta+1}{\tan^2\theta-(\sec^2\theta-2\sec\theta+1)}$

$\qquad = \dfrac{\sec^2\theta-1+2\tan\theta(\sec\theta-1)+\sec^2\theta-2\sec\theta+1}{\sec^2\theta-1-\sec^2\theta+2\sec\theta-1} = \dfrac{2\sec^2\theta-2\sec\theta+2\tan\theta(\sec\theta-1)}{-2+2\sec\theta}$

$\qquad = \dfrac{2\sec\theta(\sec\theta-1)+2\tan\theta(\sec\theta-1)}{2(\sec\theta-1)} = \dfrac{2(\sec\theta-1)(\sec\theta+\tan\theta)}{2(\sec\theta-1)} = \sec\theta+\tan\theta$

43. $\dfrac{\tan\theta-\cot\theta}{\tan\theta+\cot\theta} = \dfrac{\dfrac{\sin\theta}{\cos\theta}-\dfrac{\cos\theta}{\sin\theta}}{\dfrac{\sin\theta}{\cos\theta}+\dfrac{\cos\theta}{\sin\theta}} = \dfrac{\dfrac{\sin^2\theta-\cos^2\theta}{\cos\theta\sin\theta}}{\dfrac{\sin^2\theta+\cos^2\theta}{\cos\theta\sin\theta}} = \dfrac{\sin^2\theta-\cos^2\theta}{1} = \sin^2\theta-\cos^2\theta$

45. $\dfrac{\tan\theta-\cot\theta}{\tan\theta+\cot\theta} = \dfrac{\dfrac{\sin\theta}{\cos\theta}-\dfrac{\cos\theta}{\sin\theta}}{\dfrac{\sin\theta}{\cos\theta}+\dfrac{\cos\theta}{\sin\theta}} = \dfrac{\dfrac{\sin^2\theta-\cos^2\theta}{\cos\theta\sin\theta}}{\dfrac{\sin^2\theta+\cos^2\theta}{\cos\theta\sin\theta}} = \sin^2\theta-\cos^2\theta = \sin^2\theta-(1-\sin^2\theta) = 2\sin^2\theta-1$

47. $\dfrac{\sec\theta+\tan\theta}{\cot\theta+\cos\theta} = \dfrac{\dfrac{1}{\cos\theta}+\dfrac{\sin\theta}{\cos\theta}}{\dfrac{\cos\theta}{\sin\theta}+\dfrac{\cos\theta\sin\theta}{\sin\theta}} = \dfrac{\dfrac{1+\sin\theta}{\cos\theta}}{\dfrac{\cos\theta+\cos\theta\sin\theta}{\sin\theta}} = \dfrac{1+\sin\theta}{\cos\theta}\cdot\dfrac{\sin\theta}{\cos\theta(1+\sin\theta)} = \dfrac{\sin\theta}{\cos\theta}\cdot\dfrac{1}{\cos\theta} = \tan\theta\sec\theta$

49. $\dfrac{1-\tan^2\theta}{1+\tan^2\theta} = \dfrac{1-\tan^2\theta}{\sec^2\theta} = \dfrac{1}{\sec^2\theta}-\dfrac{\tan^2\theta}{\sec^2\theta} = \cos^2\theta-\dfrac{\sin^2\theta/\cos^2\theta}{1/\cos^2\theta} = \cos^2\theta-\sin^2\theta = \cos^2\theta-(1-\cos^2\theta) = 2\cos^2\theta-1$

51. $\dfrac{\sec\theta-\csc\theta}{\sec\theta\csc\theta} = \dfrac{\dfrac{1}{\cos\theta}-\dfrac{1}{\sin\theta}}{\dfrac{1}{\cos\theta}\cdot\dfrac{1}{\sin\theta}} = \dfrac{\dfrac{\sin\theta-\cos\theta}{\cos\theta\sin\theta}}{\dfrac{1}{\cos\theta\sin\theta}} = \sin\theta-\cos\theta$

53. $\sec\theta-\cos\theta = \dfrac{1}{\cos\theta}-\dfrac{\cos^2\theta}{\cos\theta} = \dfrac{1-\cos^2\theta}{\cos\theta} = \dfrac{\sin^2\theta}{\cos\theta} = \sin\theta\cdot\dfrac{\sin\theta}{\cos\theta} = \sin\theta\tan\theta$

55. $\dfrac{1}{1-\sin\theta}+\dfrac{1}{1+\sin\theta} = \dfrac{1+\sin\theta+1-\sin\theta}{(1+\sin\theta)(1-\sin\theta)} = \dfrac{2}{1-\sin^2\theta} = \dfrac{2}{\cos^2\theta} = 2\sec^2\theta$

57. $\dfrac{\sec\theta}{1-\sin\theta} = \dfrac{\sec\theta}{1-\sin\theta}\cdot\dfrac{1+\sin\theta}{1+\sin\theta} = \dfrac{\sec\theta(1+\sin\theta)}{1-\sin^2\theta} = \dfrac{\sec\theta(1+\sin\theta)}{\cos^2\theta} = \dfrac{1+\sin\theta}{\cos^3\theta}$

59. $\dfrac{(\sec\theta-\tan\theta)^2+1}{\csc\theta(\sec\theta-\tan\theta)} = \dfrac{\sec^2\theta-2\sec\theta\tan\theta+\tan^2\theta+1}{\dfrac{1}{\sin\theta}\left(\dfrac{1}{\cos\theta}-\dfrac{\sin\theta}{\cos\theta}\right)} = \dfrac{2\sec^2\theta-2\sec\theta\tan\theta}{\dfrac{1}{\sin\theta}\left(\dfrac{1-\sin\theta}{\cos\theta}\right)} = \dfrac{\dfrac{2}{\cos^2\theta}-\dfrac{2\sin\theta}{\cos^2\theta}}{\dfrac{1-\sin\theta}{\sin\theta\cos\theta}} = \dfrac{2-2\sin\theta}{\cos^2\theta}\cdot\dfrac{\sin\theta\cos\theta}{1-\sin\theta}$

$\qquad = \dfrac{2(1-\sin\theta)}{\cos\theta}\cdot\dfrac{\sin\theta}{1-\sin\theta} = \dfrac{2\sin\theta}{\cos\theta} = 2\tan\theta$

61. $\dfrac{\sin\theta+\cos\theta}{\cos\theta}-\dfrac{\sin\theta-\cos\theta}{\sin\theta} = \dfrac{\sin\theta(\sin\theta+\cos\theta)-\cos\theta(\sin\theta-\cos\theta)}{\cos\theta\sin\theta} = \dfrac{\sin^2\theta+\sin\theta\cos\theta-\sin\theta\cos\theta+\cos^2\theta}{\cos\theta\sin\theta}$

$\qquad = \dfrac{1}{\cos\theta\sin\theta} = \sec\theta\csc\theta$

63. $\dfrac{\sin^3\theta+\cos^3\theta}{\sin\theta+\cos\theta} = \dfrac{(\sin\theta+\cos\theta)(\sin^2\theta-\sin\theta\cos\theta+\cos^2\theta)}{\sin\theta+\cos\theta} = \sin^2\theta+\cos^2\theta-\sin\theta\cos\theta = 1-\sin\theta\cos\theta$

65. $\dfrac{\cos^2\theta-\sin^2\theta}{1-\tan^2\theta} = \dfrac{\cos^2\theta-\sin^2\theta}{1-\dfrac{\sin^2\theta}{\cos^2\theta}} = \dfrac{\cos^2\theta-\sin^2\theta}{\dfrac{\cos^2\theta-\sin^2\theta}{\cos^2\theta}} = \cos^2\theta$

67. $\dfrac{(2\cos^2\theta-1)^2}{\cos^4\theta-\sin^4\theta} = \dfrac{[2\cos^2\theta-(\sin^2\theta+\cos^2\theta)]^2}{(\cos^2\theta-\sin^2\theta)(\cos^2\theta+\sin^2\theta)} = \cos^2\theta-\sin^2\theta = (1-\sin^2\theta)-\sin^2\theta = 1-2\sin^2\theta$

69. $\dfrac{1+\sin\theta+\cos\theta}{1+\sin\theta-\cos\theta} = \dfrac{(1+\sin\theta)+\cos\theta}{(1+\sin\theta)-\cos\theta}\cdot\dfrac{(1+\sin\theta)+\cos\theta}{(1+\sin\theta)+\cos\theta} = \dfrac{1+2\sin\theta+\sin^2\theta+2(1+\sin\theta)(\cos\theta)+\cos^2\theta}{1+2\sin\theta+\sin^2\theta-\cos^2\theta}$

$\qquad = \dfrac{1+2\sin\theta+\sin^2\theta+2(1+\sin\theta)(\cos\theta)+(1-\sin^2\theta)}{1+2\sin\theta+\sin^2\theta-(1-\sin^2\theta)} = \dfrac{2+2\sin\theta+2(1+\sin\theta)(\cos\theta)}{2\sin\theta+2\sin^2\theta}$

$\qquad = \dfrac{2(1+\sin\theta)+2(1+\sin\theta)(\cos\theta)}{2\sin\theta(1+\sin\theta)} = \dfrac{2(1+\sin\theta)(1+\cos\theta)}{2\sin\theta(1+\sin\theta)} = \dfrac{1+\cos\theta}{\sin\theta}$

71. $(a\sin\theta+b\cos\theta)^2+(a\cos\theta-b\sin\theta)^2 = a^2\sin^2\theta+2ab\sin\theta\cos\theta+b^2\cos^2\theta+a^2\cos^2\theta-2ab\sin\theta\cos\theta+b^2\sin^2\theta$

$\qquad = a^2(\sin^2\theta+\cos^2\theta)+b^2(\cos^2\theta+\sin^2\theta) = a^2+b^2$

73. $\dfrac{\tan \alpha + \tan \beta}{\cot \alpha + \cot \beta} = \dfrac{\tan \alpha + \tan \beta}{\dfrac{1}{\tan \alpha} + \dfrac{1}{\tan \beta}} = \dfrac{\tan \alpha + \tan \beta}{\dfrac{\tan \beta + \tan \alpha}{\tan \alpha \tan \beta}} = (\tan \alpha + \tan \beta) \cdot \dfrac{\tan \alpha \tan \beta}{\tan \alpha + \tan \beta} = \tan \alpha \tan \beta$

75. $(\sin \alpha + \cos \beta)^2 + (\cos \beta + \sin \alpha)(\cos \beta - \sin \alpha) = (\sin^2 \alpha + 2 \sin \alpha \cos \beta + \cos^2 \beta) + (\cos^2 \beta - \sin^2 \alpha)$
$= 2 \cos^2 \beta + 2 \sin \alpha \cos \beta = 2 \cos \beta(\cos \beta + \sin \alpha)$

77. $\ln |\sec \theta| = \ln |\cos \theta|^{-1} = -\ln |\cos \theta|$

79. $\ln |1 + \cos \theta| + \ln |1 - \cos \theta| = \ln(|1 + \cos \theta| \, |1 - \cos \theta|) = \ln |1 - \cos^2 \theta| = \ln |\sin^2 \theta| = 2 \ln |\sin \theta|$

Exercise 3.2

1. $\frac{1}{4}(\sqrt{6} + \sqrt{2})$ **3.** $\frac{1}{4}(\sqrt{2} - \sqrt{6})$ **5.** $-\frac{1}{4}(\sqrt{2} + \sqrt{6})$ **7.** $\dfrac{\sqrt{3} - 1}{1 + \sqrt{3}} = 2 - \sqrt{3}$ **9.** $-\frac{1}{4}(\sqrt{6} + \sqrt{2})$

11. $\dfrac{4}{\sqrt{6} + \sqrt{2}} = \sqrt{6} - \sqrt{2}$ **13.** $\frac{1}{2}$ **15.** 0   **17.** 1   **19.** -1 **21.** $-\sqrt{3}/2$ **23. (a)** $\dfrac{2\sqrt{5}}{25}$ **(b)** $\dfrac{11\sqrt{5}}{25}$ **(c)** $\dfrac{2\sqrt{5}}{5}$ **(d)** 2

25. (a) $\dfrac{4 - 3\sqrt{3}}{10}$ **(b)** $\dfrac{-3 - 4\sqrt{3}}{10}$ **(c)** $\dfrac{4 + 3\sqrt{3}}{10}$ **(d)** $\dfrac{4 + 3\sqrt{3}}{4\sqrt{3} - 3} = \dfrac{25\sqrt{3} + 48}{39}$

27. (a) $-\dfrac{1}{26}(5 + 12\sqrt{3})$ **(b)** $\dfrac{1}{26}(12 - 5\sqrt{3})$ **(c)** $-\dfrac{1}{26}(5 - 12\sqrt{3})$ **(d)** $\dfrac{-5 + 12\sqrt{3}}{12 + 5\sqrt{3}} = \dfrac{-240 + 169\sqrt{3}}{69}$

29. $\sin\left(\dfrac{\pi}{2} + \theta\right) = \sin \dfrac{\pi}{2} \cos \theta + \cos \dfrac{\pi}{2} \sin \theta = 1 \cdot \cos \theta + 0 \cdot \sin \theta = \cos \theta$

31. $\sin(\pi - \theta) = \sin \pi \cos \theta - \cos \pi \sin \theta = 0 \cdot \cos \theta - (-1)\sin \theta = \sin \theta$

33. $\sin(\pi + \theta) = \sin \pi \cos \theta + \cos \pi \sin \theta = 0 \cdot \cos \theta + (-1)\sin \theta = -\sin \theta$

35. $\tan(\pi - \theta) = \dfrac{\tan \pi - \tan \theta}{1 + \tan \pi \tan \theta} = \dfrac{0 - \tan \theta}{1 + 0} = -\tan \theta$

37. $\sin\left(\dfrac{3\pi}{2} + \theta\right) = \sin \dfrac{3\pi}{2} \cos \theta + \cos \dfrac{3\pi}{2} \sin \theta = (-1)\cos \theta + 0 \cdot \sin \theta = -\cos \theta$

39. $\sin(\alpha + \beta) + \sin(\alpha - \beta) = \sin \alpha \cos \beta + \cos \alpha \sin \beta + \sin \alpha \cos \beta - \cos \alpha \sin \beta = 2 \sin \alpha \cos \beta$

41. $\dfrac{\sin(\alpha + \beta)}{\sin \alpha \cos \beta} = \dfrac{\sin \alpha \cos \beta + \cos \alpha \sin \beta}{\sin \alpha \cos \beta} = \dfrac{\sin \alpha \cos \beta}{\sin \alpha \cos \beta} + \dfrac{\cos \alpha \sin \beta}{\sin \alpha \cos \beta} = 1 + \cot \alpha \tan \beta$

43. $\dfrac{\cos(\alpha + \beta)}{\cos \alpha \cos \beta} = \dfrac{\cos \alpha \cos \beta - \sin \alpha \sin \beta}{\cos \alpha \cos \beta} = \dfrac{\cos \alpha \cos \beta}{\cos \alpha \cos \beta} - \dfrac{\sin \alpha \sin \beta}{\cos \alpha \cos \beta} = 1 - \tan \alpha \tan \beta$

45. $\dfrac{\sin(\alpha + \beta)}{\sin(\alpha - \beta)} = \dfrac{\sin \alpha \cos \beta + \cos \alpha \sin \beta}{\sin \alpha \cos \beta - \cos \alpha \sin \beta} = \dfrac{\dfrac{\sin \alpha \cos \beta + \cos \alpha \sin \beta}{\cos \alpha \cos \beta}}{\dfrac{\sin \alpha \cos \beta - \cos \alpha \sin \beta}{\cos \alpha \cos \beta}} = \dfrac{\dfrac{\sin \alpha \cos \beta}{\cos \alpha \cos \beta} + \dfrac{\cos \alpha \sin \beta}{\cos \alpha \cos \beta}}{\dfrac{\sin \alpha \cos \beta}{\cos \alpha \cos \beta} - \dfrac{\cos \alpha \sin \beta}{\cos \alpha \cos \beta}} = \dfrac{\tan \alpha + \tan \beta}{\tan \alpha - \tan \beta}$

47. $\cot(\alpha + \beta) = \dfrac{\cos(\alpha + \beta)}{\sin(\alpha + \beta)} = \dfrac{\cos \alpha \cos \beta - \sin \alpha \sin \beta}{\sin \alpha \cos \beta + \cos \alpha \sin \beta} = \dfrac{\dfrac{\cos \alpha \cos \beta - \sin \alpha \sin \beta}{\sin \alpha \sin \beta}}{\dfrac{\sin \alpha \cos \beta + \cos \alpha \sin \beta}{\sin \alpha \sin \beta}} = \dfrac{\dfrac{\cos \alpha \cos \beta}{\sin \alpha \sin \beta} - \dfrac{\sin \alpha \sin \beta}{\sin \alpha \sin \beta}}{\dfrac{\sin \alpha \cos \beta}{\sin \alpha \sin \beta} + \dfrac{\cos \alpha \sin \beta}{\sin \alpha \sin \beta}} =$
$\dfrac{\cot \alpha \cot \beta - 1}{\cot \beta + \cot \alpha}$

49. $\sec(\alpha + \beta) = \dfrac{1}{\cos(\alpha + \beta)} = \dfrac{1}{\cos \alpha \cos \beta - \sin \alpha \sin \beta} = \dfrac{1}{\dfrac{\cos \alpha \cos \beta - \sin \alpha \sin \beta}{\sin \alpha \sin \beta}} = \dfrac{\dfrac{1}{\sin \alpha} \cdot \dfrac{1}{\sin \beta}}{\dfrac{\cos \alpha \cos \beta}{\sin \alpha \sin \beta} - \dfrac{\sin \alpha \sin \beta}{\sin \alpha \sin \beta}} =$
$\dfrac{\csc \alpha \csc \beta}{\cot \alpha \cot \beta - 1}$

51. $\sin(\alpha - \beta) \sin(\alpha + \beta) = (\sin \alpha \cos \beta - \cos \alpha \sin \beta)(\sin \alpha \cos \beta + \cos \alpha \sin \beta) = \sin^2 \alpha \cos^2 \beta - \cos^2 \alpha \sin^2 \beta$
$= (\sin^2 \alpha)(1 - \sin^2 \beta) - (1 - \sin^2 \alpha)(\sin^2 \beta) = \sin^2 \alpha - \sin^2 \beta$

53. $\sin(\theta + k\pi) = \sin \theta \cos k\pi + \cos \theta \sin k\pi = (\sin \theta)(-1)^k + (\cos \theta)(0) = (-1)^k \cdot \sin \theta$, k any integer

55. $\sqrt{3}/2$ **57.** $-\frac{24}{25}$ **59.** $-\frac{33}{65}$ **61.** $\frac{65}{63}$ **63.** $\frac{1}{39}(48 - 25\sqrt{3})$ **65.** $u\sqrt{1 - v^2} - v\sqrt{1 - u^2}$ **67.** $\dfrac{u\sqrt{1 - v^2} - v}{\sqrt{1 + u^2}}$

69. $\dfrac{uv - \sqrt{1 - u^2}\sqrt{1 - v^2}}{v\sqrt{1 - u^2} + u\sqrt{1 - v^2}}$

71. $\dfrac{\sin(x + h) - \sin x}{h} = \dfrac{\sin x \cos h + \cos x \sin h - \sin x}{h} = \dfrac{\cos x \sin h - (\sin x)(1 - \cos h)}{h} = \cos x \cdot \dfrac{\sin h}{h} - \sin x \cdot \dfrac{1 - \cos h}{h}$

73. $\sin(\sin^{-1} u + \cos^{-1} u) = \sin(\sin^{-1} u)\cos(\cos^{-1} u) + \cos(\sin^{-1} u)\sin(\cos^{-1} u) = (u)(u) + \sqrt{1-u^2}\sqrt{1-u^2} = u^2 + 1 - u^2 = 1$

75. $\tan\dfrac{\pi}{2}$ is not defined; $\tan\!\left(\dfrac{\pi}{2} - \theta\right) = \dfrac{\sin\!\left(\dfrac{\pi}{2} - \theta\right)}{\cos\!\left(\dfrac{\pi}{2} - \theta\right)} = \dfrac{\cos\theta}{\sin\theta} = \cot\theta$ **77.** $\tan\theta = \tan(\theta_2 - \theta_1) = \dfrac{\tan\theta_2 - \tan\theta_1}{1 + \tan\theta_1\tan\theta_2} = \dfrac{m_2 - m_1}{1 + m_1 m_2}$

Exercise 3.3

1. (a) $\frac{24}{25}$ **(b)** $\frac{7}{25}$ **(c)** $\sqrt{10}/10$ **(d)** $3\sqrt{10}/10$ **3. (a)** $\frac{24}{25}$ **(b)** $-\frac{7}{25}$ **(c)** $2\sqrt{5}/5$ **(d)** $-\sqrt{5}/5$

5. (a) $-2\sqrt{2}/3$ **(b)** $\frac{1}{3}$ **(c)** $\sqrt{\dfrac{3+\sqrt{6}}{6}}$ **(d)** $\sqrt{\dfrac{3-\sqrt{6}}{6}}$ **7. (a)** $4\sqrt{2}/9$ **(b)** $-\frac{7}{9}$ **(c)** $\sqrt{3}/3$ **(d)** $\sqrt{6}/3$

9. (a) $-\frac{4}{5}$ **(b)** $\frac{3}{5}$ **(c)** $\sqrt{\dfrac{5+2\sqrt{5}}{10}}$ **(d)** $\sqrt{\dfrac{5-2\sqrt{5}}{10}}$ **11. (a)** $-\frac{3}{5}$ **(b)** $-\frac{4}{5}$ **(c)** $\frac{1}{2}\sqrt{\dfrac{10-\sqrt{10}}{5}}$ **(d)** $-\frac{1}{2}\sqrt{\dfrac{10+\sqrt{10}}{5}}$ **13.** $\dfrac{\sqrt{2-\sqrt{2}}}{2}$

15. $1 - \sqrt{2}$ **17.** $-\dfrac{\sqrt{2+\sqrt{3}}}{2}$ **19.** $\dfrac{2}{\sqrt{2}+\sqrt{2}} = (2-\sqrt{2})\sqrt{2+\sqrt{2}}$ **21.** $-\dfrac{\sqrt{2-\sqrt{2}}}{2}$

23. $\sin^4\theta = (\sin^2\theta)^2 = \left(\dfrac{1-\cos 2\theta}{2}\right)^2 = \dfrac{1}{4}(1 - 2\cos 2\theta + \cos^2 2\theta) = \dfrac{1}{4} - \dfrac{1}{2}\cos 2\theta + \dfrac{1}{4}\cos^2 2\theta = \dfrac{1}{4} - \dfrac{1}{2}\cos 2\theta + \dfrac{1}{4}\left(\dfrac{1+\cos 4\theta}{2}\right) =$

$\dfrac{1}{4} - \dfrac{1}{2}\cos 2\theta + \dfrac{1}{8} + \dfrac{1}{8}\cos 4\theta = \dfrac{3}{8} - \dfrac{1}{2}\cos 2\theta + \dfrac{1}{8}\cos 4\theta$

25. $\sin 4\theta = \sin 2(2\theta) = 2\sin 2\theta\cos 2\theta = (4\sin\theta\cos\theta)(1 - 2\sin^2\theta) = 4\sin\theta\cos\theta - 8\sin^3\theta\cos\theta =$

$(\cos\theta)(4\sin\theta - 8\sin^3\theta)$

27. $16\sin^5\theta - 20\sin^3\theta + 5\sin\theta$ **27.** $\dfrac{1-\cos\theta}{\sin\theta} = \dfrac{1-\cos 2(\theta/2)}{\sin 2(\theta/2)} = \dfrac{1 - [1 - 2\sin^2(\theta/2)]}{2\sin(\theta/2)\cos(\theta/2)} = \dfrac{2\sin^2(\theta/2)}{2\sin(\theta/2)\cos(\theta/2)} = \tan\dfrac{\theta}{2}$

29. $\cos^4\theta - \sin^4\theta = (\cos^2\theta + \sin^2\theta)(\cos^2\theta - \sin^2\theta) = \cos 2\theta$

31. $\cot 2\theta = \dfrac{1}{\tan 2\theta} = \dfrac{1}{\dfrac{2\tan\theta}{1-\tan^2\theta}} = \dfrac{1-\tan^2\theta}{2\tan\theta} = \dfrac{1 - \dfrac{1}{\cot^2\theta}}{2\left(\dfrac{1}{\cot\theta}\right)} = \dfrac{\dfrac{\cot^2\theta - 1}{\cot^2\theta}}{\dfrac{2}{\cot\theta}} = \dfrac{\cot^2\theta - 1}{\cot^2\theta} \cdot \dfrac{\cot\theta}{2} = \dfrac{\cot^2\theta - 1}{2\cot\theta}$

33. $\sec 2\theta = \dfrac{1}{\cos 2\theta} = \dfrac{1}{2\cos^2\theta - 1} = \dfrac{1}{\dfrac{2}{\sec^2\theta} - 1} = \dfrac{1}{\dfrac{2-\sec^2\theta}{\sec^2\theta}} = \dfrac{\sec^2\theta}{2 - \sec^2\theta}$ **35.** $\cos^2 2\theta - \sin^2 2\theta = \cos 2(2\theta) = \cos 4\theta$

37. $\dfrac{\cos 2\theta}{1 + \sin 2\theta} = \dfrac{\cos^2\theta - \sin^2\theta}{1 + 2\sin\theta\cos\theta} = \dfrac{(\cos\theta - \sin\theta)(\cos\theta + \sin\theta)}{\sin^2\theta + \cos^2\theta + 2\sin\theta\cos\theta} = \dfrac{(\cos\theta - \sin\theta)(\cos\theta + \sin\theta)}{(\sin\theta + \cos\theta)(\sin\theta + \cos\theta)} = \dfrac{\cos\theta - \sin\theta}{\cos\theta + \sin\theta}$

$= \dfrac{\dfrac{\cos\theta - \sin\theta}{\sin\theta}}{\dfrac{\cos\theta + \sin\theta}{\sin\theta}} = \dfrac{\dfrac{\cos\theta}{\sin\theta} - \dfrac{\sin\theta}{\sin\theta}}{\dfrac{\cos\theta}{\sin\theta} + \dfrac{\sin\theta}{\sin\theta}} = \dfrac{\cot\theta - 1}{\cot\theta + 1}$

39. $\sec^2\dfrac{\theta}{2} = \dfrac{1}{\cos^2(\theta/2)} = \dfrac{1}{\dfrac{1+\cos\theta}{2}} = \dfrac{2}{1 + \cos\theta}$

41. $\cot^2\dfrac{\theta}{2} = \dfrac{1}{\tan^2(\theta/2)} = \dfrac{1}{\dfrac{1-\cos\theta}{1+\cos\theta}} = \dfrac{1+\cos\theta}{1-\cos\theta} = \dfrac{1 + \dfrac{1}{\sec\theta}}{1 - \dfrac{1}{\sec\theta}} = \dfrac{\dfrac{\sec\theta + 1}{\sec\theta}}{\dfrac{\sec\theta - 1}{\sec\theta}} = \dfrac{\sec\theta + 1}{\sec\theta} \cdot \dfrac{\sec\theta}{\sec\theta - 1} = \dfrac{\sec\theta + 1}{\sec\theta - 1}$

43. $\dfrac{1 - \tan^2(\theta/2)}{1 + \tan^2(\theta/2)} = \dfrac{1 - \dfrac{1-\cos\theta}{1+\cos\theta}}{1 + \dfrac{1-\cos\theta}{1+\cos\theta}} = \dfrac{\dfrac{1 + \cos\theta - (1 - \cos\theta)}{1+\cos\theta}}{\dfrac{1 + \cos\theta + 1 - \cos\theta}{1+\cos\theta}} = \dfrac{2\cos\theta}{1+\cos\theta} \cdot \dfrac{1+\cos\theta}{2} = \cos\theta$

45. $\dfrac{\sin 3\theta}{\sin\theta} - \dfrac{\cos 3\theta}{\cos\theta} = \dfrac{\sin 3\theta\cos\theta - \cos 3\theta\sin\theta}{\sin\theta\cos\theta} = \dfrac{\sin(3\theta - \theta)}{\frac{1}{2}(2\sin\theta\cos\theta)} = \dfrac{2\sin 2\theta}{\sin 2\theta} = 2$

47. $\tan 3\theta = \tan(\theta + 2\theta) = \dfrac{\tan\theta + \tan 2\theta}{1 - \tan\theta\tan 2\theta} = \dfrac{\tan\theta + \dfrac{2\tan\theta}{1-\tan^2\theta}}{1 - \dfrac{\tan\theta(2\tan\theta)}{1-\tan^2\theta}} = \dfrac{\tan\theta - \tan^3\theta + 2\tan\theta}{1 - \tan^2\theta - 2\tan^2\theta} = \dfrac{3\tan\theta - \tan^3\theta}{1 - 3\tan^2\theta}$

49. $\sqrt{3}/2$ **51.** $\frac{7}{25}$ **53.** $\frac{24}{7}$ **55.** $\frac{24}{25}$ **57.** $\frac{1}{5}$ **59.** $\frac{25}{7}$ **61.** 4 **63.** $-\frac{1}{4}$ **65.** $A = \frac{1}{2}h\,(\text{base}) + \frac{1}{2}h\,(\text{base}) = S\cos\dfrac{\theta}{2}\cdot S\sin\dfrac{\theta}{2} = S^2\dfrac{\sin\theta}{2} = \frac{1}{2}S^2\sin\theta$

67.
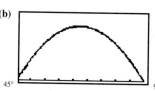
69. $\sin\dfrac{\pi}{24} = \dfrac{\sqrt{2}}{4}\sqrt{4 - \sqrt{6} - \sqrt{2}};\ \cos\dfrac{\pi}{24} = \dfrac{\sqrt{2}}{4}\sqrt{4 + \sqrt{6} + \sqrt{2}}$

71. $\sin^3\theta + \sin^3(\theta + 120°) + \sin^3(\theta + 240°) = \sin^3\theta + (\sin\theta\cos 120° + \cos\theta\sin 120°)^3 + (\sin\theta\cos 240° + \cos\theta\sin 240°)^3$

$= \sin^3\theta + \left(-\dfrac{1}{2}\sin\theta + \dfrac{\sqrt{3}}{2}\cos\theta\right)^3 + \left(-\dfrac{1}{2}\sin\theta - \dfrac{\sqrt{3}}{2}\cos\theta\right)^3$

$= \sin^3\theta + \frac{1}{8}(3\sqrt{3}\cos^3\theta - 9\cos^2\theta\sin\theta + 3\sqrt{3}\cos\theta\sin^2\theta - \sin^3\theta) - \frac{1}{8}(\sin^3\theta + 3\sqrt{3}\sin^2\theta\cos\theta + 9\sin\theta\cos^2\theta + 3\sqrt{3}\cos^3\theta)$

$= \frac{3}{4}\sin^3\theta - \frac{9}{4}\cos^2\theta\sin\theta = \frac{3}{4}[\sin^3\theta - 3\sin\theta(1 - \sin^2\theta)] = \frac{3}{4}(4\sin^3\theta - 3\sin\theta) = -\frac{3}{4}\sin 3\theta$

73. $\dfrac{1}{2}(\ln|1 - \cos 2\theta| - \ln 2) = \ln\left(\dfrac{|1 - \cos 2\theta|}{2}\right)^{1/2} = \ln|\sin^2\theta|^{1/2} = \ln|\sin\theta|$

75. **(a)** $R = \dfrac{v_0^2\sqrt{2}}{16}(\sin\theta\cos\theta - \cos^2\theta) = \dfrac{v_0^2\sqrt{2}}{16}\left(\dfrac{1}{2}\sin 2\theta - \dfrac{1 + \cos 2\theta}{2}\right)$ **(b)**   **(c)** $\theta = 67.5°$ makes R largest

$= \dfrac{v_0^2\sqrt{2}}{32}(\sin 2\theta - \cos 2\theta - 1)$

Exercise 3.4

1. $\frac{1}{2}(\cos 2\theta - \cos 6\theta)$ **3.** $\frac{1}{2}(\sin 6\theta + \sin 2\theta)$ **5.** $\frac{1}{2}(\cos 8\theta + \cos 2\theta)$ **7.** $\frac{1}{2}(\cos\theta - \cos 3\theta)$ **9.** $\frac{1}{2}(\sin 2\theta + \sin\theta)$ **11.** $2\sin\theta\cos 3\theta$

13. $2\cos 3\theta\cos\theta$ **15.** $2\sin 2\theta\cos\theta$ **17.** $2\sin\theta\sin(\theta/2)$ **19.** $\dfrac{\sin\theta + \sin 3\theta}{2\sin 2\theta} = \dfrac{2\sin 2\theta\cos(-\theta)}{2\sin 2\theta} = \cos(-\theta) = \cos\theta$

21. $\dfrac{\sin 4\theta + \sin 2\theta}{\cos 4\theta + \cos 2\theta} = \dfrac{2\sin 3\theta\cos\theta}{2\cos 3\theta\cos\theta} = \dfrac{\sin 3\theta}{\cos 3\theta} = \tan 3\theta$ **23.** $\dfrac{\cos\theta - \cos 3\theta}{\sin\theta + \sin 3\theta} = \dfrac{-2\sin 2\theta\sin(-\theta)}{2\sin 2\theta\cos(-\theta)} = \dfrac{\sin\theta}{\cos\theta} = \tan\theta$

25. $\sin\theta(\sin\theta + \sin 3\theta) = \sin\theta[2\sin 2\theta\cos(-\theta)] = 2\sin 2\theta\sin\theta\cos\theta = \cos\theta(2\sin 2\theta\sin\theta) = \cos\theta[2\cdot\frac{1}{2}(\cos\theta - \cos 3\theta)]$

$= \cos\theta(\cos\theta - \cos 3\theta)$

27. $\dfrac{\sin 4\theta + \sin 8\theta}{\cos 4\theta + \cos 8\theta} = \dfrac{2\sin 6\theta\cos(-2\theta)}{2\cos 6\theta\cos(-2\theta)} = \dfrac{\sin 6\theta}{\cos 6\theta} = \tan 6\theta$

29. $\dfrac{\sin 4\theta + \sin 8\theta}{\sin 4\theta - \sin 8\theta} = \dfrac{2\sin 6\theta\cos(-2\theta)}{2\sin(-2\theta)\cos 6\theta} = \dfrac{\sin 6\theta}{\cos 6\theta}\cdot\dfrac{\cos 2\theta}{-\sin 2\theta} = \tan 6\theta(-\cot 2\theta) = -\dfrac{\tan 6\theta}{\tan 2\theta}$

31. $\dfrac{\sin\alpha + \sin\beta}{\sin\alpha - \sin\beta} = \dfrac{2\sin\dfrac{\alpha + \beta}{2}\cos\dfrac{\alpha - \beta}{2}}{2\sin\dfrac{\alpha - \beta}{2}\cos\dfrac{\alpha + \beta}{2}} = \dfrac{\sin\dfrac{\alpha + \beta}{2}}{\cos\dfrac{\alpha + \beta}{2}}\cdot\dfrac{\cos\dfrac{\alpha - \beta}{2}}{\sin\dfrac{\alpha - \beta}{2}} = \tan\dfrac{\alpha + \beta}{2}\cot\dfrac{\alpha - \beta}{2}$

33. $\dfrac{\sin\alpha + \sin\beta}{\cos\alpha + \cos\beta} = \dfrac{2\sin\dfrac{\alpha + \beta}{2}\cos\dfrac{\alpha - \beta}{2}}{2\cos\dfrac{\alpha + \beta}{2}\cos\dfrac{\alpha - \beta}{2}} = \dfrac{\sin\dfrac{\alpha + \beta}{2}}{\cos\dfrac{\alpha + \beta}{2}} = \tan\dfrac{\alpha + \beta}{2}$

35. $1 + \cos 2\theta + \cos 4\theta + \cos 6\theta = (1 + \cos 6\theta) + (\cos 2\theta + \cos 4\theta) = 2\cos^2 3\theta + 2\cos 3\theta\cos(-\theta) = 2\cos 3\theta(\cos 3\theta + \cos\theta)$

$= 2\cos 3\theta(2\cos 2\theta\cos\theta) = 4\cos\theta\cos 2\theta\cos 3\theta$

37. **(a)** $y = 2\sin 2061\pi t\cos 357\pi t$

(b)
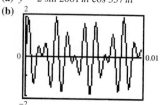

(c) $y_{\max} = 2$

39. $\sin 2\alpha + \sin 2\beta + \sin 2\gamma = 2\sin(\alpha + \beta)\cos(\alpha - \beta) + \sin 2\gamma = 2\sin(\alpha + \beta)\cos(\alpha - \beta) + 2\sin\gamma\cos\gamma$
$$= 2\sin(\pi - \gamma)\cos(\alpha - \beta) + 2\sin\gamma\cos\gamma = 2\sin\gamma\cos(\alpha - \beta) + 2\sin\gamma\cos\gamma = 2\sin\gamma[\cos(\alpha - \beta) + \cos\gamma]$$
$$= 2\sin\gamma\left(2\cos\frac{\alpha - \beta + \gamma}{2}\cos\frac{\alpha - \beta - \gamma}{2}\right) = 4\sin\gamma\cos\frac{\pi - 2\beta}{2}\cos\frac{2\alpha - \pi}{2} = 4\sin\gamma\cos\left(\frac{\pi}{2} - \beta\right)\cos\left(\alpha - \frac{\pi}{2}\right)$$
$$= 4\sin\gamma\sin\beta\sin\alpha$$

41.
$$\sin(\alpha - \beta) = \sin\alpha\cos\beta - \cos\alpha\sin\beta$$
$$\sin(\alpha + \beta) = \sin\alpha\cos\beta + \cos\alpha\sin\beta$$
$$\sin(\alpha - \beta) + \sin(\alpha + \beta) = 2\sin\alpha\cos\beta$$
$$\sin\alpha\cos\beta = \tfrac{1}{2}[\sin(\alpha + \beta) + \sin(\alpha - \beta)]$$

43. $2\cos\dfrac{\alpha + \beta}{2}\cos\dfrac{\alpha - \beta}{2} = 2 \cdot \dfrac{1}{2}\left[\cos\left(\dfrac{\alpha + \beta}{2} + \dfrac{\alpha - \beta}{2}\right) + \cos\left(\dfrac{\alpha + \beta}{2} - \dfrac{\alpha - \beta}{2}\right)\right] = \cos\dfrac{2\alpha}{2} + \cos\dfrac{2\beta}{2} = \cos\alpha + \cos\beta$

Exercise 3.5

1. $\pi/6, 5\pi/6$ **3.** $5\pi/6, 11\pi/6$ **5.** $\pi/2, 3\pi/2$ **7.** $\pi/2, 7\pi/6, 11\pi/6$ **9.** $3\pi/4, 7\pi/4$ **11.** $4\pi/9, 8\pi/9, 16\pi/9$ **13.** $0.4115168, \pi - 0.4115168$
15. $1.3734008, \pi + 1.3734008$ **17.** $2.6905658, 2\pi - 2.6905658$ **19.** $1.8234766, 2\pi - 1.8234766$ **21.** $\pi/2, 2\pi/3, 4\pi/3, 3\pi/2$
23. $\pi/2, 7\pi/6, 11\pi/6$ **25.** $0, \pi/4, 5\pi/4$ **27.** $\pi/4, 5\pi/4$ **29.** $0, \pi/3, \pi, 5\pi/3$ **31.** $\pi/2, 3\pi/2$ **33.** $0, 2\pi/3, 4\pi/3$
35. $0, \pi/3, 2\pi/3, \pi/2, 4\pi/3, \pi, 5\pi/3, 3\pi/2$ **37.** $0, \pi/5, 2\pi/5, 3\pi/5, 4\pi/5, \pi, 6\pi/5, 7\pi/5, 8\pi/5, 9\pi/5$ **39.** $\pi/6, 5\pi/6, 3\pi/2$ **41.** $\pi/3, 5\pi/3$
43. No real solutions   **45.** No real solutions   **47.** $\pi/2, 7\pi/6$ **49.** $0, \pi/3, \pi, 5\pi/3$

51. $0, -1.29$
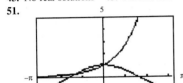

53. $-2.24, 0, 2.24$
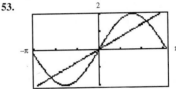

55. $-0.82, 0.82$
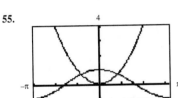

57. $-1.30, 1.97, 3.83$ **59.** 0.52 **61.** 1.25 **63.** $-1.02, 1.02$ **65.** $0, 2.14$ **67.** $0.76, 1.34$ **69.** (a) $60°$ (b) $60°$ (c) $A(60°) = 12\sqrt{3}$ sq in
71. $2.02, 4.91$ **73.** (a) $29.99°$ (b)

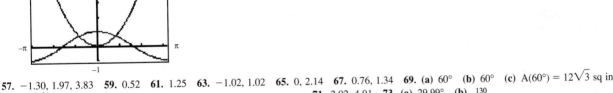

$\theta\text{max} = 60°$
Maximum Area $= 20.78$ sq in

(d) 123.6 meters   **75.** $28.9°$ **77.** Yes; it varies from 1.27 to 1.34   **79.** 1.47
81. If θ is the original angle of incidence and ϕ is the angle of refraction, then $(\sin\theta)(\sin\phi) = n_2$. The angle of incidence of the emerging beam is also ϕ, and the index of refraction is $1/n_2$. Thus, θ is the angle of refraction of the emerging beam.

Fill-in-the-Blank Items

1. identity; conditional   **2.** $-$ **3.** $+$ **4.** $\sin^2\theta$; $2\cos^2\theta$; $2\sin^2\theta$ **5.** $1 - \cos\alpha$

True/False Items

1. T   **2.** F   **3.** T   **4.** F   **5.** F   **6.** F

Review Exercises

1. $\tan\theta\cot\theta - \sin^2\theta = 1 - \sin^2\theta = \cos^2\theta$ **3.** $\cos^2\theta(1 + \tan^2\theta) = \cos^2\theta\sec^2\theta = 1$

5. $4\cos^2\theta + 3\sin^2\theta = \cos^2\theta + 3(\cos^2\theta + \sin^2\theta) = 3 + \cos^2\theta$

7. $\dfrac{1 - \cos\theta}{\sin\theta} + \dfrac{\sin\theta}{1 - \cos\theta} = \dfrac{(1 - \cos\theta)^2 + \sin^2\theta}{\sin\theta(1 - \cos\theta)} = \dfrac{1 - 2\cos\theta + \cos^2\theta + \sin^2\theta}{\sin\theta(1 - \cos\theta)} = \dfrac{2(1 - \cos\theta)}{\sin\theta(1 - \cos\theta)} = 2\csc\theta$

9. $\dfrac{\cos\theta}{\cos\theta - \sin\theta} = \dfrac{\dfrac{\cos\theta}{\cos\theta}}{\dfrac{\cos\theta - \sin\theta}{\cos\theta}} = \dfrac{1}{1 - \dfrac{\sin\theta}{\cos\theta}} = \dfrac{1}{1 - \tan\theta}$

11. $\dfrac{\csc\theta}{1 + \csc\theta} = \dfrac{\dfrac{1}{\sin\theta}}{1 + \dfrac{1}{\sin\theta}} = \dfrac{1}{1 + \sin\theta} = \dfrac{1}{1 + \sin\theta}\cdot\dfrac{1 - \sin\theta}{1 - \sin\theta} = \dfrac{1 - \sin\theta}{1 - \sin^2\theta} = \dfrac{1 - \sin\theta}{\cos^2\theta}$

13. $\csc\theta - \sin\theta = \dfrac{1}{\sin\theta} - \sin\theta = \dfrac{1 - \sin^2\theta}{\sin\theta} = \dfrac{\cos^2\theta}{\sin\theta} = \cos\theta\cdot\dfrac{\cos\theta}{\sin\theta} = \cos\theta\cot\theta$

15. $\dfrac{1 - \sin\theta}{\sec\theta} = \cos\theta(1 - \sin\theta)\cdot\dfrac{1 + \sin\theta}{1 + \sin\theta} = \dfrac{\cos\theta(1 - \sin^2\theta)}{1 + \sin\theta} = \dfrac{\cos^3\theta}{1 + \sin\theta}$

17. $\cot\theta - \tan\theta = \dfrac{\cos\theta}{\sin\theta} - \dfrac{\sin\theta}{\cos\theta} = \dfrac{\cos^2\theta - \sin^2\theta}{\sin\theta\cos\theta} = \dfrac{1 - 2\sin^2\theta}{\sin\theta\cos\theta}$

19. $\dfrac{\cos(\alpha + \beta)}{\cos\alpha\sin\beta} = \dfrac{\cos\alpha\cos\beta - \sin\alpha\sin\beta}{\cos\alpha\sin\beta} = \dfrac{\cos\alpha\cos\beta}{\cos\alpha\sin\beta} - \dfrac{\sin\alpha\sin\beta}{\cos\alpha\sin\beta} = \cot\beta - \tan\alpha$

21. $\dfrac{\cos(\alpha - \beta)}{\cos\alpha\cos\beta} = \dfrac{\cos\alpha\cos\beta + \sin\alpha\sin\beta}{\cos\alpha\cos\beta} = \dfrac{\cos\alpha\cos\beta}{\cos\alpha\cos\beta} + \dfrac{\sin\alpha\sin\beta}{\cos\alpha\cos\beta} = 1 + \tan\alpha\tan\beta$

23. $(1 + \cos\theta)\left(\tan\dfrac{\theta}{2}\right) = \left(2\cos^2\dfrac{\theta}{2}\right)\dfrac{\sin(\theta/2)}{\cos(\theta/2)} = 2\sin\dfrac{\theta}{2}\cos\dfrac{\theta}{2} = \sin\theta$

25. $2\cot\theta\cot 2\theta = 2\left(\dfrac{\cos\theta}{\sin\theta}\right)\left(\dfrac{\cos 2\theta}{\sin 2\theta}\right) = \dfrac{2\cos\theta(\cos^2\theta - \sin^2\theta)}{2\sin^2\theta\cos\theta} = \dfrac{\cos^2\theta - \sin^2\theta}{\sin^2\theta} = \cot^2\theta - 1$

27. $1 - 8\sin^2\theta\cos^2\theta = 1 - 2(2\sin\theta\cos\theta)^2 = 1 - 2\sin^2 2\theta = \cos 4\theta$ **29.** $\dfrac{\sin 2\theta + \sin 4\theta}{\cos 2\theta + \cos 4\theta} = \dfrac{2\sin 3\theta\cos(-\theta)}{2\cos 3\theta\cos(-\theta)} = \tan 3\theta$

31. $\dfrac{\cos 2\theta - \cos 4\theta}{\cos 2\theta + \cos 4\theta} - \tan\theta\tan 3\theta = \dfrac{-2\sin 3\theta\sin(-\theta)}{2\cos 3\theta\cos(-\theta)} - \tan\theta\tan 3\theta = \tan 3\theta\tan\theta - \tan\theta\tan 3\theta = 0$ **33.** $\frac{1}{4}(\sqrt{6} - \sqrt{2})$ **35.** $\frac{1}{4}(\sqrt{6} - \sqrt{2})$

37. $\frac{1}{2}$ **39.** $\sqrt{\dfrac{2 - \sqrt{2}}{2 + \sqrt{2}}} = \sqrt{2} - 1$ **41.** (a) $-\frac{33}{65}$ (b) $-\frac{56}{65}$ (c) $-\frac{63}{65}$ (d) $\frac{33}{56}$ (e) $\frac{24}{25}$ (f) $\frac{119}{169}$ (g) $5\sqrt{26}/26$ (h) $2\sqrt{5}/5$

43. (a) $-\frac{16}{65}$ (b) $-\frac{63}{65}$ (c) $-\frac{56}{65}$ (d) $\frac{16}{63}$ (e) $\frac{24}{25}$ (f) $\frac{119}{169}$ (g) $\sqrt{26}/26$ (h) $-\sqrt{10}/10$

45. (a) $-\frac{63}{65}$ (b) $\frac{16}{65}$ (c) $\frac{33}{65}$ (d) $-\frac{63}{16}$ (e) $\frac{24}{25}$ (f) $-\frac{119}{169}$ (g) $2\sqrt{13}/13$ (h) $-\sqrt{10}/10$

47. (a) $(-\sqrt{3} - 2\sqrt{2})/6$ (b) $(1 - 2\sqrt{6})/6$ (c) $(-\sqrt{3} + 2\sqrt{2})/6$ (d) $(-\sqrt{3} - 2\sqrt{2})/(1 - 2\sqrt{6}) = (8\sqrt{2} + 9\sqrt{3})/23$ (e) $-\sqrt{3}/2$
 (f) $-\frac{7}{9}$ (g) $\sqrt{3}/3$ (h) $\sqrt{3}/2$ **49.** (a) 1 (b) 0 (c) $-\frac{1}{9}$ (d) Not defined (e) $4\sqrt{5}/9$ (f) $-\frac{1}{9}$ (g) $\sqrt{30}/6$ (h) $-\sqrt{6}\sqrt{3 - \sqrt{5}}/6$

51. $\pi/3, 5\pi/3$ **53.** $3\pi/4, 5\pi/4$ **55.** $3\pi/4, 7\pi/4$ **57.** $0, \pi/2, \pi, 3\pi/2$ **59.** $1.1197695, \pi - 1.1197695$ **61.** $0, \pi$ **63.** $0, 2\pi/3, \pi, 4\pi/3$
65. $0, \pi/6, 5\pi/6$ **67.** $\pi/6, \pi/2, 5\pi/6$ **69.** $\pi/2, \pi$ **71.** 1.11 **73.** 0.86 **75.** 2.21

CHAPTER 4 Exercise 4.1

1. $a \approx 13.74, c \approx 14.62, \alpha = 70°$ **3.** $b \approx 5.03, c \approx 7.83, \alpha = 50°$ **5.** $a \approx 0.705, c \approx 4.06, \beta = 80°$ **7.** $b \approx 10.72, c \approx 11.83, \beta = 65°$
9. $b \approx 3.08, a \approx 8.46, \alpha = 70°$ **11.** $c \approx 5.83, \alpha \approx 59.0°, \beta \approx 31.0°$ **13.** $b \approx 4.58, \alpha \approx 23.6°, \beta \approx 66.4°$ **15.** 1.72 in., 2.46 in.
17. 6.10 in. or 8.72 in. **19.** 23.6° and 66.4° **21.** 70 ft **23.** 985.9 ft **25.** 137 m **27.** 20.67 ft **29.** 449.36 ft **31.** 80.5° **33.** 30 ft
35. 530 ft **37.** 555 ft **39.** (a) 112 ft/sec or 76.3 mph (b) 82.4 ft/sec or 56.2 mph (c) under 18.8° **41.** (a) 130° (b) 103.4° **43.** 14.9°
45. (a) 3.1 mi (b) 3.2 mi (c) 3.8 mi

47. (a) $\cos\dfrac{\theta}{2} = \dfrac{3960}{3960 + h}$ (b) $d = 3960\,\theta$ (c) $\cos\dfrac{d}{7920} = \dfrac{3960}{3960 + h}$ (d) 206 mi (e) 2990 miles

Exercise 4.2

1. $a = 3.23, b = 3.55, \alpha = 40°$ **3.** $a = 3.25, c = 4.23, \beta = 45°$ **5.** $\gamma = 95°, c = 9.86, a = 6.36$ **7.** $\alpha = 40°, a = 2, c = 3.06$
9. $\gamma = 120°, b = 1.06, c = 2.69$ **11.** $\alpha = 100°, a = 5.24, c = 0.92$ **13.** $\beta = 40°, a = 5.64, b = 3.86$ **15.** $\gamma = 100°, a = 1.31, b = 1.31$
17. One triangle; $\beta = 30.7°, \gamma = 99.3°, c = 3.86$ **19.** One triangle: $\gamma = 36.2°, \alpha = 43.8°, a = 3.51$ **21.** No triangle
23. Two triangles; $\gamma_1 = 30.9°, \alpha_1 = 129.1°, a_1 = 9.08$ or $\gamma_2 = 149.1°, \alpha_2 = 10.9°, a_2 = 2.21$ **25.** No triangle

27. Two triangles; $\alpha_1 = 57.7°$, $\beta_1 = 97.3°$, $b_1 = 2.35$ or $\alpha_2 = 122.3°$, $\beta_2 = 32.7°$, $b_2 = 1.28$

29. (a) Station Able is 143.3 mi from the ship; Station Baker is 135.6 mi from the ship. **(b)** Approx. 41 min

31. 1490.5 ft **33.** 381.7 ft **35. (a)** 169 mi **(b)** 161.3° **37.** 84.7°; 183.7 ft **39.** 2.64 mi **41.** 1.88 mi

43. $\dfrac{a+b}{c} = \dfrac{a}{c} + \dfrac{b}{c} = \dfrac{\sin\alpha}{\sin\gamma} + \dfrac{\sin\beta}{\sin\gamma} = \dfrac{\sin\alpha + \sin\beta}{\sin\gamma} = \dfrac{2\sin\dfrac{\alpha+\beta}{2}\cos\dfrac{\alpha-\beta}{2}}{2\sin\dfrac{\gamma}{2}\cos\dfrac{\gamma}{2}} = \dfrac{\sin\left(\dfrac{\pi}{2}-\dfrac{\gamma}{2}\right)\cos\dfrac{\alpha-\beta}{2}}{\sin\dfrac{\gamma}{2}\cos\dfrac{\gamma}{2}} = \dfrac{\cos\frac{1}{2}(\alpha-\beta)}{\sin\frac{1}{2}\gamma}$

45. $a = \dfrac{b\sin\alpha}{\sin\beta} = \dfrac{b\sin[180° - (\beta+\gamma)]}{\sin\beta} = \dfrac{b}{\sin\beta}(\sin\beta\cos\gamma + \cos\beta\sin\gamma) = b\cos\gamma + \dfrac{b\sin\gamma}{\sin\beta}\cos\beta = b\cos\gamma + c\cos\beta$

47. $\sin\beta = \sin(\text{Angle } AB'C) = b/2r$; $\dfrac{\sin\beta}{b} = \dfrac{1}{2r}$; the result follows using the Law of Sines

Exercise 4.3

1. $b = 2.95$, $\alpha = 28.7°$, $\gamma = 106.3°$ **3.** $c = 3.75$, $\alpha = 32.1°$, $\beta = 52.9°$ **5.** $\alpha = 48.5°$, $\beta = 38.6°$, $\gamma = 92.9°$
7. $\alpha = 127.2°$, $\beta = 32.1°$, $\gamma = 20.7°$ **9.** $c = 2.57$, $\alpha = 48.6°$, $\beta = 91.4°$ **11.** $a = 2.99$, $\beta = 19.2°$, $\gamma = 80.8°$
13. $b = 4.14$, $\alpha = 43.0°$, $\gamma = 27.0°$ **15.** $c = 1.69$, $\alpha = 65.0°$, $\beta = 65.0°$ **17.** $\alpha = 67.4°$, $\beta = 90°$, $\gamma = 22.6°$
19. $\alpha = 60°$, $\beta = 60°$, $\gamma = 60°$ **21.** $\alpha = 33.6°$, $\beta = 62.2°$, $\gamma = 84.3°$ **23.** $\alpha = 97.9°$, $\beta = 52.4°$, $\gamma = 29.7°$ **25.** 70.75 ft
27. (a) 12° **(b)** 220.8 mph **29. (a)** 63.7 ft **(b)** 66.8 ft **(c)** 92.5° **31. (a)** 492.6 ft **(b)** 269.3 ft **33.** 342.3 ft
35. Using the Law of Cosines:

$$L^2 = x^2 + r^2 - 2rx\cos\theta$$
$$x^2 - 2rx\cos\theta + r^2 - L^2 = 0$$
$$x = r\cos\theta + \sqrt{r^2\cos^2\theta + L^2 - r^2}$$

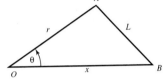

37. $\cos\dfrac{\gamma}{2} = \sqrt{\dfrac{1+\cos\gamma}{2}} = \sqrt{\dfrac{1 + \dfrac{a^2+b^2-c^2}{2ab}}{2}} = \sqrt{\dfrac{2ab + a^2 + b^2 - c^2}{4ab}} = \sqrt{\dfrac{(a+b)^2 - c^2}{4ab}} = \sqrt{\dfrac{(a+b+c)(a+b-c)}{4ab}}$

$= \sqrt{\dfrac{2s(2s-2c)}{4ab}} = \sqrt{\dfrac{s(s-c)}{ab}}$

39. $\dfrac{\cos\alpha}{a} + \dfrac{\cos\beta}{b} + \dfrac{\cos\gamma}{c} = \dfrac{b^2+c^2-a^2}{2abc} + \dfrac{a^2+c^2-b^2}{2abc} + \dfrac{a^2+b^2-c^2}{2abc} = \dfrac{b^2+c^2-a^2+a^2+c^2-b^2+a^2+b^2-c^2}{2abc} = \dfrac{a^2+b^2+c^2}{2abc}$

Exercise 4.4

1. 2.83 **3.** 2.99 **5.** 14.98 **7.** 9.56 **9.** 3.86 **11.** 1.48 **13.** 2.82 **15.** 1.53 **17.** 30 **19.** 1.73 **21.** 19.90 **23.** 19.81 **25.** $5446.38
27. $A = \dfrac{1}{2}ab\sin\gamma = \dfrac{1}{2}a\sin\gamma\left(\dfrac{a\sin\beta}{\sin\alpha}\right) = \dfrac{a^2\sin\beta\sin\gamma}{2\sin\alpha}$ **29.** 0.92 **31.** 2.27 **33.** 5.44 **35.** 0.84 **37.** $A = \frac{1}{2}r^2(\theta - \sin\theta)$

Exercise 4.5

1. Amplitude $= 2$; Period $= 2\pi$ **3.** Amplitude $= 4$; Period $= \pi$ **5.** Amplitude $= 6$; Period $= 2$ **7.** Amplitude $= \frac{1}{2}$; Period $= 4\pi/3$
9. Amplitude $= \frac{5}{3}$; Period $= 3$ **11.** F **13.** A **15.** H **17.** C **19.** J **21.** A **23.** D **25.** B
27.

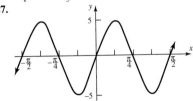

29.

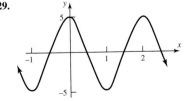

31.

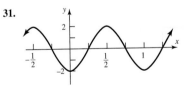

33.

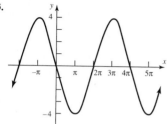

35.

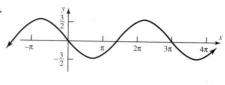

37. $y = 5 \cos \dfrac{\pi}{4}x$ **39.** $y = -3 \cos \frac{1}{2}x$ **41.** $y = \frac{3}{4} \sin 2\pi x$ **43.** $y = -\sin \frac{3}{2}x$ **45.** $y = -2 \cos \dfrac{3\pi}{2}x$ **47.** $y = 3 \sin \dfrac{\pi}{2}x$ **49.** $y = -4 \cos 3x$

51. Amplitude $= 4$
Period $= \pi$
Phase shift $= \pi/2$

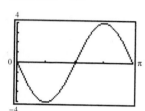

53. Amplitude $= 2$
Period $= 2\pi/3$
Phase shift $= -\pi/6$

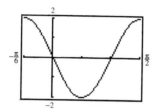

55. Amplitude $= 3$
Period $= \pi$
Phase shift $= -\pi/4$

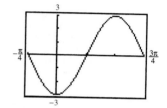

57. Amplitude $= 4$
Period $= 2$
Phase shift $= -2/\pi$

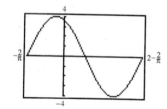

59. Amplitude $= 3$
Period $= 2$
Phase shift $= 2/\pi$

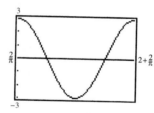

61. Amplitude $= 3$
Period $= \pi$
Phase shift $= \pi/4$

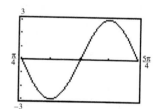

63. $d = 5 \cos \pi t$ **65.** $d = 6 \cos 2t$ **67.** $d = 5 \sin \pi t$ **69.** $d = 6 \sin 2t$
71. (a) Simple harmonic   (b) 5 m   (c) $2\pi/3$ sec   (d) $3/(2\pi)$ oscillation/sec
73. (a) Simple harmonic   (b) 6 m   (c) 2 sec   (d) $\frac{1}{2}$ oscillation/sec   **75.** (a) Simple harmonic   (b) 3 m   (c) 4π sec   (d) $1/(4\pi)$ oscillation/sec
77. (a) Simple harmonic   (b) 2 m   (c) 1 sec   (d) 1 oscillation/sec
79. Period $= \frac{1}{30}$, Amplitude $= 220$ **81.** Period $= \frac{1}{15}$, Amplitude $= 120$, Phase shift $= \frac{1}{90}$

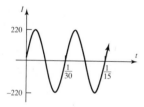

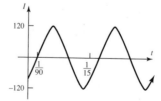

83. (a) Amplitude = 220, period = $\frac{1}{60}$
(b) & (e)

(c) $I = 22 \sin 120\pi t$
(d) Amplitude = 22, Period = $\frac{1}{60}$

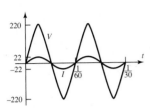

85. (a) $P = \dfrac{(V_0 \sin 2\pi ft)^2}{R} = \dfrac{V_0^2}{R} \sin^2 2\pi ft$ **(b)** $P = \dfrac{V_0^2}{R} \dfrac{1}{2}(1 - \cos 4\pi ft)$

Exercise 4.6

1.
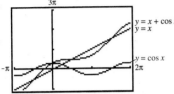

3.
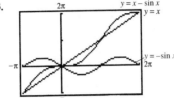

5.
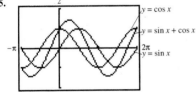

7.
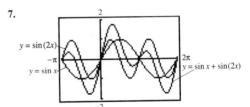

9.
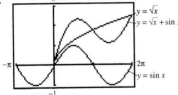

11.
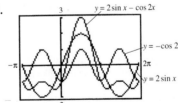

13.
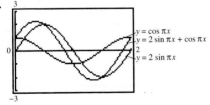

15.
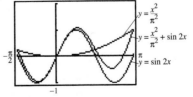

17.
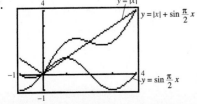

19.
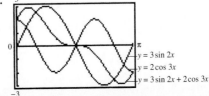

21.

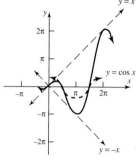

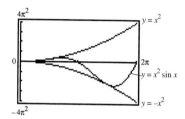

Touches: $(0, 0)$; $(\pi, -\pi)$; $(2\pi, 2\pi)$; $(3\pi, -3\pi)$
Turning Points: $(0.86, 0.56)$, $(3.42, -3.28)$, $(6.43, 6.36)$

23.

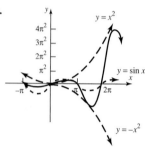

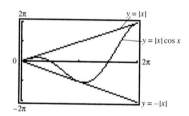

Touches: $(0, 0)$, $\left(\dfrac{\pi}{2}, \dfrac{\pi^2}{4}\right)$, $\left(\dfrac{3\pi}{2}, -\dfrac{9\pi^2}{4}\right)$
Turning Points: $(2.28, 3.94)$, $(5.08, -24.08)$

25.

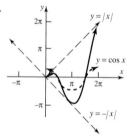

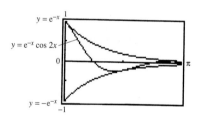

Touches: $(0, 0)$, $(\pi, -\pi)$, $(2\pi, 2\pi)$
Turning Points: $(0.86, 0.56)$, $(3.42, -3.28)$

27.

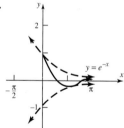

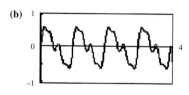

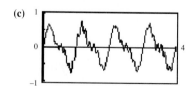

Touches: $(0, 1)$, $(1.57, -0.20)$, $(3.14, 0.04)$
Turning Points: $(1.33, -0.23)$, $(2.90, 0.04)$

29.

31. (a)

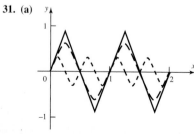

(b)

(c)

33. 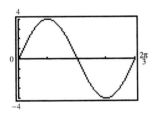 As x approaches 0, $\dfrac{\sin x}{x}$ approaches 1.   **35.** As x increases, the graphs approach the line $y = 0$.

Fill-In-The Blank Items

1. Sines   **2.** Cosines   **3.** Heron's   **4.** $y = 3 \sin \pi x$ **5.** 3; $\pi/3$ **6.** $5x$ **7.** simple harmonic motion

True/False Items

1. F   **2.** T   **3.** T   **4.** F

Review Exercises

1. $\alpha = 70°$, $b \approx 3.42$, $a \approx 9.4$ **3.** $a \approx 4.58$, $\alpha \approx 66.4°$, $\beta \approx 23.6°$ **5.** $\gamma = 100°$, $b = 0.65$, $c = 1.29$ **7.** $\beta = 56.8°$, $\gamma = 23.2°$, $b = 4.25$
9. No triangle   **11.** $b = 3.32$, $\alpha = 62.8°$, $\gamma = 17.2°$ **13.** No triangle   **15.** $c = 2.32$, $\alpha = 16.1°$, $\beta = 123.9°$ **17.** $\beta = 36.2°$, $\gamma = 63.8°$, $c = 4.56$
19. $\alpha = 39.6°$, $\beta = 18.5°$, $\gamma = 121.9°$ **21.** Two triangles: $\beta_1 = 13.4°$, $\gamma_1 = 156.6°$, $c_1 = 6.86$; $\beta_2 = 166.6°$, $\gamma_2 = 3.4°$, $c_2 = 1.02$
23. $a = 5.23$, $\beta = 46°$, $\gamma = 64°$ **25.** 1.93 **27.** 18.79 **29.** 6 **31.** 3.80 **33.** 0.32 **35.** Amplitude $= 4$; Period $= 2\pi$
37. Amplitude $= 8$; Period $= 4$
39. Amplitude $= 4$
Period $= 2\pi/3$
Phase shift $= 0$

41. Amplitude $= 2$
Period $= 4$
Phase shift $= -1/\pi$

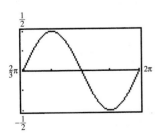

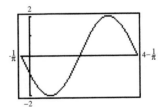

43. Amplitude $= \frac{1}{2}$
Period $= 4\pi/3$
Phase shift $= 2\pi/3$

45. Amplitude $= \frac{2}{3}$
Period $= 2$
Phase shift $= 6/\pi$

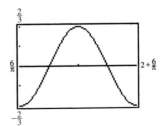

47. $y = 5 \cos \dfrac{x}{4}$ **49.** $y = -6 \cos \dfrac{\pi}{4}x$

51.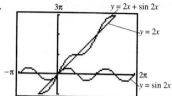

53.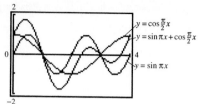

55.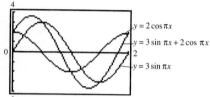

57.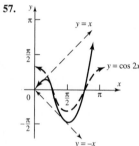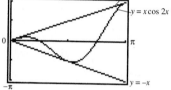

Touches: $(0, 0)$, $\left(\dfrac{\pi}{2}, -\dfrac{\pi}{2}\right)$, (π, π)

Turning Points: $(0.43, 0.28)$, $(1.71, -1.64)$

59.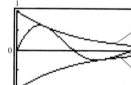

Touches: $(0.50, 0.60)$, $(1.50, -0.22)$

Turning Points: $(0.40, 0.63)$, $(1.40, -0.23)$

61.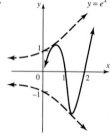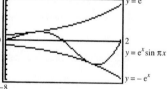

Touches: $(0.50, 1.64)$, $(1.50, -4.48)$

Turning Points: $(0.59, 1.73)$, $(1.59, -4.71)$

63. (a) Simple harmonic   **(b)** 6 ft   **(c)** π sec   **(d)** $1/\pi$ oscillation/sec

65. (a) Simple harmonic   **(b)** 2 ft   **(c)** 2 sec   **(d)** $\frac{1}{2}$ oscillation/sec

67. (a) 120 **(b)** $\frac{1}{60}$ **(c)**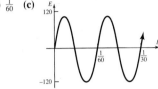

69. 839 ft   **71.** 23.32 ft   **73.** 2.15 mi   **75.** 204.1 mi

77. (a) 2.59 mi   **(b)** 2.92 mi   **(c)** 2.53 mi   **79. (a)** 131.8 mi   **(b)** 23.1°   **(c)** 0.2 hr   **81.** 8799 sq ft.   **83.** 77 in.
85. 26.70 feet of cable contact if a 'figure-eight' is used. 13.35 feet of cable contact if just one wheel is used.

CHAPTER 5   *Exercise 5.1*

1.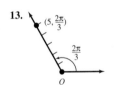

3.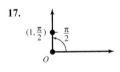

5.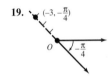

7.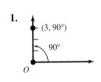

9.

11.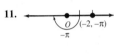

13.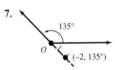

15.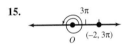

17.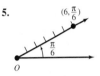

19.

13.
(a) $(5, -4\pi/3)$
(b) $(-5, 5\pi/3)$
(c) $(5, 8\pi/3)$

15.
(a) $(2, -2\pi)$
(b) $(-2, \pi)$
(c) $(2, 2\pi)$

17.
(a) $(1, -3\pi/2)$
(b) $(-1, 3\pi/2)$
(c) $(1, 5\pi/2)$

19.
(a) $(3, -5\pi/4)$
(b) $(-3, 7\pi/4)$
(c) $(3, 11\pi/4)$

21. $(0, 3)$ **23.** $(-2, 0)$ **25.** $(-3\sqrt{3}, 3)$ **27.** $(\sqrt{2}, -\sqrt{2})$ **29.** $(-\frac{1}{2}, \sqrt{3}/2)$ **31.** $(2, 0)$ **33.** $(-2.57, 7.05)$ **35.** $(-4.98, -3.86)$ **37.** $(3, 0)$
39. $(1, \pi)$ **41.** $(\sqrt{2}, -\pi/4)$ **43.** $(2, \pi/6)$ **45.** $(2.47, -1.02)$ **47.** $(9.3, 0.47)$ **49.** $r^2 = \frac{3}{2}$ **51.** $r^2 \cos^2 \theta - 4r \sin \theta = 0$ **53.** $r^2 \sin 2\theta = 1$
55. $r \cos \theta = 4$ **57.** $x^2 + y^2 - x = 0$ or $(x - \frac{1}{2})^2 + y^2 = \frac{1}{4}$ **59.** $(x^2 + y^2)^{3/2} - x = 0$ **61.** $x^2 + y^2 = 4$ **63.** $y^2 = 8(x + 2)$
65. $d = \sqrt{(r_2 \cos \theta_2 - r_1 \cos \theta_1)^2 + (r_2 \sin \theta_2 - r_1 \sin \theta_1)^2} = \sqrt{r_1^2 + r_2^2 - 2r_1 r_2 \cos(\theta_2 - \theta_1)}$

Exercise 5.2

1. Circle, radius 4, center at pole

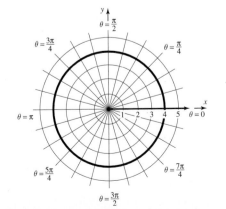

3. Line through pole, making an angle of $\pi/3$ with polar axis

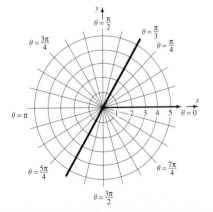

AN39

5. Horizontal line 4 units above the pole

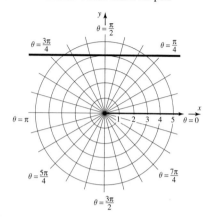

7. Vertical line 2 units to the left of the pole

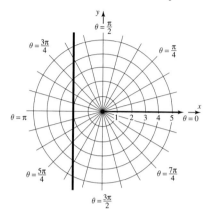

9. Circle, radius 1, center at $(1, 0)$ in rectangular coordinates

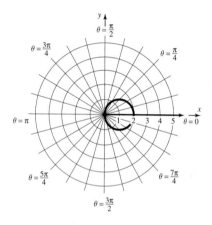

11. Circle, radius 2, center at $(0, -2)$ in rectangular coordinates

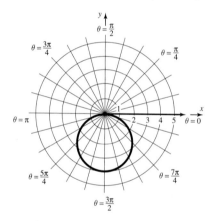

13. Circle, radius 2, center at $(2, 0)$ in rectangular coordinates

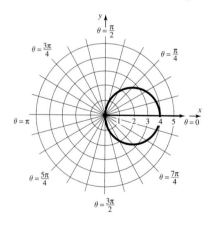

15. Circle, radius 1, center at $(0, -1)$ in rectangular coordinates

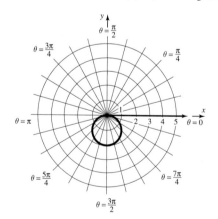

17. E **19.** F **21.** H **23.** D **25.** D **27.** F **29.** A

31. Cardioid

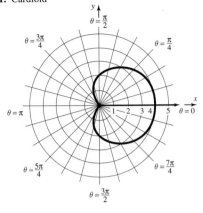

33. Cardioid

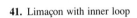

35. Limaçon without inner loop

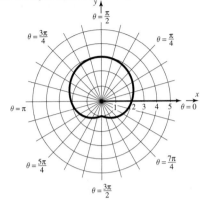

37. Limaçon without inner loop

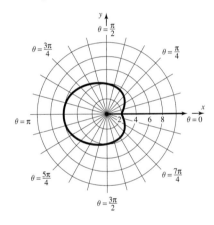

39. Limaçon with inner loop

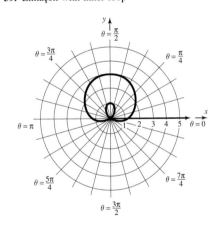

41. Limaçon with inner loop

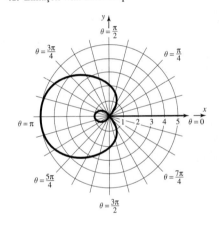

43. Rose

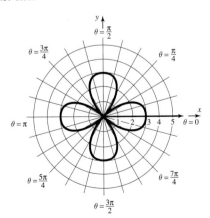

45. Rose

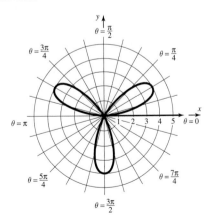

47. Lemniscate

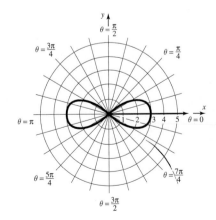

49. Spiral

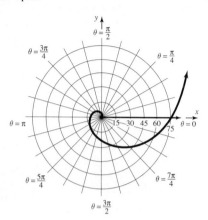

51. Cardioid

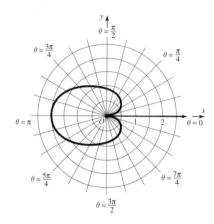

53. Lemniscate with inner loop

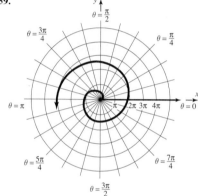

55.

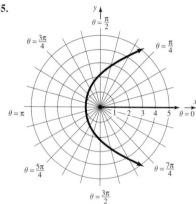

57.

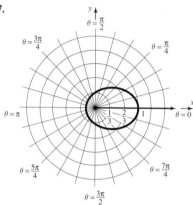

59.

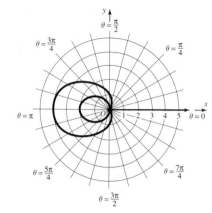

61.

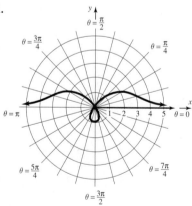

63.

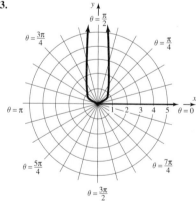

65. $r \sin \theta = a$
$y = a$

67.
$r = 2a \sin \theta$
$r^2 = 2ar \sin \theta$
$x^2 + y^2 = 2ay$
$x^2 + y^2 - 2ay = 0$
$x^2 + (y - a)^2 = a^2$
Circle, radius a, center at $(0, a)$
in rectangular coordinates

69.
$r = 2a \cos \theta$
$r^2 = 2ar \cos \theta$
$x^2 + y^2 = 2ax$
$x^2 - 2ax + y^2 = 0$
$(x - a)^2 + y^2 = a^2$
Circle, radius a, center at $(a, 0)$
in rectangular coordinates

71. (a) $r^2 = \cos \theta$: $r^2 = \cos(\pi - \theta)$ $(-r)^2 = \cos(-\theta)$
$r^2 = -\cos \theta$ $r^2 = \cos \theta$
Not equivalent; test fails   New test works

    (b) $r^2 = \sin \theta$: $r^2 = \sin(\pi - \theta)$ $(-r)^2 = \sin(-\theta)$
$r^2 = \sin \theta$ $r^2 = -\sin \theta$
Test works   New test fails

Exercise 5.3

1.
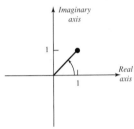
$\sqrt{2}(\cos 45° + i \sin 45°)$

3.
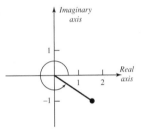
$2(\cos 330° + i \sin 330°)$

5.
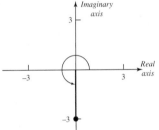
$3(\cos 270° + i \sin 270°)$

7.
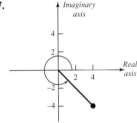
$4\sqrt{2}(\cos 315° + i \sin 315°)$

9.
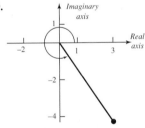
$5(\cos 306.9° + i \sin 306.9°)$

11.
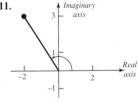
$\sqrt{13}(\cos 123.7° + i \sin 123.7°)$

13. $-1 + \sqrt{3}i$ **15.** $2\sqrt{2} - 2\sqrt{2}i$ **17.** $-3i$ **19.** $-0.035 + 0.197i$ **21.** $1.97 + 0.347i$

23. $zw = 8(\cos 60° + i \sin 60°)$; $z/w = \frac{1}{2}(\cos 20° + i \sin 20°)$ **25.** $zw = 12(\cos 40° + i \sin 40°)$; $z/w = \frac{3}{4}(\cos 220° + i \sin 220°)$

27. $zw = 4\left(\cos \frac{9\pi}{40} + i \sin \frac{9\pi}{40}\right)$; $\frac{z}{w} = \cos \frac{\pi}{40} + i \sin \frac{\pi}{40}$ **29.** $zw = 4\sqrt{2}(\cos 15° + i \sin 15°)$; $z/w = \sqrt{2}(\cos 75° + i \sin 75°)$

31. $32(-1 + \sqrt{3}i)$ **33.** $32i$ **35.** $\frac{27}{2}(1 + \sqrt{3}i)$ **37.** $\frac{25\sqrt{2}}{2}(-1 + i)$ **39.** $4(-1 + i)$ **41.** $-23 + 14.15i$

43. $\sqrt[6]{2}(\cos 15° + i \sin 15°)$, $\sqrt[6]{2}(\cos 135° + i \sin 135°)$, $\sqrt[6]{2}(\cos 255° + i \sin 255°)$

45. $\sqrt[4]{8}(\cos 75° + i \sin 75°)$, $\sqrt[4]{8}(\cos 165° + i \sin 165°)$, $\sqrt[4]{8}(\cos 255° + i \sin 255°)$, $\sqrt[4]{8}(\cos 345° + i \sin 345°)$

47. $2(\cos 67.5° + i \sin 67.5°)$, $2(\cos 157.5° + i \sin 157.5°)$, $2(\cos 247.5° + i \sin 247.5°)$, $2(\cos 337.5° + i \sin 337.5°)$

49. $\cos 18° + i \sin 18°$, $\cos 90° + i \sin 90°$, $\cos 162° + i \sin 162°$, $\cos 234° + i \sin 234°$, $\cos 306° + i \sin 306°$

51. $1, i, -1, -i$ **53.** Look at formula (8); $|z_k| = \sqrt[n]{r}$ for all k.   **55.** Look at formula (8). The z_k are spaced apart by an angle of $2\pi/n$.

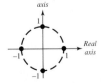

Exercise 5.4

1.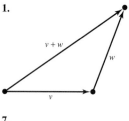

3.

5.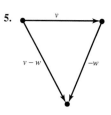

7.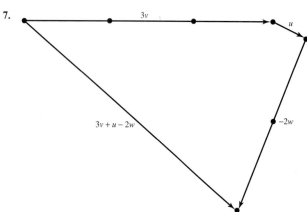

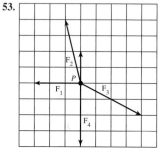

9. $x = A$ **11.** $C = -F + E - D$ **13.** $E = -G - H + D$ **15.** $x = 0$ **17.** 12   **19.** $\mathbf{v} = 3\mathbf{i} + 4\mathbf{j}$ **21.** $\mathbf{v} = 2\mathbf{i} + 4\mathbf{j}$ **23.** $\mathbf{v} = 8\mathbf{i} - \mathbf{j}$

25. $\mathbf{v} = -\mathbf{i} + \mathbf{j}$ **27.** 5   **29.** $\sqrt{2}$ **31.** $\sqrt{13}$ **33.** $-\mathbf{j}$ **35.** $\sqrt{89}$ **37.** $\sqrt{34} - \sqrt{13}$ **39.** $\mathbf{i}$ **41.** $\frac{3}{5}\mathbf{i} - \frac{4}{5}\mathbf{j}$ **43.** $\dfrac{\sqrt{2}}{2}\mathbf{i} - \dfrac{\sqrt{2}}{2}\mathbf{j}$

45. $\mathbf{v} = \dfrac{8\sqrt{5}}{5}\mathbf{i} + \dfrac{4\sqrt{5}}{5}\mathbf{j}$ or $\mathbf{v} = -\dfrac{8\sqrt{5}}{5}\mathbf{i} - \dfrac{4\sqrt{5}}{5}\mathbf{j}$ **47.** $\{-2 + \sqrt{21}, -2 - \sqrt{21}\}$ **49.** 460 kph   **51.** 218 mph

53.

![grid diagram with vectors F_1, F_2, F_3, F_4 from point P]

Exercise 5.5

1. 0; 0   **3.** 4; $\frac{4}{5}$ **5.** $\sqrt{3} - 1$; $(\sqrt{6} - \sqrt{2})/4$ **7.** 24; $\frac{24}{25}$ **9.** 0; 0   **11.** $\frac{3}{2}$ **13.** $\mathbf{v}_1 = \text{proj}_{\mathbf{w}}\,\mathbf{v} = \frac{5}{2}(\mathbf{i} - \mathbf{j})$, $\mathbf{v}_2 = -\frac{1}{2}\mathbf{i} - \frac{1}{2}\mathbf{j}$

15. $\mathbf{v}_1 = \text{proj}_{\mathbf{w}}\,\mathbf{v} = -\frac{1}{5}(\mathbf{i} + 2\mathbf{j})$, $\mathbf{v}_2 = \frac{6}{5}\mathbf{i} - \frac{3}{5}\mathbf{j}$ **17.** $\mathbf{v}_1 = \text{proj}_{\mathbf{w}}\,\mathbf{v} = \frac{7}{5}(2\mathbf{i} + \mathbf{j})$, $\mathbf{v}_2 = \frac{1}{5}\mathbf{i} - \frac{2}{5}\mathbf{j}$ **19.** 496.7 mph; 51.5° south of west

21. 8.7° off direct heading into the current; 1.5 min   **23.** 60°; 17.32 min   **25.** 2.68 ft-lb   **27.** 1732 ft-lb

29. Let $\mathbf{u} = a_1\mathbf{i} + b_1\mathbf{j}$, $\mathbf{v} = a_2\mathbf{i} + b_2\mathbf{j}$, $\mathbf{w} = a_3\mathbf{i} + b_3\mathbf{j}$. Compute $\mathbf{u} \cdot (\mathbf{v} + \mathbf{w})$ and $\mathbf{u} \cdot \mathbf{v} + \mathbf{u} \cdot \mathbf{w}$.

31. $\cos \alpha = \dfrac{\mathbf{v} \cdot \mathbf{i}}{\|\mathbf{v}\|\,\|\mathbf{i}\|} = \mathbf{v} \cdot \mathbf{i}$; if $\mathbf{v} = x\mathbf{i} + y\mathbf{j}$, then $\mathbf{v} \cdot \mathbf{i} = x = \cos \alpha$ and $\mathbf{v} \cdot \mathbf{j} = y = \cos\left(\dfrac{\pi}{2} - \alpha\right) = \sin \alpha$.

33. $\mathbf{v} = a\mathbf{i} + b\mathbf{j}$; $\text{proj}_{\mathbf{i}}\,\mathbf{v} = \dfrac{\mathbf{v} \cdot \mathbf{i}}{\|\mathbf{i}\|^2}\mathbf{i} = (\mathbf{v} \cdot \mathbf{i})\mathbf{i}$; $\mathbf{v} \cdot \mathbf{i} = a$, $\mathbf{v} \cdot \mathbf{j} = b$, so $\mathbf{v} = (\mathbf{v} \cdot \mathbf{i})\mathbf{i} + (\mathbf{v} \cdot \mathbf{j})\mathbf{j}$

35. $(\mathbf{v} - \alpha\mathbf{w}) \cdot \mathbf{w} = \mathbf{v} \cdot \mathbf{w} - \alpha\mathbf{w} \cdot \mathbf{w} = \mathbf{v} \cdot \mathbf{w} - \alpha\|\mathbf{w}\|^2 = \mathbf{v} \cdot \mathbf{w} - \dfrac{\mathbf{v} \cdot \mathbf{w}}{\|\mathbf{w}\|^2}\|\mathbf{w}\|^2 = 0$ **37.** 0

Fill-in-the-Blank Items

1. pole; polar axis   **2.** -2 **3.** $r = 2 \cos \theta$ **4.** the polar axis (x-axis)   **5.** magnitude or modulus; argument   **6.** unit   **7.** 0

True/False Items

1. F   **2.** T   **3.** F   **4.** T   **5.** T   **6.** T   **7.** T

Review Exercises

1. $(3\sqrt{3}/2, \frac{3}{2})$

3. $(1, \sqrt{3})$

5. $(0, 3)$

7. $(3\sqrt{2}, 3\pi/4), (-3\sqrt{2}, -\pi/4)$ **9.** $(2, -\pi/2), (-2, \pi/2)$ **11.** $(5, 0.93), (-5, 4.07)$ **13.** $3r^2 - 6r \sin \theta = 0$ **15.** $r^2(2 - 3 \sin^2 \theta) - \tan \theta = 0$
17. $r^3 \cos \theta = 4$ **19.** $x^2 + y^2 - 2y = 0$ **21.** $x^2 + y^2 = 25$ **23.** $x + 3y = 6$
25. Circle; radius 2, center at   **27.** Cardioid   **29.** Limaçon without inner loop
$\quad\quad$ $(2, 0)$ in rectangular coordinates

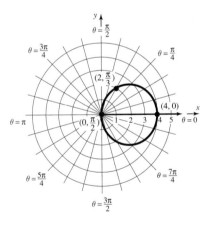

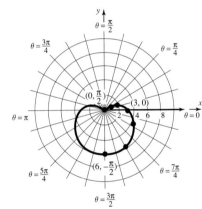

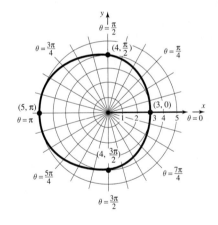

31. $\sqrt{2}(\cos 225° + i \sin 225°)$ **33.** $5(\cos 323.1° + i \sin 323.1°)$ **35.** $-\sqrt{3} + i$ **37.** $-\frac{3}{2} + (3\sqrt{3}/2)i$ **39.** $0.098 - 0.017i$
41. $zw = \cos 130° + i \sin 130°$; $z/w = \cos 30° + i \sin 30°$ **43.** $zw = 6(\cos 0 + i \sin 0)$; $\dfrac{z}{w} = \dfrac{3}{2}\left(\cos \dfrac{8\pi}{5} + i \sin \dfrac{8\pi}{5}\right)$
45. $zw = 5(\cos 5° + i \sin 5°)$; $z/w = 5(\cos 15° + i \sin 15°)$ **47.** $\frac{27}{2}(1 + \sqrt{3}i)$ **49.** $4i$ **51.** 64   **53.** $-527.1 - 335.8i$
55. $3, 3(\cos 120° + i \sin 120°), 3(\cos 240° + i \sin 240°)$ **57.** $\mathbf{v} = 2\mathbf{i} - 4\mathbf{j}$; $\|\mathbf{v}\| = 2\sqrt{5}$ **59.** $\mathbf{v} = -\mathbf{i} + 3\mathbf{j}$; $\|\mathbf{v}\| = \sqrt{10}$ **61.** $-20\mathbf{i} + 13\mathbf{j}$
63. $\sqrt{5}$ **65.** $\sqrt{5} + 5 \approx 7.24$ **67.** $\dfrac{-2\sqrt{5}}{5}\mathbf{i} + \dfrac{\sqrt{5}}{5}\mathbf{j}$ **69.** $\mathbf{v} \cdot \mathbf{w} = -11$; $\cos \theta = -11\sqrt{5}/25$ **71.** $\mathbf{v} \cdot \mathbf{w} = -4$; $\cos \theta = -2\sqrt{5}/5$
73. $\text{proj}_{\mathbf{w}} \mathbf{v} = \frac{9}{10}(3\mathbf{i} + \mathbf{j})$ **75.** 30.5°   **77.** $\sqrt{29} \approx 5.39$ mph; 0.4 mi   **79.** At an angle of 70.5° to the shore

1. B **3.** E **5.** H **7.** C **9.** F **11.** G **13.** D **15.** B

17. $y^2 = 16x$

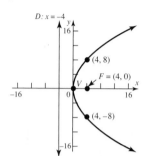

19. $x^2 = -12y$

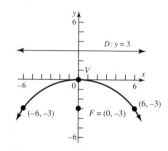

21. $y^2 = -8x$

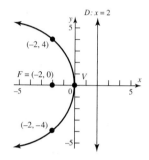

23. $x^2 = 2y$

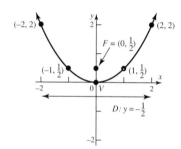

25. $(x - 2)^2 = -8(y + 3)$

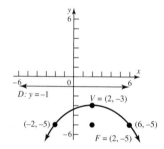

27. $x^2 = \frac{4}{3}y$

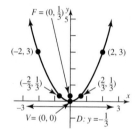

29. $(x + 3)^2 = 4(y - 3)$

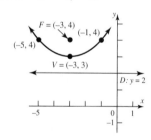

31. $(y + 2)^2 = -8(x + 1)$

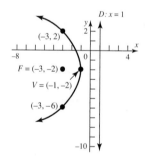

33. Vertex: (0, 0); Focus: (0, 1);
Directrix: $y = -1$

35. Vertex: (0, 0); Focus: (−4, 0);
Directrix: $x = 4$

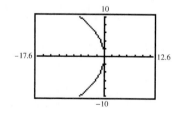

37. Vertex: (−1, 2); Focus: (1, 2);
Directrix: $x = -3$

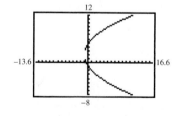

39. Vertex: $(3, -1)$; Focus: $(3, -\frac{5}{4})$;
Directrix: $y = -\frac{3}{4}$

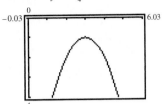

41. Vertex: $(2, -3)$; Focus: $(4, -3)$;
Directrix: $x = 0$

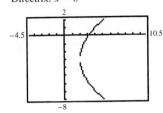

43. Vertex: $(0, 2)$; Focus: $(-1, 2)$;
Directrix: $x = 1$

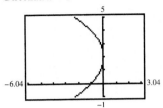

45. Vertex: $(-4, -2)$; Focus: $(-4, -1)$;
Directrix: $y = -3$

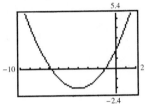

47. Vertex: $(-1, -1)$; Focus: $(-\frac{3}{4}, -1)$;
Directrix: $x = -\frac{5}{4}$

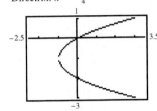

49. Vertex: $(2, -8)$; Focus: $(2, -\frac{31}{4})$;
Directrix: $y = -\frac{33}{4}$

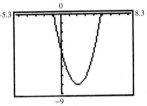

51. $(y - 1)^2 = x$ **53.** $(y - 1)^2 = -(x - 2)$ **55.** $x^2 = 4(y - 1)$ **57.** $y^2 = \frac{1}{2}(x + 2)$

59. 1.5625 ft from the base of the dish, along the axis of symmetry   **61.** 1 in from the vertex   **63.** 20 ft

65. 0.78125 ft   **67.** 4.17 ft from the base along the axis of symmetry   **69.** 24.31 ft, 18.75 ft, 7.64 ft

71. $Ax^2 + Ey = 0$

$$x^2 = -\frac{E}{A}y$$

This is the equation of a parabola with vertex at $(0, 0)$ and axis of symmetry the y-axis. The focus is at $(0, -E/4A)$; the directrix is the line $y = E/4A$. The parabola opens up if $-E/A > 0$ and down if $-E/A < 0$.

73. $Ax^2 + Dx + Ey + F = 0, A \neq 0$

$$Ax^2 + Dx = -Ey - F$$

$$x^2 + \frac{D}{A}x = -\frac{E}{A}y - \frac{F}{A}$$

$$\left(x + \frac{D}{2A}\right)^2 = -\frac{E}{A}y - \frac{F}{A} + \frac{D^2}{4A^2}$$

$$\left(x + \frac{D}{2A}\right)^2 = -\frac{E}{A}y + \frac{D^2 - 4AF}{4A^2}$$

(a) If $E \neq 0$, then the equation may be written as

$$\left(x + \frac{D}{2A}\right)^2 = -\frac{E}{A}\left(y - \frac{D^2 - 4AF}{4AE}\right)$$

This is the equation of a parabola with vertex at
$(-D/2A, (D^2 - 4AF)/(4AE))$ and axis of symmetry parallel
to the y-axis.

(b)–(d) If $E = 0$, the graph of the equation contains no points if
$D^2 - 4AF < 0$, is a single vertical line if $D^2 - 4AF = 0$,
and is two vertical lines if $D^2 - 4AF > 0$.

Exercise 6.3

1. C **3.** B **5.** C **7.** D

9. Vertices: $(-5, 0)$, $(5, 0)$
Foci: $(-\sqrt{21}, 0)$, $(\sqrt{21}, 0)$

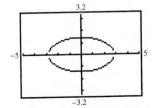

11. Vertices: $(0, -5)$, $(0, 5)$
Foci: $(0, -4)$, $(0, 4)$

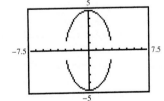

13. Vertices: $(0, -4)$, $(0, 4)$
Foci: $(0, -2\sqrt{3})$, $(0, 2\sqrt{3})$

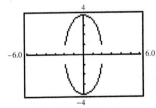

15. Vertices: $(-2\sqrt{2}, 0)$, $(2\sqrt{2}, 0)$; Foci: $(-\sqrt{6}, 0)$, $(\sqrt{6}, 0)$

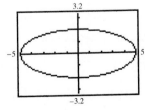

17. Vertices: $(-4, 0)$, $(4, 0)$, $(0, -4)$, $(0, 4)$; Focus: $(0, 0)$

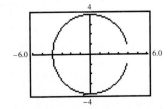

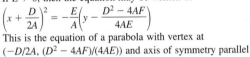

19. $\frac{x^2}{25} + \frac{y^2}{16} = 1$

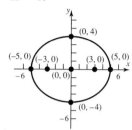

21. $\frac{x^2}{9} + \frac{y^2}{25} = 1$

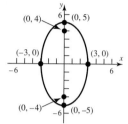

23. $\frac{x^2}{9} + \frac{y^2}{5} = 1$

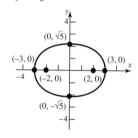

25. $\frac{x^2}{4} + \frac{y^2}{13} = 1$

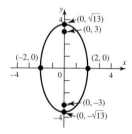

27. $x^2 + \frac{y^2}{16} = 1$

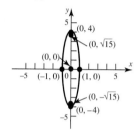

29. $\frac{(x+1)^2}{4} + (y-1)^2 = 1$

31. $(x-1)^2 + \frac{y^2}{4} = 1$

33. Center: $(3, -1)$; Vertices: $(3, -4)$, $(3, 2)$
Foci: $(3, -1 - \sqrt{5})$, $(3, -1 + \sqrt{5})$

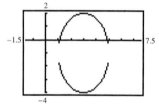

35. Center: $(-5, 4)$; Vertices: $(-9, 4)$, $(-1, 4)$
Foci: $(-5 - 2\sqrt{3}, 4)$, $(-5 + 2\sqrt{3}, 4)$

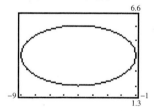

37. Center: $(-2, 1)$; Vertices: $(-4, 1)$, $(0, 1)$
Foci: $(-2 - \sqrt{3}, 1)$, $(-2 + \sqrt{3}, 1)$

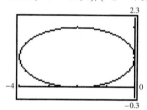

39. Center: $(2, -1)$; Vertices: $(2 - \sqrt{3}, -1)$,
$(2 + \sqrt{3}, -1)$; Foci: $(1, -1)$, $(3, -1)$

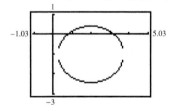

41. Center: $(1, -2)$; Vertices: $(1, -5)$, $(1, 1)$
Foci: $(1, -2 - \sqrt{5})$, $(1, -2 + \sqrt{5})$

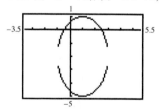

43. Center: $(0, -2)$; Vertices: $(0, -4)$, $(0, 0)$
Foci: $(0, -2 - \sqrt{3})$, $(0, -2 + \sqrt{3})$

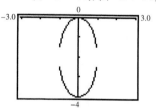

45. $\frac{(x-2)^2}{25} + \frac{(y+2)^2}{21} = 1$

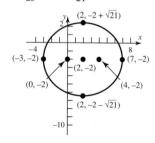

47. $\frac{(x-4)^2}{5} + \frac{(y-6)^2}{9} = 1$

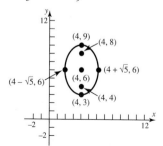

49. $\frac{(x-2)^2}{16} + \frac{(y-1)^2}{7} = 1$

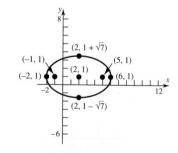

51. $\dfrac{(x-1)^2}{10} + (y-2)^2 = 1$

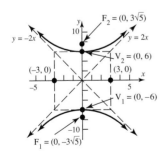

53. $\dfrac{(x-1)^2}{9} + (y-2)^2 = 1$

55.

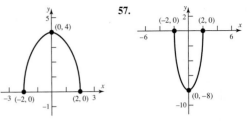

57.

59. $\dfrac{x^2}{100} + \dfrac{y^2}{36} = 1$ **61.** 43.3 ft   **63.** 24.65 ft, 21.65 ft, 13.82 ft   **65.** 0 ft, 12.99 ft, 15 ft, 12.99 ft, 0 ft

67. 91.5 million miles; $\dfrac{x^2}{(93)^2} + \dfrac{y^2}{8646.75} = 1$ **69.** perihelion: 460.6 million miles; mean distance: 483.8 million miles; $\dfrac{x^2}{(483.8)^2} + \dfrac{y^2}{233524} = 1$

71. 30 ft

73. **(a)** $Ax^2 + Cy^2 + F = 0$ $\qquad$ If A and C are of the same sign and F is of opposite sign, then the equation takes the form
$\qquad\qquad Ax^2 + Cy^2 = -F$ $\qquad$ $x^2/(-F/A) + y^2/(-F/C) = 1$, where $-F/A$ and $-F/C$ are positive. This is the equation of an ellipse
$\qquad\qquad\qquad$ with center at $(0, 0)$.

$\qquad$ **(b)** If $A = C$, the equation may be written as $x^2 + y^2 = -F/A$. This is the equation of a circle with center at $(0, 0)$ and radius equal to $\sqrt{-F/A}$.

Exercise 6.4

1. B   **3.** A   **5.** B   **7.** C

9. $x^2 - \dfrac{y^2}{8} = 1$

11. $\dfrac{y^2}{16} - \dfrac{x^2}{20} = 1$

13. $\dfrac{x^2}{9} - \dfrac{y^2}{16} = 1$

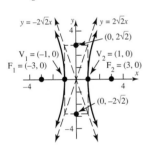

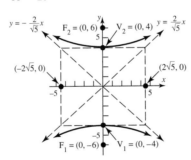

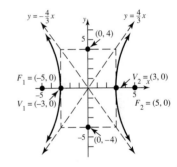

15. $\dfrac{y^2}{36} - \dfrac{x^2}{9} = 1$

17. $\dfrac{x^2}{8} - \dfrac{y^2}{8} = 1$

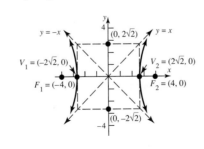

19. Center: $(0, 0)$
Transverse axis: x-axis
Vertices: $(-4, 0)$, $(4, 0)$
Foci: $(-2\sqrt{5}, 0)$, $(2\sqrt{5}, 0)$
Asymptotes: $y = \pm\frac{1}{2}x$

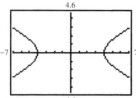

21. Center: $(0, 0)$
Transverse axis: x-axis
Vertices: $(-2, 0)$, $(2, 0)$
Foci: $(-2\sqrt{5}, 0)$, $(2\sqrt{5}, 0)$
Asymptotes: $y = \pm 2x$

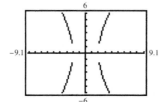

23. Center: $(0, 0)$
Transverse axis: y-axis
Vertices: $(0, -3)$, $(0, 3)$
Foci: $(0, -\sqrt{10})$, $(0, \sqrt{10})$
Asymptotes: $y = \pm 3x$

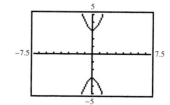

25. Center: $(0, 0)$
Transverse axis: y-axis
Vertices: $(0, -5)$, $(0, 5)$
Foci: $(0, -5\sqrt{2})$, $(0, 5\sqrt{2})$
Asymptotes: $y = \pm x$

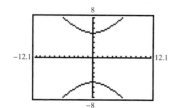

27. $x^2 - y^2 = 1$

29. $\dfrac{y^2}{36} - \dfrac{x^2}{9} = 1$

31. $\dfrac{(x-4)^2}{4} - \dfrac{(y+1)^2}{5} = 1$

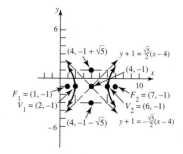

33. $\dfrac{(y+4)^2}{4} - \dfrac{(x+3)^2}{12} = 1$

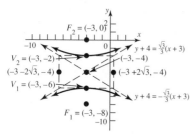

35. $(x-5)^2 - \dfrac{(y-7)^2}{3} = 1$

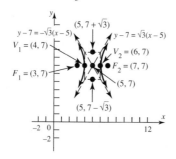

37. $\dfrac{(x-1)^2}{4} - \dfrac{(y+1)^2}{9} = 1$

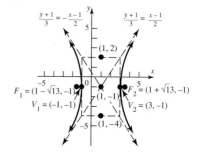

39. Center: $(2, -3)$
Transverse axis: Parallel to x-axis
Vertices: $(0, -3)$, $(4, -3)$
Foci: $(2 - \sqrt{13}, -3)$, $(2 + \sqrt{13}, -3)$
Asymptotes: $y + 3 = \pm\frac{3}{2}(x - 2)$

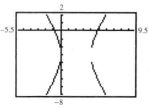

41. Center: $(-2, 2)$
Transverse axis: Parallel to y-axis
Vertices: $(-2, 0)$, $(-2, 4)$
Foci: $(-2, 2 - \sqrt{5})$, $(-2, 2 + \sqrt{5})$
Asymptotes: $y - 2 = \pm 2(x + 2)$

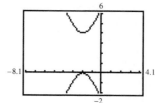

43. Center: $(-1, -2)$
Transverse axis: Parallel to x-axis
Vertices: $(-3, -2)$, $(1, -2)$
Foci: $(-1 - 2\sqrt{2}, -2)$, $(-1 + 2\sqrt{2}, -2)$
Asymptotes: $y + 2 = \pm(x + 1)$

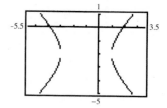

45. Center: $(1, -1)$
Transverse axis: Parallel to x-axis
Vertices: $(0, -1), (2, -1)$
Foci: $(1 - \sqrt{2}, -1), (1 + \sqrt{2}, -1)$
Asymptotes: $y + 1 = \pm(x - 1)$

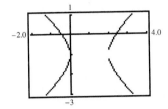

47. Center: $(-1, 2)$
Transverse axis: Parallel to y-axis
Vertices: $(-1, 0), (-1, 4)$
Foci: $(-1, 2 - \sqrt{5}), (-1, 2 + \sqrt{5})$
Asymptotes: $y - 2 = \pm 2(x + 1)$

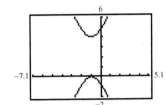

49. Center: $(3, -2)$
Transverse axis: Parallel to x-axis
Vertices: $(1, -2), (5, -2)$
Foci: $(3 - 2\sqrt{5}, -2), (3 + 2\sqrt{5}, -2)$
Asymptotes: $y + 2 = \pm 2(x - 3)$

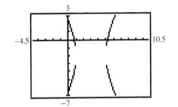

51. Center: $(-2, 1)$
Transverse axis: Parallel to y-axis
Vertices: $(-2, -1), (-2, 3)$
Foci: $(-2, 1 - \sqrt{5}), (-2, 1 + \sqrt{5})$
Asymptotes: $y - 1 = \pm 2(x + 2)$

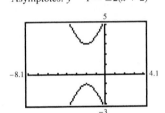

53.

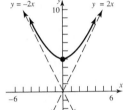

55.

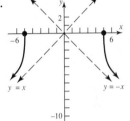

57. (a) The ship will reach shore or a point 64.66 miles from the master station. **(b)** 0.00086 second **(c)** (104, 50)
59. (a) 450 ft **61.** If e is close to 1, narrow hyperbola; if e is very large, wide hyperbola
63. $\dfrac{x^2}{4} - y^2 = 1$; asymptotes $y = \pm \dfrac{1}{2}x$ $y^2 - \dfrac{x^2}{4} = 1$; asymptotes $y = \pm \dfrac{1}{2}x$

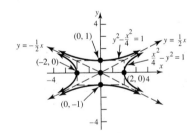

65. $Ax^2 + Cy^2 + F = 0$ If A and C are of opposite sign and $F \neq 0$, this equation may be written as $x^2/(-F/A) + y^2/(-F/C) = 1$, where
$Ax^2 + Cy^2 = -F$ $-F/A$ and $-F/C$ are opposite in sign. This is the equation of a hyperbola with center at $(0, 0)$. The transverse axis is
the x-axis if $-F/A > 0$; the transverse axis is the y-axis if $-F/A < 0$.

Exercise 6.5

1. Parabola **3.** Ellipse **5.** Hyperbola **7.** Hyperbola **9.** Circle **11.** $x = \sqrt{2}/2(x' - y'), y = \sqrt{2}/2(x' + y')$
13. $x = \sqrt{2}/2(x' - y'), y = \sqrt{2}/2(x' + y')$ **15.** $x = \frac{1}{2}(x' - \sqrt{3}y'), y = \frac{1}{2}(\sqrt{3}x' + y')$ **17.** $x = \sqrt{5}/5(x' - 2y'), y = \sqrt{5}/5(2x' + y')$
19. $x = \sqrt{13}/13(3x' - 2y'), y = \sqrt{13}/13(2x' + 3y')$

21.

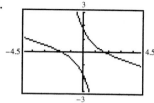

$\theta = 45°$ (see Problem 11)

$x'^2 - \dfrac{y'^2}{3} = 1$

Hyperbola
Center at origin
Transverse axis is the x'-axis
Vertices at $(\pm 1, 0)$

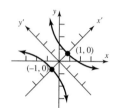

23.

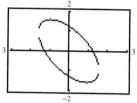

$\theta = 45°$ (see Problem 13)

$x'^2 + \dfrac{y'^2}{4} = 1$

Ellipse
Center at $(0, 0)$
Major axis is the y'-axis
Vertices at $(0, \pm 2)$

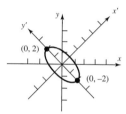

25.

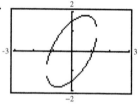

$\theta = 60°$ (see Problem 15)

$\dfrac{x'^2}{4} + y'^2 = 1$

Ellipse
Center at $(0, 0)$
Major axis is the x'-axis
Vertices at $(\pm 2, 0)$

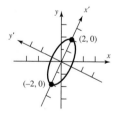

27.

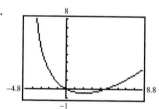

$\theta \approx 63°$ (see Problem 17)

$y'^2 = 8x'$

Parabola
Vertex at $(0, 0)$
Focus at $(2, 0)$

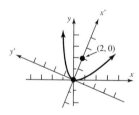

29.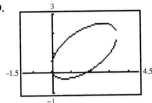

$\theta \approx 34°$ (see Problem 19)

$\dfrac{(x' - 2)^2}{4} + y'^2 = 1$

Ellipse

Center at $(2, 0)$

Major axis is the x'-axis

Vertices at $(4, 0)$ and $(0, 0)$

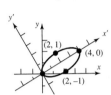

31.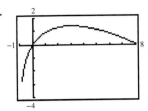

$\cot 2\theta = \frac{7}{24}$; $\theta \approx 37°'$

$(x' - 1)^2 = -6\left(y' - \frac{1}{6}\right)$

Parabola

Vertex at $\left(1, \frac{1}{6}\right)$

Focus at $\left(1, -\frac{4}{3}\right)$

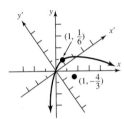

33. Hyperbola **35.** Hyperbola **37.** Parabola **39.** Ellipse **41.** Ellipse

43. Refer to equation (6): $A' = A \cos^2 \theta + B \sin \theta \cos \theta + C \sin^2 \theta$

$B' = B(\cos^2 \theta - \sin^2 \theta) + 2(C - A)(\sin \theta \cos \theta)$

$C' = A \sin^2 \theta - B \sin \theta \cos \theta + C \cos^2 \theta$

$D' = D \cos \theta + E \sin \theta$

$E' = -D \sin \theta + E \cos \theta$

$F' = F$

45. Use Problem 43 to find $B'^2 - 4A'C'$. After much cancellation, $B'^2 - 4A'C' = B^2 - 4AC$.

47. Use formulas (5) to find $d^2 = (x_2 - x_1)^2 + (y_2 - y_1)^2$. After simplifying, $(x_2 - x_1)^2 + (y_2 - y_1)^2 = (x'_2 - x'_1)^2 + (y'_2 - y'_1)^2$.

Exercise 6.6

1. Parabola; directrix is perpendicular to the polar axis 1 unit to the right of the pole.

3. Hyperbola; directrix is parallel to the polar axis $\frac{4}{3}$ units below the pole.

5. Ellipse; directrix is perpendicular to the polar axis $\frac{3}{2}$ units to the left of the pole.

7.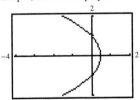

Parabola; directrix is perpendicular to
the polar axis 1 unit to the right of the pole.

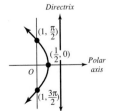

9.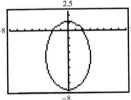

Ellipse; directrix is parallel to the
polar axis $\frac{8}{3}$ units above the pole.

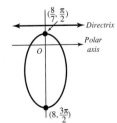

11.

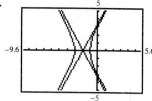

Hyperbola; directrix is perpendicular to the polar axis $\frac{3}{2}$ units to the left of the pole.

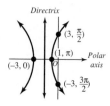

13.

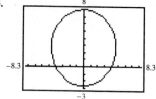

Ellipse; directrix is parallel to the polar axis 8 units below the pole; vertices are at $(8, \pi/2)$ and $(\frac{8}{3}, 3\pi/2)$.

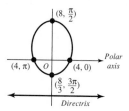

15.

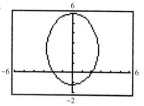

Ellipse; directrix is parallel to the polar axis 3 units below the pole; vertices are at $(6, \pi/2)$ and $(\frac{6}{5}, 3\pi/2)$.

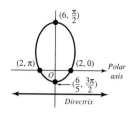

17.

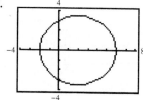

Ellipse; directrix is perpendicular to the polar axis 6 units to the left of the pole; vertices are at $(6, 0)$ and $(2, \pi)$.

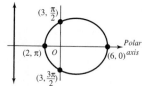

19. $y^2 + 2x - 1 = 0$ **21.** $16x^2 + 7y^2 + 48y - 64 = 0$ **23.** $3x^2 - y^2 + 12x + 9 = 0$ **25.** $4x^2 + 3y^2 - 16y - 64 = 0$
27. $9x^2 + 5y^2 - 24y - 36 = 0$ **29.** $3x^2 + 4y^2 - 12x - 36 = 0$ **31.** $r = 1/(1 + \sin\theta)$ **33.** $r = 12/(5 - 4\cos\theta)$ **35.** $r = 12/(1 - 6\sin\theta)$
37. Use $d(D, P) = p - r\cos\theta$ in the derivation of equation (a) in Table 5.
39. Use $d(D, P) = p + r\sin\theta$ in the derivation of equation (a) in Table 5.

1.

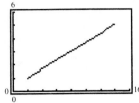

$x - 3y + 1 = 0$

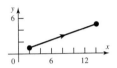

3.

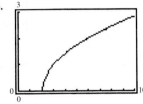

$y = \sqrt{x - 2}$

5.

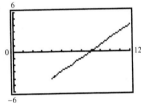

$x = y + 8$

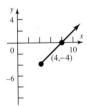

7.

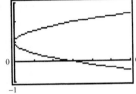

$x = 3(y - 1)^2$

9.

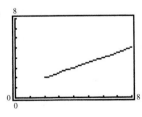

$2y = 2 + x$

11.

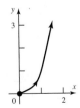

$y = x^3$

13.

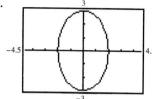

$$\frac{x^2}{4} + \frac{y^2}{9} = 1$$

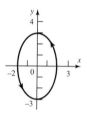

15.

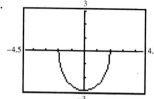

$$\frac{x^2}{4} + \frac{y^2}{9} = 1$$

17.

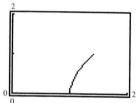

$$x^2 - y^2 = 1$$

$(\sqrt{2}, 1)$

19.

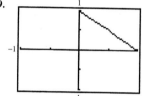

$$x + y = 1$$

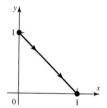

21. $x = t,\ y = t^3;\ x = \sqrt[3]{t},\ y = t$ **23.** $x = t,\ y = t^{2/3};\ x = t^{3/2},\ y = t$ **25.** $x = 2\cos \pi t,\ y = -3\sin \pi t,\ 0 \le t \le 2$
27. $x = -2\sin 2\pi t,\ y = 3\cos 2\pi t,\ 0 \le t \le 1$
29.

31. The orientation is from (x_1, y_1) to (x_2, y_2).

33.

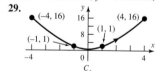

35.

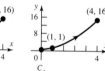

Fill-in-the-Blank Items

1. parabola **2.** ellipse **3.** hyperbola **4.** major; transverse **5.** y-axis **6.** $y/3 = x/2$; $y/3 = -x/2$ **7.** $\cot 2\theta = (A - C)/B$
8. $\frac{1}{2}$; ellipse; parallel; 4; below **9.** ellipse

True/False Items

1. T **2.** F **3.** T **4.** T **5.** T **6.** T **7.** T **8.** T **9.** T **10.** F

Review Exercises

1. Parabola; vertex $(0, 0)$, focus $(-4, 0)$, directrix $x = 4$
3. Hyperbola; center $(0, 0)$, vertices $(5, 0)$ and $(-5, 0)$, foci $(\sqrt{26}, 0)$ and $(-\sqrt{26}, 0)$, asymptotes $y = \frac{1}{5}x$ and $y = -\frac{1}{5}x$
5. Ellipse; center $(0, 0)$, vertices $(0, 5)$ and $(0, -5)$, foci $(0, 3)$ and $(0, -3)$
7. $x^2 = -4(y - 1)$: Parabola; vertex $(0, 1)$, focus $(0, 0)$, directrix $y = 2$
9. $\dfrac{x^2}{2} - \dfrac{y^2}{8} = 1$: Hyperbola; center $(0, 0)$, vertices $(\sqrt{2}, 0)$ and $(-\sqrt{2}, 0)$, foci $(\sqrt{10}, 0)$ and $(-\sqrt{10}, 0)$, asymptotes $y = 2x$ and $y = -2x$
11. $(x - 2)^2 = 2(y + 2)$: Parabola; vertex $(2, -2)$, focus $(2, -\frac{3}{2})$, directrix $y = -\frac{5}{2}$
13. $\dfrac{(y - 2)^2}{4} - (x - 1)^2 = 1$: Hyperbola; center $(1, 2)$, vertices $(1, 4)$ and $(1, 0)$, foci $(1, 2 + \sqrt{5})$ and $(1, 2 - \sqrt{5})$, asymptotes $y - 2 = \pm\, 2(x - 1)$
15. $\dfrac{(x - 2)^2}{9} + \dfrac{(y - 1)^2}{4} = 1$: Ellipse; center $(2, 1)$, vertices $(5, 1)$ and $(-1, 1)$, foci $(2 + \sqrt{5}, 1)$ and $(2 - \sqrt{5}, 1)$
17. $(x - 2)^2 = -4(y + 1)$: Parabola; vertex $(2, -1)$, focus $(2, -2)$, directrix $y = 0$
19. $\dfrac{(x - 1)^2}{4} + \dfrac{(y + 1)^2}{9} = 1$: Ellipse; center $(1, -1)$, vertices $(1, 2)$ and $(1, -4)$, foci $(1, -1 + \sqrt{5})$ and $(1, -1 - \sqrt{5})$

21. $y^2 = -8x$

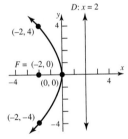
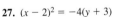
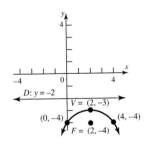

23. $\dfrac{y^2}{4} - \dfrac{x^2}{12} = 1$

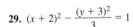
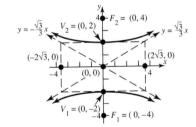

25. $\dfrac{x^2}{16} + \dfrac{y^2}{7} = 1$

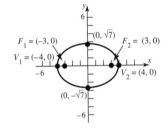

27. $(x - 2)^2 = -4(y + 3)$

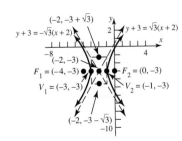

29. $(x + 2)^2 - \dfrac{(y + 3)^2}{3} = 1$

31. $\dfrac{(x + 4)^2}{16} + \dfrac{(y - 5)^2}{25} = 1$

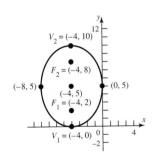

33. $\dfrac{(x + 1)^2}{9} - \dfrac{(y - 2)^2}{7} = 1$

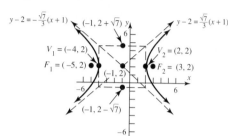

$y - 2 = -\dfrac{\sqrt{7}}{3}(x + 1)$ $(-1, 2 + \sqrt{7})$ $y - 2 = \dfrac{\sqrt{7}}{3}(x + 1)$

$V_1 = (-4, 2)$ $V_2 = (2, 2)$

$F_1 = (-5, 2)$ $F_2 = (3, 2)$

$(-1, 2)$

$(-1, 2 - \sqrt{7})$

35. $\dfrac{(x - 3)^2}{9} - \dfrac{(y - 1)^2}{4} = 1$

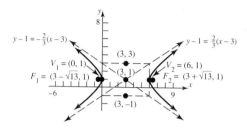

$y - 1 = -\dfrac{2}{3}(x - 3)$ $(3, 3)$ $y - 1 = \dfrac{2}{3}(x - 3)$

$V_1 = (0, 1)$ $V_2 = (6, 1)$

$F_1 = (3 - \sqrt{13}, 1)$ $(3, 1)$ $F_2 = (3 + \sqrt{13}, 1)$

$(3, -1)$

37. Parabola   **39.** Ellipse   **41.** Parabola   **43.** Hyperbola   **45.** Ellipse

47. $x'^2 - \dfrac{y'^2}{9} = 1$

Hyperbola
Center at the origin
Transverse axis the x'-axis
Vertices at $(\pm 1, 0)$

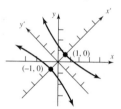

$(1, 0)$

$(-1, 0)$

49. $\dfrac{x'^2}{2} + \dfrac{y'^2}{4} = 1$

Ellipse
Center at origin
Major axis the y'-axis
Vertices at $(0, \pm 2)$

$(0, 2)$

$(0, -2)$

51. $y'^2 = -\dfrac{4\sqrt{13}}{13} x'$

Parabola
Vertex at the origin
Focus on the x'-axis at $(-\sqrt{13}/13, 0)$

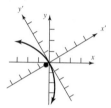

53. Parabola; directrix is perpendicular to the polar axis 4 units to the left of the pole.

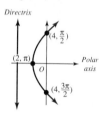

Directrix

$(4, \frac{\pi}{2})$

$(2, \pi)$ *Polar axis*

O

$(4, \frac{3\pi}{2})$

55. Ellipse; directrix is parallel to the polar axis 6 units below the pole; vertices are $(6, \pi/2)$ and $(2, 3\pi/2)$.

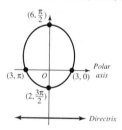

$(6, \frac{\pi}{2})$

$(3, \pi)$ O $(3, 0)$ *Polar axis*

$(2, \frac{3\pi}{2})$

Directrix

57. Hyperbola; directrix is perpendicular to the polar axis 1 unit to the right of the pole; vertices are at $(\frac{2}{3}, 0)$ and $(-2, \pi)$

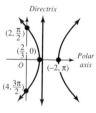

Directrix

$(2, \frac{\pi}{2})$

$(\frac{2}{3}, 0)$ *Polar axis*

O $(-2, \pi)$

$(4, \frac{3\pi}{2})$

59. $y^2 - 8x - 16 = 0$ **61.** $3x^2 - y^2 - 8x + 4 = 0$

63.

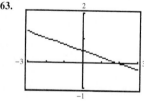

$x + 4y = 2$

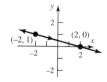

$(-2, 1)$ $(2, 0)$

65.

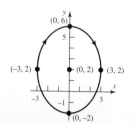

$\dfrac{x^2}{9} + \dfrac{(y - 2)^2}{16} = 1$

$(0, 6)$

$(-3, 2)$ $(0, 2)$ $(3, 2)$

$(0, -2)$

67.

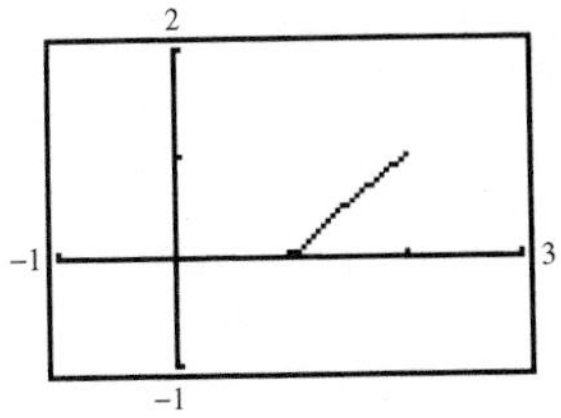

$1 + y = x$

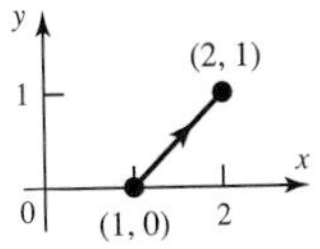

69. $\dfrac{x^2}{5} - \dfrac{y^2}{4} = 1$ **71.** The ellipse $\dfrac{x^2}{16} + \dfrac{y^2}{7} = 1$ **73.** $\frac{1}{4}$ ft or 3 in **75.** 19.72 ft, 18.86 ft, 14.91 ft
77. **(a)** 45.24 miles from the Master Station **(b)** 0.000645 second **(c)** (66, 20)

CHAPTER 7 *Exercise 7.1*

1. **(a)** 11.212 **(b)** 11.587 **(c)** 11.664 **(d)** 11.665 **3.** **(a)** 8.815 **(b)** 8.821 **(c)** 8.824 **(d)** 8.825
5. **(a)** 21.217 **(b)** 22.217 **(c)** 22.440 **(d)** 22.459 **7.** 3.320 **9.** 0.427 **11.** B **13.** D **15.** A **17.** E **19.** A **21.** E **23.** B

25.

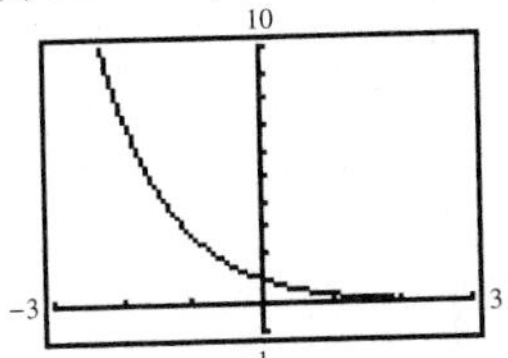

27.

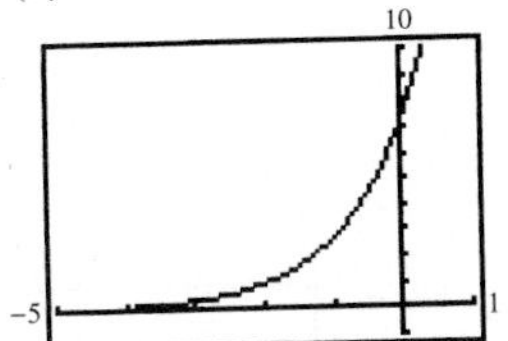

29.

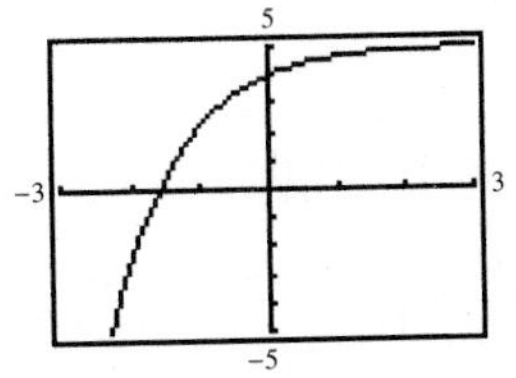

31.

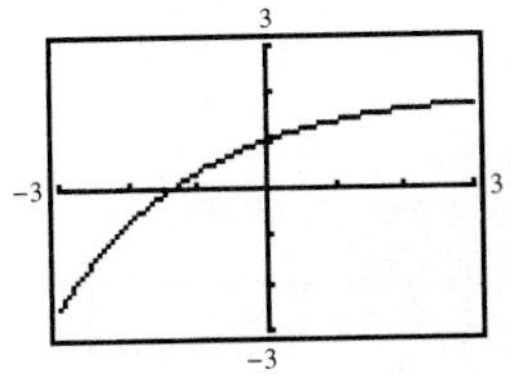

33. $\frac{1}{49}$ **35.** $\frac{1}{4}$ **37.** **(a)** 74% **(b)** 47% **39.** **(a)** 44 watts **(b)** 11.6 watts **41.** 3.35 milligrams; 0.45 milligrams
43. **(a)** 56% **(b)** 68% **(c)** 70% **(d)** $R = 40\%$ just after 6 days

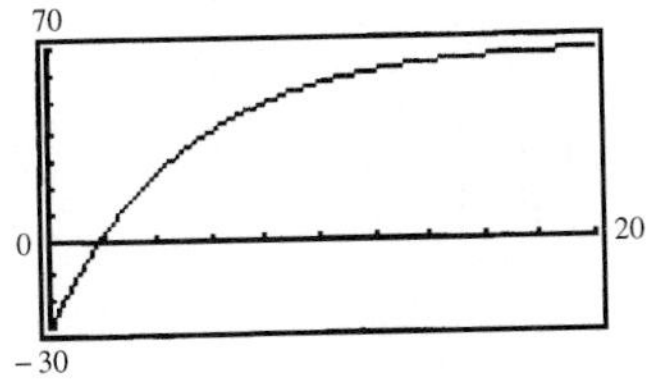

45. **(a)** 5.414 amperes, 7.5854 amperes, 10.38 amperes **(b)** 12 amperes **(c)** $I_1(t) = 12\,(1 - e^{-2t})$ **(d)** 3.343 amperes, 5.309 amperes, 9.443 amperes **(e)** 24 amperes **(f)** $I_2(t) = 24\,(1 - e^{-1/2t})$

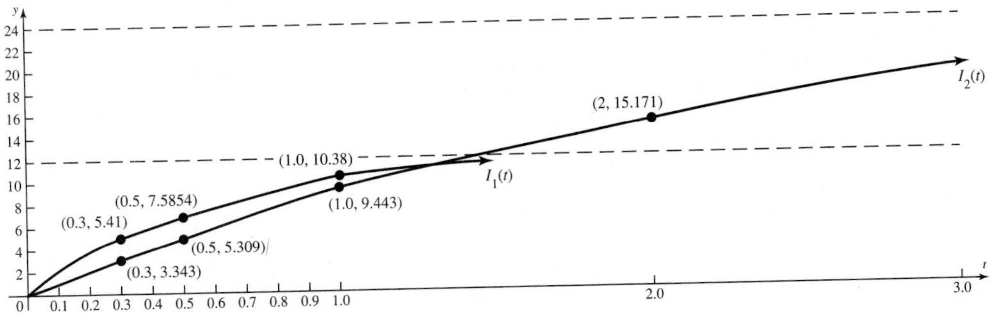

47. (a) 9.23×10^{-3} **(b)** 0.81 **(c)** 5 **(d)** 57.91°, 43.98°, 30.06°

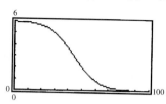

49. $n = 4$: 2.7083; $n = 6$: 2.7181; $n = 8$: 2.7182788; $n = 10$: 2.7182818

51. $\dfrac{f(x + h) - f(x)}{h} = \dfrac{a^{x+h} - a^x}{h} = \dfrac{a^x a^h - a^x}{h} = \dfrac{a^x(a^h - 1)}{h}$ **53.** $f(-x) = a^{-x} = \dfrac{1}{a^x} = \dfrac{1}{f(x)}$

55. (a) $\sinh(-x) = \frac{1}{2}(e^{-x} - e^{x})$
$= -\frac{1}{2}(e^x - e^{-x})$
$= -\sinh x$

(b)

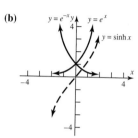

57. $f(1) = 5, f(2) = 17, f(3) = 257, f(4) = 65{,}537;$
$f(5) = 4{,}294{,}967{,}297 = 641 \times 6{,}700{,}417$

Exercise 7.2

1. $2 = \log_3 9$ **3.** $2 = \log_a 1.6$ **5.** $2 = \log_{1.1} M$ **7.** $x = \log_2 7.2$ **9.** $\sqrt{2} = \log_x \pi$ **11.** $x = \ln 8$ **13.** $2^3 = 8$ **15.** $a^6 = 3$ **17.** $3^x = 2$
19. $2^{1.3} = M$ **21.** $(\sqrt{2})^x = \pi$ **23.** $e^x = 4$ **25.** 0 **27.** 2 **29.** -4 **31.** $\frac{1}{2}$ **33.** 4 **35.** $\frac{1}{2}$ **37.** $\{x | x < 3\}$ **39.** All real numbers except 0
41. $\{x | x < -2 \text{ or } x > 3\}$ **43.** $\{x | x > 0, x \neq 1\}$ **45.** $\{x | x < -1 \text{ or } x > 0\}$ **47.** 0.511 **49.** 30.099 **51.** $\sqrt{2}$ **53.** B **55.** D **57.** A **59.** E
61. C **63.** A **65.** D

67.

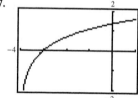

69.

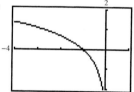

71.

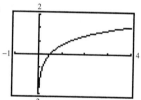

73.

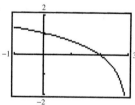

75.

77. (a) $n = 6.93$ so 7 panes are necessary **(b)** $n = 13.86$ so 14 panes are necessary
79. (a) $d = 127.7$ so it takes about 128 days **(b)** $d = 575.6$ so it takes about 576 days **81.** $h = 2.29$ so the time between injections is $2\text{-}2\frac{1}{2}$ hours
83. 0.2695 sec; 0.8959 sec **85. (a)** $k = 20.07$ **(b)** 91% **(c)** 0.175 **(d)** 0.08 **87.** $y = 20\, e^{0.023t}$; $y = 89.2$ is predicted

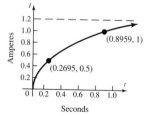

Exercise 7.3

1. $a + b$ **3.** $b - a$ **5.** $a + 1$ **7.** $2a + b$ **9.** $\frac{1}{5}(a + 2b)$ **11.** $\dfrac{b}{a}$ **13.** $2 \ln x + \frac{1}{2}\ln(1 - x)$ **15.** $3 \log_2 x - \log_2(x - 3)$

17. $\log x + \log(x + 2) - 2 \log(x + 3)$ **19.** $\frac{1}{3}\ln(x - 2) + \frac{1}{3}\ln(x + 1) - \frac{2}{3}\ln(x + 4)$ **21.** $\ln 5 + \ln x + \frac{1}{2}\ln(1 - 3x) - 3\ln(x - 4)$ **23.** $\log_5 u^3 v^4$

25. $-\frac{5}{2}\log_{1/2} x$ **27.** $-2\ln(x - 1)$ **29.** $\log_2[x(3x - 2)^4]$ **31.** $\log_a\left[\dfrac{25x^6}{(2x + 3)^{1/2}}\right]$ **33.** 2.771   **35.** -3.880 **37.** 5.615   **39.** 0.874

41. 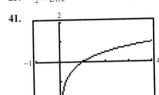   **43.** 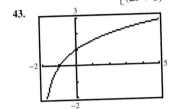   **45.**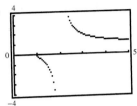

47. $\log_a(x + \sqrt{x^2 - 1}) + \log_a(x - \sqrt{x^2 - 1}) = \log_a[(x + \sqrt{x^2 - 1})(x - \sqrt{x^2 - 1})] = \log_a[x^2 - (x^2 - 1)] = \log_a 1 = 0$

49. $\ln (1 + e^{2x}) = \ln[e^{2x}(e^{-2x} + 1)] = \ln e^{2x} + \ln (e^{-2x} + 1) = 2x + \ln(1 + e^{-2x})$

51. $y = f(x) = \log_a x;\ a^y = x;\ \left(\dfrac{1}{a}\right)^{-y} = x;\ -y = \log_{1/a} x;\ -f(x) = \log_{1/a} x$ **53.** $f(AB) = \log_a AB = \log_a A + \log_a B = f(A) + f(B)$

55. $y = Cx$ **57.** $y = Cx(x + 1)$ **59.** $y = Ce^{3x}$ **61.** $y = Ce^{-4x} + 3$ **63.** $y = \dfrac{\sqrt[3]{C}(2x + 1)^{1/6}}{(x + 4)^{1/9}}$ **65.** 3   **67.** 1

69. If $A = \log_a M$ and $B = \log_a N$, then $a^A = M$ and $a^B = N$. Then $\log_a (M/N) = \log_a (a^A/a^B) = \log_a a^{A-B} = A - B = \log_a M - \log_a N$.

Exercise 7.4

1. $\frac{7}{2}$ **3.** $\{-2\sqrt{2}, 2\sqrt{2}\}$ **5.** 16   **7.** 8   **9.** 3   **11.** 5   **13.** 2   **15.** $\{-2, 4\}$ **17.** 21   **19.** $\frac{1}{2}$ **21.** $\{-\sqrt{2}, 0, \sqrt{2}\}$

23. $\left\{1 - \dfrac{\sqrt{6}}{3}, 1 + \dfrac{\sqrt{6}}{3}\right\}$ **25.** 0   **27.** 2   **29.** 0   **31.** $\frac{3}{2}$ **33.** 3.322   **35.** -0.088 **37.** 0.307   **39.** 1.356   **41.** 0

43. 0.534   **45.** 0.226   **47.** 2.027   **49.** $\frac{9}{2}$ **51.** 2   **53.** -1 **55.** 1   **57.** 16   **59.** 1.92   **61.** 2.78   **63.** -0.56 **65.** -0.70
67. 0.56   **69.** $\{0.39, 1.00\}$ **71.** 1.31   **73.** 1.30

Exercise 7.5

1. \$108.29   **3.** \$609.50   **5.** \$697.09   **7.** \$12.46   **9.** \$125.23   **11.** \$88.72   **13.** \$860.72   **15.** \$554.09   **17.** \$59.71   **19.** \$361.93

21. 5.35%   **23.** 26%   **25.** $6\frac{1}{4}$% compounded annually   **27.** 9% compounded monthly   **29.** 104.32 mo; 103.97 mo

31. 61.02 mo; 60.82 mo   **33.** 15.27 yrs or 15 yrs, 4 months   **35.** \$104,335   **37.** \$12,910.62   **39.** About \$30.17 per share or \$3017

41. 9.35%   **43.** Not quite. You will have \$1057.60. The second bank gives a better deal, since you have \$1060.62 after 1 year

45. You have \$11,632.73; your friend has \$10, 947.89

47. (a) Interest is \$30,000   (b) Interest is \$38,613.59   (c) Interest is \$37,752.73. Simple interest at 12% is best

49. (a) \$1364.62   (b) \$1353.35   **51.** \$4631.93

59. (a) 6.1 years   (b) 18.45 yrs   (c) $mP = P\left(1 + \dfrac{r}{n}\right)^{nt}$

$$m = \left(1 + \dfrac{r}{n}\right)^{nt}$$

$$\ln m = \ln \left(1 + \dfrac{r}{n}\right)^{nt} = nt \ln \left(1 + \dfrac{r}{n}\right)$$

$$t = \dfrac{\ln m}{n \ln \left(1 + \frac{r}{n}\right)}$$

1. (a) 34.7 days; 69.3 days **(b)** 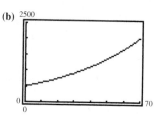 **3.** 28.4 yr **5. (a)** 94.4 yr **(b)**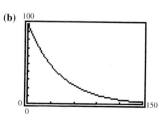

7. 5832; 3.9 days **9.** 25,198 **11.** 9.797 g **13. (a)** 9727 years ago **(b)** 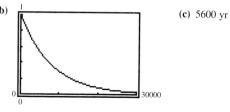 **(c)** 5600 yr

15. (a) 5:18 P.M. **(b)** 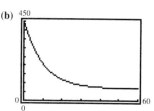 **(c)** After 14.3 minutes, the pizza will be 160°F.
(d) As time passes, the temperature of the pizza gets closer to 70°F.

17. (a) 18.63°C; 25.1°C **(b)** 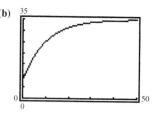 **19.** 7.34 kg; 76.6 hr **21.** 26.5 days.

1. (a) 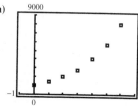 **(b)** Population = 1000.187781 (1.41414215)^t
(c) $N = 1000.187781e^{0.3465230928t}$
(d) 11,312

3. (a) 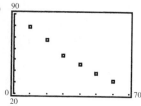 **(b)** $P = 96.0100692 (0.981490683)^Q$
(c) $Q = 25$

5. (a)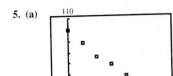

(b) $y = 100.3262508 (0.8768651017)^t$

(c) $N = 100.3262508e^{-0.1314021163t}$

(d) 5.3 weeks

(e) 0.14 grams

7. (a)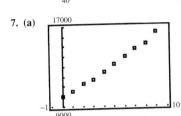

(b) Value $= 10014.4963 (1.056554737)^t$

(c) 5.655%

(d) \$68,682.99

9. (a)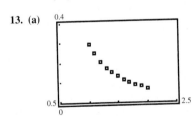

(b) $T = 173.2867183 (0.9889878268)^t$

(c) 41.04

11. (a)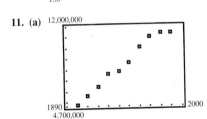

(b) $P = -1163825970 + 154808355.6 \ln t$

(c) 12,469,735 people

13. (a)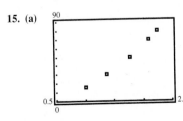

(b) $I = 0.2995056364x^{-2.012000304}$

(c) Close; -2.01 versus -2

(d) 0.056054

15. (a)

(b) $s = 15.97274236t^{2.001820627}$

(c) $s = \frac{1}{2} \cdot 31.94548472t^{2.001820627}$

$g \approx 31.9455$ feet/sec^2

(d) $t \approx 2.5$ sec

17. (a)

(b) Exponential model:

$y = 99.06645375 (1.652917358)^x$

19. (a)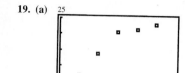

(b) Logarithmic model:
$$y = -16.55688464 + 10.31473128 \ln x$$

21. (a)

(b) Exponential model:
$$y = 346.113717\,(0.0352268374)^x$$

23. (a)

(b) Logarithmic model:
$$\text{CPI} = -69946.44499 + 9225.465596 \ln t$$
(c) 147.7

25. (a)

(b) Linear model:
$$\text{GDP} = 77.45714286t - 134670.9429$$
(c) $19,779

27. (a)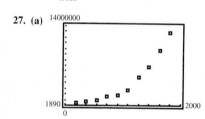

(b) Exponential model:
$$P = 2.829551 \times 10^{-25}\,(1.037329901)^t$$
(c) 16,066,684

Exercise 7.8

1. 70 decibels **3.** 111.76 decibels **5.** 10 W/m² **7.** 4.0 on the Richter scale
9. 70,794.58 mm; the San Francisco earthquake was 11.22 times as intense as the one in Mexico City.

Fill-in-the-Blank Items

1. (0, 1) and (1, a) **2.** 1 **3.** 4 **4.** sum **5.** 1 **6.** 7 **7.** $x > 0$ **8.** (1, 0) and (a, 1) **9.** 1 **10.** 7

True/False Items

1. T **2.** T **3.** F **4.** F **5.** T **6.** F **7.** F **8.** T

Review Exercises

1. -3 **3.** $\sqrt{2}$ **5.** 0.4 **7.** $\frac{25}{4}\log_4 x$ **9.** $\ln\left[\dfrac{1}{(x+1)^2}\right] = -2\ln(x+1)$ **11.** $\log\left(\dfrac{4x^3}{[(x+3)(x-2)]^{1/2}}\right)$ **13.** $y = Ce^{2x^2}$
15. $y = (Ce^{3x^2})^2$ **17.** $y = \sqrt{e^{x+C} + 9}$ **19.** $y = \ln(x^2 + 4) - C$

21.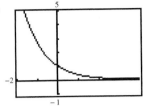

23.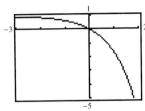

25.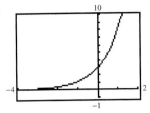

27.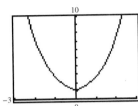

29.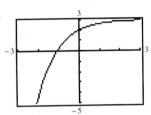

31. $\frac{1}{4}$ **33.** $\left\{\dfrac{-1-\sqrt{3}}{2}, \dfrac{-1+\sqrt{3}}{2}\right\}$ **35.** $\frac{1}{4}$ **37.** 4.301 **39.** $\frac{12}{5}$ **41.** 83 **43.** $\left\{-3, \frac{1}{2}\right\}$ **45.** -1 **47.** -0.609 **49.** -9.327

51. 3229.5 m **53.** 7.6 mm of mercury **55.** (a) 37.3 watts (b) 6.9 decibels

57. (a) 71% (b) 85.5% (c) 90% (d) About 1.6 months (e) About 4.8 months

59. (a) 9.85 years (b) 4.27 years **61.** $41,669 **63.** 80 decibels **65.** 24,203 years ago

67. 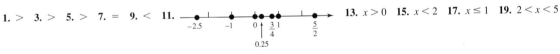 (b) Power model:
$y = 1.298985429x^{-0.2721702954}$

69. (a) 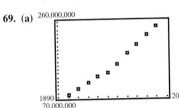 (b) Power model:
$P = 9.545211 \times 10^{-76}t^{25.29069649}$
(c) 273,930,160 people

APPENDIX *Exercise 1*

1. $>$ **3.** $>$ **5.** $>$ **7.** $=$ **9.** $<$ **11.** 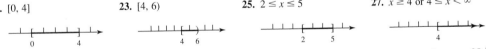 **13.** $x > 0$ **15.** $x < 2$ **17.** $x \le 1$ **19.** $2 < x < 5$

21. $[0, 4]$ **23.** $[4, 6)$ **25.** $2 \le x \le 5$ **27.** $x \ge 4$ or $4 \le x < \infty$ **29.** 1 **31.** 5 **33.** 1

35. $\frac{1}{16}$ **37.** $\frac{4}{9}$ **39.** $\frac{1}{9}$ **41.** $\frac{9}{4}$ **43.** 324 **45.** 27 **47.** 16 **49.** $4\sqrt{2}$ **51.** $-\frac{2}{3}$ **53.** y^2 **55.** $\frac{y}{x^2}$ **57.** $\frac{1}{x^3 y}$ **59.** $\frac{25y^2}{16x^2}$ **61.** $\frac{1}{x^3 y^3 z^2}$ **63.** $\frac{-8x^3}{9yz^2}$

65. $\frac{y^2}{x^2 + y^2}$ **67.** $\frac{16x^2}{9y^2}$ **69.** 13 **71.** 26 **73.** 25 **75.** 4 **77.** 24 **79.** Yes; 5 **81.** Not a right triangle **83.** Yes; 25 **85.** 9.4 in.

87. 58.3 ft **89.** 46.7 mi **91.** 3 mi **93.** $a \le b, c > 0; a - b \le 0$ **95.** $\dfrac{a+b}{2} - a = \dfrac{a+b-2a}{2} = \dfrac{b-a}{2} > 0$; therefore, $a < \dfrac{a+b}{2}$
$(a - b)c \le 0(c)$
$ac - bc \le 0$ $b - \dfrac{a+b}{2} = \dfrac{2b-a-b}{2} = \dfrac{b-a}{2} > 0$; therefore, $b > \dfrac{a+b}{2}$
$ac \le bc$

97. No; No **99.** 1 **101.** 3.15 or 3.16

103. The light can be seen on the horizon 23.3 miles distant. Planes flying at 10,000 feet can see it 99 miles away. The ship would need to be 185.8 ft tall for the data to be correct.

Exercise 2

1. -1 **3.** -18 **5.** -3 **7.** -16 **9.** 0.5 **11.** 2 **13.** 2 **15.** -1 **17.** 3 **19.** No real solution **21.** $\{0, 9\}$ **23.** $\{0, 9\}$ **25.** No real solution
27. 2 **29.** 21 **31.** $\{-2, 2\}$ **33.** $-\frac{20}{39}$ **35.** 6

1. 4 **3.** $\frac{1}{16}$ **5.** $\frac{1}{9}$ **7.** $(x-2)^2 + (y+2)^2 = 9$ **9.** $(x+3)^2 + (y-1)^2 = 9$ **11.** $(x+\frac{1}{2})^2 + (y+\frac{1}{2})^2 = 1$ **13.** $\{-7, 3\}$ **15.** $\{-\frac{1}{4}, \frac{3}{4}\}$
17. $\left\{\dfrac{-1-\sqrt{7}}{6}, \dfrac{-1+\sqrt{7}}{6}\right\}$

1. $8 + 5i$ **3.** $-7 + 6i$ **5.** $-6 - 11i$ **7.** $6 - 18i$ **9.** $6 + 4i$ **11.** $10 - 5i$ **13.** 37 **15.** $\frac{6}{5} + \frac{8}{5}i$ **17.** $1 - 2i$ **19.** $\frac{5}{2} - \frac{7}{2}i$ **21.** $-\frac{1}{2} + (\sqrt{3}/2)i$
23. $2i$ **25.** $-i$ **27.** i **29.** -6 **31.** $-10i$ **33.** $-2 + 2i$ **35.** 0 **37.** 0 **39.** $2i$ **41.** $5i$ **43.** $5i$ **45.** 6 **47.** 25
49. $z + \bar{z} = (a + bi) + (a - bi) = 2a;\ z - \bar{z} = (a + bi) - (a - bi) = 2bi$
51. $z + w = (a + bi) + (c + di) = (a + c) + (b + d)i = (a + c) - (b + d)i = (a - bi) + (c - di) = \bar{z} + \bar{w}$

1. Strong linear relation **3.** Strong linear relation **5.** Nonlinear relation
7. (a)

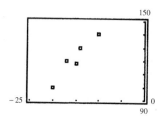

(b) $y = 2.0357x - 2.3571$

9. (a)

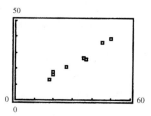

(b) $y = 2.2x + 1.2$

11. (a)

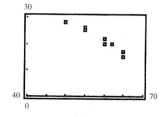

(b) $y = -0.72x + 116.6$

13. (a)

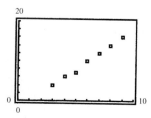

(b) $y = 3.861313869x + 180.2919708$

15. (a)

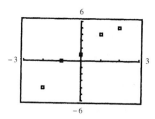

(b) $I = 0.7548890222C + 0.6266346731$

(c) As disposable income increases by $1, consumption increases by about $0.75.
(d) $32,000

17. (a)

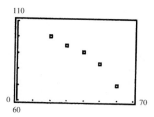

(b) $y = -0.8265306122x + 70.39030612$

(c) As speed increases by 1 mph, mpg decreases by 0.8265.
(d) 20 mpg

19. (a)

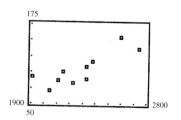

(b) NIBT $= 0.084195641$ sales $- 88.57761368$
(c) As sales increase by $1, NIBT increases by $0.08.
(d) $118.2 billion

Index

Abscissa, 2
Absolute value, 530–31
Absolute value function, 84
Acute angle, 163
Addition
  of complex numbers, 547
  of vectors, 349–50, 353, 354
Additive inverse property of vector addition, 349–50
Advertising, exponential response to, 454–55, 458
Air, velocity relative to, 355
Airspeed problem, 358
Alcohol and driving, function for risk involved in, 465–67, 470
Algebra, to solve geometry problems, 7–8
Algebraic functions, 448
Alternating current (ac) problems, 296–97, 305, 458, 459
Altitude problems, 524
Amplified sine wave, 301
Amplifying factor, 301
Amplitude
  of sinusoidal graph, 284–87, 288–90
  of vibration, 290
Analytic trigonometry, 205–48
  double-angle formulas, 221–24
    exact values using, 221–22
    for sines and cosines, 221
    for tangent, 222
    variations of, 223–24
  half-angle formulas, 224–27
  product-to-sum formulas, 230–32
  sum and difference formulas, 211–19
    for cosines, 211–14
    for sines, 214–16
    for tangent, 216–18
  sum-to-product formulas, 232–33
  trigonometric equations, 234–42
    solution of, 234
    solving, 235–42
  trigonometric identities, 206–10
Ancient tools, estimating age of, 498–99
Angle(s), 126–36, 139. *See also* Trigonometric
        functions
  acute, 163
  central, 129
  circular motion and, 132–34
  complementary, 164–65
  definition of, 126
  degrees, 127–32
  of depression, 252
  drawing, 127–30
  of elevation, 252
  of incidence, 245

  of inclination, 151
  notation for, 126
  positive and negative, 126
  quadrantal, 127, 141–42
  quadrant of, 156
  radians, 129–32
  reference, 165–68
  of refraction, 245
  right, 127, 535
  sides of, 126
  in standard position, 126–27
  straight, 127
  trigonometric functions of, 139
    45° and −45°, 142–43
    30° and 60°, 144–46
  between vectors, 360
Angular speed, 133
Annual compounding, 487, 488
Aphelion, 400, 432
Apollonius of Perga, 372
Approximations of value of inverse trigonometric
        functions, 188
Arch problems, 399, 446
Arc length, 130, 132
Area
  of circle, 62, 536
  of isosceles triangle, 229
  of lake, 283, 308
  problems involving, 539
  of rectangle, 536
  of triangle, 278–83, 536
    Heron's formula, 279–82
Argument
  of function, 61–62, 73
  of z, 341
Āryabhata the Elder, 148
Associative property of vector addition, 349
Asymptotes, 407–9
Atmospheric pressure problems, 457, 524
Axis/axes
  of conics, 372, 427
    of ellipses, 386
    of hyperbolas, 401
  coordinate, 2
  imaginary, 340
  real, 340
  rotation of, 419–24
  of symmetry of parabola, 373

Bacterial growth, 497, 501
Base, 531
  e, 453–55, 459–60
Baseball problems, 13, 275, 276

Bearing, navigational, 254–55, 258
Bell, Alexander Graham, 517
Bernoulli, Jacob, 335
Bernoulli, Johann, 440
Biology, growth in, 496–97, 510
Biorhythms, 297
BOX function, 25–26
Brachistochrone, 440
Bracket x, 84
Bridge problems, 384, 385, 387, 399, 446
Briggs, Henry, 477
Bürgi, Jobst, 477

Calculator(s). *See also* Graphing calculators; Graph-
    ing utilities
    degree, minute, second notation (D°M′S″)-to-
        decimal interconversion, 128–29
    Demoivre's Theorem with, 344
    inverse trigonometric functions on, 188
    trigonometric equations on, 237–38
    trigonometric functions on, 146–47
Calculus of variations, 440
Capacitor problems, 303
Carbon dating, 498–99, 501, 525
Cardano, Girolamo, 346
Cardioids, polar equation of, 328–30, 336
Carpentry problem, 57
Cartesian (rectangular) coordinates, 2–3
    conversion between polar coordinates and,
        315–18
    polar coordinates vs., 312, 313
Cartesian (rectangular) form of complex number,
    341
CBL Experiment problems, 512–15
Center of circle, 47
Central angle, 129
Challenger disaster problem, 459
Change-of-Base Formula, 475–76
Chebyshëv, P.L., 222
Chebyshëv polynomials, 222
Chemistry problems, 469, 511
Circle(s), 45–51, 372
    area of, 62, 536
    center of, 47
    defined, 47
    equation of, 47–50
        general form of, 49–50
        standard form of, 47–49
    inscribed, 281–82
    length of arc, 130, 132
    polar equation of, 320, 324–26, 336
    radius of, 47
    unit
        definition of, 136
        trigonometric functions and, 126, 146–52
CIRCLE function, 47
Circular functions. *See* Trigonometric functions
Circular motion, 132–34
Closed interval, 529–30
Coefficients, 534
Cofunctions, 164–65
Common logarithm (log), 474–75
Commutative property
    of dot products, 359
    of vector addition, 349
Compass heading, 366

Complementary angles, 164–65
Complete graph, 15, 19–21
Completing the square, 543–46
Complex numbers, 546–51
    conjugates, 548–50
    converting between polar and rectangular forms
        of, 342
    magnitude (modulus) of, 340–41
    operations with, 546–48
    in polar form, 341–43
    powers of i, 550–51
    rectangular (Cartesian) form of, 341
Complex plane, 340–47
    complex roots, 345–46
    Demoivre's Theorem, 343–44
    plotting point in, 341–42
Compound interest, 486–95, 525
    comparing investments using different compound-
        ing periods, 487–88
    computing, 486–87
    continuous compounding, 488–89
    defined, 486
    doubling and tripling time for investment,
        491–92, 495
    formula, 487
    interest rate required to double investment, 491
    present value and, 489–91
Conditional equation, 206
Cone, 372
Congruent triangles, 8
Conics, 372–446. *See also* Circle(s); Ellipse(s);
        Hyperbolas; Parabolas; Plane curves
    defined, 372, 427
    general form of, 418–26
    identifying, 418–25
        without completing the squares, 418–19
        with rotation of axes, 419–24
        without rotation of axes, 424–25
    major axis of, 427
    polar equations of, 426–32
        converting to rectangular equation, 431
        graphing, 428–31
    transverse axis of, 427
Conjugate axis of hyperbolas, 401
Conjugate hyperbolas, 417
Conjugates, 341, 548–50
Connected mode, 85
Constant function, 75–76, 81–82
Constants, 528
Continuous compounding, 488–89. *See also* Growth
        and decay
Cooling, Newton's Law of, 479, 499–500, 501–2,
        512–13
Coordinate axes, 2
Coordinate of point, 2
Coordinates, 2, 528. *See also* Polar coordinates;
        Rectangular (Cartesian) coordinates
Copernicus, 126
Cosecant, 138, 140, 146
    domain of, 153
    graphs of, 179–80
    inverse, 197–98
    periodic property of, 154–55
    range of, 153
    as ratios of sides of right triangle, 164
    sign of, 156

Cosine, 138, 140, 175–77
    characteristics of, 176
    domain of, 152, 153
    double-angle formulas for, 221
    graph of, 175–77
    half-angle formulas for, 224–25
    hyperbolic, 460
    inverse, 189–91
    periodic property of, 154–55
    product-to-sum formulas for, 230
    range of, 153
    as ratios of sides of right triangle, 164
    sign of, 156
    sum and difference formulas for, 211–14
    sum-to-product formulas for, 232
Cosines, Law of, 270–77
    proof of, 270
    SAS triangles, 270–71
    SSS triangles, 271–72
Cost problems, 55, 61–62, 72, 87, 93, 309
Cotangent, 138, 140, 146
    domain of, 153
    graphs of, 180
    inverse, 197–98
    periodic property of, 154–55
    range of, 153
    as ratios of sides of right triangle, 164
    sign of, 156
Counting numbers, 528
Cube function, 61, 82–83
Cube root, complex, 345–46
Cubic equations, real roots of, 346
Curve fitting. *See also* Linear curve fitting; Non-
        linear curve fitting
    exponential, 502, 503–5, 512–13
    logarithmic, 502, 505–6, 513
    power, 502, 506–8, 513–15
    problems, 512–15
Curves, plane. *See* Plane curves
Curvilinear motion, 437
Cycloids, 439–41
    inverted, 440

Daily compounding, 488, 489
Damped motion, 291
Damped vibrations, 298–303
Damping factor, 301
Decay. *See* Growth and decay
Decibels, 517–19
Decimals, 528
Decomposition of vectors, 364
Decreasing function, 75–76
Degree(s), 127–32
    of polynomial, 534
    radians and, 130–32
Demand problems, 117, 524
Demoivre, Abraham, 343
Demoivre's Theorem, 343–44
Dependent variable, 61–62
Depression, angle of, 252
Descartes, René, 2, 245
Descartes' Law, 245
Difference formulas. *See* Sum and difference
        formulas
Difference of complex numbers, 547
Difference quotient, 74

Directed line segments, 348–49, 353
Direction
  of aircraft, 361–62, 366, 367
  for crossing stream, 366, 367
  of motorboat, 367
  in navigation, 254–55
  of swimmer, 367
  of vectors, 348–49, 361–62
Directrix, 427
  of parabola, 373
Dirichlet, Lejeune, 2
Discontinuity, 85
Distance(s)
  mean, 400
  between points, 5–8, 531
  speed/distance problems, 65–66, 255, 256–59,
    268–69, 305, 307, 355–56, 358, 366,
    538–39
Distance formula, 5–7
Distributive property of dot products, 359
Domain
  of functions, 57, 58, 59, 60–61
    absolute value function, 84
    constant function, 82
    cube function, 83
    greatest integer function, 85
    identity function, 82
    inverse cosine function, 189
    inverse functions, 109
    inverse sine function, 186
    inverse tangent function, 192
    linear function, 81
    logarithmic functions, 462–63
    reciprocal function, 84
    square function, 82
    square root function, 83
    trigonometric functions, 152–53
  of variable, 528
Dot mode, 85
Dot product, 359–68
Double-angle formulas, 221–24
  exact values using, 221–22
  for sines and cosines, 221
  for tangent, 222
  variations of, 223–24
Doubling time for investment, 491–92, 495
Drug medication problems, 458, 470

e, 453–55, 459–60
Earthquake, magnitude of, 519–21
Eccentricity, 427
  of ellipses, 401
  of hyperbolas, 417
Economics problems, 510–11, 515–16, 526, 558–59
Effective rate of interest, 489
Elements (Euclid), 273
Elements of sets, 527
Elevation, angle of, 252
Elevation-weight problem, 72
Ellipse(s), 372, 386–401, 427
  applications of, 395–96
  center of, 386
  defined, 386
  eccentricity of, 401
  equations of, 386–95
  foci of, 386

graphing, 389–90, 394
  major axis of, 386
  minor axis of, 386
  vertices of, 386
Ellipsis, 528
End points, 529
Equality
  of complex numbers, 547
  of vectors, 353–54
Equation(s). *See also* Linear equations; Polar equa-
    tions; Trigonometric equations
  of circle, 47–50
    general form of, 49–50
    standard form of, 47–49
  conditional, 206
  of ellipses, 386–95
    with center at $(0, 0)$, 388–92
    with center at (h, k), 392–95
  exponential, 481–85
  of hyperbolas, 401–11
    with center at $(0,)$, 403–9
    with center at (h, k), 409–11
  logarithmic, 478–81, 484
  of parabolas, 374–80
    with vertex at $(0, 0)$, 374–77
    with vertex at (h, k), 377–80
  parametric, 433–42
    cycloids, 439–41
    finding, 438–39
    graphing, 434–35
    hypocycloid, 437
    time as parameter, 437
  quadratic, 544–45
  roots of, 540
  sides of, 539
  of sinusoid, 287–88
  solving, 539–43
    equivalent equations, 540–42
    steps for, 542
  in two variables, graphing of, 14–34
    complete, 15, 19–21
    by hand, 14–16
    intercepts, 21–26
    symmetry, 26–31
    using graphing utilities, 16–21
    $y = 1/x$, 30–31
Equilateral triangle, 13, 144
Equilibrium (rest) position, 290
Equivalent equations, 540–42
Euclid, 273
Eüler, Leonhard, 2, 126, 460
Even function, 78–81, 159–60
Even-odd identities, 206–7
Explicit form of function, 59
Exponential curve fitting, 502, 503–5, 512–13
Exponential equations, 481–85
Exponential expressions, changing logarithmic
    expressions to, 461
Exponential functions, 448–60
  base e, 453–55, 459–60
  defined, 449
  graphs of, 449–53
  inverse of. *See* Logarithmic functions
  properties of, 455
Exponential law, 496–502
EXPonential REGression option, 504

Exponents, 531–34
  laws of, 449, 481, 532–34
  rational, 533
  relating logarithms to, 461

Fermat, Pierre de, 460
Finance problems, 512
Fincke, Thomas, 126, 148
First-degree equation. *See* Linear equations
Flashlight problem, 384
Flight plan problems, 274
Focus/foci, 427
  of ellipses, 386
  of hyperbolas, 401
  of parabola, 373
Force, work done by constant, 364–65
Formulas, geometry, 536
Frequency, 291, 297
Function(s), 57–94. *See also* Domain: of functions;
    Exponential functions; Logarithmic func-
    tions; Range of functions; Trigonometric
    functions
  absolute value, 84
  algebraic, 448
  constant, 75–76, 81–82
  cube, 61, 82–83
  decreasing, 75–76
  definitions of, 2, 57–59
  dependent variable of, 61–62
  difference quotient of, 74
  even, 78–81, 159–60
  example of, 58
  explicit form of, 59
  graphs of, 62–65, 85
    product of two functions, 299–300, 301
    sum of two functions, 298–99
  greatest-integer, 84–85
  identically equal, 206
  identity, 82
  implicit form of, 59
  increasing, 75–76
  independent variable (argument) of, 61–62, 73
  inverse, 109–15
    defined, 109
    domain and range of, 109
    finding, 112–13
    geometric interpretation of, 110–11
    graph of, 111–12
    of one-to-one function, 113–15
  linear, 81–82
  local maxima and minima of, 76–78
  odd, 78–81, 159–60
  one-to-one, 107–8, 113–15
  periodic, 154
  piecewise defined, 86–88
  reciprocal, 83, 179
  as set of ordered pairs, 64–66
  square, 64, 82
  square root, 83
  step, 85
  transcendental, 448
  values of, 57, 58–59, 73
Function keys, 60
Function minimum command, 66
Function notation, 73–74
Fundamental period, 154

Gauss, Karl Friedrich, 346
General form of equation of circle, 49–50
Generators of cone, 372
Geometry formulas, 536
Geometry problems, 14, 70
  algebra to solve, 7–8
Gibbs, Josiah Willard, 357
Grade, 57
Grade computation problems, 307
Graph(s), 1–57. *See also* Linear curve fitting
  of circles, 47–51
  compressed, 97–99
  distance between two points, 5–8
  of ellipses, 389–90, 394
  of equations in two variables, 14–34
    complete, 15, 19–21
    by hand, 14–16
    intercepts, 21–26
    symmetry, 26–31
    using graphing utilities, 16–21
    $y = 1/x$, 30–31
  of functions, 62–63, 64–65, 85
    exponential, 449–53
    logarithmic, 463–67, 476
  horizontally shifting, 96–97
  of hyperbolas, 410–11
  of inequalities, 529–30
  of inverse functions, 111–12
  midpoint formula, 8–9
  of parabolas, 374, 379
  of parametric equations of plane curves,
    434–35
  of polar equations, 319–38
    cardioids, 328–30, 336
    circles, 320, 324–26, 336
    hyperbolas, 428–31
    lemniscates, 333–34, 336
    limaçons, 329–32, 336
    lines, 324, 336
    roses, 332–33, 336
    spirals, 334–35
  of product of two functions, 299–300, 301
  rectangular (Cartesian) coordinates and, 2–3
  reflection about x-axis and y-axis, 99–100
  sinusoidal, 283–98
    amplitude of, 284–87, 288–90
    characteristics of, 283–84
    equation of, 287–88
    period of, 285–87, 288–90
    phase shift of, 288–90
    simple harmonic motion, 290–92
  straight-line, 35–46
    given a point and a slope, 38–39
  stretched, 97–99
  of sum of two functions, 298–99
  of trigonometric functions, 173–204
    cosecant function, 179–80
    cosine function, 175–77
    cotangent function, 180
    inverse cosine function, 189–91
    inverse sine function, 184, 186–87
    inverse tangent function, 192–93
    secant function, 179–80
    sine function, 173–75
    tangent function, 177–79
  of vectors, 350–51
  vertically shifting, 94–95, 96–97

of $y = e^{-x} \sin x$, 301
of $y = x \sin x$, 300–301
Graphing calculators, 3
  to calculate logarithms with base other than 10 or
    e, 474–76
  CBL Experiment problems, 512–15
  e^x or $exp(x)$ key on, 453
  to evaluate powers of 2, 448
Graphing utilities, 3–5
  BOX function, 25–26
  CIRCLE function, 47
  connected and dot modes of, 85
  establishing identity and, 207
  to estimate domain of function, 60–61
  function keys, 60
  function minimum command, 66
  to graph bacterial growth, 497
  to graph logarithmic and exponential equations,
    483–85
  to graph logarithmic function with base other
    than 10 or e, 476
  to graph Newton's Law of Cooling, 499–500
  to graph radioactive decay, 498–99
  to identify functions as odd or even, 79–81
  to locate local maxima and minima, 77–78
  REGression options on, 503, 504, 506, 507
  square screens on, 37–38
  $\underset{\text{PLOT}}{\text{STAT}}$ function, 86
  TRACE function, 23–24, 30, 77–78, 79–81
    for trigonometric equations, 242
  ZOOM-IN function, 24–25
  ZOOM-OUT feature, 19–21
  ZSQR function, 37
Grassmann, Hermann, 357
Gravity problems, 70
Greater than, 529
Greater than or equal to, 529
Greatest-integer function, 84–85
Grids, polar, 319
Ground, velocity relative to, 355
Ground speed problem, 358
Growth and decay, 496–502, 503, 511, 512–13

Half-angle formulas, 224–27
Half-closed intervals, 529–30
Half-life, 126, 498
Half-open intervals, 529–30
Hamilton, William Rowan, 357
Headlight problem, 384
Healing of wounds, exponential model of, 458, 470
Heat transfer problem, 244
Height problems, 252–59, 267, 305
Height vs. weight problem, 558
Heron's (Hero's) Formula, 279–82
Highway construction problems, 258, 268, 308
Home run problems, 231
Horizontal line, equation of, 41–43
Horizontal-line test, 107–8
Huygens, Christian, 441
Hyperbolas, 372, 401–18, 427
  applications of, 412–14
  asymptotes of, 407–9
  branches of, 401
  center of, 401
  conjugate, 417
  conjugate axis of, 401

defined, 401
eccentricity of, 417
equations of, 401–11
  with center at $(0, 0)$, 403–9
  with center at (h, k), 409–11
  polar, 426–32
equilateral, 417
foci of, 401
graphing, 410–11
transverse axis of, 401
vertices of, 401
Hyperbolic cosine function, 460
Hyperbolic sine function, 460
Hypocycloid, 437
Hypotenuse, 163, 535

i, 546, 550–51
Identity(ies)
  defined, 206
  involving inverse trigonometric functions, 196–97
  polarization, 367
  solving trigonometric equations using, 238–40
  trigonometric, 206–10
    establishing, 207–10
    even-odd, 206–7
    Pythagorean, 157–58, 206
    quotient, 157–58, 206
    reciprocal, 156, 158, 206
Identity function, 82
Identity property of vector addition, 349
Image of x, 57, 59
Imaginary axis, 340
Imaginary number, pure, 546
Imaginary part of complex number, 546
Imaginary unit (i), 546, 550–51
Implicit form of function, 59
Incidence, angle of, 245
Inclination, angle of, 151
Incline problems, 257–58, 268
Income problem, 557
Increasing function, 75–76
Independent variable, 61–62, 73
Index, 533
  of refraction, 245
Individual Retirement Account (IRA), computing
    value of, 489
Inequalities
  graphing, 529–30
  nonstrict, 529
  strict, 529
Inequality signs, 529
Initial point of vector, 348
Initial side of angle, 126
Inscribed circle, 281–82
Integers, 528
Intensity
  of earthquake, 520–21
  of sound wave, 517, 518–19, 521
Intercepts, 21–26
  of line, 44
Interest
  compound. *See* Compound interest
  defined, 486
  rate of, 486
    effective, 489
  simple, 486
Intermediate Value Theorem, 484

INTERSECT command, 491
Intervals, 529–30
Inverse functions, 109–15
  defined, 109
  domain and range of, 109
  finding, 112–13
  geometric interpretation of, 110–11
  graph of, 111–12
  of one-to-one function, 113–15
Inverse trigonometric functions, 183–200
  approximating value of, 189, 198
  finding exact value of expressions involving,
      186–88, 194–96
  identity involving, 196–97
  inverse cosine, 189–91
  inverse cotangent, secant, and cosecant,
      197–98
  inverse sine, 184–88
  inverse tangent, 192–93
Investments
  doubling and tripling time for, 491–92, 495
  interest rate required to double, 491
  time to reach goal for, 495
  using different compounding periods, comparing,
      487–88
Irrational numbers, 528, 546
Isosceles triangle, 13
  area of, 229

Kepler, Johannes, 509
Keplerian Laws of Planetary Motion, 509

Ladder around corner problem, 244
Latus rectum, 374
Laws of exponents, 449, 481
Leading coefficient, 534
Learning curve function, 470
Left end point, 529
Legs of triangle, 163, 535
Leibniz, Gottfried Wilhelm von, 2
Lemniscate, polar equation of, 333–34, 336
Length
  of arc, 130, 132
  of line segment, 7
  problems involving, 257, 266, 275, 305, 538
Less than, 529
Limaçon, polar equation of, 329–32, 336
  without inner loop, 330–31
  with inner loop, 331
Limiting magnitude of telescope, 524
Line(s)
  family of, 56
  intercepts of, 44
  polar equation of, 320, 321–24, 336
  slope of, 35–37, 45–46
  tangent, 56
  y-intercept of, 45–46
Linear curve fitting, 552–59
  correlation coefficients, 555
  lines of best fit, 554–55
  prediction and accuracy, 556
  scatter diagrams, 552–53
Linear equations, 35–46
  defined, 43
  general form of, 43–44
  horizontal line, 41–43
  point-slope form of, 41–42

  slope-intercept form of, 44–46
  vertical line, 40–41
Linear function, 81–82
LINear REGression option, 504
Linear speed, 133
Line segment
  directed, 348–49, 353
  length of, 7
  midpoint of, 8–9
Local maxima and minima of functions, 76–78
Logarithm(s)
  Change-of-Base Formula, 475–76
  common (log), 474–75
  natural (ln), 474–75, 477
  properties of, 471–74, 476
  relating exponents to, 461
Logarithmic curve fitting, 502, 505–6, 513
Logarithmic equations, 478–81, 484
Logarithmic expressions, changing exponential
      expressions to, 461
Logarithmic functions, 461–71
  defined, 461
  domain of, 462–63
  graphs of, 463–67, 476
  inverse of. See Exponential functions
  natural logarithm function (ln), 463–64
  properties of, 462, 467
Logarithmic scales, 517–21
Logarithmic spiral, 335
LOng RAnge Navigation system (LORAN),
      412–14, 417, 446
Loudness of sound, 517–19, 521, 525

Magnitude(s)
  of complex number, 340–41
  of earthquake, 519–21
  of telescope, limiting, 524
  of vectors, 348, 351, 353, 361–62
Major axis of conics, 427
  ellipses, 386
Marginal propensity to consume, 557
Marginal propensity to save, 558
Marketing problems, 510–11
Mean distance, 400
Medians of triangle, 13
Menelaus of Alexandria, 126
Metrica (Heron), 280
Midpoint formula, 8–9
Mile, nautical and statute, 136
Minor axis of ellipses, 386
Minutes, 128–29
Mirror problem, 446
Mitosis, 496–97
Modulus (magnitude) of complex number, 340–41
Mollweide, Karl, 280
Mollweide's Formula, 269, 280
Monomial, 534
Monthly compounding, 488, 489
Motion
  circular, 132–34
  curvilinear, 437
  damped, 291
  describing, 437–38
  projectile, 150, 151, 205, 224, 229, 244–45, 442
  simple harmonic motion, 290–92
Mountain trail, inclination of, 251
Multiplication. See Product(s)

Napier, John, 477
Nappes, 372
Nasîr ed-dîn, 126, 273
Natural (counting) numbers, 528
Natural logarithm (ln), 474–75, 477
  function, 463–64
Nautical mile, 136
Navigation problems, 267, 272, 274, 307, 308
Negative real numbers, 528
Newton, Isaac, 499
Newton's Law of Cooling, 479, 499–500, 501–2,
      512–13
Niccolo of Brescia (Tartaglia), 346
Nonlinear curve fitting, 512–17
  choosing best model, 508–9, 515–17
  exponential, 502, 503–5, 512–13
  logarithmic, 502, 505–6, 513
  power, 502, 506–8, 514–15
  scatter diagrams of, 502, 503, 504, 506, 507,
      509
Nonstrict inequalities, 529
Notation
  for angles, 126
  function, 73–74
  for inverse trigonometric functions, 184
  quadrantal angle, 127
  set-builder, 527
  nth root, principal, 532–33
Number line, 2
  finding distance on, 531
  real numbers, 528
Numbers
  complex. See Complex numbers
  counting (natural numbers), 528
  irrational, 528, 546
  rational, 528, 546
  real, 527–30, 546

Oblique triangles, trigonometric functions to solve,
      259–62
  ambiguous cases, 262–65
  applied problems, 264–69
  Law of Cosines, 270–77
  Law of Sines, 259–69, 273
Odd functions, 78–81, 159–60
One-to-one functions, 107–8
  inverse of, 113–15
Open interval, 529–30
Optics problems, 457, 469
Ordered pair, 2
Ordinate, 2
Origin, 2, 528
  symmetry with respect to, 27–28, 29, 327
Orthogonal vectors, 362–63

Parabolas, 15, 372–85, 427
  axis of symmetry of, 373
  defined, 373
  directrix of, 373
  equations of, 374–86
  focus of, 373
  graphing, 374, 379
  reflecting property of, 380–81
  vertex of, 373
Paraboloid of revolution, 371, 372, 380
Parallel vectors, 362–63
Parameter, 433

Parametric equations, 433–42
 cycloids, 439–41
 finding, 438–39
 graphing, 434–35
 hypocycloid, 437
 time as parameter, 437
Pascal, Blaise, 440
Payment period, 486
Pendulum problems, 72, 105, 117, 507–8
Perfect roots, 533
Perihelion, 400, 432
Period
 fundamental, 154
 of sinusoid, 285–87, 288–90
 of trigonometric functions, 154–55
 of vibrating object, 290
Periodic function, 154
Periodic property of trigonometric functions, 154–55
Phase shift, 288–90
Piecewise defined functions, 86–88
Pistons and rods problems, 276
Pixels, 3
Plane
 complex, 340–47
  complex roots, 345–46
  Demoivre's Theorem, 343–44
  plotting point in, 341–42
 representing vectors in, 351–55
Plane curves, 433–42
 defined, 433
 orientation of, 434
 parametric equations of, 433–42
 rectangular equation of, 436, 438
Planetary Motion, Keplerian Laws of, 509
Point(s)
 coordinate of, 2
 distance between, 5–8, 531
 plotting, 2, 313–14, 341–42
 of tangency, 56
Polar axis, 312
Polar coordinates, 311–47
 complex plane, 340–47
  complex roots, 345–46
  Demoivre's Theorem, 343–44
  plotting point in, 341–42
 conversion between rectangular coordinates and, 315–18
 defined, 312
 plotting points using, 313–14
 rectangular coordinates vs., 312, 313
 of single point, 312–13, 314
 symmetry of, 326–28
Polar equations, 319–38
 cardioids, 328–30, 336
 circles, 320, 324–26, 336
 classification, 335, 336
 of conics, 426–32
  converting to rectangular equation, 431
  graphing, 428–31
  hyperbolas, 426–32
 defined, 319
 identifying and graphing, 320–25
 lemniscates, 333–34, 336
 limaçons, 329–32, 336
 lines, 320, 321–24, 336
 roses, 332–33, 336

 spirals, 334–35
 transforming between rectangular form and, 318
Polar form of complex number, 341–43
 products and quotients of, 342–43
Polar grids, 319
Polarization identity, 367
Pole, 312. *See also* Origin
 symmetry with respect to, 327
Polynomial(s), 534–35
 Chebyshëv, 222
 completing the square, 543–46
 degree of, 534
 terms of, 534
 zero, 534
Population growth problems, 450, 497, 501, 513, 517, 525
Position vector, 352–53
Positive real numbers, 528
Power(s), 297, 531–32. *See also* Exponents
 of *i*, 550–51
Power curve fitting, 502, 506–8, 514–15
PoWeR REGression option, 507
Present value, 489–91
Principal, 486
Principal *n*th root, 532–33
Product(s)
 of complex numbers, 547–48
  in polar form, 342–43
 scalar, 350, 354, 359
 of vectors, 350–51, 535
Product-to-sum formulas, 230–32
Profit problems, 55, 458
Projectile motion problems, 150, 151, 205, 224, 229, 244–45, 442
Projection, 291
Ptolemy, 273
Pure imaginary number, 546
Pythagorean identities, 157–58, 206
Pythagorean theorem, 535–36

Quadrantal angles, 127, 141–42
Quadratic equation, solving by completing the square, 544–45
Quarterly compounding, 488, 489
Quaternions, 357
Quotient identities, 157, 158, 206
Quotient of complex number, 549
 in polar form, 342–43

Radians, 129–32
 degrees and, 130–32
Radicals, 533–34
Radicand, 533
Radioactive decay, 498–99, 501, 502, 511
Radius of circle, 47
Rain gutter construction, 244
Range of functions, 57–58
 cube function, 83
 identity function, 82
 inverse cosine function, 189
 inverse functions, 109
 inverse sine function, 186
 inverse tangent function, 192
 reciprocal function, 84
 square function, 82
 square root function, 83
 trigonometric functions, 152–53

Range of projectile, 205, 224
RANGE setting, 3
Rate of interest, 486
 effective, 489
Rational exponents, 533
Rational numbers, 528, 546
Ray (half-line), 126
Real axis, 340
Real number line, 528
Real numbers, 527–30, 546
Real part of complex number, 546
Real roots of cubic equations, 346
Reciprocal functions, 83, 179
Reciprocal identities, 156, 158, 206
Reciprocal of complex numbers, 549
Rectangle, area of, 536
Rectangular box, volume of, 536
Rectangular (Cartesian) coordinates, 2–3
 conversion between polar coordinates and, 315–18
 polar coordinates vs., 312, 313
Rectangular (Cartesian) form of complex number, 341
Rectangular equations, transforming between polar form and, 318
Reference angle, 165–68
Reflecting property of parabolas, 380–81
Reflecting telescopes problem, 385
Reflection about *x*-axis and *y*-axis
 graphing exponential functions using, 452–53
 graphing logarithmic functions using, 464–65
Reflections, graphing using
 cosine functions, 16–77
 sine functions, 174–75
Refraction, 245
Regiomontanus, 126, 273
REGression options, 503, 504, 506, 507
Rescue at sea problems, 264–65, 266
Resultant, 355
Rhaeticus, 126
Richter, C.F., 519
Richter scale, 519
Right angle, 127, 535
Right end point, 529
Right triangle(s), 163, 535–36
 trigonometric functions to solve, 250–83
Right triangle trigonometry, 126, 163–74
 applications of, 169–72
 complementary angles, 164–65
 reference angle, 165–68
 solving right triangles, 169–72
Rise and run, 35
River width problem, 251
Rods and pistons problems, 276
Root(s)
 complex, 345–46
 of equation, 540
 square, 345, 533
Rose, polar equation of, 332–33, 336
Roster method, 527
Rumors, exponential spread of, 458, 470
Run and rise, 35

Salvage value problems, 524
Satellite dish problem, 380–81, 384
Satellites in orbit problems, 259
Sawtooth curve problems, 230, 303
Scalar product, 350, 354, 359

Scalars, 350
Scale, 528
  logarithmic, 517–21
Scatter diagrams, 502–4, 506, 507, 509
Searchlights problem, 384–85
Secant, 138, 140, 146
  domain of, 152, 153
  graphs of, 179–80
  inverse, 197–98
  periodic property of, 154–55
  range of, 153
  as ratios of sides of right triangle, 164
  sign of, 156
Seconds, 128–29
Set-builder notation, 527
Sets, 527
  solution, 540
Setting the RANGE, 3
Shifts
  graphing cosine functions using, 176–77
  graphing functions using, 452–53, 464–65
  graphing sine functions using, 174–75
Sidereal year, 508, 509
Sides of equation, 539
Signs of trigonometric functions, 155–56
Similar triangles, 13, 259
Simple harmonic motion, 290–92
Simple interest, 486
Simple pendulum, 507–8
Simplifying radicals, 533–34
Sine, 137, 140, 148, 173–75
  characteristics of, 174
  domain of, 152, 153
  double-angle formulas for, 221
  graph of, 173–75
  half-angle formulas for, 224–25
  hyperbolic, 460
  inverse, 184–88
  periodic property of, 154–55
  product-to-sum formulas for, 230
  range of, 153
  as ratios of sides of right triangle, 164
  sign of, 156
  sum and difference formulas for, 214–16
  sum-to-product formulas for, 232
Sines, Law of, 259–69, 273
  applied problems, 264–69
  proof of, 260–61
  to solve ASA triangle, 261–62
  to solve SAA triangle, 261
  to solve SSA triangle, 262–65
Sine wave, amplified, 301
Sinusoidal graphs, 283–98
  amplitude of, 284–87, 288–90
  characteristics of, 283–84
  equation of, 287–88
  period of, 285–87, 288–90
  phase shift of, 288–90
  simple harmonic motion, 290–92
Slope
  of line, 35–37, 45–46
  straight-line graph given a point and, 38–39
Slope-intercept form of equation of line, 44–46
Snell, Willebrod, 245
Snell's Law of Refraction, 245
Solar heat problem, 385
Solution set, 540

Sound, loudness of, 517–19, 521, 525
Sound amplification problems, 524
Sound wave, intensity of, 517, 518–19, 521
Space satellites problems, 457, 469
Speed, angular and linear, 133
Speed/time/distance problems, 65–66, 124, 255,
  256–59, 268–69, 305, 307, 355–56, 358,
  366, 538–39, 558
Spiral, polar equation of, 334–35
Square function, 64, 82
Square root(s), 533
  complex, 345
Square root function, 83
Standard form
  complex numbers in, 546, 549
  of equation of circle, 47–49
  polynomials in, 535
Standard position, angle in, 126–27
$\frac{\text{STAT}}{\text{PLOT}}$ function, 86
Statute miles, 136
Step function, 84–85
Straight angle, 127
Straight-line graphs, 35–46
  given a point and a slope, 38–39
Strict inequalities, 529
Subscript, 534
Subsets, 527
Subtraction of vectors, 354
Sum. See Addition
Sum and difference formulas, 211–19
  for cosines, 211–14
  for sines, 214–16
  for tangent, 216–18
Sum-to-product formulas, 232–33
Surveillance satellites, 259
Surveying problems, 266, 274, 308
Symmetry, 26–31
  of polar coordinates, 326–28
  with respect to origin, 27–28, 29, 327
  with respect to x-axis, 26–27, 28, 29
  with respect to y-axis, 27, 28, 29
  tests for, 28–29

Tangency, point of, 56
Tangent, 138, 140, 148
  characteristics, 178
  domain of, 152, 153
  double-angle formulas for, 222
  graphs of, 177–79
  half-angle formulas for, 224–25, 227
  inverse, 191–94
  periodic property of, 154–55
  range of, 153
  as ratios of sides of right triangle, 164
  sign of, 156
  sum and difference formulas for, 216–18
Tangent line, 56
Tangents, Law of, 269, 273
Tartaglia (Niccolo of Brescia), 346
Tautochrone, 441
Telescope
  limiting magnitude of, 524
  reflecting, 385
Temperature problems, 55, 105, 117, 124, 479,
  499–500, 501–2, 512–13
Terminal point of vector, 348

Terminal side of angle, 126
Terms of polynomial, 534
Time as parameter, 437
Time value of money (present value), 489–91
Touch-tone phone sound problem, 234, 302
TRACE function, 30
  to identify functions as odd or even, 79–81
  to locate intercepts, 23–24
  to locate local maxima and minima, 77–78
Transcendental functions, 448. See also Exponential
    functions; Logarithmic functions
Transverse axis of conics, 427
  hyperbolas, 401
Triangle(s)
  area of, 278–83, 536
    Heron's formula, 279–82
  circumscribing, 269
  congruent, 8
  equilateral, 13, 144
  inscribed circle of, 281–82
  isosceles, 13, 229
  legs of, 163, 535
  medians of, 13
  oblique, 259–69
    ambiguous cases, 262–65
    applied problems, 264–69
    Law of Cosines and, 270–77
    Law of Sines and, 259–69, 273
  right, 163, 535–36. See also Right triangle
      trigonometry
  trigonometric functions to solve, 250–83
  similar, 13, 259
Trigonometric equations, 234–42
  solution of, 234
  solving, 235–42
    with calculator, 237–38
    linear in sin θ and cos θ, 241–42
    other methods for, 240–42
    quadratic in form, 238
    using graphing utility, 242
    using identities, 238–40
Trigonometric functions, 125–204, 249–309. See
    also Analytic trigonometry
  of angles, 126–36, 139
    circular motion and, 132–34
    degrees, 127–32
    45° and −45°, 142–43
    radians, 129–32
    30° and 60°, 144–46
  area of triangle, 278–83
    Heron's formula, 279–82
  damped vibrations, 298–303
  domain and range of, 152–53
  evaluating, 139–44
  even-odd properties of, 159–60
  finding value of, 138–39
    exact value, 140–44
    from right triangle, 164, 168
    using calculator, 146–47
    using identities, 157–58
    using odd-even properties, 160
    using periodic properties, 155
    using reference angles, 167
  fundamental (basic) identities, 156–59
    Pythagorean, 157–58
    quotient, 157, 158
    reciprocal, 156, 158

Trigonometric functions, *(cont.)*
  graphs of. *See under* Graph(s)
  period of, 154–55
  right triangle trigonometry, 126, 163–74
    applications of, 169–72
    complementary angles, 164–65
    reference angle, 165–68
    solving right triangles, 169–72
  signs of, 155–56
  sinusoidal graphs, 283–98
    amplitude of, 284–87, 288–90
    characteristics of, 283–84
    equation of, 287–88
    period of, 285–87, 288–90
    phase shift of, 288–90
    simple harmonic motion, 290–92
  to solve oblique triangles, 259–62
    ambiguous cases, 262–65
    applied problems, 264–69
    Law of Cosines, 270–77
    Law of Sines, 259–69, 273
  to solve right triangles, 250–83
  unit circle approach to, 126, 136–52
Trigonometric identities, 206–10
  establishing, 207–10
  even-odd, 206–7
  Pythagorean, 206
  quotient, 206
  reciprocal, 206
Tripling time for investment, 491–92, 495
TV dish problem, 384

Uninhibited decay, law of, 502
  radioactive decay, 498–99, 501, 502, 511
Uninhibited growth, law of, 496–97, 502, 510
Unit, 528
Unit circle
  definition of, 136
  trigonometric functions and, 126, 146–47
    of angles, 139

evaluating, 139–44
  types of, 137–38
Unit vector, 351, 354–55

Values of functions, 57, 58–59, 73
Variable(s), 528
  dependent, 61–62
  domain of, 528
  independent, 61–62, 73
Vector(s), 348–70
  adding, 349–50, 354
  angle between, 360
  applications of, 355–56
  components of, 352, 354
  decomposition of, 364
  defined, 348
  directed line segments and, 348–49, 353
  direction of, 348–49, 361–62
  dot product, 359–68
  equality of, 353
  graphing, 350–51
  history of, 357
  initial point of, 348
  magnitudes of, 348, 351, 354, 361–62
  multiplying by numbers, 350–51, 354
  parallel and orthogonal, 362–63
  plane representation of, 351–55
  position, 352–53
  projection onto another vector, 363–64
  subtracting, 354
  terminal point of, 348
  unit, 351, 354–55
  word done by constant force, 364–65
  zero, 348, 349
Velocity relative to air and ground, 355
Vertex/vertices, 126
  of cone, 372
  of ellipses, 386
  of hyperbola, 401
  of parabola, 373

Vertical line
  equation of, 40–41
  slope of, 35
Vertical-line test, 63
Vibration, 290–91
  damped, 298–303
Viète, François, 273
Viewing rectangle, 3
Volume of rectangular box, 536

Wallis, John, 346
Weather satellites problem, 56
Whispering galleries, 395–96, 399, 446
Width problems, 256, 305
Window, 3
Work
  computing, 367
  done by constant force, 364–65

x-axis, 2
  projection of *P* on, 291
  reflection about, 99–100
  symmetry with respect to, 26–27, 28, 29, 327
  x-coordinate, 2
  x-intercept, 21

y-axis, 2
  projection of *P* on, 291
  reflection about, 99–100
  symmetry with respect to, 27, 28, 29, 327
y-coordinate, 2
y-intercept, 21
y-intercept of line, 45–46

Zero-coupon bond, computing value of, 490–91, 494–95
Zero-level earthquake, 519
Zero polynomial, 534
Zero vector, 348, 349
ZOOM-IN function, 24–25
ZOOM-OUT feature, 19–21
ZSQR function, 37

Conics

Parabola

$$y^2 = 4ax \qquad\qquad y^2 = -4ax \qquad\qquad x^2 = 4ay \qquad\qquad x^2 = -4ay$$

Ellipse

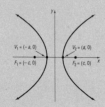

$$\frac{x^2}{a^2} + \frac{y^2}{b^2} = 1, \quad c^2 = a^2 - b^2 \qquad\qquad \frac{x^2}{b^2} + \frac{y^2}{a^2} = 1, \quad c^2 = a^2 - b^2$$

Hyperbola

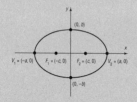 

$$\frac{x^2}{a^2} - \frac{y^2}{b^2} = 1, \quad c^2 = a^2 + b^2 \qquad\qquad \frac{y^2}{a^2} - \frac{x^2}{b^2} = 1, \quad c^2 = a^2 + b^2$$

$$\text{Asymptotes:} \quad y = \frac{b}{a}x, \quad y = -\frac{b}{a}x \qquad\qquad \text{Asymptotes:} \quad y = \frac{a}{b}x, \quad y = -\frac{a}{b}x$$

Properties of Logarithms

$$\log_a MN = \log_a M + \log_a N$$

$$\log_a\left(\frac{M}{N}\right) = \log_a M - \log_a N$$

$$\log_a M^r = r \log_a M$$

$$\log_a M = \frac{\log M}{\log a} = \frac{\ln M}{\ln a}$$